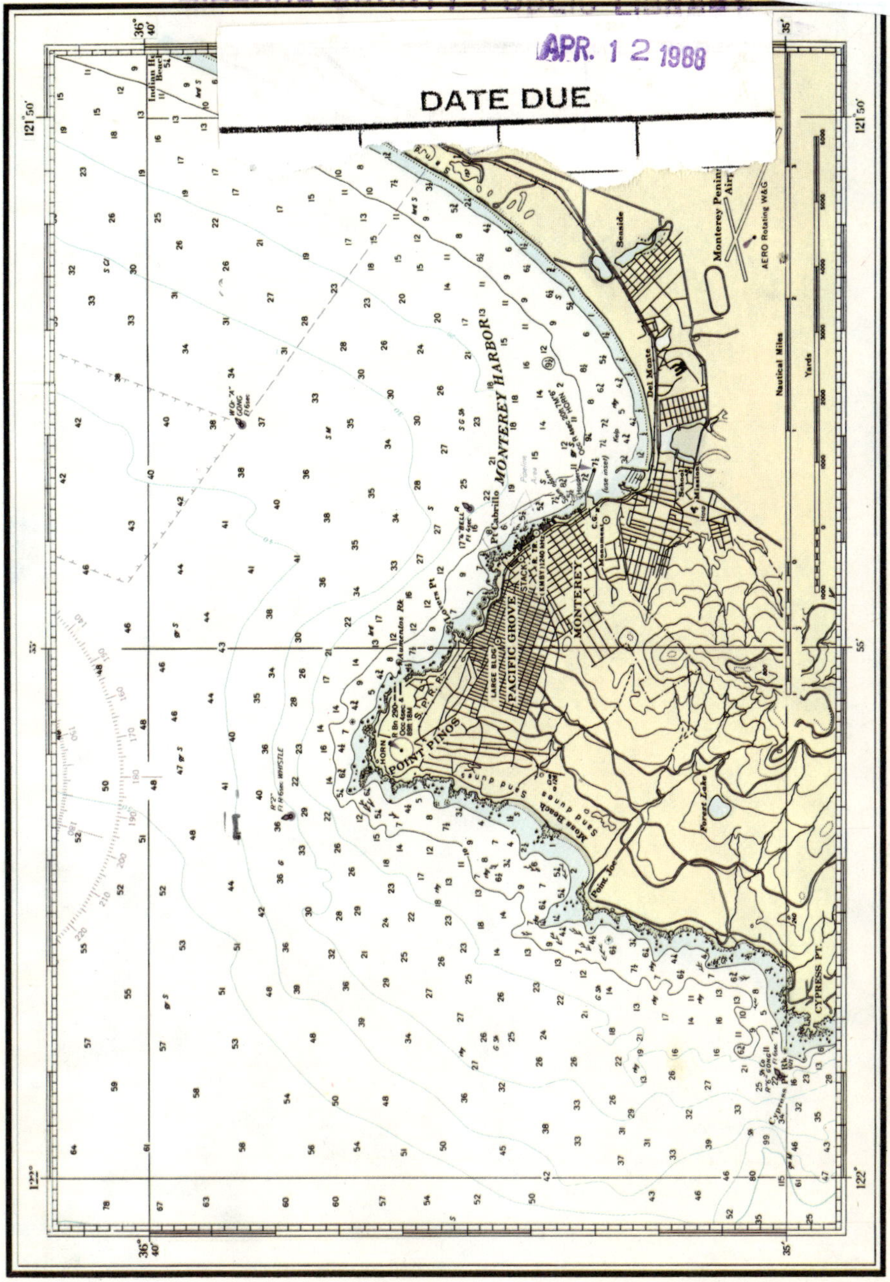

Hydrographic chart. (*Courtesy National Ocean Survey.*)

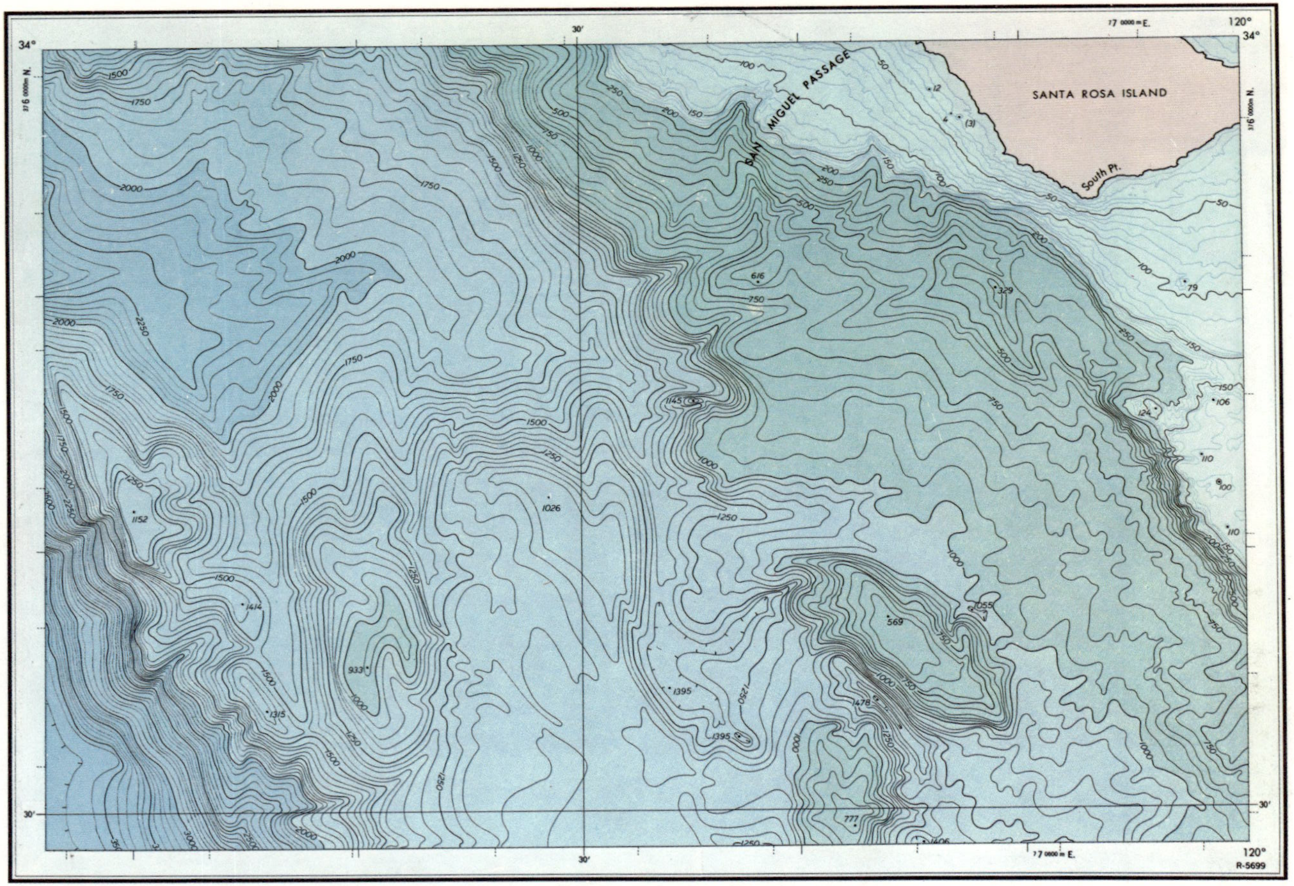

Bathymetrio map. (*Courtesy National Ocean Survey.*)

Introduction to SURVEYING

Introduction to SURVEYING

James M. Anderson, Ph.D.
Professor of Civil Engineering
University of California, Berkeley

Edward M. Mikhail, Ph.D.
Professor of Civil Engineering
Purdue University

A textbook based on
Surveying: Theory and Practice
now in its 6th edition, by
Raymond E. Davis
Francis S. Foote
Late Professors of Civil Engineering
University of California
James M. Anderson
Edward M. Mikhail

McGraw-Hill Book Company

New York St. Louis San Francisco
Auckland Bogotá Hamburg Johannesburg
London Madrid Mexico Montreal New Delhi
Panama Paris São Paulo Singapore
Sydney Tokyo Toronto

Library of Congress Cataloging in Publication Data

Anderson, J. M. (James McMurry), date
Introduction to surveying.

"A textbook based on Surveying, theory and practice" 6th ed.

Includes bibliographies and index.

1. Surveying. T. Mikhail, Edward M. II. Surveying, theory and practice. c1981. III. Title.
TA545.A545 1985 526.9 83-19949
ISBN 0-07-001653-4

INTRODUCTION TO SURVEYING

1 2 3 4 5 6 7 8 9 0 D O C D O C 8 9 8 7 6 5 4

ISBN 0-07-001653-4

This book was set in Times Roman by Science Typographers, Inc.
The editors were Kiran Verma and Madelaine Eichberg;
the production supervisor was Marietta Breitwieser.
New drawings were done by J & R Services, Inc.
The cover was designed by Suzanne Haldane;
cover art courtesy of the Lietz Company, Inc.
R. R. Donnelley & Sons Company was printer and binder.

Contents

PART II BASIC SURVEY MEASUREMENTS

Contents of field and office problems

Preface

The sixth edition of *Surveying: Theory and Practice* by Davis, Foote, Anderson, and Mikhail came out in 1981. Within a very few months from publication, the publisher and the last two authors received excellent feedback from those who adopted it as a textbook as well as those who acquired it for a reference. At the same time several professors indicated that they would like to see the same two authors produce an elementary textbook to go with the more comprehensive book. In response to these comments, a detailed questionnaire was designed and distributed to a large number of professors who teach surveying. The response was favorable and therefore this book was written to satisfy this indicated need.

This book is produced in such a manner as to accommodate, as much as possible, the specific suggestions and desires of the respondents to the questionnaire. It is designed to cover fully the material for at least one college level course in surveying. The style is such that it can be used for either a two-year technician program or a four-year technology course of study. For instance, matrix algebra, which was used extensively in the sixth edition of the large book, is not employed at all in this elementary textbook. Instead, mathematical concepts are kept as simple as possible. Please note that this simplification is *not* done at the price of sacrificing rigor. One of the most distinguishing features of this book is that it avoids superficiality. In other words, whenever a topic is addressed, it is covered in sufficient depth to provide the student with solid, useful information. A common complaint frequently heard among students is that many textbooks only "skim" the surface, thus providing coverage of many topics but leaving the reader with little information usable for practical problems. A concerted effort was made to avoid this problem throughout the writing of this book. Naturally, when a certain level is reached beyond which information is deemed too advanced for the beginning student, the reader is referred to other sources of information.

Another feature of this book is the inclusion in Appendix B of material on elementary mathematical concepts. It is felt that such an appendix will provide the elementary student with an extremely valuable tool. College and university students, such as those majoring in engineering, forestry, etc., would in general have covered these concepts in their courses in mathematics. Even so, this Appendix B will provide them with an excellent reference when needed. For technology and technician schools, Appendix B may actually be used by the instructor to provide students with basic information on the topics covered. Outstanding features of Appendix B include a direct and uncomplicated presentation of the material and an abundance of worked numerical examples which will be helpful not only to students enrolled in formal courses but also for self-study. As a matter of fact, numerical examples are found throughout the book since the authors are of the opinion that worked examples are indispensable for books in general and in particular for an introductory textbook.

The book contains 15 chapters and two appendices. Chapter 1 is introductory, containing basic definitions and brief descriptions of operations in surveying. Chapter 2 describes in a simplified manner survey measurements and their associated errors. It includes a brief nonmathematical introduction to probability concepts as they relate to survey measurements. Further discussion of the related topics of error propagation and adjustment of survey measurements is deferred to Appendix A. In some situations, the instructor may wish to continue with this appendix immediately following Chap. 2. However, in general, it will probably be better to wait until more specific knowledge of surveying operations is acquired before discussing the material in Appendix A. Chapter 3 completes a three-chapter part on general matters. In it are discussed the important subjects of field work, field notes, office work, etc., along with a description of significant figures and modern computational aids.

Basic survey measurements are covered in Chaps. 4 through 7. This overall topic includes: distance measurement (Chap. 4), leveling (Chap. 5), angle and direction measurement (Chap. 6), and stadia and tacheometry (Chap. 7). The sequence is logical and the coverage emphasizes the fundamental measurements in surveying, without encumbering the reader with details on techniques, methodology, and operations all at once.

Chapters 8 and 9 constitute a part on survey operations. Traverse is covered in detail in Chap. 8. In Chap. 9, the student is introduced briefly to other survey operations such as triangulation, trilateration, intersection, resection, and modern positioning systems. This last item makes up almost half the chapter since it is felt that the student should be aware of the power and direction of the field of surveying due to modern technological developments. This segment is up-to-date but is presented without too much detailed development of theory and operations, much of which is considered beyond the scope of beginning students using this introductory textbook.

Mapping is covered in Chaps. 10 and 11. Mapping and map drafting are included in Chap. 10, while a simple treatment of map projections and specifically state plane coordinate systems is provided in Chap. 11. The remaining chapters cover distinct types of surveys: control and topographic surveys (Chap. 12), photogrammetric surveys (Chap. 13), route and construction surveys (Chap. 14), and land surveys (Chap. 15). Throughout the text, each chapter is followed by problems and references. Where appropriate, field and/or office problems are included at the ends of certain chapters.

In summary, this book provides a logically arranged sequence of topics that are in tune with current practices and technological developments in the field of surveying.

It is a timely and up-to-date introductory textbook designed to lead into the more extensive *Surveying: Theory and Practice*, 6th edition, and will be useful for a variety of elementary one-semester (or single-quarter) courses in surveying, including short courses.

James M. Anderson
Edward M. Mikhail

PART I
CONCEPTS

CHAPTER 1
Surveying and mapping

1.1. Surveying Surveying has to do with the determination of the relative spatial location of points on or near the surface of the earth. It is the art of measuring horizontal and vertical distances between objects, of measuring angles between lines, of determining the direction of lines, and of establishing points by predetermined angular and linear measurements.

Accompanying the actual measurements of surveying are mathematical calculations. Distances, angles, directions, locations, elevations, areas, and volumes are thus determined from data of the survey. Also, much of the information of the survey is portrayed graphically by the construction of maps, profiles, cross sections, and diagrams.

The equipment available and methods applicable for measurement and calculation have changed tremendously in the past decade. Aerial photogrammetry, satellite observations, remote sensing, inertial surveying, and electronic distance measurement and laser techniques are examples of modern systems utilized to collect data usable in the surveying process. The relatively easy access to electronic computers of all sizes facilitates the rigorous processing and storage of large volumes of data.

With the development of these modern data acquisition and processing systems, the duties of the surveyor have expanded beyond the traditional tasks of the field work of taking measurements and the office work of computing and drawing. Surveying is required not only for conventional construction engineering projects, mapping, and property surveys, but is also used increasingly by other physical sciences, such as geology and geophysics; biology, including agriculture, forestry, grasslands, and wildlife; hydrology and oceanography; and geography, including human and cultural resources. The tasks in these physical science operations need to be redefined to include design of the surveying procedure and selection of equipment appropriate for the project; acquisition of data in the field or by way of remote stations; reduction or analysis of data in the office or in the field; storage of data in a form compatible with future retrieval; preparation of maps or other displays in the graphical (including photographic) or numerical form needed for the purpose of the survey; and setting of monuments and boundaries in the field as well as control for construction layout. Performance of these tasks requires a familiarity with the uses of surveying, knowledge of the fundamentals of the surveying process, and a knowledge of the various means by which data can be prepared for presentation.

1.2. Uses of surveys The earliest surveys known were for the purpose of establishing the boundaries of land, and such surveys are still the important work of many surveyors.

Every construction project of any magnitude is based to some degree upon measurements taken during the progress of a survey and is constructed about lines and points established by the surveyor. Aside from land surveys, practically all surveys of a private nature and most of those conducted by public agencies are of assistance in the conception, design, and execution of engineering works.

For many years the government, and in some instances the individual states, have conducted surveys over large areas for a variety of purposes. The principal work so far accomplished consists of the fixing of national and state boundaries; the charting of coastlines and navigable streams and lakes; the precise location of definite reference points throughout the country; the collection of valuable facts concerning the earth's magnetism at widely scattered stations; the establishment and observation of a greater network of gravity stations throughout the world; the establishment and operation of tidal and water level stations; the extension of hydrographic and oceanographic charting and mapping into the approximately three-fourths of the world which is essentially unmapped; and the extension of topographic mapping of the land surfaces of the earth, for which the United States has achieved coverage of nearly one-half of its surface area to map scales as large as 1 : 24,000 or 1 in to 2000 ft.

Observations of a worldwide net of satellite triangulation stations were made during the decade 1964–1974. Results of the computations are expected to be completed by 1985. They will allow determination of the shape of the earth from one to two orders of magnitude better than has heretofore been known. Consequently, surveys of global extent have been performed and will become common in the future.

Thus surveys are divided into three classes: (1) those for the primary purpose of establishing the boundaries of land, (2) those providing information necessary for the construction of public or private works, and (3) those of large extent and high precision conducted by the government and to some extent by the states. There is no hard and fast line of demarcation between surveys of one class and those of another as regards the methods employed, results obtained, or use of the data of the survey.

1.3. The earth as a spheroid The earth has the approximate shape of an oblate spheroid of revolution, the length of its polar axis being somewhat less than that of its equatorial axis. For example, the Clarke (1866) ellipsoid has a polar (or minor) axis of 12,713,168 m and an equatorial (or major) axis of 12,756,413 m. It is the figure generally accepted in the United States and has been used in government land surveys. It is seen that the polar axis is shorter than the equatorial axis by about 27 mi, or about 43 km. Relative to the diameter of the earth this is a very small quantity, less than 0.34 percent.

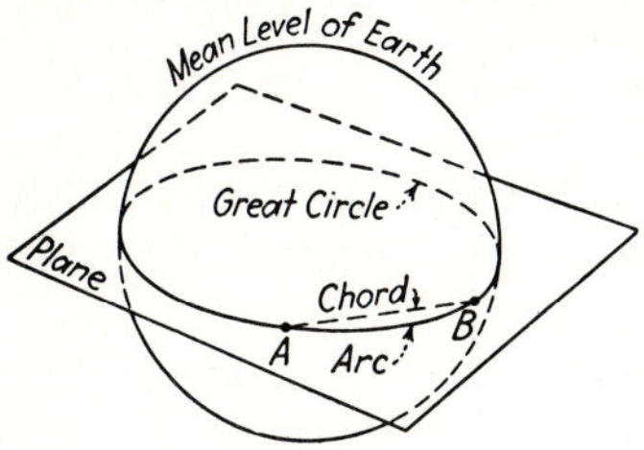

Fig. 1.1

Let us consider that the irregularities of the earth have been removed. The surface of this imaginary spheroid is a curved surface every element of which is normal to the plumb line. Such a surface is termed a *level surface*. The particular surface at the average sea level is termed *mean sea level*.

The student's understanding of the various concepts regarding the earth is greatly facilitated by assuming it to be a sphere. Imagine a plane as passing through the center of the earth, which is assumed to be spherical, as in Fig. 1.1. Its intersection with the level surface forms a continuous line around the earth. Any portion of such a line is termed a *level line*, and the circle defined by the intersection of such a plane with the mean level of the earth is termed a *great circle* of the earth. The distance between two points on the earth, as A and B (Fig. 1.1), is the length of the arc of the great circle passing through the points, and is always more than the chord intercepted by this arc. The arc is a level line; the chord is a mathematically straight line.

If a plane is passed through the poles of the earth and any other point on the earth's surface, as A (Fig. 1.2), the line defined by the intersection of the level surface and plane is called a *meridian*. Imagine two such planes as passing through two points as A and B (Fig. 1.2) on the earth, and the section between the two planes removed like the slice of an orange, as in Fig. 1.3. At the equator the two meridians are parallel; above and below the equator they converge, and the angle of convergency increases as the poles are approached. No two meridians are parallel except at the equator.

Imagine lines, normal to the meridians, drawn on the two cut surfaces of the slice. If the earth is as a perfect sphere, these lines (AO and BO, Fig. 1.3) converge at a point O at the center of the earth. Considering the lines on either or both of the cut surfaces, no two are parallel. The radial lines may be considered as vertical or plumb

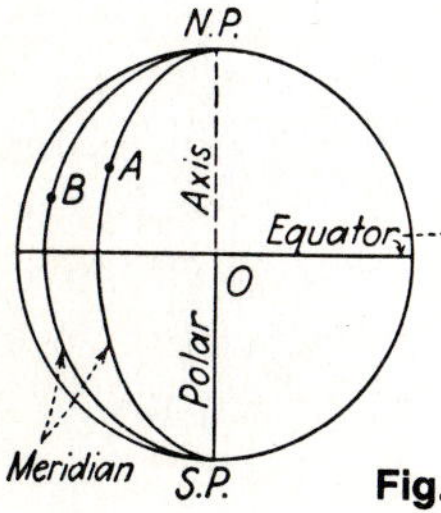

Fig. 1.2

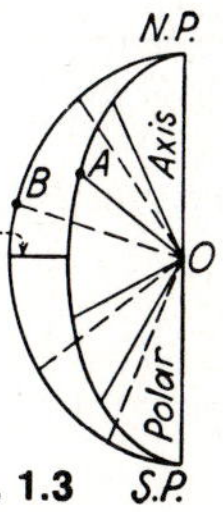

Fig. 1.3

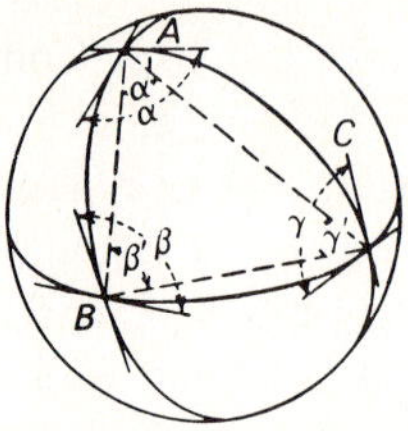

Fig. 1.4

lines, and hence we arrive at the deductions that all plumb lines converge at the earth's center and that no two are parallel. (Strictly speaking, this is not quite true, owing to the unequal distribution of the earth mass and to the fact that normals to an oblate spheroid do not all meet at a common point.)

Consider three points on the mean surface of the earth. Let us make these three points the vertices of a triangle, as in Fig. 1.4. The surface within the triangle *ABC* is a curved surface, and the lines forming its sides are arcs of great circles. The figure is a spherical triangle. In the figure the dashed lines represent the plane triangle whose vertices are points *A*, *B*, and *C*. (Actually, the "auxiliary plane triangle" has sides equal in length to the *arcs* of the corresponding spherical triangle.) Lines drawn tangent to the sides of the spherical triangle at its vertices are shown. The angles α, β, and γ of the spherical triangle are seen to be greater than the corresponding angles α', β', and γ' of the plane triangle. The amount of this excess would be small if the points were close together, and the surface forming the triangle would not depart far from a plane passing through the three points. If the points were far apart, the difference would be considerable. Evidently, the same conditions would exist for a figure of any number of sides. Hence, we see that angles on the surface of the earth are spherical angles.

In everyday life we are not concerned with these facts, principally because we are dealing with only a small portion of the earth's surface. We think of a line passing along the surface of the earth directly between two points as being a straight line, we think of plumb lines as being parallel, we think of a level surface as a flat surface, and we think of angles between lines in such a surface as being plane angles.

As to whether the surveyor must regard the earth's surface as curved or may regard it as plane (a much simpler premise) depends upon the character and magnitude of the survey and upon the precision required. In this book, consideration is given mainly to *plane surveying* for areas of limited extent.

1.4. Geodetic surveying That type of surveying which takes into account the true shape of the earth is defined as *geodetic surveying*. Surveys employing the principles of geodesy are of high precision and generally extend over large areas. Where the area involved is not great, as for a state, the required precision may be obtained by assuming that the earth is a perfect sphere. Where the area is large, as for a country, the true spheroidal shape of the earth is considered. Surveys of the latter character have been conducted primarily through the agencies of governments. In the United States such surveys have been conducted principally by the U.S. National Geodetic Survey and the U.S. Geological Survey. Geodetic surveys have also been conducted by the Great Lakes Survey, the Mississippi River Commission, several boundary commissions, and others. Surveys conducted under the assumption that the earth is a perfect sphere have been made by such large cities as Washington, Baltimore, Cincinnati, and Chicago.

Though relatively few engineers and surveyors are employed full time in geodetic work, the data of the various geodetic surveys are of great importance in that they furnish precise points of reference to which the multitude of surveys of lower precision may be tied. For each state, a system of plane coordinates has been devised, to which all points in the state can be referred without an error of more than one part in 10,000 in distance or direction arising from the difference between the reference surface and the actual mean surface of the earth.

1.5. Plane surveying That type of surveying in which the mean surface of the earth is considered as a plane, or in which its spheroidal shape is neglected, is generally defined as *plane surveying.* With regard to horizontal distances and directions, a level line is considered as mathematically straight, the direction of the plumb line is considered to be the same at all points within the limits of the survey, and all angles are considered to be plane angles.

By far the greater number of all surveys are of this type. When it is considered that the length of an arc 11.5 mi or 18.5 km long lying in the earth's surface is only 0.02 ft or 0.007 m greater than the subtended chord, and further that the difference between the sum of the angles in a plane triangle and the sum of those in a spherical triangle is only 1 second for a triangle at the earth's surface having an area of 75.5 mi^2 or 196 km^2, it will be appreciated that the shape of the earth must be taken into consideration only in surveys of precision covering large areas.

Surveys for the location and construction of highways, railroads, canals, and, in general, the surveys necessary for the works of human beings are plane surveys, as are the surveys made for the purpose of establishing boundaries, except state and national. The U.S. system of subdividing the public lands employs the methods of plane surveying but takes into account the shape of the earth in the location of certain of the primary lines of division.

The operation of determining *elevation* is usually considered as a division of plane surveying. Elevations are referred to a spheroidal surface, a tangent at any point in the surface being normal to the plumb line at that point. The curved surface of reference, usually mean sea level, is called a "datum" or, curiously and incorrectly, a "datum plane." The procedure ordinarily used in determining elevations automatically takes into account the curvature of the earth, and elevations referred to the curved surface of reference are secured without extra effort on the part of the surveyor. In fact, it would be more difficult for the surveyor to refer elevations to a true plane than to the imaginary spheroidal surface which was selected. Imagine a true plane, tangent to the surface of mean sea level at a given point. At a horizontal distance of 10 mi from the point of tangency, the vertical distance (or elevation) of the plane above the surface represented by mean sea level is 67 ft, and at a distance of 100 mi from the point of tangency the elevation of the plane is 6670 ft above mean sea level. In metric units at a horizontal distance of 1 km from the point of tangency, the vertical distance above mean sea level is 78.5 mm; at 10 km the vertical distance is 7.85 m; and at 100 km it is 785 m. Evidently, the curvature of the earth's surface is a factor that cannot be neglected in obtaining even very rough values of elevations.

For the reasons enumerated above, this book deals chiefly with the methods of plane surveying.

1.6. Operations in surveying A *control survey* consists of establishing the horizontal and vertical positions of arbitrary points. A *land*, *boundary*, or *property survey* is performed to determine the length and direction of land lines and to establish the position of these lines on the ground. A *topographic survey* is made to secure data from which may be made a *topographic map* indicating the configuration of the terrain and the location of natural and human-made objects. *Hydrographic surveying* refers to surveying bodies of water for the purposes of navigation, water supply, or subaqueous construction. *Mine surveying* utilizes the principles for control, land, geologic, and topographic surveying to control, locate, and map underground and surface works related to mining operations. *Construction surveys* are performed to lay out, locate, and monitor public and private engineering works. *Route surveying* refers to those control, topographic, and construction surveys necessary for the location and construction of lines of transportation or communication, such as highways, railroads, canals, transmission lines, and pipelines. *Photogrammetric surveys* utilize the principles of aerial and terrestrial photogrammetry, in which measurements made on photographs are used to determine the positions of photographed objects. Photogrammetric surveys are applicable in practically all the operations of surveying and in a great number of other sciences.

1.7. Summary of definitions A *level surface* is a curved surface every element of which is normal to a plumb line. Disregarding local deviations of the plumb line from the vertical, it is parallel with the mean spheroidal surface of the earth. A body of still water provides the best example.

A *horizontal plane* is a plane tangent to a level surface at a particular point.

A *horizontal line* is a line tangent to a level surface. In surveying, it is commonly understood that a horizontal line is straight.

A *horizontal angle* is an angle formed by the intersection of two lines in a horizontal plane.

A *vertical line* is a line perpendicular to the horizontal plane. A plumb line is an example.

A vertical line in the direction toward the center of the earth is said to be in the direction of the nadir. A vertical line in the direction away from the center of the earth and above the observer's head is said to be directed toward the zenith.

A *vertical plane* is a plane in which a vertical line is an element.

A *vertical angle* is an angle between two intersecting lines in a vertical plane. In surveying, it is commonly understood that one of these lines is horizontal, and a vertical angle to a point is understood to be the angle in a vertical plane between a line to that point and the horizontal plane.

A *zenith angle* is an angle between two lines in a vertical plane where it is understood that one of the lines is directed toward the zenith. A *nadir angle* is an angle between two lines in a vertical plane where it is understood that one of the lines is directed toward the nadir.

In plane surveying, distances measured along a level line are termed *horizontal distances*. The distance between two points is commonly understood to be the horizontal distance from the plumb line through one point to the plumb line through the other. Measured distances may be either horizontal or inclined, but in practically all cases the inclined distances are reduced to equivalent horizontal lengths.

The *elevation* of a point is its vertical distance above (or below) some arbitrarily assumed level surface, or datum.

A *contour* is an imaginary line of constant elevation on the ground surface. The corresponding line on the map is called a *contour line*.

The vertical distance between two points is termed the *difference in elevation*. It is the distance between an imaginary level surface containing the high point and a similar surface containing the low point. The operation of measuring difference in elevation is called *leveling*.

The *grade*, or *gradient*, of a line is its slope, or rate of ascent or descent.

1.8. Units of measurement The operations of surveying entail both angular and linear measurements.

The sexagesimal units of angular measurement are the *degree*, *minute*, and *second*. A plane angle extending completely around a point equals 360° (degrees); $1° = 60'$ (minutes); 1 minute = $60''$ (seconds). In Europe, the centesimal unit, the *grad*, or *grade*, is the angular unit; 400 grads$^{(g)}$ equals 360°; 1 grad$^{(g)}$ = 100 centesimal minutes$^{(c)}$ = 0.9°; 1 centesimal minute$^{(c)}$ = 100 centesimal seconds$^{(cc)}$ = $0°00'32.4''$. Grads are usually expressed in decimals. For example, $100^{(g)}42^{(c)}88^{(cc)}$ is expressed as 100.4288 grads.

The international unit of linear measure is the metre. Originally, the metre was defined as 1/10,000,000 of the earth's meridional quadrant. The metre as used in the United States, where it has always been employed for geodetic work, was initially based on an iron metre bar standardized at Paris in 1799. In 1866, by act of Congress, the use of metric weights and measures was legalized in the United States and English equivalents were specified. Among these equivalents, the metre was given as 39.37 in, or the ratio of the foot (12 in) to the metre was specified as 1200/3937. This ratio yields a yard (3 ft) equal to 0.914401829 m.

In 1875, a treaty was signed in Paris by representatives of seventeen nations (the United States included) to establish a permanent International Bureau of Weights and Measures. As a direct result of this treaty, the standard for linear measure was established as the International Metre, defined by two marks on a bar composed of 90 percent platinum and 10 percent iridium. The original International Metre Bar was deposited at Sèvres, near Paris, France. Two copies of this metre bar (called *prototype metres*) were sent to the United States in 1889. These bars were first acquired by the U.S. Coast and Geodetic Survey (now called the National Ocean Survey or NOS) and later were transferred to the National Bureau of Standards in

Washington, D.C. By executive order in 1893, these new standards were established as *fundamental standards* for the nation. In this executive order the foot-to-metre ratio was defined to be 1200/3937, the same as specified by Congressional action in 1866.

Also established by the Treaty of 1875 was a General Conference on Weights and Measures (Conference General de Poids et Mesures, abbreviated to CGPM) that was to meet every 6 years. In 1960 the CGPM modernized the metric system and established the *Système International d'Unités* (SI). The standard unit for linear measure, as fixed by this conference, was the metre, defined as a length equal to 1,650,763.73 wavelengths of the orange-red light produced by burning the element krypton at a specified energy level. Such a standard has the advantage of being reproducible and is immune to inadvertent or purposeful damage.

The metre is subdivided into the following units:

$$1 \text{ decimetre (dm)} = 0.1 \text{ metre (m)}$$
$$1 \text{ centimetre (cm)} = 0.01 \text{ metre}$$
$$1 \text{ millimetre (mm)} = 0.001 \text{ metre}$$
$$1 \text{ micrometre } (\mu\text{m}) = 0.001 \text{ mm} = 10^{-6} \text{ m}$$
$$1 \text{ nanometre (nm)} = 0.001\ \mu\text{m} = 10^{-9} \text{ m}$$

In the United States by executive order on July 1, 1959, the yard (3 ft) was redefined to be 0.9144 m. This order changed the foot-to-metre ratio as previously established so as to shorten the foot by 1 part per 500,000. The relationships between the English system of linear units of measure (the foot and the inch) as agreed upon in 1959 are:

1 U.S. inch = 1 British inch = 2.54 cm
1 U.S. foot = 30.48 cm

The English system was used by the United States and the United Kingdom and is now employed officially only in the United States, where conversion to the International system is slowly but surely occurring.

The difference between the U.S. *survey foot* (1200/3937) and the International foot is very small and amounts to 0.2 m/100,000 m or 0.66 ft/60 mi. When six digits or fewer are used in surveying computations, no significant difference will result. However, note that many hand-held calculators contain automatic conversion factors based on the International system. The surveyor must be aware of this possibility and recognize that when six to ten digits are used in computations, the proper conversion factor of 1200/3937 must be calculated or significant differences will occur.

In the United States, the *rod* and *Gunter's chain* were units used in land surveying. Gunter's chain is still employed in the subdivision of U.S. public lands. The Gunter's chain is 66 ft long and is divided into 100 *links* each 7.92 in in length. One mile (mi) = 80 chains = 320 *rods*, *perches*, or *poles* = 5280 ft.

The *vara* is a Spanish unit of measure used in Mexico and several other countries that were under early Spanish influence. In portions of the United States formerly belonging to Spain or Mexico, the surveyor will frequently have occasion to rerun property lines in which lengths are given in terms of the vara. Commonly, 1 vara equals 32.993 in (Mexico), 33 in (California), or $33\frac{1}{3}$ in (Texas); but other somewhat different values have been used for many surveys.

The units of area, as used in the United States, are the *square foot* (ft^2), *acre*, and *square mile* (mi^2). Formerly, the *square rod* and *square Gunter's chain* were also used. One acre = 10 square Gunter's chains = 160 square rods = 43,560 square feet. One square mile equals 640 acres.

In the metric system the unit of area is the square metre (m^2). A European metric unit of area is 1 hectare = 10,000 m^2 = 107,639.10 ft^2 = 2.471 acres. One acre equals 0.4047 hectare. The unit of volume is the *cubic metre* (m^3), where 1 cubic metre = 35.315 ft^3 = 1.308 yd^3.

The units of volume measurement in the United States have been the *cubic foot* (ft^3) and the *cubic yard* (yd^3). One cubic foot = 0.0283 m^3 and 1 cubic yard = 0.7646 m^3.

1.9. The drawings of surveying The drawings of surveying consist of maps, profiles, cross sections, and to a certain extent graphical calculations. The usefulness of these drawings is largely dependent upon the accuracy with which the points and lines are projected on paper. For the most part, few dimensions are shown, and the person who makes use of the drawings must rely on distances measured with a scale or angles measured with a protractor. Moreover, the drawings of surveying are so irregular and the data upon which the drawings are based are of such a nature that the use of conventional drafting tools for the construction of these drawings is the exception rather than the rule. The drawings may be constructed manually by the application of traditional drafting methods or may be compiled from numerical data stored in the central memory of an electronic computer, on magnetic tape, or in some other peripheral storage unit and plotted by a computer-controlled automatic flat-bed plotter.

An important concept involved in the construction of these drawings is that of *scale*. Scale is defined as the ratio of a distance on the drawing to the corresponding distance on the object. Scale can be expressed as a ratio: for example, 1 : 1000, where the units of measure are the same or by stating that 1 cm on the drawing represents 1000 cm on the object. Another way is to use different units; for example, 1 in on the drawing representing 100 ft on the object, which is equivalent to the scale ratio of 1 : 1200.

1.10. Map projections A map shows graphically the location of certain features on or near the surface of the earth. Since the surface of the earth is curved and the surface of the map is flat, no map can be made to represent a given territory without some distortion. If the area considered is small, the earth's surface may be regarded as a plane, and a map constructed by orthographic projection, as in mechanical drawing, will represent the relative locations of points without measurable distortion. The maps of plane surveying are constructed in this manner, points being plotted by rectangular coordinates or by horizontal angles and distances.

As the size of the territory increases, this method becomes inadequate, and various forms of projection are employed to minimize the effect of map distortion. Control points are plotted by spherical coordinates, or the latitude and longitude of the points. *Latitude* is the angular distance above or below the plane of the equator and *longitude* is the angular distance in the plane of the equator east or west of the Greenwich Meridian. It is also customary to show meridians and parallels of latitude on the finished map. The maps of states and countries, as well as those of some smaller areas, such as the U.S. Geological Survey topographic quadrangle maps, are constructed in this manner. Map projection, or the projection of the curved surface of the earth onto the plane of the map sheet, is usually *conformal*, which means that angles measured on the earth's surface are preserved on the map. There are several different conformal map projection methods, each suited for a particular purpose or area of specific configuration. Each method is based on minimizing one or more of the distortions inherent in map production, which include distortions in length, distortions in azimuth and angle, and distortions in area. The various methods of map projection are discussed in Chap. 11.

State plane coordinate systems have been designed for each state in the United States, whereby, even over large areas, points can be mapped accurately using a simple orthogonal coordinate system without the direct use of spherical coordinates.

1.11. Maps Maps may be divided into two classes: those that become a part of public records of land division, and those that form the basis of studies for private and public works. The best examples of the former are the plans filed as parts of deeds in the county registry of deeds in most states of the United States. Good examples of the latter are the preliminary maps along the proposed route of a highway. It is evident that any dividing line between these two classes is indistinct, since many maps might serve both purposes.

Maps that are the basis for studies may be divided into two types: (1) historical *line-drawn maps* conventionally produced by a draftsman, although at the present time an increasing proportion of such maps is being drawn on a so-called flat-bed plotter under the control of an electronic computer; and (2) *pictorial maps* produced from aerial photographs (Chap. 13).

Maps are further divided into the following two general categories: (1) *planimetric maps*, which graphically represent in plan such natural and artificial features as streams, lakes, boundaries, condition and culture of the land, and public and private works; and (2) *topographic maps*, which include not only some or all of the preceding features but also represent the relief or contour of the ground.

Maps of large areas, such as a state or country, which show the locations of cities, towns, streams, lakes, and the boundary lines of principal civil divisions are called *geographic maps*. Maps of this character, which show the general location of some kind of the works of human beings, are designated by the name of the works represented. Thus we have a *railroad map of the United States* or an *irrigation map of California*. Maps of this type, which emphasize a single topic and where the entire map is devoted to showing the geographic distribution or concentration of a specific subject, are called *thematic maps*.

Topographic maps indicate the relief of the ground in such a way that elevations may be determined by inspection. Relief is usually shown by irregular lines, called *contour lines*, drawn through points of equal elevation (Chap. 10). General topographic maps represent topographic and geographic features, public and private works, and are usually drawn to a small scale. The quadrangle maps of the U.S. Geological Survey are good examples.

Hydrographic maps show shorelines, locations and depths of soundings or lines of equal depth, bottom conditions, and sufficient planimetric or topographic features of lands adjacent to the shores to interrelate the positions of the surface with the underwater features. Charts of the U.S. National Ocean Survey are good examples of hydrographic maps. Such maps form an important part of the basic information required for the production of environmental impact statements.

Maps may be constructed using data acquired in the field by traditional field survey methods. Under these conditions, the map is usually drawn manually using traditional drafting techniques. Maps are more frequently compiled photogrammetrically using data from aerial photography or other so-called *remote sensors* supported by either aircraft or satellite. Maps of this type are constructed using a combination of manual, photographic, and automatic compilation techniques. Note that for a map compiled by photogrammetric methods, a sufficient number of ground control points of known positions must be identifiable in the photographic or sensor record to allow scaling the map and selecting the proper datum.

An increasing number of maps is also being produced from map data obtained from a wide range of sources. Thus, data from old maps, from aerial photography, and data obtained by traditional field methods can be digitized and stored in the memory of an electronic computer or on magnetic tape. Information stored electronically in this way is frequently referred to as a *data bank*. The data bank so formed is called a *digital map*. When desired, these data can be retrieved and plotted automatically on a flat-bed plotter (see Chap. 10) under the control of an electronic computer. The overall procedure involved in acquiring and processing data in this way is called *automatic mapping*.

1.12. Precision of measurements In dealing with abstract quantities, we have become accustomed to thinking in terms of exact values. The student of surveying should appreciate that the physical measurements acquired in the process of surveying are correct only within certain limits, because of errors that cannot be eliminated. The degree of precision of a given measurement depends on the methods and instruments employed and upon other conditions surrounding the survey. It is desirable that all measurements be made with high precision, but unfortunately a given increase in precision is usually accompanied by more than a directly proportionate increase in the time and effort of the surveyor. It therefore becomes the duty of the surveyor to maintain a degree of precision as high as can be justified by the purpose of the survey, but not higher. It is important, then, that the surveyor have a thorough knowledge of (1) the sources and types of errors, (2) the effect of errors upon field measurements, (3) the instruments and methods to be employed to keep the magnitude of the errors within allowable limits, and (4) the intended use of the survey data.

A complete discussion of error analysis as related to surveying procedures is presented in Chap. 2.

1.13. Principles involved in surveying Plane surveying requires a knowledge of geometry and trigonometry, and to a lesser degree of physics and astronomy, and of the behavior of random variables and the adjustment of survey data. Some knowledge of error propagation, which, in turn, is needed for design of survey procedures and selection of equipment is quite valuable.

Data processing with electronic computers, desktop electronic calculators, and pocket electronic calculators is so much a part of the recording, storage, and retrieval of survey data that familiarity with these systems is essential to the surveyor. No attempt is made to cover these subjects in this book because such tools are being changed continuously and newer models introduced. It is assumed that the student will have the opportunity to acquire some of these skills in other courses and from other textbooks.

1.14. The practice of surveying The practice of surveying is complex. No amount of theory will make a good surveyor unless the requisite skill in the art and science of measuring and in field and office procedures is obtained. The importance of the practical phases of the subject cannot be overemphasized.

In addition to those who will major in the field, surveying is frequently one of the first courses taught to students in engineering and the earth sciences. Even though students in the latter disciplines may not expect to major in surveying, they should understand that the education received in the art and science of measuring and computing, and in the practice of surveying and mapping, will contribute directly to success in other subjects, regardless of their chosen fields.

References

1. Bomford, G., *Geodesy*, 2d ed., Oxford at the Clarendon Press, London, 1962.
2. Ewing, C. E., and Mitchell, M. M., *Introduction to Geodesy*, American Elsevier Publishing Company, Inc., New York, 1970.
3. Gergen, J. G., "New Adjustment of the North American Horizontal Datum, Article 26: Interim Status Report," *ACSM Bulletin*, No. 76, February 1982.
4. Hamilton, A. C., "Surveying: 1976–2176," *Surveying and Mapping*, Vol. 37, No. 2, June 1977, pp. 119–126.
5. Moffitt, F. H., and Bouchard, H., *Surveying*, 7th ed., Harper & Row, Publishers, New York, 1981.
6. National Bureau of Standards Publication, *The International System of Units (SI)*, U.S. Government Printing Office, Washington, D.C., 1974 ed.
7. Strasser, Georg, "The Toise, the Yard, and the Metre—The Struggle for a Universal Unit of Length," *Surveying and Mapping*, Vol. 35, No. 1, March 1975, pp. 25–46.
8. Whitten, C. A., "Surveying 1776 to 1976," *Surveying and Mapping*, Vol. 37, No. 2, June 1977, pp. 109–118.
9. Whitten, C. A., "Metric Corner," *ACSM Bulletin*, No. 54, August 1976.

CHAPTER 2
Survey measurements and errors

2.1. Introduction Measurements are essential to the functions of the surveyor. The surveyor's task is to design the survey; plan its field operations; designate the amount, type, and acquisition techniques of the measurements; and then adjust and analyze these measurements to arrive at the required survey results. It is important, then, that the individual studying surveying understand the basic idea of a measurement or an observation (the two terms are used alternatively as having the same meaning).

Except for counting, measuring entails performing a physical operation which usually consists of several more elementary operations, such as preparations (either instrument setup or calibration or both), centering, pointing, matching, setting, comparing, and reading. Yet, it is the result of these physical operations that is assigned a numerical value and called "measurement." Therefore, it is important to note that a measurement is really an indirect thing, even though in some simple instances it may appear to be direct. Consider, for example, the simple task of determining the length of a line using a measuring ruler. This operation involves several steps: setting up the ruler, aligning the zero mark to the left end of the line, and observing the reading on the ruler opposite to the right end. The value of the distance as obtained by subtracting zero from the second reading (40.1 cm in Fig. 2.1) is what we call the measurement, although actually two alignments have been made. To make this point clear, visualize the case when the ruler is simply aligned next to the line and readings are taken opposite to its ends. In this case, the reading on the left end may be 11.9 and the one on the right end 52.0, with the net length measurement of $52.0 - 11.9 = 40.1$ cm.

Another more common situation regarding observations involves the determination of an angle. This raises the question of distinguishing between directions and angles. Directions are more fundamental than angles, since an angle can be derived from two directions. In the cases where both directions and angles may be used, care must be exercised in treating each group, since an angle is the difference between two directions. Similar to the case of measuring a line, an angle is directly obtained

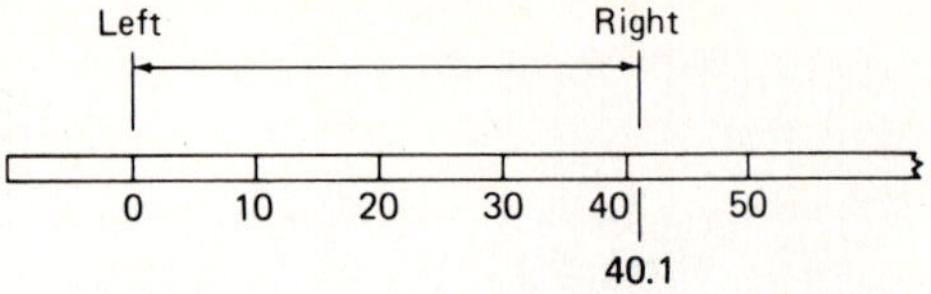

Fig. 2.1 Length of line using a tape (in centimetres).

if the first pointing (or direction) is taken specifically at zero; then the reading at the second direction will be the value of the angle and may be considered as the observation. However, if the surveying process is basically in terms of directions, the values read for these directions should then be the observations. In this case, any angle determination will be a simple linear combination (or function) of the observed directions and its properties can be evaluated from the properties of the directions and the known function.

2.2. Observations and errors Taking measurements is an operation which is subject to variations that will occur even if all the conditions remain the same during the period of repeated measurements. These variations are caused by the fact that no observation can be repeated exactly (except by sheer chance) because of instrument limitations and human weaknesses in the ability to center, point, match, and read. All these variations in the elementary operations, however small, produce corresponding variations in the resulting measurement. Therefore, an observation or a measurement is a variable, but a special kind of variable. It is called a *random variable* because there is some randomness in its value when the measurement operation is repeated under similar conditions. Consequently, it is impossible to say, with *absolute certainty*, what a value of a measurement should be. Such a value has been called *true value*, which is unattainable in practice. What we get in reality is an estimate of the true value. Some estimates may be better than others, and the surveyor should strive to obtain as good an estimate as possible at a reasonable cost and effort.

Admitting to the fact that measurement variation is a natural phenomenon, then a measurement will usually differ from its true value, whatever that true value may be. The difference between a measurement and its true value is called the measurement *error*. Thus, if x is a given measurement and x_t is the (unknown) true value, then the error e is given by

$$e = x - x_t \tag{2.1}$$

or

$$\text{error} = \text{measured value} - \text{true value}$$

Since x_t is never known, e is also unknown. What we often do know is an estimate for x_t denoted by $\hat{x}$. Therefore, an *estimate* for the error can be calculated as

$$\hat{e} = x - \hat{x} \tag{2.2}$$

It is not customary in practice to work with estimates of errors such as $\hat{e}$. Instead, over the years people have been more comfortable with the quantity having the same magnitude as $\hat{e}$ but with the opposite sign. This is as a result of the notion of a *correction* which is the quantity added to the measurement x to "improve" it and bring it to the (better) estimate of $\hat{x}$. Rearranging Eq. (2.2) yields

$$(-\hat{e}) = \hat{x} - x \tag{2.3}$$

or

correction = estimated value − measured value

The quantity $(-\hat{e})$ in Eq. (2.3) is denoted by v and is called a *residual*, representing the *variation* of the observation x with respect to its estimate $\hat{x}$. Hence Eq. (2.3) becomes

$$v = \hat{x} - x \tag{2.4}$$

or

$$\hat{x} = x + v \tag{2.5}$$

which may be stated in words as

estimated value = measured value + residual

The choice of the English word "error" may have been unfortunate in that it suggests that somehow something wrong has been committed. As noted above, errors will *always* occur no matter how carefully planned and executed the measurement operation is. The only exception to that is the so-called *gross error*, also known as a *blunder* or a *mistake*. Occurrence of blunders or mistakes does imply that something wrong has happened, as explained in Art. 2.3.

2.3. Gross errors Gross errors such as blunders or mistakes are not actually errors, because they are usually gross in magnitude compared to the other types of errors. One of the most common reasons for mistakes is simple carelessness on the part of the observer, who may take the wrong reading of a scale or a dial, or if the proper value was read, may record the wrong one by, for example, transposing numbers (e.g., reordering 14.76 as 14.67). If the operation of collecting the observations is performed through an automatic recording technique, mistakes may still occur through failure of the equipment, although they may be less frequent in this case. Another cause of blunders is failure in technique, as in the case of selecting the wrong whole degree in the measurement of angles that are very close to an integer of degrees (e.g., 71°59′58″ instead of 72°00′02″). Finally, a mistake may also occur because of misinterpretation, such as sighting to the wrong target.

Once the possibility that mistakes and blunders will occur is recognized, observational procedures and methodology must be designed to allow for their detection and elimination. Here, a variety of ways can be employed, such as taking multiple independent readings (not mere replications) and checking for reasonable consistency; careful checking of both sighting on targets and recording; using simple and quick techniques for verification, applying logic and common sense; checking and verifying the performance of equipment, particularly that with automatic readout; repeating the experiment with perhaps slightly different techniques or adopting a different datum or indexes; in the case of relatively complex models, applying simplified geometric or algebraic checks to detect the mistakes; and simply noting that most mistakes have a large magnitude, which may lead directly to their detection. For example:

1. Measuring an angle or a distance several times and computing the average. Any single measurement deviating from that average by an amount that is larger than a preset value can be assumed to contain a blunder.
2. Taking two readings on the horizontal circle of a transit differing by $180°$.
3. Using two types of units, such as metres and feet, on a level rod and converting one into the other for comparison.
4. Realizing simple facts such as that a tape is usually 100 ft long; vertical angles must lie between $\pm 90°$; and so on.
5. Making a quick check such as in a plane triangle where the sum of the interior angles is $180°$.
6. Using small portable calculators or even slide rules to check surveying computations to a low number of significant digits.

It is quite important that the surveyor should not tolerate mistakes and must safeguard against their occurrence. An observation with a mistake is not useful unless the mistake is removed; otherwise, that observation must be discarded.

After mistakes have been detected and eliminated from the measurements, the remaining errors are usually classified as *systematic* or *random*. Each type will be treated separately.

2.4. Systematic errors Systematic errors or effects occur according to a *system* which, if known, can always be expressed by functional formulation. They follow a defined pattern, and if the experiment is repeated while maintaining the same conditions, the same pattern will be duplicated and the systematic errors will reoccur. The system causing the pattern can be dependent on the observer, the instrument used, or the physical environmental conditions of the observational experiment. Any change in one or more of the elements of the system will cause a change in the character of the systematic effect if the observational process is repeated. It must be emphasized, however, that repetition of the measurement under the same conditions will not result in the elimination of the systematic error. One must find the system causing the systematic errors before they can be eliminated from the observations.

Systematic effects take on different forms depending on the value and sign of each of the effects. If the value and sign remain the same throughout the measuring process, so-called *constant error* is present. Making distance measurements with a tape that is either too short or too long by a constant value is an example of a constant error. All lengths measured by that tape will undergo the same systematic effect due to the tape alone. If the sign of the systematic effect changes, perhaps because of the personal bias of an observer, the resulting systematic errors are often called *counteracting*.

In surveying, systematic errors occur because of natural causes, instrumental factors, and the observer's human limitations. Temperature, humidity, and barometric pressure are examples of natural sources that will affect angle measurements and distance measurements either by tapes or electronic distance measuring equipment. Instrumental factors are caused by either imperfections in construction or lack of adequate adjustment of equipment before their use in data acquisition. Examples include unequal graduations on linear and circular scales; lack of centering of different components of the instrument; compromise in optical design, thus leaving certain amounts of distortions and aberrations; and physical limitations in machining parts, such as straight ways and pitch of screws.

Although automation has been considered and in some cases introduced (with its own sources of systematic effects!) to several tasks, the human observer remains an important element in the activities of surveying. The observer relies mostly on the natural senses of vision and hearing, both of which have limitations and vary because of circumstances and from one individual to another. Although some personal systematic errors are constant and some are counteracting, many others may be erratic. In a way, the set of errors committed by an observer will depend on the precise physical, psychological, and environmental conditions that exist during the particular observational experiment.

2.5. Examples of systematic errors and their compensation Examples of a variety of systematic errors abound in surveying operations. Consider the operation of taping for the determination of distances between points on the earth's surface. The length of a given tape may be physically different from the values indicated by the numbers written on its graduations, owing to some or all of the following factors:

1. Change in temperature between that used for tape standardization (calibration) and the temperature actually recorded in the field during observation [Eq. (4.6), Art. 4.18].
2. The tension or pull applied to the tape during measurement is different from that used during calibration [Eq. (4.7), Art. 4.19].
3. The method of tape support is different during measurement from that used during calibration [Eq. (4.8), Art. 4.20].
4. The end points of the distance to be measured are at different elevations and the horizontal distance is desired. In this case a correction is needed because of the fact that one would be measuring the slope distance instead of the horizontal distance [Eq. (4.4), Art. 4.14].

Electronic (and electro-optical) distance measuring techniques are also subject to a number of systematic effects whose sources and characteristics must be determined and alleviated. Of these sources the following are mentioned here: a change in the density of air through which the signal travels can cause a change in the signal frequency (due to variations in wave propagation velocity); the instrument (and sometimes the reflector or remote unit) may not be properly centered on the ends of the line to be measured; and the path of propagation of the signal may not conform to the straight-line assumption and may be deviated because of environmental and other factors.

Another surveying operation for which systematic effects must be determined is leveling, or the observation of differences in elevation between points on the earth's surface. The "level" is the instrument used for acquisition of data by placing it at one point and sighting on a graduated rod on another point. In the instrument there is a spirit bubble which must be centered to ensure that the line of sight is horizontal. But if the axis of the telescope is not parallel to the axis of the bubble, the line of sight will be inclined to the horizontal even when the bubble is centered. This error, however, may be compensated for by a relatively simple observational procedure of selecting equal backsight and foresight distances. In this fashion, the systematic effects will be counteracting and cancel each other.

In addition to the level itself, the rod may also cause systematic errors. For example, the rod may not be held vertical, or it may change in dimensions because of thermal expansion (being in the sun as opposed to being in the shade). The environment also contributes its share as a source of systematic effects. One aspect is the fact that the earth is not a plane and the line of sight is not a straight line, because of atmospheric refraction.

Another instrument that is used extensively in surveying to measure horizontal and vertical angles is the transit (see Chap. 6). The following are some of the sources of systematic effects that may be associated with such instruments:

1. Horizontal circle may be off center.
2. Graduations on either or both horizontal and vertical circles may not be uniform.
3. The horizontal axis of the telescope (about which it rotates) may not be perpendicular to the vertical axis of the instrument.
4. The longitudinal (or optical) axis of the telescope may not be normal to the horizontal axis of rotation. (In this case, the axis of the telescope would describe a cone instead of a plane as it rotates through a complete revolution.)
5. When the optical axis of the telescope is horizontal, the reading on the vertical circle is different from zero.
6. The telescope axis and the axis of the leveling bubble may not be parallel.
7. This source does not pertain to the transit but to the target on which sightings are taken. If the natural illuminating conditions are such that part of the target is in shadow, the observer will tend to center the transit's cross hair so as to bisect the illuminated portion of the target. This type of error is often referred to as *phase error*.

These are only some examples of systematic sources in surveying. Once the systematic errors are recognized they can often be accounted for, by:

1. Actual formulation and computation of corrections, which are then applied to the raw observations
2. Careful calibration and adjustment of equipment and measuring under the same conditions specified by calibration results
3. Devising observational procedures which will result in the elimination of systematic errors that would otherwise occur

2.6. Treatment of systematic errors Classically, if an "error" is removed from a measurement, the value of that measurement should improve. This idea is applicable to *systematic errors*. Thus, if x is an observation and e_s is a systematic error in that observation, then the quantity

$$x_c = x - e_s \tag{2.6}$$

is the value of the observation "corrected" for the systematic error. Again, in the classical error theory, a *correction* c_s for systematic error was taken equal in magnitude but opposite in sign to the systematic error; thus with $c_s = -e_s$ we have

$$x_c = x + c_s \tag{2.7}$$

In the reduction of survey measurements, it is imperative that the surveyor detect and correct for all possible systematic errors.

2.7. Random errors After mistakes are detected and eliminated, and all sources of systematic errors are identified and corrected, the observations will be free of any biases. However, there will still remain some variations in the measurements which result from observational errors that are random in behavior. These errors are called *random errors* and are the consequence of the fact that the observations themselves are *random variables* as introduced in Art. 2.2. In order to understand the behavior of random errors, and equivalently the observations, more must be learned about the properties of random variables. The behavior of random variables in turn depends on *probability*, which is simply the area dealing with variations due to chance. The following article is a very simple introduction. Students interested in more details are advised to consult basic books on probability and statistics.

2.8. Elementary concepts in probability Suppose that we wanted to determine, in a fair manner, who out of five persons should be selected for a particular task or chore. First write the names of the five individuals on identical pieces of paper, place them in a box, and mix them thoroughly. Next another person, different from the five under consideration, is blindfolded and draws one paper from the box. The name on this paper would be the selected person. This operation is a simple *experiment* and the name on the slip of paper is an *outcome*. The selection of that person was done at *random*, and each person of the five had an *equal chance* of being selected. In other words, the *probability* is the same for any one of them to be selected. Each person had one chance in five of being selected, or the probability is $\frac{1}{5} = 0.2$ for any one person to be selected. This number is obtained by dividing one by the total number of outcomes which are equally likely, in this case five. Another simple example involves tossing a fair die. There are six equally likely outcomes and the probability of three dots on the top face is $\frac{1}{6}$ or 0.166.

In many cases we may be interested not only in *one* outcome, but in a *set of outcomes* which are referred to as the *event*. For example in tossing a fair die, the *event* "that less than four dots appear on the top face" is of interest. Therefore, any one of the outcomes one, two, or three dots would be a *favorable* outcome. Hence there are three possible favorable outcomes out of the total of six possible outcomes leading to a probability of $\frac{3}{6} = \frac{1}{2}$. Let E be the stated event and $P(E)$ its probability;

then, in this case, $P(E) = \frac{1}{2}$. In general the probability statement is formulated as

$$P(E)\,[\text{probability of a favorable outcome}] = \frac{f\,[\text{number of favorable outcomes}]}{n\,[\text{number of all possible outcomes}]}$$

or more simply,

$$P(E) = \frac{f}{n} \tag{2.8}$$

A survey measurement, such as a distance or an angle, after mistakes are eliminated and systematic errors are corrected, is a random variable just as the number of dots on the top face of a die is. In order to understand the random nature of a survey measurable quantity, such as an angle, it may be measured repeatedly and its frequency distribution describing its random variation is then studied. This operation is described, in relatively simple terms, in Art 2.9.

2.9. Frequency distributions—histograms Consider a horizontal angle measured a large number of times (the number is essentially arbitrary), as for example $n = 177$ (Table 2.1), by a Wild T2 theodolite (an instrument used to observe directions or angles; see Chap. 6). Further, assume that all measurements are free of gross errors and corrected for all systematic errors. Table 2.1 lists the seconds of arc parts of the observed angles. A scanning of this table reveals that the range in values is between 2.9 and 13.1″. It also shows that the observations change within this range, with very few close to either the low value (in the 2 to 3″ range) or to the high value (in the 12 to 13″ range). On the other hand, there is a preponderance of values close to the 8″ of arc value.

It is rather difficult to study the *distribution* of the observed values by examining a listing such as that given in Table 2.1. Therefore, a diagram is devised which will make it easier for one to evaluate the data. In such a diagram, instead of plotting *each observed value*, a small *interval* of values is selected and the number of observations within each interval is counted. For the data in Table 2.1, an interval of 1″ is selected as an example. Starting with the interval 2.0 to 2.9″ inclusive, there is one observation (2.9″); in the interval 3.0 to 3.9″ inclusive there are three observations (3.1″, 3.2″, and 3.6″); and so on. This interval grouping is shown in Table 2.2. It shows the *frequency* of occurrence of certain value ranges (second line in Table 2.2). In order to remove the effect of actual number of observations, all the frequency numbers in the second line of Table 2.2 are divided by 177, the total number of observations. The new numbers are called *relative frequencies* and are shown in the third line of Table 2.2. The *sum* of all relative frequencies must by necessity be equal to unity.

The relative frequency for a given interval is an indication of *probability*, as implied by Eq. (2.8). For example, for the inclusive interval 6.0 to 6.9″, there are $f = 25$ observations. The corresponding relative frequency is $f/n = 25/177 = 0.14$.

With the given class intervals and corresponding relative frequencies, the diagram referred to above may be constructed (Fig. 2.2) to describe the distribution of frequencies. It is called a *frequency distribution*, or more frequently, a *histogram*. It is

Table 2.1 177 Micrometer Readings Taken with the Wild T2 Theodolite†

Reading	Value, ″ of arc	Reading	Value, ″ of arc	Reading	Value, ″ of arc	Reading	Value, ″ of arc
1	10.3	45	08.5	89	09.5	133	03.2
2	10.2	46	07.9	90	10.5	134	04.8
3	10.3	47	09.1	91	08.0	135	06.1
4	10.2	48	08.3	92	09.0	136	06.5
5	10.4	49	08.3	93	10.5	137	07.0
6	11.9	50	11.0	94	07.5	138	06.2
7	09.2	51	10.0	95	08.2	139	07.5
8	10.2	52	08.2	96	07.0	140	08.2
9	10.4	53	09.1	97	09.9	141	06.5
10	10.9	54	11.6	98	07.4	142	07.2
11	04.0	55	07.9	99	06.8	143	06.9
12	04.9	56	07.0	100	08.9	144	05.5
13	13.1	57	08.1	101	09.2	145	06.0
14	12.7	58	08.8	102	08.5	146	06.2
15	07.0	59	08.5	103	09.5	147	05.0
16	07.9	60	05.0	104	07.2	148	08.0
17	08.0	61	07.0	105	11.9	149	06.2
18	07.8	62	11.1	106	09.1	150	07.5
19	08.8	63	12.0	107	10.0	151	06.9
20	05.1	64	08.8	108	12.1	152	07.2
21	07.3	65	06.2	109	07.0	153	07.2
22	06.4	66	07.0	110	06.0	154	05.0
23	06.2	67	11.0	111	07.3	155	03.6
24	10.9	68	07.5	112	08.5	156	04.8
25	03.1	69	07.8	113	07.0	157	05.1
26	07.3	70	10.2	114	07.2	158	04.8
27	07.1	71	11.0	115	08.4	159	07.2
28	08.0	72	07.8	116	10.2	160	08.0
29	07.8	73	10.1	117	08.8	161	07.8
30	08.3	74	09.2	118	06.9	162	08.0
31	07.0	75	05.7	119	10.1	163	08.9
32	07.2	76	07.9	120	08.2	164	08.0
33	08.3	77	06.2	121	06.4	165	08.0
34	07.5	78	05.0	122	10.4	166	07.2
35	07.2	79	08.5	123	07.8	167	08.0
36	08.0	80	11.0	124	06.9	168	05.2
37	08.1	81	11.9	125	05.6	169	07.0
38	07.8	82	06.9	126	04.6	170	07.2
39	10.0	83	09.8	127	02.9	171	09.3
40	08.2	84	08.3	128	06.5	172	06.8
41	08.5	85	08.5	129	06.1	173	06.1
42	07.9	86	07.2	130	08.9	174	06.9
43	08.9	87	09.5	131	11.3	175	08.2
44	09.2	88	08.9	132	06.9	176	08.2
		Mean value = 1413.6/177 = 8.0″				177	09.1

†Degrees and minutes omitted to simplify presentation.

Table 2.2

Interval range, ″of arc	2.0–2.9	3.0–3.9	4.0–4.9	5.0–5.9	6.0–6.9	7.0–7.9	8.0–8.9	9.0–9.9	10.0–10.9	11.0–11.9	12.0–12.9	13.0–13.9	Sum
Number of observations in each interval	1	3	6	10	25	43	41	15	19	10	3	1	177
Relative frequency	0.01	0.02	0.03	0.06	0.14	0.24	0.23	0.08	0.10	0.06	0.02	0.01	1.00

drawn such that over each class interval a rectangle is constructed the area of which is equal to the corresponding frequency. Since the selected class interval in this case is a unit value, the height of each rectangle is equal to the relative frequency.

The frequency distribution shown in Fig. 2.2 is centered on or very near the value 8.0″. The highest frequencies of occurrence are at or near the central value. If all the 177 readings are added together, and the resulting sum is divided by 177, we get the mean value of 8.0″. It is interesting to note that this mean value has the highest relative frequency. More will be said about this later on.

2.10 Probability density function If the number of repeated angle measurements is increased from 177 to a very large number and the class interval is made very

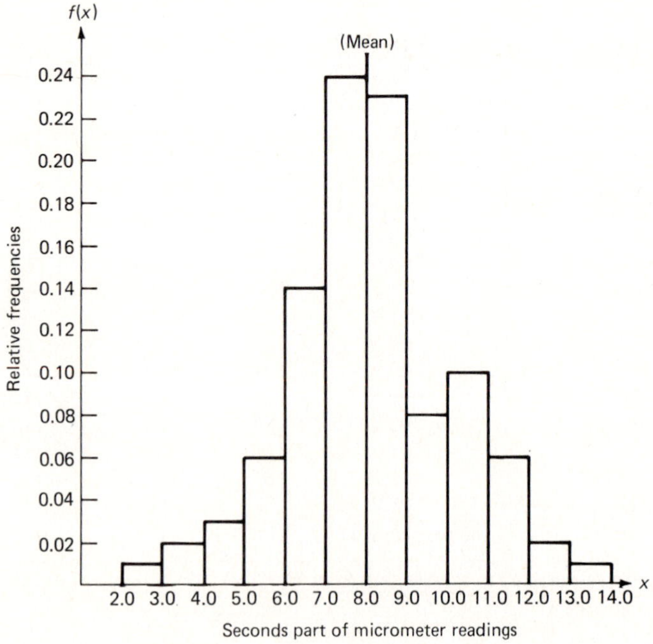

Fig. 2.2 Histogram of micrometer readings.

narrow, it will be found that each relative frequency approaches a stable limit which is the *probability*. Correspondingly, the histogram shown in Fig. 2.2 approaches the appearance of an idealized curve, called the *probability distribution curve*, as shown in Fig. 2.3*a*. In practice, histograms are frequently not used because their construction requires large numbers of observations. Furthermore, it is not convenient to use histograms to represent probabilities. Instead, the probability distribution curves are used.

The x axis in Fig. 2.3*a* represents the values that the measured angle (as a random variable) may take, while the ordinate, given the symbol $f(x)$, is called *probability density*. The curve itself is called the *probability density function*.

Since in the case of a histogram, the probability is expressed by the area of the rectangles, in Fig. 2.3*a* probability is represented by the area under the curve. For example, the probability that the measured angle falls between the two values x_1 and x_2 is given by the shaded area in Fig. 2.3*a*. The central value, μ, in this figure represents the *distribution mean value* of the angle measurements. The distribution mean, μ, designates the "true" value.

Since the sum of the areas of all the rectangles in a histogram is equal to 1, the total area under the density function curve is also equal to 1. Furthermore, the numerical value of a probability, being equal to the area of a region under the curve,

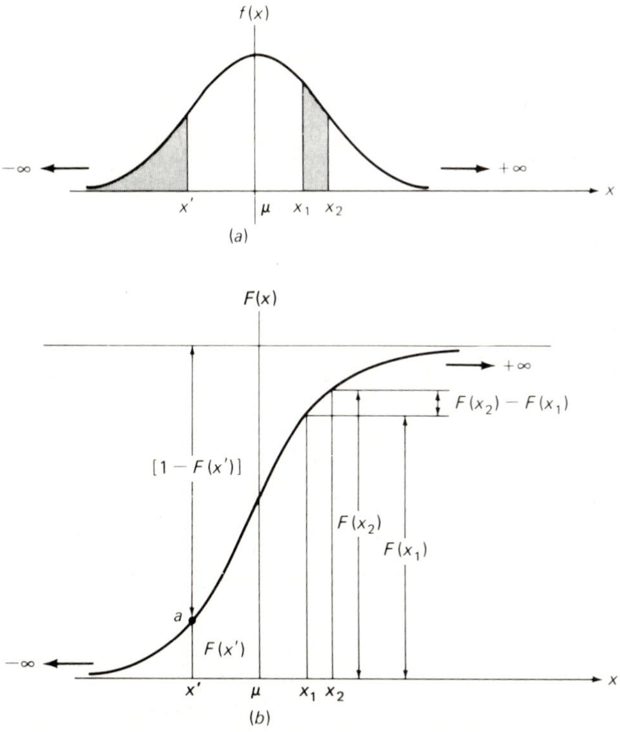

Fig. 2.3 (*a*) Density function curve. (*b*) Distribution function curve.

will have to be less than or equal to 1. More specifically, any probability P will have the necessary limits

$$0 \le P \le 1.0 \tag{2.9}$$

Further examination of Fig. 2.3*a* indicates that the density curve approaches the x axis, in either direction, as a tangent (i.e., asymptotically). Consequently, the curve will meet the x axis at $+\infty$ and $-\infty$. This indicates that the idealized density curve implies an *infinite* number of observations. Such observations are said to form the *population* which is the totality of all possible values the random variable (the angle) may take. The symbol $\tilde{x}$ is used to represent a random variable which may assume different values such as x', x_1, x_2, etc., as described subsequently.

2.11 Cumulative distribution function The density function is not the only kind of function associated with a random variable. Another function is called the *cumulative distribution function* or simply the *distribution function*. It is shown in Fig. 2.3*b* directly below the corresponding density function. The distribution function is given the symbol $F(x)$. Each value of the function gives the *probability that the random variable, $\tilde{x}$, takes a value that is less than or equal to x*, or

$$F(x) = P(\tilde{x} \le x) \tag{2.10}$$

Therefore, at the abscissa $x = x'$, the probability that $\tilde{x} < x'$ is represented by the shaded area under the density function to the left of x'. In the distribution function it is represented by the ordinate at $x = x'$.

Also, the probability that $\tilde{x} > x'$ would be represented by the area under the density function to the right of x'. Since the total area under the density function equals 1, then $P(\tilde{x} > x') = 1 - P(\tilde{x} < x')$. Therefore, on the distribution curve, the probability that $\tilde{x} > x'$ would be represented by the vertical distance from the point a on the curve (Fig. 2.3*b*) to the maximum value possible (i.e., $+1$), or $1 - F(x')$.

The probability that $\tilde{x}$ lies between x_1 and x_2 is represented by the shaded area under the density function between $x = x_1$ and $x = x_2$. Correspondingly, on the distribution function, it is the difference between the two ordinates, or $F(x_2) - F(x_1)$.

Finally, as shown in Fig. 2.3*b*, the distribution function goes to zero as x approaches $-\infty$, and to $+1$ as x approaches $+\infty$.

Summarizing, the following are the general properties of the distribution function:

$$P(x_1 < \tilde{x} \le x_2) = F(x_2) - F(x_1) \tag{2.11}$$

$$0 \le F(x) \le 1 \qquad \text{for all } x \tag{2.12}$$

$$F(x_1) \le F(x_2) \qquad \text{for } x_1 < x_2 \tag{2.13}$$

$$F(-\infty) = 0 \tag{2.14}$$

$$F(+\infty) = +1 \tag{2.15}$$

The probability distribution expresses the whole *population* and is characterized by a number of variables called *parameters*. Knowing the parameters totally specifies the distribution function. There are a number of probability functions which are useful in the reduction of survey data. The most pertinent and commonly used density function is the normal distribution density function, which is discussed in Art. 2.12.

2.12. The normal distribution The normal distribution is the most frequently used distribution in the statistical analysis of all types of data, including survey measurements. The normal density function is characterized by two *parameters*, the distribution mean, μ_x, and the standard deviation, σ_x. The significance of μ_x and σ_x is graphically depicted in Fig. 2.4. The value of $f(x)$ is a maximum at $x = \mu_x$, about which the normal curve is symmetric (Fig. 2.4*a*). The distribution function is shown in Fig. 2.4*b*.

The points of inflection of the curve lie at a distance σ_x on either side of μ_x, as shown in Fig. 2.4*a*. The mean, μ_x, is called a parameter of location; other location parameters include the median and mode. The *median* is the value along the x axis (i.e., the abscissa) at which a vertical line (or ordinate) divides the area under the

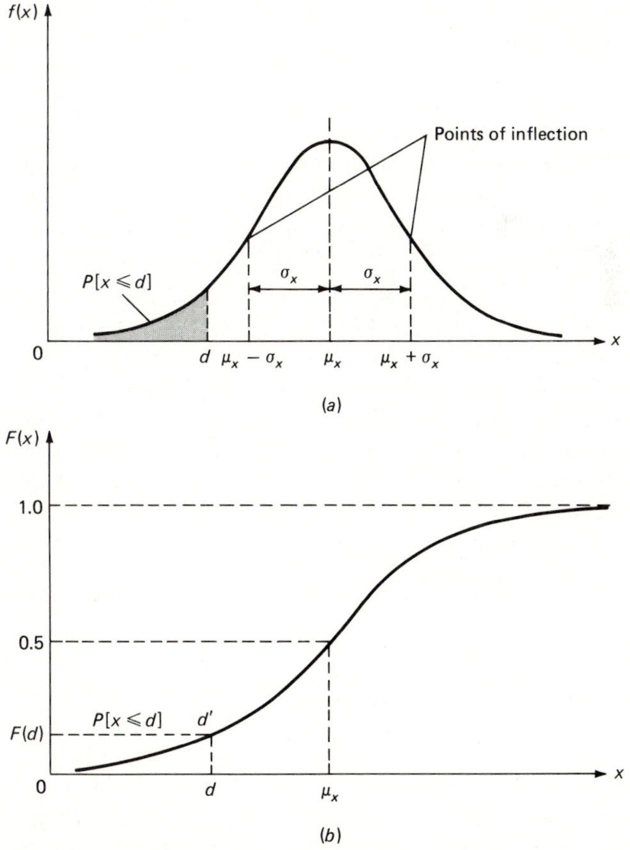

Fig. 2.4 (*a*) Density function curve. (*b*) Distribution function curve.

curve into two equal parts. The *mode* is the value of the random variable at which the density function has a relative maximum. Because of the symmetry of the normal distribution, the median and mode are equal to the distribution mean, μ_x.

The *standard deviation*, σ_x, is a measure of the spread or dispersion of the probability distribution (of the population), as indicated in Fig. 2.4*a*. It reflects the degree of variation in the measurements. Thus a set of measurements with larger variation will have a larger value of σ_x compared with another set with less variation. The square of the standard deviation, σ_x^2, is called the *variance*.

If $\tilde{x}$ is the random variable with a normal distribution, then the probability that $\tilde{x}$ is less than or equal to d, or $P(\tilde{x} \leq d)$ is represented by the shaded area in Fig. 2.4*a*, and by the ordinate $F(d)$, or the line dd', in Fig. 2.4*b*. As shown in Fig. 2.4*a*, the normal density function approaches zero as x goes to $+\infty$ and to $-\infty$.

The normal distribution is used extensively in surveying and geodesy, and in many other engineering fields. Although many of the random physical phenomena involved may not precisely obey the normal law, the normal distribution often provides a good approximation for other distributions. It also has many properties that make it easy to manipulate and apply.

2.13. Sampling The discussion so far has concentrated on probability distributions which pertain to the total *population*. To say that we *know* the distribution is taken to mean that its type (as for example normal) and all its parameters are known. Although in a few cases it is possible theoretically to construct a probability distribution, it is more often the case that one cannot be so constructed. This is the situation with survey measurements. The alternative is to make observations and calculate *estimates* for the distribution parameters. The observations obtained constitute a *sample*, which is assumed to be randomly drawn from the population. For a survey measurement the population is assumed to be the infinite number of all possible values of the measurement. On the other hand, the sample will have a specific *number* of measurements, which is referred to as the *sample size*. It is logical to see that in general the larger the sample size the better are the estimates calculated from it. These estimates of the distribution parameters are called *sample statistics*.

From the sample data, particularly from samples of larger size, frequency diagrams, such as histograms, may be constructed as was described in Art. 2.9. In addition, two other groups of sample statistics may be calculated:

1. Sample statistics for location (sample mean, median, mode, and midrange)
2. Sample statistics for dispersion (range, mean deviation, standard deviation, and variance)

These are described in the following articles.

2.14. Sample statistics for location

The sample mean The first and most commonly used measure of location is the *mean* of a sample, which is defined as

$$\bar{x} = \frac{1}{n} \sum_{i=1}^{n} x_i \tag{2.16}$$

where $\bar{x}$ is the mean, x_i are the observations, and n is the total number of observations in the sample. This is nothing more than the average value of the observations. The arithmetic mean of a set of independent observations is a good estimate of the distribution mean, μ, of the population describing the random variable for which observations are made. The sample mean of the 177 micrometer readings listed in Table 2.1 is $\bar{x} = 1413.6/177 = 8.0''$.

The sample median Another measure of location is the median, x_m, which is obtained by arranging the values in the sample in their order of magnitude. The median is the value in the middle if the number of observations n is odd, and is the mean of the middle two values if n is even. Thus, the number of observations larger than the median equals the number smaller than the median.

The sample median from Table 2.1 is $8.0''$.

The sample mode The sample mode is the value that occurs most often in the sample.

Midrange If the observation of *smallest magnitude* is subtracted from that of the *largest magnitude*, a value called the *range* is obtained. The value of the observation that is midway along the range is called the *midrange* and may be used, though infrequently, as a measure of location for a given sample. It is simply the arithmetic mean of the largest and smallest observations. The range of the 177 values in Table 2.1 is $13.1 - 2.9 = 10.2''$. The midrange is $2.9 + 10.2/2 = 8.0''$.

2.15. Sample statistics for dispersion

The range The simplest of dispersion (scatter or spread) measures is the range as defined in the preceding article. It is not, however, as useful a measure as others.

The mean (or average) deviation This is a more useful measure of dispersion and has been conventionally called the *average error*. It is the arithmetic mean of the absolute values of the deviations from any measure of position (usually the mean). Thus, the mean deviation from the mean for a sample of n observations would be given by

$$\text{mean deviation} = \frac{1}{n}\sum_{i=1}^{n} |(x_i - \bar{x})| = \frac{1}{n}\sum_{i=1}^{n} |v_i| \tag{2.17}$$

where v_i represents the deviation of observation x_i from the mean $\bar{x}$.

Sample variance and standard deviation The mean deviation, although useful in certain cases, does not reflect the dispersion or scatter of the measured values as effectively as the standard deviation. In Art. 2.12, the standard deviation was discussed first because of its geometric significance in a distribution. Then the variance was described as the square of the standard deviation. In reality, when working with numerical sample data, the sample variance, $\hat{\sigma}_x^2$, is calculated first, and then its square root is taken as the sample standard deviation. Thus, for a sample of n independent observations x_i, whose sample mean is $\bar{x}$, the sample variance is

$$\hat{\sigma}_x^2 = \frac{1}{n-1} \sum_{i=1}^{n} (x_i - \bar{x})^2 = \frac{1}{n-1} \sum_{i=1}^{n} v_i^2 \tag{2.18}$$

Thus $\hat{\sigma}_x^2$ is an unbiased estimate of σ_x^2. It is often called the estimate of the variance of any one observation. On the other hand, the estimate of the variance of the mean $\bar{x}$ is given by

$$\hat{\sigma}_{\bar{x}}^2 = \frac{\hat{\sigma}_x^2}{n} \tag{2.19}$$

Obviously, the corresponding standard deviations are $\hat{\sigma}_x$ and $\hat{\sigma}_{\bar{x}}$, respectively. For surveying measurements, the sample standard deviation $\hat{\sigma}_x$ is often called the *standard error*.

Example 2.1 Twenty measurements of an angle are given in Table 2.3. Compute the mean of the angle $\bar{x}$, the standard deviation $\hat{\sigma}_x$, and the standard deviation of the mean $\hat{\sigma}_{\bar{x}}$.

SOLUTION Applying Eq. (2.16), the sample mean is

$$\bar{x} = 31°02'\frac{532.4''}{20} = 31°02'26.6''$$

First, the deviations from the mean are calculated from

$$v_i = x_i - \bar{x}$$

For example,

$$v_1 = x_1 - \bar{x} = 31°02'29.3'' - 31°02'26.6'' = 2.7''$$

$$v_2 = x_2 - \bar{x} = 31°02'24.0'' - 31°02'26.6'' = -2.6''$$

and so forth. Then, the sample variance $\hat{\sigma}_x^2$ is equal to the sum of v_i^2 divided by $(n-1)$ or

Table 2.3

Observation number	Value of angle	Observation number	Value of angle
1	31°02′ 29.3″	11	31°02′ 24.1″
2	31°02′ 24.0″	12	31°02′ 26.2″
3	31°02′ 27.9″	13	31°02′ 30.1″
4	31°02′ 26.8″	14	31°02′ 29.7″
5	31°02′ 26.1″	15	31°02′ 24.1″
6	31°02′ 25.9″	16	31°02′ 26.2″
7	31°02′ 26.1″	17	31°02′ 27.1″
8	31°02′ 27.8″	18	31°02′ 24.9″
9	31°02′ 27.2″	19	31°02′ 25.7″
10	31°02′ 28.0″	20	31°02′ 25.2″

(20 − 1) = 19. Thus, by Eq. (2.18), the sample variance is

$$\hat{\sigma}_x^2 = \frac{61.22}{19} = 3.22 \text{ sec}^2$$

and the standard deviation of a single observation is

$$\hat{\sigma}_x = \sqrt{3.22} = 1.8''$$

Also, the standard deviation of the mean may be obtained from Eq. (2.19) as

$$\hat{\sigma}_{\bar{x}} = \frac{\hat{\sigma}_x}{\sqrt{n}} = \frac{1.8}{\sqrt{20}} = 0.4''$$

Example 2.2 Refer to the sample of 177 micrometer observations illustrated in Table 2.1. Compute the mean of observations $\bar{x}$, the estimated standard deviation $\hat{\sigma}_x$, and the estimated standard deviation of the mean $\hat{\sigma}_{\bar{x}}$.

SOLUTION By Eq. (2.16) the sample mean is

$$\bar{x} = \frac{1413.6}{177} = 8.0''$$

The variance can be obtained using Eq. (2.18):

$$\hat{\sigma}_x^2 = \frac{1}{177-1} \sum_{1}^{177} (x_i - 8.0'')^2 = 3.703 \text{ sec}^2$$

$$\hat{\sigma}_x = 1.9''$$

The standard deviation of the mean is calculated using Eq. (2.19) as:

$$\hat{\sigma}_{\bar{x}} = \frac{\hat{\sigma}_x}{\sqrt{n}} = \frac{1.9}{\sqrt{177}} = 0.14''$$

2.16 Accuracy, precision, and relative precision The term *accuracy* refers to the closeness of the mean of a sample to the population or distribution mean μ, i.e., its closeness to the true value. The same applies to a single measurement; the farther it is from the distribution mean μ, the less accurate it is. *Precision*, on the other hand, pertains to the closeness to one another of a set of repeated observations of a random variable. Thus, if such observations are clustered together, i.e., they have small deviations from their sample mean, then the observations are said to have been obtained with high precision. It should be apparent, then, that observations may be precise but not accurate if they are closely grouped together about their sample mean which is different from the population mean μ by a significant amount. Also, observations may be accurate but not precise if they are well distributed about the population mean μ but are significantly disbursed from one another. Finally, observations will be both precise and accurate if they are closely grouped around the distribution mean μ.

One of the examples used most often to demonstrate the difference between the two concepts of accuracy and precision is that of rifle shot groupings. Figure 2.5 shows three different types of groupings that it is possible to obtain. From the discussion above, group (a) is both accurate and precise, group (b) is precise but not accurate, and group (c) is accurate but not precise. One of the harder notions to

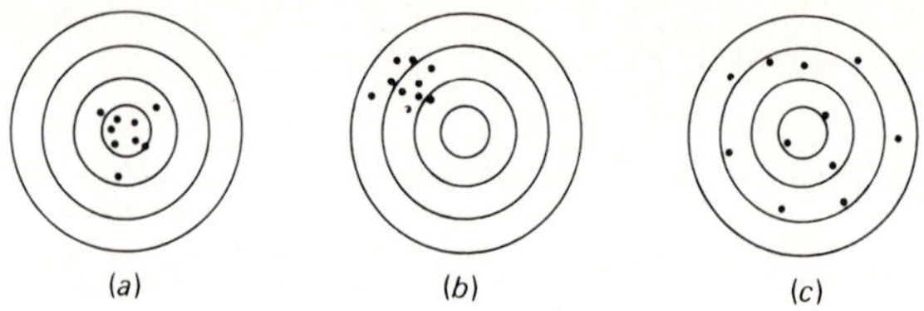

Fig. 2.5 (*a*), (*b*), and (*c*) Rifle shot groupings to illustrate accuracy and precision.

accept is that case (c) is in fact *accurate*, even though the scatter between the different shots is rather large. A justification which may help is that we can visualize that the center of mass (which is equivalent to the sample mean of the different shots) turns out to be very close to the target center (which is the true value).

Another term which is used quite frequently in conjunction with survey measurements is *relative precision*. As the term implies, it refers to a ratio (or proportion) of the precision of a given measurement and the value of the measurement itself. Thus, if d is a measured distance, and σ_d is the standard deviation of the measurement, then the relative precision is σ_d/d. Obviously, relative precision is a dimensionless quantity. It may be expressed as a percent, a fractional ratio such as $1/500$ or $1/10{,}000$, or parts per million (ppm). The last convention, ppm, is quite frequently used in conjunction with very high precision distance measurements with electro-optical distance measuring (EDM) equipment, as discussed in Art. 4.29.

Relative precision may be evaluated after a measurement and its precision are determined, or it may be given at the beginning and used to determine the required tolerance before the measurements are obtained. In the latter case, it is used to decide on both equipment and observational techniques. This will be discussed in more detail in Arts. 3.5, 3.6, and 3.7 of Chap. 3.

Precision can also be expressed by the shape (i.e., spread) of the probability (density) distribution of the measurement. A small spread indicates high precision while a large spread implies low precision, as demonstrated by the two density functions shown in Fig. 2.6. In other words, the higher the precision, the smaller is the value of the standard deviation σ or the variance σ^2, and vice versa. Since the area under the density curve is a constant value of 1, the low-precision curve shown in Fig. 2.6 peaks well below the peak of the high-precision curve, and correspondingly its point of inflection (at σ_2) is farther away from the center point (at μ).

In addition to accuracy and precision, the term *uncertainty* is another useful measure of the measurement reliability. For normally distributed observations, uncertainty is expressed in reference to the standard deviation σ of the measurement. Figure 2.7 shows that the area under the normal density curve between ordinates at $(\mu - \sigma)$ and $(\mu + \sigma)$ is 0.6826. Therefore the probability is 0.6826 that a measurement falls between $(\mu - \sigma)$ and $(\mu + \sigma)$, where μ is the distribution mean or true value. This is expressed as

$$P[(\mu - \sigma) < \tilde{x} < (\mu + \sigma)] = 0.6826 \tag{2.20}$$

Other probabilities for deviations from the mean are shown in Fig. 2.7 for multiples of σ, or

$$\begin{aligned} P[(\mu - 2\sigma) < \tilde{x} < (\mu + 2\sigma)] &= 0.954 \\ P[(\mu - 3\sigma) < \tilde{x} < (\mu + 3\sigma)] &= 0.997 \end{aligned} \tag{2.21}$$

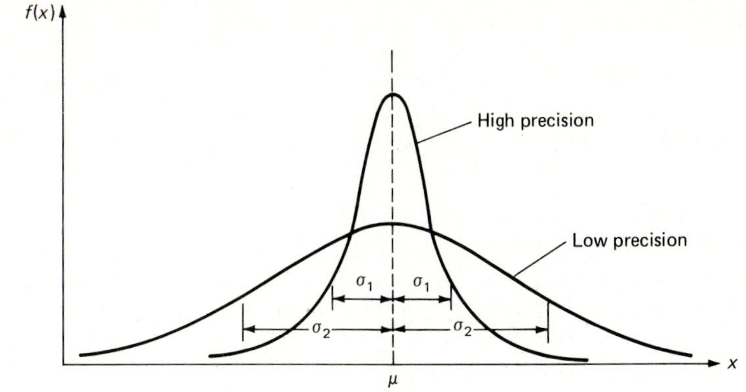

Fig. 2.6

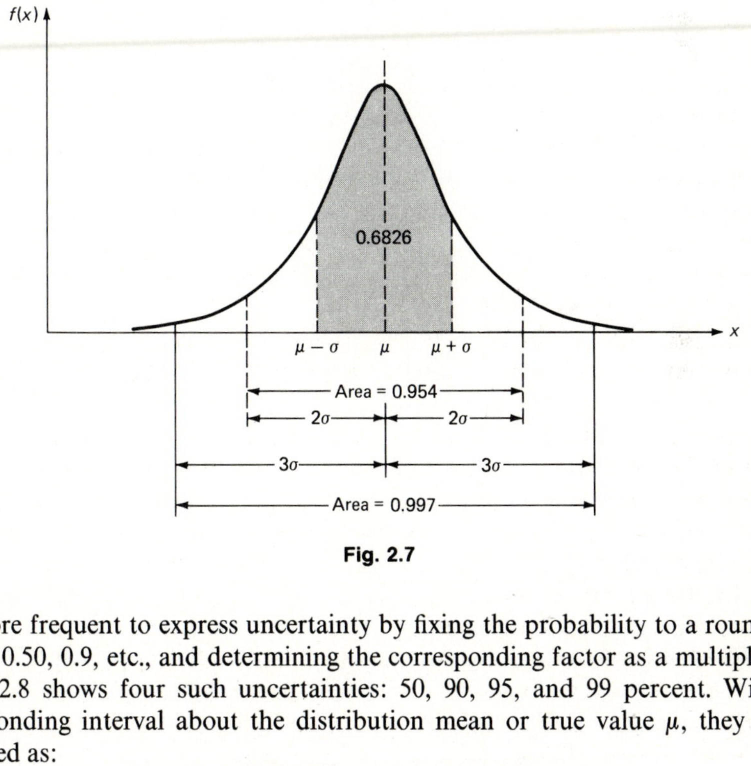

Fig. 2.7

It is more frequent to express uncertainty by fixing the probability to a round figure such as 0.50, 0.9, etc., and determining the corresponding factor as a multiplier of σ. Figure 2.8 shows four such uncertainties: 50, 90, 95, and 99 percent. With their corresponding interval about the distribution mean or true value μ, they may be expressed as:

$$
\begin{aligned}
P[(\mu - 0.674\sigma) < \tilde{x} < (\mu + 0.674\sigma)] &= 0.500 \\
P[(\mu - 1.645\sigma) < \tilde{x} < (\mu + 1.645\sigma)] &= 0.900 \\
P[(\mu - 1.960\sigma) < \tilde{x} < (\mu + 1.960\sigma)] &= 0.950 \\
P[(\mu - 2.576\sigma) < \tilde{x} < (\mu + 2.576\sigma)] &= 0.990
\end{aligned}
\qquad (2.22)
$$

The range for the *50 percent uncertainty*, between -0.674σ and $+0.674\sigma$, used to be called the *probable error*. This is now obsolete and should not be used. Instead the

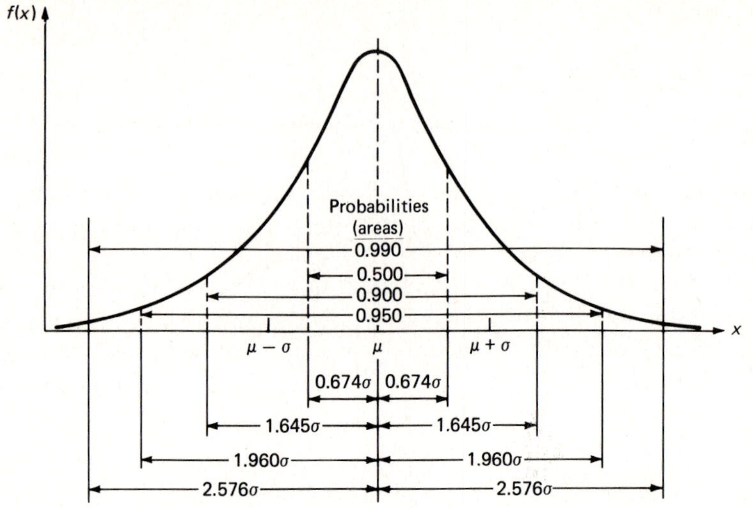

Fig. 2.8

uncertainty term should be used. As another example, the *95 percent uncertainty* is the range -1.960σ to $+1.960\sigma$. It implies that the probability is 0.95 that the measurement is within $\pm 1.960\sigma$ of its population mean or true value, μ.

Example 2.3 An angle is measured with a standard deviation $\sigma = 20''$. Calculate the 50, 90, and 95 percent uncertainty intervals.

SOLUTION The standard deviation represents the 68.26 percent uncertainty according to Eq. (2.20) and Fig. 2.7. Therefore, the interval about the population mean of the angle is smaller than $\pm\sigma$ for the 50 percent uncertainty, and larger for both the 90 and 95 percent uncertainties. By Eq. (2.22),

$$0.674\sigma = 13.5'' \qquad 1.645\sigma = 32.9'' \qquad 1.96\sigma = 39.2''$$

and then,

the 50% uncertainty interval $= \mu \pm 13.5''$
the 90% uncertainty interval $= \mu \pm 32.9''$
the 95% uncertainty interval $= \mu \pm 39.2''$

Example 2.4 A distance is measured a large number of times and its 95 percent error is calculated and found to be 0.98 cm. What is the standard deviation, or error, σ, of this distance? Also, what is the uncertainty for interval $\mu \pm 1.0$ cm for this distance?

SOLUTION The magnitude of the 95 percent error is equivalent to 1.96σ according to Eq. (2.22). Hence,

$$0.98 = 1.96\sigma \qquad \text{or} \qquad \sigma = \frac{0.98}{1.96} = 0.5 \text{ cm}$$

Next, the given interval is

$$\mu \pm 1.0\ \text{cm} = \mu \pm \frac{1.0}{0.5}\sigma = \mu \pm 2\sigma$$

and therefore by Eq. (2.21) it corresponds to an uncertainty of 95.4 percent.

2.17. Concept of weight In Art. 2.16, precision and its relationship to the standard deviation σ were described. It was noted that the *higher* the precision the *smaller* is the standard deviation σ or the variance σ^2. Since this is an inverse or opposite relationship, another *direct* relation is possible by introducing a measure called *weight*. The *higher* the value of the weight of a measurement, the *higher* is its precision, and vice versa. Consequently, the weight must be inversely related to the variance. If the weight of a single observation is denoted by w, and its variance by σ^2, then

$$w = \frac{k}{\sigma^2} \tag{2.23}$$

where k is simply a constant of proportionality. Given more than one measurement, the weights of these measurements would be proportional to the reciprocals of their variances, *only if these measurements* are assumed to be uncorrelated. The case when these measurements are correlated requires a different treatment which is considered beyond the scope of this discussion. Those interested in pursuing this further should consult Refs. 9 and 11 at the end of this chapter.

Given several *independent* measurements with standard deviations $\sigma_1, \sigma_2, \sigma_3, \ldots$, their *relative* weights are calculated as the reciprocals of their variances, or:

$$w_1 = \frac{1}{\sigma_1^2} \qquad w_2 = \frac{1}{\sigma_2^2} \qquad w_3 = \frac{1}{\sigma_3^3} \qquad \cdots \tag{2.24}$$

In this case, $k = 1$. Sometimes, it is convenient to select one measurement and assign it a unit weight. This would fix another value for k and all the other weights would be adjusted accordingly so that *their ratios remain the* same. For example, if the first measurement is given unit weight, then

$$w_1 = \frac{k}{\sigma_1^2} = 1 \qquad \text{or} \qquad k = \sigma_1^2$$

and

$$w_2 = \frac{k}{\sigma_2^2} = \frac{\sigma_1^2}{\sigma_2^2} \qquad w_3 = \frac{\sigma_1^2}{\sigma_3^2} \qquad \cdots \tag{2.25}$$

The following example demonstrates this technique.

Example 2.5 Two angle measuring instruments have standard deviations $\sigma_1 = 5'$ and $\sigma_2 = 1'$.

(*a*) Which instrument is more precise?

(*b*) Find the relative weights of an angle measured by both instruments.

SOLUTION

(*a*) The higher the precision, the smaller is the standard deviation. Therefore, instrument 2 (with $\sigma_2 = 1'$) is more precise than instrument 1 (with $\sigma_1 = 5'$).

(*b*) Weights are inversely proportional to the variances. According to Eq. (2.24)

$$w_1 = \frac{1}{\sigma_1^2} = \frac{1}{(5)^2} = \frac{1}{25} = 0.04$$

$$w_2 = \frac{1}{\sigma_2^2} = \frac{1}{(1)^2} = 1.00$$

Sometimes it is easier to avoid small numerical values for weights. Therefore, if we made $w_1 = 1$, then

$$k = \sigma_1^2 = 25$$

and according to Eq. (2.25)

$$w_2 = \frac{\sigma_1^2}{\sigma_2^2} = 25$$

It is important to note that the ratio between the weights is the same, that is,

$$\frac{0.04}{1.00} = \frac{1}{25}$$

2.18. Problems

2.1 A survey measurement, corrected for blunders and systematic errors, is considered as a random variable; so is its associated random error. Discuss the similarities and differences between the two.

2.2 Describe the three types of errors and their influence on survey measurements.

2.3 Explain several methods by which one can compensate for recognized systematic errors.

2.4 A linear distance is assumed to be normally distributed with a mean $\mu = 113.15$ m and variance $\sigma^2 = 0.20$ m^2. Sketch the distribution showing the 0.954 probability region. Give the range in distance corresponding to this probability.

2.5 Three independent distance measurements have the following standard deviations: 2 cm, 4 cm, and 6 cm. What are their relative weights?

2.6 The following are 25 independent measurements of a distance:

731.68 (m)	731.64	731.60	731.58	731.65
731.59	731.56	731.61	731.63	731.59
731.61	731.61	731.69	731.60	731.61
731.64	731.62	731.60	731.55	731.67
731.62	731.60	731.63	731.64	731.62

(*a*) Construct a histogram from these sample data.

(*b*) Calculate the sample mean, sample median, sample mode, and sample midrange.

(*c*) Calculate the sample variance and standard deviation, and the sample mean deviation.

(*d*) Calculate the variance and standard deviation of the sample mean.

2.7 An angle is measured by three different instruments with the given standard deviations: theodolite, $\sigma_t = 10''$; transit, $\sigma_r = 30''$, and a compass, $\sigma_c = 10'$.

(*a*) Rank the measurements in order of increasing precision.

(*b*) Calculate the relative weights of the measurements such that the weight of the theodolite measurement is $w_t = 1$.

(*c*) Calculate the 90 percent uncertainty for each measurement.

2.8 A line about 1000 ft long is measured with three different instruments with the given standard deviations: electro-optical measuring equipment, EDM, $\sigma_E = 0.015$ ft; a steel tape, $\sigma_t = 0.03$ ft; and a Gunter's chain, $\sigma_c = 0.05$ ft.

(*a*) Rank these measurements in order of increasing weight.

(*b*) Calculate the relative weights of the measurements such that the weight of the measurement with the steel tape, $w_t = 1$.

(*c*) Calculate the 95 percent uncertainty for each measurement.

References

1. Aguilar, A. M., "Principles of Survey Error Analysis and Adjustment," *Surveying and Mapping*, September 1973.

2. Allman, J. S., et al., "Angular Measurement in Traverses," *Australian Surveyor*, December 1973.

3. Barry, A., *Engineering Measurements*, John Wiley & Sons, Inc., New York, 1964.

4. Barry, A., *Errors in Practical Measurement in Science, Engineering, and Technology*, John Wiley & Sons, Inc., New York, 1978.

5. Cooper, M. A. R., *Fundamentals of Survey Measurements and Analysis*, Crosby Lookwood, London, 1974.

6. Cross, P. A., "The Effect of Errors in Weights," *Survey Review*, July 1972.

7. Cross, P. A., "Ellipse Problem—More Solutions," *Survey Review*, January 1972.

8. Cross, P. A., "Error Ellipse: A Simple Example," *Survey Review*, October 1971.

9. Davis, R. E., Foote, F. S., Anderson, J. M., and Mikhail, E. M., *Surveying Theory and Practice*, 6th ed., McGraw-Hill Book Co., New York, 1981.

10. Mikhail, E. M., *Observations and Least Squares*, Harper & Row Publishers, New York, 1976.

11. Mikhail, E. M., and Gracie, G., *Analysis and Adjustment of Survey Measurements*, Van Nostrand Reinhold Company, New York, 1980.

12. Moffitt, F. H., and Bouchard, H., *Surveying*, 7th ed., Harper & Row Publishers, New York, 1982.

13. Richardus, P., *Project Surveying*, North-Holland Publishing Company, Amsterdam, 1966.

14. Thompson, E. H., "A Note on Systematic Errors," *Photogrammetria*, Vol. 10, 1953–1954.

15. Uotila, U. A., "Useful Statistics for Land Surveyors," *Surveying and Mapping*, March 1973.

CHAPTER 3
Field and office work

3.1. General The nature of surveying measurements has already been indicated. Much of the field and office work involved in the acquisition and processing of measurement is performed concurrently. Field and office work for a complete survey consists of:

1. Planning and design of the survey; adoption of specifications; adoption of a map projection and coordinate system, and of a proper datum; selection of equipment and procedures
2. Care, handling, and adjustment of the instruments
3. Fixing the horizontal location of objects or points by horizontal angles and distances
4. Determining the elevations of objects or points by one of the methods of leveling
5. Recording field measurements
6. Field computations for the purpose of verifying the data
7. Office computations in which data are reduced, adjusted, and filed or stored for current utilization or for use in the near or distant future
8. The setting of points in the field to display land property location and to control construction layout (as may be necessary)
9. Performing the final *as built survey*, in which all structures built as a part of the project are located with respect to the basic control network and/or established property lines

Discussions in this chapter are restricted to general comments related to planning and design of the survey, care and adjustment of instruments, methods for recording data, computational methods, and computational aids.

3.2. Planning and design of the survey There are many types of surveys related to an almost infinite variety of projects. Thus, the planning of the survey can be discussed only in very general terms at this point. Assuming that the nature of the project is established and the results desired from the survey are known, the steps involved in planning the survey are:

1. Establishing specifications for horizontal and vertical control accuracies
2. Location and analysis of all existing control, maps, photographs, and other survey data
3. Preliminary examination of existing maps and photographs of the site in the office to locate existing control points, and to set new control points in the field

4. Selection of equipment and surveying procedures appropriate for the task
5. Selection of computational procedures and the method for presentation of the data in final form

To understand the procedure, consider an example. Assume that a private surveying firm has been engaged to prepare a topographic map of a 2000-acre (809.4-ha) tract at a scale of 1 : 1500 (see Art. 1.9 for a definition of scale) and with a contour interval of 5 ft (~ 2 m).† Control surveys are to be performed using traditional surveying methods while the major portion of the topography is to be compiled using techniques employing aerial photographs (see Art. 1.6). The tasks to be performed by the surveying firm are to select a proper coordinate grid and datum plane and to establish a horizontal and vertical control network throughout the area, including sufficient points to control aerial photographs for the photogrammetric mapping; and to map heavily wooded valleys not accessible to the photogrammetric methods. Another firm has been engaged to do the photogrammetric work.

The map is to be used by the client for presentation of plans of a proposed industrial park to the county planning commission. If plans are approved, earthwork quantities will be calculated using measurements from the map, and construction surveys will be made from control points set in the survey. Consider some of the problems involved in planning for this type of a survey.

3.3. Specifications The accuracy of the measurements should be consistent with the purpose of the survey. Each survey is a problem in itself, for which the surveyor must establish the limits of error using a knowledge of the equipment involved, the procedure to be employed, error propagation, and judgment based on practical experience. The best survey is one that provides data of adequate accuracy without wasting time or money.

The National Geodetic Survey publishes specifications for first-, second-, and third-order horizontal and vertical control surveys (Chaps. 5 and 9). These specifications provide a starting point for establishing standards on most jobs that require basic control surveys.

For the example given, the work can be divided into three categories: establishing basic control, establishing supplementary control, and performing topographic surveys.

Basic control consists of a primary network of rather widely spaced horizontal and vertical control points which will coordinate with local, state, and national control and will be used to control all other surveys throughout the entire area to be studied. Second-order specifications are generally recommended for basic control networks. Since photogrammetric surveys will follow, particular attention should be paid to establishing basic horizontal control points around the perimeter of the area to be mapped, plus several vertical control points more or less evenly distributed throughout the area of the project. Construction surveys are a definite possibility so

†A contour is the line of intersection of a level surface with the terrain surface; a contour interval is the vertical distance between two successive such level surfaces (see Art. 10.5 for more details).

that a sufficient number of basic control points should be established in the interior of the area to allow tying all subsequent construction surveys to second-order control points.

Supplementary control is set to provide additional control points for photogrammetric mapping and for topographic mapping by field methods. Third-order specifications or lower (down to a relative precision of 1 part in 3000) are usually considered adequate for supplementary horizontal control. On the example project, most of the supplementary control will be used for controlling the photogrammetric mapping, and a specification for error in position is appropriate. Thus, it could be specified that supplementary control points should be to within ± a tolerance based on the requirements of the photogrammetric mapping system. Vertical control is usually established to satisfy third-order specifications. Close cooperation is required between the surveyor and photogrammetrist in this phase of the work. Under certain circumstances, supplementary control can be established by photogrammetric methods. This could be a topic for discussion between the surveyor and the photogrammetrist.

Topographic measurements in the field are made from supplementary control points. Accuracies in these measurements are usually based on map accuracies.† For the example, since the map scale is 1 : 1500, then 0.5 mm on the map is equivalent to 0.75 m on the ground or $\frac{1}{50}$ in represents 2.5 ft. Thus, the accuracy in position for mapping is ±0.8 m or ±2.5 ft. Elevations are governed by the contour interval. National map standards for accuracy require that 90 percent of all spot elevations determined from the map be within ±one-half contour interval of the correct elevation. Experience has shown that this accuracy of contour interpolation can normally be achieved for maps if spot elevations in the field are determined to within one-tenth of the contour interval. Heights of topographic points should be located to within ±0.2 m for the example project. Since it is possible to determine elevations well within this interval by both field and photogrammetric methods, this specification presents no problem for either the field or the photogrammetric survey method.

The specifications for photogrammetric mapping are largely a function of the map scale and contour interval. As the contour interval decreases, the cost of the photogrammetric map increases at a more rapid rate. Thus, it is worth studying the desired contour interval to be sure that this interval is *really* necessary. For this example, earthwork will be calculated from measurements made on the map so that both contour interval and map scale represent optimum values that must be maintained. This being the case, aerial photography must be flown at an altitude and to specifications compatible with the stated scale and contour interval.

Unique situations on the project may require special treatment. If construction is to be concentrated in a particular area of the site, second-order control can be clustered in that region and mapping of that portion can be at a larger scale and with a smaller contour interval. For a relatively small area such as the example

† National Map Accuracy Standards require that for maps on publications scales larger than 1 : 20,000, not more than 10 percent of the points tested shall be in error by more than $\frac{1}{30}$ in (0.8 mm), measured on publication scale; and for maps on publication scales of 1 : 20,000 or smaller, $\frac{1}{50}$ in (0.5 mm).

problem of 2000 acres, mapping at two scales would probably not be practical. Such an approach could be feasible under special circumstances in larger areas. The requirements for construction surveys may be the limiting factor. When structures for the transport of water or other liquids by gravity flow are involved, first-order leveling is required. Surveys for bridges, complicated steel structures, tunnels, and areas of very high land values are a few examples where first-order control surveys may be necessary.

3.4. Existing maps and control points A thorough search should be performed to locate all survey data, existing maps, and aerial photographs of the area to be surveyed.

The National Geodetic Survey in Rockville, Maryland 20852, maintains records of all first-, second-, third-, and selected fourth-order and photogrammetrically derived control point locations in the horizontal and vertical control networks of the United States. Control point data plus descriptions will be provided for a fee when requested. The Bureau of Land Management keeps a file on property survey data related to public lands survey. State, county, city, and town engineering and surveyor's offices should also be consulted for useful survey data.

The U.S. Geological Survey is responsible for compiling maps at various scales of the United States. The U.S.G.S. *quadrangle* topographic maps at a scale of 1 : 24,000 or 1 : 25,000 in the metric series are the maps most commonly used for engineering studies. Information can be obtained from the National Cartographic Information Center (NCIC) in Reston, Virginia 22092. The National Ocean Survey compiles and sells nautical and aeronautical charts. State, county, city, and town engineering and survey offices should also be checked for maps at scales larger than the 1 : 24,000.

Aerial photography is very useful for preliminary planning purposes and can be acquired from the following government agencies: U.S. Geological Survey; U.S. Department of Agriculture, Stabilization and Conservation Service, and Soil Conservation Service; U.S. Forest Service; Bureau of Land Management; National Ocean Survey; and the Tennessee Valley Authority (TVA). Many of these agencies have local or regional offices which maintain files of photography of that particular area. State highway departments generally have photography along the right of way for highways. Private photogrammetric firms usually keep files of photography for the areas that they have mapped.

Data from all sources should be gathered and studied so as to exploit all previous survey work in the area. Preliminary studies can be made in the office using existing maps and/or aerial photographs to choose possible locations for control points. These studies are followed by field reconnaissance to locate all existing control points and to set new horizontal and vertical control points in the basic and supplementary networks to be established. At this point, close cooperation between the surveyor and photogrammetrist is absolutely essential when photogrammetric mapping is involved.

Whenever possible, the basic control network should be tied to at least two existing government monuments so as to provide a basis for performing the entire survey in a proper state or national coordinate system.

3.5. Selection of equipment and procedures The specifications for horizontal and vertical control provide an allowable tolerance for horizontal and vertical positions. Using estimates for the capabilities inherent in various types of instruments, errors in distance and direction can be estimated for a given procedure. These estimated errors are then compared with the allowable tolerance and modification in instrument or procedure can be made if necessary. Various instruments and procedures for use of these instruments are described in subsequent chapters. Considerable judgment and practical experience in surveying are required in this phase of the survey design.

3.6. Selection of computational procedures and method for data presentation Systematic procedures for gathering, filing, processing, and dissemination of the survey data should be developed. Field notes and methods for recording data are described in Arts. 3.12 to 3.14.

The prevalence of modern, efficient electronic computing equipment virtually dictates that rigorous methods and adjustment techniques be utilized in all but the smallest of surveys. Computational aids are described in Art. 3.19.

Methods for filing or storing survey information should be standardized for all jobs and ought to allow current or future retrieval without complication.

The method for presenting the data must be carefully considered. For the example case, a conventional line map with an overlay of contours would be appropriate for presentation to a planning commission.

3.7. Relation between angles and distances In Art. 3.5 reference was made to the estimated errors in distance and in direction. These errors are used to determine the uncertainty in position for a point. For example, in Fig. 3.1 a point is located by distance r and angle α. Suppose that $r = 100$ m, $\alpha = 60°00'$, and the standard deviations σ_r and σ_α are 0.5 m and $0°17' = 0.00500$ radian, respectively. The angular error produces a linear displacement at right angles to the line of (100.0)(0.0050) = 0.5 m. Since the error in the distance is also 0.5 m, the area of uncertainty for the point defined by α and r is a circle with a radius of 0.5 m and the coordinates of the point

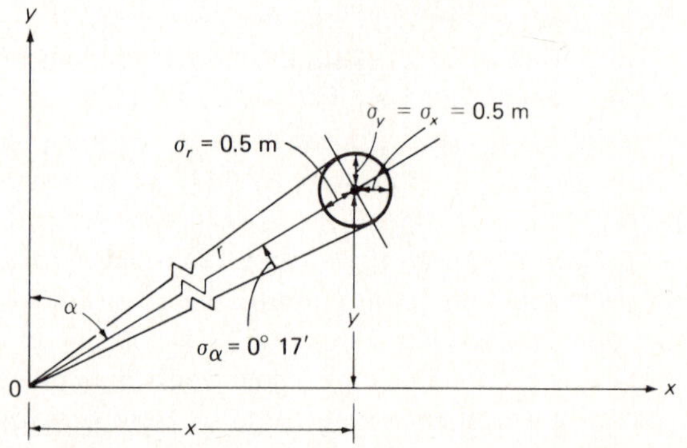

Fig. 3.1 Error ellipse for $r = 100$ m, $\alpha = 60°$, and $\sigma_r = r\sigma_\alpha$.

have standard deviations of $\sigma_x = \sigma_y = 0.5$ m as illustrated by Fig. 3.1. If the errors in distance and direction are unequal, then the area of uncertainty becomes an ellipse in which the parameters must be determined by rigorous error propagation (see App. A, Arts. A.6 and A.7).

The preceding analysis illustrates the relationship between uncertainties in direction and distance and emphasizes the desirability of maintaining consistent accuracy in the two measurements. The error in distance is normally expressed as a relative precision or ratio of the error to the distance. Thus, in the example, the relative accuracy is 0.5/100, or 1 part in 200. Similarly, the linear distance subtended by σ_α in a distance r equals 0.5 and the tangent or sine of the error or its value in radians is 1 part in 200. Accordingly, a consistent relation between accuracies in angles and distances will be maintained if the estimated standard deviation in direction equals σ_r/r radians or the relative precision in the distance.

It is impossible to maintain an exact equality between these two relative accuracies; but with some exceptions, to be considered presently, surveys should be conducted so that the difference between angular and distance accuracies is not great. Table 3.1 shows for various angular standard deviations the corresponding relative precision and the linear errors for lengths of 1000 ft and 300 m. For a length other than 1000 ft or 300 m, the linear error is in direct proportion. A convenient relation to remember is that an angular error of 01′ corresponds to a linear error of about 0.3 ft in 1000 ft or 3 cm in 100 m.

To illustrate the use of the table, suppose that distances are to be taped with a relative precision of 1/10,000. From the table the corresponding permissible angular error is 20″. As another example, suppose that the distance from the instrument to a

Table 3.1 Corresponding Angular and Linear Errors

Standard deviation in angular measurement	Linear error in:		Relative precision
	1000 ft	300 m	
10′	2.9089	0.87267	$\frac{1}{344}$
5′	1.4544	0.43633	$\frac{1}{688}$
1′	0.2909	0.08727	$\frac{1}{3440}$
30″	0.1454	0.04363	$\frac{1}{6880}$
20″	0.0970	0.02909	$\frac{1}{10,300}$
10″	0.0485	0.01454	$\frac{1}{20,600}$
5″	0.0242	0.00727	$\frac{1}{41,300}$
1″	0.004848	0.001453	$\frac{1}{206,300}$

desired point is determined as 250 m with a standard deviation of 0.8 m. For an angular error of 10′ the corresponding linear error is $(\frac{250}{300})(0.87) = 0.73$ m. Therefore, the angle needs to be determined only to the nearest 10′.

The prevalence of EDM equipment for measuring distances creates a situation where an exception occurs. Distances can be measured using EDM with very good relative precision without additional effort. For example, suppose that a distance of 3000 m is observed with an estimated standard deviation of 0.015 m, producing a relative precision of 1/200,000. The corresponding angular standard deviation is 01″. For an ordinary survey this degree of angular accuracy would be entirely unnecessary. At the other end of the spectrum, distances may be roughly determined by taping or pacing and angles measured with more than the required accuracy. For example, in rough taping, the relative accuracy in distance might be $\frac{1}{1000}$, corresponding to a standard deviation in angles of 03′. Even using an ordinary 01′ transit, angles could be observed as easily to the nearest 01′ as to the nearest 03′.

Often, field measurements are made on the basis of computations involving the trigonometric functions, and it is necessary that the computed results be of a required precision. If the values of these functions were exactly proportional to the size of the angles—in other words, if any increase in the size of an angle were accompanied by a proportional increase or decrease in the value of a function—the problem of determining the precision of angular measurements would resolve itself into that explained in the preceding section. However, the rates of change of the sines of small angles, of the cosines of angles near 90°, and the tangents and cotangents of small and large angles are relatively large. Thus, it is evident that the degree of precision with which an angle is determined should be made to depend upon the size of the angle and upon the function to be used in the computations. It is not practical to measure each angle with exactly the precision necessary to ensure sufficiently accurate computed values, but at least the surveyor should have a sufficiently comprehensive knowledge of the purpose of the survey and of the properties of the trigonometric functions to keep the angles within the required precision.

The curves of Figs. 3.2 and 3.3 show the relative precision corresponding to various standard deviations in angles from 05″ to 01′ for sines, cosines, tangents, and cotangents. For the function under consideration these curves may be used as follows:

1. To determine the relative precision corresponding to a given angle and error
2. To determine the maximum or minimum angle that for a given angular error will furnish the required relative accuracy
3. To determine the precision with which angles of a given size must be measured to maintain a required relative precision in computations

The following examples illustrate the use of the curves:

1. An angle measured with a 1′ transit is recorded as 32°00′. It is desired to know the relative precision of a computation involving the tangent of the angle if the error of the angle is 30″. In Fig. 3.3 it will be seen that the relative precision opposite the intersection of the curve $\sigma_\alpha = 30''$ and a line corresponding to 32° is $\frac{1}{3000}$.

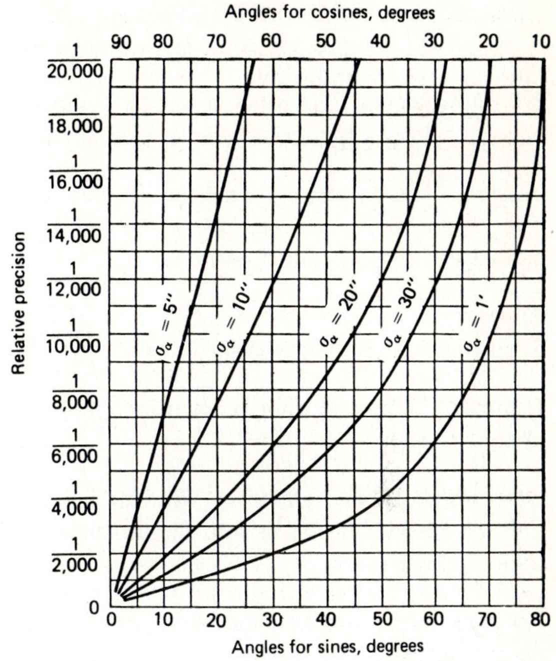

Fig. 3.2 Relative precision for sines and cosines.

2. In a triangulation system the angles can be measured with σ_α not exceeding 05″. Computations involving the use of sines must maintain a precision not lower than 1/20,000. It is desired to determine the minimum allowable angle. In Fig. 3.2 (sines) the angle corresponding to a relative precision of 1/20,000 and σ_α of 05″ is about 26°.
3. In computations involving the use of cosines a relative precision of 1/10,000 is to be maintained. It is desired to know with what precision angles must be measured. In Fig. 3.2

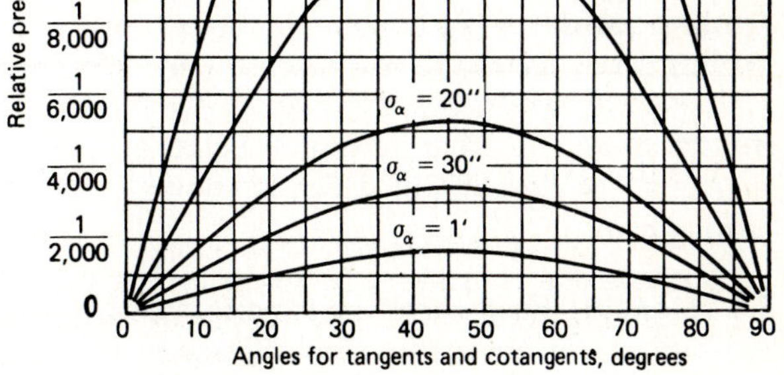

Fig. 3.3 Relative precision for tangents and cotangents.

(cosines) opposite 1/10,000 it will be seen that for angles of about 76° the σ_α cannot exceed 05″; for angles of about 64° the σ_α cannot exceed 10″; and so on.

3.8. Definitions For a better understanding of the following sections, brief definitions of a few of the terms of surveying are appropriate.

Taping The operation of measuring horizontal or inclined distances with a tape. The people who make such measurements are called "tapemen." Historically, this procedure has been called *chaining*.

Flagman A person whose duty it is to hold the range pole, at selected points, as directed by the transit operator or other person in charge.

Rodman A person whose duty it is to hold the rod and to assist the level operator or topographer.

Backsight (1) A sight taken with the level to a point of known elevation. (2) A sight or observation taken with the transit along a previous line of known direction to a reference point.

Foresight (1) A sight taken with the level to a point the elevation of which is to be determined. (2) A sight taken with the transit to a point (usually in advance), along a line whose direction is to be determined.

Grade or gradient The slope, or rate of regular ascent or descent, of a line. It is usually expressed in percent; for example, a 4 percent grade is one that rises or falls 4 ft in a horizontal distance of 100 ft or 4 m in 100 m. The term *grade* is also used to denote an established line on the profile of an existing or a proposed roadway. In such expressions as "at grade" or "to grade" it denotes the elevation of a point either on a grade line or at some established elevation as in construction work.

Hub A transit station, or point over which the transit is set, in the form of a heavy stake set nearly flush with the ground, with a tack in the top marking the point.

Line The path or route between points of control along which measurements are taken to determine distance or angle. To *give line* is to direct the placing of a range pole, pin, or other object on line.

Turning point A fixed point or object, often temporary in character, used in leveling where the rod is held first for a foresight, then for a backsight.

Bench mark A fixed reference point or object, more or less permanent in character, the elevation of which is known. A bench mark may also be used as a turning point.

3.9. Signals Field surveying is performed by *field parties* composed of from two to seven people. Except for short distances, a good system of hand signals between different members of the party makes a more efficient means of communication than is possible by word of mouth. A few of the more common hand signals are as follows.

Right or left The corresponding arm is extended in the direction of the desired movement. A long, slow, sweeping motion of the hand indicates a long movement; a short, quick motion indicates a short movement. This signal may be given by the transit operator in directing the tapeman on line, by the level operator in directing the rodman for a turning point, by the chief of the party to any member, or by one tapeman to another tapeman.

Up or down The arm is extended upward or downward, with wrist straight. When the desired movement is nearly completed, the arm is moved toward the horizontal. The signal is given by the level operator.

All right Both arms are extended horizontally and the forearms waved vertically. The signal may be given by any member of any party.

Plumb the range pole or plumb the rod The arm is held vertically and moved in the direction in which the flagpole or rod is to be plumbed. The signal is given by the transit operator or level operator.

Give a foresight The instrument operator holds one arm vertically above the head.

Establish a turning point or set a hub The instrument operator holds one arm above the head and waves it in a circle.

Turning point or bench mark In profile leveling the rodman holds the rod horizontally above the head and then brings it down on the point.

Give line The flagman holds the range pole horizontally in both hands above the head and then brings it down and turns it to a vertical position. If a hub is to be set, the range pole is waved (with one end of it on the ground) from side to side.

Wave the rod The level operator holds one arm vertically and moves it from side to side.

Pick up the instrument Both arms are extended outward and downward, then inward and upward, as they would be in grasping the legs of the tripod and shouldering the instrument. The signal is given by the chief of the party or by the head tapeman when the transit is to be moved to another point.

When distances are sufficiently long to render hand signals inadequate, the use of a citizens' band radio provides an excellent method for communication among members of a field party.

3.10. Care and handling of instruments As the use of the various surveying instruments is discussed in the following chapters, suggestions for the care and manipulation of these instruments are given.

Surveying instruments The following suggestions apply to such instruments as the theodolite, transit, level, and plane table. More detailed information is given in the references at the end of the chapter.

1. Handle the instrument with care, especially when removing it from or replacing it in its case.
2. See that it is securely fastened to the tripod head.
3. Normally, carry the instrument mounted on the tripod and over one shoulder with the tripod legs forward and held together by the hand and forearm of that shoulder.
4. Avoid carrying the instrument on the shoulder while passing through doorways or beneath low-hanging branches; carry it under the arm, with the head of the instrument in front.
5. When walking along a side hill always carry the instrument on the downhill shoulder to leave the uphill arm free to catch the body in case of tripping or stumbling and to prevent the instrument from slipping off the shoulder.
6. Before climbing over a fence or similar obstacle, place the instrument on the other side, with the tripod legs well spread.
7. Whenever the instrument is being carried or handled, the clamp-screws should be clamped very lightly so as to allow the parts to move if the instrument is struck.
8. Protect the instrument from impact and vibration.
9. If the instrument is to be shipped, pack paper, cloth, or styrofoam padding around it in the case; pack the case, well padded, in a larger box.
10. Never leave the instrument while it is set up in the street, on the sidewalk, near construction work, in fields where there are livestock, or in any other place where there is a possibility of accident.
11. Just before setting up the instrument, adjust the wing nuts controlling the friction between tripod legs and head so that each leg when placed horizontally will barely fall under its own weight.
12. Do not set the tripod legs too close together, and see that they are firmly planted. Push *along* the leg, not vertically downward. As far as possible, select solid ground for instrument stations. On soft or yielding ground, do not step near the feet of the tripod.
13. While an observation is being made, do not touch the instrument except as necessary to make a setting; and do not move about.
14. In tightening the various clamp-screws, adjusting screws, and leveling screws, bring them only to a firm bearing. *The general tendency is to tighten these screws far more than necessary*. Such a practice may strip the threads, twist off the screw, bend the connecting parts, or place undue stresses in the instrument, so that the setting may not be stable. Special care should be taken not to strain the small screws that hold the cross-hair ring.
15. For the plumb-bob string, learn to make a sliding bowknot that can be easily undone. Hard knots in the string indicate an inexperienced or careless instrument operator.
16. Before observations are begun, focus the eyepiece on the cross hairs and (by moving the eye slightly from side to side) see that no parallax is present.
17. *When the magnetic needle is not in use, see that it is raised off the pivot*. While the needle is resting on the pivot, impact is apt to blunt the point of the pivot or to chip the jewel, thus causing the needle to be sluggish.

18. Always use the sunshade. Attach or remove it by a clockwise motion, in order not to unscrew the objective.
19. If the instrument is to be returned to its box, put on the dust cap if one was supplied by the manufacturer, and wipe the instrument clean and dry.
20. Never rub the coated lenses of a telescope with the fingers or with a rough cloth. Use a camel's-hair brush to remove dust, or use clean chamois or lint-free soft cloth if the dust is caked or damp. Occasionally, the lenses may be cleaned with a mixture of equal parts of alcohol and water. Keep oil off the lenses.
21. Never touch the graduated circles, verniers, micrometer lenses, or prisms with the fingers. Do not wipe them more than necessary, and particularly do not rub the edges.
22. Do not touch the level vials or breathe on them while in use, as unequal heating of the level tube will cause the bubble to move out of its correct position.
23. In cold weather, the instrument should not be exposed to sudden changes in temperature (as by bringing it indoors); and the observer should be careful not to breathe on the eyepiece.
24. Return the instrument regularly to a manufacturer's representative or qualified instrument repair shop for cleaning, maintenance, and repair.

Suggestions regarding the adjustment of instruments are given in Art. 3.11.

Taping equipment Keep the tape straight when in use; most tapes will break when kinked and subjected to a steady pull. Steel tapes rust readily and for this reason should be wiped dry after being used.

Use special care when working near electric power lines. Fatal accidents have resulted from throwing a metallic tape over a power line.

Do not use the range pole as a bar to loosen stakes or stones, such use bends the steel point and soon renders the point unfit for lining purposes.

To avoid losing pins, tie a piece of colored cloth (preferably bright red) through the ring of each. Plastic flagging is also available in various bright colors and may be used for this purpose.

Leveling rod Do not allow the metal shoe on the foot of the rod to strike against hard objects, as this, if continued, will round off the foot of the rod and thus introduce a possible error in leveling. Keep the foot of the rod free of dirt. When not in use, long rods should be either placed upright or supported for their entire length; otherwise, they are likely to warp. When not in use, jointed rods should have all clamps loosened to allow for possible expansion of the wood.

EDM equipment EDM instruments are precision surveying devices which require the same care and handling as a theodolite or level. These instruments should always be transported in the carrying case especially designed for that purpose. All EDM devices have a power unit. Excessive shock can damage this unit. Care must be exercised when transporting the power unit. Most EDM instruments require little service other than the normal cleaning of transmitter lens and other parts. The power unit does require periodic recharging and checking to ensure that power is not lost at a critical moment. EDM instruments should be calibrated and tested periodically to verify instrument constants and check for frequency drift (Art. 4.34).

3.11. Adjustment of instruments By "adjustment" of a surveying instrument is meant the bringing of the various fixed parts into proper relation with one another, as distinguished from the ordinary operations of leveling the instrument, aligning the telescope, and so on.

The ability to perform the adjustments of the ordinary surveying instruments is an important qualification of the surveyor. Although it is a fact that the effect of instrumental errors may largely be eliminated by proper field methods, it is also true that instruments in good adjustment greatly expedite the field work. It is important that the surveyor:

1. Understand the principles upon which the adjustments are based.
2. Learn the method by which nonadjustment is discovered.
3. Know how to make the adjustments.
4. Appreciate the effect of one adjustment upon another.
5. Know the effect of each adjustment upon the use of the instrument.
6. Learn the order in which adjustments may most expeditiously be performed.

The frequency with which adjustments are required depends upon the particular adjustment, the instrument and its care, and the precision with which measurements are to be taken. Often in a good instrument, well cared for, the adjustments will be maintained with sufficient precision for ordinary surveys over a period of months or even years. On the other hand, blows that may pass unnoticed are likely to disarrange the adjustments at any time. On ordinary surveys, it is good practice to test the critical adjustments once each day, especially on long surveys where frequent checks on the accuracy of the field data are impossible. Failure to observe this simple practice sometimes results in the necessity of retracing lines, which may represent the work of several days. Testing the adjustments with reasonable frequency lends confidence to the work and is a practice to be strongly commended. The instrument operator should, if possible, make the necessary tests at a time that will not interfere with the general progress of the survey party. Some adjustments may be made with little or no loss of time during the regular progress of the work.

The adjustments are made by tightening or loosening certain screws. Usually, these screws have capstan heads which may be turned by a pin called an *adjusting pin*. Following are some general suggestions:

1. The adjusting pin should be carried in the pocket and not left in the instrument box. Disregard of this rule frequently leads to loss of valuable time.
2. The adjusting pin should fit the hole in the capstan head. If the pin is too small, the head of the screw is soon ruined.
3. Preferably make the adjustments with the instrument in the shade.
4. Before adjusting the instrument, see that no parts (including the objective) are loose. When an adjustment is completed, always check it before using the instrument.
5. When several interrelated adjustments are necessary, time will be saved by first making an approximate or rough series of adjustments and then by repeating the series to make finer adjustments. In this way, the several disarranged parts are gradually brought to their correct position. This practice does not refer to those adjustments which are in no way influenced by others.

6. Most of the more precise instruments, such as theodolites, precise levels, and EDM equipment, require more complicated tools, more stable or constant environmental conditions for the test and adjustment phase, and more specialized personnel to achieve the more refined adjustments. Consequently, such instruments are rarely adjusted in the field, but the operator does need to test the instrument periodically to determine if adjustment is necessary.

3.12. Field notes No part of the operations of surveying is of greater importance than the field notes. The competence of a surveyor is reflected with great fidelity by the character of the notes recorded in the field. These notes should constitute a permanent record of the survey with data in such form as to be interpreted with ease by anyone having a knowledge of surveying. Unfortunately, this is often not the case. Many surveyors seem to think that their work is well done if the field record, reinforced by their own memories, is sufficiently comprehensive to make the field data of immediate use for whatever purpose the survey may have. On most surveys, however, it is impossible to predict to what extent the information gathered may become of value in the remote future. Often court proceedings involve surveys made long before. Often, it is desirable to rerun, extend, or otherwise make use of surveys made years previously. In such cases it is quite likely that the old field notes will be the only visible evidence, and their value will depend largely upon the clarity and completeness with which they are recorded.

The notes consist of numerical data, explanatory notes, and sketches. Also, the record of every survey or student problem should include a title indicating the location of the survey and its nature or purpose; the equipment used, including manufacturer and number for major instruments; the date, the weather conditions; and the names and duties of the members of the party.

All field notes should be recorded *in the field book* at the time the work is being done. In the case of court proceedings involving a survey, the field notebook may become a legal document. Notes made later from memory or copied from temporary notes, may be useful, but they are not field notes and are not admissible as evidence in court. Field notes are generally recorded neatly in pencil and they should be regarded as a permanent record and not as memoranda to be used only in the immediate future.

It is not easy to take good notes. The recorder should realize that the notes may be used by persons not familiar with the locality, who must rely entirely upon what has been recorded. Not only should the notebook contain all necessary information, but the data should be recorded in a form which will allow only the correct interpretation. A good sketch will help to convey a correct impression, and sketches should be used freely. The use to be made of the notes will guide the recorder in deciding which data are necessary and which are not. To make the notes clear, recorders should put themselves in the place of one who is not in the field at the time the survey is made. Before any survey is made, the necessary data to be collected should be considered carefully, and in the field all such data should be obtained, but no more. The ABCs of field notes, where A = accurate, B = brief, and C = concise, provides an excellent, easily remembered guide for good field notes.

Although convenient forms of notes such as those shown in this book are in common use, it will generally be necessary to supplement these, and in many cases it will be necessary for surveyors to devise their own form of record. A code of symbols is desirable.

In some cases, as in locating details for mapping, the field notes may be supplemented by photographs taken with an ordinary camera. Frequently, in field completion surveys for photogrammetric mapping, an enlargement of the aerial photograph that covers the area to be mapped is annotated in the field, thus providing a record of additional information.

Other methods for recording field data are described in Art. 3.14.

3.13. Notebook In practice the field notebook should be of good-quality rag paper, with a stiff board or leather cover, made to withstand hard usage, and of pocket size. Treated papers are available which will shed rain; some of these can be written on when wet.

Special field notebooks are sold by engineering supply companies for particular kinds of notes, such as cross sections or earthwork. For general surveying or for students in field work, where the problems to be done are general in character, an excellent form of notebook has the right-hand page divided into small rectangles with a red line running up the middle, and has the left-hand page divided into six columns; both pages have the same horizontal ruling. In general, tabulated numerical values and other field-measured data are written on the left-hand page, sketches and explanatory notes on the right. This type is called a *field book* (see, e.g., Fig. 8.12). Another common form, used in leveling, has both pages ruled in columns and has wider horizontal spacing than the field book; this is called a *level book*.

The field notebook may be bound in any of three ways: conventional sewn binding, spiral ring, or loose-leaf. The ring type, which consists of many metal rings passing through perforations in the pages, is not loose-leaf; it has the advantage over the conventional binding that the book opens quite flat and that the covers can be folded back against each other.

Loose-leaf notebooks are increasing in use because of the following advantages:

1. Only one book need be carried, as in it may be inserted blank pages of various rulings, together with notes and data relating to the current field work.
2. Sheets can be withdrawn for use in the field office while the survey is being continued.
3. Carbon copies can be made in the field, for use in the field or headquarters office. Carbon copies are also a protection against loss of data. Duplicating books are available.
4. Notes of a particular survey can be filed together. Files can be made consecutive and are less bulky than for bound books.
5. The cost of binders is less than that for bound books.

The disadvantages are:

1. Sheets may be lost or misplaced.
2. Sheets may be substituted for other sheets—an undesirable practice.
3. There may be difficulty in establishing the identity of the data in court, as compared with a bound book. (This feature is important in land surveying.) When loose-leaf books are used, *each sheet* should be fully identified by date, serial number, and location.

Loose leaves are furnished in either single or double sheets. Single sheets are ruled on both sides and are used consecutively. Double sheets comprising a left-hand and a right-hand page joined together are ruled on one side only. Sheets for carbon copies need not be ruled.

3.14. Recording data A 4H pencil, well pointed, should be used. Lines made with a harder pencil are not as distinct: lines made with a softer pencil may become smeared. Office entries of reduced or corrected values should be made in red ink, to avoid confusion with the original data.

The figures used should be plain; one figure should never be written over another. In general, *numerical data should not be erased*; if a number is in error, a line should be drawn through it, and the corrected value written above.

In tabulating numbers, the recorder should place all figures of the tens column, etc., in the same vertical line. Where decimals are used, the decimal point should never be omitted. The number should always show with what degree of precision the measurement was taken; thus, a rod reading taken to the nearest 0.01 ft should be recorded not as 7.4 ft but as 7.40 ft. Recordings of values less than unity should always contain a zero to the left of the decimal point to indicate that nothing is missing. Notes should not be made to appear either more precise or less precise than they really are.

Sketches are rarely made to exact scale, but in most cases they are made approximately to scale. They are made freehand and of liberal size. The recorder should decide in advance just what the sketch is to show. A sketch crowded with unnecessary data is often confusing, even though all necessary features are included. Large detailed sketches may be made of portions having much detail. Many features may be most readily shown by conventional symbols (Art. 10.17); special symbols may be adopted for the particular organization or job.

Explanatory notes are employed to make clear what the numerical data and sketches fail to do. Usually, they are placed on the right-hand page in the same line with the numerical data that they explain. If sketches are used, the explanatory notes are placed where they will not interfere with other data and as close as possible to that which they explain.

If a page of notes is abandoned, either because it is illegible or because it contains erroneous or useless data, it should be retained and the word "void" written in large letters diagonally across the page. The page number of the continuation of the notes should be indicated.

3.15. Other media for recording data The development of compact, portable, programmable electronic calculators, solid-state memory units, and magnetic tape cassette recorders, all of which have low power requirements, has caused the recording of field data to enter a new era.

Data observed in the field can be keyed into a portable, programmable calculator, partially processed to detect blunders, and stored on magnetic tape cassettes. At the end of the day's work, the tape cassette is taken to the field office, where the data are read into a desktop computer which produces a *hard-copy* listing of the observed data. Then, using a remote terminal at the field office, the data are transmitted to a centrally located electronic computer for adjustment and analysis. This computer produces another hard-copy listing that contains all the observed data, processed results, and pertinent statistics. The data stored in the central computer can then be filed on disk or tape for use in subsequent adjustment of the complete job of which this day's work was only a portion.

Another option is available in digitized EDM instruments and theodolites (Arts. 6.31 and 9.13). In these instruments, digitized distances and/or angles are partially processed by an on-board computer and stored in a portable solid-state memory unit or magnetic tape recorder fastened to the instrument tripod. These data are then transmitted to a central computer, by way of a remote terminal, for further processing. The computer makes the necessary calculations and furnishes a listing of the observations and processed data.

In each case, the computer listing constitutes the permanent record of the day's work and would be placed in the field notebook along with sketches and other information pertinent to the survey. The legality of such a process, particularly with respect to land surveys, is by no means certain at the present time. When the inevitable disputes arise, a court decision will no doubt set a precedent.

The examples cited represent only two of many similar possible schemes using available components. Obviously, a substantial investment must be made to initiate such methods and time will be required to standardize equipment and techniques for optimum efficiency. The National Ocean Survey is currently involved in utilizing the first option discussed, for first-order leveling (see Ref. 9).

Another method for recording data consists of entering the data into computer memory by voice using a microphone connected to a microcomputer which controls the measuring system.

When large amounts of survey information are stored systematically in the solid-state memory of an electronic computer or on other peripheral devices such as magnetic tapes and disk storage, this information is frequently said to be filed in a *data bank* (see also Art. 1.11).

3.16. Computations Calculations of one kind or another form a large part of the work of surveying, and the ability to compute with speed and accuracy is an important qualification for the surveyor. To become an expert computer, one must not only have practice but must also (1) possess a knowledge of the precision of measurements and the effects of errors in the given data upon the precision of values calculated therefrom, and (2) be familiar with the algebraic and graphical processes and computational aids which help reduce the labor of computing to a minimum.

Computations are made *algebraically* by the use of simple arithmetical procedures and trigonometric functions, *graphically* by accurately scaled drawings, or *electronically* by pocket or desk calculators or high-speed electronic computers.

Before making calculations of importance, the surveyor should plan a clear and orderly arrangement, using tabular forms wherever possible. Such steps save time, prevent mistakes, make the calculations legible to others, allow proper checks, and facilitate the work of the checker. The work should not be crowded.

Particular care should be exercised in preparing data for computation by a program on an electronic computer. The presence of undetected blunders or data out of order punched by an unsuspecting keypunch operator or otherwise entered through a keyboard can lead to the wasting of inestimable amounts of valuable computer time.

All computations should be preserved in a notebook maintained for that purpose. Since a large portion of the computations will be output from an electronic computer, a loose-leaf notebook is advisable so that sheets from the computer printout can easily be included.

Office computations are generally a continuation of some field work. They should be easily accessible for future reference; for this reason, the pages should be numbered and a table of contents included. Parts of problems separated by other computations should be cross-referenced. Each problem should have a clear heading, which should include the name of the survey, the kind of computations, the field book number and page of original notes, the name of the computer, the name of the checker, and the dates of computing and checking. Usually, enough of the field notes should be transcribed to make computations possible without further reference to the field notebook. All transcripts should be checked.

3.17. Checking In practice, *no confidence is placed in results that have not been checked*, and important results are preferably checked by more than one method. Rarely is a lengthy computation made without a mistake. A good habit to develop is that of checking one's work until certain that the results are correct. These checks should be performed independently. Comparing each step with that of other students is not a true check and would not be accepted in practice.

Many problems can be solved by more than one method. Since the use of the same method in checking may cause the same error to occur, results should be checked by a different method when this is feasible. Approximate checks to discover large arithmetical mistakes may be obtained by applying approximate solutions to the problem with a pocket calculator or checking the answer by scaling from a map or drawing with protractor and scale. Graphical methods may be used as an approximate check; they take less time than arithmetical solutions, and possible incorrect assumptions in the precise solution may be detected.

Each step in a long computation that cannot be verified otherwise should be checked by repeating the computation.

When work is being checked and a difference is found, the computation should be repeated before a correction is made, since the check itself may be incorrect.

In many cases, large mistakes such as faulty placing of a decimal point, can be located by inspection of the value to see if it looks reasonable.

The checking procedures outlined in the previous paragraphs are primarily directed toward manual computations performed with a pocket or desk calculator. When a computer program, known to be operating properly, is used for the solution, checking should be concentrated on the input and output listings for and from the computer. Exhaustive checking should be applied to input data for a computer program to eliminate tabulation and punching errors. It is also good practice to list the input data with the program output to provide another verification of the information used for the solution.

3.18. Significant figures The expression *significant figures* is used to designate those digits in a number which have meaning. A significant figure can be any one of the digits 1, 2, 3, . . . , 9; and 0 is a significant figure except when used to fix a decimal point. Thus, the number 0.00456 has three significant figures and the number 45.601 has five significant figures.

In making computations you will have occasion to use numbers which are mathematically exact and numbers determined from observations (directly or indirectly) which inevitably contain observational errors. Mathematically exact numbers are absolute and have as many significant figures as are required. For example, if you multiply a given quantity by 2, the number 2 is absolute and is exact.

An observed value is *never* exact. A distance measured with a steel tape is an example of a directly observed quantity. If the distance is measured roughly, it may be recorded as 52 ft or as 16 m. If greater precision is required, the distance is observed as 52.3 ft (15.95 m) or if still more refinement is desired 52.33 ft (15.950 m) may be the value recorded. Note that none of these three numbers expresses the correct distance exactly, but that they contain two, three, four, and five significant figures, respectively. Thus, the number of significant figures in a directly observed quantity is related to the precision or refinement employed in the observation.

An indirectly measured quantity is obtained when the observed value is determined from several related observations. For example, a total distance determined by a summation of a number of directly measured quantities is an indirectly observed distance.

The number of significant figures in a directly observed quantity is evident. Thus, if you measure the width of a room using a 50-ft steel tape (graduated to hundredths of a foot), your answer can contain up to four significant figures. Assume that the width is recorded as 32.46 ft. This is an example of a directly observed measurement in which the number of significant figures is related to the precision of the equipment and the method of observation. Note that *precision* of the method of observation is a function of the *skill* and *experience of the observer*, as well as the *least count* of the instrument being used.

When numbers are determined indirectly from observed quantities, the number of significant figures is less easily determined. Suppose that the length of the same room is slightly over 50 ft and that two directly observed quantities 25.12 and 27.56 ft, yield a sum of 52.68 ft. This distance could also have been found by utilizing the full 50-ft length of tape and a short increment, resulting in measurements of 50.00 ft and 2.68 ft, to yield a total of 52.68 ft. The total distance is correctly expressed to four significant figures even though one distance contained only three significant digits. This example illustrates the fact that the number of significant figures in individual directly observed quantities does not control the precision of the quantity calculated by addition of these quantities.

Consistency in computations requires that one distinguish between exact and observed quantities and that the rules of significant figures be followed. First consider the act of rounding numbers.

Rounding of numbers Assume that 27 and 13.1 are exact numbers. The quotient of $27/13.1 = 2.061068702 \cdots$ never terminates. To use this quotient, superfluous digits are removed from the right and the quotient is retained as 2.06 or 2.0611. This process is called *rounding off*.

To round a number, retain a certain number of digits counted from the left and drop the others. If it is desired to round π to three, four, and five significant figures, the results are 3.14, 3.142, and 3.1416, respectively. Rounding is performed to cause the least possible error and should be done according to the following rule:

To round a number to n significant figures, discard all digits to the right of the nth place. If the discarded digit in the $(n+1)$st place is less than 5, leave the nth digit unchanged; if the discarded digit is greater than 5, add one to the nth digit. If the discarded digit is 5, leave the nth digit unaltered if it is even and increase the nth digit by one if it is odd.

A number rounded according to this rule is said to be correct to n significant figures. Examples of numbers rounded to four significant figures are:

31.68234	becomes	31.68
45.6874	becomes	45.69
2.3465	becomes	2.346
10.475	becomes	10.48

Rules of significant numbers applied to arithmetic operations

ADDITION

Find the sum of approximate numbers 561.32, 491.6, 86.954, and 3.9462, where each number is correct to its last digit.

These four numbers contain from one to four significant decimal places. The least number of decimals is the controlling factor. The practice is to carry one or more decimal place than occurs in the number with the least number of decimals. Thus in the example, round all the numbers to two decimals and add:

$$\begin{array}{r} 561.32 \\ 491.6 \\ 86.95 \\ 3.95 \\ \hline 1143.82 \end{array} \quad \text{or} \quad 1143.8$$

Retention of the extra significance in the more accurate numbers eliminates errors inherent in these numbers and reduces the total error in the sum. Round the final answer to tenths or in this case to the least number of decimals in the original data.

The average of a set of measurements provides a more reliable result than any of the individual measurements from which the average is computed. Retention of one more figure in the average than in the numbers themselves is justified.

SUBTRACTION

To subtract one approximate number from another, first round each number to the same decimal place before subtracting. Consider the difference between 821.8 and 10.464:

$$821.8 - 10.5 = 811.3$$

Errors resulting from subtraction of approximate numbers are most significant and serious when the numbers are very nearly equal and the leading significant figures are lost. Such errors in subtraction of nearly equal figures can cause computations to be worthless.

MULTIPLICATION

Round the numbers with the most significant figures to one more significant figure than the number with the least significant figures. Answer should be given to the same number of significant figures as are found in the number with the least significant figures. For example:

$$(349.1)(863.4) = 301{,}412.94$$

which should be rounded to

$$301{,}400 = 3.014 \times 10^5$$

which has four significant figures.

DIVISION

Same rules as for multiplication. For example, assume that 5 is an exact number. Then

$$\frac{56.3}{\sqrt{5}} = \frac{56.3}{2.236}$$

where $\sqrt{5} = 2.236$ is rounded to one more significant figure than the numerator and the quotient, $56.3/2.236 = 25.2$, is rounded to the same number of significant figures as are found in the least significant number in the calculation.

POWERS (see Scarborough, Ref. 8)

If k is the value of the first significant figure of a number having n significant digits, its pth power is correct to

$n - 1$	significant figures	if $p \le k$
$n - 2$	significant figures	if $p \le 10k$

For example, raise 0.3862 to the fourth power. Thus, $k = 3$, $n = 4$, and $p = 4$, so that $p < 10k$ or $p < 30$ and $(0.3862)^4 = 0.02225$ should be rounded to $n - 2$ significant figures, or to 0.022.

ROOTS (see Scarborough, Ref. 8)

The rth root of a number having n significant digits and in which k is the first significant digit is correct to

n	significant figures	if $rk \geq 10$
$n-1$	significant figures	if $rk < 10$

As examples, consider roots of the following numbers where the given numbers are correct to the last digit:

$$(25)^{1/2} = 5 \qquad \text{where } r = 2, k = 2, n = 2, \text{ and } rk = 4 < 10$$
$$(615)^{1/4} = 4.98 \qquad \text{where } r = 4, k = 6, n = 3, \text{ and } rk = 24 > 10$$
$$(32{,}768)^{1/5} = 8.0000 \qquad \text{where } r = 5, k = 3, n = 5, \text{ and } rk = 15 > 10$$

3.19. Computational aids Electronic calculators and high-speed electronic computers are the major computational aids available to the surveyor and students of surveying. The various types of electronic devices listed according to increasing speed, capacity, and memory are: pocket calculators, including those which are programmable; programmable desktop calculators; microcomputers; minicomputers; and large-capacity high-speed main-frame computers.

Pocket calculators burst upon the market in the early 1970s and have virtually eliminated mechanical desk calculators, slide rules, and logarithms from surveying schools, offices, and field operations. The simplest of these devices allows addition, subtraction, multiplication, and division. Other models permit taking roots, powers, exponential functions, trigonometric functions, fixed- and floating-point arithmetic, and regression analysis. Many of these latter types have numerous storage registers, so that matrices of a limited size can be manipulated. Programmable pocket calculators are available in various models which can be programmed according to the following methods: internally by keying the program in from the keyboard, by means of programmed magnetic cards, and by plugging in preprogrammed solid-state modules for special routines. Many of these models have storage capacity of from 10 words to 1k words ($k = 2^{10} = 1024$) and several can be used with portable printing and plotting devices so that hard-copy output is available. One of the more recent models has input/output ports which permit an interface with a micro- or minicomputer. All pocket calculators can be run on rechargeable batteries or may be plugged into ac current. These moderately priced calculators are extremely powerful tools for students and the practicing surveyor in the field and office.

Microcomputers vary in size from the so-called *pocket* microcomputer with 1.9k words of memory, a full alphanumeric keyboard, and hard-copy output, to full-size desktop models with video display units. Desktop microcomputers have random access memory (RAM) capacity of from 1.9k to 256k 8-bit words† (some systems use a 16-bit word). Instructions and data can be input by keyboard, floppy disks, hard

†A bit is a binary digit, i.e., 1 or 0. An 8-bit word contains a decimal number with three digits; a 16-bit word has a decimal number with five digits. A word represents an addressable storage location in memory. Word lengths vary from 4 bits in the smaller microcomputers to 64 bits in some large main-frame computers.

disks (disks in a sealed unit sometimes called *Winchester disks*), and magnetic tape cassettes. Output is on a printer, disks, or cassettes. Programming can be in Basic, Fortran, or other languages. Programs are available for many routine surveying calculations such as for horizontal curves, earthwork computations, traverse, triangulation, and many others.

Minicomputers are high-speed machines of moderate size (about the size of a five-drawer file cabinet) which have RAMs of from $32k$ to $1000k$ 16-bit words. There are also the so-called "super minicomputers" with *virtual* (almost unlimited) RAMs and a 32-bit word length. Most minicomputers accept input from punched cards, paper tape, a terminal keyboard, and magnetic tape. Output is by line printer, magnetic tape, or punched cards. With the proper type of peripheral hardware, such as disk files and magnetic tape units, these machines can do jobs that formerly required large-capacity computers. These systems operate at much higher speeds than microcomputers, can be used by several people simultaneously, are permanently installed, and are suitable for organizations of moderate size which do not need a large computer but want the independence of having their own machine.

Large-capacity, main-frame computers have RAMs of at least 2×10^6 words and are equipped with additional disk and/or magnetic tape storage. Numerical calculations are performed at extremely high speeds using word lengths of from 32 to 64 bits. These systems are designed for very large numerical calculations and in particular are suitable for servicing a large number of remote terminals.

Remote terminals are input/output stations which allow communication with a large central computer by way of a special cable or a telephone line. These terminals are usually equipped with a teletypewriter, a punched-card reader, and a line printer. Instructions and data can be typed on the teletypewriter or may be read in the form of punched cards by the card reader. When suitable instructions are typed on the teletypewriter, the program and data are processed by the central computer. Answers are printed by the line printer in the remote terminal or, on request, may be output at the computer center on the line printer or on punched cards.

Interactive systems consist of remote terminals equipped with a keyboard for input and a cathode ray tube (CRT) for observing in real time the manipulation of the program and data. Thus, the microcomputers described previously are self-contained interactive systems. One way in which a remote terminal can be operated interactively is to store an operating program on disk at the central computer center. Then when desired, data are input at the remote terminal using the keyboard. Also entered are the instructions requesting use of the desired program, which is stored on disk. Data and the program can be displayed on the CRT for checking and solution of the problem. When computations are completed, the answers may be displayed on the CRT or printed on the line printer. If changes are needed during the solution, they can be made immediately by entering the appropriate instructions at the keyboard. Data can be displayed on the CRT in numerical, alphanumeric, or graphical form. The potential for such systems in surveying and mapping is enormous.

The trend toward large centralized systems with many remote terminals was especially strong during the late 1960s. The emergence of high-speed minicomputers with very adequate core storage and well-designed peripheral hardware, all at a fairly reasonable cost, has reversed this trend at the time of this writing. In large government agencies for surveying and mapping, minicomputers are interfaced with a large central computer in a distributive array manner. At the time of this writing, the rapid development and intense marketing of microcomputers is now causing further confusion among users as to what is the best computational system.

Technological advances in the computer industry have been so rapid during the past few years that it is very difficult for the practitioner to stay abreast of the most recent developments. Rational decisions as to what constitutes the best system for a particular organization are difficult even for those well acquainted with computer equipment and operations. Members of the surveying profession must utilize the best in computational aids, since computations are their business. To do this, they need to continually familiarize themselves with the various options and become well-enough acquainted with the technicalities of different systems to allow optimum use of the best facilities available.

3.20. Problems

3.1 A point is to be established on the ground at a distance of about 200 m from a given point by means of one linear and one angular measurement. It is desired to establish the point within 3 cm of its true location. With what precision need the angle be measured?

3.2 An angle of 40° is measured with a transit having a vernier with a least count of 30″. The maximum error is half the least count of the vernier. What is the ratio of precision if the sine of the angle is to be used in computations? The cosine? The tangent?

3.3 If tangents or cotangents are involved, what is the highest precision corresponding to single measurements (of angles of any size) with the transit of the preceding problem?

3.4 What is the minimum allowable angle that for computations involving sines will permit a ratio of precision of 1/10,000 to be maintained if the angular error is 10″?

3.5 Round off the following numbers to four significant figures: 72.8432; 10.285; 0.0082495; 93,821; 521,242; 0.0800189; 6490.8.

3.6 The following distances are added: 167.8084 ft, 228.01 ft, 73.133 m, and 167.8084 ft. What is the sum of the distances? (1 m = 3.280833 ft.)

3.7 Perform the following subtractions: 779.9 − 36.543; 65.3956 − 65.214; 18,230 − 18,228.

3.8 The hypotenuse of a right triangle is 384.79 m long and makes an angle of 25°11′ 30″ with the long leg of the triangle. Calculate, to five significant figures, the lengths of both legs of the triangle.

3.9 In a 30°-60°-90° triangle, the long leg is 100.0 m in length. Calculate the length of the hypotenuse of this triangle.

3.10 In an oblique triangle, the two sides and the included angle are 100 m, 130 m, and 120°, respectively. Calculate the third side of the triangle.

3.11 Calculate the following values correct to the proper number of significant figures: $\sqrt{12} - 4$; $\sqrt{324}$; $\sqrt{1359}$; $(12.5)^2$; $(2.845)^3$; $(225)^2$; $(34.662)^{1/2}$; $(625)^{1/4}$.

3.12 How many significant figures are there in each of the following numbers?

(*a*) 0.208 (*d*) 30.0×10^6 (*g*) 28.46 ± 0.08

(*b*) 76248.0 (*e*) 0.006

(*c*) 1.4800 (*f*) 1.04 ± 0.02

3.13 Find the product of each of the following series of numbers, to the proper number of significant figures:

(*a*) $2.64 \times 79.18 \times 0.0767 \times 1.0028$

(*b*) $(3.415)^2 \times \sqrt[3]{72.849}$

(*c*) $(6.618)^{1/3} \times (68.627)^{1/2}$

3.14 Find the quotient for each of the following series of numbers, to the proper number of significant figures:

(*a*) 76.21 ÷ 49.20
(*b*) 0.4092 ÷ 0.006314
(*c*) 1.4823 ÷ 16.4917
(*d*) 8.472 ÷ tan 42°21′

3.15 How many feet and how many metres are there in $3\frac{1}{2}$ rods? In 6 chains and 8 links?

3.16 How many acres are there in a rectangular field 18.25 rods wide and 920.0 ft long? A rectangular field 100.55 by 424.8 m? How many hectares are in each field?

3.17 If the angles of an approximately equilateral triangle are measured with a standard deviation of 30″ and the distances with corresponding precision, how many significant figures will there be in the area computed by sines and cosines?

References

1. Beek, B., Broglie, J., and Vonusa, R., "Voice Data Entry for Cartographic Applications," *Proceedings, Fall Convention, ACSM,* 1977, p. 161.
2. Chapman, W. H., "Programmable Desk Calculators—Their Use in Topographic Mapping," *Proceedings, Fall Convention, ACSM,* 1972, p. 238.
3. Kimberly, P. C., *Micro-Processors; An Introduction,* McGraw-Hill Book Co., New York, 1981.
4. Mayfield, G. M., "Engineering Surveying, Chapters 2, 5, 6, 8, 16, and 20," *Journal of the Surveying and Mapping Division, ASCE,* Vol. 102, No. SU1, Proceedings Paper 12648, December 7, 1976, pp. 59–94.
5. Meade, B. K., "Solution of Surveying Problems by Mini-Programmable Computers," *Proceedings, 38th Annual Meeting, ACSM,* 1978, p. 312.
6. Miller, M. L., "Perpetuation and Consolidation of Federal Survey Notes," *Proceedings, Fall Convention, ACSM,* 1974, p. 287.
7. *National Geodetic Survey Data: Availability, Explanation, and Application,* NOAA Technical Memorandum NOS NGS-5, National Geodetic Survey, Rockville, Md., June 1976.
8. Scarborough, J. B., *Numerical Mathematical Analysis,* The Johns Hopkins Press, Baltimore, Md., 1962.
9. Whalen, C. T., and Balacz, E. I., "Test Results of First-Order Class III Leveling," *Surveying and Mapping,* Vol. 37, No. 1, March 1977, p. 45.

PART II
BASIC SURVEY MEASUREMENTS

CHAPTER 4
Distance measurement

4.1. Distance The distance between any two points in three-dimensional space is a spatial distance. This concept is particularly pertinent considering the advanced stage of electronic distance measurement (EDM) techniques, which automatically provide spatial or slope distances in the range of from 1 m or ~ 3.3 ft to 60 km or 36 mi. Although slope distances are frequently observed in the surveying operation, these distances are then reduced to a horizontal projection for more convenient use in subsequent calculations and field layout. Horizontal distance is further reduced to its equivalent at sea level for geodetic surveys (Art. 9.7). In plane surveying, horizontal distances are reduced to sea level only when it is desired to convert them into equivalent distances at another elevation, such as that of the state plane-coordinate system or the average elevation of a survey for which the variation in elevation over the area is large.

There are several methods of determining distance, the choice of which depends on the accuracy required, the cost, and other conditions. The methods in ascending order of accuracy are estimation, scaling from a map, pacing, odometer, tacheometry, taping, photogrammetry, inertial systems, and electronic distance measurement. For example, on rough reconnaissance, a relative precision of 1 part in 100 (i.e., a standard deviation of 1/100 of the distance) or less may be adequate for the survey; on the other hand, specifications for first-order surveys (refer to Table 9.2) call for base-line measurement to have a standard deviation of 1 part in 1 million of the distance measured. On certain surveys, a combination of methods may be warranted.

Table 4.1 classifies the principal methods of measuring distance according to the usual relative precisions obtained. The various methods are discussed further in the following sections.

METHODS OF MEASUREMENT

4.2. Pacing Pacing furnishes a rapid means of approximately checking more precise measurements of distance. It is used on reconnaissance surveys and, in small-scale mapping, for locating details and traversing with the plane table. Pacing over rough country has furnished a relative precision of 1/100; under average

Table 4.1 Methods of Measuring Distance

Method	Relative precision[a]	Use	Instruments required	Instruments for measuring angles with corresponding accuracy
Pacing odometer or mileage recorder	1/100	Reconnaissance, small-scale mapping, checking tape measurement, quantity surveys	Pedometer, odometer	Hand compass, peepsight alidade
Tacheometry				
Stadia	1/300–1/1000	Location of details for topographic mapping, rough traverse, checking more accurate measurements	Level rod or stadia board, calibrated optical line of sight with stand	Transit, telescopic alidade and plane table, surveyor's compass
Distance wedge	1/5000–1/10,000	Traverse for land surveys, control of route and topographic surveys and construction work	Horizontal graduated rod and support, calibrated optical line of sight equipped with a distance wedge	Transit or theodolite
Subtense bar	1/1000–1/9000	Hydrographic surveys, traverse	Calibrated subtense bar and tripod; 1″ theodolite	Transit or theodolite
Ordinary taping	1/3000–1/5000	Traverse for land surveys and for control of route and topographic surveys and construction	Steel tape, chaining pins, plumb bobs	Transit

Precise taping	1/10,000–1/30,000	Traverses for city surveys, base lines for triangulation of low accuracy, and construction surveys requiring high accuracies	Calibrated steel tape, thermometer, tension handle, hand level, plumb bobs	Transit or theodolite
Photogrammetry	1/10,000–1/100,000	Location of detail for topographic mapping, second- and third-order ground control surveys	Stereoplotters, mono and stereo comparators, electronic computer	
Inertial systems	Up to 1/50,000	Rapid, reconnaissance surveying: large area surveys, second- and third-order ground control surveys	Inertial positioning system	
Base-line taping	1/100,000–1/1,000,000	First-, second-, and third-order triangulation for large areas, city surveys, long bridges, and tunnels	Calibrated steel tape, thermometers, tension handle, taping supports, level, level rod	Theodolite (1″)
EDM	0.02 or 6 mm +1 ppm	Traverse, triangulation, and trilateration for control surveys of all types and for construction surveys	EDM equipment	Theodolite (1″) which can be built into the EDM

[a]Relative precision may be defined as the ratio of the allowed standard deviation to the distance measured (see Art. 2.16, Chap. 2).

conditions, a person of experience will have little difficulty in pacing with a relative precision of 1/200.

Each two paces or double step is called a *stride*. Thus, the stride would be approximately 5 ft, or there would be roughly 1000 strides per mile.

Paces or strides are usually counted by means of a tally register operated by hand or by means of a *pedometer* attached to the leg. In hilly country, rough corrections for slope can be applied.

Beginners should standardize their pace by walking over known distances on level, sloping, and uneven ground.

4.3. Mileage recorder, odometer, and other methods Distance may be measured by observing the number of revolutions of the wheel of a vehicle. The *mileage recorder* attached to the ordinary automobile speedometer registers distance to 0.1 mi and may be read by estimation to 0.01 mi. Special speedometers are available reading to 0.01 or 0.002 mi. The *odometer* is a simple device that can be attached to any vehicle and directly registers the number of revolutions of a wheel. With the circumference of the wheel known, the relation between revolutions and distance is fixed. The distance indicated by either the mileage recorder or the odometer is somewhat greater than the true horizontal distance, but in hilly country a rough correction based on the estimated average slope may be applied.

Distances are sometimes roughly estimated by *time interval of travel* for a person at walk, a saddle animal at walk, or a saddle animal at gallop.

Unknown distances may be determined by *mathematical* or *graphical methods* through their relation to one or more known distances. These methods are used in triangulation and plane-table work.

4.4 Tacheometry Tacheometry includes stadia with transit and stadia rod; stadia with alidade, plane table, and rod; distance wedge and horizontal rod; and subtense bar and theodolite. The *stadia* method, described in detail in Chap. 7, offers a rapid indirect means of determining distances. It is used extensively and is particularly useful in topographic surveying. The telescope of the transit, level, or plane-table alidade is equipped with two horizontal hairs, one above and the other an equal distance below the horizontal cross hair. The distance from the instrument to a given point is indicated by the interval between these stadia hairs as shown on a graduated rod held vertically at the point. The precision of the stadia method depends upon the instrument, the observer, the atmospheric conditions, and the length of sights. Under average conditions the stadia method will yield a relative precision between 1/300 and 1/1000. The use of a self-reducing tacheometer or self-reducing alidade for the stadia method simplifies the operation substantially.

The *distance wedge* is used in conjunction with a horizontally supported rod placed perpendicular to the optical line of sight.

Another method for determining distance by tacheometry consists of using a *subtense bar* and precise transit or theodolite. The subtense bar is a 2-m calibrated bar mounted horizontally on a tripod and placed perpendicular to the line of sight. The subtended angle is then measured with a 1″ theodolite placed at the other end of the line to be measured so that the horizontal distance between the points can be calculated. Further details regarding this procedure can be found in Chap. 7.

4.5. Taping Taping involves direct measurement of the distance with *steel tapes* varying in length from 3 ft (1 m) to 1000 ft (300 m). Graduations are in feet, tenths, and hundredths, or metres, decimetres, centimetres, and millimetres. Formerly, on surveys of ordinary precision it was the practice to measure the length of lines with the *engineer's chain* or *Gunter's chain*; for measurements of the highest precision, special bars were used.

The *engineer's chain* is 100 ft long and is composed of 100 links each 1 ft long. At every 10 links brass tags are fastened, notches on the tags indicating the number of 10-link segments between the tag and the end of the tape. Distances measured with the engineer's chain are recorded in feet and decimals.

The *surveyor's* or *Gunter's chain* is 66 ft long and is divided into 100 links each 0.66 ft = 7.92 in long; distances are recorded in chains and links. Thus,

$$\text{1 (Gunter's) chain} = 66 \text{ ft} = 100 \text{ links} = 4 \text{ rods}$$
$$\text{80 (Gunter's) chains} = 1 \text{ mi}$$
$$\text{10 square (Gunter's) chains} = 1 \text{ acre} = 43{,}560 \text{ ft}^2$$

Measuring with chains was originally called "chaining." The term has survived and is often applied to the operation of measuring lines with tapes.

The precision of distance measured with tapes depends upon the degree of refinement with which measurements are taken. On the one hand, rough taping through broken country may be less accurate than the stadia. On the other hand, when extreme care is taken to eliminate all possible errors, measurements have been taken with a relative precision of better than 1/1,000,000. In ordinary taping over flat, smooth ground, the relative precision is about 1/3000 to 1/5000.

4.6. Electronic distance measurement Recent scientific advances have led to the development of electro-optical and electromagnetic instruments which are of great value to the surveyor for accurate measurements of distances. Measurement of distance with electronic distance measuring (EDM) equipment is based on the invariant speed of light or electromagnetic waves in a vacuum. EDM equipment which can be used for traverse, triangulation, and trilateration as well as for construction layout is rapidly supplanting taping for modern surveying operations except for short distances and certain types of construction layout. A detailed explanation of EDM systems and procedures is given in Arts. 4.29 to 4.38.

4.7. Choice of methods Most boundary, control, and construction surveys involving long lines and large areas can be performed most accurately and economically using modern EDM equipment. Where the distances involved are relatively short or specific construction layout requirements are present, taping the distances can be more practical. Stadia is still unsurpassed for small topographic surveys and preliminary surveys for projects of limited extent.

Each of the methods mentioned in the preceding sections has a field of usefulness. On the survey for a single enterprise, the surveyor may find occasion to employ a combination of methods to advantage.

4.8. Tapes Tapes are made in a variety of materials, lengths, and weights. Those more commonly used by the surveyor and for engineering measurements are the steel tapes, sometimes called the engineer's or surveyor's tape, and woven nonmetallic and metallic tapes.

The woven *metallic tape* (Fig. 4.1) is a ribbon of waterproofed fabric into which are woven small brass or bronze wires to prevent its stretching. It is usually 50, 100, or 150 ft long; is graduated in feet, tenths, and half-tenths; and is $\frac{5}{8}$ in wide. The metric metallic tape is 10, 20, 30, or 50 m long; is graduated in metres, centimetres, and 2 mm; and is usually 13 mm wide. Metallic tapes are used principally in earthwork cross sectioning, in location of details, and in similar work where a light, flexible tape is desirable and where small errors in length are not of consequence. Nonmetallic glass fiber tapes which are nonconductors of electricity have been developed for use near power lines.

A tape of phosphor bronze is rustproof and is particularly useful for work in the vicinity of salt water.

For very precise measurements, such as those for base lines and in city work, the *invar tape* has come into general use. Invar is a composition of nickel and steel with

Fig. 4.1 Metallic tapes graduated in feet and/or metric units. (*Courtesy Lietz Instruments, Inc.*)

a very low coefficient of thermal expansion, sometimes as small as one-thirtieth that of steel, and is affected little by temperature changes. Since the compositions having the lowest coefficients of expansion may not remain constant in length over a period of time, it is customary to use a composition having a large coefficient; the thermal expansion of commercial invar tapes is about one-tenth that of steel tapes. Invar is a soft metal, and the tape must be handled very carefully to avoid bends and kinks. This property and its high cost make it impractical for ordinary use.

Steel tapes for which the foot is the unit of length are graduated as follows: Lightweight tapes and some engineer's tapes are graduated to hundredths of feet throughout the length; normally, the heavier tapes have graduations with numbers every foot, with only the end feet graduated in tenths and/or hundredths of feet.

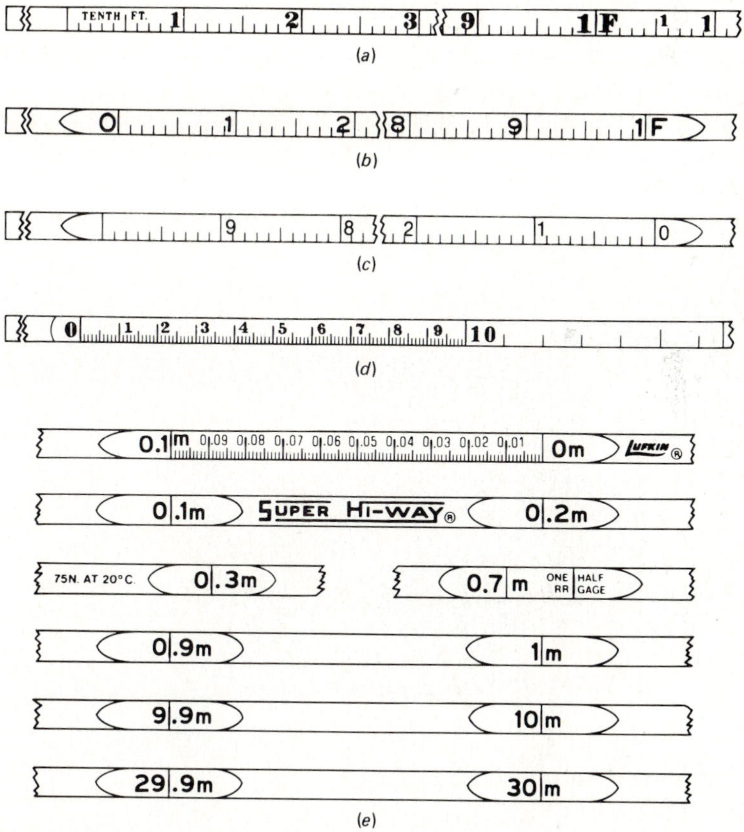

Fig. 4.2 Various styles of steel tapes (*a*) Tape graduated in feet, tenths, and hundredths throughout. (*b*) First foot graduated to hundredths, tape graduated throughout to feet. A "cut" tape. (*c*) Added foot graduated to hundredths of a foot, tape graduated to feet throughout. An "add" tape. (*d*) Metric tape. First decimetre graduated to millimetres, first metre graduated to centimetres, graduated throughout to decimetres. A "cut" tape. (*e*) Metric tape. Added decimetre graduated to millimetres. Tape graduated throughout to decimetres. An "add" tape. [*Courtesy the Cooper Group (Lufkin).*]

Metric tapes are 15, 25, 30, or 50 m long. The light box tapes are graduated throughout in metres, decimetres, and centimetres. Heavier tapes are graduated as follows: throughout to metres and decimetres; first and last metre in centimetres; and first and last decimetres in millimetres; or throughout to metres and half metres with end metres to decimetres. Various styles of steel tapes graduated in feet and metres are shown in Fig. 4.2*a* through *e*.

Fiberglass tapes are also available in units of feet and metres and may be obtained in lengths of 50, 100, and 150 ft or 15, 30, and 50 m.

Tapes for which the Gunter's chain is the unit of length are 1, 2, and 2.5 chains in length and are graduated in links throughout with end links in tenths.

Ordinarily, rawhide thongs serving as handles are fastened to the rings at each end of the chain tape. Wire handles are sometimes used, but they are objectionable when the tape must be dragged through grass or brush. Detachable clamp handles are available for grasping the tape at any point. Figure 4.3 illustrates the use of a tape clamp.

The tape should be kept straight when in use; any tape will break when kinked and subjected to a straight pull. Steel tapes rust readily and should be wiped dry after being used. Special care is required when working near power lines. Fatal accidents have occurred from throwing a steel or metallic tape over a power line.

Fig. 4.3 Use of the tape clamp.

4.9. Equipment for taping Additional equipment employed for determining the lengths of lines by direct measurement with a tape consists of plumb bobs, the hand level, the tension handle, chaining pins, and range poles.

The *plumb bob* is a pointed metal weight used to project the horizontal location of a point from one elevation to another.

The *hand level* is described in Art. 5.19 and can be used to keep the two ends of the tape at the same elevation when measuring over irregular terrain. A *tension handle* is a spring scale which can be attached to the end of the tape and allows applying the proper tension.

Steel taping pins, also called *chaining pins*, *taping arrows*, or *surveyor's arrows*, are commonly employed to mark the ends of the tape during the process of taping between two points more than a tape length apart. They are usually 10 to 14 in (25 to 35 cm) long. A set consists of 11 pins (see Fig. 4.4*a*).

For more precise taping or for future reference, nails may be driven into the earth. Marking tags, which can be stamped with identification as desired, may be threaded over the nails before they are driven. On paved surfaces, tape lengths and other points may be marked with keel (carpenter's chalk), pencil, or spray paint; for precise work a short piece of opaque adhesive tape may be stuck on the pavement and the point marked with a pencil or ball-point pen.

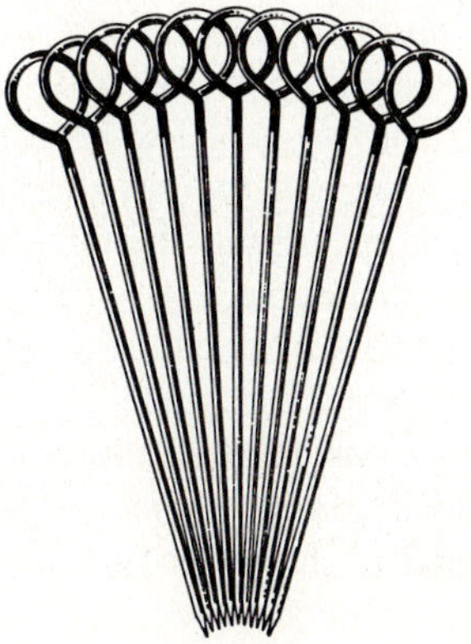

Fig. 4.4 *a* Taping pins.

Fig. 4.4 *b* Range pole.

Wooden stakes or hubs are driven into the ground to mark the significant points of profile-level and transit-traverse lines. Taping tripods and wooden or concrete posts are used to mark tape ends in base-line taping (Chap. 9).

Metal, wooden, or fiberglass *range poles*, also called *flags*, *flagpoles*, or *lining rods*, are used as temporary signals to indicate the location of points or the direction of lines. They are of octagonal or circular cross section and are pointed at the lower end. Wooden and fiberglass range poles are shod with a steel point. The common length is 6 or 8 ft, or 2 or 3 m. Usually, the pole is painted with alternate bands of red and white 1 ft or $\frac{1}{2}$ m long (see Fig. 4.4*b*).

4.10. Taping on smooth, level ground The procedure followed in measuring distances with the tape depends to some extent upon the required accuracy and the purpose of the survey. The following description represents the usual practice when the measurements are of ordinary relative precision of, say, 1/5000. Errors in taping are discussed in Art. 4.26 and mistakes in Art. 4.28.

The tape is supported throughout its length. If only the distance between two fixed points (as the corners of a parcel of land) is to be determined, the equipment will consist of one or more range poles, 11 chaining pins, and a 100-ft or 30-m heavy steel tape graduated as described in Art. 4.8. One range pole is placed behind the distant point to indicate its location.

The rear tapeman *with one pin* is stationed at the point of beginning. The head tapeman, with the zero (graduated) end of the tape and 10 pins, advances toward the distant point. When the head tapeman has gone nearly 100 ft, or 30 m, the rear tapeman calls "chain" or "tape," a signal for the head tapeman to halt. The rear tapeman holds the 100-ft or 30-m mark at the point of beginning and, by hand signals or by voice, lines in a taping pin (held by the head tapeman) with the range pole (or other signal) marking the distant point. During the lining-in process, the rear tapeman is in a kneeling position on the line and facing the distant point; the head tapeman is in a kneeling position to one side of and facing the line so that the tape can be held steady and so that the rear tapeman will have a clear view of the signal marking the distant point. The head tapeman with one hand sets the pin vertically on line and a short distance to the rear of the zero mark. With the other hand the tapeman then pulls the tape taut and, making sure that it is straight, brings it in contact with the pin. The rear tapeman, upon observing that the 100-ft mark is at the point of beginning, calls "good" or "mark." The head tapeman pulls the pin and sticks it at the zero mark of the tape, with the pin sloping away from the line (Fig. 4.5). As a check, the head tapeman again pulls the tape taut and notes that the zero point coincides with the pin at its intersection with the ground. The head tapeman then calls "good," the rear tapeman releases the tape, the head tapeman moves forward as before, and so the process is repeated.

As the rear tapeman leaves each intermediate point, the pin is pulled. Thus, there is always one pin in the ground, and the number of pins held by the rear tapeman at any time indicates the number of hundreds of feet, or *stations*, or multiples of 30 m, from the point of beginning to the pin in the ground.

Fig. 4.5 Taping on smooth level ground.

At the end of each 1000 ft (10 full stations) or 300 m, the head tapeman has placed the last pin in the ground. The head tapeman signals for pins, the rear tapeman comes forward and gives the 10 pins to the head tapeman, and both check the total, which is recorded. The procedure is then repeated. The count of pins is important, as the number of tape lengths is easily forgotten, owing to distractions.

When the end of the course is reached, the head tapeman halts and the rear tapeman comes forward to the last pin set. The head tapeman holds the zero mark at the terminal point. The rear tapeman pulls the tape taut and observes the number of *whole feet*, metres, or decimetres between the last pin and the end of the line (Fig. 4.6*a*). The rear tapeman then holds the next *larger* foot, metre, or decimetre mark at the pin, and the head tapeman pulls the tape taut and reads the decimal by means of the finer graduations at the end of the tape (Fig. 4.6*b*). For tapes graduated in feet and having fine graduations between the 0 and 1-ft points, (a "cut" tape), the distance is the whole number of feet held by the rear tapeman minus the fractional increment observed by the head tapeman. For example, if the rear tapeman holds 87 ft and the head tapeman observes 0.32 ft directly, the distance is $87 - 0.32 = 86.68$ ft as illustrated in Fig. 4.6*b*.

For tapes having an extra graduated foot beyond the zero point (an "add" tape), the rear tapeman holds the next *smaller* foot mark at the pin, and the head tapeman reads the decimal. Figure 4.6*c* and *d* illustrates this procedure for the same distance of 86.68 ft.

Metric tapes have the first decimetre graduated to millimetres and the first metre to centimetres (a "cut" tape; see Fig. 4.2*d*) or a decimetre graduated to millimetres is added to the zero end of the tape (an "add" tape; see Fig. 4.2*e*). Both styles are usually graduated throughout to decimetres.

Cut and add metric tapes are used in the same way as tapes graduated in feet. Suppose that when the zero end of a metric cut tape is at the terminal point, the rear tapeman observes that the rear pin falls just past 25.3 m. The rear tapeman "gives" a decimetre and holds 25.400 m at the rear taping pin. If the head tapeman then reads

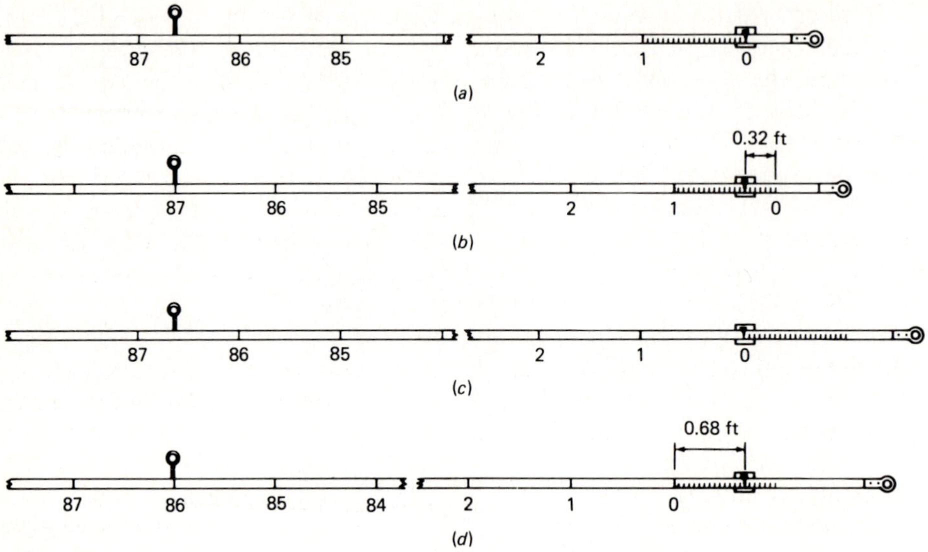

Fig. 4.6 (*a*) and (*b*) Measuring with a "cut" tape. (*c*) and (*d*) Measuring with an "add" tape. Plan view of tape graduated in feet.

0.075 m on the first graduated decimetre at the terminal point, the distance is 25.400 − 0.075 = 25.325 m.

In the same situation, using a metric add tape, the rear tapeman would hold 25.300 m at the rear taping pin. The head tapeman would then read 0.025 m on the added decimetre at the terminal point, yielding a distance of 25.300 + 0.025 = 25.325 m.

When the transit is set up on the line to be measured, the transit operator usually directs the head tapeman in placing the pins on line.

On some surveys it is required that stakes be set on line at short intervals, usually 100 ft or 30 m. Sometimes stakes are driven by the rear tapeman after the pin is pulled. On surveys of low precision, no pins are used and the head tapeman sets the stakes in the manner previously described for pins, measuring the distance between centers of stakes at their junction with the ground. On more precise surveys, measurements are carried forward by setting a tack in the head of each stake. When the tack has been driven, it is tested for line and distance.

4.11. Horizontal taping over sloping ground The process of taping over uneven or sloping ground or over grass and brush is much the same as that just described for smooth, level ground, except that the plumb bob is used. The tape is held horizontal, and a plumb bob is used by either, or at times by both, tapemen for projecting from tape to pin, or vice versa. For rough work, plumbing can be accomplished with the range pole.

To secure accuracy comparable to taping over level ground, considerable skill is required. Some experience is necessary to determine when the tape is nearly horizontal; the tendency is to hold the downhill end of the tape too low. A hand level is useful to estimate the proper height to hold the tape ends so as to have a horizontal tape. The tape is unsupported between its two ends, and either the pull must be increased to eliminate the effect of sag or a correction for sag must be applied. A firm stance is important; the tapemen should place the planes of their bodies parallel to the tape, with legs well apart. The forearms should be in line with the tape, and should be steadied or snubbed against the respective legs or bodies, as required by the height at which the tape must be held.

Where the slope is less than 5 ft in 100 ft, or 1.5 m in 30 m, the head tapeman advances a full tape length at a time. If the course is downhill, the head tapeman holds the plumb line at the zero point of the tape, with the tape horizontal and the plumb bob a few inches from the ground, then pulls the tape taut and is directed to line by the rear tapeman. When the plumb bob comes to rest, it is lowered carefully to the ground (see Fig. 4.7) and a pin is set in its place. As a check, the measurement is repeated. If the course is uphill, the rear tapeman holds a plumb line suspended from the 100-ft or 30-m mark and signals the head tapeman to give or take until the rear tapeman's plumb bob comes to rest over the pin. The head tapeman sets a pin, and the measurement is repeated.

Fig. 4.7 Plumbing at the downhill end of a horizontal tape.

Where the course is steeper and it is still desired to establish 100-ft or 30-m stations, as for a route survey, the following procedure is recommended. Assuming that the slope is downhill, the head tapeman advances a full tape length and then returns to an intermediate point from which the tape can be held horizontal. The head tapeman suspends the plumb line at a foot mark, is lined in by the rear tapeman, and sets a pin at the indicated point. The rear tapeman comes forward, *gives the head tapeman a pin*, and at the pin in the ground holds the tape at the foot mark, from which the plumb line was previously suspended. The head tapeman proceeds to another point from which the tape can be held horizontal, and so the process is repeated until the head tapeman reaches the zero mark on the tape. At each *intermediate* point of a tape length the rear tapeman gives the head tapeman a pin, but not at the point marking the full tape length. In this manner the tape is always advanced a full length at a time; the number of pins held by the rear tapeman at each 100-ft or 30-m point indicates the number of hundreds of feet or multiples of 30 m from the last tally, and the rear tapeman's count of pins is not confused. The process is called "breaking tape."

To illustrate, Fig. 4.8 represents the profile of a line to be measured in the direction of *A* to *D*, and *A* represents a pin marking the end of a 30-m interval from the point of beginning. The head tapeman goes forward until the 30-m mark is at *A*, where the rear tapeman is stationed. The head tapeman then returns to *B*, where the tape is held horizontal and is plumbed from the 25-m mark to set a pin at *B*. The rear tapeman gives the head tapeman a pin and holds the 25-m mark at *B*. The head tapeman plumbs from the 15-m mark and sets a pin at *C*. The rear tapeman gives the head tapeman a pin and holds the 15-m mark at *C*. The head tapeman plumbs from the zero mark to set a pin at *D* at the end of the full tape length. The rear tapeman goes forward but keeps the pin that was pulled at *C*.

Some surveyors prefer to measure distances less than the tape length individually and add these measurements. However, this practice requires recording and may lead to mistakes in addition.

Usually, the tape is estimated to be horizontal by eye. This practice commonly results in the downhill end's being too low, sometimes causing a significant error in horizontal measurement. The safe procedure in rough country is to use a hand level.

In horizontal measurements over uneven or sloping ground, the tape sags between supports and becomes effectively shorter. The effect of sag can be eliminated by standardizing the tape, by applying a computed correction, or by using the normal tension (Art. 4.21). In breaking tape or when the tape is supported for part of its

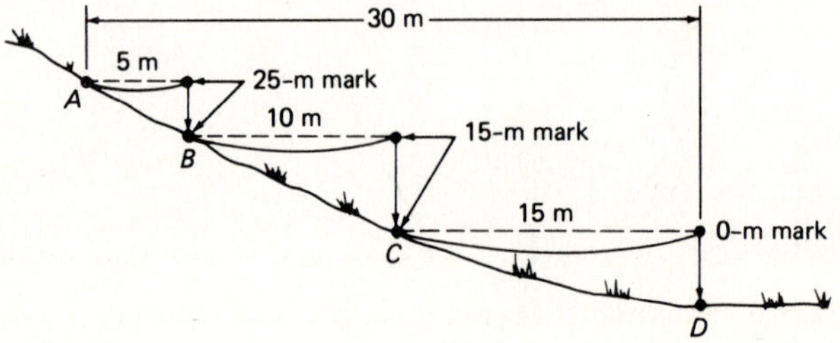

Fig. 4.8 Horizontal measurements on a steep slope ("breaking tape").

length, the difference in effect of sag as between a full tape length and the unsupported length can be taken into account roughly by varying the pull on the tape.

In measurements over low obstructions, if the tape is partially supported between the tapemen, account should be taken of the effect this support makes on sag corrections.

4.12. Slope taping Where the ground is fairly smooth, slope measurements are generally preferred because they are usually made more accurately and quickly than horizontal measurements. Some means of determining either the slope or the difference in elevation between successive tape ends and/or breaks in slope is required. For surveys of ordinary accuracy, either the clinometer or transit (for measuring vertical angles) or the hand level (for measuring difference in elevation) may be used to advantage. If only the distance between the ends of the line is required, the procedure of taping is the same as on level ground, but a record is kept either of the slope or of the difference in elevation of each tape length (or less at breaks in slope). The horizontal distances are then computed from the distances measured on the slope.

Where stakes are to be placed at regular intervals (e.g., 100 ft, 20 m, 30 m), corrections to the slope distances may be applied as the taping progresses.

4.13. Systematic errors in taping Systematic errors in taping linear distances are those attributable to the following causes: (1) slope taping (tape not horizontal), (2) tape not of standard length, (3) variations in temperature, (4) variations in tension, (5) sag, (6) incorrect alignment of tape, and (7) tape not straight. Each systematic error will be designated by the symbol e with the appropriate subscript; corresponding corrections are given the symbol C.

4.14. Correction for slope Where the slopes are measured with sufficient accuracy to warrant, the corresponding horizontal distances may be computed by exact trigonometric relations. In Fig. 4.9 let s represent the slope distance between two points A and B, h the difference in elevation, and H the horizontal distance, all in the same units. θ is the vertical angle observed with a clinometer or transit. If θ has been observed, then

$$H = s \cos \theta \tag{4.1}$$

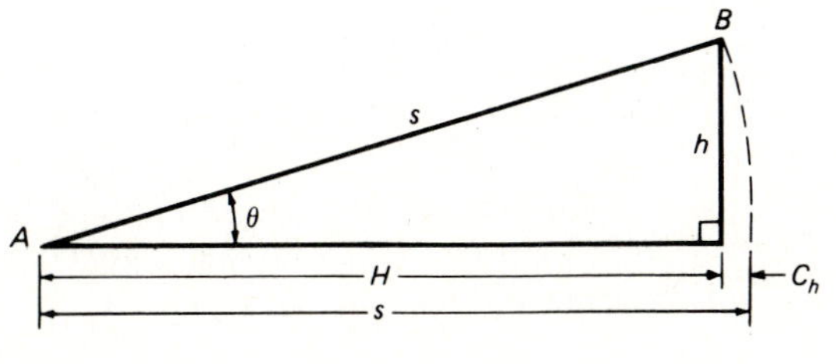

Fig. 4.9 Slope correction.

and the slope correction is

$$C_h = H - s$$

such that $s + C_h = H$.

Example 4.1 A slope measurement of 29.954 m was made between two points where θ is 4°30′. Determine the horizontal distance.

SOLUTION Using Eq. (4.1) yields

$$H = (29.954)(\cos 4°30') = (29.954)(0.996917) = 29.862 \text{ m}$$

Example 4.2 It is desired to set point D a horizontal distance of 195.00 ft from point E along a line which has a slope angle of 5°30′. What slope distance should be measured in the field?

SOLUTION Using Eq. (4.1), we obtain

$$s = \frac{H}{\cos\theta} = \frac{195.00}{0.995396} = 195.90 \text{ ft}$$

Reduction of slope to horizontal distances can also be determined by using the difference in elevation between the two ends of the line. Referring to Fig. 4.9 yields

$$H^2 = s^2 - h^2 \tag{4.2}$$

from which

$$s = (H^2 + h^2)^{1/2} \tag{4.3}$$

and

$$H = (s^2 - h^2)^{1/2} \tag{4.4}$$

which are the direct relations for the slope and horizontal distances, respectively.

Expanding the right side of Eq. (4.4) with the binomial theorem yields

$$H = s + \left(-\frac{h^2}{2s} - \frac{h^4}{8s^3} - \cdots \right) \tag{4.4a}$$

Traditionally, the quantity enclosed by parentheses in Eq. (4.4*a*) is designated the *slope correction*, labeled C_h in Fig. 4.9. The same equation can be used to calculate the slope distance necessary to lay out a given horizontal distance.

For moderate slopes the first term within the parentheses of Eq. (4.4*a*), $h^2/2s$, is usually adequate. The error introduced by Eq. (4.4*a*) using only the first term is negligible for ordinary slopes.

Thus, for slopes not exceeding 15 percent, the second term is not significant unless relative precisions of 1:15,000 or better are required.

Example 4.3 A distance of 130.508 m was measured over terrain with a constant slope along a sloping line which has a difference in elevation between the two ends of 5.56 m. Calculate the horizontal distance between the two points.

SOLUTION By Eq. (4.4),

$$H = [(130.508)^2 - (5.56)^2]^{1/2} = 130.390 \text{ m}$$

Example 4.4 Point R is to be set at a horizontal distance of 98.25 ft from point Q along a sloping line where the difference in elevation between R and Q is 4.35 ft. Calculate the slope distance to be measured in the field.

SOLUTION Using Eq. (4.3) yields

$$s = [(98.25)^2 + (4.35)^2]^{1/2} = 98.35 \text{ ft}$$

Example 4.5 A distance was measured over irregularly sloping terrain. Slope distances and differences in elevation are tabulated in the two columns on the left of the following table. Calculate the horizontal distance.

Slope distance, ft	Difference in elevation, ft	Horizontal distance, ft
100.00	3.50	99.94
100.00	5.30	99.86
80.50	4.20	80.39
100.00	8.05	99.68
62.35	5.25	62.13
442.85		442.00
−0.85		
442.00		

SOLUTION H is calculated using Eq. (4.4), individually for each length and tabulated in the third column. The total horizontal distance is the sum of the calculated H's.

4.15. Precision required for θ and *h* The relative precision of a measured line is usually expressed as the ratio of the allowable standard deviation to the distance measured. Thus, a relative precision of 1 in 10,000 implies a standard deviation of 1 unit in 10,000 or 0.01 unit in 100 units. Required relative precisions in measured distances for various classes of work are given in Chaps. 8 and 9. Ordinary taping is generally said to have a relative precision of 1 part in 3000 to 1 part in 5000.

The final precision in horizontal distance H will depend on the precision with which s and either θ or h are measured. Application of the principles of error propagation to Eqs. (4.1) and (4.2) will yield the relationship between standard deviation in slope angle and difference in elevation and the horizontal distance. These relationships are displayed graphically for various relative precisions in Figs. 4.10 and 4.11. Figure 4.10 shows the standard deviation in slope angle θ for various vertical angles and relative precisions in measurement of distances of 1:5000, 1:10,000, and 1:25,000. Using these curves one can determine the precision required in a vertical angle to achieve a certain relative precision. For example, if a

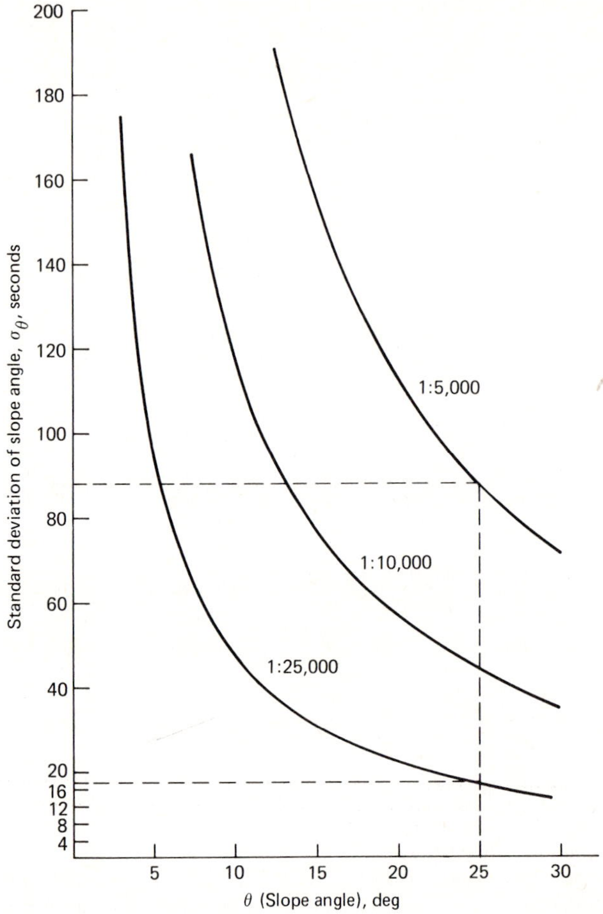

Fig. 4.10 Standard deviation in vertical angle for various angles and relative precisions.

relative precision of 1 : 5000 is required, the slope angle needs to be measured only to the nearest 1.5′ assuming the steepest angle does not exceed 25° (see Fig. 4.10). Under the same conditions, slope angles must be measured with a standard deviation of 18″ to achieve a relative precision of 1 : 25,000.

Figure 4.11 shows the standard deviation in difference in elevation for various values of h and relative precisions when using a 30-m tape length. Thus, if a relative precision of 1 : 5000 is required when elevation differences do not exceed 3 m, then these elevation differences should be determined with a standard deviation of 0.06 m or about 0.2 ft.

4.16. Tape not of standard length The nominal length of a tape, as stated by the manufacturer, rarely corresponds exactly with the true length. The true length is taken as the length determined by comparison with a known standard length under given conditions of temperature, tension, and support. The National Bureau of Standards in Washington, D.C., will standardize tapes and provide a certificate of

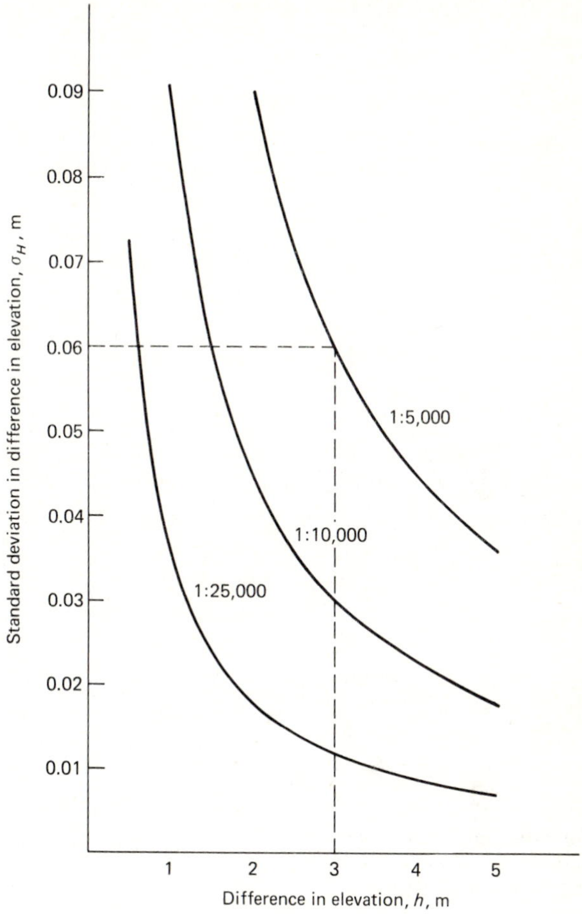

Fig. 4.11 Standard deviations in elevation difference for various differences in elevation and relative precisions for a 30-m tape length.

standardization upon payment of a fee. The absolute value for the tape correction C_d is

$$C_d = \text{true length} - \text{nominal length} \tag{4.5}$$

This discrepancy is normally assumed to be distributed uniformly throughout the tape and is directly proportional to the fractional portion of the tape used for a measurement.

Tapes are usually standardized at 20 °C (68 °F) and with tensions varying from 4 to 14 kg or 10 to 30 lb. Tapes may be supported throughout, at specified points, or at the two ends only. Desired conditions of standardization are specified when requesting the standardization.

The normal use and abuse of a tape in the field over a period of time can invalidate the standard length. Consequently, the standard tape should be kept in the office and used only for purposes of comparison. At regular intervals, field tapes should be compared with the standard tape or to a standard length measured with the standardized tape. Such comparisons are extremely important, if specified accuracies are to be maintained in the field.

4.17. Tape not horizontal When horizontal taping is being performed over sloping ground, discrepancies frequently occur because the tape is not truly horizontal. Slopes are often deceptive, even to experienced surveyors; the tendency is to hold the downhill end of the tape too low. Inexperienced tapemen have been observed keeping very careful alignment, yet taping what they thought were horizontal distances on a slope of perhaps 10 percent. The resulting discrepancy is 0.5 ft/100 ft or 26 ft/mi (0.15 m/30 m or $\simeq$ 5 m/km). In ordinary taping, this is one of the largest of contributing errors. It will not be eliminated by repeated measurements, but it can be reduced to a negligible amount by leveling the tape with either a hand level or a clinometer.

4.18. Variations in temperature The tape expands as the temperature rises and contracts as the temperature falls. Therefore, if the tape is standardized at a given temperature and measurements are taken at a higher temperature, the tape is too long. For a change in temperature of 15 °F, a 100-ft steel tape will undergo a change in length of about 0.01 ft, introducing an error of about 0.5 ft/mi. Similarly, a change of 15 °C produces a change in length of 5 mm/30 m or 0.17 m/km. Under a change of 50 °F the error would be 1.5 ft/mi or 0.58 m/km. It is seen that, even for measurements of ordinary precision, the error due to thermal expansion becomes of consequence when the measurements are taken during cold or hot weather.

The coefficient of thermal expansion of steel, α, is approximately 0.00000645 (ft/ft)/1° F or 0.0000116 (m/m)/1° C. If the tape is standard at a temperature of T_0 degrees and measurements are taken at a temperature of T degrees, the correction C_t for change in length is given by the formula

$$C_t = \alpha L(T - T_0) \tag{4.6}$$

in which L is the measured length and α the coefficient of thermal expansion.

Errors due to variations in temperature are greatly reduced by using an invar tape. If a steel tape is used for precise taping, one or more tape thermometers should be taped to it.

4.19. Variations in tension If the tension or pull is greater or less than that for which the tape is standardized, the tape is elongated or shortened accordingly. The correction for variation in tension in a steel tape is given by the formula

$$C_p = \frac{(P - P_0)L}{aE} \tag{4.7}$$

where C_p = correction per distance L, ft or m
P = applied tension, lb or kg
P_0 = tension for which the tape is standardized, lb or kg
L = length, ft or m
a = cross-sectional area, in^2 or cm^2
E = elastic modulus of the steel, lb/in^2 or kg/cm^2

The modulus of elasticity is taken as 28 to 30 million lb/in^2 or 2.1×10^6 kg/cm^2. The cross-sectional area of the tape can be computed from the weight and dimensions, since steel weighs approximately 490 lb/ft^3 or 7.85×10^{-3} kg/cm^3. Light (1-lb) and heavy (3-lb) 100-ft tapes have cross-sectional areas of approximately 0.003 and 0.009 in^2, respectively. Light and heavy 30-m tapes have respective cross-sectional areas of about 0.019 and 0.058 cm^2.

Some idea of the effect of variation in tension can be obtained from the following examples.

Example 4.6 Assume that a light 100-ft tape is standardized under a tension of 10 lb, $E = 30{,}000{,}000$ lb/in^2, and the cross-sectional area of the tape is 0.003 in^2. Determine the elongation for an increase in tension from 10 to 30 lb.

SOLUTION Using Eq. (4.7), we obtain

$$C_p = \frac{(30 - 10)(100)}{(30{,}000{,}000)(0.003)} = 0.0222 \text{ ft}$$

Example 4.7 A heavy 30-m tape having a cross-sectional area of 0.06 cm^2 has been standardized at a tension of 5 kg. If $E = 2.1 \times 10^6$ kg/cm^2, calculate the elongation of the tape for an increase in tension from 5 to 15 kg.

SOLUTION Using Eq. (4.7) yields

$$C_p = \frac{(15 - 5)(30)}{(2{,}100{,}000)(0.06)} = 0.0024 \text{ m}$$

Corrections for variation in tension are seen to be significant for ordinary taping when light tapes are used and the differences in tension are substantial. For measurements of high accuracy such as base lines, tension variations are significant regardless of the weight of tape.

4.20. Correction for sag When the tape sags between points of support, it takes the form of a catenary. The correction to be applied is the difference in length between the arc and the subtending chord. Since the required distance is the chord which is always shorter than the arc, the correction C_s evaluated by Eq. (4.8) is always subtracted from the observed value (an arc). For the purpose of determining the correction, the arc may be assumed to be a parabola, and the correction is then given with sufficient precision for most purposes by the formula

$$C_s = \frac{w^2L^3}{24P^2} = \frac{W^2L}{24P^2} \tag{4.8}$$

where C_s = correction between points of support, ft or m
w = weight of tape, lb/ft or kg/m
W = total weight of tape between supports, lb or kg = wL
L = distance *between supports*, ft or m
P = applied tension, lb or kg

The correction is seen to vary directly as the cube of the unsupported length and inversely as the square of the pull. Although the equation is intended for use with a *level* tape, it may be applied without error or consequence to a tape held on a slope up to approximately 10%.

The changes in correction for sag due to variation in tension, weight of tape, and distance between supports are illustrated in Fig. 4.12 for 100-ft and 25-ft spans and in Fig. 4.13 for 10-m and 30-m tapes.

Analysis of Figs. 4.12 and 4.13 reveals the hazard of measuring with a heavy tape without using a tension handle. A variation of −0.5 kg, from a standard tension of

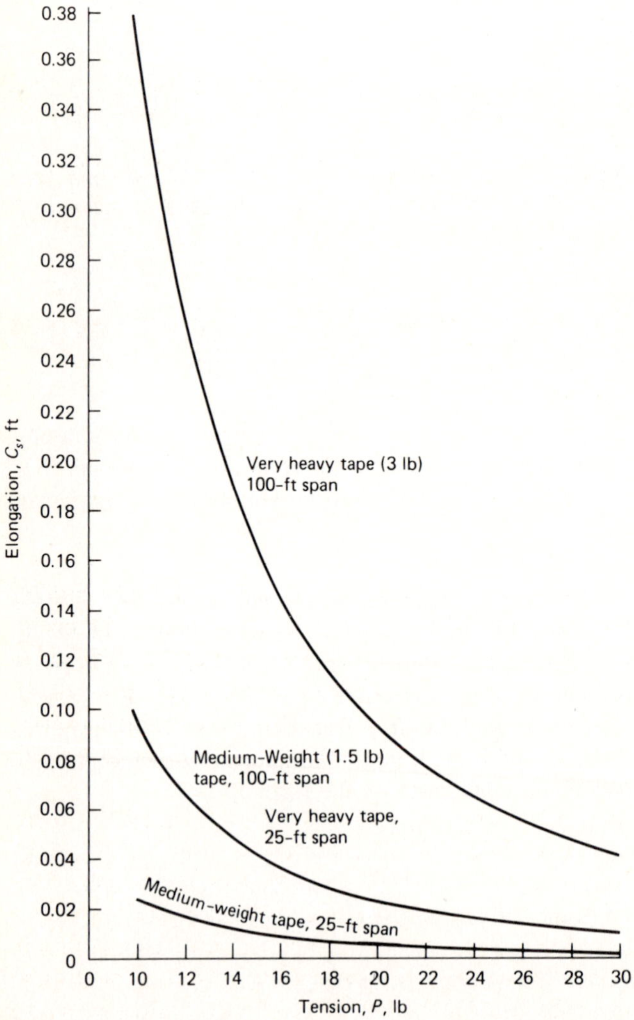

Fig. 4.12 Sag correction versus tension, 100- and 25-ft spans.

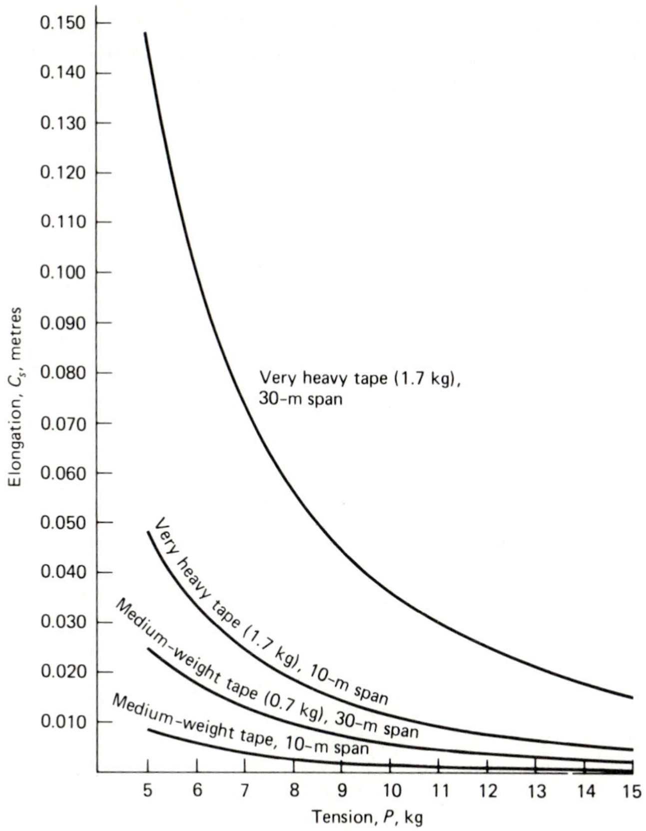

Fig. 4.13 Sag correction versus tension, 30- and 10-m spans.

8 kg, with a very heavy tape over a 30-m span results in a discrepancy of 8 mm in the sag correction or a relative precision of 1 part in 3750. A similar variation in tension using a medium-weight 30-m tape causes discrepancies of 1.5 mm for a relative precision of 1 in 20,000. Estimation of tension to within ± 0.5 kg or about ± 1 lb is very difficult, especially with a heavy tape. Consequently, even for ordinary taping where a relative precision of 1 part in 5000 is sought, a tension handle should be used for measurements with heavy tapes.

The effect of reducing span length is also illustrated in Figs. 4.12 and 4.13, where discrepancies in C_s are substantially reduced for the shorter 10-m and 25-ft spans. This factor should not be interpreted to indicate that shorter tape length should be used, as the error propagation due to setting and measuring additional points would probably offset any reduction in sag correction errors. Reduction of span length is most useful when using long heavy tapes for precise base-line measurement (Art. 9.6). Note that a 25-ft unsupported span of a very heavy 100-ft tape has the same sag corrections as a medium-weight 100-ft unsupported span.

4.21. Normal tension By equating the right-hand members of Eqs. (4.7) and (4.8), the elongation due to increase in tension is made equal to the shortening due to sag; thus, the effect of sag can be eliminated. The pull that will produce this condition, called *normal tension* P_n, is given by the formula

$$P_n = \frac{0.204W\sqrt{aE}}{\sqrt{P_n - P_0}} \tag{4.9}$$

This equation can be solved using successive approximations for P_n. Normal tensions for very heavy ($W = 3$ lb or 1.7 kg; $a = 0.01$ in^2 or 0.06 cm^2) and medium-weight ($W = 1.5$ lb or 0.7 kg; $a = 0.005$ in^2 or 0.03 cm^2) 100-ft and 30-m tapes assuming that $P_0 = (10$ lb, 5 kg) and $E = (30{,}000{,}000$ lb/in^2, 2,100,000 kg/cm^2), respectively, are as follows:

	Very heavy tape		Medium-weight tape	
	100 ft	30 m	100 ft	30 m
Normal tension	51.8 lb	26.5 kg	28.0 lb	12.8 kg

It can be seen that the normal tension for the very heavy tape over the long span is not practical.

4.22. Imperfect alignment of tape The head tapeman is likely to set the pin sometimes on one side and sometimes on the other side of the correct line. This produces a variable systematic error, since the horizontal angle that the tape makes with the line is not the same for one tape length as for the next. The error cannot be eliminated, but it can be reduced to a negligible quantity by care in lining. Generally, it is the least important of the errors of taping, and extreme care in lining is not justified. The linear error when one end of the tape is off line a given amount can be computed by Eq. (4.4) in the same manner as the slope correction. For a 100-ft tape, the error amounts to 0.005 ft when one end with respect to the other is off line 1 ft and to only 0.001 ft when the error in alignment is 0.5 ft. Displacements at the ends of a 30-m tape of 3 dm and 1.5 dm produce discrepancies of 1.5 mm and 0.4 mm, respectively. Many surveyors use unnecessary care in securing good alignment without paying much attention to other more important sources of error. Errors in alignment tend to make the measured length between fixed points greater than the true length and hence are positive.

4.23. Tape not straight In taping through grass and brush or when the wind is blowing, it is impossible to have all parts of the tape in perfect alignment with its ends. The error arising from this cause is systematic and variable and is of the same sign (positive) as that from measuring with a tape that is too short. Care must be exercised to stretch the tape taut and to observe that it is straight by sighting over it, so that the error is not of consequence.

Table 4.2 Systematic Errors in Taping

Source	Amount	Error of 0.01 ft/100 ft or 3 mm/30 m tape length caused by:	Makes tape too:	Importance in 1:5000 taping	Procedure to eliminate or reduce	
Tape not of standard length	~ 0.05 ft/100 ft or 1.5 mm/30 m	—	Long or short	Usually small but must be checked	Standardize tape and apply computed correction—cumulates ∝ to length used	
Temperature	$C_t = \alpha L(T - T_0)$	15°F or 9°C	Long or short	Of consequence only in hot or cold weather	Measure temperature and apply computed correction; for precise work tape at favorable times and/or use invar tape	
Change in pull or tension	$C_p = \frac{(P - P_0)L}{aE}$	15 lb or 4.2 kg	Long or short	Negligible	Apply computed correction; in precise work use a tension handle	Use normal tension $P_n = \frac{0.204W\sqrt{aE}}{\sqrt{P_n - P_0}}$ or use standard tension P_0
Sag	$C_s = \frac{w^2L^3}{24P^2} = \frac{W^2L}{24P^2}$	$\Delta P = 0.6$ lb or[a] 1 kg too small	Short	Large, especially with heavy tape	Apply computed correction; use tape supported fully	
Slope	$H = (s^2 - h^2)^{1/2}$ $C_h = -\frac{h^2}{2s}$ (approx.) $H = s\cos\theta$	1.4 ft or[b] 0.42 m in h, 0°48′ in θ (both from horizontal)	Short	—	At breaks in slope determine difference in elevation or slope angle; apply computed correction	
Imperfect horizontal alignment	Same as slope	1.4 ft or 0.42 m	Short	Not serious	Use reasonable care in aligning tape; keep tape taut and reasonably straight	

Note: In measuring a distance between fixed points with a tape that is too long, add the correction (assume a tape of medium weight).

[a]Error affected nonlinearly by the tension P.

[b]Nonlinear relationship, error increases nonlinearly as the slope increases.

Table 4.3

Taping operation	Sign of correction to be applied	
	Tape too long	Tape too short
Measurement	+	−
Layout	−	+

4.24. Summary of systematic errors in taping The various systematic errors discussed in Arts. 4.13 through 4.23 are summarized in Table 4.2. The apparently simple task of linear measurement is affected by a remarkable number of factors.

When corrections are applied to the observed length of a line measured between fixed points (designated *measurement*) with a tape that is too long, the correction is *added*. When layout of a distance is performed (i.e., a second point is established at a specified distance from a starting point) with a tape that is too long, the correction is subtracted from the specified distance to determine the distance to be read on the tape. For a tape that is too short, the corrections are opposite in direction to those just stated. Table 4.3 provides a convenient summary of the signs for these corrections.

Refer to the sample problems in Art. 4.25 for examples of how corrections are applied.

4.25. Combined corrections Whenever corrections for several effects such as slope, tension, temperature, and sag are to be applied, for convenience they may be combined as a single net correction per tape length. Since the corrections are relatively small, the value of each is not appreciably affected by the others and each may be computed on the basis of the nominal tape length. For example, even if the verified length of a tape were 100.21 ft, the correction for temperature (within the required precision) would be found to be the same whether computed for the exact length or for a nominal length of 100 ft. The combined correction per tape length can be used as long as conditions remain constant; however, the temperature is likely to vary considerably during the day, and the ground may be such that it may be necessary to support the tape at various intervals of length and with different slopes so that corrections must be calculated for individual tape lengths (see Example 4.5). Several examples to illustrate applying corrections follow.

Example 4.8 A 30-m tape weighing 0.55 kg and with a cross-sectional area of 0.02 cm^2 was standardized and found to be 30.005 m at 20°C, with 5-kg tension and supported at the 0 and 30-m points. This tape was used to measure a distance of about 89 m over terrain of a uniform 5 percent slope. The temperature was constant at 30°C, the tape was fully supported throughout, and tension of 5 kg was applied to each tape length. The observed distances were 30.000 m, 30.000 m, and 29.500 m. Calculate the horizontal distance between the points.

SOLUTION Slope, temperature, and standardization are in direct proportion to the length and can be calculated for the total distance. On the other hand, the sag correction must be calculated for each segment.

Distance corrected for slope [Eq. (4.4)]:

$$H = [(s^2 - h^2)]^{1/2} = [(89.500)^2 - (0.05 \times 89.500)^2]^{1/2}$$
$$= [(1 - 0.0025)(89.500)^2]^{1/2}$$
$$= 89.388 \text{ m}$$

Sag correction [Eq. (4.8)] for two 30.000-m and one 29.500-m tape lengths is

$$C_s = (2)\frac{(0.55)^2(30.005)}{(24)(5)^2} + \frac{[0.55(29.5/30)]^2(29.500)(30.005/30.000)}{(24)(5)^2} = 0.045 \text{ m}$$

Since the tape was fully supported the tape is long and C_s is positive.

Temperature correction [Eq. (4.6)]:

$$C_t = (0.0000116)(89.50)(30 - 20) = 0.0104 \text{ m} \qquad \text{tape is long}$$

Standardization correction [Eq. (4.5)]:

$$C_d = (30.005 - 30.000)\left(\frac{89.50}{30}\right) = 0.0149 \text{ m} \qquad \text{tape is long}$$

Corrected distance = 89.388 + 0.045 + 0.010 + 0.015 = 89.458 m.

Example 4.9 A 100-ft tape has a standardized length of 100.000 ft at 68°F, with 20-lb tension and with the tape fully supported. The observed distance between two points, with the tape held horizontal, supported at the 0 and 100-ft points only, and 30-lb tension is 100.452 ft. The tape weighs 3.0 lb and has a cross-sectional area of 0.01 in^2. Calculate the true distance between the two points. Assume that $E = 3 \times 10^7$ lb/in^2.

SOLUTION First calculate the sag correction at the standard tension with Eq. (4.8):

$$C_s = \frac{(3)^2(100.452)}{(24)(30)^2} = 0.042 \text{ ft}$$

The tape is shorter, so the correction is negative when measuring between two fixed points. Next, compute the correction due to the increase in tension using Eq. (4.7):

$$C_p = \frac{(30 - 20)(100.452)}{(0.01)(3)(10^7)} = 0.003 \text{ ft}$$

The increased tension makes the tape long, so that the correction is positive. The corrected distance is

$$100.452 - 0.042 + 0.003 = 100.413 \text{ ft}$$

Note that the foregoing corrections can be calculated with adequate accuracy with a slide rule and using $L = 100$ ft without introducing significant errors:

$$C_s = \frac{(3)^2(100)}{(24)(30)^2} = 0.042 \text{ ft}$$

$$C_p = \frac{(30 - 20)(100)}{(0.01)(3)(10^7)} = 0.003 \text{ ft}$$

4.26. Random errors in taping Random errors in taping occur as a result of human limitations in observing measurements and in manipulating the equipment. There are several causes for random errors in taping. For each source the same

Table 4.4 Random Errors in Taping

Designation	Source of random variation	Governing conditions, causes, and manner of cumulation	Estimated value per tape length
σ_v	Plumbing and uncertainty in marking tape ends	Rugged terrain, breaking tape frequently; cumulates randomly $\propto \sqrt{n}$	0.03–0.10 ft (10–30 mm)
σ_m	Uncertainty in marking tape ends with tape fully supported	Tape graduated to hundredths of ft or mm; cumulates randomly $\propto \sqrt{n}$	0.01 ft (3 mm)
σ_p	Uncertainty in applying tension	Change in sag correction due to variations in tension of ± 2 lb or 0.9 kg from standard tension; cumulates $\propto \sqrt{n}$	0.01 ft (3 mm)
σ_h	Uncertainty in determining elevation difference or slope angle (assume a maximum 6 percent slope)	In $h = \pm 0.3$ ft; ± 0.09 m In $\theta = \pm 0°10'$; cumulates $\propto \sqrt{n}$	0.02 ft (7 mm)
σ_t	Uncertainty in reading temperature	Observe t to $\pm 1°$F; cumulates $\propto n$	0.001 ft (0.3 mm)

symbol, σ, will be used to designate the standard error, but with the appropriate subscript. Thus, σ_v represents the standard error due to plumbing the tape ends, σ_m designates the standard error due to marking the supported tape ends, and so on as summarized in Table 4.4.

Corrections for most systematic errors depend on observing or otherwise determining quantities such as temperature t, tension or pull p, slope angle θ, or

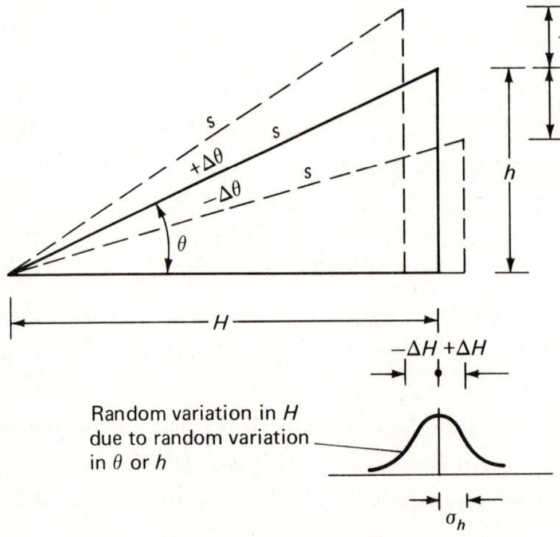

Fig. 4.14

elevation difference h. For given specific values of these quantities, *systematic* corrections are calculated from specified formulas and applied to the measured distance. However, each of these quantities is subject to *random* errors which when propagated through the formulas result in random errors in the distance. The latter random errors are due to random variation in the distance corrected for systematic errors. Figure 4.14 shows one example for the slope angle θ. The measured slope distance is s and for the given slope angle θ, the horizontal distance is H [calculated by Eq. (4.1)]. A small positive change $+\Delta\theta$ in the angle θ would result in a shortening of H, while a negative change $-\Delta\theta$ causes an increase in the value of H. The same results are obtained with changes in the elevation difference h, as shown in Fig. 4.14. Since the variations in θ or h are random, the corresponding variation in H will also be random. This is schematically shown in Fig. 4.14, with the corresponding standard error designated by σ_h.

The rest of the standard errors as listed in Table 4.4 are:

σ_p is the standard error due to uncertainty in the value of the applied pull or tension.
σ_h is the standard error in determining either the slope angle θ or the elevation difference h, as explained above.
σ_t is the standard error in determining the temperature.

Also listed in Table 4.4 are estimates of each error *per tape length*, assuming experienced personnel, conditions as given in the table, and medium-weight (1.5-lb or 0.7-kg) 100-ft or 30-m tapes having cross-sectional areas of 0.006 in^2 and 0.040 cm^2, respectively. Because all these errors are random, they accumulate as the *square root of the number of tape lengths*, n, as indicated in Art. A.2.

Note that random errors can occur as a result of (1) direct observations, such as marking tape ends, reading graduations, and so on; or (2) indirect observation of the difference in elevation or slope between the tape ends, applying tension, and so on.

Systematic errors are of a much greater magnitude than are random errors. When every device is employed to eliminate systematic errors using the methods described in Arts. 4.13 through 4.23, random errors of observation become more significant; for this reason long tapes (200 ft, 50 m) are frequently employed for base-line measurement.

4.27. Effects of errors on taping It would be valuable if a definite outline of procedure could be established to produce any desired degree of precision in taping. Unfortunately, the conditions are so varied, and so much depends upon the skill of the individual, that the surveyor must be guided largely by experience and by knowledge of the errors involved. However, by making certain assumptions concerning the errors involved and procedures employed, a reasonably methodical approach can be developed which is very useful in establishing procedures related to measurements required for specific tasks.

The usual practice in rough taping through broken country (maximum slopes 5 percent) is to take measurements with the tape horizontal, plumbing at the downhill end, breaking tape where necessary, applying tension by estimation, and making no corrections for sag, temperature, or tension. The tape is usually 100 ft or 30 m long and weighs about 2 lb or 1 kg. For this case the most significant errors and assumed magnitudes are:

Cause of error	Assigned error (σ) ft	m	Effect on measured line
e_h, tape not level	0.05	0.015	Systematic
e_s, sag in tape	0.03	0.009	Systematic
σ_v, plumbing and marking	0.03	0.010	Random

For a distance of 1000 ft (10 tape lengths) the total error would be $10 \times (0.05 + 0.03) = 0.80$ ft due to the tape not being level and sag plus $0.03\sqrt{10} = 0.10$ ft due to plumbing, for a total of 0.90 ft. Therefore, under these assumptions the relative accuracy is only about 1 part in 1100.

In order to achieve a relative accuracy of 1 part in 5000, much more control must be exercised over the taping procedure. For a distance of 1000 ft the total allowable error is $(1000)(1/5000) = 0.2$ ft. A breakdown of this total error, which is often referred to as the *error budget*, is necessary so that one can employ the proper procedures to meet the stated requirements. The following table gives the typical magnitudes for the most significant errors:

Cause of error	Assigned errors ft	m	Effect on measured line
Plumbing, σ_v	0.03	0.006	Random
Tension, σ_p	0.01	0.003	Random
Temperature, σ_t	0.004	0.001	Random
Tape not level, σ_h	0.02	0.001	Random

Thus, the total error is $(0.03 + 0.01 + 0.004 + 0.02)\sqrt{10} = 0.202$ ft, yielding the desired relative accuracy of 1 part in 5000.

To meet the stated conditions, the following specifications should govern the taping operation (assume a 100-ft medium-weight tape having a cross-sectional area of 0.006 in^2, and maximum slopes of 5 percent):

1. The tape shall be standardized with the tape supported at two ends at 68°F and with 12 lb of tension.
2. A tension handle shall be used with tension applied to within ± 2 lb of the standardized tension.
3. End points of each tape length shall be plumbed and marked to within ± 0.02 ft.
4. A hand level shall be used to ensure that the two tape ends of a full tape length are within ± 0.5 ft of horizontal, assuming the maximum 5 percent slope.
5. Temperatures at the site of the measurement shall be observed to within ± 7°F.
6. In general, increased precision is attained when measuring with the tape fully supported.

Note that some rather stringent specifications are necessary to ensure a relative accuracy of 1 part in 5000. Note also that more than half the total error occurs in plumbing or marking the end points of each tape length. Results can be improved by using a longer tape with fewer tape lengths to be marked.

4.28. Mistakes in taping Some of the mistakes commonly made by individuals inexperienced in taping are:

1. *Adding or dropping a full tape length* This is not likely to occur if both tapemen count the pins, or when numbered stakes are used, if the rear tapeman calls out the station number of the rear stake in response to which the head tapeman calls out the number of the forward stake while marking it. A tape length may be added through failure of the rear tapeman to give the head tapeman a pin at breaks marking fractional tape lengths. A tape length may be dropped through failure of the rear tapeman to take a pin at the point of beginning.
2. *Adding a foot or a decimetre* This usually happens in measuring the fractional part of a tape length at the end of the line. This distance should be checked by the head tapeman holding the zero mark on the tape at the terminal point and the rear tapeman noting the number of feet or decimetres and approximate fraction at the last pin set.
3. *Other points incorrectly taken as* 0 *or* 100-*ft* (*or* 0 *and* 30-*m*) *marks on tape* The tapeman should note whether these marks are at the end of rings or on the tape itself, also whether there is an extra graduated unit at one end of the tape (i.e., whether it is an add or cut tape).
4. *Reading numbers incorrectly* Frequently "68" is read "89" or "6" is read as "9." As a check, good practice is to observe the number of the foot or decimetre marks on each side of the one indicating the measurement, especially if the numbers are dirty or worn. Also, the tape should be read with the numbers right side up.
5. *Calling numbers incorrectly or not clearly* For example, 50.3 might be called "fifty, three" and recorded as 53.0. If called as "fifty, point, three" or preferably "five, zero, point, three," the mistake would not be likely to occur. When a decimal point or a zero occurs in a number, it should be called. When numbers are called to a recorder, the note keeper should repeat them as they are recorded.

Often, large mistakes will be prevented or discovered if those involved in the work form the habit of pacing distances or of estimating them by eye. If a transit is being used to give line, distances can be checked by reading the approximate stadia interval on a range pole.

4.29. Electronic distance measurement Electronic distance measuring (EDM) equipment includes electro-optical (light waves) and electromagnetic (microwaves) instruments. The first generation of electro-optical instruments, typified by the Geodimeter, was originally developed in Sweden in the early 1950s. The basic principle of electro-optical devices is the indirect determination of the time required for a light beam to travel between two stations. The instrument, set up on one station, emits a modulated light beam to a passive reflector set up on the other end of the line to be measured. The reflector, acting as a mirror, returns the light pulse to the instrument, where a phase comparison is made between the projected and reflected pulses. The velocity of light is the basis for computation of the distance. A clear line of sight is required, and observations cannot be made if conditions do not permit intervisibility between the two stations. The maximum range for first-generation instruments was from 3 to 5 km (2 to 3 mi) in daylight and 25 to 30 km (15 to 20 mi) at night, depending on atmospheric conditions.

Electromagnetic EDM devices, first developed in South Africa during the late 1950s, utilize high-frequency microwave transmission. A typical electromagnetic device, such as the Tellurometer, consists of two interchangeable instruments, one being set up on each end of the line to be measured. The sending instrument transmits a series of microwaves which are run through the circuitry of the receiving unit and are retransmitted to the original sending unit, which measures the time required. Distances are computed on the basis of the velocity of the radio waves. An unobstructed measuring path between the two instruments is necessary. However, intervisibility is not required, and therefore observations can be made in fog and other unfavorable weather conditions. Distances up to 66 to 80 km (40 to 50 mi) can be measured under favorable conditions.

First-generation EDM equipment, although very precise for measuring long distances, was bulky, heavy, and expensive. The practicing surveyor needed a lighter, portable device useful for measuring distances of from a few metres or feet up to 1 km (0.6 mi).

During the mid-1960s miniaturization of electronics and development of small light-emitting diodes allowed the design of a second generation of electro-optical EDM instruments. These instruments were smaller, lighter, required less power, and were easier to operate but had substantially shorter range. Included in this group are instruments that have ranges of from 0.2 m to 5 km (0.7 ft to 3 mi) with inherent standard deviations of ± 2 to ± 10 mm + 0.5 to 5 ppm or ± 0.006 to ± 0.032 ft + 0.5 to 5 ppm.[†] These instruments are well adapted to the needs of the practicing surveyor, and those with ranges of up to 5 km (3 mi) are also applicable to high-order control surveys of fairly large extent, such as city or county surveys.

Incorporation of coherent laser light into the light-wave EDM devices gave rise to a third generation of electro-optical instruments. These instruments have fairly good portability, low power requirements, and ease of readout, coupled with longer range of from 15 m to 60 km (50 ft to 36 mi) and high accuracy of ± 5 mm + 1 ppm (± 0.016 ft + 1 ppm).

Short to medium- and long-range electro-optical EDM instruments of modern design are illustrated in Figs. 4.15 and 4.16 respectively. Concurrent with the rapid development of electro-optical instruments, microwave equipment was also improved, becoming more portable and being provided with direct readout. Figures 4.17 and 4.18 respectively, illustrate long-range and moderate-range microwave equipment. However, none of the aforementioned instruments was really adaptable

[†] ppm = parts per million. Thus, if a distance of 1000 m is measured, the standard deviation would be ± 2 to ± 10 mm + 0.5 to 5 mm.

Fig. 4.15 Short- to medium-range electro-optical EDM. (*Courtesy Keuffel & Esser Co.*)

to use for simultaneously measuring the length and direction of a line. In the mid-1970s EDM instruments were reduced in size sufficiently to be mounted on theodolites. Continued development and refinement of EDM equipment can be expected in the future. As discussed in Chap. 9, EDM equipment has been combined with a theodolite equipped with horizontal and vertical axis encoders enabling real-time reduction of slope distance to positional data using an attached microprocessor. Another device under development is the laser range pole, which will allow proper orientation of a measuring path when the line of sight between two stations is obstructed.

4.30. Basic principles of electro-optical EDM instruments A continuous-wave carrier beam of light, generated in the transmitter, is modulated by an electronic shutter before entering the aiming optics and being transmitted to the other end of

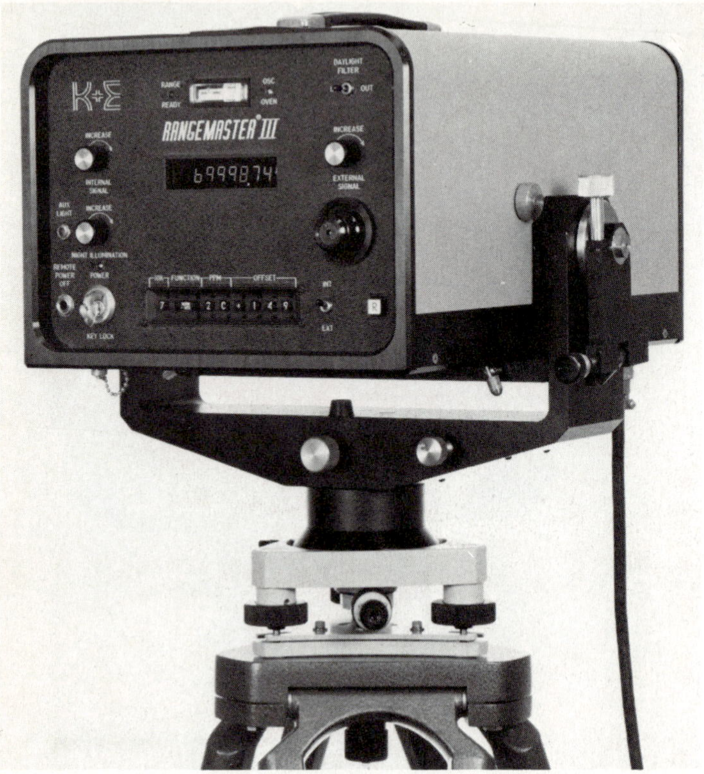

Fig. 4.16 Long-range electro-optical EDM, Rangemaster III. (*Courtesy Keuffel & Esser Co.*)

the line, where a reflector is placed. The original light-wave instruments, such as the Geodimeter, utilized incandescent or mercury light and a Kerr cell modulator to achieve the high-frequency modulation of the beam. More recent instruments use infrared or laser light as carrier beams and ultrasonic modulators to impose the intensity modulation on the continuous-wave beam. This intensity modulation is analogous to turning a light off and on using a switch. In the EDM device, the modulator chops the beam into wavelengths that are proportional to the modulating frequency, where the wavelength is given by

$$\lambda = \frac{V}{f} \tag{4.10}$$

where λ = wavelength
V = velocity of light through the atmosphere, m/s or ft/s
f = modulating frequency, Hz (hertz = cycles per second)

The velocity of light in the atmosphere varies with temperature, humidity, and the partial pressure of water vapor. Corrections for these factors are described in Art. 4.33. The intensity of the modulated light varies from zero or no light at 0° to

Fig. 4.17 Long-range Tellurometer MRA5. (*Courtesy Tellurometer— U.S.A., Division of Plessey, Inc.*)

maximum light at 90° to a second zero at 180°, to a second maximum at 270° and back to zero at 360°, as illustrated graphically in Fig. 4.19. Measurement of a distance is accomplished by placing the electro-optical transmitter at one end and a reflector at the other end of the line to be measured. The reflector is a corner cube of glass in which the sides are perpendicular to within a very close tolerance. It has the characteristic that incident light is reflected parallel to itself and is called a retrodirective prism or *retro-reflector*. The EDM instrument transmits an intensity-modulated light beam to the reflector, which reflects the light beam back to the transmitter, where the incoming light is converted to an electrical signal, allowing a phase comparison between transmitted and received signals. The amount by which transmitted and received wavelengths are out of phase can be measured electronically and registered on a meter to within a few millimetres or hundredths of feet.

Fig. 4.18 Moderate-range Tellurometer CA1000. (*Courtesy Tellurometer— U.S.A., Division of Plessey, Inc.*)

To visualize the operation, consider an electro-optical EDM instrument operating on one modulating frequency and with a null meter to record positive or negative maximum or zero intensity of modulation. This relationship is illustrated schematically in Fig. 4.20. Assume that the instrument centered over A in Fig. 4.20 is an even number of wavelengths λ from the reflector at C and the meter registers zero. Now, move the instrument slowly toward C so that the meter will reach a maximum on the right at 90°, return to zero at 180°, a second maximum on the left at 270°, and a second zero at 360°. This procedure is repeated as the instrument is moved toward C, with the needle on the meter moving from zero to one maximum to zero to

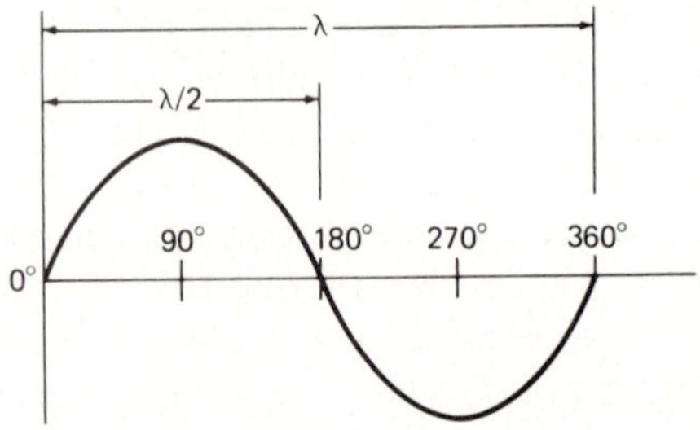

Fig. 4.19 Wavelength of modulated light.

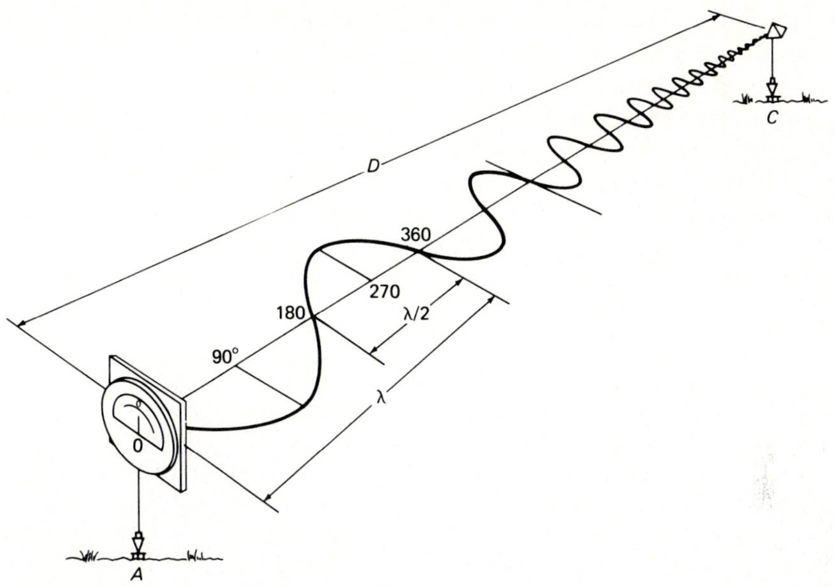

Fig. 4.20 Modulation frequency of EDM. (*Courtesy AGA, Ref. 1*)

another maximum until the instrument is at a distance of $\lambda/2$ from the reflector. Note that the constant or uniform "tape length" between successive zero points is $\lambda/2$ for the double distance.

In practice, the instrument is rarely an even number of wavelengths from the reflector and is not moved along the line of sight toward the reflector. Assume that distance AC is not an integral number of wavelengths. Figure 4.21 shows a section through AC with the transmitter at A and the reflector at C. The modulated beam transmitted at A travels the double distance from A to C and back, where the received distance is out of phase with the transmitted signal by an amount d. The increment d is measured electronically by adjusting some type of phase meter in the instrument, and the distance from A to C is

$$D = \tfrac{1}{2}(n\lambda + d) \tag{4.11}$$

in which λ is the wavelength of the modulated beam and n the integral number of wavelengths in the double path of the light. Note that if the reflector were moved $\lambda/2$ or any even number of $\lambda/2$'s, the increment d does not change. Thus, if one

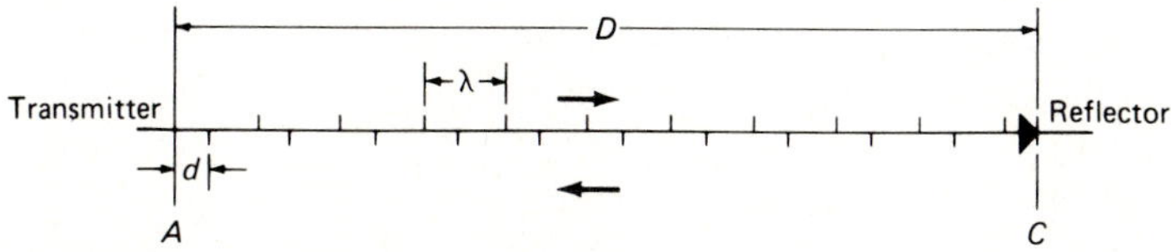

Fig. 4.21 Measurement principle of EDM.

knows the double distance to within $\lambda/2$ or the distance to within $\lambda/4$, the total distance can be resolved.

Knowledge of the distance to within $\pm\lambda/4$ is an impractical constraint on an EDM device. The ambiguity in n is resolved by using some type of multiple-frequency technique. Light is transmitted on one frequency and d_1 is observed. Light is then transmitted on a second but slightly different frequency and d_2 is observed. Two equations (4.11) are formed and solved for n. The multiple-frequency technique is built into the EDM device, permitting electronic determination of n and a direct readout of the distance. Decade modulation is one of several techniques applicable and is used for the following example.

Example 4.10 Assume that a modulation frequency of 15 MHz (megahertz $= 10^6$ cps), giving $\lambda/2$ of 10 m, is used.[†] A full sweep of the phase meter represents this 10-m distance and allows reading the unit and decimal parts of a meter from 0 to 9.999 m. If a given distance is 5495.800 m, the phase meter records 5.800 m on this frequency. This operation is analogous to measuring the distance with a steel tape graduated at every 10-m point (but not labeled), and with one added 10-m section graduated to metres and millimetres, where the increment 5.800 is observed. Now switch to a second frequency with $f = 1.5$ MHz, so that $\lambda/2 = 100$ m. A full sweep of the phase meter allows reading 10 to 90 m and permits resolving the number of tens of metres, or 90 m for this example. Next, switch to 0.15 MHz, yielding $\lambda/2 = 1000$ m, permitting resolution of the number of hundreds of metres, in this case 400 m. Finally, turn to a frequency of 15 kHz (kilohertz $= 10^3$ cps) with $\lambda/2 = 10{,}000$ m, which will allow resolving the number of 1000 m in the distance, 5000 m for the example. The total distance is $5.800 + 90 + 400 + 5000 = 5495.800$ m.

4.31. Electromagnetic or microwave EDM principles Microwave EDM transmitters generate electromagnetic carrier beams having frequencies of from 3 to 35 GHz (gigahertz $= 10^9$ cps) and modulated with frequencies varying from 10 to 75 MHz. The wavelength is given by Eq. (4.10), in which λ is the wavelength of modulation in metres or feet, V the velocity of microwaves through the atmosphere in metres m/s or ft/s, and f the modulating frequency in Hz. The velocity V is a function of temperature, atmospheric pressure, and the partial pressure of water vapor. Details of corrections for these factors are given in Art. 4.33.

The fundamental concepts of microwave instrument operations are similar to those involved in electro-optical EDM devices. The major difference is that two similar instruments and two operators, one for each end of the line, are required to measure a line using microwave equipment. The master unit transmits to the remote unit, which acts as a reflector for the transmitted microwaves. Voice communication is available for the operators at each unit. The operator selects a frequency at the master unit and by voice signal transmits this information to the operator at the other station, who sets the remote unit to correspond to the transmitted frequency. The signal is received at the remote unit and retransmitted to the master station, where the odd increment is measured with a phase meter.

[†]Assume $V \simeq 300{,}000$ km/s.

The number of integral measuring units n between the two ends of the line is determined by the method of multiple frequencies similar to that employed in the electro-optical EDM instrument. An example of a typical long-range microwave instrument is illustrated in Fig. 4.17. The multiple frequencies employed in this device are:

A pattern: 10,000 MHz
B pattern: 9.990 MHz
C pattern: 9.900 MHz
D pattern: 9.000 MHz

The *A* pattern resolves the nearest 50 ft and decimal parts of 50 ft. Combinations of the *A* and *D*, *A* and *C*, and *A* and *B* patterns enable resolving the nearest 500, 5000, and 50,000 ft, respectively.

4.32. Systematic errors in EDM equipment Effects of atmospheric conditions on wave velocity, uncertainties in the position of the electrical center of the transmitter, uncertainties in the effective center(s) of the reflectors(s), and transmitter nonlinearity all contribute to the systematic errors found in microwave instruments.

4.33. Effects of atmospheric conditions on wave velocity The velocity V of electromagnetic waves in air is a function of the speed of light in a vacuum V_0 and the refractive index of air n, yielding

$$V = \frac{V_0}{n} \tag{4.12}$$

The constant V_0 is 299,792.5 km/s. The refractive indices of light waves and microwaves in air are functions of air temperature, atmospheric pressure, and the partial pressure of water vapor, which in turn depend on temperature and relative humidity. Thus, a knowledge of these atmospheric conditions is required to determine the refractive index and the consequent effects on the velocity of the propagated waves in air. Since light waves and microwaves react somewhat differently to varying atmospheric conditions, each is treated separately.

For light waves, it is first necessary to calculate the index of refraction n_g of standard air given by the Barrell and Sears equation for an atmosphere at 0°C, 760 mmHg pressure, and 0.03 percent carbon dioxide:

$$n_g = 1 + \left(287.604 + \frac{4.8864}{\lambda^2} + \frac{0.068}{\lambda^4}\right)10^{-6} \tag{4.13}$$

in which λ is the wavelength of the carrier beam of light in micrometres. Values of λ for various sources of light are:

Carrier	λ, μm
Mercury vapor	0.5500
Standard lamp	0.5650
Red laser	0.6328
Infrared	0.900–0.9300

Owing to changes in temperature, pressure, and humidity, the refractive index of air becomes n_a, is given by

$$n_a = 1 + \frac{0.359474(n_g - 1)p}{273.2 + t} - \frac{1.5026e(10^{-5})}{273.2 + t} \tag{4.14}$$

where p = atmospheric pressure, mmHg
t = air temperature, °C
e = vapor pressure, mmHg

The humidity represented by water vapor pressure in the second term of Eq. (4.14) has little effect on light waves and usually can be ignored. For measurements of high precision an average value of 0.5 ppm can be assumed for humidity correction and is substituted for the second term in Eq. (4.14).

Example 4.11 An electro-optical instrument utilizing infrared light with a wavelength of 0.9100 μm has a modulation frequency at 24.5 MHz. At the time of measurement the temperature was 27.0°C and atmospheric pressure was 755.1 mmHg. Calculate the modulated wavelength of the light under the given atmospheric conditions.

SOLUTION Using Eq. (4.13), the refractive index of air under standard conditions is

$$n_g = 1 + \left[287.604 + \frac{4.8864}{(0.91)^2} + \frac{0.068}{(0.91)^4}\right]10^{-6} = 1.0002936$$

The refractive index of air under the given conditions by Eq. (4.14), neglecting the second term, is

$$n_a = 1 + \frac{(0.359474)(1.0002936 - 1)(755.1)}{273.2 + 27.0} = 1.00026547$$

The velocity of the infrared light through the atmosphere by Eq. (4.12), where $n = n_a$, is

$$V_a = \frac{299{,}792.5}{1.00026547} = 299{,}712.9 \text{ km/s}$$

and by Eq. (4.10) the modulated wavelength in the given atmosphere is

$$\lambda = \frac{299{,}712.9}{(24.5)(10^6)} = 0.01223318 \text{ km} = 12.23318 \text{ m}$$

Microwaves are more sensitive than light waves to humidity. The refractive index of the atmosphere for microwaves is

$$(n_r - 1)10^6 = \frac{103.49}{273.2 + t}(p - e) + \frac{86.26}{273.2 + t}\left(1 + \frac{5748}{273.2 + t}\right)e \tag{4.15}$$

where n_r = refractive index
p = atmospheric pressure, mmHg
e = vapor pressure, mmHg
t = temperature, °C

The velocity of microwaves in air is given by Eq. (4.12), where $n = n_r$.

Example 4.12 The modulation frequency for microwave transmission is exactly 10 MHz at a temperature of 15.4°C, atmospheric pressure of 645 mmHg, and vapor pressure of 3.8 mmHg. Calculate the modulated wavelength under these conditions.

SOLUTION Using Eq. (4.15), we obtain

$$(n_r - 1)(10^6) = \frac{103.49}{288.6}(645 - 3.8) + \frac{86.26}{288.6}\left(1 + \frac{5748}{288.6}\right)(3.8)$$

$$(n_r - 1)(10^6) = 253.69$$

$$n_r = 1.00025369$$

By Eq. (4.12),

$$V = \frac{299{,}792.5}{1.00025369} = 299{,}716.5 \text{ km/s}$$

and by Eq. (4.10) the modulated wavelength in the atmosphere is

$$\lambda = \frac{299{,}716.5}{(10)(10^6)} = 0.02997165 \text{ km} = 29.97165 \text{ m}$$

The examples indicate that as atmospheric conditions vary, the velocity of the modulated wave is altered, resulting in a corresponding change in the modulated wavelength and hence in the basic measuring unit of the EDM instrument.

The method of correcting the observed slope distance to account for varying atmospheric conditions is achieved differently for each type of EDM instrument. For light-wave instruments the temperature and atmospheric pressure are recorded at each end of the line. Corrections are then calculated using the average of the observed meteorological data, or the instrument circuitry is modified to permit automatic compensation as measurements progress. Most recent electro-optical EDM systems allow dialing in temperature and atmospheric pressure for this automatic compensation. In recording meteorological data, care should be exercised to obtain air temperatures. Estimating temperature to $\pm 10°\text{F}$ ($\simeq \pm 6°\text{C}$) and atmospheric pressure to $\simeq \pm 1$ in or 25 mmHg will introduce a relative error of 10 ppm in light-wave equipment. Thermometers should be placed above the ground and in the shade. Atmospheric pressure can be determined from readings on an aneroid barometer (Art. 5.6) converted to the appropriate pressure reading in mmHg or inHg using a chart provided by the instrument manufacturer.

The partial pressure of water vapor as determined by wet- and dry-bulb thermometer readings must be determined for microwave instruments. An error of 1°C between wet- and dry-bulb temperatures at normal conditions (temperature of 20°C, atmospheric pressure of 760 mmHg) produces a relative error of 7 ppm in the distance. This relative error at 45°C is 17 ppm. Consequently, wet- and dry-bulb temperatures must be carefully observed using calibrated equipment. Once the meteorological data are recorded, corrections to observed slope distances are made using charts and nomographs provided with the instrument.

4.34. Systematic Instrumental errors in EDM systems Systematic instrumental errors occurring in electro-optical systems include uncertainties in the position of the electrical center of the transmitter, uncertainties in the effective center of the reflectors, frequency drift, and instrument nonlinearity. The first two sources of error must be taken into account in all survey measurements, the third requires constant monitoring, and the fourth is critical only for measurements of high precision.

Microwave systems are affected by uncertainties in the electrical centers of the master and remote units and by a phenomenon called *ground swing* or *reflection*.

In EDM systems properly adjusted at the factory, the errors noted above will be very small and in a practical sense may be insignificant. However, it is important that users of these systems realize that periodic calibration of the instruments against a known distance is absolutely necessary to assure consistent results. First, consider calibration of an electro-optical system.

Uncertainties in the effective center of the reflector are illustrated by reference to Fig. 4.22, which shows a cross section through a corner-cube retro-reflector. The distance from the face of the cube to the back corner is t. The ray path within the cube is $a + b + c = 2t$. Owing to the refractive properties of the glass in the cube, the equivalent travel in air of the ray path within the cube is $1.517 \times 2t$. Point D represents the effective center of the corner cube and is literally the end of the line being measured. If it were possible to mount the reflector so that D coincided with the plumb line, the reflector offset would be zero. However, D is so far behind the face of the prism that such an arrangement would be unbalanced and difficult to manage in the field. Prisms are usually mounted so that c_R is from 28 to 40 mm. Some manufacturers eliminate c_R by making an adjustment in the transmitter to absorb the offset.

When slope distances are measured, light rays striking the reflector are not perpendicular to the front face of the reflector, as indicated in Fig. 4.22, thus altering

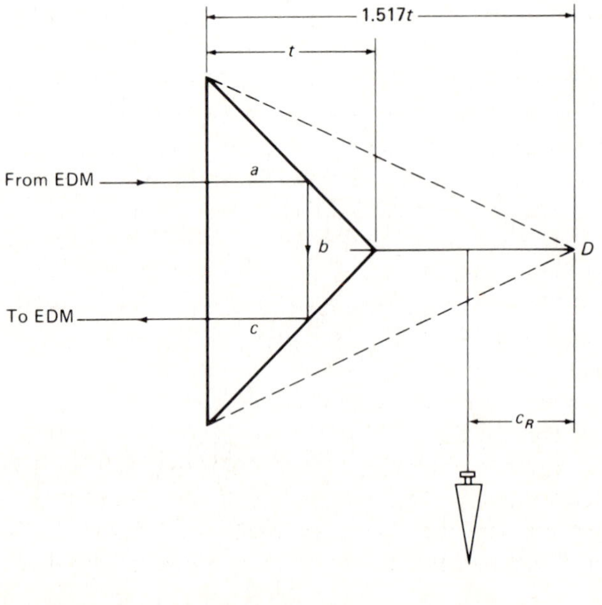

Fig. 4.22 Reflector offset.

the path of the rays within the reflector and changing the position of the effective center (Kivioja, Ref. 8). The amount of this change is a function of the degree of slope and varies from a few tenths of a millimetre at slopes of 4° to 6° to 7 to 14 mm at slopes from 30° to 40°. Given the angle of slope, corrections can be calculated to compensate for this error. Another option is to design the reflector with an adjustment allowing the front face of the reflector to be placed perpendicular to the incoming rays of light compensating for the error instrumentally. Details concerning the formulas for making corrections can be found in Ref. 8. Corrections of this type would be necessary on surveys of high precision where slope angles are consistently large.

At the other end of the line, if the plumb line does not coincide with the exact electrical center of the electro-optical transmitter, an instrument offset exists.

Similar errors exist in microwave systems where the instrument offsets in the master and remote transmitters are analogous to the instrument and reflector offsets of the light-wave system.

A precisely taped base line is required to calibrate an EDM system if high-precision results are to be obtained. In establishing this base line, all the procedures recommended for base-line taping should be followed (Art. 9.6). Preferably, the length of the base line should conform to the average length of line to be measured with the equipment.

The calibration procedure is the same for both the electro-optical and microwave systems. A series of EDM distances should be recorded (ten, for example). Meteorological data should be gathered with extreme care. The average of the EDM distances corrected for meteorological conditions and for slope of the line should agree with the taped distance for the base line. The difference between the two measurements represents the system constant, Δ_s. In electro-optical systems a system constant should be determined for each reflector or group of reflectors to be used. Reflectors should be numbered and constants recorded for use in subsequent surveys.

Once a base line has been established for calibration purposes, the instrument should be recalibrated by remeasuring the base line at periodic intervals to guard against frequency drift. The time between check calibrations is a function of the age of the equipment and the user's confidence in it. With new equipment a check every week would not be unreasonable until confidence in the system stability is established. Thereafter, once per month would be sufficient. The same equipment should be used for each check calibration. The mean of the observations should be plotted versus the date of measurement and a record of meteorological conditions should be maintained. Ideally, the graph should be a horizontal straight line within the specification range for the system.

If a known base line is not available, calibration on a line of unknown length is possible. This type of calibration is accomplished by measuring a line of unknown length in several sections. Let the distance D in Fig. 4.23 be divided into arbitrary increments $d_1, d_2, \ldots, d_n$. Measure the total length of line using the EDM system and then measure the n sections separately, where $n = 2$ represents the minimum number of sections. For electro-optical systems the same reflector should be used for

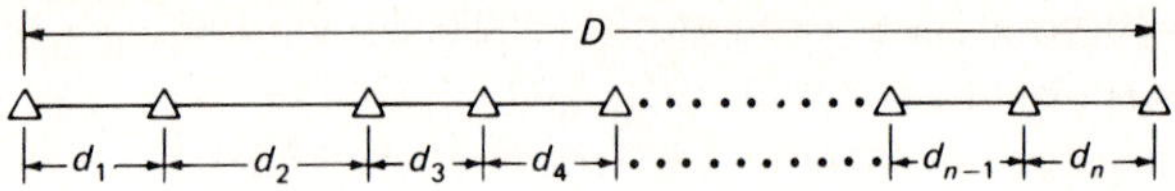

Fig. 4.23 Base-line calibration using a line of unknown length.

the entire set of measurements. All distances should be corrected for meteorological conditions and slope. Let the system constant be Δ_s so that the total distance is $D + \Delta_s$, each of the increments is $d_1 + \Delta_s, d_2 + \Delta_s, \ldots, d_n + \Delta_s$ and the following equation can be formed:

$$D + \Delta_s = (d_1 + \Delta_s) + (d_2 + \Delta_s) + \cdots + (d_n + \Delta_s)$$

from which

$$\Delta_s = \frac{-(d_1 + d_2 + \cdots + d_n) + D}{n - 1} \tag{4.16}$$

If the distance is divided into sections such that each section contains a distance with a different unit digit (e.g., for five sections $d_1 = 152$ m, $d_2 = 153$ m, $d_3 = 241$ m, $d_4 = 305$ m, $d_5 = 206$ m, and $D = 1057$ m), then errors due to nonlinearity (discussed in the next section) will be averaged and the net effect of this error on Δ_s will be small.

4.35. Nonlinearity in electro-optical systems A nonlinear periodic bounded error occurs in all electro-optical instruments. As the distance between the transmitter and reflector is changed, the error will rise to a maximum, fall to a minimum, rise to a maximum, fall to a minimum, and so forth, repeating over and over again within the same boundary. To achieve maximum accuracy from a given electro-optical system, this cyclic error must be evaluated.

The nonlinearity of an instrument can be detected by taking a series of readings, from a single instrument setup, to a reflector set successively on a series of points located at intervals the sum of which should equal one full period of the nonlinearity. It is not necessary to know the total distance from the transmitter to this span of intervals, but the distances between the points must be determined with the highest possible precision.

For example, assume that the instrument to be calibrated has a basic measuring unit of 10 m. Figure 4.24 illustrates the configuration for the calibration base line

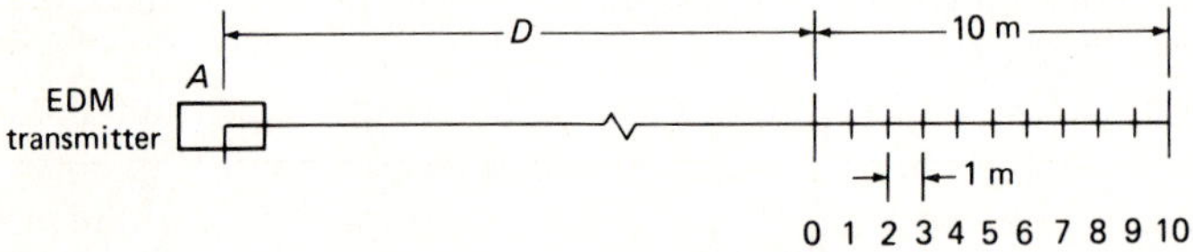

Fig. 4.24 Setup for calculating nonlinear errors in EDM.

where points are set at 1-m intervals over a span of 10 m, resulting in 11 stations. The series of measurements observed will be:

$$
\begin{array}{ll}
D + e_0 & D + 6 + e_6 \\
D + 1 + e_1 & D + 7 + e_7 \\
D + 2 + e_2 & D + 8 + e_8 \\
D + 3 + e_3 & D + 9 + e_9 \\
D + 4 + e_4 & D + 10 + e_{10} \\
D + 5 + e_5 &
\end{array}
$$

where D is the distance from instrument to the first station of the 10-m span and $e_1, e_2, \ldots, e_{10}$ are the errors in each reading due to nonlinearity. Subtract D, $D + 1$, $D + 2, \ldots, D + 10$ from each successive measurement to yield $e_1, e_2, \ldots, e_{10}$. Figure 4.25 shows the corrections or plot of $-e_0, -e_1, \ldots, -e_{10}$ versus distance for a short-range electro-optical system. Thus, if the system constant, as determined by one of the methods outlined in Art. 4.34, is -0.005 m and the observed slope distance is 342.541 m, the correct slope distance is $342.541 - 0.005 - 0.004 = 342.532$ m, where -0.004 m is taken at the intersection of the ordinate for a distance of 2.5 m with the calibration curve from the graph in Fig. 4.25. The correction is determined for the value 2.5 m because it represents the units and decimal parts of a metre which were measured.

In most electro-optical instruments factory calibration is performed so that the average nonlinearity and the true distance coincide. In other words, errors due to nonlinearity are uniformly distributed about the calibration point so that the total peak to peak error is within specified limits. From the practical standpoint, calibration for nonlinearity is rarely needed unless the instrument is to be used for projects in which very high accuracies are required, such as earth crustal movement and dam or structure deformation studies.

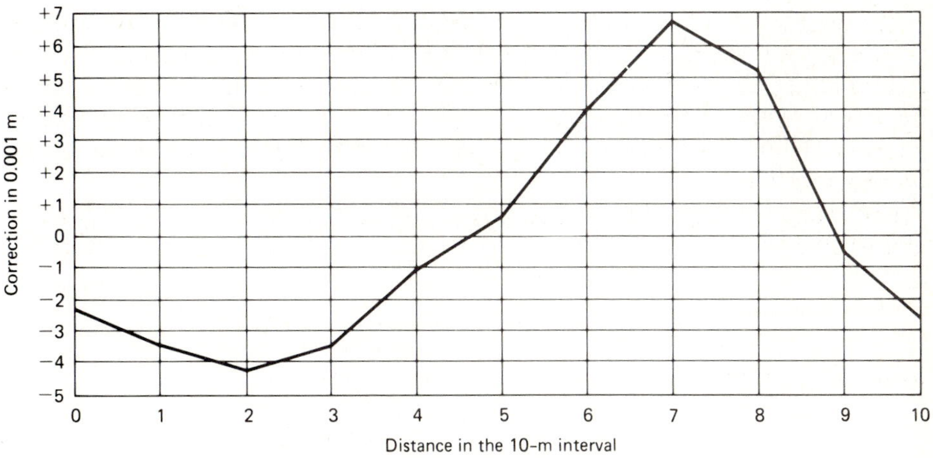

Fig. 4.25 Typical calibration curve for nonlinearity. (*After Moffitt, Ref. 11.*)

4.36. Ground reflections of microwave EDM instruments The beam transmitted by microwave instruments is relatively wide. Thus, when transmitting over smooth level terrain or water, reflections can occur which yield erroneous distances. Occurrence of errors due to reflections, or "ground swing" as it is called, can be reduced by elevating the master and remote units above the surface over which the line is to be measured. Microwave instruments of recent design operate on higher frequencies with a more narrow beam width ($1\frac{1}{2}°$), so that errors due to ground swing have been reduced. It is recommended that a series of fine readings (the *A* pattern for the Tellurometer) be taken, each at a different frequency, from both ends of the line at the beginning and ending of each measurement. Strong reflections will cause these readings to vary in a cyclic manner and if the readings are plotted versus frequency, the resulting curve will ideally take the shape of a sine wave. When swing is present, considerable experience and judgment are required to interpret the results properly. Usually a straight-line average of the fine readings will be adequate. The National Ocean Survey recommends that the two measurements from the two ends of the line agree to within 1 part in 100,000 and that the spread of the fine readings not exceed 4 cm.

4.37. Reduction of slope distances obtained by EDM For short lines less than 2 mi or 3.3 km in length and/or vertical angles of less than 5°, EDM slope distances corrected for meteorological conditions and system constants can be reduced to horizontal with the usual slope correction equations as outlined in Art. 4.14. Thus, either the vertical angle or zenith angle or the difference in elevation between the two ends of every measured line must be obtained. When vertical angles are observed, using earlier EDM instruments separate from the theodolite, additional data required for the reduction include the heights of the EDM transmitter and the reflector or remote unit above the ground at respective stations, and the heights of the theodolite used to measure the angle and the target sighted in this measurement above the ground at their respective stations.

Figure 4.26 illustrates a typical situation for distance measurement by separate EDM equipment accompanied by observation of the vertical angle. The EDM transmitter is at E while the reflector or remote unit is at E', with respective heights above the ground of EA and $E'B$. The theodolite and target, located at T and T', have heights above the ground (h.i.'s) of AT and BT', respectively. Angle α_T is the observed vertical angle and Δ_α is the correction to be calculated to determine the vertical angle α_E of the measured line EE'. The difference in heights of instruments required for this calculation is

$$\begin{aligned}\Delta\text{h.i.} &= (BE' - BT') - (AE - AT)\\ &= (\text{h.i. reflector} - \text{h.i. target}) - (\text{h.i. EDM} - \text{h.i. theodolite}) \qquad (4.17)\end{aligned}$$

and

$$\Delta_\alpha = \frac{\Delta\text{h.i.}\cos\alpha}{EE'} \quad \text{in radians} \qquad (4.18a)$$

$$= \frac{\Delta\text{h.i.}\cos\alpha}{EE'}\rho \quad \text{in seconds} \qquad (4.18b)$$

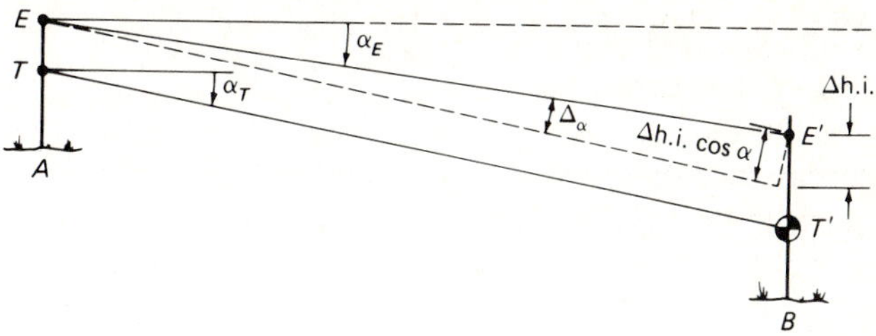

Fig. 4.26 Reduction of slope distances by EDM using angles.

where $\rho = 206{,}264.8$ seconds/radian (see Art. B.1, App. B), leading to

$$\alpha_E = \alpha_T + \Delta_\alpha \tag{4.19}$$

The sign of Δ_α is a function of the sign of Δh.i., which can be negative or positive. Care should be exercised in calculating α_E so as to reflect the proper signs of α, Δh.i., and Δ_α.

Example 4.13 The slope distance from A to B corrected for meteorological conditions and EDM system constants is 920.850 m. The EDM transmitter is 1.840 m above the ground and the reflector is 2.000 m above the ground. the observed vertical angle is $-4°30'00''$ with theodolite and target 1.740 m and 1.800 m above the ground, respectively. Calculate the horizontal distance.

SOLUTION From Eq. (4.17), calculate Δh.i.:

Δh.i. = (2.000 − 1.800) − (1.840 − 1.740) = 0.100

Using Eq. (4.18*b*), determine Δ_α:

$$\Delta_\alpha = \frac{(0.100)(\cos 4°30'00'')(206{,}264.8)}{920.850} = 22''$$

and the corrected vertical angle by Eq. (4.19) is

$$\alpha_E = \alpha_T + \Delta_\alpha = -4°30'00'' + 0°00'22'' = -4°29'38''$$

so that the horizontal distance by Eq. (4.1) is

$$H = s\cos\alpha_E = (920.850)(0.996926) = 918.019 \text{ m}$$

Example 4.14 A slope distance from C to D corrected for EDM constants and meteorological conditions is 5005.45 ft with a vertical angle of $+3°30'00''$ observed from a theodolite having an h.i. of 4.40 ft to a target with h.i. of 4.60 ft. The h.i.'s of EDM and reflector are 5.00 and 4.80 ft, respectively. Compute the horizontal distance.

SOLUTION By Eq. (4.17)

Δh.i. = (4.80 − 4.60) − (5.00 − 4.40) = −0.40 ft

Using Eq. (4.18*b*), we obtain

$$\Delta_\alpha = \frac{(-0.40)(\cos 3°30'00'')(206{,}264.8)}{5005.45} = -16''$$

so the corrected angle using Eq. (4.19) is

$$\alpha_E = \alpha_T + \Delta_\alpha = 3°29'44''$$

and the horizontal distance by Eq. (4.1) is

$$H = s \cos \alpha_E = 4996.14 \text{ ft}$$

It is recommended that vertical or zenith angles be observed with the telescope direct and reversed using a 1″ instrument (see Art. 6.43), and that vertical or zenith angles be observed from both ends of every line if first-order horizontal distances are desired.

Several types of electro-optical systems now have the EDM transmitter built into a theodolite, so that vertical angles can be observed simultaneously with the slope distance measurement. In these instruments there is a vertical offset between the electrical center of the transmitter and horizontal axis of the telescope. Also, the target sighted to observe the vertical angle may not be the same height above the ground as the reflector. These vertical offsets must be taken into account with such systems using the method outlined previously. For even more advanced systems with combined distance and direction measurement, refer to Chap. 9.

When slope distances are to be reduced using differences in elevation between the two ends of the line, the heights of the EDM instrument and reflector (or remote unit) above the ground at respective stations must be observed and recorded. In Fig. 4.27 the EDM instrument is at E with a height above the ground of AE and the reflector or remote unit is at E' at a height of BE' above the ground. Then the elevation difference between E and E' is

$$\Delta h_{EE'} = \Delta h_{A-B} - AE + BE'$$

or

$$\Delta h_{EE'} = \Delta h_{A-B} - \text{h.i. EDM} + \text{h.i. reflector or remote unit} \qquad (4.20)$$

4.38. Accuracies possible with EDM systems Short-range EDM systems (maximum range of 500 to 5000 m) have constant uncertainties which vary from 1.5 to 10 mm. Long-range systems (maximum range of 15 to 65 km) have constant uncertainties of from 0.5 to 1.5 cm. It is reasonable to assume that these figures represent

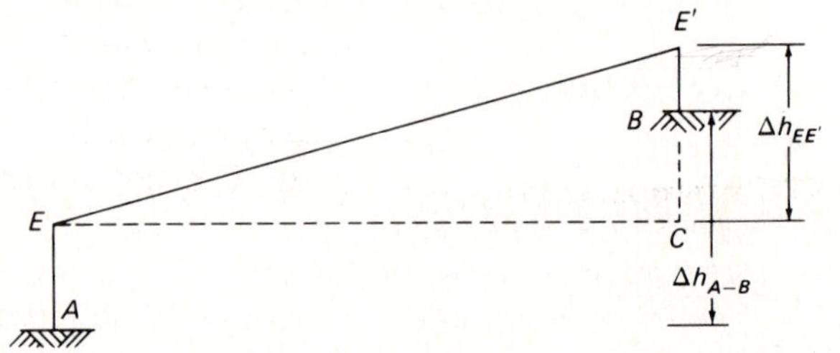

Fig. 4.27 Reduction of short slope distances by EDM using differences in elevation.

legitimate estimates of the standard deviation of a measurement when corrections for meteorological conditions and instrumental systematic errors have been applied as completely as possible. Then the worst of short-range instruments would provide a relative precision of 1 part in 25,000 or better for lines equal to or in excess of 250 m. The worst of the long-range systems would provide similar relative precisions for lines equal to or in excess of 375 m. Achievement of relative precisions of 1 part in 25,000 over comparable distances by taping would require use of taping tripods in precise taping procedures. Hence, EDM equipment has a clear superiority over traditional taping for surveys with distances of these magnitudes. For longer distances relative precisions of 1 part in 100,000 to 1 part in 1,000,000 are attainable. It is important to recognize that relative precisions of these magnitudes are possible only when corrections for meteorological conditions are very carefully evaluated and the EDM system is in good adjustment and calibrated against a legal standard.

When very long lines and high precision are involved, acquisition and correction for meteorological conditions represent the limiting factors in relative precisions possible with EDM systems. For maximum accuracy meteorological conditions should always be recorded at both ends of the line. For best results temperatures should be recorded 30 ft or about 10 m above terminal stations. The next step would be to acquire meteorological data at specified intervals along the line so as to obtain a more realistic estimate for the index of refraction. An approach in this direction has been taken on long lines measured in conjunction with fault-line movements in California. In this case an aircraft was flown between the two terminal points of the line to acquire additional meteorological data along the line while observations were being performed.

The ultimate solution probably lies in the instrumental determination of the index of refraction simultaneously with the observations for measurement by utilizing a combination of light and microwaves. A system now under development uses modulated red and blue laser light in addition to simultaneous modulated microwave transmission. The difference in velocities of the two light-wave signals permits a real-time determination of the dry air density, leading to an average air density over the path. From the average air density, the average group index of refraction can be computed. The microwave index of refraction is much more sensitive to water vapor pressure than light waves are, so that simultaneous transmission of microwaves permits determination at the average water vapor pressure from the optical-microwave dispersion. A general-purpose microcomputer is built into the instrument to allow real-time processing of the data and output consisting of slope distances corrected for meteorological conditions.

4.39. Problems

4.1 The length of a line measured with a 30-m tape is 308.550 m. When the tape is compared to the standard it is found to be 0.009 m too long under the same conditions of support, tension, and temperature as existed during measurement of the line. Compute the length of the line.

4.2 A building 80.00 ft by 160.00 ft is to be laid out with a 50-ft tape which is 0.016 ft too long. What ground measurements should be made?

4.3 The slope measurement of a line is 243.840 m. The differences in elevation between successive 30-m points are 0.50, 0.46, 0.75, 1.20, 1.40, 1.50, 2.00, and 0.20 m. Calculate the horizontal distance along the line.

4.4 The slope measurement of a line is 1246.5 ft. Slope angles measured with a clinometer are as shown below. Determine the horizontal distance.

Cumulated distance, ft	0		300		800		1000		1246.5
Slope angle, deg		$\frac{1}{2}$		$1\frac{1}{4}$		$2\frac{1}{2}$		4	

4.5 Two points at a slope distance of about 50 m apart have a difference in elevation of 4.00 m. Calculate the slope distance to be set off to establish a horizontal distance of 50 m.

4.6 Two points at a slope distance of about 25 m apart have a difference in elevation of 2.5 m. Determine the slope distance to be laid off to set the second point 25.000 m from the first.

4.7 The distance measured over smooth level ground between two monuments was recorded as 95.00 m. Measurements consisted of three full tape lengths and one partial tape length. If the first taping point was misaligned 0.4 dm left, the second 9.2 dm right, and the third 5.0 dm left, compute the correct distance between the points.

4.8 A tape that has a standardized length of 100.010 at 68°F with the tape fully supported is to be used to set the corner points of a foundation for a building that is 100.00 ft by 200.00 ft. What distances should be measured between the building corners with the tape fully supported over level terrain when the temperature is 80°F?

4.9 The tape in Prob. 4.8 was used to measure the distance between two monuments set at the corners of a city block. The tape was fully supported over level terrain and the temperature was 48°F. The distance recorded was 610.86 ft. What is the correct horizontal distance between the two monuments?

4.10 The distance along a property line from the front lot corner to the rear lot corner is 45.720 m. If the front pin is located and it is desired to set the rear corner, what distance should be measured in the field using a 30-m tape which is 5 mm too short?

4.11 A line is measured along sloping terrain and the following data are recorded:

Slope distance, ft	Difference in elevation between tape ends, ft
100.00	4.55
50.00	3.62
60.00	2.54
100.00	5.32
25.00	4.50

Compute the horizontal distance between the two ends of the line.

4.12 A standardized 30-m steel tape weighing 0.7 kg measures 29.9935 m supported at the two ends only, with 7.5 kg of tension and at 20°C.

(*a*) A horizontal distance of 3248.835 m is measured with the tape fully supported, with 7.5 kg of tension and at a temperature recorded at 38°C. Determine the correct horizontal distance;

(*b*) A horizontal distance of 200.843 m is to be measured from a monument to set an adjacent property corner. What horizontal distance must be measured under standardization conditions of support and tension to set this property corner where the temperature is 11°C?

4.13 A 100-ft tape weighing 2 lb is of standard length under a tension of 12 lb, supported for the full length of the tape. A line on smooth level ground is measured with the tape under a tension of 35 lb and found to be 4863.5 ft long. $E = 30{,}000{,}000\ \text{lb/in}^2$ and $3.53\ \text{in}^3$ of steel weighs 1 lb. Make the correction for increase in tension.

4.14 A second line is measured with the tape of Prob. 4.13, the tape being supported at intervals of 50 ft and the pull being 20 lb. The measured length is 1823.00 ft. Compute the corrections for sag and variation in tension and determine the corrected length of the line.

4.15 A 30-m steel tape weighing 0.7 kg has a cross-sectional area of $0.03\ \text{cm}^2$ and a standard length of 30.005 m when fully supported, at 20°C, and with 5 kg of tension.

(*a*) A line on a smooth level ground is measured with the tape under a tension of 10 kg at a temperature of 20°C and is recorded as 1242.823 m long. Compute the correction for the increase in tension. $E = 2{,}100{,}000\ \text{kg/cm}^2$.

(*b*) This same tape is used to measure a distance recorded as 985.423 m over level terrain with the tape supported at the two ends and 5 kg of tension (temperature = 20°C). Calculate the correction due to sag for this distance.

4.16 Compute the normal tension for the tape in (*a*) Prob. 4.13; (*b*) Prob. 4.15.

4.17 For the purpose of establishing monuments in a city, a line along a paved street having a grade of 2.5 percent is measured on the slope. The applied tension is 7.5 kg, and observations of temperature are made at each application of the tape. The measured length on the slope is 402.351 m, and the mean of the observed temperatures is 29.3°C. The 30-m steel tape used for the measurements is standardized at 20°C, supported for its full length, and is found to be 0.001 m too short under a tension of 7.5 kg. Determine the horizontal length of the line.

4.18 A 100-ft tape weighing 2 lb and having a cross-sectional area of 0.01 in^2 measures 100.020 ft fully supported and with 12-lb tension.

(*a*) Compute the effect of sag on a 100-ft tape length with the tape supported at the two ends and with 12-lb tension.

(*b*) Calculate the length measured with this tape supported at the two ends and with 19-lb tension ($E = 30{,}000{,}000\ \text{lb/in}^2$).

4.19 A tape has a standardized length of 30.0027 m at 20°C with the tape fully supported. A distance measured with the tape fully supported, in a horizontal position, and at a temperature of 15°C is recorded in the field book as 150.089 m.

(*a*) Compute the correct horizontal distance.

(*b*) What distance should be measured with the tape fully supported over a level surface in order to lay out a distance of 100,000 m? Assume that in both cases the proper standard tension was applied.

4.20 Two independent measurements were made of a line 10,000 ft long. The ground was sloping and measurements were taken with the tape horizontal. The tape is 100 ft long, and weighs 3 lb. One of the measurements of the line was made by two tapemen supporting the tape at the 0- and 100-ft marks; the second measurement was made by three tapemen supporting the tape at the 0-, 50-, and 100-ft points. The discrepancy between the two measurements was 11.7 ft. Several tests with a spring balance indicated that the average pull exerted by the tapemen was 20 lb. How much of the above discrepancy might be attributed to the different modes of supporting the tape?

4.21 A line roughly 2 mi long along a railroad track is measured with a steel tape, and corrections are made for observed temperatures. What error will be introduced if the actual temperature of the tape is 2°F higher than the observed temperature? State the error in fractional form with 1 as the numerator.

4.22 Assume that an invar tape having a coefficient of thermal expansion of 0.00000083 (ft/ft)/1°F is used under the conditions of Prob. 4.21. Compute the error introduced.

4.23 A hedge along the line *AB* makes direct measurement impossible. A point *C* is established at an offset distance of 6.10 m from the line *AB* and roughly equidistant from *A* and *B*. The distances *AC* and *CB* are then chained; *AC* = 392.34 m and *CB* = 412.39 m. Compute the length of the line *AB*.

4.24 Describe and explain the fundamental differences between electro-optical and electromagnetic distance measuring instruments.

4.25 Describe and explain the major systematic errors that affect EDM systems.

4.26 An electro-optical EDM instrument using infrared light with a wavelength of 0.9050 μm has a modulation frequency of 30 MHz. Calculate the modulated wavelength of light at 26°C and an atmospheric pressure of 752.9 mmHg. Compare this wavelength with the wavelength of the modulation frequency in a standard atmosphere at 0°C and 760 mmHg.

4.27 Compute the velocity of red laser light at a temperature of 30°C and an atmospheric pressure of 758.2 mmHg.

4.28 Compute the modulation wavelength of microwaves having a modulation frequency of 25 MHz at a temperature of 19.5°C, an atmospheric pressure of 758.2 mmHg, and a vapor pressure of 5.2 mmHg.

4.29 The vertical angle from point 1 to 2 is +2°27′01″ and the slope distance (corrected for atmospheric conditions and system constants) is 1770.897 m, as measured by an EDM instrument. Compute the horizontal distance from 1 to 2.

4.30 The slope distance from *E* to *G* corrected for atmospheric conditions and EDM system constants is 1524.500 m with a vertical angle of $+2°55'$ measured from a theodolite at E with an h.i. of 1.452 m above E to a target at G having an h.i. of 1.524 m above G. The respective h.i.'s of the EDM instrument at E and reflector at G are 1.676 and 1.825 m. Compute the horizontal distance from E to G.

4.31 The slope distance from H to J corrected for atmospheric conditions and EDM system constants is 4234.285 ft. If the difference in elevation between the two points is 67.056 m, compute the horizontal distance from H to J.

4.32 The vertical angle between stations A and B is $3°02'05''$. The slope distance AB corrected for atmospheric conditions and system constants is 2345.85 m. Compute the horizontal distance AB.

4.33 The zenith angle from C to D is $87°32'59''$ and the slope distance corrected for atmospheric conditions and EDM system constants is 5810.31 ft. Calculate the horizontal distance from C to D.

4.34 The difference in elevation between points A and B is 77.854 m and the slope distance (corrected for atmospheric conditions and EDM system constants) is 1642.183 m. Determine the horizontal distance.

4.40. Field problems

PROBLEM 1 TAPING OVER UNEVEN GROUND

Object To standardize the 100-ft or 30-m steel tape; to find, by two methods, the horizontal length of an assigned course about 800 ft or 250 m long over uneven ground; and to correct for the error in length of tape as determined by the standardization tests.

TAPING OVER UNEVEN GROUND

Sta.	Length in Feet Forward	B'kward	Mean	Correct-ion	Corr. Length		
						Chaining Equipment	J. Pratt, H.C.
						Locker No. 35	S. Fulton, R.C.
						Chicago 100 ft. Steel	Oct. 3, 1984 (2 hrs.)
						Tape	Cold and Windy
351						Tk. in Stk. Ctr. Line M.C. R.R. at Haskell St.	
	311.73	311.65	311.69	+0.07	311.76	East on Ctr. Line Haskell St. up Steep Hill	
352						Nail in Pavement	
	198.60	198.62	198.61	+0.04	198.65	North over Creek and R.R. Embankment	
353						S.B.S.E. Corner Jones' Lot	
	213.84	213.78	213.81	+0.05	213.86	North up Hill (Rough)	
354						1" Iron Pipe N.E. Corner Jones' Lot	

Standardization		
Actual length of standard	100.038'	
Observed length of standard		Tape compared with Standard at City Hall
Before	100.017'	
After	100.015'	
Mean	100.016'	
Actual length of tape	100.022'	
Correction per tape length	+0.022'	

Fig. 4.28 Notes for taping over uneven ground with tape horizontal.

Procedure (1) Standardize the tape before and after the field work, by comparing it with the official standard of length. Maintain the required pull by means of a spring balance. Determine the error with a finely divided scale. (2) Tape the course by horizontal measurement after the manner described in Art. 4.11. Maintain a pull estimated to be equal to that used during standardization. For measurements in which the line slopes more than 5 or 6 ft in 100, or 1.5 to 2 m in 30 m, "break tape." (3) Correct for the error in length of tape by adding to or subtracting from the observed distance the observed error multiplied by the number of tape lengths in the course. Record the data in a form similar to that of Fig. 4.28. (4) Tape the course by measurement on the slope, recording the difference in elevation (to the nearest foot or nearest 2 dm) of each tape length as determined by a hand level. If sharp breaks in slope occur between stations, treat each distance between breaks in the same manner as a full station. Correct the measurements for error in length of tape and for slope (Arts. 4.14 and 4.16). Compare results obtained under (3) and (4).

PROBLEM 2 PRECISE TAPING OVER UNEVEN GROUND

Object To measure a distance of a line using a standardized steel tape, tension handle, and tape thermometer, and to make corrections for incorrect tape length and temperature.

Procedure Obtain the standardized length of the steel tape and standardization conditions from the instructor or compare the steel tape with a standardized tape using a tension handle and tape thermometer attached to the tape. Set two hubs with tacks at the end points of the line to be measured. Measure the line five times as follows: (1) Mark the ends of each tape length by pencil lines on the top of a hub; (2) apply the standard tension with a tension handle; (3) use a hand level to check the

MEASUREMENT OF A DISTANCE BY TAPING

Station	Dist., m	Temp., C°	Temp. Corr., m	Calibr. Length, m	Corrected Dist., m
A_7	30.000	9.7	−0.0036	29.9934	29.9898
	30.000	9.7	−0.0036	29.9934	29.9898
	5.825	12.6	−0.0005	5.8237	5.8232
					65.8033
B_7	30.000	6.3	−0.0048	29.9934	29.9886
	30.000	4.7	−0.0053	29.9934	29.9881
	5.822	3.8	−0.0011	5.8207	5.8196
A_7					65.7963
	30.000	6.9	−0.0046	29.9934	29.9888
	30.000	4.1	−0.0055	29.9934	29.9879
	5.807	3.5	−0.0011	5.8057	5.8046
B_7					65.7813
	30.000	5.7	−0.0050	29.9934	29.9884
	30.000	8.1	−0.0041	29.9934	29.9893
	5.839	7.8	−0.0008	5.8377	5.8369
A_7					65.8146
	30.000	6.2	−0.0048	29.9934	29.9886
	30.000	7.1	−0.0045	29.9934	29.9889
	5.820	6.5	−0.0009	5.8187	5.8178
B_7					65.7953
Average of 5 measurements				=	65.7982

	Oct. 7, 1984
30-m tape No. 6	Clear, Warm
Hand Level	Morgan
Line Rod	Roberts
Thermometer No. 9	Pental
Spring Balance No. 4	Mennick
2 Plumb Bobs	

Tape 6 measures 29.9934 at 20°C, supported at the two ends, and with 7-kg tension

$$\hat{\sigma} = \left[\frac{\sum_{i=1}^{5} \bar{x} - x_i}{n-1}\right]^{1/2} = \left[\frac{0.0005926}{4}\right]^{1/2}$$

$$\hat{\sigma} = 0.012\ m$$

Fig. 4.29 Notes for precise taping over uneven ground.

horizontal position of the tape; (4) with a tape thermometer attached to the tape, observe and record the temperature of the tape after each measurement; (5) estimate distances to thousandths of a foot or to millimetres.

Compute the (1) corrections for incorrect length of tape and temperature and apply to the measured distances; (2) average of the five measurements; (3) standard deviation of the measurements (Art. 2.15). Figure 4.29 illustrates a suggested form of notes.

References

1. AGA Operating Manuals, AGA Corporation Geodimeter Division, S-181 20 Lidingö, Sweden.
2. Dracup, J. F., Kelley, C. F., Lesley, G. B., and Tomlinson, R. W., *National Geodetic Survey Lecture Notes for Use at Workshops on Surveying Instrumentation and Coordinate Computation*, National Ocean Survey, Rockville, Md.
3. Davis, R. E., Foote, F. S., Anderson, J. M., and Mikhail, E. M., *Surveying, Theory and Practice*, 6th ed., McGraw-Hill Book Co., New York, 1981.
4. Huggett, G. R., and Slater, S. L., "Precision Electromagnetic Distance Measuring Instrument for Determining Secular Strain and Fault Movement," Presented paper, International Symposium on Recent Crustal Movements, Eidgonössische Technische Hochschule, Zurich, Switzerland, August 26–31, 1974.
5. Huggett, G. R., "High Accuracy Techniques Using the HP 3800 Distance Meter," *Insight*, Hewlett-Packard Engineering Products, Vol. 2, No. 1, November 1973.
6. Huggett, G. R., "Instrument Non-linearity," *Insight*, Hewlett-Packard Engineering Products, Vol. 2, No. 4, August 1974.
7. Kester, J. M., "EDM Slope Reduction and Trigonometric Leveling," *Surveying and Mapping*, Vol. 33, No. 1, March 1973, p. 61.
8. Kivioja, L. A., "The EDM Corner Reflector Is Not Constant," *Surveying and Mapping*, Vol. 38, No. 2, June 1978, pp. 142–155.
9. Laurilla, S. H., *Electronic Surveying*, John Wiley & Sons, Inc., New York, 1976.
10. Mead, B. D., "Precision in Electronic Distance Measuring," *Surveying and Mapping*, Vol. 32, No. 1, March 1972, pp. 69–78.
11. Moffitt, F. H., "Field Evaluation of the Hewlett-Packard Model 3800 Distance Meter," *Surveying and Mapping*, Vol. 31, No. 1, March 1971.
12. Moffitt, F. H., "Calibration of EDM's for Precision Measurement," *Surveying and Mapping*, Vol. 35, No. 2, June 1975, pp. 147–154.
13. Moffitt, F. H., and Bouchard, H. B., *Surveying*, 7th ed., Harper & Row, Publishers, New York, 1982.
14. Romaniello, C. G., "EDM 1976," *Surveying and Mapping*, Vol. 37, No. 1, March 1977, p. 25.
15. Saxena, N. K., "Electro-optical Short Range Surveying Instruments," *Journal of the Surveying and Mapping Division, ASCE*, Vol. 101, No. SU1, October 1975, pp. 137–147.
16. Tomlinson, R. W., "Electronic Distance Measuring Instruments Calibration—Why, When, Where, and How," *Proceedings, Fall Convention, ACSM*, 1977, p. 56.

CHAPTER 5
Vertical distance measurement: leveling

5.1. Definitions The *elevation* of a point near the surface of the earth is its vertical distance above or below an arbitrarily assumed *level surface*. A level surface is a curved surface every element of which is perpendicular to the plumb line. The level surface (real or imaginary) used for reference is called the *datum*. A *level line* is a line in a level surface.

The datum most commonly used is mean sea level, particularly the Mean Sea Level Datum established by the United States government (Art. 10.2). If merely the *relative* elevations of points for a particular survey are desired, the datum may bear no known relation to sea level. For example, the initial point in a survey may be assumed to have given elevation and the elevations of all succeeding points computed accordingly. To avoid negative values of elevation, either the datum is taken below the lowest point of the survey or a positive value is assigned to it. However, the use of a local datum is to be discouraged and should not be considered good surveying practice. The presence of vertical control surveys, each related to its own unrelated datum, contributes to chaotic conditions and inevitably leads to complications and increased costs for subsequent surveys as well as for planning, construction, and operation of engineering and scientific projects.

A *horizontal line* is a line, in surveying taken as straight, tangent to a level surface.

A *vertical angle* is an angle between two intersecting lines in a vertical plane. In surveying, it is commonly understood that one of these lines is horizontal.

The *difference in elevation* between two points is the vertical distance between the two level surfaces in which the points lie. *Leveling* is the operation of measuring vertical distances, either directly or indirectly, to determine differences in elevation. This operation is depicted in Fig. 5.1, where A represents a point of known elevation and B is a point whose elevation is desired. An instrument known as a *level* (refer to Fig. 5.14) is set up at L between and along the route from A to B. A *level rod* (refer to Fig. 5.21*a*) is held at A and the reading h_a is taken, by sighting on the rod through the telescope of the level (line of sight assumed horizontal). The rod is then moved from A to B and the reading h_b is taken. The *observed* difference in elevation

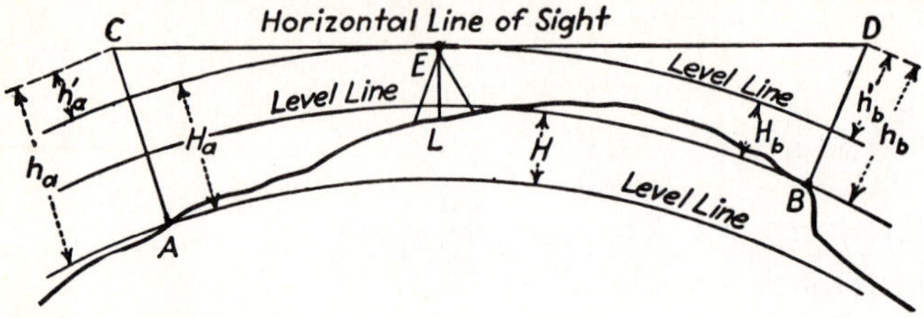

Fig. 5.1 Direct differential or spirit leveling. (Owing to refraction, line of sight is slightly curved.)

between points A and B is $(h_a - h_b)$. The readings h_a and h_b include systematic errors h_a' and h_b' (Fig. 5.1) because of the curvature of the earth and refraction of light rays in the atmosphere, as discussed in Art. 5.2. Depending on the length of sight, neglecting curvature and refraction may cause significant error.

5.2. Curvature and refraction In leveling, it is necessary to consider the effects of (1) the curvature of the earth and (2) atmospheric refraction, which affects the line of sight. Usually, these two effects are considered together.

In Fig. 5.2 is shown a horizontal line tangent at A to a level line near the surface of the earth. The vertical distance between the horizontal line and the level line is a measure of the earth's curvature. It varies approximately as the square of the distance from the point of tangency. In Fig. 5.2, let $OA = R$, the average radius of the earth. Also, let $c = ED$, the correction for earth curvature. Then

$$R^2 + \overline{AE}^2 = (R + c)^2 = R^2 + 2Rc + c^2$$

$$\overline{AE}^2 = c(2R + c)$$

$$c = \frac{\overline{AE}^2}{2R + c} \tag{5.1}$$

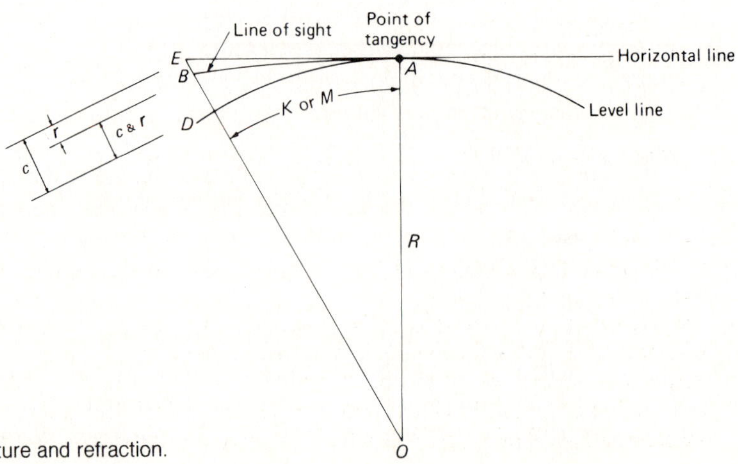

Fig. 5.2 Earth curvature and refraction.

Since c is very small compared to R, a reasonable approximation for earth curvature is

$$c = \frac{\overline{AE}^2}{2R} \tag{5.2}$$

Assuming a mean radius of the earth of 3959 mi or 6371 km, the curvature correction is

$$c = 0.667M^2 \text{ feet} \tag{5.3}$$

in which M is the distance from the point of tangency (station of the observer) in miles. In the metric system, the earth curvature correction c_m, in metres, is

$$c_m = 0.0785K^2 \tag{5.4}$$

in which K is the distance from the point of tangency in kilometres. Thus, the curvature correction for a distance of 1 mi is 0.67 ft and for a distance of 1 km is 7.9 cm. For distances of 100 ft or 30 m the respective corrections would be 0.00024 ft and 0.07 mm.

Owing to the phenomenon of atmospheric refraction, rays of light are refracted, or bent downward slightly. This bending of the rays of light towards the center of the earth tends to diminish the effect of earth curvature by approximately 14 percent. In Fig. 5.2, AB is the refracted line of sight and the distance BD represents the combined effect of curvature and refraction. Let $(c \,\&\, r) = BD$ be computed by the following equations:

$$(c \,\&\, r) = 0.574M^2 \text{ feet} \tag{5.5}$$

$$(c \,\&\, r) = 0.0206D^2 \text{ feet} \tag{5.6}$$

$$(c \,\&\, r) = 0.0675K^2 \text{ metres} \tag{5.7}$$

in which M is in miles, K is in kilometres as above, and D is in thousands of feet.

In most ordinary spirit leveling operations, the line of sight is rarely more than 6 ft or approximately 2 m above the ground, where variations in temperature cause substantial uncertainties in the refractive index of air. Fortunately, most lines of sight in leveling are relatively short (about 100 ft or 30 m) and backsight and foresight distances are balanced (see Art. 5.3). Consequently, curvature and refraction corrections are rarely significant except for precise leveling (see Ref. 3).

5.3. Methods Difference in elevations may be measured by the following methods:

1. *Direct differential* or *spirit leveling*, by measuring vertical distances directly. Direct leveling is the most precise method of determining elevations and is the one more commonly used (Art. 5.4).
2. *Indirect* or *trigonometric leveling*, by measuring vertical angles and horizontal or slope distances (Art. 5.5) and computing the elevation difference.
3. *Stadia leveling*, in which vertical distances are determined by tacheometry using the engineer's transit and level rod; plane table and alidade and level rod; or self-reducing tacheometer and level rod. Details for these procedures are given in Chap. 7.

4. *Barometric leveling*, by measuring the differences in atmospheric pressure at various stations by means of a barometer (Art. 5.6).
5. A ground "elevation meter," developed originally by the U.S. Corps of Engineers but now commercially available. This device consists of a vehicle on which is mounted an electromechanical device that integrates the vertical component of any longitudinal movement. Its accuracy is within 0.1 ft times the square root of the traveled distance in miles and it can travel up to 30 mi/h. This is one of the inertial system instruments to appear in the mid-1970s.

Precise leveling is a precise form of differential or spirit leveling.

Profile leveling is the operation—usually by direct leveling—of determining elevations of points at short measured intervals along a definitely located line, such as the center line for a highway or a sewer (Art. 5.42).

Direct leveling is also employed for determining elevations for cross sections, grades, and contours.

Reciprocal leveling is a form of differential leveling by which elevations can be transferred across wide ravines and rivers.

5.4. Direct differential or spirit leveling In Fig. 5.1, A represents a point of known elevation and B represents a point the elevation of which is desired. In the method of direct or spirit leveling, the level is set up at some intermediate point as L, and the vertical distances AC and BD are observed by holding a leveling rod first at A and then at B, the line of sight of the instrument being horizontal. (Owing to refraction, the line of sight is slightly curved, as explained previously.)

If the difference in elevation between the points A and E is designated as H_a and between E and B as H_b.

$$H_a = h_a - h'_a \qquad \text{and} \qquad H_b = h_b - h'_b$$

in which h_a and h_b are the vertical distances read at A and B, respectively, and $h'_a = (c \;\&\; r)_a$ and $h'_b = (c \;\&\; r)_b$ are the effects of curvature and refraction for the horizontal distances LA and LB, respectively, calculated using one of Eqs. (5.5), (5.6), or (5.7).

The difference in elevation H between A and B is then

$$\begin{aligned} H &= H_a - H_b = (h_a - h'_a) - (h_b - h'_b) \\ &= h_a - h_b - h'_a + h'_b \end{aligned} \tag{5.8}$$

If the backsight distance LA is equal to the foresight distance LB, then $h'_a = h'_b$ and

$$H = h_a - h_b \tag{5.9}$$

Thus, if backsight and foresight distances are balanced, the difference in elevation between two points is equal to the difference between the rod readings taken to the two points, and no correction for curvature and refraction is necessary. In direct leveling, the work is usually conducted so that the effect of curvature and refraction is reduced to a negligible amount (Art. 5.35).

On lines of direct levels the usual procedure is as follows (refer to Fig. 5.31). Let B.M.$_1$ and B.M.$_2$ be two established points some distance apart whose difference in elevation is desired. With the level in some convenient location, not necessarily on a line joining B.M.$_1$ and B.M.$_2$, a backsight is taken to B.M.$_1$ and a foresight is taken to some convenient point T.P.$_1$. The level is then moved ahead, a backsight is taken to T.P.$_1$, and a foresight is taken to some accessible point as T.P.$_2$. And so the process is repeated until the terminal point B.M.$_2$ is reached.

5.5. Trigonometric leveling Trigonometric leveling involves observing the vertical angle and either the horizontal or slope distance between two points. The difference in elevation can then be calculated. In Fig. 5.3, A represents a point of known elevation and B a point the elevation of which is desired. Within the limits of ordinary practice, triangle ECD can be assumed a right triangle and the distance $ED = EF$. If the horizontal distance ED is known, then

$$DC = ED \tan \alpha$$
$$DF = (c \,\&\, r)_{ED}$$

in which the correction for earth curvature and refraction (c & r) can be calculated using one of Eqs. (5.5), (5.6), or (5.7). The difference in elevation, ΔH_{AB}, is

$$\Delta H_{AB} = AE + DF + DE \tan \alpha - BC \tag{5.10}$$

where AE and BC are instrument height above A and target height above B, respectively. These heights are measured using a level rod (Art. 5.21) or by means of the telescopic plumbing rod if the instrument tripod is so equipped.

A more common situation occurs when the slope distance EC is measured very accurately using EDM equipment (Art. 4.34). In this case $DC = EC \sin \alpha$ and ΔH_{AB} becomes

$$\Delta H_{AB} = AE + DF + EC \sin \alpha - BC \tag{5.11}$$

Uses In ordinary surveying, indirect trigonometric leveling furnishes a rapid means of determining the elevations of points in rolling or rough country. On

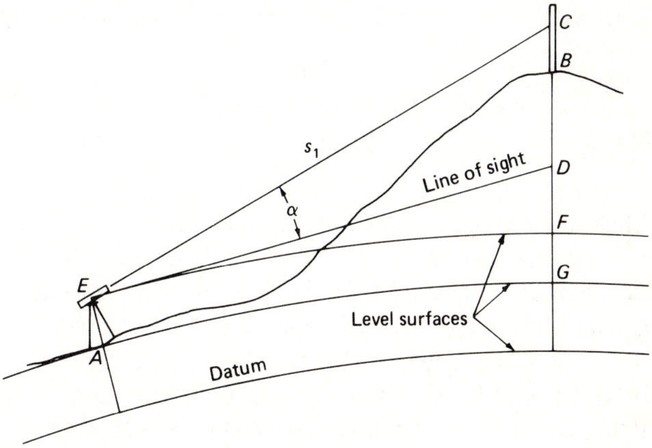

Fig. 5.3 Trigonometric leveling.

reconnaissance surveys, angles may be measured with the clinometer and distances may be obtained by pacing. On more precise surveys, angles are measured with the transit and distances by the stadia. Indirect or trigonometric leveling is used extensively in plane-table work (Chap. 7).

Example 5.1 Referring to Fig. 5.3, the slope distance s_1 from *E* to *C* is 332.791 m with an average vertical angle of 9°46′29″ from a height of instrument at *E* of 1.558 m above *A* to a target height *C* of 1.372 m above *B*. If the elevation at *A* is 21.935 m above mean sea level, calculate the difference in elevation from *A* to *B* and the elevation of *B*.

SOLUTION Use Eq. (5.11), where the correction for curvature and refraction is calculated with Eq. (5.7):

$$\Delta H_{AB} = 1.558 + (0.067)(0.332791)^2 + (332.791)(\sin 9°46'29'') - 1.372$$
$$= 56.693 \text{ m}$$
$$\text{elevation of } B = 21.935 + 56.693 = 78.628 \text{ m}$$

5.6. Barometric leveling Since the pressure of the earth's atmosphere varies inversely with the elevation, the barometer may be employed for making observations of difference in elevation. Barometric leveling is employed principally on exploratory or reconnaissance surveys where differences in elevation are large, as in hilly or mountainous country. Since atmospheric pressure may vary over a considerable range in the course of a day or even an hour, elevations determined by one ordinary barometer carried from one elevation to another may be several feet in error. However, by means of sensitive barometers and special techniques, elevations can be determined within a foot or so.

For field use an aneroid barometer is generally used because it is light and easily transported. Extremely sensitive barometers called *altimeters* (Fig. 5.4) have been developed with which elevations can be determined to within a foot or so.

In one procedure, barometric observations are taken at a fixed station during the same period that observations are made on a second barometer which is carried from point to point in the field. The carried barometer is either returned to the starting point or taken to another point of known elevation; the computation of the elevation of each point then takes account of the corresponding change in atmospheric pressure at the fixed base.

In another procedure used in topographic surveying, two of the instruments are employed at fixed bases and one or more instruments are carried from point to point over the area being surveyed. One fixed instrument is located at a point of known elevation near the highest elevation of the area, and one near the lowest elevation; these instrument stations are called the *upper base* and *lower base*, respectively. Other altimeters are carried to points whose elevations are desired, and readings are taken. Readings on the fixed instruments are taken either simultaneously (as determined by signaling) or at fixed intervals of time; in the latter case the readings at the desired instant are determined by proportion. The elevation of a portable instrument is then determined by interpolation. The horizontal location of each point at which a reading is taken is determined by conventional methods.

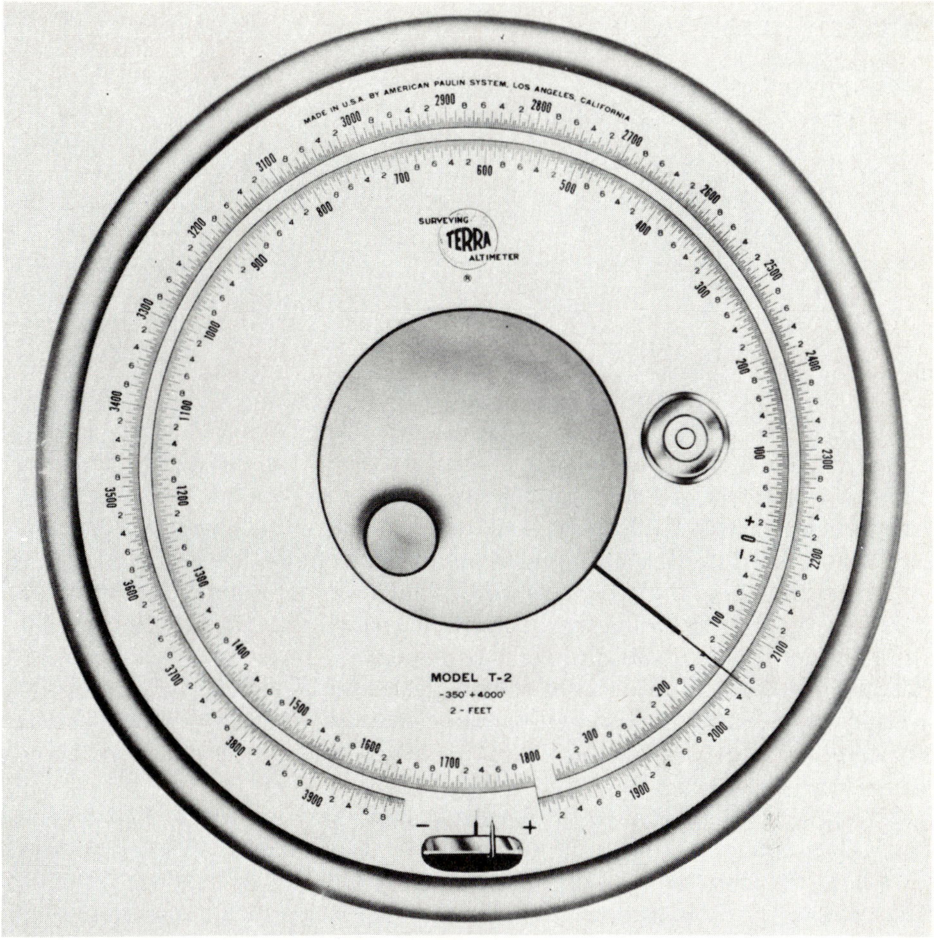

Fig. 5.4 Altimeter. (*Courtesy American Paulin System.*)

INSTRUMENTS FOR DIRECT LEVELING

5.7. Kinds of levels Any instrument commonly used for direct leveling has as its essential features a *line of sight* and *spirit level tube* or some other means of making the line of sight horizontal. The level tube is so mounted that its axis is parallel to the line of sight. The instrument used principally in the United States is the *engineer's level.* Figure 5.5*a* is the diagram of the principal parts of an engineer's level. The level consists of the telescope *A* mounted upon the level bar *B* which is rigidly fastened to the spindle *C*. Attached to the telescope or the level bar and parallel to the telescope is the level tube *D*. The spindle fits into a cone-shaped bearing of the leveling head *E*, so that the level is free to revolve about the spindle *C* as an axis. The leveling head is attached to a tripod *F*. In the tube of the telescope are cross hairs at *G*, which appear on the image viewed through the telescope, as

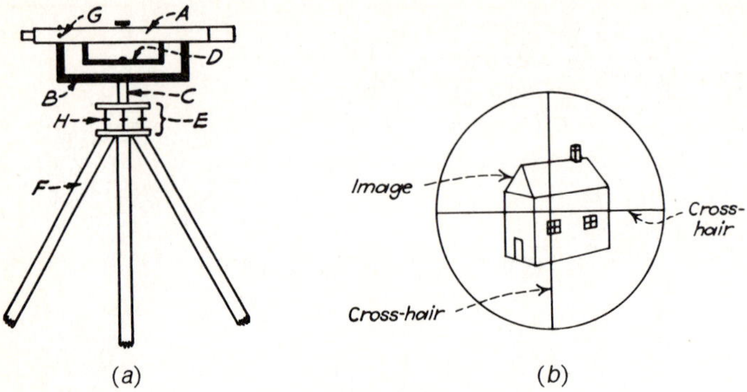

Fig. 5.5 Principal parts of an engineer's level.

illustrated by Fig. 5.5*b*. The bubble of the level is centered by means of the leveling screws *H*. The two distinct types of engineer's levels are the *dumpy level* and *self-leveling* or *automatic level*. The *architect's level* or *construction level*, a modified form of the engineer's level but with a telescope of lower magnifying power and with a less sensitive level tube, is used in establishing grades for buildings. The *hand level* (Arts. 5.19 and 5.20) is a simple, hand-held, useful device for determining differences in elevation roughly. Instruments which are frequently used for direct leveling but which are not primarily designed for this purpose are the *engineer's transit* (Art. 6.22) and the telescopic alidade used with the *plane table* (Chap. 7).

The measurements of difference in elevation are made by sighting on a graduated rod, called a *leveling rod* (Arts. 5.23 to 5.25). Other accessories sometimes used are a *rod level* (Art. 5.27) and a *turning point* (Art. 5.28).

Usually the leveling head of the engineer's level is equipped with four leveling screws, but the three-screw type is favored by many engineers and surveyors, particularly for instruments of high precision. Reasons for this preference are that the three-screw type can be leveled rapidly, requires the use of only one hand, and is relatively stable as compared with the four-screw type when the latter is not perfectly set. With the three-screw leveling head, on the other hand, when one of the screws is turned, the elevation of the telescope may change slightly.

5.8. Telescope Modern levels are equipped with internally focusing telescopes that have features as illustrated in Fig. 5.6*a*. Rays of light emanating from an object point are caught by the objective lens *A* and are brought to a focus and form an image in the plane of cross hairs *B*. The lenses of the eyepiece *C* form a microscope which is focused on the image at the cross hairs. The objective lens is fixed in the end of the telescope tube and a negative lens *L* is attached to a slide *D* within the telescope tube. This negative lens can be moved parallel to the line of sight by rotating the focusing knob *F* and allows focusing the telescope on objects at different distances from the instrument. The eyepiece *C* is held in position by threaded sleeve *G*, which may be moved in a longitudinal direction for focusing. By means of adjusting screws, the cross-hair reticule may be moved transversely so that the intersection of the cross hairs will appear in the center of the field of view.

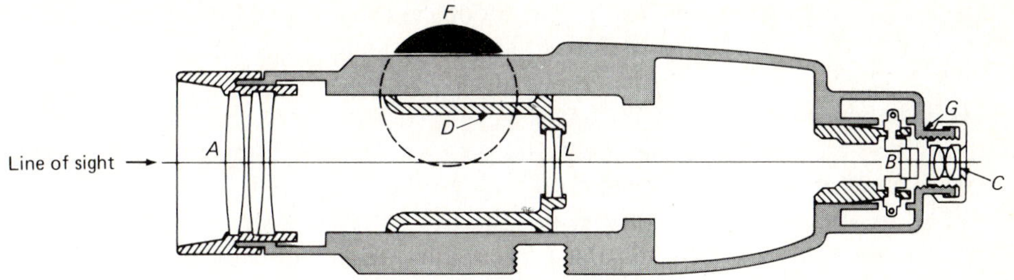

Fig. 5.6*a* Longitudinal section of an internal focusing telescope.
(*Courtesy the Lietz Co.*)

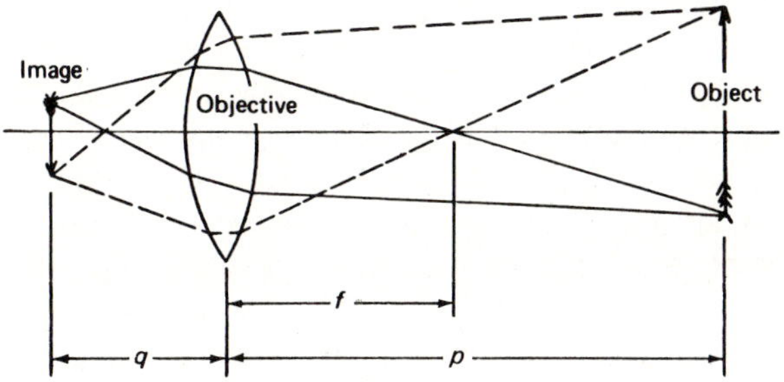

Fig. 5.6*b*

The line of sight is defined by the intersection of the cross hairs and the optical center of the objective lens. The instrument is so constructed that the optical axis of the objective lens coincides (or practically coincides) with the axis of the negative or objective slide (*D*, Fig. 5.6*a*); in other words, a given ray of light passing through the optical center of the objective always occupies the same position in the telescope tube regardless of the longitudinal position of the negative lens. The cross hairs can be adjusted so that the line of sight and the optical axis coincide.

5.9. Objective lens The principal function of the telescope objective lens is to form an image for sighting purposes. For accuracy of measurements, the objective should produce an image that is well lighted, accurate in form, distinct in outline, and free of discolorations. Such an image is provided by using an outer double-convex lens of crown glass and an inner concavo-convex lens of flint glass. The outer surface of the lens, in contact with air, is coated to prevent loss of light by reflection.

The basic principle (lens equation) which relates object distance p, image distance q, and focal length f (for externally focusing telescopes) is

$$\frac{1}{p}+\frac{1}{q}=\frac{1}{f} \tag{5.12}$$

A similar relationship exists for internally focusing telescopes. Figure 5.6*b* illustrates the manner in which rays from an object are deviated by the objective and brought to a focus to form the image. Note that the image is inverted.

5.10. Focusing When the telescope is to be used, the eyepiece is first moved in or out until the cross hairs appear sharp and distinct. This adjustment of the eyepiece should be tested frequently, as the observer's eye becomes tired.

When an object is sighted, the negative lens or objective slide is moved in or out until the image appears clear. At this point the image should be in the plane of the cross hairs. If a slight movement of the eye from side to side or up or down produces an apparent movement of the cross hairs over the image, the plane of the image and the plane of the cross hairs do not coincide, and *parallax* is said to exist. Since parallax is a source of error in observations, it should be eliminated by refocusing the eyepiece first and then the objective until further trial shows no apparent movement. To aid this process, the observer's hand or piece of white paper should be held a few inches in front of the objective lens so that the observer can concentrate on focusing the cross hairs rather than other objects viewed through the telescope. Figures 5.7*a* through *c* provide a graphical illustration of the concept of parallax. The negative or objective lens must be focused for each distance sighted.

The telescope cannot be focused on objects closer than about 6 ft or 2 m from the center of the instrument, unless special short-focus lenses are employed.

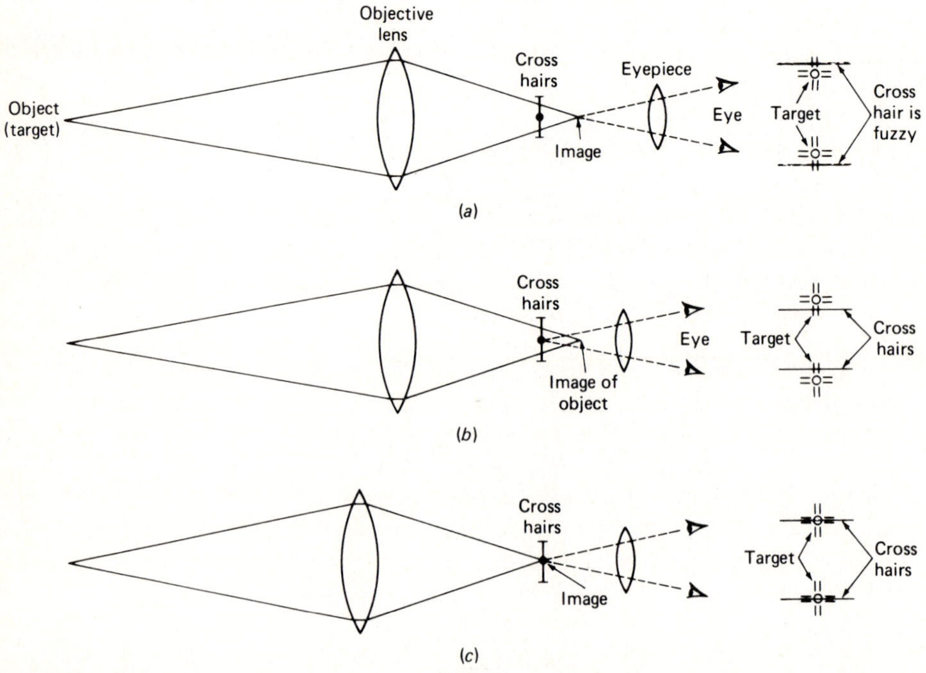

Fig. 5.7 Presence of parallax. (*a*) Objective not focused, eyepiece not focused, parallax is present. (*b*) Objective not focused, eyepiece focused on cross hairs, parallax present. (*c*) Objective focused, eyepiece focused, no parallax.

Most telescopes are equipped with peep sights on top or alongside the telescope. After the cross hairs have been focused, the observer should point the telescope approximately at the object by first sighting along the peep sights at the targets. Then looking through the telescope, the observer should focus the objective lens so that there is a clear, distinct image of the target at which the telescope is pointed. A subsequent movement of the pointing of the telescope by a degree or two one way or the other should point the line of collimation (cross hairs) precisely on the target. Now the observer should check for parallax and if it is present remove the parallax using the procedure described above.

The strain on the eyes will be reduced if the observer tries keeping both eyes open while sighting.

5.11. Cross hairs The cross hairs, which define the line of sight, used to be made of threads from the cocoon of the brown spider and later of very fine platinum wire. In modern instruments the *cross-hair ring* or *reticule* consists of a glass plate on which are etched fine vertical and horizontal lines which serve as cross hairs. Special patterns of additional lines are used on some instruments: for example, stadia hairs, or double horizontal and vertical lines closely spaced for sighting between them precisely. The spacing between a double line can be pointed more precisely than can a single line.

As shown in Fig. 5.8*a* and *b*, the cross-hair ring is held in position by four capstan-headed screws which pass through the telescope tube and tap into the ring.

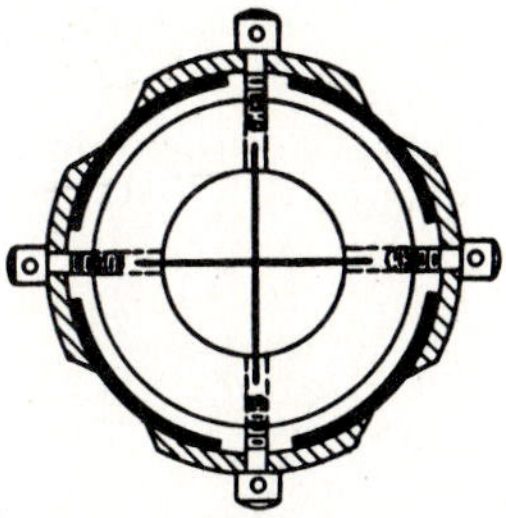

(*a*)

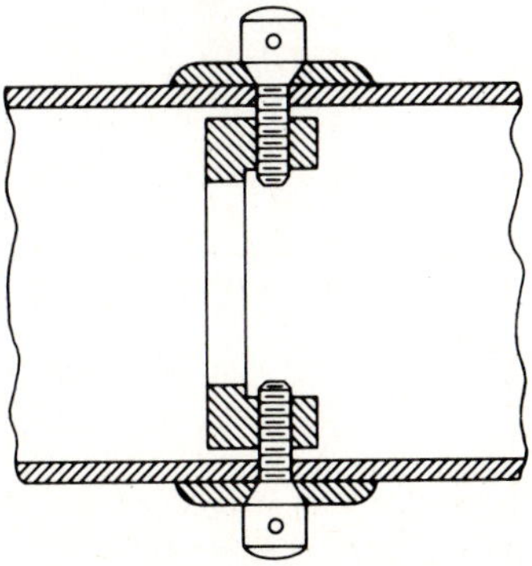

(*b*)

Fig. 5.8 Cross-hair reticule.

The holes in the telescope tube are slotted so that when the screws are loosened, the ring may be rotated through a small angle about its own axis. To rotate the ring without disturbing its centering, two adjacent screws are loosened; and the same two screws are tightened after the ring has been rotated. The ring is smaller than the inside of the tube, and it may be moved either horizontally or vertically by means of the screws. Thus, to move it to the left, the left-hand screw is loosened and the right-hand screw is tightened. If the movement is to be large, first the top or bottom screw is loosened slightly; and after the movement to the left has been completed, the same (top or bottom) screw is tightened again.

5.12. Eyepiece Attention has previously been drawn to the fact that the image formed by the objective is inverted. Eyepieces are of two general types:

The *erecting* or *terrestrial* eyepiece, the more common of the two types, reinverts the image so that the object appears to the eye in its normal position. Usually, it consists of four plano-convex lenses placed in a metal tube called the eyepiece slide (Fig. 5.9*a*). In the figure *A* represents the object, *B*, the inverted image in the plane of the cross hairs, *C* the image which is magnified by the lens nearest the eye, and *D* the magnified image as it appears to the eye.

The *inverting* or *astronomical* eyepiece simply magnifies the image without reinverting it. It is composed of two plano-convex lenses generally arranged as shown in Fig. 5.9*b*. The arrangement is seen to be identical with that of the two lenses farthest apart in the erecting eyepiece. The magnified image *D* is seen to be inverted, and the object, as viewed through the telescope, is upside down.

For either eyepiece the ratio of the angle at the eye subtended by the magnified image, to that subtended by the object itself, is the magnifying power of the telescope. If, in either Fig. 5.9*a* or Fig. 5.9*b*, *D* is the apparent length of the magnified image and *E* is the apparent length of the object as seen by the naked eye, the ratio of *D* to *E* is the magnifying power.

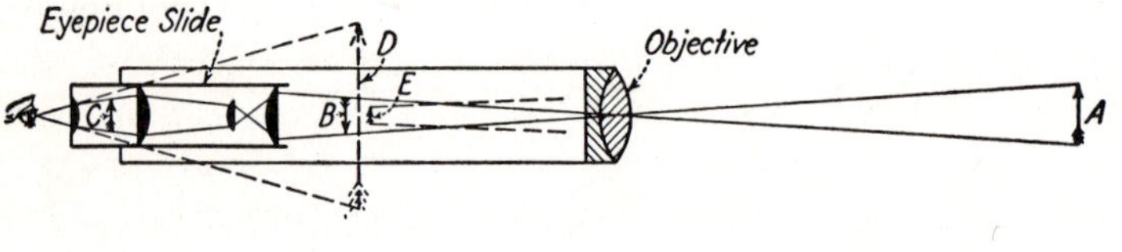

Fig. 5.9*a* Erecting eyepiece.

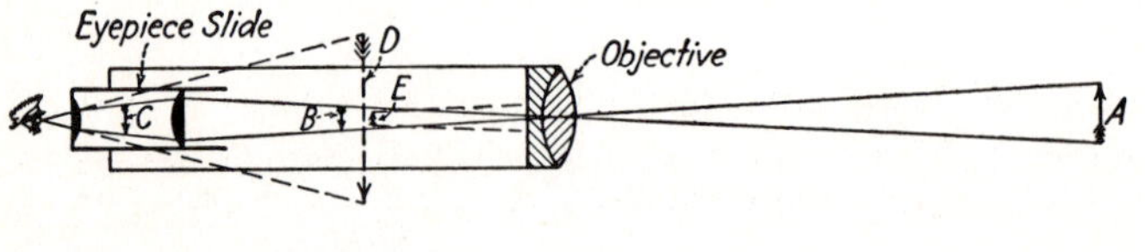

Fig. 5.9*b* Inverting eyepiece.

Each lens which is interposed between the object and the eye absorbs some of the light that strikes it and each lens surface reflects some of the entering and departing light. Hence, other things being equal, the object is more brilliantly illuminated when viewed through the inverting eyepiece; this is a great advantage, particularly when observations are made during cloudy days or near nightfall. Another important advantage of the inverting eyepiece is that the telescope is shorter and the instrument lighter in weight. The beginner experiences some inconvenience on viewing things apparently upside down, but this difficulty is overcome with a little practice. The single advantage of the erecting eyepiece is that objects appear in their natural position, and this is the reason why its use is so common.

5.13. Properties of the telescope The *illumination* of the image depends upon the effective size of the objective, the quality and number of lenses, and the magnifying power. Other conditions being the same, either the larger the objective or the smaller the magnifying power, the better the illumination; that is, the better lighted appears the object.

Distortion of the field of view so that it does not appear flat is mainly caused by what is termed the *spherical aberration* of the eyepiece. Although this introduces no appreciable error in ordinary measurements, it is not desirable when two points in the field are to be observed at the same time, as in stadia measurements.

The *definition* of a telescope is its power to produce a sharp image. It depends upon the quality of the glass, the accuracy with which the lenses are ground and polished, and the precision with which they are spaced and centered. Light rays passing through the lenses near their edges are particularly troublesome, and to improve the definition these rays are intercepted by diaphragms or screens placed between the lenses of the eyepiece and in the rear of the objective. The effect of these screens is to decrease the illumination somewhat.

The angular width of the field of view is the angle subtended by the arc, whose center is nearly at the eye and whose length is the distance between opposite points of the field viewed through the telescope. It is independent of the size of the objective. In general, the larger the telescope and the greater the magnifying power, the less the angle of the field of view. For most surveying of moderate precision, the work is greatly retarded if the instrument does not have a fairly large field of view, and this is one of the reasons why the telescopes are not usually made of high magnifying power. Usually, the angular width of the field ranges from about 1°30′ for a magnifying power of 20, to 45′ for a magnifying power of 40.

The magnifying power of the telescope for the better grade of engineer's level is about 30 diameters. The U.S. National Ocean Survey type of precise level has a magnification of about 40 diameters.

5.14. Spirit level tube The level tube used in surveying instruments (Fig. 5.10) is a glass vial with the inside ground barrel-shaped, so that a longitudinal line on its inner surface is the arc of a circle (Fig. 5.11). The tube is nearly filled with sulfuric ether or with alcohol. The remaining space is occupied by a bubble of air which takes up a location at the high point in the tube. The tube is usually graduated in both directions from the middle; thus, by observation of the ends of the bubble it may be "centered," or its center brought to the midpoint of the tube.

A longitudinal line tangent to the curved inside surface at its upper midpoint is called the *axis of the level tube* or *axis of the level*. When the bubble is centered, the axis of the level tube is horizontal.

The tube is set in a protective metal housing, usually with plaster of paris. The housing is attached to the instrument by means of screws which permit vertical adjustment at one end and lateral movement at the other end of the tube, as shown in Fig. 5.10.

Some leveling instruments are equipped with a prismatic viewing device by means of which one end of the bubble appears reversed in direction and alongside the other end. Figure 5.12*a* illustrates such a *coincidence* bubble when it is not centered. The bubble is centered by matching its ends as shown in Fig. 5.12*b*. Settings can be made more accurately by such coincidence than by matching the ends of the bubble with

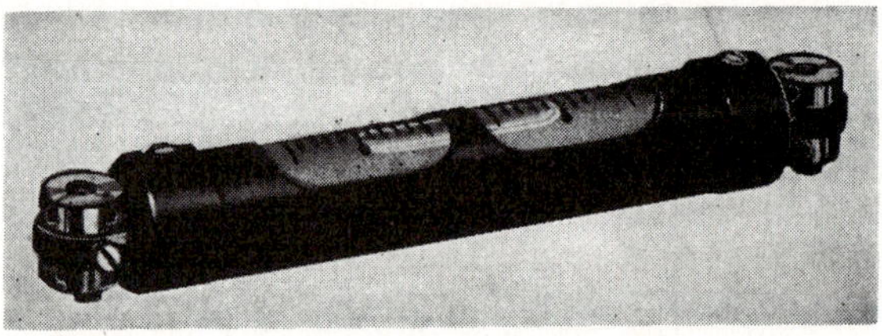

Fig. 5.10 Level tube.

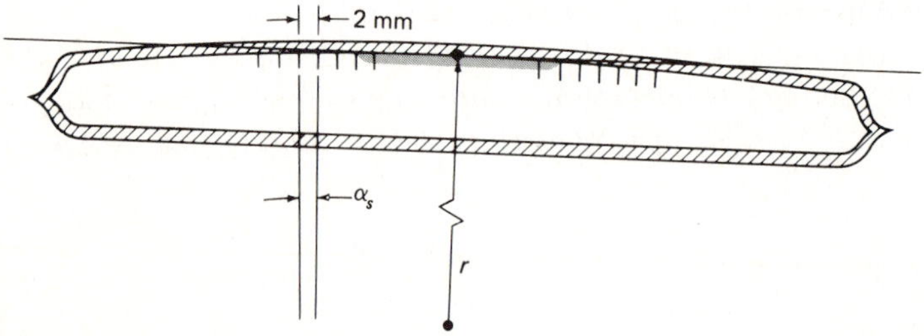

Fig. 5.11 Cross section of spirit level tube.

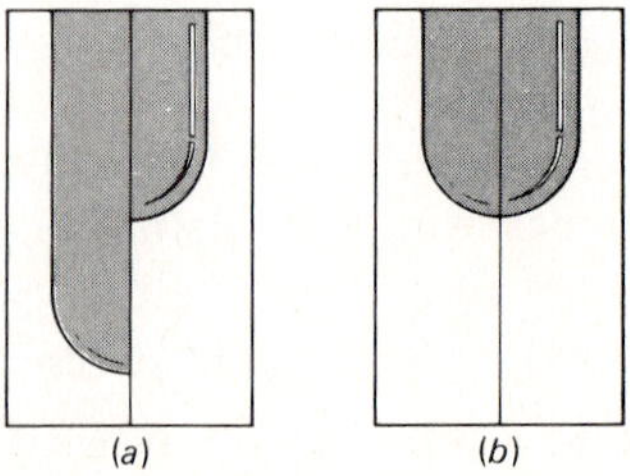

5.12 Coincidence bubble. (*a*) Bubble not centered. (*b*) Bubble centered. (*Courtesy Kern Instruments, Inc.*)

respect to the graduations on the level tube. Tilting levels (Art. 5.7) are usually equipped with coincidence viewing of the spirit level tube.

5.15. Sensitivity of level tube If the radius of the circle to which the inner chamber level tube is ground is large, a small vertical movement of one end of the tube will cause a large displacement of the bubble; if the radius is small, the displacement will be small. Thus, the radius of the tube is a measure of its sensitivity. The sensitivity is generally expressed in seconds of the central angle, α_s (Fig. 5.11), whose arc is one division of the tube. For most instruments the length of a division is 2 mm. The sensitivity expressed in seconds of arc is not a definite measure unless the spacing of graduations is known.

Sensitivities of bubble tubes, expressed in seconds per 2-mm division, vary from 1 to 2″ for precise levels up to 10 to 30″ for engineer's levels.

Should determination of bubble tube sensitivity be necessary, proceed as follows: (1) Align the bubble tube axis with a pair of diagonally opposite level screws; (2) hold a rod in a vertical position at a measured distance from the level; (3) observe the rod reading; (4) tilt the telescope by manipulating the level screws, moving the bubble tube through n divisions; and (5) observe the rod reading. Repeat steps (3) and (4) at least five times and calculate the average increment on the rod subtended by the angle caused by movement of the bubble through n divisions. From Fig. 5.13, the sensitivity α_s in seconds per division is

$$\alpha_s = \frac{d}{(\mathrm{a})(n)}\rho \tag{5.13}$$

in which d is the average increment on the rod, a the distance from level to rod in compatible units, n the number of divisions the bubble tube moved, and $\rho = 206{,}265.4$ seconds of arc per radian (see App. B).

5.16. Dumpy level Figure 5.14 shows the details of a conventional American dumpy level with erecting eyepiece. The telescope A is rigidly attached to the level

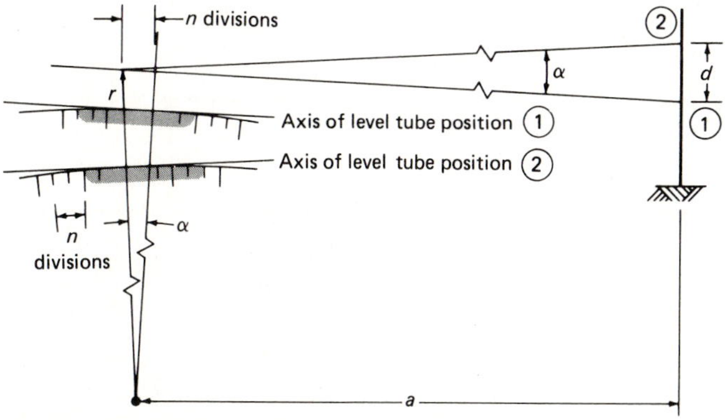

Fig. 5.13 Determination of level-tube sensitivity.

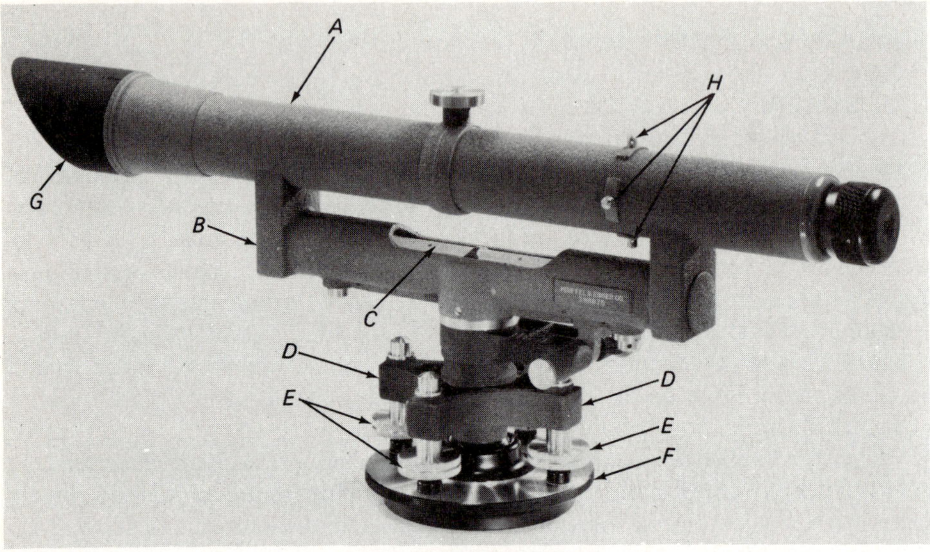

Fig. 5.14 Engineer's dumpy level. (*Courtesy Keuffel & Esser Co.*)

bar *B*, and the instrument is so constructed that the optical axis of the telescope is perpendicular to the axis of the center spindle. The level tube *C* is permanently placed so that its axis lies in the same vertical plane as the optical axis, but it is adjustable in the vertical direction by means of a capstan-headed screw at one end. The spindle revolves in the socket of the leveling head *D*, which is controlled in position by the four leveling screws *E*. At the lower end of the spindle is a ball-and socket joint which makes a flexible connection between the instrument proper and the foot plate *F*. When the leveling screws are turned, the level is moved about this joint as a center. The sunshade *G* protects the objective from the direct rays of the sun. The adjusting screws *H* for the cross-hair ring are near the eyepiece end of the telescope.

The telescope of the dumpy level usually has a magnifying power of about 30 diameters, and the level tube usually has a sensitivity of 20 seconds of arc per graduation (2 mm).

The name "dumpy level" originated from the fact that formerly this level was usually equipped with an inverting eyepiece and therefore was shorter than other American manufactured levels of the same magnifying power. The current primary advantage of the dumpy-style level is that its weight lends stability under conditions where vibration or wind render the lighter automatic levels less stable.

5.17. Tilting levels The distinctive feature of a *tilting level* (Fig. 5.15) is that the telescope is mounted on a transverse fulcrum at the vertical axis and on a micrometer screw at the eyepiece end of the telescope. After the instrument has been leveled approximately in the usual manner, perhaps by use of a circular spirit level, the telescope is pointed in the direction desired and is then "tilted," or rotated slightly in the vertical plane of its axis by turning the micrometer screw, until the sensitive

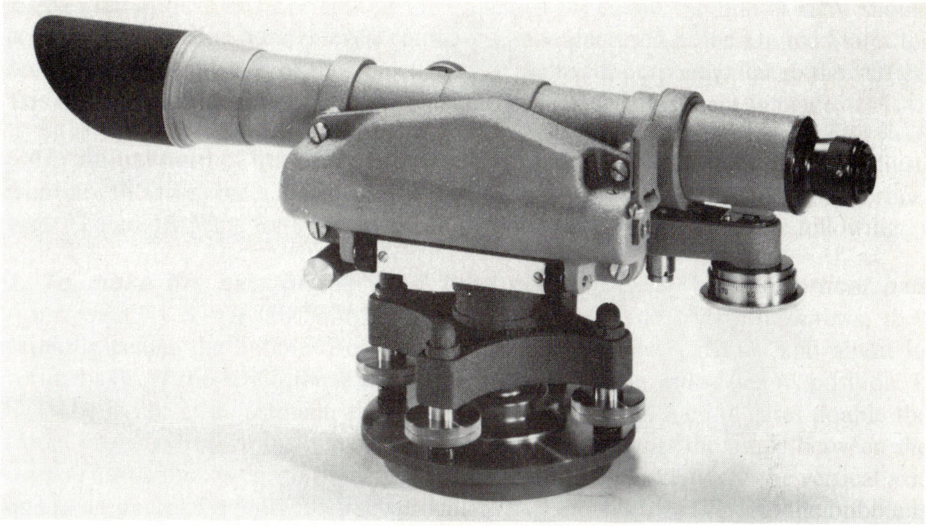

Fig. 5.15 Tilting level. (*Courtesy Keuffel & Esser Co.*)

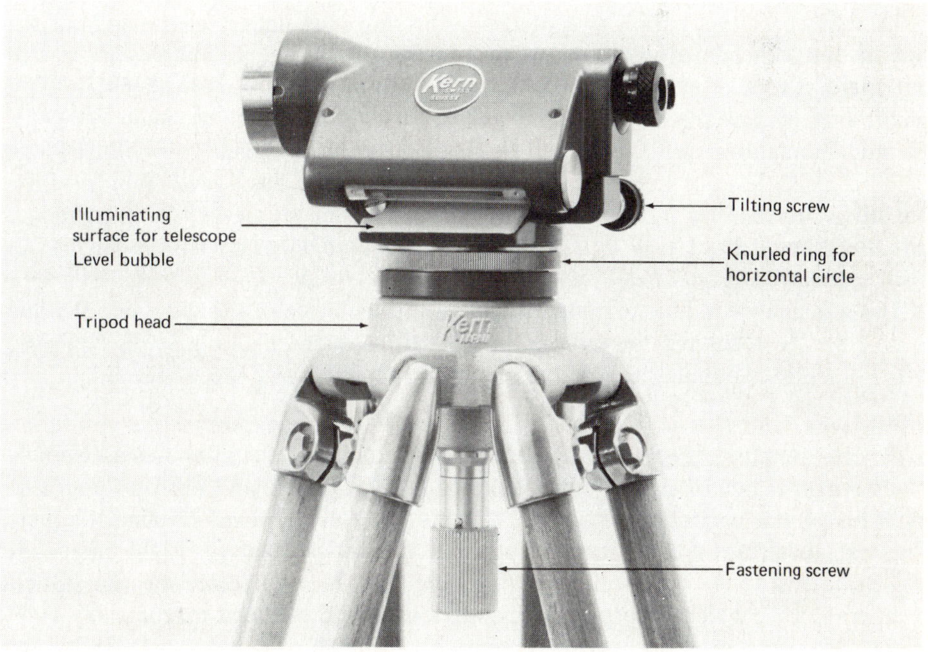

Fig. 5.16*a* European-designed tilting level. (*Courtesy Kern Instruments, Inc.*)

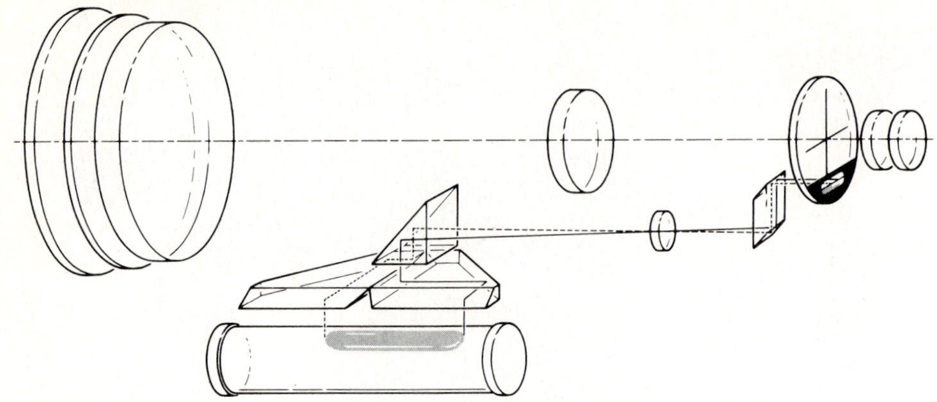

Fig. 5.16*b* Optical train for viewing coincidence bubble. (*Courtesy Kern Instruments, Inc.*)

telescope-level bubble is centered. The line of sight is then horizontal, even though the instrument as a whole is not exactly level.

Modern tilting levels, although of the dumpy type, incorporate features which make them quite different from the dumpy-style tilting level. They are small, light in weight, and permit rapid setting up and observing. They are equipped with a circular spirit level and have three leveling screws. The telescope is internally focusing, usually has an inverted image, and has the cross hairs, including stadia lines, etched on a glass diaphragm.

The telescope-level bubble is centered by bringing the images of the ends of the bubble into coincidence as shown in Fig. 5.12. These coincidence bubbles are viewed through a system of prisms from the eyepiece end of the telescope either through a separate parallel eyepiece or off to one side of the field of view of the main telescope. Electric illumination can be provided for night observation. Figure 5.16*a* illustrates a modern tilting level of European design. This instrument has a level bubble with a sensitivity of 18″ per 2-mm division. The bubble can be viewed for coincidence leveling in the field of view of the telescope (see Fig. 5.16*b*). Horizontal angles can also be observed optically to 1′ of arc.

The coincidence bubble permits more accurate centering of the spirit bubble than is possible by observing the bubble directly. Consequently, tilting levels equipped with coincidence bubbles are used for first- and second-order precise leveling.

5.18. Self-leveling levels A *self-leveling* or automatic level allows establishment of a horizontal line of sight by means of a system of prisms and mirrors supported by wires as in a pendulum. Figure 5.17 shows one form of a self-leveling level and Fig. 5.18 illustrates a cross-sectional view of the optics involved. Light enters the objective lens at *A*, passes through focusing lens *B*, is reflected by the optical compensator *C*, which is suspended by wires at *D* and clamped magnetically at *E*. When the compensator *C* is freely supported, the line of sight defined by the objective lens, compensator, and eyepiece is automatically horizontal. Most auto-

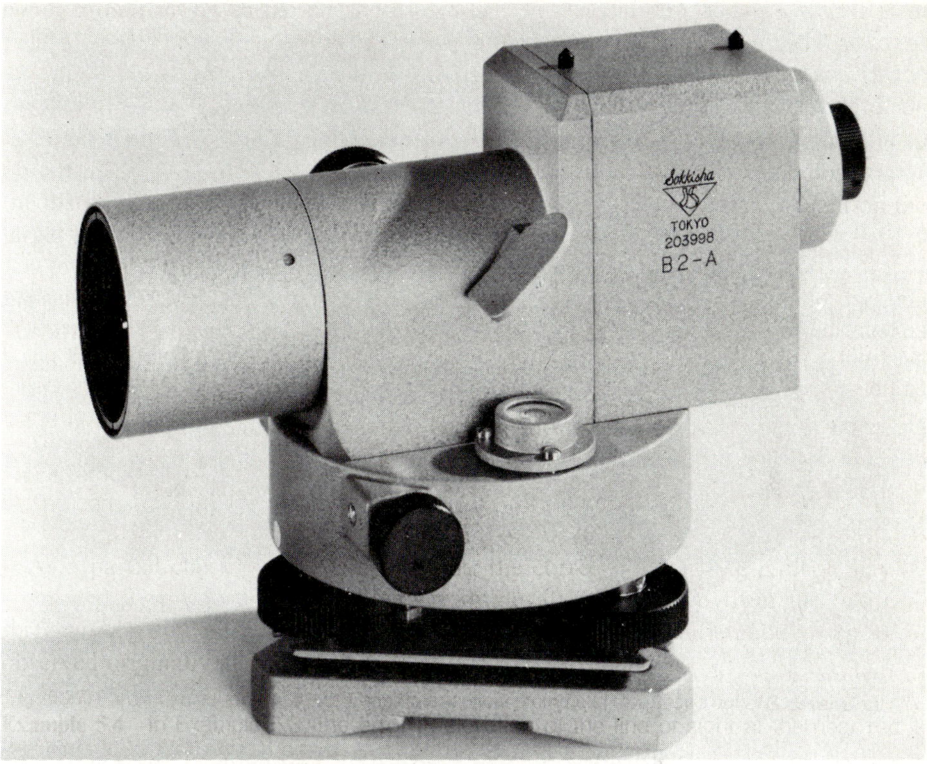

Fig. 5.17 Lietz-Sokkisha B-2A self-leveling level. (*Courtesy of the Lietz Co.*)

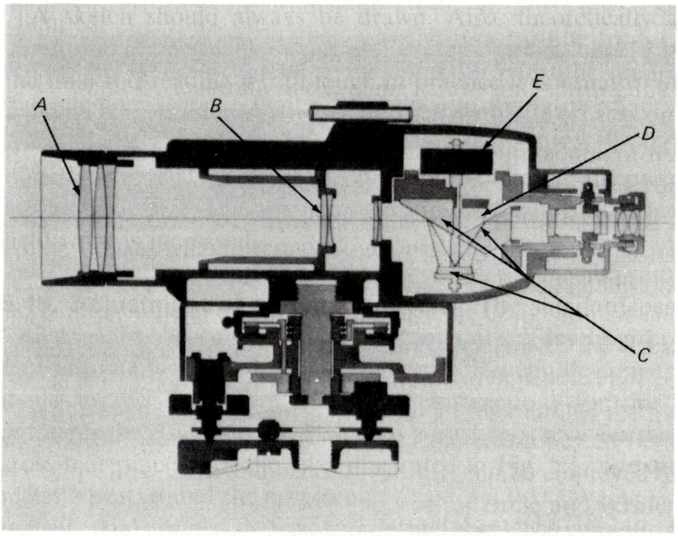

Fig. 5.18 Section view of a Lietz-Sokkisha B-1 self-leveling level.

matic levels are designed so that the line of sight is horizontal when the telescope barrel is within $\pm 10'$ of being horizontal. At each setup, the instrument is leveled approximately by use of a circular spirit or bull's-eye level (Fig. 5.17) and the pendulum maintains a horizontal line of sight. The instrument is light, easy to handle, and its operation is quick and accurate. Note that the random error in centering the bubble is absent in this style of instrument.

5.19. Locke hand levels The Locke hand level is widely used for rough leveling. It consists of a metal sighting tube about 6 in long on which is mounted a level vial (Fig. 5.19). In the tube beneath the vial is a prism which reflects the image of the bubble to the eye end of the level. Just beneath the level vial is a cross wire which is adjustable by means of a pair of screws, the heads of which protrude from the case; one screw is loosened and the other is tightened. The eyepiece consists of a peephole mounted in the end of a slide which fits inside the tube and is held in a given position by friction. Mounted on the right half of the inner end of the slide is a semicircular convex lens which magnifies the image of the bubble and cross wire as reflected by the prism. Both the object and the eye ends of the tube are closed by disks of plain glass so that dust will not collect on the prism and lens. The magnifying lens is focused by moving the eyepiece slide in or out.

In using the level the object is viewed directly through the left half of the sighting tube, without magnification, while with the same eye and at the same time the position of the bubble with respect to the cross wire is observed in the right half of the field of view. The level is held with the level vial uppermost and is tipped up or down until the cross wire bisects the bubble, when the line of sight is horizontal. After a little practice one may make observations with greater facility by keeping both eyes open. Some observers steady the hand level by holding it against, or fastening it to, a staff. Hand levels equipped with stadia hairs are available.

5.20. Abney hand level and clinometer As its name indicates, this level is suitable both for direct leveling and for measuring the angles of slopes. The

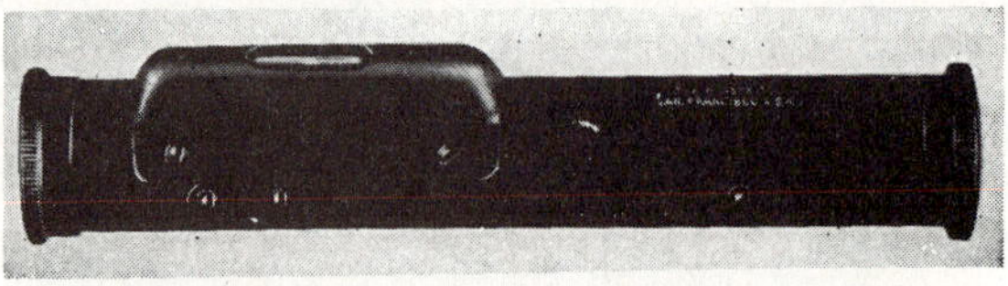

Fig. 5.19 Locke hand level. (*Courtesy Keuffel & Esser Co.*)

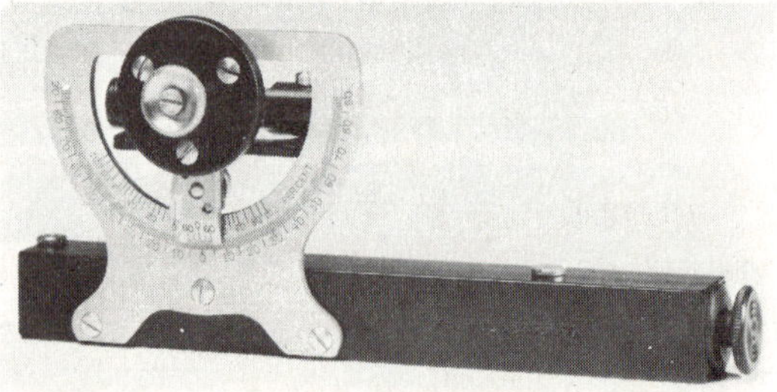

Fig. 5.20 Abney hand level and clinometer. (*Courtesy the Lietz Co.*)

instrument shown in Fig. 5.20 is graduated both in degrees and in percentage of slope, or grade. When it is used as a level, the index of the vernier is set at zero, and it is then used in the same way as the Locke hand level. When it is used as a clinometer, the object is sighted, and the level tube is caused to rotate about the axis of the vertical arc until the cross wire bisects the bubble as viewed though the eyepiece. Either the slope angle or the slope percentage is then read on the vertical arc.

5.21. Leveling rods These are graduated wooden rods of rectangular cross section by means of which difference in elevation is measured. The lower or ground end of the rod is shod with metal to protect it from wear and is usually the point of zero measurement from which the graduations are numbered. Aluminum alloy and fiberglass rods are also available in certain specific styles.

The rod is held vertically, and hence the reading of the rod as indicated by the horizontal cross hair of the level is a measure of the vertical distance between the point on which the rod is held and the line of sight.

Rods are obtainable in a variety of types, patterns, and graduations and are either in single pieces or in sections which are jointed together or slide past each other and are clamped together. Common lengths are 12 and 13 ft. In the United States, rods are graduated in hundredths of a foot and in metric units. On some government surveys the rods are graduated in decimals of the metre or the yard.

The two general classes of leveling rods are (1) *direct-reading* rods which may be read directly by the leveler while looking through the telescope, and (2) *target* rods on which a target sliding on the rod is set by the rodman as directed by the leveler. A rod such as the *Philadelphia* rod (Art. 5.22), illustrated in Fig. 5.21*a*, is a combination direct-reading and target rod. Under ordinary conditions, observations with the direct-reading rod without the target can be made with nearly the same precision and much more rapidly than when using the target. The direct-reading rod is the one commonly employed, even for precise leveling.

The San Francisco rod (Fig. 5.21*b*) is an example of a direct-reading rod designed for use without a target. It is constructed in three sections each about 4.4 ft or 1.34 m in length and is more portable than the Philadelphia rod.

Another less commonly used style of rod is the *direct elevation* rod (Fig. 5.21*c*) with an endless specially treated fabric or metal graduated ribbon that can be set by the rodman to the elevation of the leveled optical line of sight. Since the scale is graduated in numbers which increase downward, subsequent rod readings from the same instrument setup provide elevations directly with no need for additional computations. Such rods are used primarily on construction layout surveys or stadia and topographic mapping surveys.

5.22. Direct-reading rods The direct-reading rod is held vertically; the leveler observes the graduation at which the line of sight intersects the rod and records the reading. Observations closer than the smallest division on the rod are made by estimation.

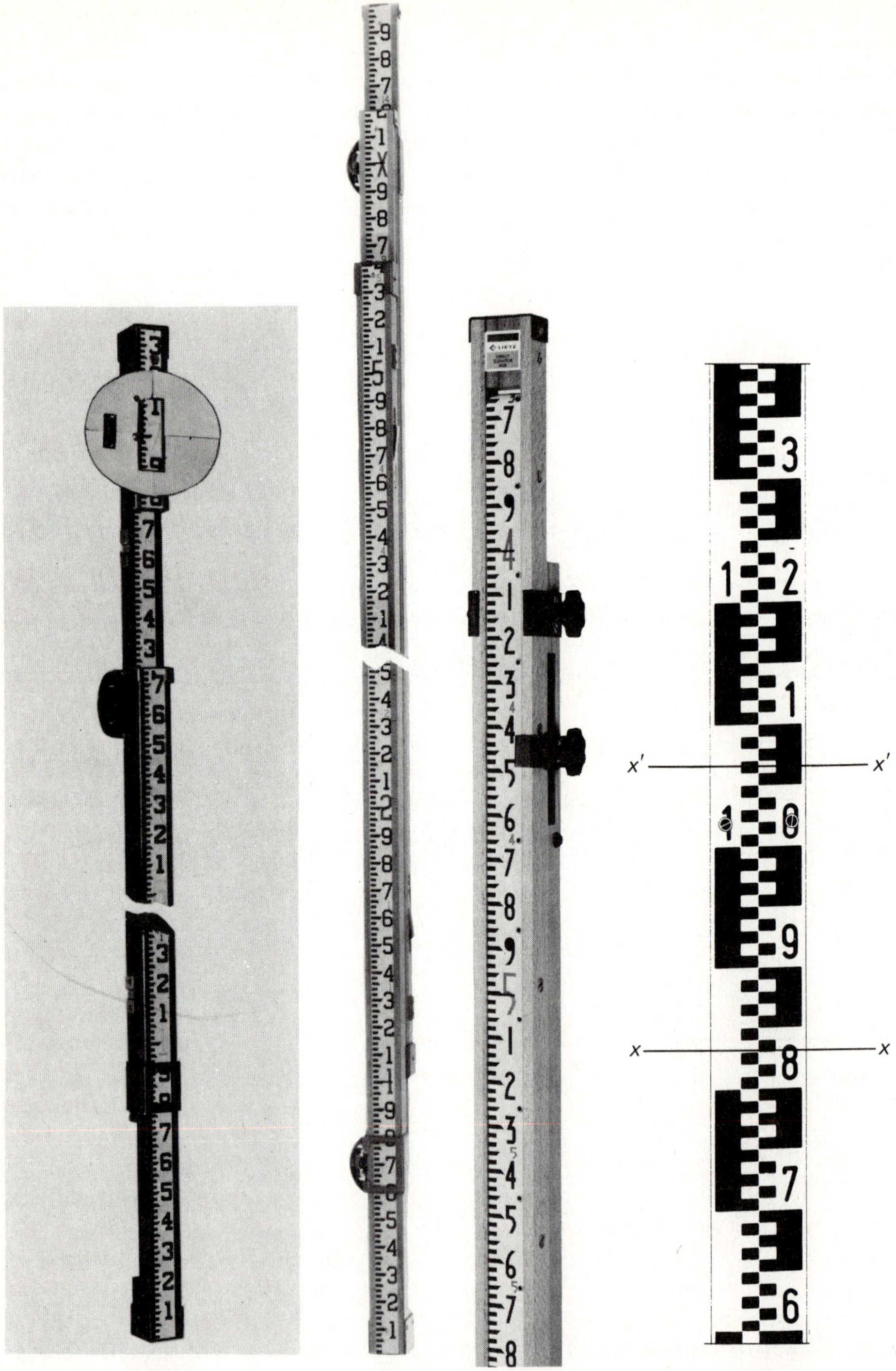

Fig. 5.21 (*a*) Philadelphia rod graduated in feet, tenths, and hundredths of feet, partially extended. (*Courtesy the Lietz Co.*) (*b*) San Francisco rod, graduated in feet, tenths, and hundredths of feet, partially extended. (*Courtesy the Lietz Co.*) (*c*) Direct elevation rod graduated in feet, tenths, and hundredths of feet. (*Courtesy the Lietz Co.*) (*d*) Portion of a rod graduated in metric units. (*Courtesy Wild Heerbrugg Instruments, Inc.*)

The direct-reading rod should be so marked that the graduations appear sharp and distinct for any normal distance between level and rod. Normally, the background is white or yellow with graduations 0.01 ft or 0.010 m wide painted in black as shown in Fig. 5.21. The readings to 0.01 ft and to 0.010 m are made on the edges of the graduations. Usually, the rod is read to half-hundredths of a foot or to millimetres, by estimation. The numbers indicating feet or metres are in red, and those indicating tenths of feet or decimetres are in black. This style of graduation is satisfactory for direct-reading when the length of sight is less than 400 to 500 ft or 120 to 150 m. When using a target equipped with a vernier (Arts. 5.24, 5.25, and 5.26), readings to smaller increments can be made.

The *Philadelphia* rod is the most widely used rod. It is made in two sliding sections held in contact by two brass sleeves. A screw attached to the upper sleeve permits clamping the two sections together in any desired relative position. For readings of 7 ft (2 m on the metric rod) or less, the back section is clamped in its normal collapsed position. For greater readings, the rod is extended to its full length so that graduations on the front face of the back section are a continuation of those on the lower front strip. When thus extended, the rod is called a "high" rod. Figure 5.21*a* shows a partially extended Philadelphia rod graduated in feet, tenths, and hundredths of feet. A portion of the face of a European-style rod graduated in metric units is shown in Fig. 5.21*d*. This rod comes in 3- and 4-m lengths, is hinged for folding to simplify carrying, and contains a built-in circular level bubble used to keep the rod vertical.

Direct observation of a direct-reading rod involves estimation to obtain the third decimal place in metres and feet. In Fig. 5.21*d*, graduations are in metres numbered on the left, decimetres numbered on the right, and centimetres designated by alternating black and white squares. These squares are arranged in a checkerboard pattern so that the cross hair will always appear on a white square. Again referring to Fig. 5.21*d*, readings at lines *xx* and *x′x′* to the nearest centimetre yield 0.83 m and 1.07 m, respectively. Estimation to the nearest millimetre at these same lines gives readings of 0.831 m and 1.066 m, respectively.

Figures 5.22*a* and *b* show the faces of Philadelphia-style rods graduated in feet and metres, respectively. In Fig. 5.22*a* the largest crosshatched (red) numerals indicate the whole numbers of feet, the black numerals tenths of feet, and the alternating black and white graduations represent hundredths of feet. Thus, the readings at lines *aa′*, *bb′*, and *cc′* to the nearest hundredth are 1.01, 1.54, and 1.98 ft, respectively. Estimation to the nearest thousandth of a foot yields respective readings of 1.006, 1.545, and 1.980. In Fig. 5.22*b* the light gray numbers are metres, black numerals are decimetres, and alternating black and white graduations are centimetres. Readings to the nearest centimetre at lines *dd′*, *ee′*, and *ff′* are 0.94, 1.02, and 1.15 m, respectively. If millimetres are desired, they must be estimated yielding respective readings of 0.940, 1.024, and 1.155 m. Note that in Fig. 5.22*a* the whole number of feet is indicated by the small crosshatched (red) number between tenth foot numerals and the whole number of metres in Fig. 5.22*b* is given by the number of dots above each decimetre mark on the metric rod. Consequently, even

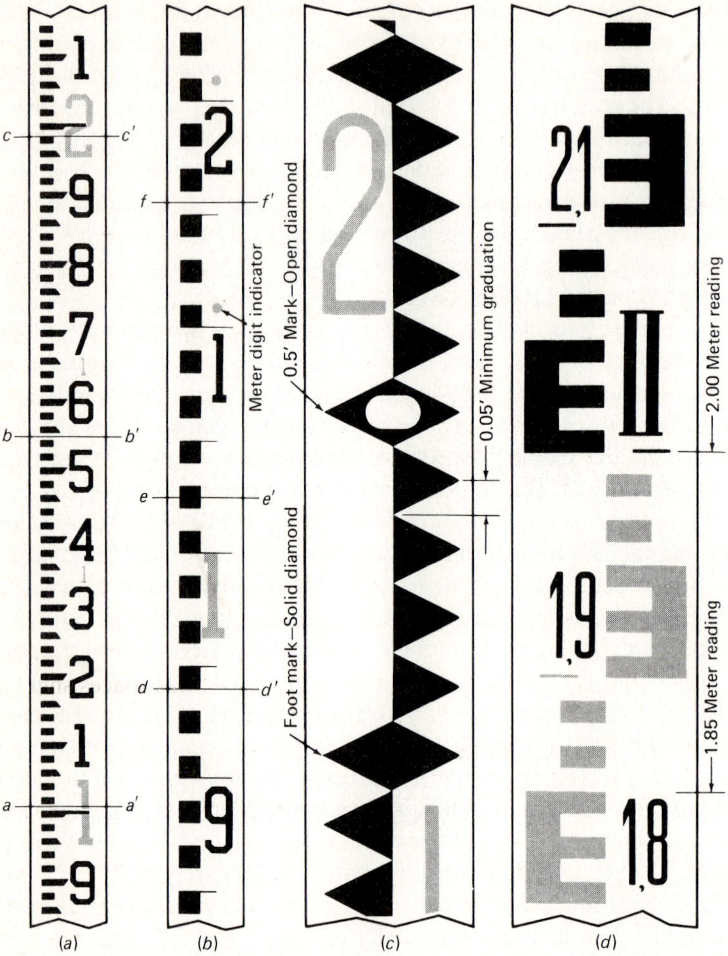

Fig. 5.22 Direct reading of: (*a*) rod graduated in feet, tenths, and hundredths of feet; (*b*) rod graduated in metres, decimetres, and centimetres; (*c*) stadia rod graduated in feet, tenths of feet, and 0.05 ft; (*d*) stadia rod graduated in metres, decimetres, and centimetres. (*Courtesy the Lietz Co.*)

though the field of view of the telescope may fall within full unit graduations, determination of the full unit is possible.

When using a Philadelphia rod, high rod readings are made with the rod fully extended. On these rods there is a metal stop against which the lower brass sleeve makes contact when the rod is fully extended. Occasionally, this metal stop becomes loose and out of adjustment, so that the rod may not be extended to the proper length, introducing errors in high rod readings. Accordingly, this metal stop and the vernier must be examined frequently and adjusted periodically.

5.23. Stadia rods This type of rod is used for stadia surveying as described in Chap. 7. Any type of self-reading rod may be used as a stadia rod, but the leveling rod graduated in hundredths of feet or in centimetres as in Fig. 5.22*a* and *b* is suitable only for sights of less than 400 ft or 120 m. For longer sights a rod with larger, heavier graduations arranged in a different pattern is necessary. Figures 5.22*c* and *d* illustrate two such patterns graduated in feet, tenths, and 0.05 ft; and metres, decimetres, and centimetres, respectively. These graduations are suitable for distances of up to about 700 ft or 210 m. Readings can be estimated to the nearest 0.01 ft and nearest millimetre on these two rods. These rods are equally suitable for use in stadia surveying and leveling and are sometimes called *stadia boards*.

5.24. Verniers A vernier, or vernier scale, is a short auxiliary scale placed alongside the graduated scale of an instrument, by means of which fractional parts of the least division of the main scale can be measured precisely; the length of one space on the vernier scale differs from that on the main scale by the amount of one fractional part. The precision of the vernier depends on the fact that the eye can determine more closely when two lines coincide than it can estimate the distance between two parallel lines. The scale may be either straight (as on a leveling rod) or curved (as on the horizontal and vertical circles of a transit). The zero of the vernier scale is the index for the main scale.

Verniers are of two types: (1) the *direct* vernier, which has spaces slightly shorter than those of the main scale; and (2) the *retrograde* vernier, which has spaces slightly longer than those of the main scale. The use of the two types is identical, and they are equally sensitive and equally easy to read. Since they extend in opposite directions, however, one or the other may be preferred because it permits a more advantageous location of the vernier on the instrument. Both types are in common use.

Direct vernier Figure 5.23*a* represents a scale graduated in hundredths of feet, and a direct vernier having each space 0.001 ft shorter than a 0.01-ft space on the main scale; thus, each vernier space is equal to 0.009 ft, and 10 spaces on the vernier are equal to 9 spaces on the scale. The index, or zero, of the vernier is set at 0.400 ft on the scale. If the vernier were moved upward 0.001 ft, its graduation numbered 1 would coincide with a graduation (0.41 ft) on the scale, and the index would be at 0.401 ft; and so on. It is thus seen that the position of the index is determined to thousandths of feet without estimation, simply by noting which graduation on the vernier coincides with one on the scale. Note that the coinciding graduation on the main scale does *not* indicate the main-scale reading.

The *least count*, or fineness of reading of the vernier, is equal to the difference between a scale space and a vernier space. For a direct vernier, if s is the length of a space on the scale and if n is the number of vernier spaces of total length equal to that of $(n-1)$ spaces on the scale, then the least count is s/n.

Retrograde vernier On the retrograde vernier shown in Fig. 5.23*b*, each space on the vernier is 0.001 ft *longer* than a 0.01-ft space on the main scale, and 10 spaces on

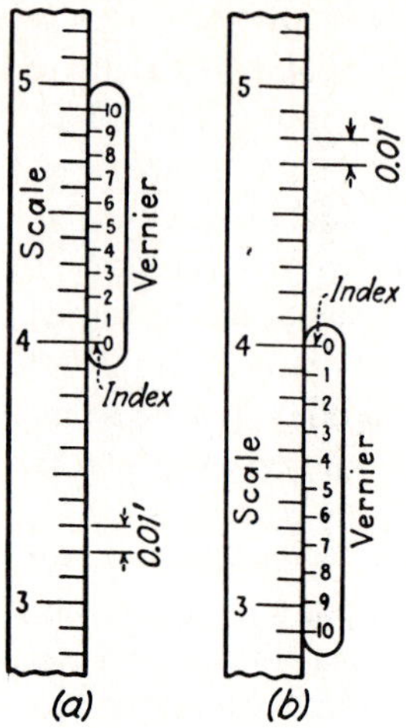

Fig. 5.23 (*a*) Direct vernier. (*b*) Retrograde vernier.

the vernier are equal to 11 spaces on the scale. As before, the index is set at 0.400 ft on the scale. If the vernier were moved upward 0.001 ft, its graduation numbered 1 would coincide with a graduation (0.39 ft) on the scale; and so on. It is seen that, from the index, the retrograde vernier extends backward along the main scale, and that the vernier graduations are also numbered in reverse order; however, the retrograde vernier is read in the same manner as the direct vernier.

The least count of the retrograde vernier is equal to the difference between a scale space and a vernier space, as in the case of the direct vernier. For the retrograde vernier, if s is the length of a space on the scale and if n is the number of vernier spaces of total length equal to that of $(n+1)$ spaces on the scale, then the least count is s/n.

Reading with vernier Figure 5.24 illustrates settings of direct verniers on the target (Fig. 5.24*a*) and on the back (Fig. 5.24*b*) of a Philadelphia leveling rod (Art. 5.22). The rod reading indicated by the target [4.347 ft (Fig. 5.24*a*)] is determined by first observing the position of the vernier index on the scale to hundredths of feet (4.34 in the figure), next by observing the number of spaces *on the vernier* from the index to the coinciding graduations (seven spaces in the figure), and finally by adding the vernier reading (0.007 ft in the figure) to the scale reading (4.34 ft). On the back of the rod (Fig. 5.24*b*) both the main scale and the direct vernier read *down* the rod. The scale reading is 9.26 ft and the vernier reading is 0.004 ft; hence the rod reading is 9.264 ft.

A helpful check in reading the vernier is to note that the lines on either side of the coinciding line should depart from coincidence by the same amount, in opposite directions. As a check against possible mistakes, it is advisable to estimate the fractional part of the main-scale division by reading the index directly.

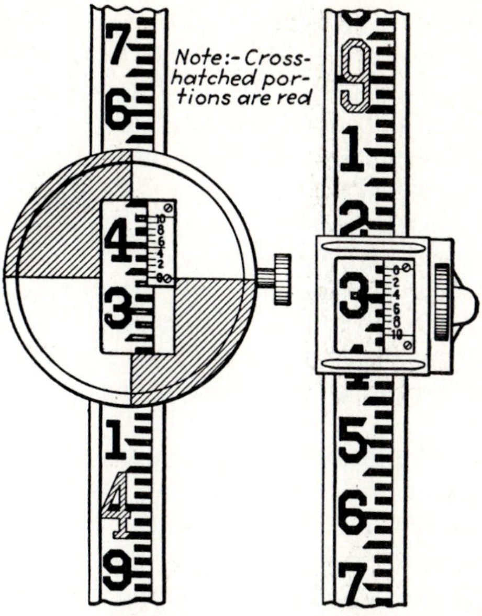

Fig. 5.24 (*a*) and (*b*) Direct vernier settings.

5.25. Targets The usual target (Figs. 5.21*a* and 5.24*a*) is a circular or elliptical disk about 5 in in diameter, with horizontal and vertical lines formed by the junction of alternate quadrants of white and red. A rectangular opening in the front of the target exposes a portion of the rod to view so that readings can be taken. The attached vernier (Art. 5.24) fits closely to the rod, its zero point or index being at the horizontal line of the target. In Fig. 5.24*a* a direct vernier is shown, but both retrograde and direct verniers are in common use.

5.26. Target rods With the target rod, the leveler signals the rodman to slide the target up or down until it is bisected by the line of sight. The target is then clamped, and the rodman, leveler, or both observe the indicated readings. Usually, the target is equipped with a vernier or other device by means of which fractional measurements of the rod graduations can be read without estimation.

The Philadelphia rod (Fig. 5.21*a*), previously described, is a combined direct-reading rod and target rod. Lugs on the target engage in a groove on either side of the front strip. For readings on the lower half of the rod, the reading is made to thousandths of feet by means of a vernier attached to the target. For example, the reading in Fig. 5.24*a* is 4.347 ft using the target.

Graduations on the back of the rear strip are a continuation of those on the front strip and read downward. On the back of the top sleeve is a vernier employed for observations with the rod extended. For readings greater than can be taken with a "short" rod, the target in Fig. 5.21*a* is clamped on the 7.000-ft graduation (the 2.000-m graduation on the metric rod shown in Fig. 5.22*b*) on the front face of the upper section. The vertically held rod is then extended until the target is bisected by the line of sight. The vertical distance from the foot of the rod is then indicated by the reading of the vernier on the back of the rod (Fig. 5.24*b*). Note that the back of the rod is a continuation of graduations on the front, with numbers increasing from top to bottom. For the example illustrated in Fig. 5.24*b* with the target set at 7.00 ft on the front of the rod, the vertical distance from the bottom of the rod to the target is 9.264 ft.

The principal advantage of the target rod is that mistakes are less likely to occur particularly if both rodman and leveler read the rod. Under certain conditions its use materially facilitates the work: for example, where very long sights are taken, where the rod is partly obscured from view, or where it is necessary to establish a number of points all at the same elevation. However, under ordinary conditions its use retards progress without adding much, if anything, to the precision.

5.27. Rod levels The rod level is an attachment for indicating the verticality of the leveling rod. One type (Fig. 5.25) consists of a circular or "bull's-eye" level vial mounted on a metal angle or bracket which is either attached by screws to the side of the rod or is held against the rod, as desired. As noted for the level rod shown in Fig. 5.21*b*, the rod level may be built into the rod.

Another type consists of a hinged casting on each wing of which is mounted a level tube that is held parallel to a face of the rod. When both of the bubbles are centered, the rod is plumb. The hinge makes it possible to fold the level tubes compactly when it is not in use.

Fig. 5.25 Rod level.

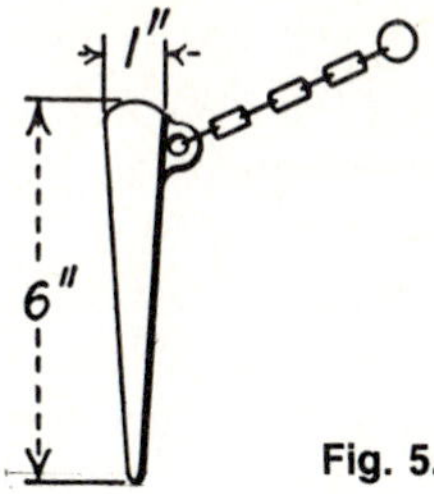

Fig. 5.26 Turning point.

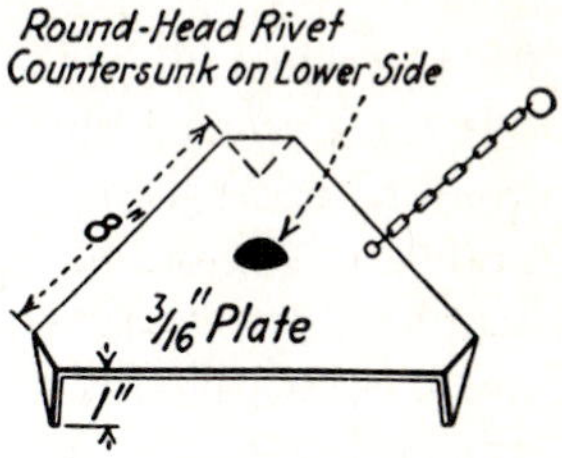

Fig. 5.27 Turning plate.

5.28. Turning points A metal plate or pin that will serve temporarily as a stable object on which the leveling rod may be held at turning points is a useful part of the leveling equipment for lines of differential levels. The iron pin shown in Fig. 5.26 is adapted for use in firm ground. Often a railroad spike or rivet pin is used.

In soft ground the steel plate of Fig. 5.27 makes a satisfactory turning point for ordinary leveling. The plate is also adapted for use where the ground is so solid as to make driving the pin impossible or at least impractical, as along highways. Under these conditions the plate with the dogs at its corners acts as a tripod, no special attempt being made to secure bearing between the lower surface of the plate and ground.

5.29. Setting up engineer's levels The engineer's level is placed in a desired location, with the tripod legs well spread and firmly pressed into the ground, with the tripod head nearly level, and with the telescope at a convenient height for sighting. If the setup is on a slope, it is preferable to orient the tripod so that one of its legs extends up the slope.

When operating three-screw tilting or automatic levels which have bull's-eye bubbles as illustrated in Fig. 5.28*a*, opposite rotation of level screws *A* and *B* moves the bubble along axis xx', as shown in Fig. 5.28*b*, while rotation of screw *C* moves the bubble in the direction of axis yy' (Fig. 5.28*c*). When the bubble is centered

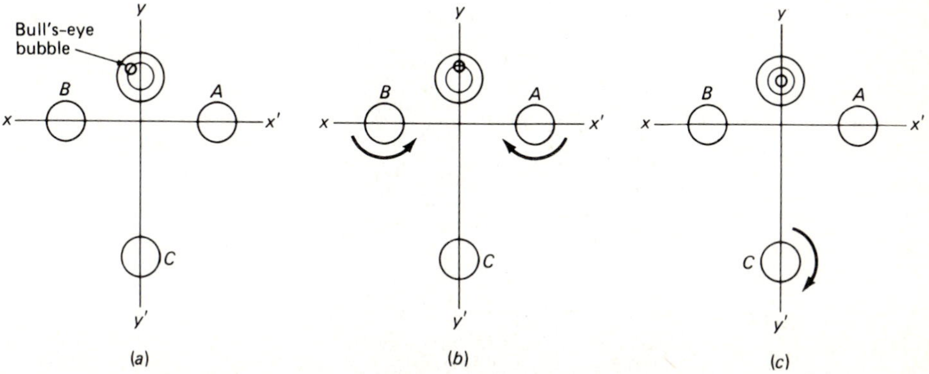

Fig. 5.28 (*a*), (*b*), and (*c*) Leveling a three-screw instrument.

within the circle, the instrument is sufficiently leveled to be within the range of the spirit level in a tilting instrument or the pendulum device in the automatic level.

If the leveling head has four screws, the telescope is brought over one pair of opposite leveling screws, and the bubble is centered approximately by turning the level screws in *opposite directions*. The bubble will follow the direction of the left thumb. Then the process is repeated with the telescope over the other pair. By repetition of this procedure the leveling screws are manipulated until the bubble remains centered. After completion of this process, the telescope is rotated 180° and the bubble is checked along both axes. If the bubble remains centered, the instrument is in adjustment and the line of sight is horizontal for any position of the telescope. If the bubble does not remain centered, a correction of one-half the amount by which the bubble is out of center is made along each axis. Following this correction, the bubble should be off center by the same amount for any direction in which the telescope is pointed. If this situation occurs, the level bubble is obviously out of adjustment and should be adjusted as soon as possible (see Art. 5.47).

5.30. Reading the rod For observations to millimetres and to hundredths or thousandths of feet, the rod is held on some well-defined point of a stable object. The rodman holds the rod vertical either by observing the rod level or by estimation. First, the cross hairs are carefully focused to suit the eye of the observer. This can be done by holding an open field book in front of the objective lens of the telescope and rotating the eyepiece adjustment knob until the cross hairs are clear, sharp, and black. Next, the level operator revolves the telescope about the vertical axis until the rod is about in the middle of the field of view, focuses the objective for distinct vision, checks for presence of parallax (Arts. 5.10 and 5.38), and carefully centers the bubble. The leveler observes and records the reading indicated by the line of sight, that is, the apparent position of the horizontal cross hair on the rod. Checking, the leveler again observes the bubble and the rod. If the target is used, the procedure is identical except that the target is set by the rodman as directed by the leveler.

For leveling of lower precision, as when rod readings for points on the ground are determined to the nearest 0.1 ft or nearest cm, the observations usually are not checked, and proportionally less care is exercised in keeping the rod vertical and the bubble centered, always bearing in mind the errors involved and the precision with which measurements are desired.

If no rod level is used, in calm air the rodman can plumb the rod accurately by balancing it upon the point on which it is held. By means of the vertical cross hair the leveler can determine when the rod is held in a vertical plane passing through the instrument, but cannot tell whether it is tipped forward or backward in this plane. If it is in either of these positions, the rod reading will be greater than the true vertical distance, as illustrated by Fig. 5.29. To eliminate this error, the rodman *waves the rod*, or tilts it forward and backward as indicated by the figure, and the leveler takes the least reading, which occurs when the rod is vertical. The larger the rod reading, the larger the error due to the rod's being held at a given inclination; hence, it is more important to wave the rod for large readings than for small readings. Further, whenever the rod is tipped backward about any support other than the front edge of

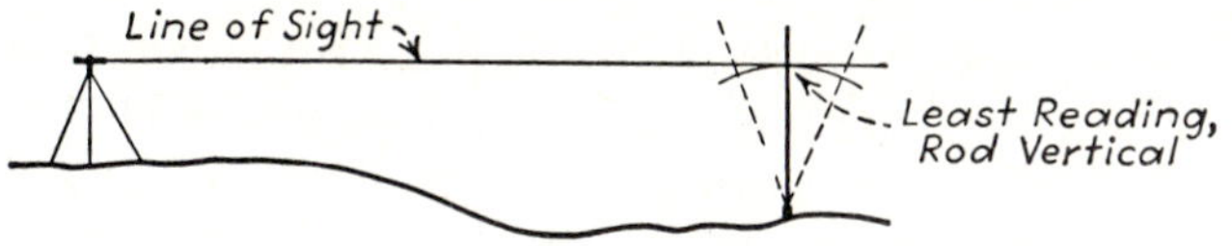

Fig. 5.29 Waving the rod.

its base, the graduated face rises and an error is introduced; for small readings this error is likely to be greater than that caused by not waving the rod.

5.31. Differential leveling Differential leveling is the operation of determining the elevations of points some distance apart. Differential leveling requires a series of setups of the instrument along the general route and, for each setup, rod readings are taken back to a point of known elevation and forward to a point of unknown elevation.

5.32. Bench marks A *bench mark* (B.M.) is a definite point on an object, the elevation and location of which are known. It may be permanent (P.B.M.) or temporary (T.B.M.). Bench marks serve as points of reference for levels in a given locality. Their elevations are established by differential leveling, except that the elevation of the initial bench mark of a local project may be assumed.

Throughout the United States permanent bench marks are established by the U.S. Geological Survey and the U.S. National Ocean Survey. Similarly, bench marks have been established by various other federal, state, and municipal agencies and by such private interests as railroads and water companies, so that the surveyor has not far to go before finding some point of known elevation.

Fig. 5.30 United States National Ocean Survey (formerly U.S. Coast & Geodetic Survey) bench mark.

The National Ocean Survey bench marks (Fig. 5.30) consist of bronze plates set in stone or concrete. Some are marked with the elevation to the nearest foot above mean sea level but most are not marked with any elevation. Such a marking would of course be invalid if the bench mark had been disturbed. Those of other agencies are similar. Other objects frequently used as bench marks are stones, pegs or pipes driven in the ground, nails or spikes driven horizontally in trees or vertically in pavements, and marks painted or chiseled on street curbs.

For any survey or construction enterprise, levels are run from some initial bench mark of known or assumed elevation to scattered points in desirable locations for future reference as bench marks.

In some areas, the elevation of bench marks may be altered by earth movements such as those caused by earthquakes, slides, lowering of water tables, pumping from oil fields, mining, or construction.

5.33. Definitions A *turning point* (T.P.) is an intervening point between two bench marks upon which point foresight and backsight rod readings are taken. It may be a pin or plate (see Art. 5.28) which is carried forward by the rodman after observations have been made, or it may be any stable object such as a street curb, railroad rail, or stone. The nature of the turning point is usually indicated in the notes, but no record is made of its location unless it is to be reused. A bench mark may be used as a turning point.

A *backsight* (B.S.) is a rod reading taken on a point of known elevation, as a bench mark or a turning point. Usually, it will be taken with the level sighting back along the line, hence the name. A backsight is sometimes called a *plus sight*. The horizontal distance from level to rod on a B.S. is called *backsight distance*.

A *foresight* (F.S.) is a rod reading taken on a point the elevation of which is to be determined, as on a turning point or on a bench mark the elevation of which is to be established. A foresight is sometimes called a *minus sight*. The horizontal distance from the level to the rod on a F.S. is called *foresight distance*.

The *height of instrument* (H.I.) is the elevation of the line of sight of the telescope above the datum when the instrument is leveled.

In surveying with the transit, the terms backsight, foresight, and height of instrument have meanings different from those defined here.

5.34. Procedure In Fig. 5.31, B.M.$_1$ represents a point of known elevation (bench mark), and B.M.$_2$ represents a bench mark to be established some distance away. It is desired to determine the elevation of B.M.$_2$. The rod is held at B.M.$_1$, and the level is set up in some convenient location, as L_1, along the general route B.M.$_1$ to B.M.$_2$. The level is placed in such a location that a clear rod reading is obtainable, but no attempt is made to keep on the direct line joining B.M.$_1$ and B.M.$_2$. A backsight is taken on B.M.$_1$. The rodman then goes forward and, as directed by the leveler, chooses a turning point T.P.$_1$ at some convenient spot within the range of the telescope along the general route B.M.$_1$ to B.M.$_2$. It is desirable, but not necessary, that each foresight distance, such as L_1-T.P.$_1$, be approximately equal to its corresponding backsight distance, such as B.M.$_1$-L_1. The chief requirement of a

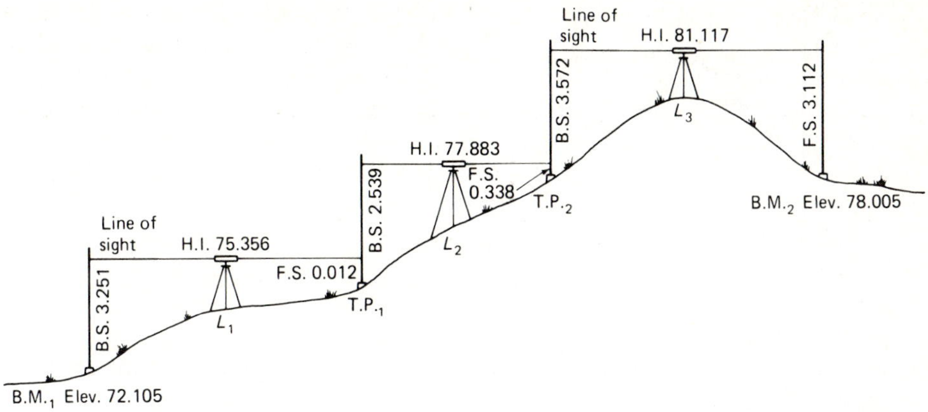

Fig. 5.31 Differential leveling.

desirable turning point is that it shall be a stable object at an elevation and in a location favorable to a rod reading of the required precision. The rod is held on the turning point, and a foresight is taken. The leveler then sets up the instrument at some favorable point, as L_2, and takes a backsight to the rod held on the turning point; the rodman goes forward to establish a second turning point T.P.$_{,2}$; and so the process is repeated until finally a foresight is taken on the terminal point B.M.$_{,2}$.

It is seen in Fig. 5.31 that a *backsight added to the elevation* of a point on which the backsight is taken gives the height of instrument, and that a *foresight subtracted from the height of instrument* determines the elevation of the point on which the foresight is taken. Thus if the elevation of B.M.$_{,1}$ is 72.105 m and the B.S. is 3.251 m, the H.I. with the instrument set at L_1 is $72.105 + 3.251 = 75.356$ m. If the following F.S. is 0.012 m, the elevation of T.P.$_{,1}$ is $75.356 - 0.012 = 75.344$ m. Also, the difference between the backsight taken on a given point and the foresight taken on the following point is equal to the difference in elevation between the two points. It follows that the difference between the sum of all backsights and the sum of all foresights gives the difference in elevation between the bench marks.

Since the level is normally at a higher elevation than that of the points on which rod readings are taken, the backsights are often called "plus" sights and the foresights "minus" sights and are so recorded in the field notes. Sometimes, however, in leveling for a tunnel or a building it is necessary to take rod readings on points which are at a higher elevation than that of the H.I. In such cases the rod is held inverted, and in the field notes each such backsight is indicated with a minus sign and each foresight with a plus sign.

When several bench marks are to be established along a given route, each intermediate bench mark is made a turning point in the line of levels. Elevations of bench marks are checked sometimes by rerunning levels over the same route but more often by "tying on" to a previously established bench mark near the end of the line or by returning to the initial bench mark. A line of levels that ends at the point of beginning is called a *level circuit*. The final observation in a level circuit is therefore a foresight on the initial bench mark. If each bench mark in a level circuit is also a turning point and the circuit checks within the prescribed limits of error (Art. 5.39), it is regarded as conclusive evidence that the elevations of all turning points in the circuit, including all bench marks used as turning points, are correct within prescribed limits.

5.35. Balancing backsight and foresight distances In Art. 5.4 it has been shown that, if a foresight distance were equal to the corresponding backsight distance, any error in readings due to earth's curvature and to atmospheric refraction (under uniform conditions) would be eliminated. In ordinary leveling no special attempt is made to balance *each* foresight distance against the preceding backsight distance. Whether or not such distances are approximately balanced between bench marks will depend upon the desired precision. The effect of the earth's curvature and atmospheric refraction is slight unless there is an abnormal difference between backsight and foresight distances. The effect of instrumental errors is likely to be of considerably greater consequence with regard to the balancing of these distances. The chances are that there is not absolute parallelism between the line of sight and the axis of the level tube, so that if the instrument were perfectly leveled, the line of sight would be inclined always slightly upward or always slightly downward. The error in a rod reading due to this imperfection of adjustment is proportional to the distance from the instrument to the rod and is of the same sign for a backsight as for a foresight. Since backsights are added and foresights are subtracted, this instrumental error is eliminated if, between bench marks, the *sum* of the foresight distances is made equal to the *sum* of the backsight distances.

In ordinary leveling no special attempt is made to equalize these distances *if there is assurance that the instrument is in good adjustment*. Normally, for levels run over flat or gently rolling ground, the line of sight will fall within the length of the rod regardless of the position of the instrument, and the distance between instrument and rod is governed by the optical qualities of the telescope. While moving forward, the leveler generally paces or estimates by eye a distance about the proper maximum length of sight; the rodman similarly estimates the proper distance from the instrument.

For leveling of moderately high precision it is necessary to equalize backsight and foresight distances between bench marks; hence these distances are recorded. In less refined leveling, distances are usually determined by pacing; in precise leveling, they are usually measured by stadia.

In leveling uphill or downhill the length of sight is usually governed by the slope of the ground. In order that maximum distances between turning points may be obtained and hence progress be most rapid, the leveler sets up the instrument in a position such that the line of sight will intersect the rod near its top if the route is uphill, or near its bottom if the route is downhill; the leveler directs the rodman to a similarly favorable location for the turning point. In leveling uphill or downhill, a balance between foresight and backsight distances can be obtained with a minimum number of setups by following a zigzag course.

When bench marks are at roughly the same elevation, the backsight and foresight distances will tend to balance in the long run, regardless of the character of the terrain. However, for levels between two points having a large difference in elevation, a very small inclination of the line of sight will produce a marked error unless some attempt is made to equalize backsight and foresight distances.

5.36. Differential-level notes For ordinary differential leveling when no special effort is made to equalize backsight and foresight distances between bench marks, usually the record of field work is kept in the form indicated by Fig. 5.32, in which the levels from B.M.$_1$ to B.M.$_2$ are the same as shown by Fig. 5.31. The left page is divided into columns for numerical data, and the right page is reserved for descriptive notes concerning bench marks and turning points. In the same horizontal line with each turning point or bench mark shown in the first column are all data concerning that point. The heights of instrument and the elevations are computed as the work progresses. Thus, when the backsight (3.251) has been taken on B.M.$_1$ it is added to the elevation (72.105) to determine the H.I. (75.356). The height of

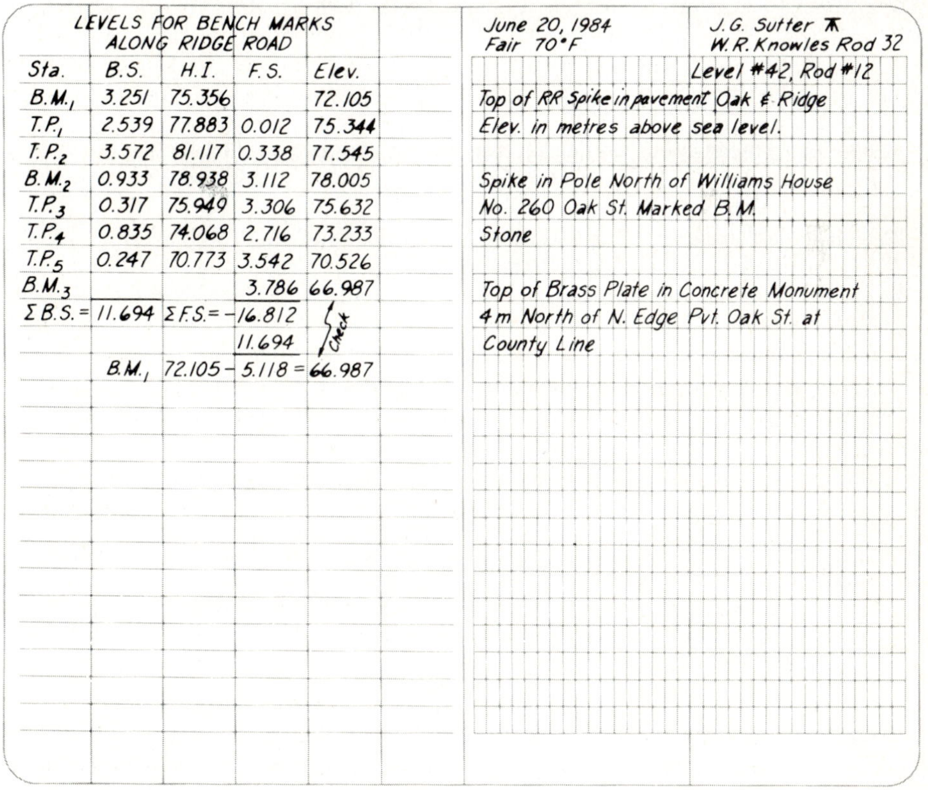

Fig. 5.32 Differential-level notes.

instrument is recorded on the same line with the backsight by means of which it is determined. When the first foresight (0.012) is observed, it is recorded on the line below and is subtracted from the preceding H.I. (75.356) to determine the elevation of T.P.$_1$ (75.344). And so the notes are continued. Usually, at the foot of each page of level notes the *computations* are checked by comparing the difference between the sum of the backsights and the sum of the foresights with the difference between the initial and the final elevation, as illustrated at the bottom of Fig. 5.32. Agreement between these two differences signifies that the additions and subtractions are correct but does not check against mistakes in observing or recording.

Bench marks should be briefly but definitely described and should be so marked in the field that they can be readily identified. They are usually marked with paint or with crayon that will withstand the effects of the weather. When the bench mark is on stone or concrete the position is often indicated by a cross cut with a chisel. A bench mark may or may not be marked with its elevation. Whenever there might arise any question as to the exact location of the point on which the rod was held, its nature should be clearly indicated in the notes. A description of turning points is of no particular importance unless the points are on objects that can be identified and might therefore become of some value in future leveling operations. Such points are usually marked with crayon and briefly described in the notes.

5.37. Mistakes in leveling Some of the mistakes commonly made in leveling are:

1. Confusion of numbers in reading the rod, as, for example, reading and recording 4.92 when it should be 3.92. The mistake is not likely to occur if the observer notices the numbers on both sides of the observed reading.
2. Recording backsights in the foresight column and vice versa.
3. Faulty additions and subtractions; adding foresights and subtracting backsights. As a check, the difference between the sum of the backsights and the sum of the foresights should be computed for each page or between bench marks.
4. Rod not held on same point for both foresight and backsight. This is not likely to occur if the turning points are marked or otherwise clearly defined.
5. Not having the Philadelphia rod fully extended when reading the long rod. Before a reading on a turning point is taken, the clamp should be inspected to see that it has not slipped.
6. Wrong reading of vernier when the target rod is used.
7. When the target is used on the extended rod, not having the vernier on the target set to read exactly the same as the vernier on the back of the rod when the rod is not extended.

5.38. Errors in leveling In leveling, errors are due to some or all of the following causes (see also Table 5.1):

1. Imperfect adjustment of the instrument The essential relation is that the line of sight should be parallel to the axis of the level tube. Any inclination between these lines causes a systematic error, for then if the bubble were perfectly centered, the line of sight would be inclined always slightly upward or downward. The error in a rod reading due to this imperfection would be proportional to the distance from the instrument to the rod, and for a given distance would be of the same magnitude and sign for a backsight as for a foresight. Since backsights are added and foresights are subtracted, it is clear that the error in elevations will be eliminated to the extent that, between bench marks, the sum of the backsight distances is made equal to the sum of the foresight distances. Conversely, a systematic error will result to the extent that these distances are not equalized between any two bench marks. Often, these distances will be sufficiently balanced in the long run, regardless of the terrain, to yield a satisfactory final result; but that fact does not ensure a corresponding accuracy for the bench marks established along the line.

The effect of imperfect adjustment of the instrument is minimized by adjusting the instrument and by balancing backsight and foresight distances.

2. Parallax This condition is present when either or both of the following occur: (*a*) The objective lens is not focused on the object; (*b*) the observer's eye is not focused on the plane of the cross hairs (see Fig. 5.7). The effect of parallax is to cause relative movement between the image of the cross hairs and image of the object when the eye is moved up and down. Parallax causes a random error and can be practically eliminated by careful focusing as described in Art. 5.10.

3. Earth's curvature This produces an error only when backsight and foresight distances are not balanced. Under ordinary conditions these distances do not tend to vary greatly, and whatever resultant error arises from this source in ordinary leveling is so small as to be of no consequence. When backsight distances are consistently made greater than foresight distances, or vice versa, a systematic error of considerable magnitude is produced, particularly when the sights are long. The effect is the same as that due to the line of sight being inclined. The error varies as the square of the distance from instrument to rod and hence will be eliminated not merely by equalizing the sum total of backsight and foresight distances between bench marks but rather by balancing *each* length of backsight by a corresponding length of foresight.

4. Atmospheric refraction This varies as the square of the distance, but under normal conditions is only about one-seventh of that due to the earth's curvature, and its effect is opposite in sign. It is usually considered together with the earth's curvature, but though the effect of the latter will be entirely eliminated if each backsight distance is made equal to the following foresight distance, the atmospheric refraction often changes rapidly and greatly in a short distance. It is particularly uncertain when the line of sight passes close to the ground. Hence, it is impossible to eliminate entirely the effect of refraction even though the backsight and foresight distances are balanced. In ordinary leveling its effect is negligible. In leveling of greater precision the change in refraction can be minimized by keeping the line of sight well above the ground (at least 2 ft or ~ 0.7 m) and by taking the backsight and foresight readings in quick succession. So-called "heat waves" are evidence of rapidly fluctuating refraction. Errors from this source can be reduced by shortening the length of sight until the rod appears steady.

5. Variations in temperature The sun's rays falling on top of the telescope, or on one end and not on the other, will produce a warping or twisting of its parts and hence may influence rod readings through temporarily disturbing the adjustments. Although this effect is not of much consequence in leveling of ordinary precision, it may produce an appreciable error in more refined work. The error is usually random, but under certain conditions it may become systematic. It is practically eliminated by shielding the instrument from the rays of the sun.

6. Rod not standard length If the error is distributed over the length of the rod, a systematic error is produced which varies directly as the difference in elevation and bears no relation to the length of the line over which levels are run. The error can be eliminated by comparing the rod with a standard length and applying the necessary corrections. The case is analogous to measurement of distance with a tape that is too long or too short. If the rod is too long, the correction is added to a measured difference in elevation; if the rod is too short, the correction is subtracted.

Most manufactured rods are nearly of standard length, but where large differences in elevation are to be determined, few rods are near enough to the standard that corrections can be ignored in precise work.

If the rod is worn uniformly at the bottom, an erroneous height of instrument is shown at each setup, but the error in backsight is balanced by that in the following foresight, and no error results in the elevation of the foresight point.

7. Rod not held plumb This condition produces rod readings which are too large. The error varies directly with the first power of the rod reading and directly as the square of the inclination. Thus, if a 3-m rod is 0.06 m out of plumb, the error amounts to 0.0006 m for a 3-m reading and 0.0002 m for a 1-m reading; but if the rod is 0.12 m out of plumb, the corresponding errors are 0.002 and 0.0008 m, respectively. It is therefore evident that appreciable inclinations of the rod must be avoided, particularly for high rod readings. The error can be eliminated by waving the rod, or by using a rod level.

8. Faulty turning points A random error results when turning points are not well defined. Even a flat, rough stone, for example, does not make a good turning point for precise leveling for the reason that no definite point exists on which to hold the rod, which is not likely to be held in the same position for both backsight and foresight.

9. Settlement of tripod or turning points If the tripod settles in the interval that elapses between taking a backsight and the following foresight, the foresight will be too small and the observed elevation of the forward turning point will be too large. Similarly, if a turning point settles in the interval after the previous foresight but before the current backsight readings, the height of instrument as computed from the backsight reading will be too great. It is thus seen that by the normal leveling procedure, if either the level tripod or the turning point settles, as may occur to some extent when leveling over soft ground, an error will occur and the resulting elevations will always be too high by an unknown amount.

Table 5.1 Errors in Leveling

Source	Type	Cause	Remarks	Procedure to eliminate or reduce
Instrumental	Systematic	Line of sight not parallel to axis of level tube	Error of each sight proportional to distance[a]	Adjust instrument; balance sum of backsight and foresight distances
		Rod not standard length (throughout length)[b]	May be due to manufacture, moisture, or temperature; error usually small	Standardize rod and apply corrections, same as for tape
Personal	Random	Parallax	. . .	Focus carefully
		Bubble not centered at instant of sighting	Error varies as length of sight	Check bubble before making each sight
		Rod not held plumb	Readings are too large; error of each sight proportional to square of inclination[a]	Wave the rod, or use rod level
		Faulty reading of rod or setting of target	. . .	Check each reading before recording; for self-reading rod, use fairly shot sights
		Faulty turning points	. . .	Choose definite and stable points
Natural	Random	Temperature variation	May disturb adjustment of level	Shield level from sun
	Systematic	Earth's curvature	Error of each sight proportional to square of distance[a]	Balance each backsight and foresight distance; or apply computed correction
	Random	Variations in atmospheric refraction	Error of each sight proportional to square of distance[a]	Same as for earth's curvature; also take short sights, well above ground, and take backsight and foresight readings in quick succession
	Pseudo-systematic[c]	Settlement of tripod or turning points	Observed elevations are too high	Choose stable locations; take backsight and foresight readings in quick succession preferably alternating order of sights

[a]The error of each sight is systematic, but the resultant error is the difference between the systematic error for foresights and that for backsights; hence, the resultant error tends to be random.

[b]Uniform wear of the bottom of the rod causes no error.

[c]The sign is always the same but the error is of an unknown magnitude.

Few occasions arise when turning points cannot be so selected or established as to eliminate the possibility of settlement, but care should be taken not to strike the bottom of the rod against the turning point between sights or to exert any pressure whatsoever on the turning point.

On the other hand, some settlement of the instrument is nearly certain to occur when leveling over muddy, swampy, or thawing ground or over melting snow. The errors due to such settlement can be greatly reduced by employing two rods and two rodmen, one rodman setting the turning point ahead while the other remains at the turning point in the rear. Backsight and foresight readings can then be made in quick succession. Small errors remaining from this source can be minimized by reversing the order of sights at alternate setups.

10. Bubble not exactly centered at instant of sighting This condition produces a random error which tends to vary as the distance from instrument to rod. Hence, the longer the sight, the greater is the care that should be observed in leveling the instrument.

11. Inability of observer to read the rod exactly or to set the target exactly on the line of sight This inability causes a random error of a magnitude depending upon the instrument, weather conditions, length of sight, and observer. It can be confined within reasonable limits through proper choice of length of sight.

5.39. Error propagation in leveling The sources of errors discussed in Art. 5.38 are summarized in Table 5.1. Obviously, a great many factors affect the differential leveling operation. Although the final accuracy achieved is influenced by the instrument used, a great deal depends on the skill of the leveler and the degree of refinement with which the work is executed. Assuming that proper leveling procedures are employed and care is given to detail, systematic errors can be nearly eliminated. Thus, the remaining errors are random and can be attributed to centering the spirit level, reading the rod, and variations in refraction in air. The first two errors are proportional to the length of sight. Although refraction varies as the square of the distance, the exact relationship cannot be determined, since these variations may occur anywhere between the instrument and the rod. Consequently, refraction is also assumed proportional to the length of sight, and these three sources of random error act together.

By using statistical error propagation techniques similar to but more advanced than those developed in App. A, it is possible to show that the propagated error in a line of levels varies as the square root of the number of instrument setups or as the square root of the distance leveled and the square root of the lengths of sights. Thus, most specifications governing closures in lines or circuits of levels are expressed as a function of the square root of the distance leveled and the maximum length of sights is specified.

Levels to establish vertical control are classified as first, second, and third order, depending on the equipment and procedures employed. Table 5.2 shows the general specifications and standards of accuracy for first-, second-, and third-order vertical control. These standards were prepared by the Federal Geodetic Control Committee and are published by the National Ocean Survey.

5.40. Reciprocal leveling Occasionally, it becomes necessary to determine the relative elevations of two widely separated intervisible points between which levels cannot be run in the ordinary manner. For example, it may be desired to transfer levels from one side to the other of a deep canyon, or from bank to bank of a wide stream.

If *A* and *B* are two such points, then the level is set up near *A* and one or more rod readings are taken on both *A* and *B* (Fig. 5.33*a*). Then the level is set up in a similar location near *B*, and rod readings to near and distant points are taken as before. The mean of the two differences in elevation thus determined is taken to be the difference between the two points. Usually, the distance between points is large (often a half mile or a kilometre) so that it is necessary to use a target on the distant rod. If more accurate results are desired, a series of foresights is taken on the distant rod and sometimes also a series of backsights on the near rod, the bubble being recentered and the target reset after each observation. The difference in elevation is then computed by using the mean of the backsights and the mean of the foresights at each setup.

Two factors that may appreciably affect the results are variations in temperature, causing unequal expansion of parts of the instrument, and variations in atmospheric refraction. To minimize these effects, the instrument should be shaded and corrections for curvature and refraction applied (Art. 5.2). A more satisfactory approach is to use two levels, each having about the same magnifying power and sensitiveness of level bubble. With one level set up near *A* and the other near *B*, sets of observations are obtained simultaneously from both locations. Thus, the effects of curvature and refraction cancel. The two instruments are interchanged and the operation is repeated. The mean of the differences in elevations is taken as the best estimate of actual difference in elevation.

5.41. Adjustment of intermediate bench marks When a line of levels makes a complete circuit, almost invariably the final elevation of the initial bench mark as computed from the level notes will not agree with the initial elevation of this point. The difference is the error of running the circuit and is called the *error of closure*. It is evident that elevations of intermediate bench marks established while running the

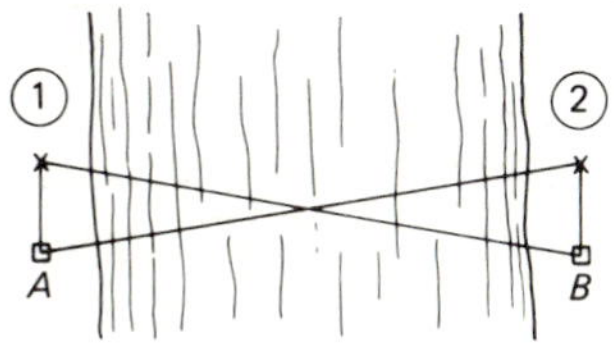

Fig. 5.33 Reciprocal leveling. (*a*) Plan view. (*b*) Section.

Table 5.2 Classification, Standards of Accuracy, and General Specifications for Vertical Control

Classification	First-order: Class I, Class II	Second-order		Third-order
		Class I	Class II	
Principal uses Minimum standards; higher accuracies may be used for special purposes	Basic framework of the National Network and of metropolitan area control Extensive engineering projects Regional crustal movement investigations Determining geopotential values	Secondary control of the National Network and of metropolitan area control Large engineering projects Local crustal movement and subsidence investigations Support for lower-order control	Control densification, usually adjusted to the National Net; local engineering projects Topographic mapping Studies of rapid subsidence Support for local surveys	Miscellaneous local control may not be adjusted to the National Network; small engineering projects Small-scale topographic mapping Drainage studies and gradient establishment in mountainous areas
Recommended spacing of lines National Network	Net A; 100–300 km *Class I* Net B; 50–100 km *Class II*	Secondary net; 20–50 km	Area control; 10–25 km	As needed
Metropolitan control; other purposes	2–8 km As needed	0.5–1 km As needed	As needed As needed	As needed As needed
Spacing of marks along lines	1–3 km	1–3 km	Not more than 3 km	Not more than 3 km
Gravity requirement	0.20×10^{-3} gpu[d]	. . .	. . .	. . .
Instrument standards	Automatic or tilting levels with parallel-plate micrometers; invar scale rods	Automatic or tilting levels with optical micrometers or three-wire levels; invar scale rods	Geodetic levels and invar scale rods	Geodetic levels and rods

Field procedures	Double-run; forward and backward, each section	Double-run; forward and backward, each section	Double- or single-run	Double- or single-run
Section length	1–2 km	1–2 km	1–3 km for double-run	1–3 km for double-run
Maximum length of sight	50 m *class I*; 60 m *Class II*	60 m	70 m	90 m
Field procedures[a]				
Max. difference in lengths				
Forward and backward sights per setup	2 m *class I*; 5 m *Class II*	5 m	10 m	10 m
per section (cumulative)	4 m *Class I*; 10 m *Class II*	10 m	10 m	10 m
Maximum length of line between connections	Net A; 300 km Net B; 100 km	50 km	50 km double-run 25 km single-run	25 km double-run 10 km single-run
Maximum closures[b]				
Section; forward and backward	3 mm $\sqrt{K}$ Class I; 4 mm $\sqrt{K}$ Class II	6 mm $\sqrt{K}$	8 mm $\sqrt{K}$	12 mm $\sqrt{K}$
Loop or line	4 mm $\sqrt{K}$ Class I; 5 mm $\sqrt{K}$ Class II	6 mm $\sqrt{K}$ 0.017 ft $\sqrt{M}$	8 mm $\sqrt{K}$ 0.035 ft $\sqrt{M}$	12 mm $\sqrt{K}$ 0.05 ft $\sqrt{M}$[c]

[a]The maximum length of line between connections may be increased to 100 km for double-run for second-order Class II, and to 50 km for double-run for third-order in those areas where the first-order control has not been fully established.

[b]Check between forward and backward runnings, where K is the distance in kilometres.

[c]M is distance leveled in miles.

[d]One geopotential unit (gpu) = 1 km (10^5 cm^2 s^2)

circuit will also be in error, and there arises the problem of determining the errors for these intermediate points and of adjusting their elevations accordingly.

For level circuits of lower order, it is sufficient to assume that the appropriate correction to the observed elevation of a given bench mark in the circuit is directly proportional to its distance from the point of beginning. Thus, if E_c is the error of closure of a level circuit of length L, and if $C_a, C_b, \ldots, C_n$ are the respective corrections to be applied to observed elevations of bench marks $A, B, \ldots, N$ whose respective distances from the point of beginning are $a, b, \ldots, n$, then

$$C_a = -\frac{a}{L}E_c; \quad C_b = -\frac{b}{L}E_c; \cdots \qquad C_n = -\frac{n}{L}E_c \tag{5.14}$$

Example 5 The accepted elevation of the initial bench mark B.M.$_i$ of a level circuit is 470.46 ft. The length of the circuit is 10 mi. The final elevation of the initial bench mark as calculated from the level notes is 470.76. The observed elevations of bench marks established along the route and the distances to the bench marks from B.M.$_i$ are as shown in the third and second columns of the accompanying tabulation. The elevations of these intermediate points are required.

$E_c = 470.76 - 470.46 = +0.30$ ft

By Eq. (5.14),

$C_a = -\frac{2}{10} \times 0.30 = -0.06$ ft
$C_b = -\frac{5}{10} \times 0.30 = -0.15$ ft
$C_c = -\frac{7}{10} \times 0.30 = -0.21$ ft

Point	Distance from B.M.$_i$, mi	Observed elevation, ft	Correction, ft	Adjusted elevation, ft
B.M.$_i$	0	470.46	0.0	
B.M.$_a$	2	780.09	−0.06	780.03
B.M.$_b$	5	667.41	−0.15	667.26
B.M.$_c$	7	544.32	−0.21	544.11
B.M.$_i$	10	470.76	−0.30	470.46

These corrections added algebraically from the corresponding observed elevations give the adjusted elevations as tabulated. It is to be noted that, if the error of closure is positive, all corrections are to be subtracted, and vice versa.

If desired, corrections may be made directly proportional to the number of setups instead of the distances.

The principles described in this section apply also to the adjustment of elevations of bench marks on a line of levels run between two points whose difference in elevation has previously been determined by more accurate methods and is assumed to be correct.

PROFILE LEVELING; GRADES

5.42. Profile leveling The process of determining the elevations of points at short measured intervals along a fixed line is called *profile leveling*. During the location and construction of highways, railroads, canals, and sewers, stakes or other marks are placed at regular intervals along an established line, usually the center line. Ordinarily, the interval between stakes is 100 ft, 50 ft, or 25 ft, with intervals of 100 m, 50 m, 20 m, and 10 m being utilized in the metric system. The 100-ft points or 100-m points, reckoned from the beginning of the line, are called *full stations*, and all other points are called *plus stations*. Each stake is marked with its station and plus. Thus, a stake set at 1600 ft from the point of beginning is numbered "16" or "16 + 00," and one set at 1625 ft from the point of beginning is numbered "16 + 25." Similarly, a point 450 m from the origin is station 4 + 50 and a point set 1410 m from the origin is number 14 + 10. Elevations by means of which the profile may be constructed are obtained by taking level-rod readings on the ground at each stake and at intermediate points where marked changes in slope occur.

Figure 5.34 illustrates in plan and elevation the steps in leveling for profile. In this case stakes are set every 100 ft, according to the common practice in highway and railway location. The instrument is set up in some convenient location not necessarily on the line (as at L_1), the rod is held on a bench mark (B.M. 28, elevation 564.31 ft above mean sea level), a backsight (1.56) is taken, and the height of instrument (565.87) is obtained as in differential leveling. Readings (in ft) are then taken with the rod held on the ground at successive stations along the line. These rod readings, being for points of unknown elevation, are foresights regardless of whether they are back or ahead of the level. They are frequently designated as *intermediate foresights* to distinguish them from foresights taken on turning points or bench marks. The intermediate foresights (0.7, 2.9, ..., 11.9) subtracted from the H.I. (565.87) give ground elevations of stations. When the rod has been advanced to a

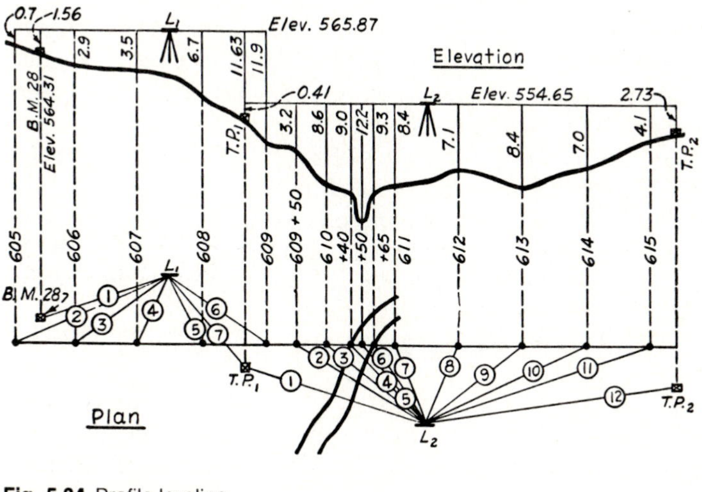

Fig. 5.34 Profile leveling.

point beyond which further readings to ground points cannot be observed, a turning point (T.P.$_1$) is selected, and a foresight (11.63) is taken to establish its elevation. The level is set up in an advanced position (L_2), and a backsight (0.41) is taken on the turning point (T.P.$_1$) just established. Rod readings on ground points are then continued as before. The rodman observes where changes of slope occur (as $609 + 50, 610 + 40, \ldots, 610 + 65$), and readings are taken to these intermediate stations. The "plus," or distance from the preceding full station to the intermediate point, is measured by pacing or with a tape or the rod according to the precision required.

The care exercised in taking observations on turning points depends upon the distance between bench marks, the elevations of which have been determined previously, and upon the required precision of the profile. For a ground profile the backsights and foresights are usually read to hundredths of feet (mm) and no particular attention is paid to balancing backsight and foresight distances; the intermediate foresights to ground points are read to tenths of feet (cm) only. Occasions arise when it is desirable or necessary to determine intermediate foresights to hundredths of feet (~ 3 mm), for example, in securing the profile of railroad track or of the water grade in a canal; rod readings on turning points are then generally taken to thousandths of feet or to millimetres, and backsight and foresight distances are often balanced.

As the work of leveling for profile progresses, bench marks are generally established to facilitate later work; these are made turning points wherever possible. To check the elevation of turning points, it is necessary either to run levels back to the point of beginning or to run short lines of differential levels connecting with bench marks previously established by some other survey. The effect of an occasional error in the elevations of intermediate ground points on the profile is usually not of sufficient importance to justify the additional work which checking would make necessary, and if turning points are checked, it is regarded as sufficient.

5.43. Profile-level notes The notes for profile leveling may be recorded as shown in Fig. 5.35, in which foresights to turning points and bench marks are in a separate column from intermediate foresights to ground points. The notes for turning points are kept in the same manner as for differential leveling.

The computations shown at the foot of the notes of Fig. 5.35 check all computations for H.I.'s and elevations of T.P.'s on the page and thus for the notes shown the difference between the sum of all backsights and the sum of all foresights is equal to the error of closure on B.M. 30. Elevations of ground points are recorded only to the *number of decimal places contained in the intermediate foresights*, regardless of the number of places in the H.I.

The right-hand page is reserved for concise descriptions of bench marks and for other pertinent items. Occasionally, as shown in Fig. 5.35, simple sketches are employed in conjunction with the explanatory notes.

5.44. Plotting the profile To provide a useful product for the designer, the profile is plotted on profile paper for specified horizontal and vertical scales. Profile paper is a cross-sectioned paper of good quality with a specified number of lines per

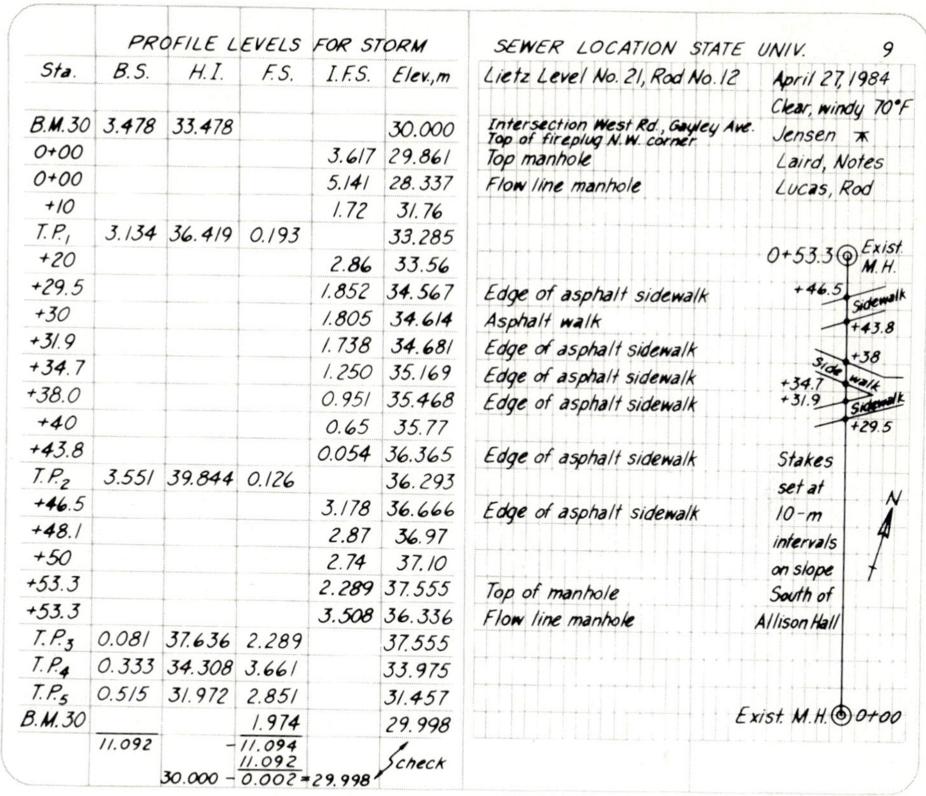

PROFILE LEVELS FOR STORM SEWER LOCATION STATE UNIV. 9

Sta.	B.S.	H.I.	F.S.	I.F.S.	Elev., m	Lietz Level No. 21, Rod No. 12
B.M.30	3.478	33.478			30.000	Intersection West Rd., Gayley Ave. Top of fireplug N.W. corner
0+00				3.617	29.861	Top manhole
0+00				5.141	28.337	Flow line manhole
+10				1.72	31.76	
T.P.$_1$	3.134	36.419	0.193		33.285	
+20				2.86	33.56	
+29.5				1.852	34.567	Edge of asphalt sidewalk
+30				1.805	34.614	Asphalt walk
+31.9				1.738	34.681	Edge of asphalt sidewalk
+34.7				1.250	35.169	Edge of asphalt sidewalk
+38.0				0.951	35.468	Edge of asphalt sidewalk
+40				0.65	35.77	
+43.8				0.054	36.365	Edge of asphalt sidewalk
T.P.$_2$	3.551	39.844	0.126		36.293	
+46.5				3.178	36.666	Edge of asphalt sidewalk
+48.1				2.87	36.97	
+50				2.74	37.10	
+53.3				2.289	37.555	Top of manhole
+53.3				3.508	36.336	Flow line manhole
T.P.$_3$	0.081	37.636	2.289		37.555	
T.P.$_4$	0.333	34.308	3.661		33.975	
T.P.$_5$	0.515	31.972	2.851		31.457	
B.M.30			1.974		29.998	
	11.092		−11.094			
			11.092			
		30.000 −	0.002 = 29.998		✓check	

Fig. 5.35 Profile notes in metric units.

horizontal and vertical unit. Common scales are 1 : 600 (50 ft/in) horizontally and 1 : 120 (10 ft/in) vertically. Profile paper has 20 lines per inch vertically and 10 lines per inch horizontally. In the metric system comparable horizontal and vertical scales are 1 : 500 and 1 : 100 in the horizontal and vertical scales, respectively. Figure 5.36 illustrates a profile plotted using the elevations in Fig. 5.35. The horizontal and vertical scales in Fig. 5.36 were 1 : 500 and 1 : 100, respectively before reduction of the drawing to fit the book. Note that a certain balance between horizontal and vertical scale is necessary in order to portray the ground profile in a realistic manner. The plotted profile is utilized by the designer to establish grade lines for highways, roads, sewer lines, canals, and so on. Frequently, a plan view of the area through which the profile runs is placed above or below the profile on the same sheet.

5.45. Adjustment of levels Regardless of the precision of manufacture, levels in process of use require certain field adjustments from time to time. It becomes an important duty of the surveyor to test the instrument at short intervals and to make with facility such adjustments as are found necessary.

In some instances one adjustment is likely to be altered by, or depends upon, some other adjustment made subsequently. For example, lateral movement of the

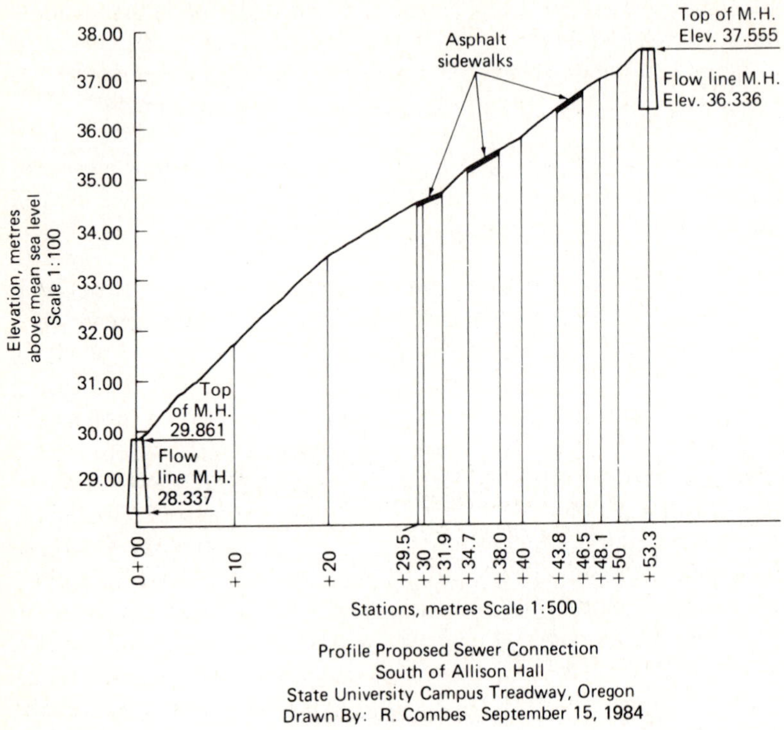

Fig. 5.36 Plotted profile (not to scale).

cross-hair ring may likewise produce a small rotation, and the lateral adjustment of the level tube depends upon the vertical adjustment. Hence, if an instrument is badly out of adjustment, related adjustments must be repeated until they are gradually perfected.

5.46. Desired relations in the dumpy level For a dumpy level in perfect adjustment the following relations should exist (see Fig. 5.37):

1. The axis of the level tube should be perpendicular to the vertical axis.
2. The horizontal cross hair should lie in a plane perpendicular to the vertical axis, so that it will lie in a horizontal plane when the instrument is level.
3. The line of sight should be parallel to the axis of the level tube.

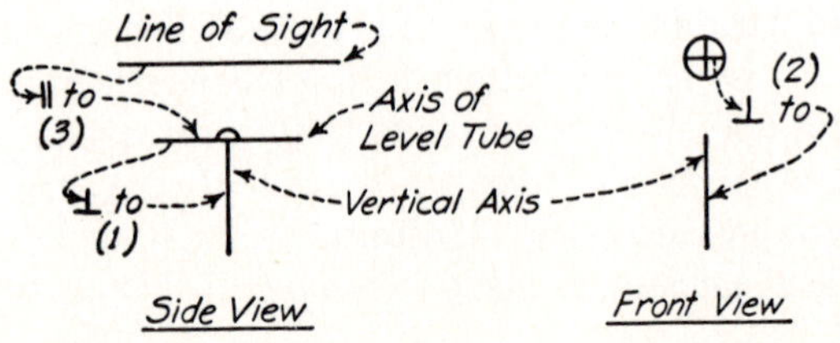

Fig. 5.37 Desired relations among principal lines of a dumpy level.

Also the optical axis, the axis of the objective slide, and the line of sight should coincide; but for the dumpy levels commonly manufactured in the United States the optical axis and the axis of the objective slide are fixed, perpendicular to the vertical axis by the manufacturer, and no provision for further adjustment is made.

5.47. Adjustment of the dumpy level The parts capable of and requiring adjustment are the cross hairs and the level tube. The basis for adjustments is the vertical axis. The adjustments for a level with four foot-screws are given in the following.

1. To make the axis of the level tube perpendicular to the vertical axis Approximately center the bubble over each pair of opposite leveling screws; then carefully center the bubble over one pair. Rotate the level end for end about its vertical axis. If the level tube is in adjustment, the bubble will retain its position. If the tube is not in adjustment, the displacement of the bubble indicates double the actual error, as shown by Fig. 5.38. If $(90° - \alpha)$ represents the angle between the vertical axis and axis of level tube, then when the bubble is centered the vertical axis makes an angle of α with the true vertical. When the level is reversed, the bubble is displaced through the arc whose angle is 2α. Hence, the correction is the arc whose angle is α. Make the correction by bringing the bubble halfway back to the center by means of the capstan nuts at one end of the tube. Relevel the instrument with the leveling screws, and repeat the process until the adjustment is perfected. Usually, three or four trials are necessary. As a final check, the bubble should remain centered over each pair of opposite leveling screws. This adjustment involves the principle of *reversion*.

2. To make the horizontal cross hair lie in a plane perpendicular to the vertical axis (and thus horizontal when the instrument is level) Sight the horizontal cross hair on some clearly defined point (as A, Fig. 5.39) and rotate the instrument slowly about its vertical axis. If the point appears to travel along the cross hair, no adjustment is needed.

If the point departs from the cross hair and takes some position as A' on the opposite side of the field of view, loosen two adjacent capstan screws and rotate the cross-hair ring until by further trial the point appears to travel along the cross hair. Tighten the same two screws. The instrument need not be level when the test is made.

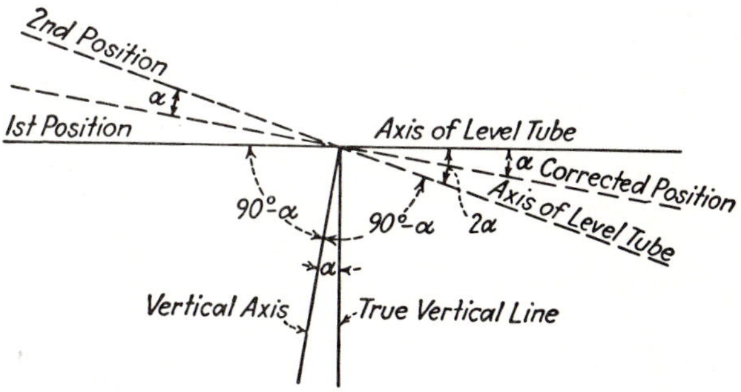

Fig. 5.38 Adjustment of axis of level tube of dumpy level.

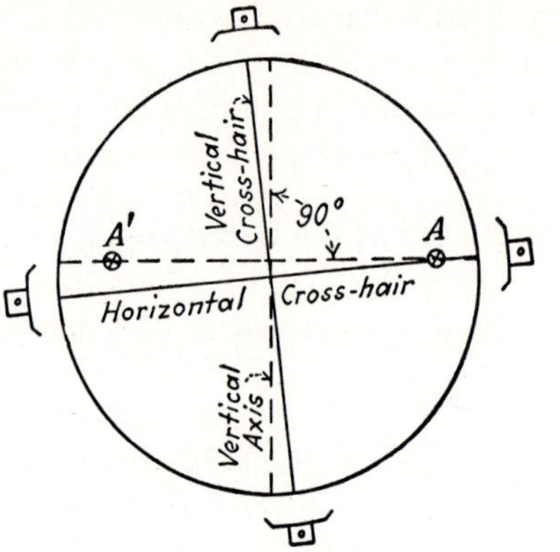

Fig. 5.39 Adjustment of horizontal cross hair.

3. To make the line of sight parallel to the axis of the level tube (two-peg test) Set two pegs 200 to 300 ft (60 to 90 m) apart on approximately level ground, and designate as A the peg near which the second setup will be made (Fig. 5.40); call the other peg B. Set up and level the instrument at any point M equally distant from A and B, that is, in a vertical plane bisecting the line AB. Take rod readings a on A and b on B; then $(a-b)$ will be the true difference in elevation, since any error would be the same for the two equal sight distances L_m. Due account must be taken of signs throughout the test.

Move the instrument to a point P near A, preferably but not necessarily on line with the pegs; set up as before, and measure the distances L_a to A and L_b to B. Take rod readings c on A and d on B. Then $(c-d)$, taken in the same order as before, is the indicated difference in elevation; if $(c-d)=(a-b)$, the line of sight is parallel to the axis of the level tube, and the instrument is in adjustment. If not, $(c-d)$ is called the "false" difference in elevation, and the inclination (error) of the line of sight in the net distance $(L_b - L_a)$ is equal to $(a-b)-(c-d)$. By proportion, the

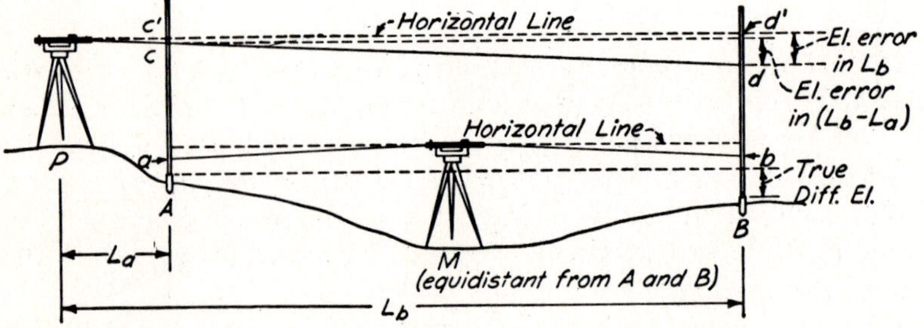

Fig. 5.40 Two-peg test.

error in the reading on the far rod is

$$e_{fr} = \frac{L_b}{L_b - L_a}[(a-b)-(c-d)] \tag{5.15}$$

Subtract algebraically the amount of this error from the reading d on the far rod to obtain the correct reading d' at B for a horizontal line of sight with the position of the instrument unchanged at P. Set the target at d' and bring the line of sight on the target by moving the cross-hair ring vertically.

Example 5.3 With the level at M, the rod reading a is 0.970 and b is 2.986; the true difference in elevation $(a-b)$ is then $0.970 - 2.986 = -2.016$ ft, with B thus indicated as being lower than A. With level at P, the rod reading c is 5.126 and d is 7.018; the false difference in elevation $(c-d)$ is then $5.126 - 7.018 = -1.892$, with B again indicated as being lower than A. The distance L_a is observed to be 30 ft and L_b to be 230 ft. The inclination of the line of sight in $(230 - 30 = 200)$ ft is $(-2.016) - (-1.892) = -0.124$ ft. The error in elevation of the line of sight at the far rod is $(230/200) \times (-0.124) = -0.143$ ft. The correct rod reading d' for a horizontal line of sight is $7.018 - (-0.143) = 7.161$ ft.

As a partial check on the computations, the correct rod reading c' at A may be computed by proportion; the difference in elevation computed from the two corrected rod readings c' and d' should be equal to the true difference in elevation observed originally at M.

Example 5.4 In Example 5.3, the error in elevation of the line of sight at the near rod is $(30/200) \times (-0.124) = -0.019$ ft. The correct rod reading c' is $5.126 - (-0.019) = 5.145$ ft. The "false" difference in elevation is $5.145 - 7.161 = -2.016$ ft, which is equal to the true difference in elevation; hence, the computations are checked to this extent.

A sketch should always be drawn. Also, theoretically a correction for earth's curvature and atmospheric refraction (see Art. 5.2) should be added numerically to the final rod reading d', although in practice it is usually considered negligible.

Some surveyors prefer to set up at P within 2 to 3 m or 6 to 10 ft of A and to consider $[(a-b)+(c-d)]$ as being the *total* error in elevation, to be subtracted directly from d. This serves as a first approximation; the procedure for the setup at P is then repeated. The amount of computation is thus reduced, but the amount of field work is increased.

5.48. Adjustment of automatic levels The standard peg test described in Art. 5.47 can also be employed with automatic, self-leveling levels. If the slope of the optical line of sight is excessive, the cross hair can be adjusted. Reference should be made to the instrument manual of instructions for this operation, since most automatic levels do not have exposed adjusting screws for the cross-hair reticule. The cross-hair reticule for the level illustrated in Fig. 5.17 is exposed by unscrewing the metal cover around the eyepiece.

Prior to leveling, the circular bubble and activity of the compensator, which automatically makes the line of sight horizontal, should be checked. Center the

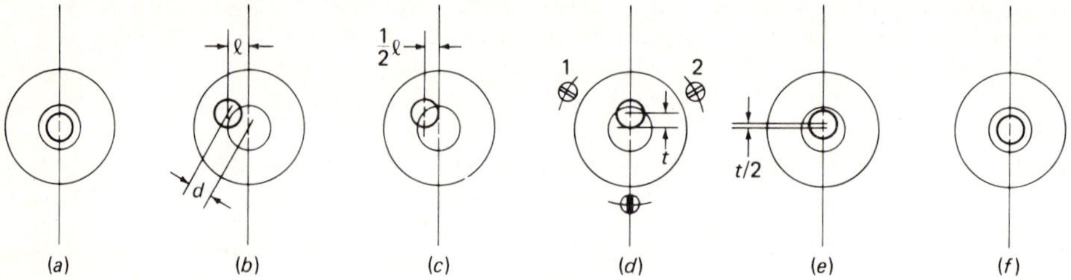

Fig. 5.41 (*a*) through (*f*) Adjustment of circular level bubble.

circular bubble in the engraved circle (Fig. 5.41*a*) and rotate the instrument through 180° or to the direction of maximum bubble displacement (Fig. 5.41*b*). When the displacement d exceeds one-half the bubble diameter, adjustment is necessary. Adjustment is performed in two steps by reducing the two components of the displacement, ℓ from left to right and t from top to bottom. First, remove $\frac{1}{2}\ell$ using the appropriate level screws (Fig. 5.41*c*). Second, remove the remaining $\frac{1}{2}\ell$ by manipulating the adjusting screws (Fig. 5.41*d*). One-half the top to bottom displacement t (Fig. 5.41*e*) is then removed by the level screws and the remainder $t/2$ is compensated using the adjusting screws to bring the bubble to the center of the engraved circle. Turn 180° and check the bubble. Repeat the entire procedure until the bubble remains centered in all positions.

To check the compensator, sight a target about 30 m away. Center the bull's-eye bubble. Next, tap one tripod leg with a hand. The image of the target will appear to swing in the field of view but the target will return to its original position. Next, turn a level screw, causing the line of sight to slope slightly. Once again the target will appear to swing but will return to the original position.

5.49. Problems

5.1 What is the combined effect of the earth's curvature and mean atmospheric refraction in a distance of 300 ft? In a distance of 3000 ft? In a distance of 6 mi? In a distance of 60 mi?

5.2 Calculate the combined effect of the earth's curvature and refraction for the following distances: (*a*) 100 m; (*b*) 300 m; (*c*) 1 km; (*d*) 12 km; (*e*) 100 km.

5.3 An observer standing on the shoreline of a lake can just see the top of a tower on an island. If the eye of the observer is 1.65 m above lake level and the top of the tower is 15 m above lake level, how far is the tower from the observer?

5.4 The backsight to a bench mark is 2.125 m and the foresight from the same instrument setup is 4.587 m. If the distance from the level to the bench mark is 5 m and from the level to the turning point is 400 m, compute the difference in elevation between the bench mark and the turning point.

5.5 Two points, *A* and *B*, are each 2000 ft distant from a third point *C*, from which the measured vertical angle to *A* is +3°21′ and that to *B* is +0°32′. What is the difference in elevation between *A* and *B*?

5.6 Let *A* be a point of elevation 100.00 ft, and let *B* and *C* be points of unknown elevation. By means of an instrument set 4.00 ft above *B*, vertical angles are observed, that to *A* being −1°55′ and that to *C* being +3°36′. If the horizontal distance *AB* is 1500 ft and the horizontal distance *BC* is 5000 ft, what are the elevations of *B* and *C*, making due allowance for the earth's curvature and atmospheric refraction?

5.7 Two points, *A* and *B*, are 1000 ft apart. The elevation of *A* is 615.03 ft. A level is set up on the line between *A* and *B* and at a distance of 250 ft from *A*. The rod reading on *A* is 9.15 ft and that on *B* is 2.07 ft. Making the due allowance for curvature and refraction, what is the elevation of *B*? What would be the magnitude and sign of the error introduced if the correction for curvature and refraction were omitted?

5.8 The zenith angles between stations *A* and *B* are 86°57′55″ from *A* to *B* and 93°12′55″ from *B* to *A*. The slope distance measured by EDM from *A* to *B* is 23,458.487 m and the elevation of station *A* above the datum is 322.073 m. The heights of theodolite and target above stations *A* and *B* are 1.55 m and 1.38 m, respectively. Similarly, the heights of theodolite and target above *B* and *A* for observing the zenith angle from *B* are 1.43 m and 1.51 m, respectively. The heights of EDM and reflector above stations *A* and *B* are 1.62 m and 1.41 m, respectively. Compute the elevation of station *B*.

5.9 What is the radius of curvature of a level tube graduated to 0.1 in and having a sensitivity of 30″ per division?

5.10 What is the sensitivity of a level tube graduated to 2 mm having a radius of curvature of 10 m?

5.11 A sight is taken with an engineer's level at a rod held 100 m away, and an initial reading of 1.927 m is observed. The bubble is then moved through five spaces on the level tube, when the rod reading is 2.008 m. What is the sensitivity of the level tube in seconds of arc? What is the radius of curvature of the level tube if one space is 2 mm?

5.12 Design (*a*) a direct vernier and (*b*) a retrograde vernier reading to thousandths of feet, each space on the rod being equal to 0.005 ft.

5.13 Design a direct vernier and retrograde vernier both reading to 1 mm for a rod graduated to centimetres. Draw a neat sketch of each vernier and a portion of the rod for a reading of 1.215 m with graduations shown and labeled.

5.14 On a rod graduated to 5 mm, a direct-reading vernier is to read to millimetres. State the following: (*a*) length of one vernier space; (*b*) number of spaces on the vernier; (*c*) number of spaces on the main scale corresponding to the full length of vernier scale; (*d*) least count of the vernier.

5.15 Design a direct vernier reading to 1 mm, applied to a scale graduated to 0.5 cm. For the same scale, design a retrograde vernier with least count of 0.2 mm. Sketch the verniers in relation to the scale.

5.16 What is the least count of a retrograde vernier which is to be graduated with 9/64-in spaces, when used with a main scale graduated to 1/8 in? How many spaces are there on the vernier scale? Make a neat sketch of the vernier applied to the scale.

5.17 If the rod were inclined 0.4 ft forward in a length of 13 ft, what error would be introduced in a rod reading of 5.0 ft?

5.18 The rod placed on a pointed turning point is inclined backward 15 cm in a length of 3 m. What error is introduced in a rod reading of (*a*) 0.500 m; (*b*) 2.50 m?

5.19 A rod that is 50 mm square at the base is placed on a flat surface for a turning point. If the top of the 3-m rod is inclined backward 20 cm, what error is introduced in a reading of 0.300 m?

5.20 A line of differential levels was run between two bench marks 20 mi apart, and the measured difference in elevation was found to be 2163.4 ft. Later the rod whose nominal length was 13 ft was found to be 0.003 ft too short, the error being disturbed over its full length. Correct the measured difference in elevation for erroneous length of rod.

5.21 Suppose that the levels of Prob. 5.20 had been run by using a rod which was 0.003 ft too short owing to wear on the lower end. What would have been the error?

5.22 Differential levels were run from B.M.$_1$ (el. 143.277 m) to B.M.$_2$, a distance of 60 km. On the average, the backsight distances were 100 m in length and the foresight distances were 50 m in length. The elevation of B.M.$_2$ as computed from the level notes was 1113.355 m. Compute the error due to earth's curvature and atmospheric refraction, and correct the elevation of B.M.$_2$.

5.23 The levels of Prob. 5.22 were rerun using an average backsight distance of 50 m and an average foresight distance of 25 m. The elevation of B.M.$_2$ as deduced from the level notes was 1112.941 m. Compute the error due to curvature and refraction and correct the elevation of B.M.$_2$.

5.24 Suppose that the instrument used in running the levels of Probs. 5.22 and 5.23 was out of adjustment, so that when the bubble was centered the line of sight was inclined 0.001 m upward in a distance of 100 m. Correct the observed results of Probs. 5.22 and 5.23 for inclination of line of sight.

5.25 If in running levels between two points the rod were inclined 10 cm forward in a height of 3.5 m, what error would be introduced per setup when backsight readings averaged 3 m and foresight readings averaged 0.3 m?

5.26 If levels are run from B.M.$_1$ (el. 600 m) to B.M.$_2$ (observed el. 900 m) and the rod is on the average 6 cm out of plumb in a height of 3 m, what error is introduced because of the rod's not being plumb? What is the correct elevation of B.M.$_2$?

5.27 Complete the differential-level notes shown below, putting the entire set of notes in proper field note form and making all the customary checks. All units are in metres.

Station	B.S.	H.I.	F.S.	Elevation
B.M.$_A$	0.232		...	45.272
T.P.$_1$	0.503		2.892	
T.P.$_2$	0.212		3.056	
T.P.$_3$	1.246		3.302	
T.P.$_4$	2.169		1.257	
T.P.$_5$	2.895		0.678	
T.P.$_6$	3.004		0.202	
B.M.$_B$	...		1.423	

5.28 Complete the differential-level notes shown below, putting the entire set of notes in proper field note form. Determine the error of closure of the level circuit and adjust the elevations of B.M.$_2$ and B.M.$_3$, assuming that the error is a constant per setup.

Station	B.S.	H.I.	F.S.	Elevation
B.M.$_1$	4.127		...	100.000
T.P.$_1$	3.831		9.346	
T.P.$_2$	4.104		10.725	
T.P.$_3$	2.654		12.008	
B.M.$_2$	4.368		7.208	
T.P.$_4$	6.089		6.534	
T.P.$_5$	8.863		4.736	
B.M.$_3$	12.356		2.100	
T.P.$_6$	10.781		3.662	
T.P.$_7$	12.365		4.111	
B.M.$_1$	...		9.059	

5.29 Complete the differential-level notes shown below, performing the customary checks. All units are in feet. Put the entire set of notes in proper field note form.

Station	B.S.	H.I.	F.S.	Elevation
B.M.$_1$	6.11		...	416.23
T.P.$_1$	9.25		7.36	
T.P.$_2$	11.48		3.12	
T.P.$_3$	8.30		2.98	
B.M.$_2$	12.29		4.37	
T.P.$_4$	7.73		5.16	
T.P.$_5$	8.24		3.38	
T.P.$_6$	10.66		0.47	
B.M.$_3$	...		4.33	

5.30 Complete the differential-level notes shown below. Determine the error of closure of the level circuit and adjust the elevations of B.M.$_{10}$ and B.M.$_{11}$ assuming that the error is a constant per instrument setup. The elevation of the closing B.M.$_{12}$ is 98.991 m. Put the entire set of notes in proper field note form.

Station	B.S.	H.I.	F.S.	Elevation
B.M.$_{9}$	0.215			100.814
T.P.$_{1}$	0.156		3.045	
T.P.$_{2}$	0.352		2.821	
B.M.$_{10}$	0.506		0.145	
T.P.$_{3}$	2.468		0.523	
T.P.$_{4}$	3.062		0.106	
B.M.$_{11}$	1.423		0.321	
T.P.$_{5}$	2.156		3.042	
B.M.$_{12}$			2.142	

5.31 The data for profile levels in the order taken are as follows: B.S. on B.M.$_{10}$ = 3.845, F.S. on T.P.$_{1}$ = 1.234, B.S. on T.P.$_{1}$ = 3.005, I.F.S. on 0 + 00 = 2.54, I.F.S. on 0 + 10 = 1.52, I.F.S. on 0 + 20 = 0.80, I.F.S. on 0 + 30 = 0.12, F.S. on T.P.$_{2}$ = 0.064, B.S. on T.P.$_{2}$ = 3.945, I.F.S. on 0 + 35 = 3.80, I.F.S. on 0 + 40 = 3.00, I.F.S. on 0 + 50 = 2.35, I.F.S. on 0 + 60 = 2.00, I.F.S. on 0 + 70 = 1.50, I.F.S. on 0 + 80 = 1.00, I.F.S. on 0 + 90 = 2.42, I.F.S. on 1 + 00 = 2.68, I.F.S. on 1 + 08.50 = 2.785 (hub), F.S. on T.P.$_{3}$ = 3.002, B.S. on T.P.$_{3}$ = 0.105, F.S. on T.P.$_{4}$ = 3.421, B.S. on T.P.$_{4}$ = 0.062, F.S. on B.M.$_{10}$ = 3.231 m. Prepare profile-level notes for these data, make all necessary checks, adjust the H.I.'s in proportion to the number of setups, and calculate elevations for all stations. The elevation of B.M.$_{10}$ is 50.508 m above the datum.

5.32 Plot the profile for the data of Prob. 5.31 at a horizontal scale of 1 : 500 and a vertical scale of 1 : 100.

5.33 The data for profile levels in the order taken in the field are as follows: B.S. on B.M.$_{22}$ = 12.31, I.F.S. on 0 + 00 = 11.02 (hub), I.F.S. on 0 + 25 = 5.2, I.F.S. on 0 + 50 = 0.4, F.S. on T.P.$_{1}$ = 0.31, B.S. on T.P.$_{1}$ = 11.34, I.F.S. 0 + 75 = 7.9, I.F.S. on 0 + 85 = 8.60 (edge of sidewalk), I.F.S. on 0 + 97 = 6.42, I.F.S. 1 + 00 = 5.2, F.S. on T.P.$_{2}$ = 2.57, B.S. on T.P.$_{2}$ = 11.16, I.F.S. on 1 + 25 = 9.5, I.F.S. 1 + 44 = 6.8, I.F.S. 1 + 50 = 6.2, I.F.S. 1 + 52 = 6.1, I.F.S. 1 + 75 = 3.5, F.S. on T.P.$_{3}$ = 3.49, B.S. on T.P.$_{3}$ = 2.67, F.S. on T.P.$_{4}$ = 12.73, B.S. on T.P.$_{4}$ = 1.88, F.S. on T.P.$_{5}$ = 12.42, B.S. on T.P.$_{5}$ = 3.24, F.S. on B.M.$_{22}$ = 11.14, all units in feet. Prepare profile-level notes for these data, make all necessary checks, adjust the H.I.'s in proportion to the number of instrument setups, and calculate elevations for all stations. The elevation of B.M.$_{22}$ is 155.25 ft above the datum.

5.34 Plot the profile of the data in Prob. 5.33 using a horizontal scale of 1 : 240 and a vertical scale of 1 : 120. Include all the information considered necessary for a profile on the drawing.

5.35 Reciprocal leveling across a canyon from station 1 to station 2, with simultaneous readings using two tilting levels and two rods on opposite sides of the canyon, yielded the following average readings:

Instrument station	Average near reading, m	Average distant readings, m
1	2.685	3.241
2	1.521	0.973

The distance from station 1 to 2 is approximately 1 km. Compute the difference in elevation between stations 1 and 2.

5.36 In the two-peg test of a level, the following observations are taken:

	Instrument at M	Instrument at P
Rod reading on A	3.612	1.862
Rod reading on B	3.248	0.946

M is equidistant from A and B; P is 40 ft from A and 240 ft from B. What is the true difference in elevation between the two points? With the level in the same position at P, to what rod reading on B should the line of sight be adjusted? What is the corresponding rod reading on A for a horizontal line of sight? Check these two rod readings against the true difference in elevation, previously determined.

5.37 Complete the differential-level notes shown below. Determine error of closure and adjust the elevations of B.M.$_X$ and B.M.$_Y$. Assume the error is constant per instrument setup. Put notes in a proper field note form. Units are metres.

Station	B.S.	H.I.	F.S.	Elevation
B.M.$_Z$	1.893			310.456
T.P.$_1$	2.618		0.862	
B.M.$_X$	3.138		0.377	
T.P.$_2$	2.198		0.973	
B.M.$_Y$	1.958		1.747	
T.P.$_3$	0.169		2.010	
T.P.$_4$	0.952		3.354	
T.P.$_5$	0.789		2.481	
B.M.$_Z$			1.919	

5.38 Data for a differential-level circuit in the order readings were taken B.S., F.S.,..., etc., starting with the initial reading on B.M.$_8$ (elevation 248.924 m) are as follows: 1.209, 1.861, 0.128, 1.918, 0.399, 2.617, 0.882, 1.762, 1.961, 0.920, 2.599, 0.918, 2.256, 0.517, 2.308, 1.172. The final reading closes the circuit on B.M.$_8$. Put the data in a complete field note form, making all the necessary calculations and checks. All units are metres.

5.39 Rod readings in metres for differential levels run from B.M.$_K$ to B.M.$_L$ in the order taken (e.g., B.S., F.S.) are the following: 3.118, 0.245, 3.122, 0.355, 2.719, 1.195, 1.138, 2.681, 0.353, 3.283, 0.376, 3.084. Put these notes in field note form, make all necessary checks, and determine the elevation of B.M.$_L$ if B.M.$_K$ has an elevation of 50.152 m above the datum.

5.40 The data for profile levels in the order taken in the field are: B.S. on B.M.$_{20}$ = 2.702, I.F.S. on 0 + 00 (hub) = 3.003, I.F.S. on 0 + 10 = 1.18, F.S. on T.P.$_1$ = 0.379, B.S. on T.P.$_1$ = 3.608, I.F.S. on 0 + 20 = 2.82, I.F.S. on 0 + 30 = 1.55, I.F.S. on 0 + 40 = 0.40, I.F.S. on 0 + 41.7 = 0.248 (edge sidewalk), I.F.S. on 0 + 42.7 = 0.083 (edge sidewalk), F.S. on T.P.$_2$ = 0.304, B.S. on T.P.$_2$ = 2.463, I.F.S. on 0 + 46.1 = 1.792 (edge sidewalk), I.F.S. on 0 + 50 = 1.611 (edge sidewalk), F.S. on 0 + 54.13 = 0.711 (top hub), B.S. on 0 + 54.13 = 0.106 (top hub), F.S. on T.P.$_3$ = 3.572, B.S. on T.P.$_3$ = 0.071, F.S. on T.P.$_4$ = 3.468, B.S. on T.P.$_4$ = 1.197, F.S. on B.M.$_{20}$ = 1.772, all units being in metres. Prepare profile-level notes for these data, make all necessary checks, adjust H.I.'s in proportion to the number of instrument setups, and calculate elevations for all stations. The elevation of B.M.$_{20}$ = 245.500 m above the datum.

5.41 Plot the profile of the data given in Prob. 5.40 using a horizontal scale of 1 : 500 and a vertical scale of 1 : 100. Include all the information necessary for a profile on the drawing.

5.50. Field problems

PROBLEM 1 RADIUS OF CURVATURE AND SENSITIVITY OF LEVEL TUBE

Object To determine, in the field without the use of special apparatus, the radius of curvature of the level tube of transit or level and the sensitivity of the bubble (Art. 5.15).

Procedure (1) Hold the rod on a solid point 300 ft or 100 m from the instrument. With one end of the bubble at a division near the end of the level tube, take a careful rod reading to the nearest 0.001 ft or 0.001 m. Note the exact position of each end of the bubble. (2) Manipulate the leveling screws until the other end of the bubble falls near the other end of the tube. Take another rod reading, and measure the exact distance traversed by each end of the bubble. (3) Determine the bubble movement expressed as the number of divisions moved times the length of a single graduation on the bubble tube = $(n)(i)$. On most levels, $i = 2$ mm. Also determine the difference between the two rod readings d (see Fig. 5.13), expressed in the same units as the bubble tube graduation i. (4) In this manner obtain a series of five bubble movements and their corresponding rod readings. (5) Compute the radius of curvature by the formula $r = (ni/d)a$, in which r is the radius of curvature, a is the distance from the instrument to the rod, ni is the mean of the five bubble movements, and d is the mean of the five differences in rod readings. (6) Compute the sensitivity of the level tube in seconds of arc per division using Eq. (5.13).

PROBLEM 2 DIFFERENTIAL LEVELING WITH ENGINEER'S LEVEL AND SELF-READING ROD

Object Given the elevation of an initial bench mark, to determine the elevations of points in an assigned level circuit.

Procedure Follow the procedure outlined in Art. 5.34. Keep notes in the form of the sample notes shown in Fig. 5.36, but estimate rod readings to thousandths of feet or to millimetres. Check each rod reading by taking a second observation. Close the circuit, and compute the error of closure.

Hints and precautions (1) When using *self-leveling instruments* (Art. 5.18): (*a*) Try to keep the base plate of the tripod as close to a horizontal position as possible when setting the tripod, or it may be impossible to center the bull's eye bubble, owing to insufficient run on the level screws. (*b*) Center the bull's-eye bubble as directed in Art. 5.29 and then turn the instrument through 180°. The bull's-eye bubble should stay within the circle (see Art. 5.48). (*c*) To check the compensator, sight the rod, note the reading at the middle cross hair, and then tap the leg of the tripod. If the compensator is working properly, the image will appear to move in the field of vision but the middle cross hair will return to its original position. (*d*) Always double-check to make sure the rod reading was taken at the middle and not at one of the stadia cross hairs. (2) When using *spirit level instruments* of the dumpy level type: (*a*) Sight in the direction of the rod and focus before centering the bubble exactly; (*b*) center the bubble before each reading and check its position immediately afterward. For each type of instrument, (3) test for parallax (Art. 5.38). (4) Choose the turning points with an eye to simplicity of field operations, but roughly balance the backsight and foresight distances between bench marks; keep no record of these distances. (5) Use signals (Art. 3.9). (6) Keep the foot of the rod free from dirt. (7) Be sure that the rod is held vertical while a sight is being taken. When the "long rod" is used, wave the rod, and before each reading check the position of the back vernier. (8) Describe each bench mark in the field notes. (9) Check the computations.

References

1. Berry, R. M., "History of Geodetic Leveling in the United States," *Surveying and Mapping*, Vol. 36, No. 2, June 1976, pp. 137–154.
2. Berry, R. M., "Observational Techniques for Use with Compensator Leveling Instruments for First-Order Levels," *Surveying and Mapping*, Vol. 37, No. 1, March 1977, p. 17.
3. Davis, R. E., Foote, F.S., Anderson, J.M., and Mikhail, E. M., *Surveying: Theory and Practice*, 6th ed., McGraw-Hill Book Co., New York, 1981.
4. Federal Geodetic Control Committee, *Specifications to Support Classification, Standards of Accuracy, and General Specifications of Geodetic Control Surveys*, U.S. Department of Commerce, National Oceanic and Atmospheric Administration, Rockville, Md., 1975, 1976.
5. Hou, G.-Y., Veress, S. A., and Colcord, J.E., "Refraction in Precise Leveling," *Surveying and Mapping*, Vol. 32, No. 2, June 1972, p. 231.
6. Hradilek, L., "Refraction in Trigonometric and Three-Dimensional Terrestrial Networks," *The Canadian Surveyor*, Vol. 26, No. 1, March 1972, pp. 59–70.
7. Huether, G., "New Automatic Level NI 002 of the Jean Optical Works for First and Second Order Leveling," *Proceedings, Fall Convention, ACSM*, 1974, p. 365.
8. Moffitt, F. H., and Bouchard, H., *Surveying*, 7th ed., Harper & Row Publishers, New York, 1982.

9. Rappleye, H. S., *Manual of Leveling Computation and Adjustment*, Spec. Publ. No. 240, U.S. Department of Commerce, Coast and Geodetic Survey (now National Ocean Survey, NOAA), Washington, D.C., 1948; reprinted 1975.

10. Rappleye, H.S., *Manual of Geodetic Leveling*, Spec. Publ. No. 239, U.S. Department of Commerce, Coast and Geodetic Survey (now National Ocean Survey, NOAA), Washington, D.C., 1948; reprinted 1963/1976.

11. Selley, A. D., "A Trigonometric Level Crossing of the Strait of Belle Isle," *The Canadian Surveyer*, Vol. 31, No. 3, September 1977.

12. Whalen, C. T., and Balacz, E. I., "Test Results of First-Order Class III Leveling," *Surveying and Mapping*, Vol. 37, No. 1, March 1977, pp. 45–58.

CHAPTER 6
Angle and direction measurement

6.1. Location of points As previously stated, the purpose of a survey is to determine the relative locations of points below, on, or above the surface of the earth. Since the earth is three-dimensional, the most logical reference framework with which to locate points is a three-dimensional rectangular coordinate system. Figure 6.1 shows an earth-centered, three-dimensional rectangular coordinate system $X'Y'Z'$ called the *geocentric coordinate system*. It is a right-handed system, with X' passing through Greenwich in England and Z' passing through the North Pole (Y' is uniquely fixed because $X'Y'Z'$ is specified as right-handed). In addition to the $X'Y'Z'$ coordinates, any point may also be located by a set of spherical coordinates consisting of latitude ϕ, longitude λ, and the distance $R + h$ along the normal to the earth ellipsoid. The latter system is used mainly for geodetic surveys and is not discussed in further detail in this text.

A reference more suitable for plane surveying is a local XYZ right-handed rectangular coordinate system similar to the one illustrated in Fig. 6.2. The origin of such a system is usually chosen near the center of the area to be surveyed. Its XY plane is tangent to the reference ellipsoid at the point of origin, and the Y axis is generally directed toward the North Pole. Any point P may be located either by XYZ or by the two angles, horizontal angle α (in the XY plane) and vertical angle β, and the distance r from O to P, as shown in Fig. 6.2.

Visualization of the local XYZ system is more convenient when the Z axis coincides with the reader's local vertical or the plumb line, as illustrated in Fig. 6.3. In this figure, point A, which is assumed to be of known position, is also introduced. With respect to Fig. 6.3, the position of point P in space may be determined in one of the following measurement procedures:

1. The direction of line OP (angles α, β) and the distance $r = OP$
2. The directions of lines OP (angles α, β) and AP (angles α' in the XY plane and β' in plane $APP'A'$) from two known points O and A
3. The distances r (from O to P), r' (from A to P), and h' (height of P above the XY plane)

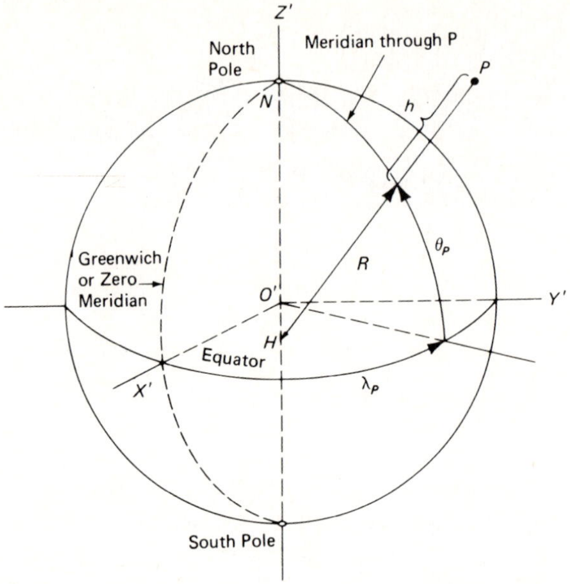

Fig. 6.1 Position of P in geocentric system.

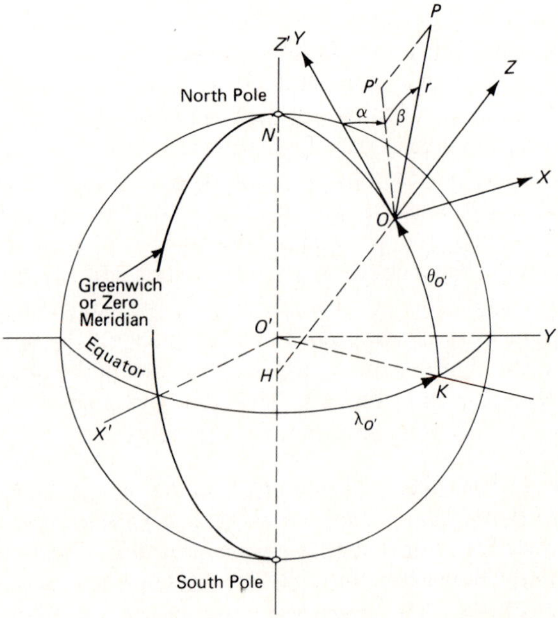

Fig. 6.2 Local coordinate system.

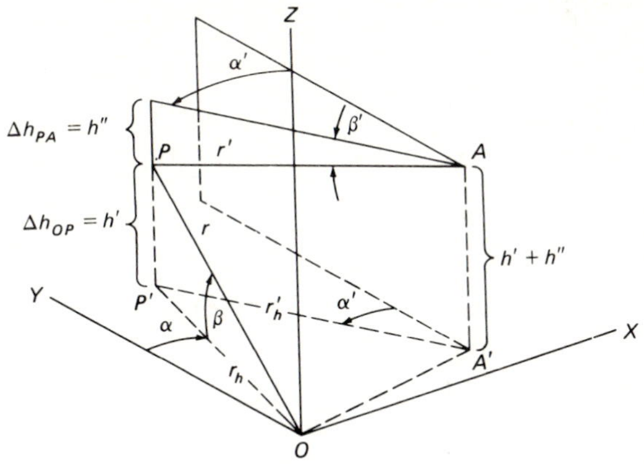

Fig. 6.3 Position of a point in local coordinate system.

For most surveys of small extent, the positions of P and A and the slope distances r and r' are projected onto the horizontal XY plane. Still referring to Fig. 6.3, the position of P' in the horizontal XY plane is defined when measurements in one of the following cases are given:

1. Distance and direction from one known point, such as r_h and α from known point O
2. Directions from two known points such as α and α' from known points O and A', respectively
3. Distances from two known points such as r_h and r'_h from points O and A'

Note that the direction of spatial line OP is specified by angles α in the horizontal XY plane and β in the vertical plane OPP'. Similarly, the direction of line AP is determined by α' and β' in the horizontal XY plane and the vertical plane $APP'A'$, respectively. The directions of horizontal lines OP' and $A'P'$ are specified only by horizontal angles α and α', respectively. Thus, in general, the angular measurements of surveying are either horizontal or vertical.

The angle between two points is understood to mean the horizontal angle or angle between projections in the horizontal plane of two lines passing through the two points and converging at a third point. Thus, at P in Fig. 6.3, the angle between A and O is the horizontal angle $OP'A'$.

The vertical angle to a point is the angle of elevation or depression measured from the horizontal. In Fig. 6.3, β is an elevation vertical angle from O to P and β' is a depression vertical angle from A to P. Vertical angles can also be measured from the zenith (the OZ axis). For example, in Fig. 6.3 the zenith angle from O to P is angle ZOP.

6.2. Meridians The relative directions of lines connecting survey points may be obtained in a variety of ways. Figure 6.4 shows lines intersecting at a point. The direction of any line (as OB) with respect to an adjacent line (as OA) is given by the horizontal angle between the two lines (as α_2) and the direction of rotation (as clockwise). The direction of any line (as OC) with respect to a line not adjacent (as

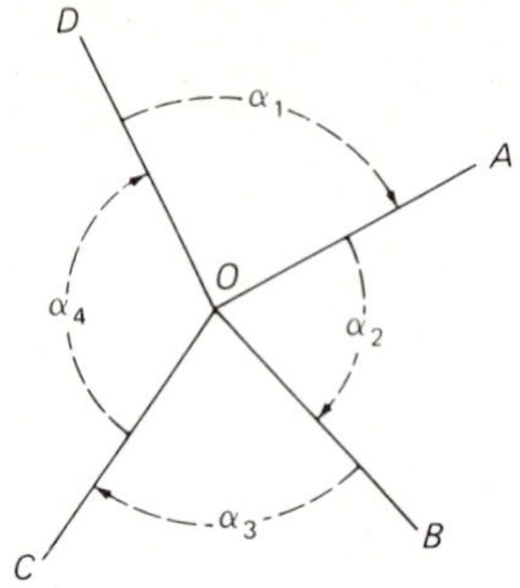

Fig. 6.4 Directions by angles.

OA) is not given by any of the measured angles but may be computed by adding the intervening angles (as $\alpha_2 + \alpha_3$).

Figure 6.5 shows the same system of lines but with all angles measured from a line of reference *OM*. The direction of any line (as *OA*) with respect to the line of reference (as *OM*) is given by the angle between the lines (as β_1) and its direction of rotation (as clockwise). The angle between any two lines (as *AOC*) is not given directly, but may be computed by taking the difference between the direction angles of the two lines (as $\angle\beta_3 - \angle\beta_1 = \angle AOC$).

The fixed line of reference may be any line in the survey, or it may be purely imaginary. It is termed a *meridian*. If it is arbitrarily chosen, it is called an *assumed meridian*; if it is a north-and-south line passing through the geographical poles of the earth, it is called a *true* or *astronomic meridian*; if it is a line parallel to a central true meridian, it is called a *grid meridian*; or if it lies parallel with the magnetic lines of force of the earth as indicated by the direction of a magnetized needle, it is called a *magnetic meridian*.

6.3. True meridian The *true meridian*, or *astronomic meridian*, is determined by astronomical observations. In Fig. 6.2, arc *NOK* is a true meridian. For any given point on the earth its direction is always the same, and hence directions referred to the true meridian remain unchanged regardless of time. The lines of most extensive surveys and usually the lines marking the boundaries of property are referred to the true meridian.

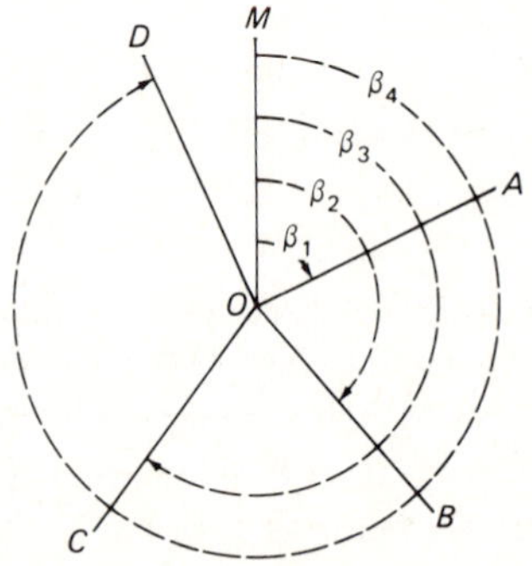

Fig. 6.5 Directions referred to meridian.

At many triangulation stations established throughout the United States by the National Ocean Survey, reference lines of known true direction have been established for use by surveyors.

6.4. Grid meridian Examination of Fig. 6.1 shows that true meridians converge at the poles. In plane surveys of limited extent, it is convenient to perform the work in a rectangular *XY* coordinate system in which one central meridian coincides with a true meridian. All remaining meridians are parallel to this central true meridian, eliminating the need to calculate the convergence of meridians when determining positions of points in the system. These parallel meridians are called *grid meridians*.

Each state in the United States has plane coordinate projections established by the National Ocean Survey. These projections are designed so that surveys can be performed as on a horizontal surface within tolerable levels of accuracy. The two projections used for this purpose in the United States are the *Lambert conformal conic projection* and the *transverse Mercator projection*, which are described in Chap. 11.

6.5. Magnetic meridian The direction of the *magnetic meridian* is that taken by a freely suspended magnetic needle. The magnetic poles are at some distance from the true geographic poles; hence, in general the magnetic meridian is not parallel to the true meridian. The location of the magnetic poles is constantly changing; therefore, the direction of the magnetic meridian is not constant. However, the magnetic meridian is employed as a line of reference on rough surveys where a magnetic compass is used and often is employed in connection with more precise surveys in which angular measurements are checked approximately by means of the compass. It was formerly used extensively for land surveys and still is for retracement surveys and surveys in hilly or mountainous terrain in areas of low land values. Details concerning magnetic compasses and meridians can be found in Arts. 6.13 to 6.20.

6.6. Angles and directions Angles and directions may be defined by means of *bearings*, *azimuths*, *deflection angles*, *angles to the right*, or *interior angles*, as described in the following sections. These quantities are said to be *observed* when obtained directly in the field, and *calculated* when obtained indirectly by computation. Conversion from one means of expressing angles and directions to another is a simple matter if a sketch is drawn to show the existing relations.

6.7. Bearings The direction of any line with respect to a given meridian may be defined by the *bearing*. Bearings are called *true* (*astronomic*) *bearings*, *magnetic bearings*, or *assumed bearings* depending on whether the meridian is true, magnetic, or assumed. The bearing of a line is indicated by the quadrant in which the line falls and the acute angle which the line makes with the meridian in that quadrant. Thus, in Fig. 6.6 the bearing of the line *OA* is read north 37° east and is written N37°E. The bearings of *OB*, *OC*, and *OD* are, respectively, S62°E, S50°W, and N20°W. In all cases, values of bearing angles lie between 0° and 90°. If the direction of the line

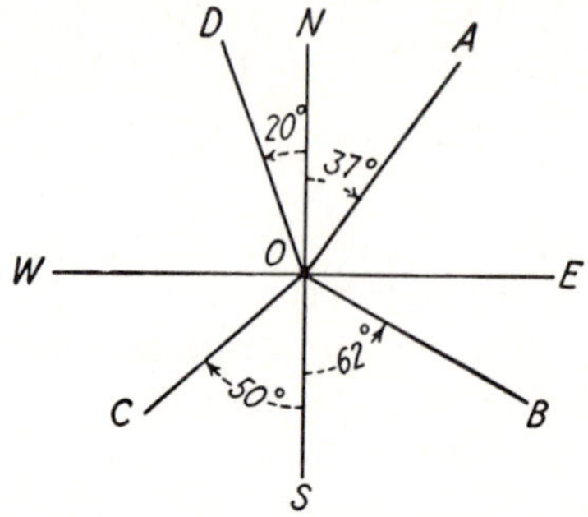

Fig. 6.6 Bearings.

is parallel to the meridian and north, it is written as N0°E, N0°W or *due North*; if perpendicular to the meridian and east, it is written as N90°E, N90°W, or *due East*.

If the bearings of two lines are known, the angle between them can be determined. If the bearing of one line is known and the angle that it makes with a second line is known, the bearing of the second line can be determined. For example in Fig. 6.7, if the observed bearing of *OA* is N37°E and the angle $AOB = 81°$, the calculated bearing of *OB* is S62°E.

6.8. Azimuths The *azimuth* of a line is its direction as given by the angle between the meridian and the line measured in a clockwise direction usually from the north branch of the meridian. In astronomical observations azimuths are generally reckoned from the true south; in surveying, some surveyors reckon azimuths from the south and some from the north branch of whatever meridian is chosen as a reference, but on any given survey the direction of zero azimuth is either always south or always north. Thus, it is necessary to designate whether the azimuth is from the north or from the south. Azimuths are called *true* (*astronomic*) *azimuths*, *magnetic azimuths*, or *assumed azimuths*, depending on whether the meridian is true, magnetic, or assumed. Azimuths may have values between 0° and 360°.

In Fig. 6.8, azimuths measured from the south point are $A_{OA} = 217°$, $A_{OB} = 298°$, $A_{OC} = 50°$, and $A_{OD} = 160°$; or in Fig. 6.9, in which are shown the same lines with azimuths measured from the north point, $A_{OA} = 37°$, $A_{OB} = 118°$, $A_{OC} = 230°$, and $A_{OD} = 340°$. In Fig. 6.8 if the observed azimuth of *OA* as reckoned from the south is 217° and the observed azimuth of *OB* is 298°, the calculated angle *AOB* is 81°.

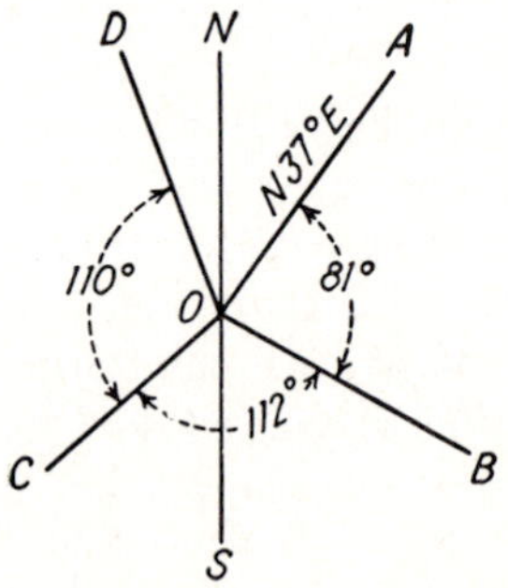

Fig. 6.7 Angles and bearings.

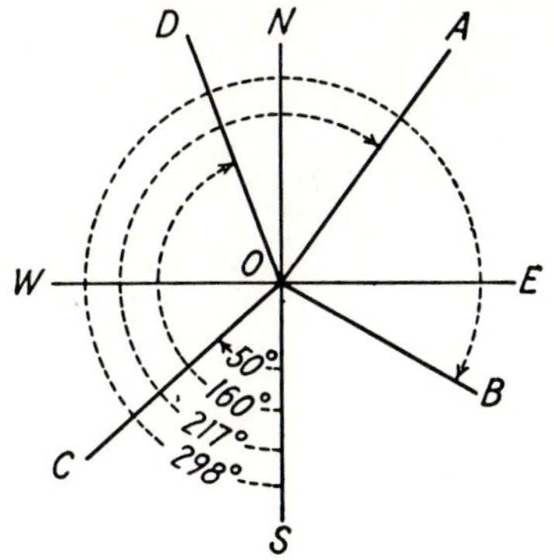

Fig. 6.8 Azimuths from the south.

Azimuths may be calculated from bearings, or vice versa, preferably with the aid of a sketch. For example, if the bearing of a line is N16°E, its azimuth (from south) is $180 + 16 = 196°$; and if the azimuth (from south) of a line is 285°, its bearing is $360 - 285 =$ S75°E.

It is assumed that when an azimuth is given as being from O to A, as in Fig. 6.9, where the azimuth from north $A_{OA} = 37°$, this specifies the direction of the line from an origin at O to a terminal point A and the azimuth is called a *forward azimuth*. Conversely, the azimuth from A to O is called the *back azimuth* of OA. When the azimuth of a line is less than 180°, the back azimuth of the line is the forward azimuth plus 180°. For example, the back azimuth of OA in Fig. 6.9 is $37° + 180° = 217°$. When the forward azimuth of a line is greater than 180°, the back azimuth equals the forward azimuth minus 180°. In Fig. 6.9, the back azimuth of line $OC = 230° - 180° = 50°$. The concept of forward and back azimuth is further illustrated in Fig. 6.10, where the azimuth from the north of line $CD = 60°$. Now, suppose that the initial point was D. Then the azimuth of DC is 240° and the back azimuth of DC is $240° - 180° = 60°$. Note that the idea of forward and back azimuth as developed here is valid only for plane surveys of limited extent or where grid north is used as the reference meridian.

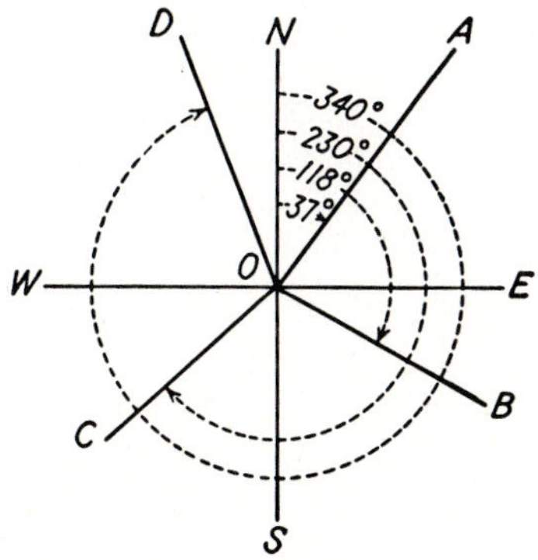

Fig. 6.9 Azimuths from the north.

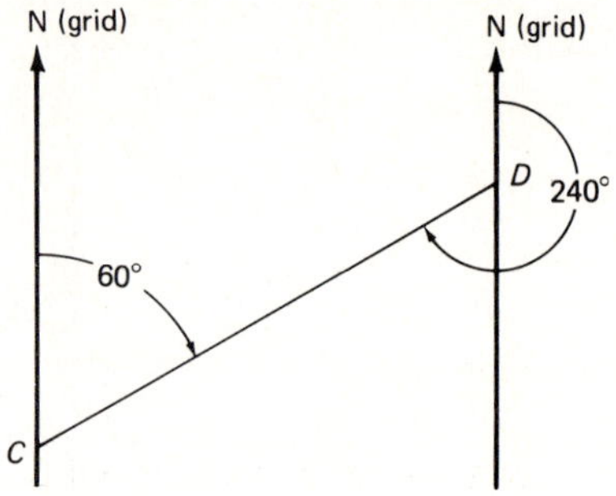

Fig. 6.10 Forward and back azimuths.

6.9. Interior angles In a closed polygon, angles inside the figure between adjacent lines are called *interior angles*. Figure 6.11 illustrates interior angles $\alpha_1, \alpha_2, \ldots, \alpha_6$ in polygon *ABCDEF*. If n equals the number of sides in a closed polygon, the sum of the interior angles is $(n-2)(180°)$. In Fig. 6.11, $\sum_{i=1}^{6} \alpha_i = (4)(180°) = 720°$.

6.10. Deflection angles The angle between a line and the prolongation of the preceding line is called a *deflection angle*. Deflection angles are recorded as *right* or *left* depending on whether the line to which measurement is taken lies to the right (clockwise) or left (counterclockwise) of the prolongation of the preceding line. Thus, in Fig. 6.12 the deflection angle at B is 23°R, and at C is 33°L. Deflection angles may have values between 0° and 180°, but usually they are not employed for angles greater than 90°. In any closed polygon the algebraic sum of the deflection angles (considering right deflections as of sign opposite to left deflections) is 360°. Figure 6.13 shows polygon *ABCDE* with deflection angles $\alpha_1, \alpha_2, \ldots, \alpha_5$. In this example $\alpha_1 + \alpha_2 - \alpha_3 + \alpha_4 + \alpha_5 = 360°$.

6.11. Angles to the right Angles may be determined by clockwise measurements from the preceding to the following line, as illustrated by Fig. 6.14. Such angles are called *angles to right* or *azimuths from back line*.

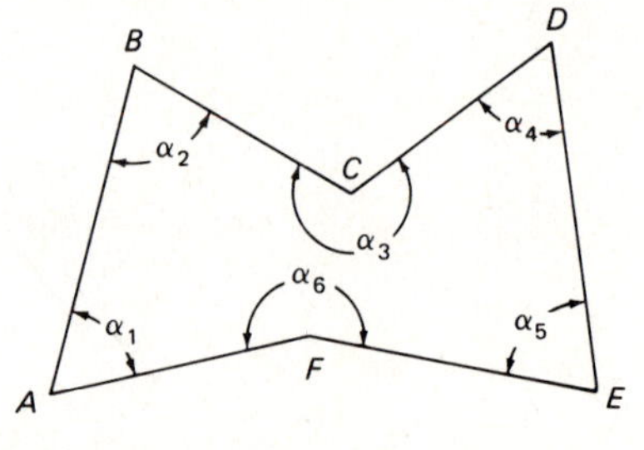

Fig. 6.11 Interior angles in polygon.

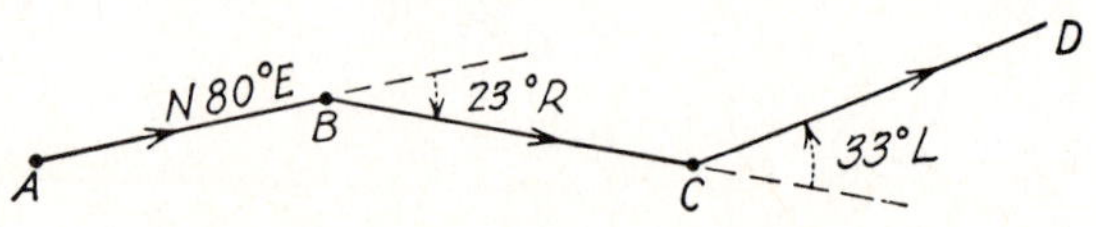

Fig. 6.12 Deflection angles.

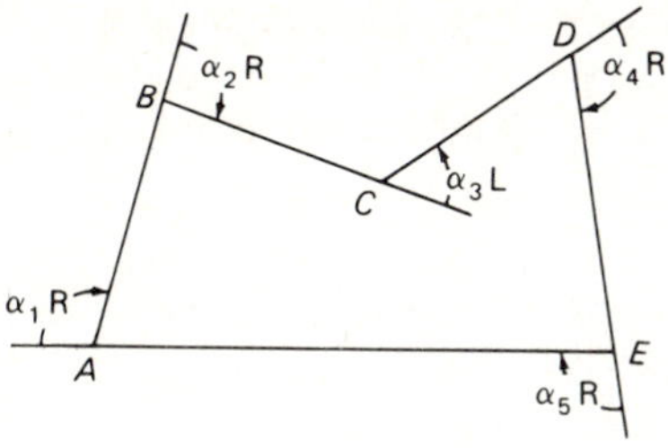

Fig. 6.13 Deflection angles in polygon.

6.12. Methods of determining angles and directions Angles are normally measured with a transit but can also be determined by means of a tape, plane-table alidade, sextant, or compass. Directions are observed with a direction theodolite or with a magnetic compass.

1. Transit The engineer's transit is designed to enable observing horizontal and vertical angles. Most transits allow measuring the horizontal angle to the nearest minute of arc but on some instruments the horizontal angle can be observed to the nearest 10″ of arc. Vertical angles can be read to the nearest minute of arc. A detailed description of the engineer's transit and its use can be found in Arts. 6.22 to 6.29.

2. Tape If the sides of a triangle are measured, sufficient data are obtained for computing the angles in the triangle. The error in the computed values of the angles depends on the care with which points are established and the precision with which measurements are taken. For acute angles on level ground the error need not exceed 05′ to 10′. For angles greater than 90°, the corresponding acute angle should be observed. The method is slow and is generally used only as a check.

3. Plane table and alidade Using a plane table, which is essentially a drawing board mounted on a tripod, and an alidade, which is a straight edge with parallel line of sight, angles can be determined graphically. The plane table and alidade and its use are described in Arts. 7.18 and 7.19.

4. Sextant Although the sextant is used primarily by the navigator, it is also employed by the surveyor, principally on hydrographic surveys. Using the sextant, an angle may be measured while the observer is moving; hence, angles may be read from a boat from which soundings are taken. The angle measured by the sextant is in the plane defined by the two points sighted and the telescope and therefore is not

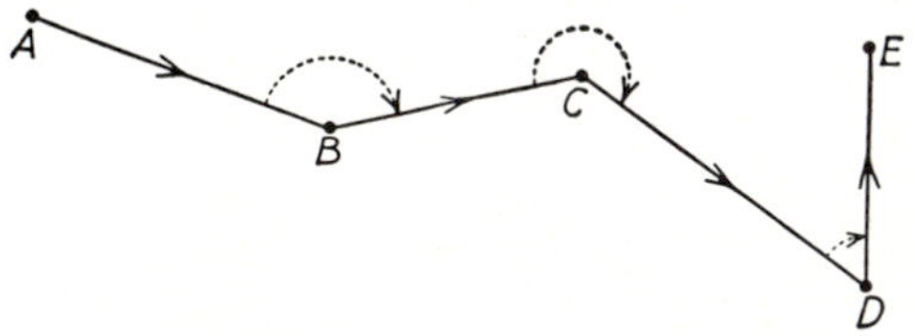

Fig. 6.14 Angles to right.

in general a horizontal angle. The sextant is not an instrument of high accuracy for small angles of less than 15° or for short distances of less than 300 m or 1000 ft. The sextant and its use are described in Arts. 6.58 and 6.59.

5. Direction theodolite The terms "transit" and "theodolite" are frequently used interchangeably. In the United States, *direction theodolite* refers to an extremely precise transit which has only one horizontal motion. With this type of instrument, directions are observed and the angle is then computed as the difference between the two directions. Direction theodolites usually have a nominal accuracy of $\pm 1''$ of arc. The characteristics and use of these instruments are given in Arts. 6.30, 6.40, and 6.41.

6. Magnetic compass The use of the magnetic compass is described in the following sections. By itself the compass is useful in making rough surveys and retracing early land surveys. Mounted on the transit, the compass is useful as a means of approximately checking horizontal angles measured by more precise methods.

6.13. Magnetic compass Any slender symmetrical bar of magnetized iron when freely suspended at its center of gravity takes up a position parallel with the lines of magnetic force of the earth. In horizontal projection these lines define the magnetic meridians. In elevation, the lines are inclined downward toward the north in the northern hemisphere, and downward toward the south in the southern hemisphere. Since the bar takes a position parallel with the lines of force, it becomes inclined with the horizontal. This phenomenon is called the *magnetic dip*. The angle of dip varies from 0° at or near the equator to 90° at the magnetic poles. The needle of the magnetic compass rests on a pivot. To counteract the effect of dip, so that the needle will take a horizontal position when directions are observed, a counterweight is attached to one end (the south end in the northern hemisphere). The counterweight usually consists of a short piece of fine brass wire wound around the needle and held in place by spring action. As long as the needle is used in a given locality and loses none of its magnetism, it will remain balanced. When for any reason it becomes unbalanced, it is adjusted to the horizontal by sliding the counterweight along the needle. At the midpoint of the needle is a jewel which forms a nearly frictionless bearing for the pivot.

The essential features of the magnetic compass used by the surveyor are (1) a compass box with a circle graduated from 0° to 90° in both directions from the N and S points and usually having the E and W points interchanged, as illustrated in Fig. 6.15; (2) a line of sight in the direction of the SN points of the compass box; and (3) a magnetic needle. When the line of sight is pointed in a given direction, the compass needle (when pivoted and brought to rest) gives the magnetic bearing. Thus, in Fig. 6.15 the bearing of *AB* is N60°E. If the N point of the compass box is nearest the object sighted, the bearing is read by observing the north end of the needle.

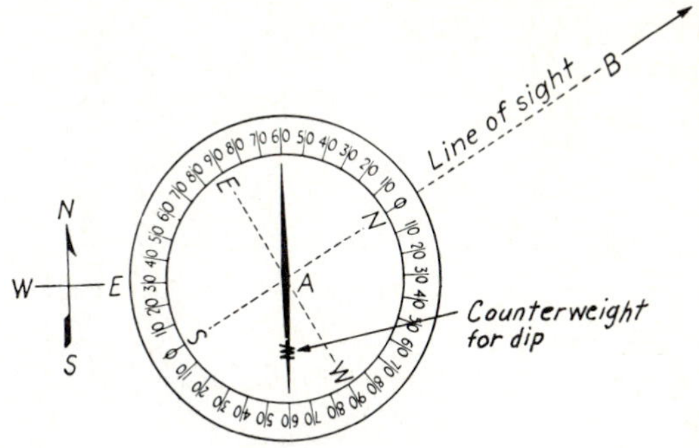

Fig. 6.15 Features of magnetic compass used in surveying.

The varieties of compasses exhibiting the features just mentioned are:

1. Various *pocket compasses*, which are generally held in the hand when bearings are observed and are used in reconnaissance or other rough surveys. Figure 6.16 illustrates one type of pocket compass. Another style, the Brunton pocket transit, is shown in Fig. 6.17. It is designed primarily as a hand instrument but may be mounted on a tripod or a Jacob's staff [a pointed stick about 5 ft (1.52 m) long].

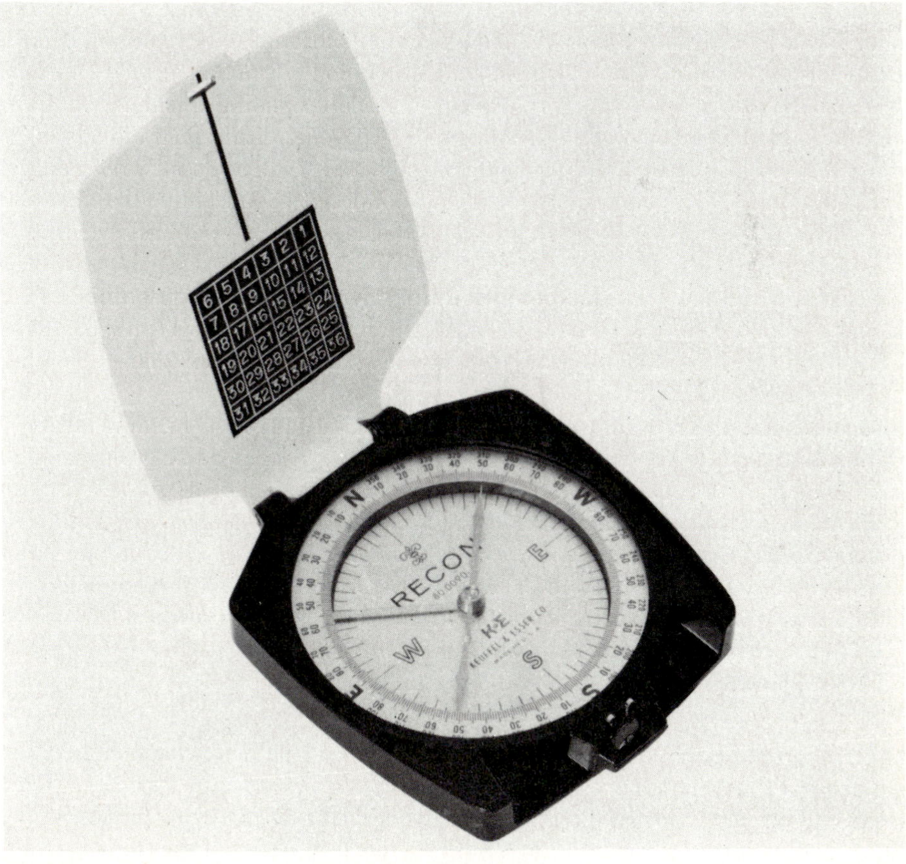

Fig. 6.16 Pocket compass. (*Courtesy Keuffel & Esser Co.*)

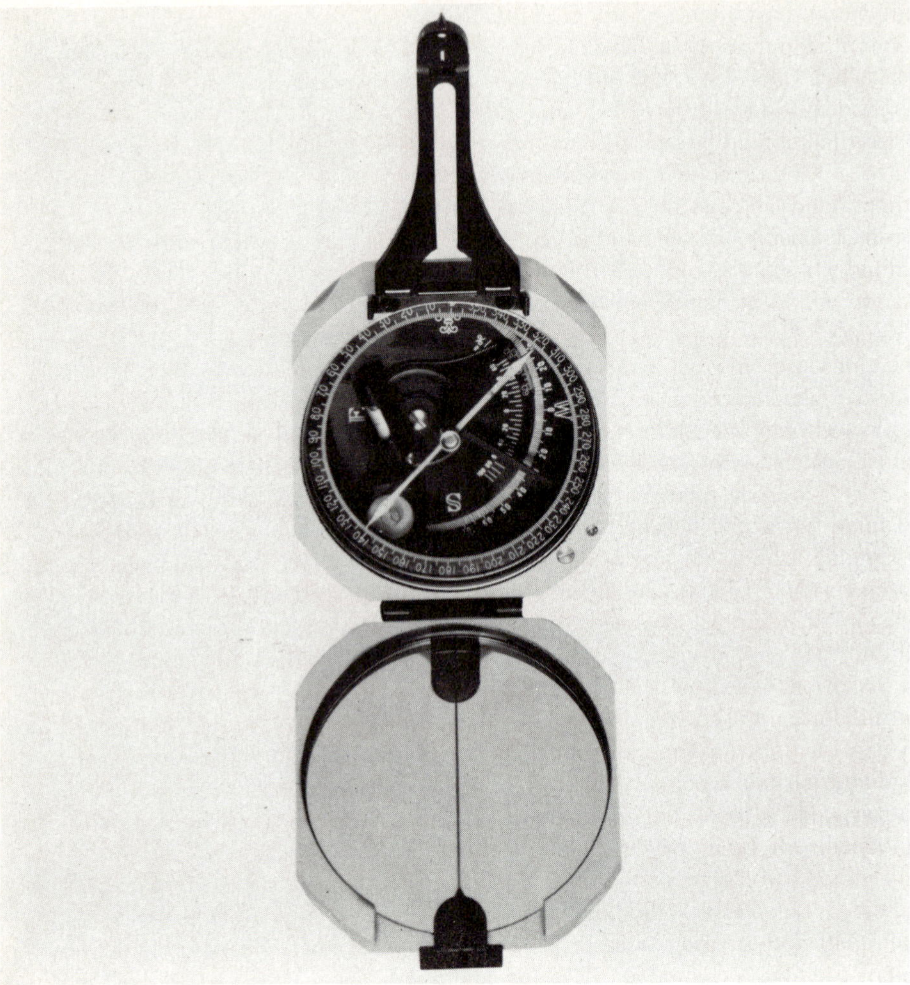

Fig. 6.17 Brunton pocket transit. (*Courtesy Keuffel & Esser Co.*)

2. The *surveyor's compass*, which is usually mounted on a light tripod or sometimes on a Jacob's staff, is shown in Fig. 6.18. This type of compass is now used only for forest surveys, retracing old land surveys or for land surveys in areas of low property values.
3. The *transit compass*, a compass box similar to that of the surveyor's compass, mounted on the upper or vernier plate of the engineer's transit (Art. 6.22) and often used to check horizontal angles.

6.14. Magnetic declination The angle between the true or astronomic meridian and the magnetic meridian is called the *magnetic declination*, or *variation*. If the north end of the compass needle points to the east of the true meridian, the declination is said to be east (Fig. 6.19); if it points to the west of the true meridian,

Fig. 6.18 Surveyor's compass. (*Courtesy Keuffel & Esser Co.*)

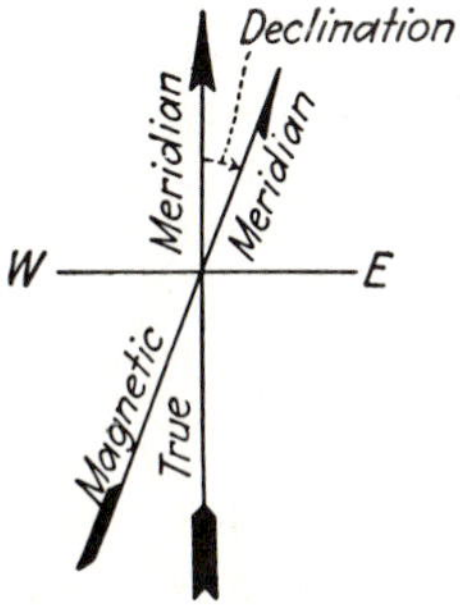

Fig. 6.19 Declination east.

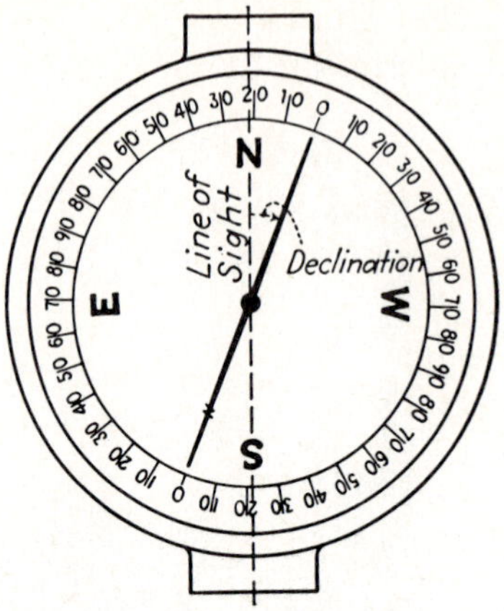

Fig. 6.20 Declination set off on compass circle.

the declination is said to be west. Declination may be set off on the compass as shown in Fig. 6.20.

If a true north-and-south line is established, the mean declination of the needle for a given locality can be determined by compass observations extending over a period of time. The declination may be estimated with sufficient precision for most purposes from an *isogonic chart* published by the U.S. National Ocean Survey; specific values for a particular locality can be obtained from the Survey.

6.15. Isogonic chart The isogonic chart of the continental United States shown in Fig. 6.21 applies to January 1, 1980. It is based upon observations made by the U.S. National Ocean Survey at stations widely scattered throughout the country. The solid lines are lines of equal magnetic declination, or *isogonic lines*. East of the heavy solid line of zero declination, or *agonic line*, the north end of the compass needle points west of north; west of that line it points east of north. The north end of the compass needle is moving eastward over the area of eastward annual change and westward elsewhere over the chart at an annual rate indicated by the lines of equal annual change.

6.16. Variations in magnetic declination The magnetic declination changes more or less systematically in cycles over periods of (1) approximately 300 years, (2) 1 year, and (3) 1 day, as follows:

1. Secular variation Like a pendulum, the magnetic meridian swings in one direction for perhaps a century and a half until it gradually comes to rest and then swings in the other direction, and as with a pendulum the velocity of movement is greatest at the middle of the swing. The rate of change per year, however, varies irregularly. The causes of this secular variation are not well understood. In the United States, it amounts to several degrees in a half-cycle of approximately 150 years. In Fig. 6.21 the annual rates of change in the secular variation for the year 1980 are shown by dashed lines. Because of its magnitude, the secular variation is of considerable importance to the surveyor, particularly in retracing lines the directions of which are referred to the magnetic meridian as it existed years previously. When *variation* is mentioned without further qualification, it is taken to mean the secular variation.

2. Annual variation This is a small annual swing distinct from the secular variation. For most places in the United States, it amounts to less than 01′.

3. Daily variation This variation, also called *solar-diurnal variation*, is a periodic swing of the magnetic needle occurring each day. For points in the United States the north end of the needle reaches its extreme easterly swing at about 8 or 9 A.M. and its extreme westerly swing at about 1 or 2 P.M. The needle usually reaches its mean position between 10 and 11 A.M. and between 7 and 11 P.M. In general, the higher the latitude, the greater the range in the daily variation. The average range for points in the United States is less than 08′, a quantity so small as to need no consideration for most of the work for which the compass needle is employed. However, in the United States in summer, a line run 1000 ft by compass at 8 A.M. would end as much as 2.3 ft or 0.7 m to the right of the point where it would end if run at 1 P.M.

4. Irregular variations Irregular variations are due to magnetic disturbances usually associated with sunspots. They cannot be predicted but are most likely to occur during magnetic storms, when auroral displays occur and radio transmission is disturbed. They may amount to a degree or more, particularly at high latitudes.

6.17. Correction for declination When magnetic directions are used to obtain coarse estimates for bearings or when an old survey must be retraced, it is necessary to reduce the magnetic directions to true bearings or azimuths. Conversion from magnetic directions to true directions, or vice versa, is most easily accomplished by using azimuths. Consider the following examples.

Example 6.1 A magnetic azimuth of 54°30′ was observed along line *AD* in June 1977. The declination for the area surveyed is found by interpolation from an isogonic chart dated 1970 to be 17°30′E with an annual change of 1′ westward. Compute the true azimuth of line *AD*.

SOLUTION First draw a careful sketch of the relationship among true north, magnetic north, and the direction of the line as illustrated by Fig. 6.22*a*.

Magnetic azimuth, 1977	54°30′
Declination 1970	17°30′
Change = (7 yr)(1′/yr)	−0°07′
True azimuth	71°53′

Example 6.2 A magnetic bearing of N34°30′W is recorded on an old survey plan dated August 20, 1950. It is desired to reestablish this direction on the site in 1980. The 1970 isogonic chart shows a declination of 10°W for the area, with an annual change of 2′

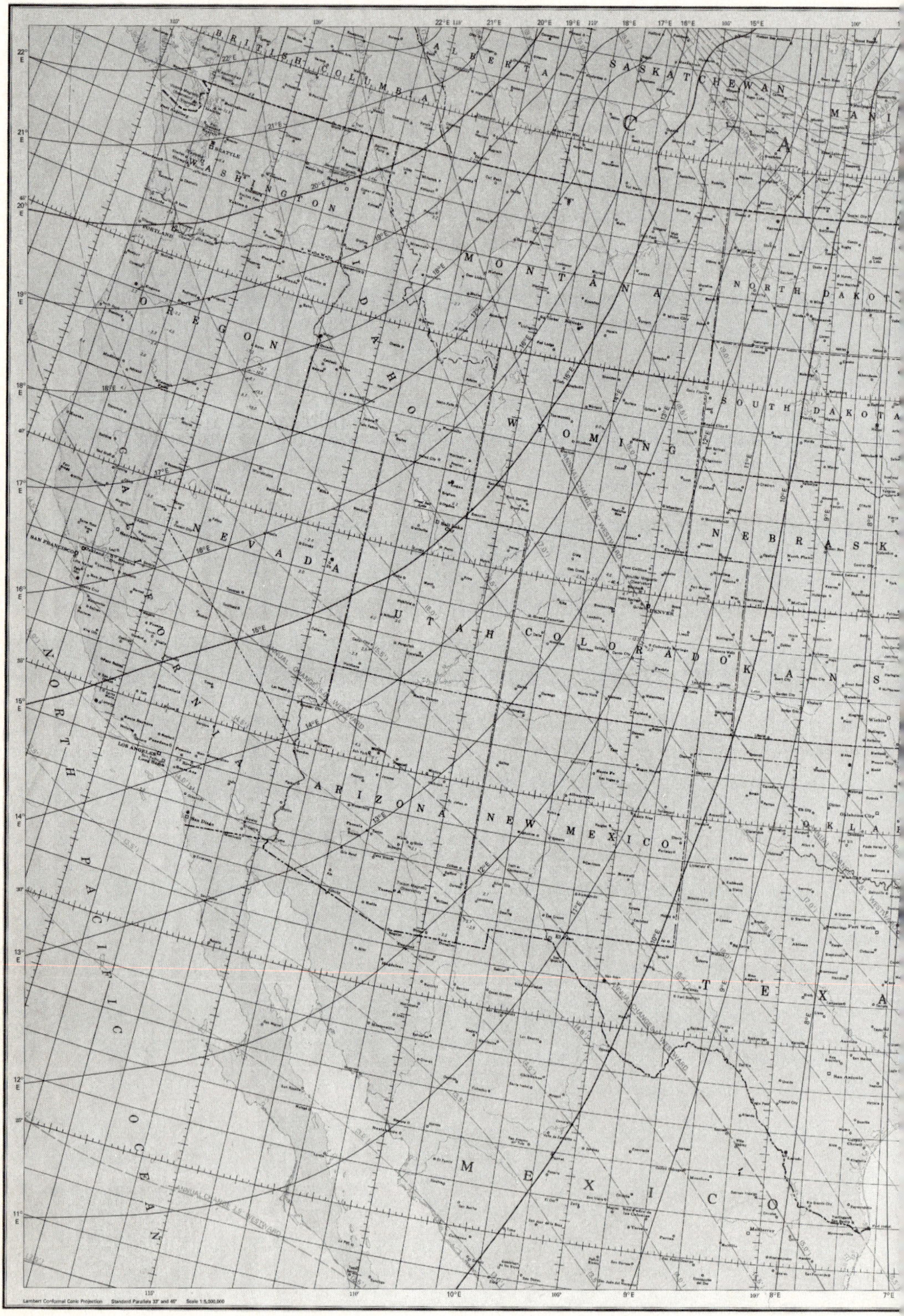

Fig. 6.21 Isogonic chart for the United States, showing magnetic declination in the United States for 1980. (*Courtesy U.S. Geological Survey.*)

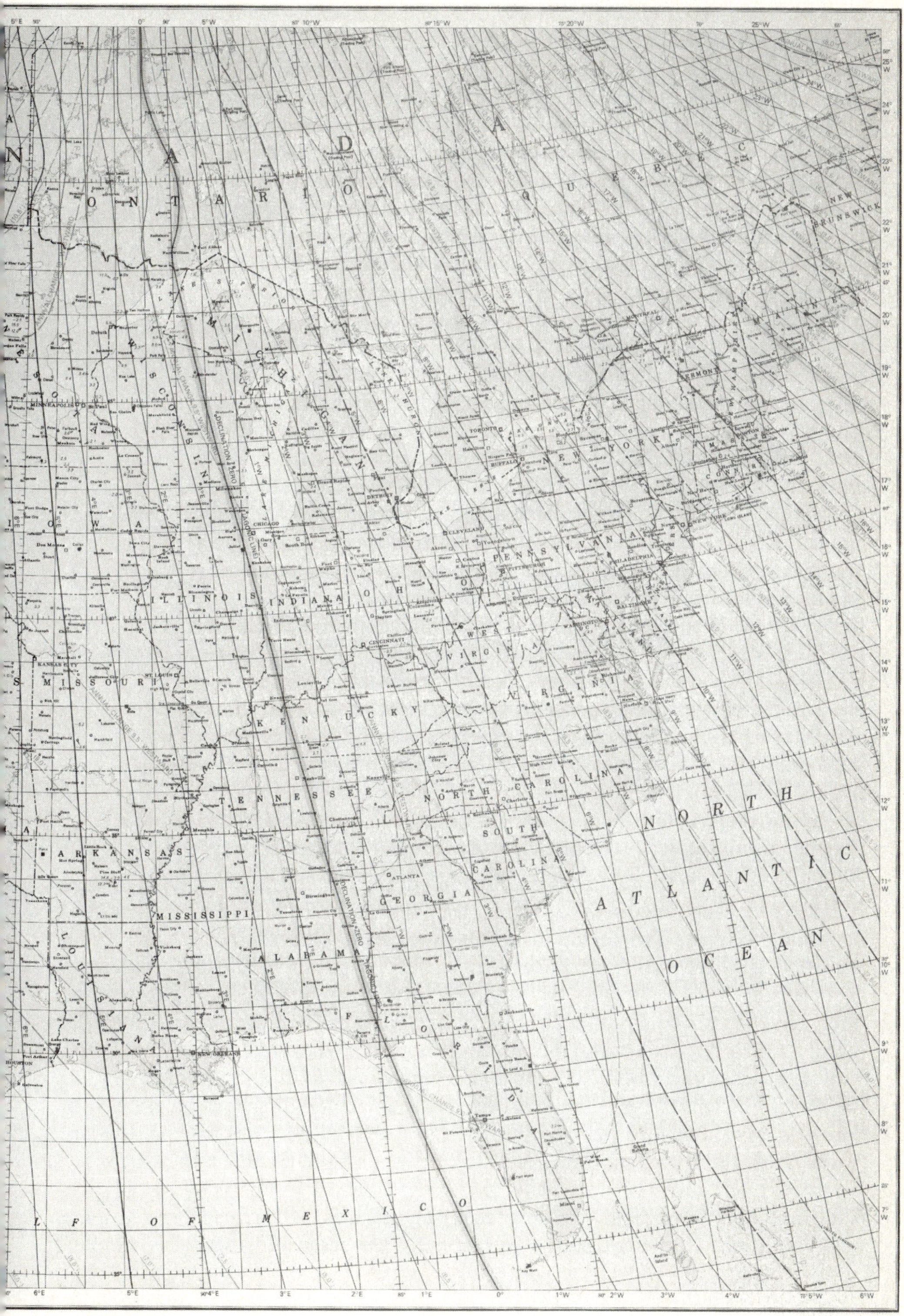
O N T A R I O
Q U E B E C
NEW BRUNSWICK
M A I N E
VERMONT
N E W Y O R K
P E N N S Y L V A N I A
M I C H I G A N
W I S C O N S I N
I O W A
I L L I N O I S
I N D I A N A
O H I O
W E S T V I R G I N I A
V I R G I N I A
K E N T U C K Y
T E N N E S S E E
N O R T H C A R O L I N A
S O U T H C A R O L I N A
G E O R G I A
A L A B A M A
M I S S I S S I P P I
A R K A N S A S
M I S S O U R I
F L O R I D A
N O R T H A T L A N T I C O C E A N
M E X I C O
MINNEAPOLIS
CHICAGO
DETROIT
CLEVELAND
PITTSBURGH
PHILADELPHIA
NEW YORK
BALTIMORE
WASHINGTON
CINCINNATI
ST LOUIS
KANSAS CITY
ATLANTA
NEW ORLEANS
TORONTO
BUFFALO
MONTREAL
DECLINATION ZERO
AGONIC LINE

Fig. 6.21 Cont.

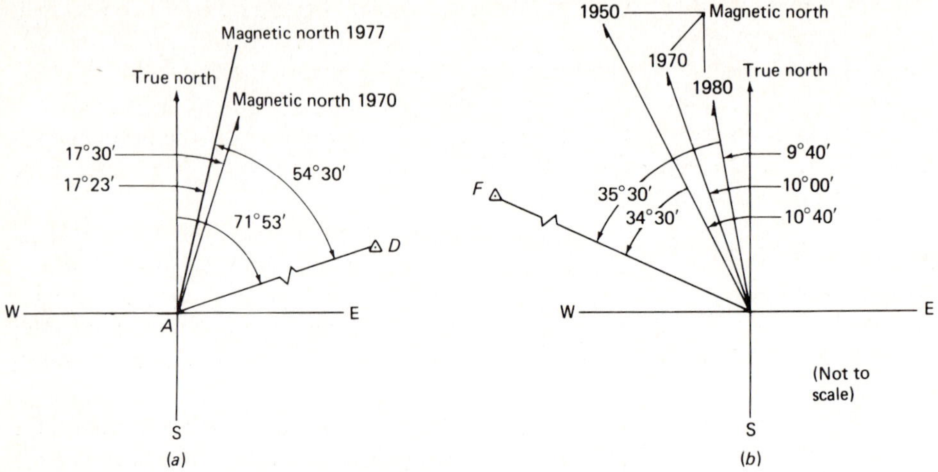

Fig. 6.22 (*a*) and (*b*) Corrections for declination.

eastward. Determine the magnetic bearing that must be used to relocate the direction of the line in the field.

SOLUTION As before, draw a careful sketch of the lines involved, as shown in Fig. 6.22*b*.

Declination in 1970	10°00′W
Change in 20 yr = 20 × 2′	0°40′
Declination 1950	10°40′W
Magnetic bearing 1950	N34°30′W
True bearing of line	N45°10′W
Declination 1980 = 10°W − (10 yr)(2′E)	9°40′W
Magnetic bearing, 1980	N35°30′W

6.18. Local attraction Objects of iron or steel, some kinds of iron ore, and currents of direct electricity alter the direction of the lines of magnetic force in their vicinity and hence are likely to cause the compass needle to deviate from the magnetic meridian. The deviation arising from such local sources is called *local attraction* or *local disturbance*. In certain localities, particularly in cities, its effect is so pronounced as to render the magnetic needle of no value for determining directions. It is not likely to be the same at one point as at another, even though the points may be but a short distance apart. It is even affected by such objects as the steel tape, chaining pins, axe, and small objects of iron or steel that are on the person. Usually, its magnitude can be determined, and directions observed with the compass can be corrected accordingly. Local attraction can usually be detected by observing the compass bearing of a line at two or more points on the line.

6.19. Use of the compass In order that true bearings may be read directly, some compasses, such as the one shown in Fig. 6.20, are designed so that the compass circle may be rotated with respect to the box in which it is mounted. When the circle is in its normal position, the line of sight as defined by the vertical slits in the sight vanes is in line with the N and S points of the compass circle, and the observed bearings are magnetic. If the magnetic declination is set off by means of the circle, the observed bearings will be true, as is evident from Fig. 6.20. If the declination is east, as in the figure, the circle is rotated clockwise with respect to the plate; if the declination is west, counterclockwise.

When the direction of a line is to be determined, the compass is set up on line and is leveled. The needle is released, and the compass is rotated about its vertical axis until a range pole or other object on line is viewed through the slits in the two sight vanes. The eye is placed next to the sight vane at the edge of the compass box marked "S" and when the needle comes to rest, the bearing is read from the north end of the needle.

The following suggestions apply to compass observations: At each observation the compass box should be tapped lightly as the needle comes to rest, so that the needle may swing freely. In order not to confuse the north and south ends of the needle when taking bearings, the observer should always note the position of the counterbalancing wire (which is on the south end in the northern hemisphere). Since the *precision* with which angles may be read depends on the delicacy of the needle, special care should be taken to avoid jarring between the jewel bearing of the needle and the pivot point. *Never move the instrument without making certain that the needle is lifted and clamped.*

Sources of magnetic disturbances such as chaining pins and axe should be kept away from the compass while a reading is being taken. Care should be taken not to produce static charges of electricity by rubbing the glass; a moistened finger pressed against the glass will remove such charges. Ordinarily the amount of metal about the person of the instrumentman is not large enough to deflect the needle appreciably, but a change of position between two readings should be avoided.

Surveying with the compass is usually by traversing. Only alternate stations need be occupied, but a check is secured and local attraction is detected if both a backsight and a foresight are taken from each station. Unlike a transit traverse, in which an error in any angle affects the observed or computed directions of all following lines, an error in the observed bearing of one line in a compass traverse has no effect upon the observed directions of any of the other lines. This is an important advantage, especially in the case of a traverse having many angles. Another advantage of the compass is that obstacles such as trees can be passed readily by offsetting the instrument a short measured distance from the line. The procedure for performing and adjusting a compass traverse is outlined in Art. 8.11.

6.20. Sources of error in compass work Systematic errors in compass observations consist of: (1) bent needle, pivot or sighting vanes; (2) graduated circle not horizontal; and (3) undetected magnetic variations from a great variety of causes. The last type represents the largest and most important source of systematic errors in compass work. Largely because of such variations, the compass, no matter how finely constructed is not a suitable instrument for any except rough surveys.

Deviations due to local attraction can be detected and corrections applied as described in Arts. 6.18 and 8.11.

Common sources of random errors are a sluggish needle and errors in reading the point on the graduated circle where the needle comes to rest. Other conditions being equal, the longer the needle, the smaller the error of observing. With the 6-in needle found on many surveyor's compasses, the estimated standard deviation need not exceed $\pm 10'$; with the $3\frac{1}{2}$- or 4-in needle on the American-designed engineer's transit, the standard deviation is likely to be as much as $15'$.

6.21. Theodolite and transit A theodolite or transit (see Figs. 6.23 and 6.24) is an instrument designed to measure horizontal and vertical angles. It consists of a telescope mounted so as to rotate vertically on a horizontal axis supported by a pair of vertical standards attached to a revolvable circular plate containing a graduated circle for reading horizontal angles. Another graduated arc is attached to one standard so that vertical angles can be observed. Strictly speaking, "theodolite" is the correct word to be used in describing such an instrument. However, through common usage in the United States, the term "transit" has come to be applied to the conventional instrument such as the one shown in Fig. 6.24.

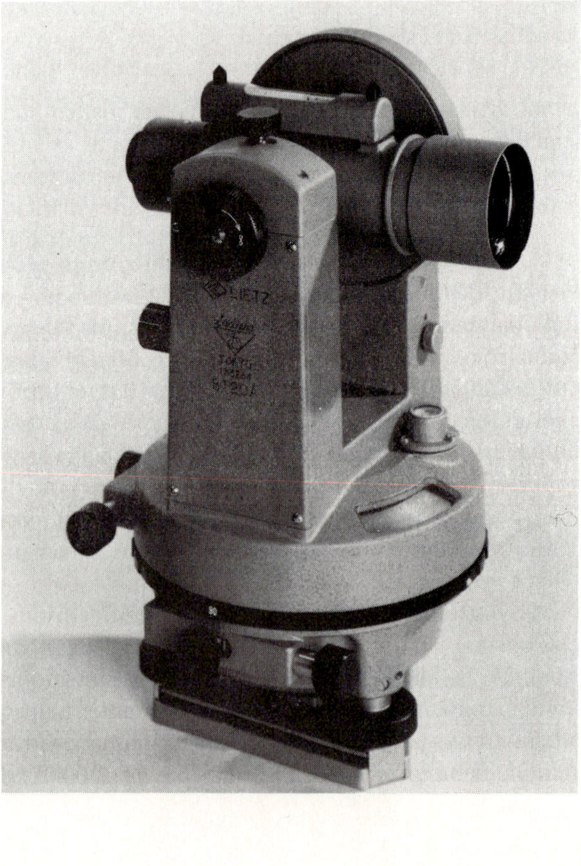

Fig. 6.23 Engineer's transit with three-screw leveling and optical plummet. (*Courtesy Lietz/Sokkisha Co.*)

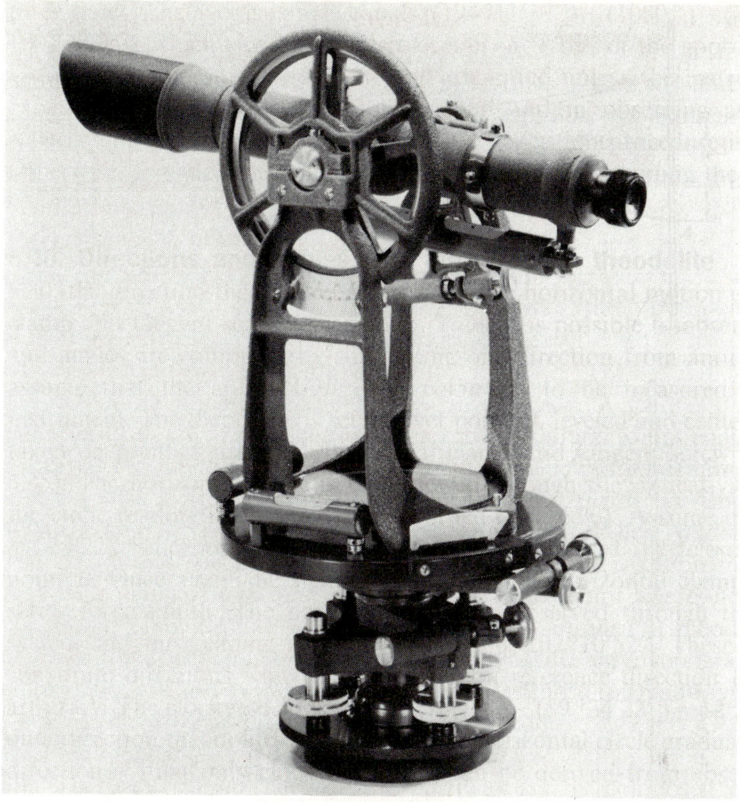

Fig. 6.24 Traditional-style engineer's transit. (*Courtesy Keuffel & Esser Co.*)

Emphasis in the American design has been concentrated on retaining the traditional configuration of the instrument (with four-screw leveling) equipped with metal graduated scales and verniers for direct reading of horizontal and vertical angles using a pocket magnifier. European designers concentrated on developing instruments equipped with graduated scales etched on glass which are viewed by way of optical trains that contain optical micrometers for measuring horizontal and vertical angles. These instruments have three-screw leveling and optical plummets. Regardless of the differences cited, both groups of instruments are designed according to the same fundamental principles.

Instruments of the first group are normally called *engineer's transits* and have been used in the United States for property and construction surveying. In general, instruments of the second group possess higher nominal angular accuracies approaching 1″ of arc. These instruments have been used primarily in control and construction surveys. They are usually designated as *repeating* or *direction theodolites*.

In this chapter all angle-measuring instruments in which the telescope can be reversed will be classified as follows: (1) Repeating instruments with direct-reading verniers. These instruments will be designated *transits* or *engineer's transits*. (2) Double-center instruments with optical scales. These instruments will be referred to as *repeating theodolites* or *optical transits*. (3) Single-center instruments with optical scales. These instruments will be called *direction theodolites*.

6.22. Engineer's transit The engineer's transit is sometimes called the "universal surveying instrument" by reason of the wide variety of uses for which it is adapted. It may be employed for measuring and laying off horizontal angles, directions, vertical angles, differences in elevation, distances by the stadia principle (Art. 7.4), and for prolonging lines.

Some of the modern types of transit differ considerably in design and construction from those long in use, but their essential features do not differ greatly, and their use not at all. Some differences do exist and are typified by the transit illustrated in Fig. 6.23, which has three-screw leveling and an optical plummet. The optical plummet can be found on other engineer's transits, but three-screw leveling is not typical of this style of instrument. In this discussion, the more traditional engineer's transit with four-screw leveling will be discussed first. Figure 6.24 shows such an instrument and Fig. 6.25 illustrates the three main components of a transit. The upper part, called the *upper plate*, *vernier plate*, or *alidade* (not to be confused with the alidade used with the plane table, see Art. 7.19) contains the horizontal circle vernier, the compass box, and the standards supporting the telescope and vertical circle and vernier. The middle part, called the *lower plate* or *horizontal limb*, contains the horizontal graduated circle. The upper and lower plates are fastened, respectively, to vertical inner and outer spindles, the two axes of rotation being coincident with and at the geometric center of the graduated circle. The outer spindle is seated in the tapered socket of the leveling head or lower part in Fig. 6.25. The leveling head assembly contains the leveling screws mentioned before. Near the bottom of the leveling head is a ball-and-socket joint which secures the instrument to the foot plate yet permits rotation of the instrument about the joint as a center.

The outer spindle carrying the lower plate can be clamped in any position with respect to the leveling head by means of the lower *clamp-screw*. Similarly, the inner spindle carrying the upper plate can be clamped in any position with respect to the outer spindle by means of the upper clamp-screw. After either clamp has been tightened, small movements of the spindle can be made by turning the corresponding *tangent screw*. The axis about which the spindles revolve is called the *vertical axis*.

The upper clamp and corresponding tangent screw permit rotating the inner spindle with respect to the outer spindle. This rotation is called the *upper motion* which allows changing the reading on the horizontal circle and the direction of the line of sight. Similarly, the lower clamp and tangent screw permit rotating the outer spindle in the tapered socket of the leveling head. This rotation is called the *lower motion*. Use of the lower motion changes only the direction of the line of sight.

Level tubes, called *plate levels*, are mounted at right angles to each other on the upper plate. Four *leveling screws*, or foot screws, are threaded into the leveling head and bear against the foot plate; when the screws are turned, the instrument is tilted

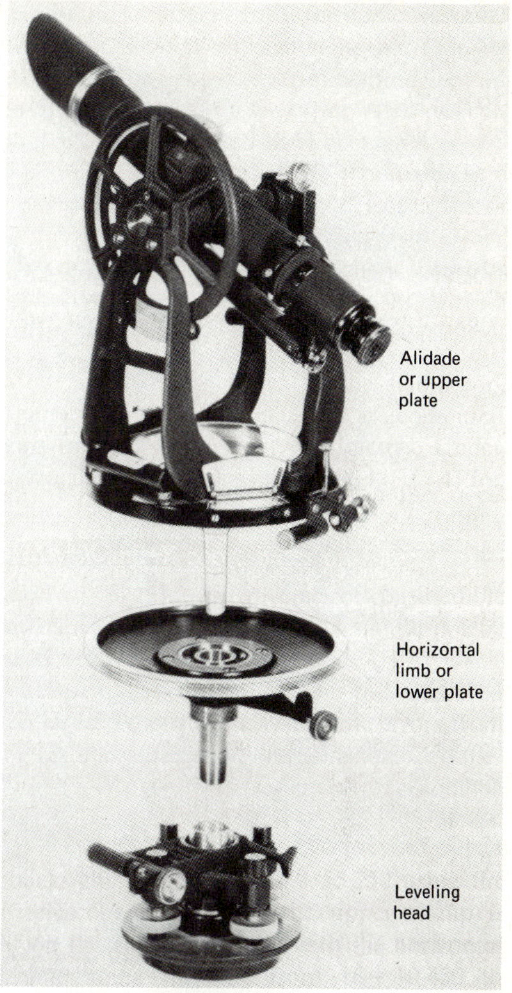

Fig. 6.25 Disassembled view of engineer's transit. (*Photograph by Anderson, Moffit, and Thomas.*)

about the ball-and-socket joint. When all four screws are loosened, pressure between the sliding plate and the foot plate is relieved, and the transit may then be shifted laterally with respect to the foot plate. From the end of the spindle and at the center of curvature of the ball-and-socket joint is suspended a chain with hook for the plumb line. The instrument is mounted on a tripod by screwing the foot plate onto the tripod head.

The spindle design is modified in the transit equipped with an optical plummet. Figure 6.26 shows a vertical section of an instrument similar to the transit shown in Fig. 6.24 but with an optical plummet. Figure 6.27 illustrates the cross-slide arrangement on the tripod designed to accommodate a transit with the optical centering.

The *telescope* is fixed to a transverse *horizontal axis* which rests in bearings on the standards. The telescope may be rotated about this horizontal axis and may be fixed

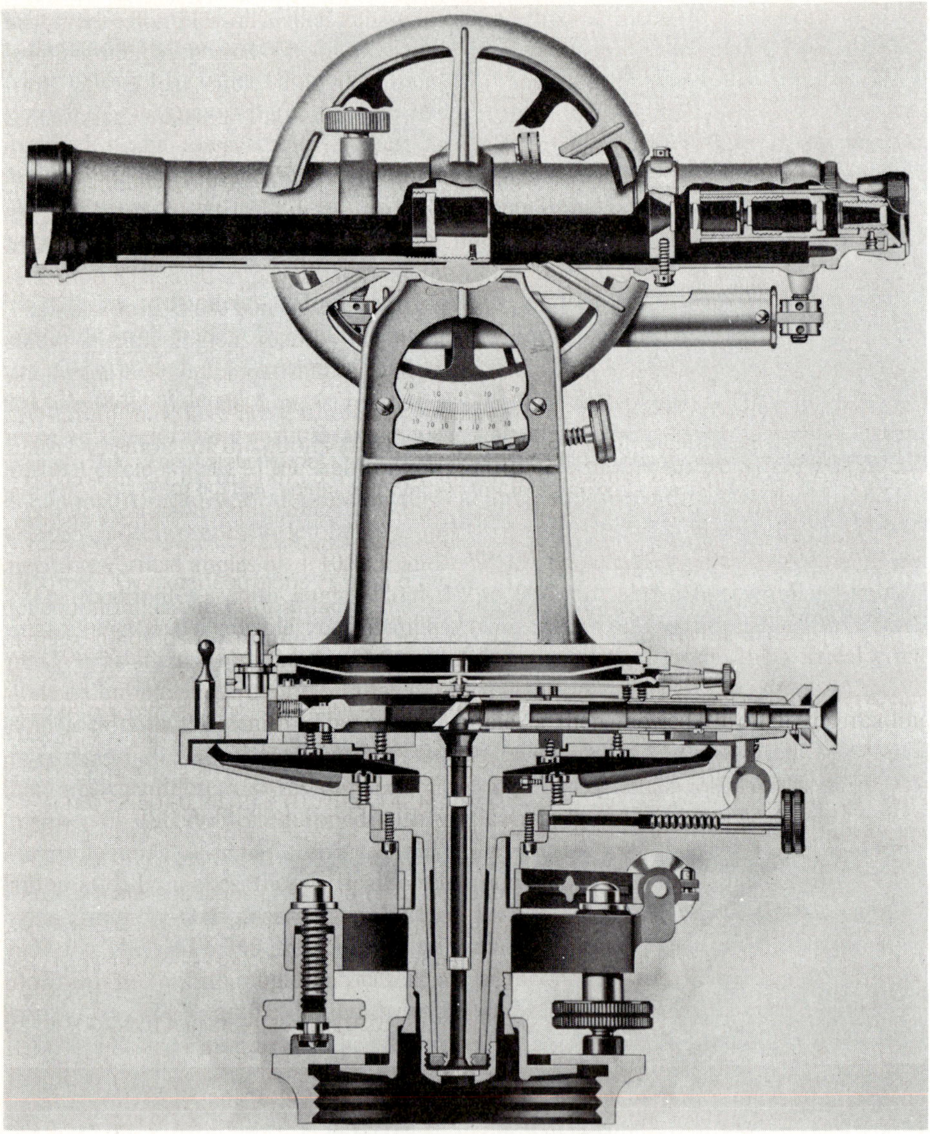

Fig. 6.26 Vertical section, transit with an optical plummet. (*Courtesy Keuffel & Esser Co.*)

in any position in a vertical plane by means of the telescope clamp-screw; small movements about the horizontal axis may then be secured by turning the telescope tangent screw. Fixed to the horizontal axis is the *vertical circle*, and attached to one of the standards is the vertical vernier. Beneath the telescope is the *telescope level tube*.

Attached to the upper plate is the *compass box*. If the compass circle is fixed, its N and S points are in the same vertical plane as the line of sight of the telescope. The compass boxes of some transits are so designed that the compass circle may be

Fig. 6.27 Parallel-shift tripod head for transit with optical plummet. (*Courtesy Keuffel & Esser Co.*)

rotated with respect to the upper plate, so that the magnetic declination may be laid off and true bearings may be read. At the side of the compass box is a screw, or needle lifter, by means of which the magnetic needle may be lifted from its pivot and clamped.

Summing up the main features: (1) The center of the transit can be brought over a given point by loosening the leveling screws and shifting the transit laterally; (2) the instrument can be leveled by means of the plate levels and the leveling screws; (3) the telescope can be rotated about either the horizontal or the vertical axis; (4) when the upper clamp-screw is tightened and the telescope is rotated about the vertical axis, there is no relative movement between the verniers and the horizontal circle; (5) when the lower clamp-screw is tightened and the upper one is loose, a rotation of the telescope about the vertical axis causes the vernier plate to revolve but leaves the horizontal circle fixed in position hence the designation *double-center* instrument; (6) when both upper and lower clamps are tightened, the telescope cannot be rotated about the vertical axis; (7) the telescope can be rotated about the horizontal axis and can be fixed in any direction in a vertical plane by means of the telescope clamp-screws and tangent screws; (8) the telescope can be leveled by means of the telescope level tube, and hence the transit can be employed for direct leveling; (9) by means of direct observation of the vertical circle and vernier, vertical angles can be determined, and hence the transit is suitable for indirect leveling; (10) by means of direct observation of the horizontal circle and vernier, horizontal angles can be measured; and (11) by means of the compass, magnetic bearings can be determined.

6.23. Level tubes The sensitivity of the several spirit levels of the transit should be such as to produce a well-balanced instrument and hence should correspond to the fineness of graduation of the circles and the optical properties of the telescope. If the levels are more sensitive than necessary to maintain this balance, time is wasted in centering the bubbles; if less sensitive than necessary, the precision of measurements is less than it should be for the transit as otherwise designed.

The plate levels of the ordinary transit reading to 01′ usually are alike in sensitivity and have a value of about 60″ per 0.1-in graduation, or about 75″ per 2-mm graduation, of the level tube. When horizontal angles are measured between points nearly in the same horizontal plane, it can be shown that no appreciable error is introduced even if the bubbles are some distance off center. On the other hand, where there is a large difference in vertical angle between the points sighted, a small displacement of the bubble in the tube that is parallel to the horizontal axis causes a relatively large error in the horizontal angle. Some instruments are equipped with a 20″ or 30″ striding level which is employed for leveling the horizontal axis whenever sights are sharply inclined.

The telescope level has a sensitivity of about 30″ per 2-mm graduation, depending upon the magnifying power of the telescope. The sensitivity of the vertical-vernier control bubble should depend upon the least reading of the vernier; for a vertical circle reading to 01′, a level tube having a sensitivity of 40″ per 2-mm graduation is commonly employed. For further details concerning level tubes, see Arts. 5.14 and 5.15.

6.24. Telescope The telescope of the transit is similar to that of the engineer's level (Art. 5.8). Assuming a perfectly adjusted instrument, when the transit is used for either direct or trigonometric leveling, any point on the horizontal cross hair may be used in sighting; when the transit is used for establishing lines, measuring angles, or taking bearings, any point on the vertical cross hair may be used. However, in practice the instrument may not be in perfect adjustment and it is always wise to use the intersection of the horizontal and vertical cross hairs for sighting in all situations.

Most instruments are equipped with stadia hairs (Chap. 7), which are usually mounted in the same plane with the cross hairs. The magnifying power is usually 20 to 28 diameters. For the transit, as for the level, usually the erecting eyepiece is employed; but the superior optical properties of the inverting eyepiece make it the favorite of some surveyors and it is the type used in instruments of precision. The instruments illustrated in Figs. 6.23 and 6.24 have erecting eyepieces. For the relative merits of inverting and erecting eyepieces, see Art. 5.12. Telescopes on modern transits are usually of the internal focusing type (Art. 5.8).

6.25. Graduated circles The vertical circle has two opposite zero points and is graduated usually in half-degrees, the numbers increasing to 90° in both directions from the zero points, as illustrated in Fig. 6.28*a*. When the telescope is level, the index of the vernier is at 0°.

The horizontal circle is likewise usually graduated in half-degrees, but may be graduated in third-degrees or quarter-degrees. It may be numbered from 0° to 360°

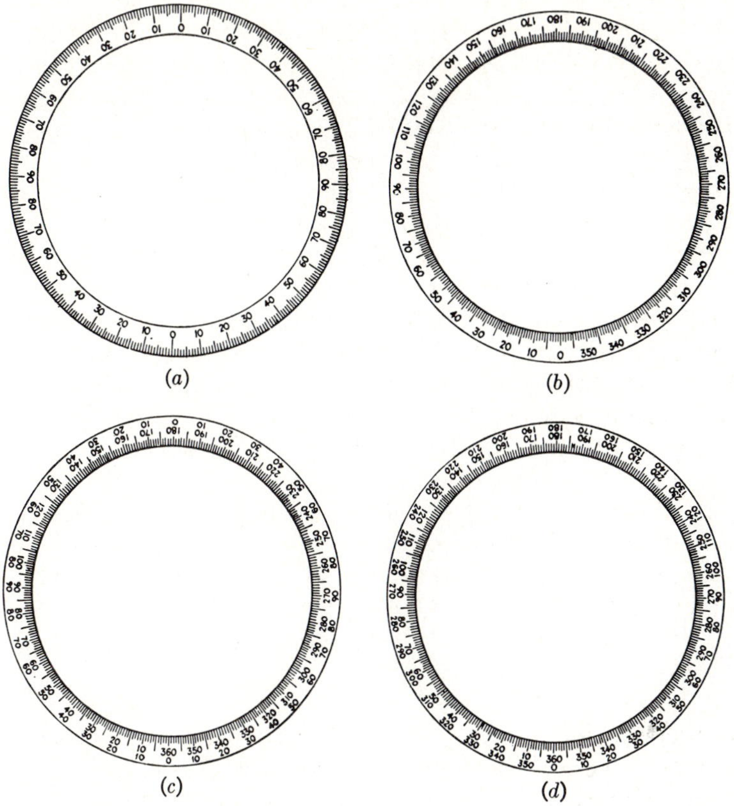

Fig. 6.28 Numbering of circles. (*a*) Vertical circle numbered in quadrants. (*b*) Horizontal circle numbered 0 to 360. (*c*) Horizontal circle numbered 0 to 360 and in quadrants. (*d*) Horizontal circle numbered 0 to 360 and 360 to 0.

clockwise (Fig. 6.28*b*), 0° to 360° clockwise and 0° to 90° in quadrants (Fig. 6.28*c*), or 0° to 360° in both directions (Fig. 6.28*d*). Most surveyors prefer the numbering system illustrated in Fig. 6.28*d*. Usually, the numbers slope in the direction of reading.

The horizontal circles of transits designed for work of moderately high precision are graduated to 20′ or to 15′. Those for repeating theodolites are often graduated to 10′. In special cases, circles may be graduated in divisions based on the *mil* or the *grad*.

6.26. Verniers The verniers employed for reading the horizontal and vertical circles of the transit are identical in principle with those for the level rod target, described in Art. 5.26. Practically all transit verniers are of the direct type.

Figure 6.29*a* shows the usual type of double direct vernier reading to minutes. The circle is graduated to half-degrees and the vernier contains 30 divisions on either side of the index. One division on the vernier is $\frac{1}{30}$ less than the least division (half-degree

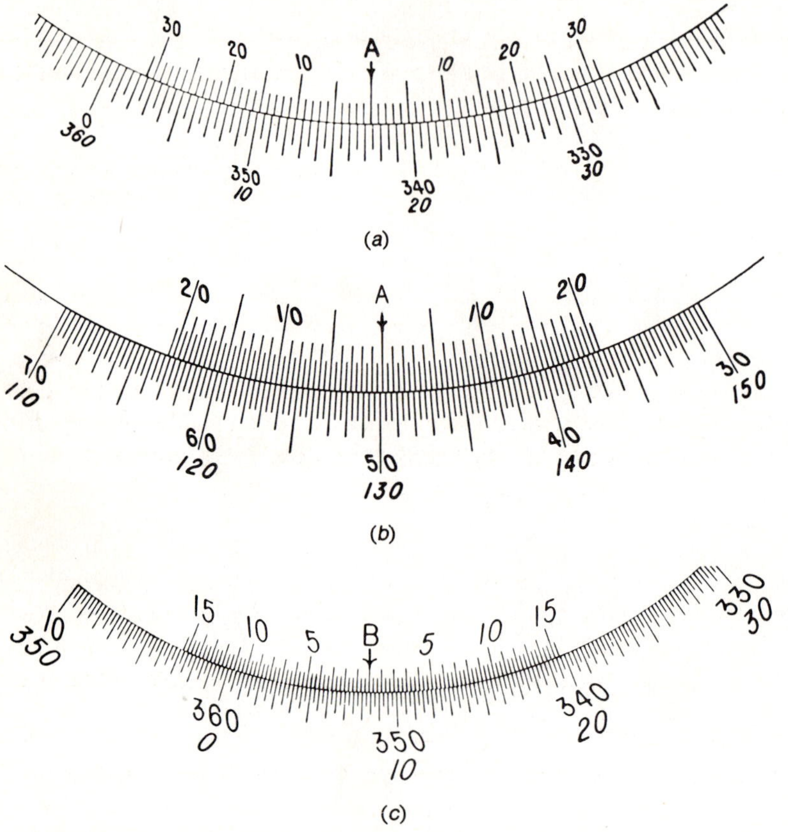

Fig. 6.29 Vernier styles. (*a*) One-minute double direct vernier. (*b*) Thirty-second double direct vernier. (*c*) Twenty-second double direct vernier. (*Courtesy Keuffel & Esser Co.*)

interval) on the main scale, so that 30 divisions on the vernier cover 29 divisions on the main scale. Thus, the least count is 30′/30, or 1′. In Fig. 6.29*a*, the clockwise angle is 342°30′ + 05′ (on the vernier reading to the left of the index) = 342°35′. Reading the counterclockwise circle and the vernier to the right of the index yields an angle of 17°25′.

Figure 6.29*b* shows a double direct vernier on a circle with a least division of 20′. The vernier contains 40 divisions which cover 39 divisions on the circle. The least count is $20'/40 = \frac{1}{2}'$ or 30″. The clockwise angle is 49°40′ + 10′30″ = 49°50′30″. The counterclockwise angle is 130° + 9′30″ = 130°09′30″. The reader should attempt to determine the least count and the clockwise and counterclockwise angles illustrated by Fig. 6.29*c*.

6.27. Eccentricity of verniers and centers Transits have two verniers for reading the horizontal circle, their indexes being 180° apart. The one nearest the upper clamp and tangent screw is known as the *A* vernier, and the one opposite is known as the *B* vernier. The verniers are attached to the upper plate and are adjusted by the instrument maker so that they are much nearer to being truly 180° apart than their least count. Failure of the two verniers to register readings exactly 180° apart on the circle may be due to either or both of the two following causes:

1. Eccentricity of verniers The verniers may have become displaced, so that a line joining their indexes does not pass through the center of rotation of the upper plate. The error will be the same for all parts of the graduated circle.

2. Eccentricity of centers The spindles may have become worn or otherwise damaged, so that the center of rotation of the upper plate does not coincide with the geometrical center of the graduated horizontal circle. The error varies according to the position of the verniers with respect to the horizontal plate. As indicated by Fig. 6.30, there will be one setting on the graduated circle for which the indexes are exactly 180° apart (first position), and 90° therefrom there will be another setting for which the verniers fail to register 180° apart by a maximum amount (second position).

To correct either of these defects requires the services of an instrument maker, but neither defect limits the precision with which angles can be measured. Taking the mean of the two vernier readings (*A* and *B* verniers) eliminates errors due to either or both types of eccentricity. Furthermore, if the verniers only are eccentric, no error is introduced in an angle as long as the same vernier is used for making the final reading as for making the initial setting.

6.28. Transit with three level screws The engineer's transit illustrated in Fig. 6.23 is fundamentally similar to the transit described in Arts. 6.22 to 6.27 with certain specific exceptions. There is no vertical spindle as such. Double centering is achieved by means of the black positioning ring attached to the horizontal circle, which permits the horizontal circle to be rotated with respect to the alidade that contains the verniers. A lower lock and lower tangent screw serve the same function as on a conventional transit. A circular or bull's eye level bubble is provided for approximate leveling and a single-level tube is used for precise leveling. The

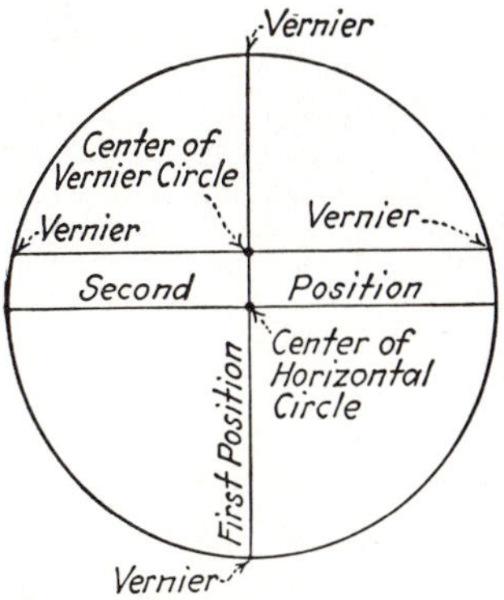

Fig. 6.30 Eccentricity of verniers.

three-screw leveling and lack of a vertical spindle mean that an optical plummet is very easily adapted. Consequently, this instrument contains a combination of features found on both the traditional engineer's transit and the optical theodolites.

6.29. Geometry of the transit A knowledge of the geometry of the transit is necessary for a thorough understanding of its operation. Figure 6.31 shows the geometry of the transit in isometric, schematic form. The intersecting lines labeled A define the plane of the plate level tubes; the vertical axis is designated as B; the horizontal axis is C; the optical line of sight or collimation axis is D; and the axis of the spirit level attached to the telescope is E. In a perfectly constructed and adjusted instrument, the following relationships should exist among these axes: B is perpendicular to A; C is perpendicular to B; D is perpendicular to C; E is parallel to D; and B, C, and D pass through a single point. Achievement of these relationships by adjustment is detailed in Arts. 6.50 and 6.51.

6.30. Theodolites Repeating and direction theodolites have features that are common to both types of instruments. Characteristics of modern optical theodolites will be discussed before pointing out the differences between repeating and direction models.

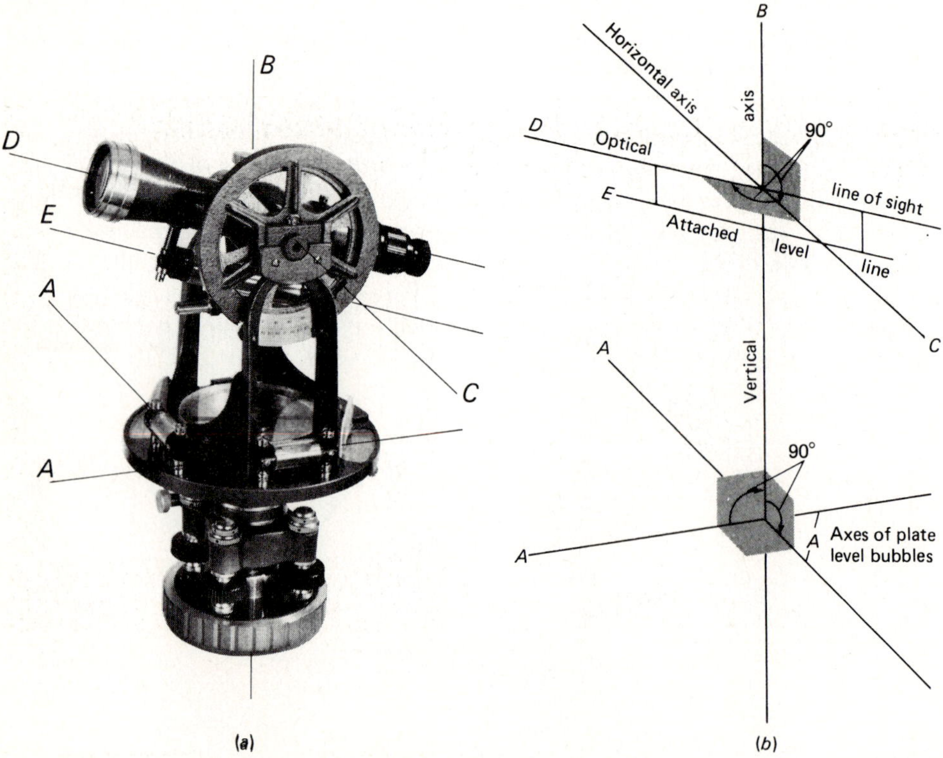

Fig. 6.31 (*a*) and (*b*) Geometry of the transit.

Although there are differences as among the various manufacturers, certain general features of theodolites are fairly common. The weight (without tripod) is about 10 lb or 5 kg as compared with about 15 lb or 7 kg for the conventional American transit; the difference is due both to compact size and to the use of light materials in manufacture. The finish is light in color to minimize temperature effects of sunlight. There is no vernier plate, and the center is cylindrical and rotates on ball bearings. A circular bubble of moderate sensitivity is used for approximate leveling, while a single plate level bubble of high sensitivity serves for precise leveling. The telescope level is tubular and of high sensitivity. There are three leveling screws. An optical plummet is used instead of a plumb line. All motions are enclosed, so that the instrument is dustproof and moistureproof. Interior lighting is accomplished by means of mirrors and prisms, and provision is made for night lighting. The only field adjustments are those of the level bubble. These adjustments are similar to those for the conventional American transit.

The telescope of the theodolite is short and can be rotated completely about the horizontal axis. It is of the internal-focusing type, so that the stadia constant (Art. 7.5) is zero. The reticule is of glass with etched crosslines, and the stadia interval factor (Art. 7.6) is fixed at 100.0. For European use the eyepiece is of the inverting type; for American use it is usually of the erecting type.

Figure 6.32 shows a repeating theodolite; Figs. 6.33 and 6.34 illustrate direction theodolites.

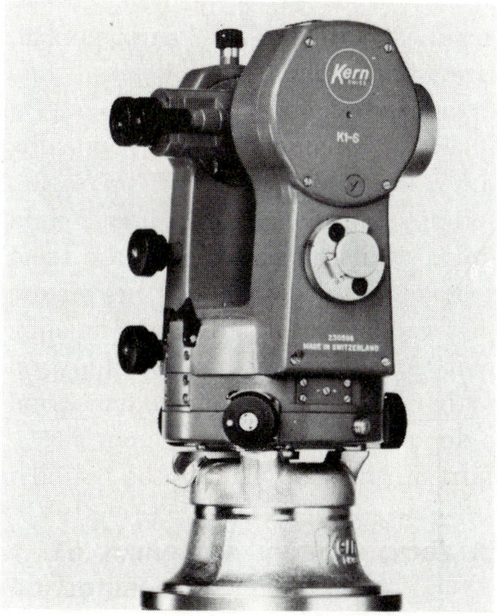

(*a*)

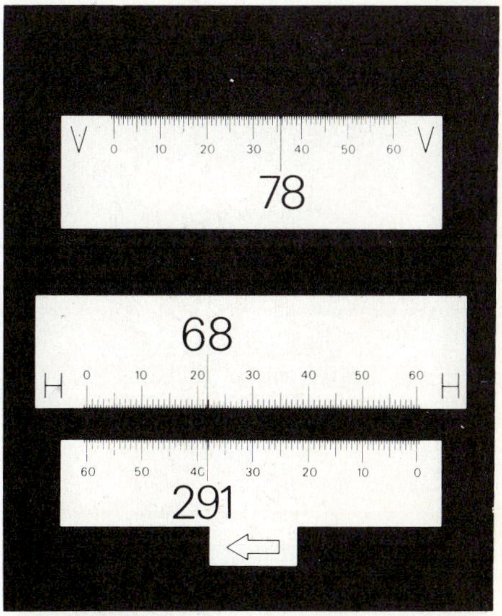

Vertical circle reading	78°35.7′
Horizontal clockwise circle	68°21.8′
Horizontal counterclockwise circle	291°38.2′

(*b*)

Fig. 6.32 (*a*) Repeating theodolite. (*b*) Vertical and horizontal circle readings. (*Courtesy Kern Instruments, Inc.*)

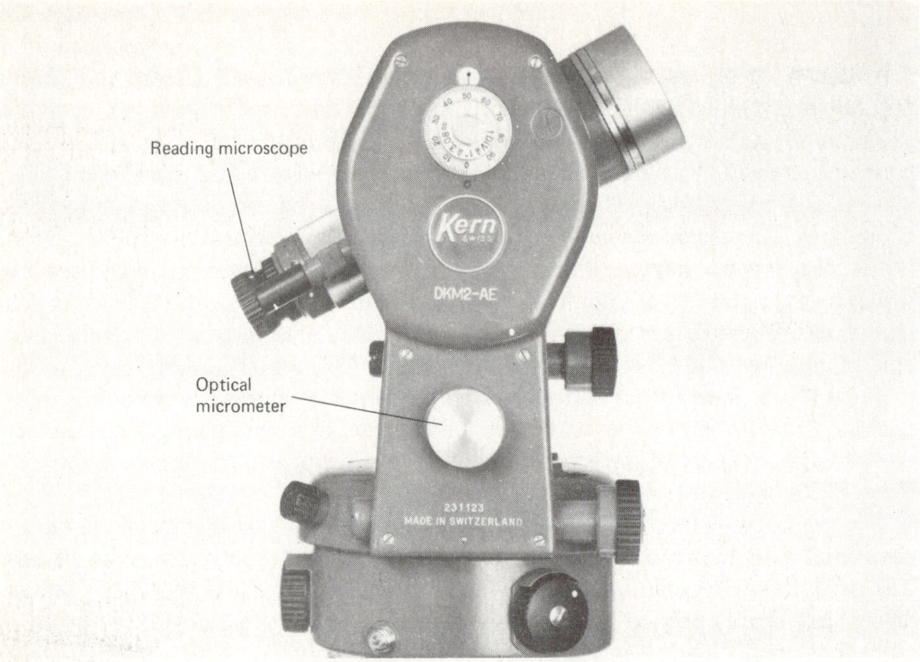

(*a*)

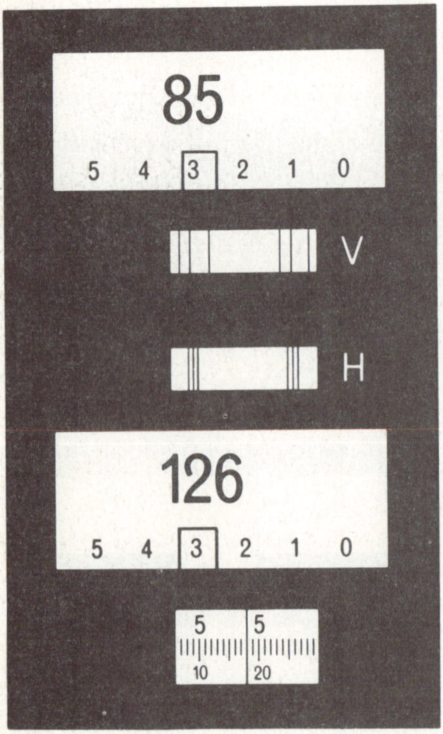

(*b*)

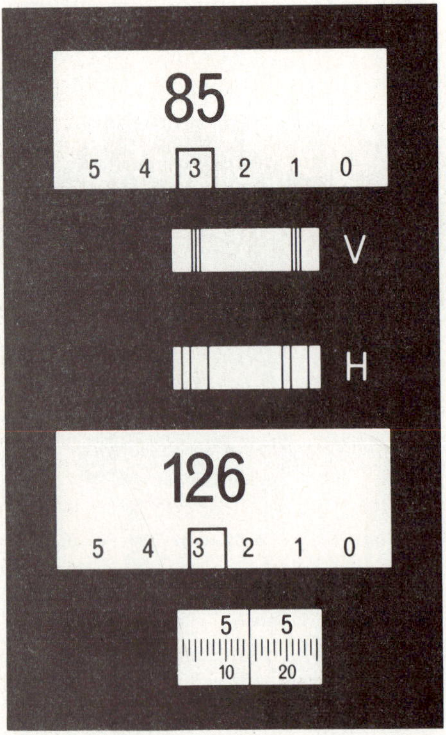

(*c*)

Fig. 6.33 (*a*) Direction theodolite, Kern DKM2-AE. (*b*) Horizontal circle reading. (*c*) Vertical circle reading yielding a zenith angle in degrees. (*Courtesy Kern Instruments, Inc.*)

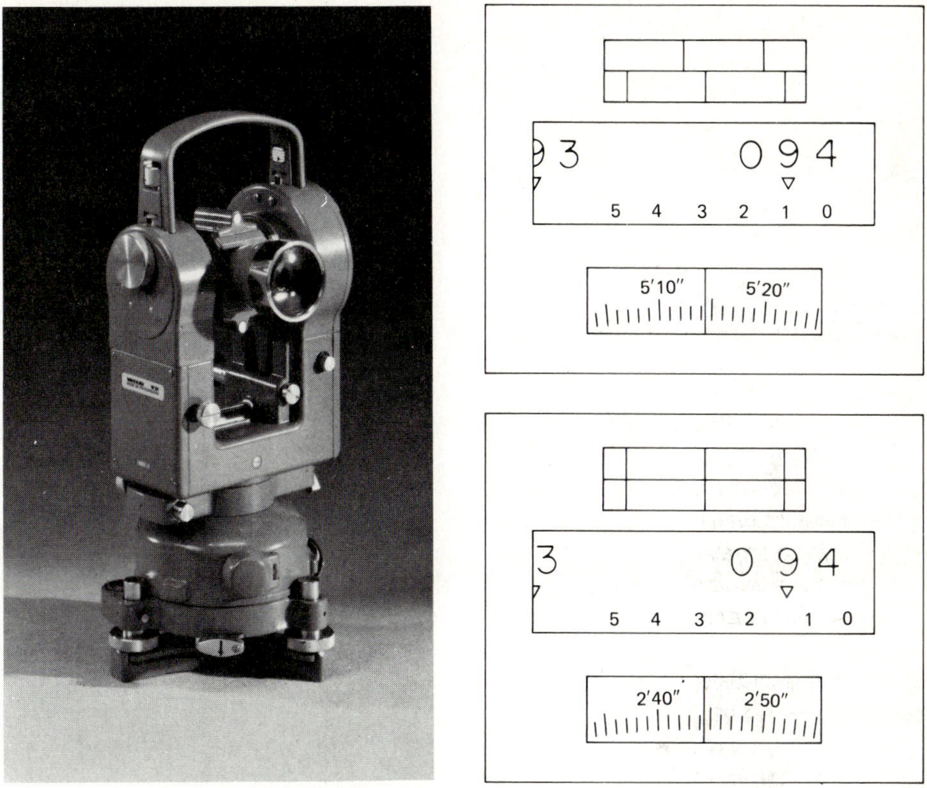

(*a*) (*b*)

Fig. 6.34 (*a*) Wild theodolite, T2. (*b*) Horizontal circle reading. (*Courtesy Wild Heerbrugg Instruments, Inc.*)

The graduated horizontal and vertical circles are of glass and are relatively small; they are viewed simultaneously from the eye end of the telescope by means of a system of microscopes, prisms, and mirrors. Figure 6.35 shows the optical train of a direction theodolite. The horizontal circle is read by means of an optical micrometer, in accordance with systems devised by the various manufacturers. Each reading is obtained by one observation which is the average of two readings at opposite sides of the circle and which is therefore free of errors due to eccentricity. Direction theodolites are usually read directly to 01″. Repeating theodolites are read directly to 20″ or 01′ and by estimation to one-tenth the corresponding direct reading.

Most manufacturers produce mounting brackets or tribrachs on which the theodolite, a sighting target, an EDM instrument, or subtense bar may be mounted interchangeably.

The repeating theodolite has all the features described above plus a lower lock and tangent screw which allow angles to be set and repeated. Thus, it is a double-center instrument with the same capabilities as the engineer's transit. The repeating theodolite illustrated in Fig. 6.32*a* allows reading horizontal and zenith angles directly to minutes and estimating to tenths of minutes.

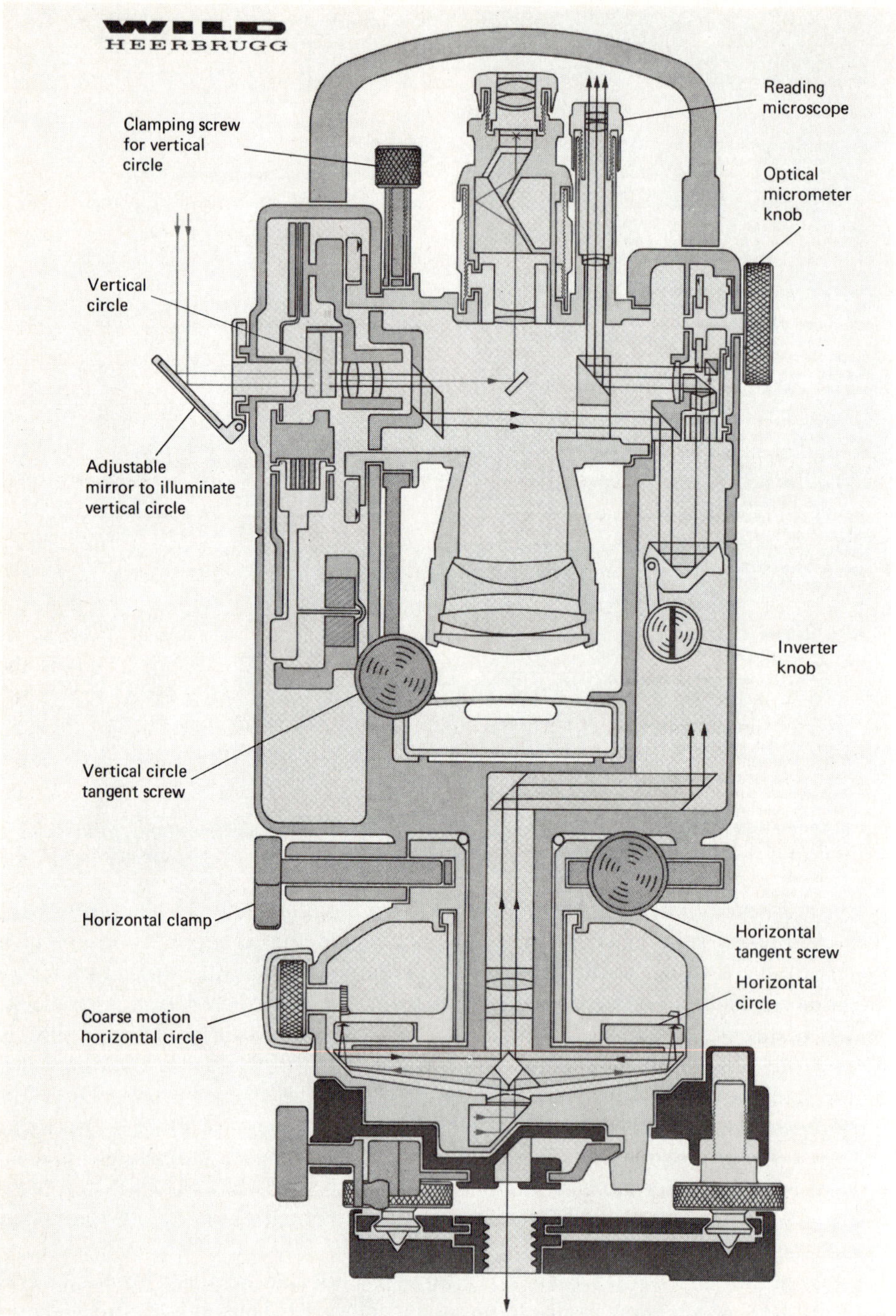

Fig. 6.35 Optical train of a direction theodolite, Wild T2. (*Courtesy Wild Heerbrugg Instruments, Inc.*)

Direction theodolites have only a single horizontal motion in contrast to the double-center instruments. *Directions* are observed with these theodolites and the angle is the difference between the two directions (Art. 6.40). In general, direction instruments are more precise than are repeating theodolites. The plate level bubble has a sensitivity of 20″ per 2-mm division, and the vertical circle, on instruments of recent design, has an automatic compensator setting accuracy of from 0.3′ to 0.5′. The direction theodolite illustrated in Fig. 6.33*a* has a glass horizontal circle divided into 20′ intervals. The micrometer permits reading both horizontal and zenith angles directly to 1″ and estimating to 0.2″.

The method of observing horizontal and vertical angles on both repeating and direction theodolites involves viewing through a reading microscope mounted parallel to the telescope (Fig. 6.33*a*). By means of an optical train (Fig. 6.35), the observer's eye can be focused on the horizontal or vertical circle. The optical train is illuminated by adjustable mirrors (Fig. 6.35) for daylight operations. These mirrors can be replaced by attachments with light bulbs for night observations.

The procedure for reading angles varies with the make of instrument. First, the eyepiece of the reading microscope should be focused to the observer's eye so that a sharp image free of parallax is presented. The field of view visible when looking through the reading microscope on the repeating theodolite in Fig. 6.32*a* is illustrated in Fig. 6.32*b*. Three illuminated scales in separate frames are visible: the scale at the top, labeled *V*, shows the zenith angle and can be read directly to minutes and estimated to tenths of minutes; the second scale, labeled *H*, is the horizontal clockwise circle; and the bottom scale, with the arrow to the left, is the horizontal counterclockwise circle reading. Each of the three scales can be read directly to degrees and minutes, and estimated to tenths of minutes. The circle readings for this example are indicated on Fig. 6.32*b*. On theodolites of older design, the control spirit level attached to the vertical circle must be centered before observing the vertical or zenith angle. The instrument illustrated in Fig. 6.32*a* has automatic vertical circle indexing, so that this operation is not necessary.

A direction theodolite in which the horizontal and vertical circles can be read directly to seconds is illustrated in Fig. 6.33. The scales visible when viewing through the reading microscope are shown in Fig. 6.33*b* and *c*. The uppermost frame contains the vertical circle reading in degrees and tens of minutes. The large numbers are degrees and the small numerals are the number of tens of minutes. In Fig. 6.33*b*, the approximate vertical circle reading is 85°30′. Immediately below is a much smaller frame, labeled *V*, which contains a pair of split graduations and two index marks which are controlled by the optical micrometer. These index marks must be centered between respective split graduations in order to observe vertical circle readings to seconds. Below this *V* coincidence scale is a similar coincidence scale, labeled *H*, for horizontal circle observations. Note that index marks are centered between their respective split graduations. Thus, Fig. 6.33*b* illustrates a horizontal circle reading. The large frame at the bottom contains horizontal circle readings in degrees and tens of minutes. The smaller scale at the very bottom contains horizontal or vertical circle readings to minutes and seconds. Readings can be estimated to tenths of seconds. First, coincidence of the index mark and the appropriate split graduations is obtained by rotating the optical micrometer. The horizontal circle reading is 126°30′ plus 05′18″, or 126°35′18.0″. The 05′18.0″ is read directly from the lowest scale. A similar array is shown in Fig. 6.33*c*, where the index marks are centered on respective split graduations in the *V* scale. Here the vertical circle reading is 85°30′ from the topmost frame plus 05′14.0″ from the lowest scale, or 85°35′14.0″. This reading is a zenith angle. In each case the reading to minutes and seconds read from the lowest scale is obtained by rotating the optical micrometer and centering the index marks between split graduations on either the *V* or *H* scale, depending on which circle observation is desired.

Another style of direction theodolite with a different scheme for obtaining circle observations is shown in Fig. 6.34. The field of view visible when looking through the reading microscope is illustrated before and after coincidence in Fig. 6.34*b*. In the top diagram, before coincidence, the top frame contains a scale that shows degrees and minutes to the nearest 10′, while the bottom frame has a scale graduated in minutes and seconds. The horizontal and vertical circles are divided into 20′ intervals and the optical micrometer is employed to measure directly to seconds. Before coincidence, the horizontal circle reading is estimated as 94°10′. Coincidence is obtained by rotating the micrometer knob until the vertical marks in the top scale coincide. This rotation causes the pointer in the middle scale to indicate the nearest 10′, while the bottom scale provides minutes and seconds. The setting shown on the bottom of Fig. 6.34*b* yields a horizontal circle reading of 94°10′ plus 02′44.4″, or 94°12′44.4″. Horizontal circle readings are made with the indentation on the inverter knob (Fig. 6.35) in a horizontal position. Vertical circle readings are obtained in exactly the same way but with the inverter knob rotated 90° so that the indentation is in a vertical position. Turning this knob rotates a reflective prism so that the optical path is directed toward either the horizontal or vertical circle. In Fig. 6.35, illustrating the optical train, this knob is set to permit viewing the vertical circle.

When making vertical circle observations, due regard must be given to taking into account instrumental vertical circle index error. On engineer's transits, where zero corresponds to the horizontal, the index error is noted and applied to observed vertical angles (see Art. 6.45). The procedure is different in theodolites in which vertical circles are oriented with zero at the zenith. In order that vertical circle readings be referred to the true zenith, theodolites of the most recent design have automatic vertical circle indexing. One method of achieving automatic indexing is to have a prism or pairs of prisms suspended by a pendulum apparatus incorporated into the ray path of the optical-reading microscope for vertical circle observations. Figures 6.36 and 6.37 show in schematic form, how typical V types and X types of pendulum compensators operate.

Another style of theodolite utilizes a special colorless liquid in a sealed glass container as a compensator. This unit is mounted at the top of the telescope standard, where it reflects light in the optical path for reading the vertical circle. When the theodolite is perfectly leveled, and zero on the vertical circle coincides with the zenith, the optical path of the reading rays forms an isosceles triangle as shown in Fig. 6.38*a*. In this case the vertical circle reading yields the correct angle referred to the zenith. When the instrumental vertical axis does not coincide with the zenith, the free surface of the liquid in the compensator remains horizontal and the rays in

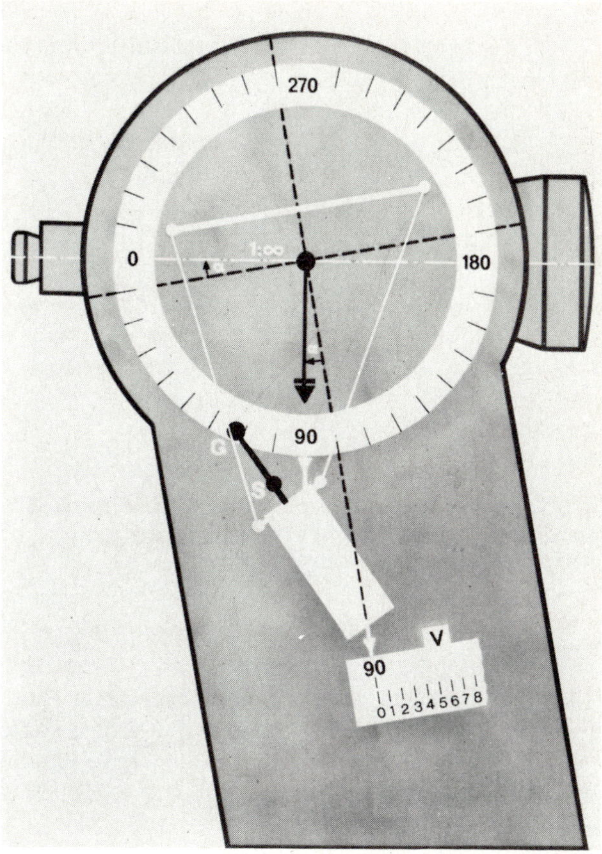

Fig. 6.36 Pendulum prism V-type compensator. (*Courtesy Keuffel & Esser Co.*)

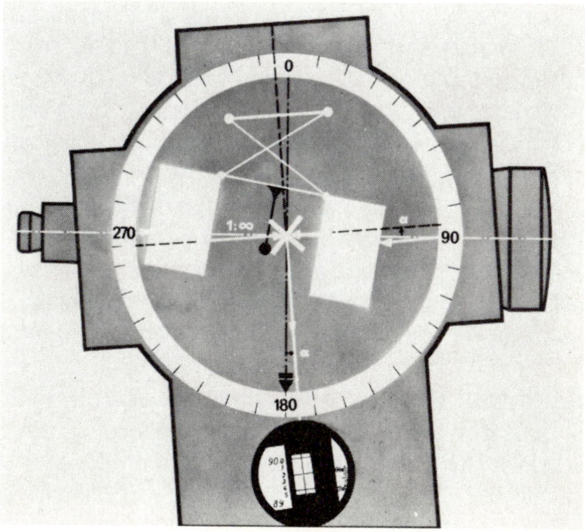

Fig. 6.37 Pendulum prism X-type compensator. (*Courtesy Keuffel & Esser Co.*)

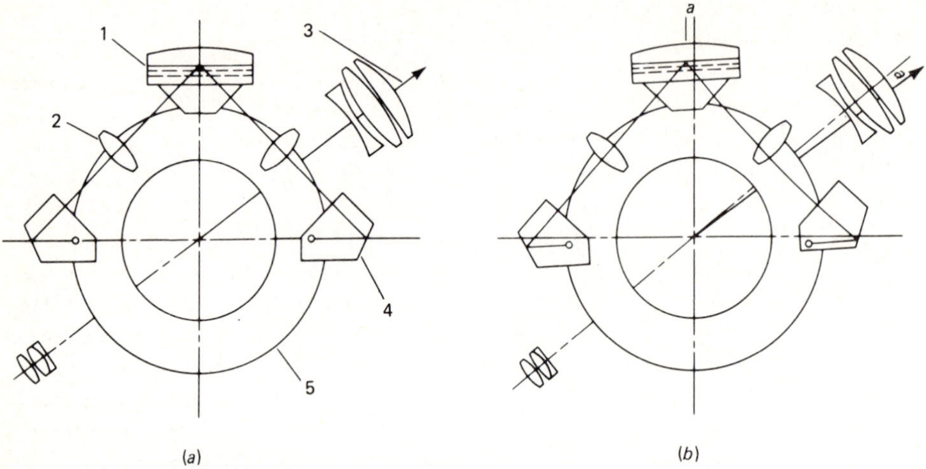

Fig. 6.38 Optical path of automatic index compensator, Kern DKM2A. (*a*) Vertical axis plumb: (1) liquid reservoir; (2) objective; (3) optical line of sight; (4) deviating prism; (5) vertical circle. (*b*) Vertical axis not plumb. (*Courtesy Kern Instruments, Inc.*)

the optical path are deviated, as illustrated in Fig. 6.38*b*. The optics of the reading system are designed so that this deviation is observed as an equal amount acting in the opposite direction, as indicated by Fig. 6.38*b*. Thus, the vertical circle reading is referred to the true zenith even though the vertical axis is not truly vertical.

All of the several types of compensators have the same function, to refer vertical circle readings to the direction of gravity regardless of small inclinations of the vertical axis of the instrument.

Theodolites of earlier design do not contain automatic vertical circle indexing but do have sensitive spirit levels attached to the vertical circle so that zero on this circle can be made to correspond with the zenith. This bubble usually has a sensitivity of 20″ per 2-mm division and is viewed by a prism arrangement which allows split bubble coincidence. Assuming that the bubble is in adjustment, no index error results if it is centered each time a vertical circle observation is made.

Accuracies possible using either automatic index compensation or the level bubble are comparable assuming an instrument in good adjustment and careful centering of the spirit level in the latter case. Automatic index compensation does reduce the time required for observation and also relieves the strain on the operator.

All optical theodolites have optical plummets, which consist of a vertical collimation line that coincides with the instrumental vertical axis. The eyepiece, which allows viewing along this collimation line, is usually located near the base of the instrument. To center the theodolite over a station with the optical plummet, the instrument must first be leveled, then centered and again releveled. This process is repeated until the plate level bubble is centered, making the instrumental vertical axis vertical so that the cross in the vertical collimation line coincides with the point over which the theodolite is being centered. The optical plummet permits centering to within ± 0.5 mm and is immune to the effects of wind.

Another method for centering which utilizes a telescopic centering rod attached directly to the leveling head is shown in Fig. 6.39. Note that the centering tripod does not support the instrument directly but is attached to a shifting tripod head, the upper portion of which has a spherical surface (Fig. 6.39*a*). This spherical surface supports a plate on which the instrument is clamped. The plate is perpendicular to and is rigidly connected to the centering rod. An adjustable circular level bubble is attached to the centering rod. A clamping grip holds the centering rod against the lower surface of the tripod head. In order to make a setup, the tripod is set over the station and with the clamping grip loose, the lower pointed end of the rod is placed on the station point. Next, the legs of the tripod are adjusted or moved to roughly center the circular level bubble on the centering rod. Force the legs of the tripod firmly into the ground to complete this rough centering. Now, shift the head on the tripod plate until the circular bubble is accurately centered. The centering rod is vertical and the tripod plate is horizontal. Tighten the clamping grip and place the theodolite on the support plate, fastening it securely with the lever provided for that purpose. The instrument is then leveled accurately using the level screws to center the plate level bubble and the setup is completed. The theodolite can be released from the support plate simply by releasing the clamping lever. Thus, it is very

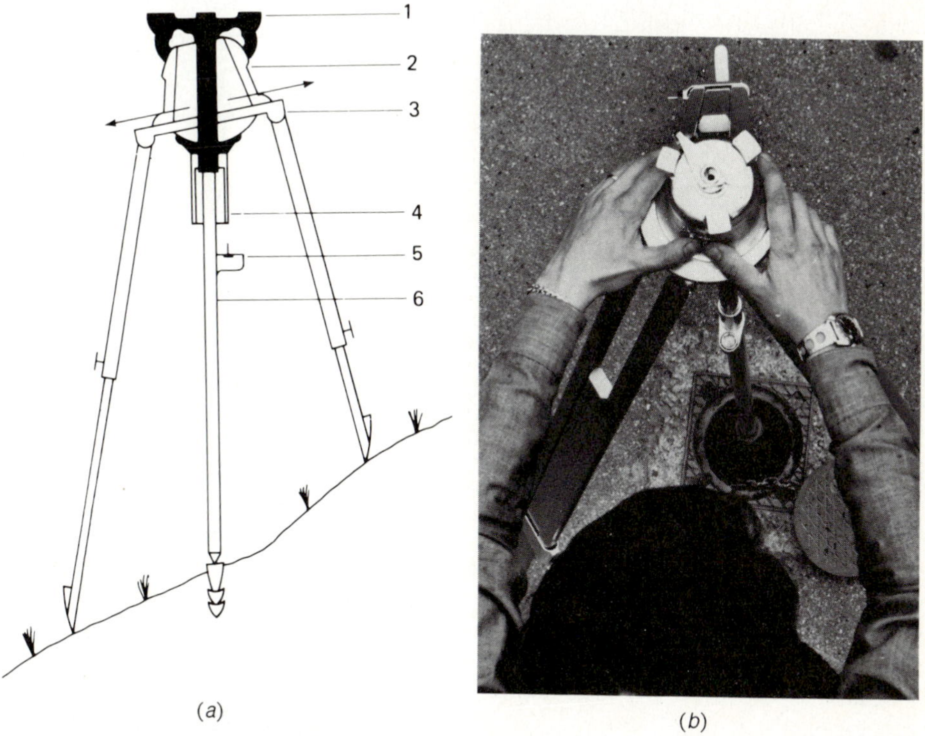

Fig. 6.39 (*a*) Centering tripod with telescopic plumbing rod: (1) supporting plate; (2) tripod head with spherical surface; (3) tripod plate; (4) clamping grip; (5) circular level; (6) centering rod. (*b*) Centering with telescoping plumbing rod. (*Courtesy Kern Instruments, Inc.*)

convenient to replace the theodolite with a target (see Fig. 8.43) or EDM reflector (Figs. 9.10*a*, *b*) without disturbing the setup. This procedure is called *forced centering*.

6.31. Digitized theodolites Theodolites containing digitized horizontal and vertical scales are now on the market. Figure 6.40 illustrates a digitized direction theodolite. Horizontal and vertical angles are output from this instrument with a least count of 3″.

One method of digitizing angular output is to use a glass encoder disk such as the one shown in Fig. 6.41. This disk contains a metallic film pattern deposited on the surface. Illuminators are placed on one side of the disk, which are opposite to photodiodes on the other side of the disk.

Several sets of illuminators and diodes are placed at specific positions about the circle. Each diode senses the collimated illumination transmitted through the pattern. The amount of current generated by each photodiode is a function of the amount of light passing through the circle and hence depends on the position of the circle. The impulses generated by the photodiodes are converted by an on-board

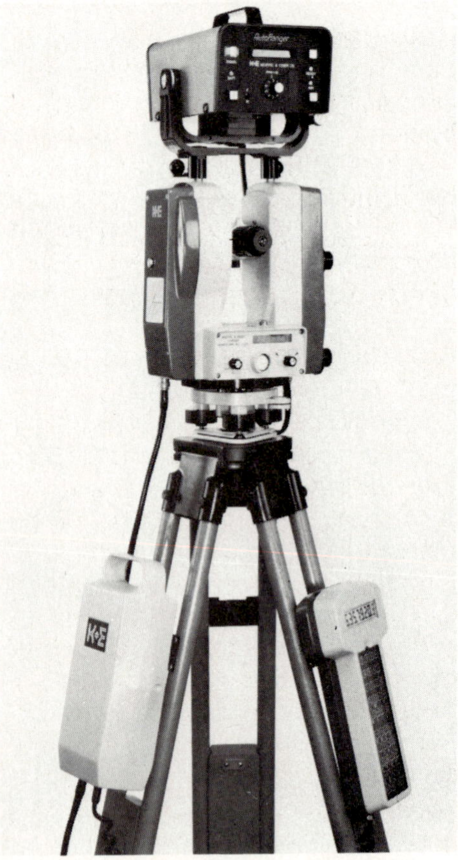

Fig. 6.40 Digitized theodolite. (*Courtesy Keuffel & Esser Co.*)

Fig. 6.41 Glass encoder disk. (*Courtesy Hewlett Packard, Inc.*)

microcomputer to angular values. In the most elementary form of such an instrument, angles in degrees, minutes, and seconds; grads, centesimal minutes, and centesimal seconds; or mils are output on small nixie tubes. The raw data can also be stored in a memory unit connected to the on-board computer or on magnetic tape. On the more sophisticated models, repeated angles can be stored and processed by the on-board computer so that output consists of the average angle. Further discussions concerning electronic surveying systems incorporating digitized theodolites can be found in Art. 9.13. Technical details about these devices are outlined in Refs. 3 and 4.

6.32. Use of the engineer's transit Succeeding sections describe operations with a transit for measuring horizontal and vertical angles as well as for running lines. Transit surveys are described in detail in Chap. 8.

The process of taking magnetic bearings with the transit is the same as with the magnetic compass (Art. 6.13). The transit may be employed for running direct levels in the same manner as with the engineer's level, the telescope level bubble being centered each time a reading is taken.

The primary function of the transit is for observing horizontal and vertical angles. One of the major assets of the transit in this application is that the telescope can be reversed. *Reversing* or plunging the telescope consists of rotating it about the horizontal axis. The telescope is said to be in the *direct* or *normal* position when the focusing knob is upright or on top; the telescope is said to be in the *reversed* or *inverted* position when the focus knob is upside down. As will be explained subsequently, observations repeated with the telescope direct and reversed permit compensation of most of the systematic instrumental errors.

First, consider the procedure for setting a transit over a station point.

6.33. Setting up the engineer's transit Ordinarily, the transit is set over a definite point, such as a tack in a stake. For centering the transit with four level screws, a plumb line is suspended from the hook and chain beneath the instrument. First the transit is placed approximately over the point, with the telescope near the observer's eye level. On side-hill setups, one tripod leg should be up the hill, with two on the downhill side of the point. Each tripod leg is then moved as required to bring the plumb bob within about 0.02 ft or 6 mm of being over the tack, with the foot plate nearly level and with the shoe of each tripod leg pressed firmly into the ground. The instrument is leveled approximately by means of the leveling screws.

The procedure for manipulating the level screws is similar to that described for the four-screw dumpy level (Art. 5.29). Each plate level is aligned with a pair of diagonally opposite level screws, as illustrated in Fig. 6.42*a*, where the axis of tube *A* is in line with level screws 1 and 2. The position of the bubble in tube *A* is controlled by screws 1 and 2 and the bubble in tube *B* is controlled by screws 3 and 4. Level bubble *A* is brought to center by turning screws 1 and 2, keeping both screws lightly in contact with the foot plate. It is convenient to remember that the bubble travels in the *same direction as the left thumb*. The motions indicated in Fig. 6.42*a* will cause the bubble in tube *A* to move from left to right. Next, the bubble in tube *B* is centered in similar fashion using screws 3 and 4. With the two bubbles approximately centered in this position, two adjacent screws (2 and 4, for example) are loosened by turning them in the same direction. This action releases the pressure of the foot-screws on the foot plate and permits lateral movement of the leveling head so that the plumb bob can be centered accurately over the tack. The length of plumb bob line is changed as necessary to make the plumb bob just clear the tack. Tighten the two adjacent screws to a firm but not tight bearing and recenter the bubbles in tubes *A* and *B* very carefully. Rotate the instrument through 90°, as shown in Fig. 6.42*b*, and center the bubbles in tubes *A* and *B* as before, using the foot-screws. The instrument is now rotated back to the original position (Fig. 6.42*a*), where the positions of the level bubbles are checked. The bubbles are recentered if necessary. When the plate level bubbles remain centered in both positions, the transit is rotated

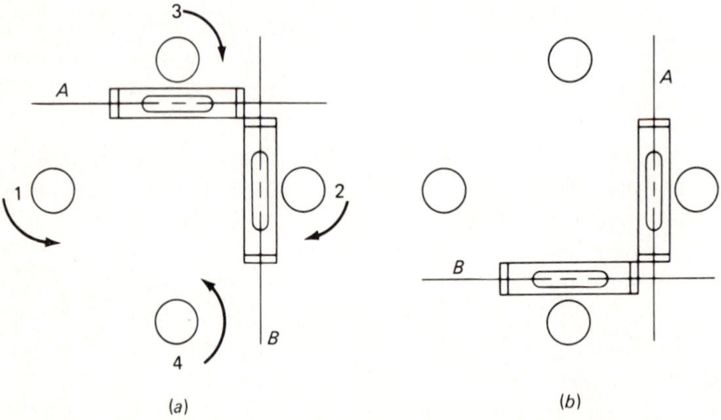

Fig. 6.42 (*a*) and (*b*) Leveling the plate level bubbles, four-screw instrument.

180° from the original position and the centering of the level bubbles is checked again. If the bubbles remain centered and the plumb bob is over the tack, the setup is complete. If the bubbles move off center a small amount (about one 2-mm division), bring each bubble halfway to center using the appropriate pair of diagonally opposite level screws. If the deviation is large, adjustment of the bubbles is necessary (Art. 6.51). The telescope should be tested for parallax before observations begin.

The traditional-style engineer's transit equipped with a vertical optical collimator or optical plummet (Figs. 6.26 and 6.27) is set up in a similar fashion except for centering over the point. Initially, a plumb bob is attached to the vertical spindle and the transit is approximately leveled and centered as described in the preceding paragraph. Then the plumb bob is removed and the instrument is centered by loosening the knurled screw which secures the cross-slide apparatus that supports the transit on the tripod (Fig. 6.27). Centering with the optical collimator is then possible. Releveling is necessary, followed by recentering, since adjustment of the level screws alters the vertical collimation line. When the optical collimation line is centered over the point and the plate level bubbles are centered, the instrument is turned through 90° and the entire procedure is repeated. The process of alternately leveling and recentering is repeated in the original position and with the instrument rotated through 90° until the bubbles remain centered and the optical plumb coincides with the point. As before, the instrument should be rotated through 180° from the original position as a check.

Just before the transit is moved, the instrument is centered on the foot plate, the leveling screws are roughly equalized, the upper motion is clamped, and the lower motion is either unclamped or clamped lightly, and the telescope is pointed vertically with the vertical motion lightly clamped.

6.34. Setting up instruments with three leveling screws These instruments are equipped with a circular level bubble, only one plate level bubble, and an optical plummet. Consequently, certain significant differences in the setting up procedure exist.

The instrument is set approximately over the point using a plumb bob attached to a hook which can be inserted into the screw that secures the instrument to the top of the tripod. Care should be exercised to ensure that the foot plate is nearly level, so as to simplify succeeding operations. Next, center the circular level bubble by manipulating the three level screws as described in Art. 5.29. Remove the plumb bob and focus the optical plummet on the point, so as to be free of parallax. After slightly loosening the central screw, the instrument is moved laterally until the cross mark in the optical plummet coincides with the station point. The circular level bubble is checked and if recentering is required, centering of the optical plummet should be checked. When the optical plummet is centered over the station point and the circular level bubble remains centered, the instrument is rotated so that the axis of the plate level bubble is aligned with any two level screws, as shown in Fig. 6.43*a*. The bubble is centered by rotating screws 1 and 2 in opposite directions. The rotations indicated in Fig. 6.43*a* will cause the bubble to move from left to right. Next, rotate the instrument so that one end of the bubble tube is aligned with the remaining screw (Fig. 6.43*b*). Center the bubble in this position by rotating this remaining screw. The rotation indicated in Fig. 6.43*b* will cause the bubble to move away from level screw 3. Return to the original position (Fig. 6.43*a*) and check centering of the bubble. Rotate through 180° so that end *A* of the bubble tube is on line with level screw 1. If the bubble moves off center, bring the bubble halfway to center using level screws 1 and 2. Rotate 90° so as to be in the position of Fig. 6.43*b* and check the level bubble. Rotate through 180° so that end *B* of the bubble tube coincides with level screw 3 (Fig. 6.43*b*). If the bubble moves off center, bring halfway to center with level screw 3. If this deviation from center is large, the bubble tube must be adjusted (see Art. 6.51). Finally, check the centering of the optical plummet.

If the theodolite is mounted on a tripod with adjustable legs, the procedure for setting up may be modified as follows. Set the instrument approximately over the survey point as previously described (Fig. 6.44*a*). Then looking through the optical plummet, focus on the cross hairs and then on the point. At this stage the center cross should be within 2 to 3 cm (0.07 to 0.1 ft) of the point (Fig. 6.44*b*). Still looking through the optical plummet, center the cross hairs on the survey point by manipulating one or two of the level screws (Fig. 6.44*c*). Now the optical plummet is centered (Fig. 6.44*d*) but the circular level bubble is off-center (Fig. 6.44*e*). Loosen the clamp on the tripod leg which is in the direction you want the bubble to move to bring it to center. Adjust the leg until the bubble moves toward the ring. The bubble will move toward a leg being lengthened and away from a leg being shortened. In Fig. 6.44*e* and *g*, leg 2 should be shortened. If the bubble moves to one side during this adjustment, clamp the leg when the bubble is radially opposite another leg as in Fig. 6.44*h*. In Fig. 6.44*j*, leg 3 is shortened to yield a centered circular level bubble (Fig. 6.44*k*). Under certain conditions, these operations may need to be repeated using the same two or possibly the third tripod leg. Next, level the theodolite precisely using the plate level bubble as previously described and shown graphically in Fig. 6.43*a* and *b*. When the theodolite is leveled, look through the optical plummet and check the centering. If the cross hairs are not centered on the point,

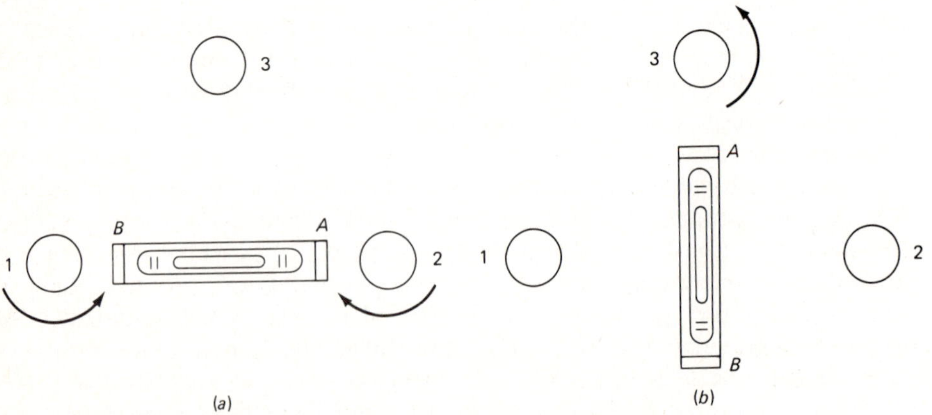

Fig. 6.43 (*a*) and (*b*) Leveling the plate level bubble, three-screw instrument.

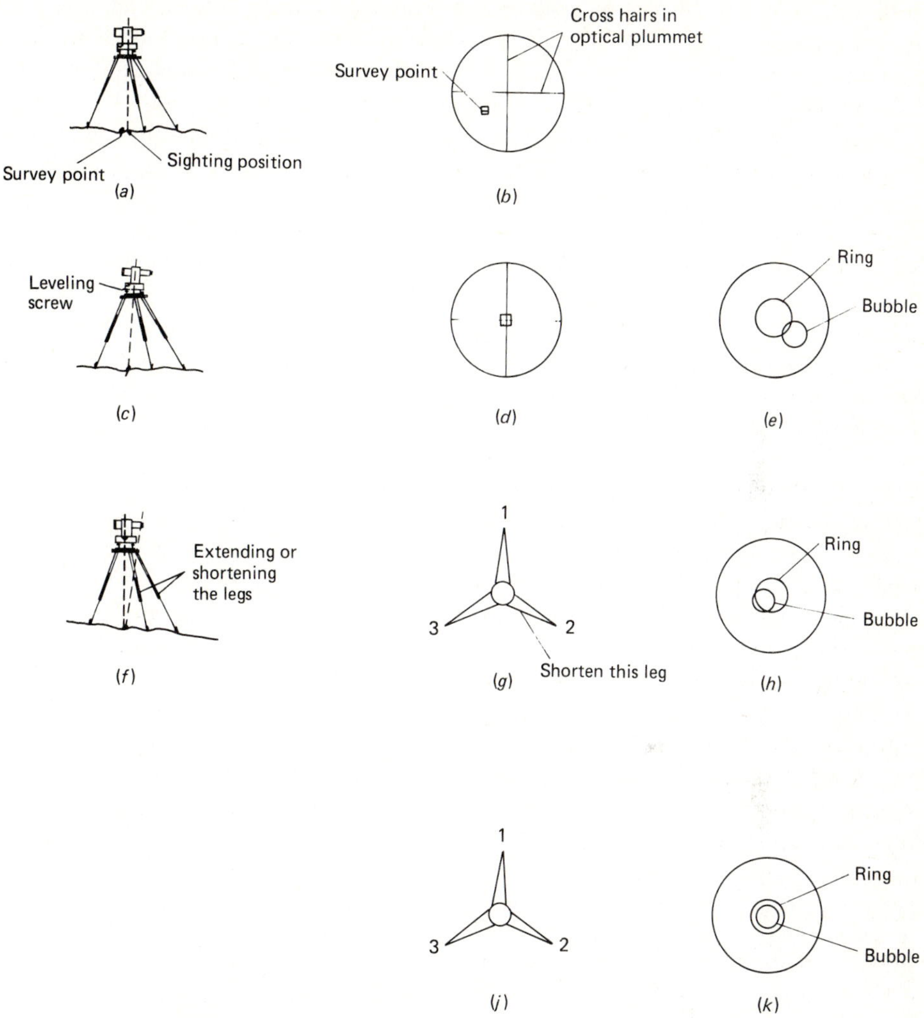

Fig. 6.44 (*a*) through (*k*) Setup for three-screw theodolite with adjustable tripod legs.

loosen the clamp that holds the theodolite to the tripod, and while sighting through the optical plummet, shift the instrument laterally on the top of the tripod until the cross hairs are centered on the point. Reclamp the theodolite to the tripod and check the plate level bubble, centering as before if necessary. These last two steps of centering and leveling should be repeated until the cross hairs in the optical plummet remain centered on the survey point and the plate level bubble is centered in the tube throughout a 360° revolution of the theodolite about its vertical axis.

The plate level bubble on most direction theodolites is quite sensitive (about 20″ per 2-mm division) and centering the bubble throughout a 360° revolution is a challenging task unless the theodolite is protected from the rays of the sun.

The process of sequentially leveling and centering is made much easier when the top of the tripod is close to being level initially. Care should be exercised to ensure that this condition is satisfied. Some tripods are equipped with level bubbles built into the supporting plate to facilitate this operation.

Setup of an engineer's transit or theodolite equipped with a tripod having a telescopic centering rod was described in the last paragraph of Art. 6.30. When using the centering tripod, the support plate on which the instrument rests is automatically level after the rod is plumbed. Thus, leveling the instrument requires only a few turns of the level screws. One precaution should be observed. After the circular level bubble on the centering rod has been centered, it should be rotated 180° and checked to see if the bubble remains in the center. If the bubble deviates from center, loosen the clamping grip (Fig. 6.39*a*), bring the bubble halfway to center, and then tighten the clamping grip. If the discrepancy is excessive, the circular level bubble must be adjusted as described in Art. 5.48.

6.35. Horizontal angles with the engineer's transit If a horizontal angle such as *AOB* in Fig. 6.45*a* is to be measured, the transit is set over *O*. Using the upper clamp, the upper motion is clamped with one of the verniers set near zero, and by means of the upper tangent screw one vernier is set to zero. With the lower clamp loose, the telescope is sighted approximately to *A* (telescope in the direct position) by first sighting over the top of the telescope to get near the point and then sighting through the telescope. The lower clamp-screw is locked, the telescope is raised or lowered so that the cross-hair intersection is near the point, and the telescope is focused on point *A*. Care should be exercised to detect and remove parallax. By turning the lower tangent screw, the vertical cross hair is set exactly on the plumbed line, range pole, or other target which marks point *A*. At this stage the optical line of sight is on line between *O* and *A* and the vernier is set to zero on the horizontal circle. Next, the upper clamp is loosened and the telescope is turned until the line of sight is approximately on *B*. The upper clamp is tightened and the line of sight is set

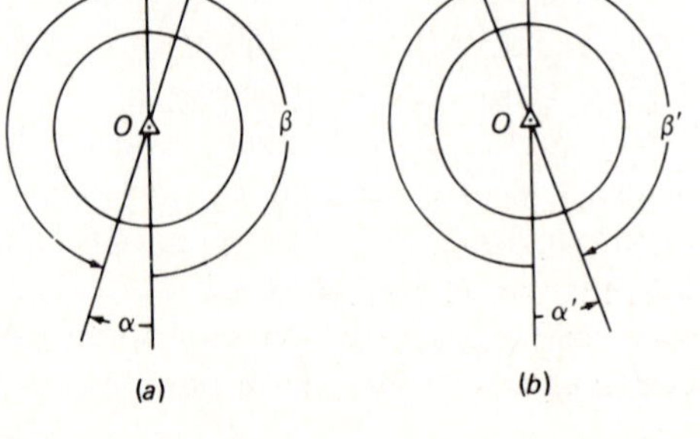

Fig. 6.45 (*a*) Horizontal angle to the right. (*b*) Horizontal angle to the left.

exactly on B by turning the upper tangent screw. The reading of the vernier that was set to zero on point A gives the value of the angle. If point B falls to the right of point A, as shown in Fig. 6.45*a*, the objective lens of the telescope has moved to the right and the angle AOB is said to be *turned to the right* or *clockwise*. Assume that the horizontal circle setting when sighted on B is as shown in Fig. 6.29*b*. The horizontal angle $\alpha = AOB$ turned to the right is read on the clockwise circle as being $49°40'$ plus $10'30'' = 49°50'30''$, where the $10'30''$ is observed on the vernier to the left of the index mark. From this same setup, the counterclockwise angle from A to B or β (Fig. 6.45*a*) is observed on the counterclockwise horizontal circle and the vernier is read on the right side of the index. This counterclockwise angle $\beta = AOB$ is $130°09'30'' + 180° = 310°09'30''$.

If B falls to the left of A, the objective lens of the telescope is turned to the left and angle $\alpha' = AOB$ is said to be turned to the left, as shown in Fig. 6.45*b*. Assume that the horizontal circle settings are as illustrated in Fig. 6.29*a*. An angle turned to the left is observed on the counterclockwise circle. Observation of the counterclockwise circle and the right-hand side of the vernier yields $\alpha' = 17°25'00''$. The clockwise angle from A to B or β', observed on the clockwise circle, is $342°35'00''$.

The following is a list of suggestions for turning horizontal angles with an engineer's transit:

1. Make reasonably close settings by hand so that the tangent screws will not need to be turned through more than one or two revolutions.
2. Make the last movement of the tangent screw clockwise, thus compressing the opposite spring.
3. When reading the vernier, have the eye directly over the coinciding graduation, to avoid parallax. It is also helpful to observe that the graduations on both sides of those coinciding fail to concur by the same amount.
4. As a check on the reading of one vernier, the other vernier may be read also. Or, check readings may be taken at each end of the vernier scale; these differ from the vernier reading by a value which is constant for the given type of vernier.
5. The plate bubbles should be centered before measuring an angle, but between initial and final settings of the line of sight the leveling screws should not be disturbed. When an angle is being measured by repetition (Art. 6.39), the plate may be releveled after each turning of the angle before again sighting on the initial point.
6. Long sights may be made on a range pole plumbed over the point. The person giving the sight should stand directly behind the pole, holding it lightly with the fingers of both hands, and balancing it on the tack or other mark indicating the point. Sights of moderate length are best taken on a plumbed line.
7. In sighting at a range pole the bottom of which is not visible, particular care should be taken to see that it is held vertical. When the view is obstructed for a considerable distance above the point to which the sight is taken, use a plumb line behind which a white card is held. For short sights a pencil or ruler held on the point makes a satisfactory target. Where the lighting is poor, the sight may be taken on a flashlight.
8. When a number of angles are to be observed from one point without moving the horizontal circle, the instrument operator should sight at some clearly defined object that will serve as a reference mark and should observe the angle. If occasionally the angle to the reference mark is read again, any accidental movement of the horizontal circle will be detected.

9. Whenever an angle is doubled, if the instrument is in adjustment, the two readings should not differ by more than the least count of the vernier. A greater discrepancy, if confirmed by repeating the measurement, will indicate that the instrument is out of adjustment.

6.36. Horizontal angles with the repeating theodolite Figure 6.32 shows a repeating theodolite. Assume that this instrument is set over point *O* in order to observe horizontal angle *AOB*. The initial setting of the horizontal circle is accomplished by viewing through the reading microscope and setting the horizontal circle reading and index mark to zero using the upper clamp and tangent screw. A sight is then taken on *A* using the lower clamp and tangent screw. At this point the optical line of sight coincides with the line *OA* and the horizontal circle reading is zero. The upper clamp is loosened and the telescope is turned to point *B*, where a sight is taken with the upper clamp locked and using the upper tangent screw. The angle is then observed by viewing through the reading microscope and following the instructions given in Art. 6.35. Figure 6.32*b* illustrates a horizontal circle reading of 68°21.8′. These scales are read directly. Note that both the clockwise and counterclockwise angles can be observed on this instrument.

6.37. Laying off horizontal angles If an angle *AOB* is to be laid off from line *OA*, the transit is set up at *O*, one vernier (or index and micrometer scale in optical-reading instruments) is set at zero, and the line of sight is set on *A*. The upper clamp is loosened, and the plate is turned until the index of the vernier or optical-reading system is approximately at the required angle. The upper clamp is tightened, and the vernier (or index and micrometer scale in an optical-reading instrument) is set exactly on the given angle by means of the upper tangent screw. Point *B* is then established on the line of sight.

6.38. Common mistakes In measuring horizontal angles, mistakes often made are:

1. Turning the wrong tangent screw
2. Failing to tighten the clamp
3. Reading numbers on the horizontal scale from the wrong row
4. Reading angles in the wrong direction
5. Dropping 30′ or 20′ by failure to take the full-scale reading before reading the vernier, for example, with a circle graduated to 30′ calling an angle 21°14′ when it is actually 21°44′, the vernier reading being 14′
6. Reading the vernier in the wrong direction
7. Reading the wrong vernier

6.39. Measuring horizontal angles by repetition By means of the engineer's transit or repeating theodolite, a horizontal angle may be mechanically cumulated and the sum can be read with the same precision as the single value. When this sum is divided by the number of repetitions, the resulting angle has a precision that exceeds the nominal least count of the instrument. Thus, with a transit having verniers reading to single minutes, an angle for which the true value is between the limits 30°00′30″ and 30°01′30″ will be read as 30°01′ and the limits of possible error will be $\pm 30''$. If the angle is cumulated six times on the horizontal circle, the sum, also read to the nearest minute, might be $180°04'$, its true value being within the limits $180°03'30''$ and $180°04'30''$; the limits of possible error, as far as reading the horizontal circle is concerned, will also be $\pm 30''$. Dividing the observed sum $180°04'$ by 6, the single value becomes $30°00'40''$ for which the limits of possible "reading" error are $\pm 30''/6 = \pm 05''$. Similarly, a repeating theodolite (Fig. 6.32) with a least count of $30''$ (estimating to $15''$) would yield an angle having a possible reading error of about $\pm 03''$ in six repetitions. This method of determining an angle is called *measurement by repetition*. The precision with which an angle can be observed by this method increases directly with the number of times the angle is cumulated or repeated up to between six and twelve repetitions; beyond this number the precision is not appreciably increased by further repetition because of errors in graduations of the horizontal circle; eccentricity of instrument centers; play in the instrument; and random errors such as those due to setting the line of sight, setting the angle on the horizontal circle, and reading angles on the horizontal circle.

When angles are observed by repetition, the principle of *double centering* is also employed. As noted in Art. 6.29, on geometry of the transit, the lines and planes defined by vertical axis, horizontal axis, optical line of sight, and so on, theoretically have exact relationships with respect to one another. In practice, these relationships are not exact, even in the best-adjusted transit or theodolite, so that systematic instrumental errors are present (see Art. 6.55). One of the excellent features of the transit or theodolite is that the telescope can be transited or rotated about the horizontal axis so that observations can be made with the telescope in the direct and reversed positions (Art. 6.32). When operations on the instrument include a direct and reversed observation or setting, systematic instrumental errors will occur in opposite directions. Thus, if the average of the observations or settings is used, the effect of the systematic errors cancels. Implementation of this process is called *double centering* or *double sighting*.

Another reason for repeating an angle is to decrease the possibility of mistakes in observations. Thus, the primary reasons for observing an angle by repetition with the telescope direct and reversed are to increase the accuracy of reading the angle, compensate for systematic errors, and eliminate mistakes. Consider some examples.

To repeat an angle once direct and once reversed, as for angle AOB in Fig. 6.45*a*, the transit is set up at O, where it is very carefully centered and leveled. A sight is taken on point A with the telescope direct and a single value of the angle is observed as described in Art. 6.35 or 6.36. Next, with this angle still set on the plate, the lower motion is unlocked, the telescope is reversed, the instrument is turned on its lower motion, and a second sight is taken on A using the lower clamp and lower tangent screw. Note that on this sighting from O to A the value of the angle AOB is on the horizontal circle. The upper clamp is loosened and the telescope, now in the reversed position, is again sighted on point B. At this stage, approximately twice the original value of the first angle appears at the index of the horizontal circle, and the angle has been repeated with the telescope direct and reversed. The accompanying notes show how data can be recorded for observation of angle AOB by repetition once direct and once reversed using an engineer's transit with a least count of $30''$.

Station	From/to	Rep.	Tel.	Circle reading	Angle
O	A	0	Direct	0°00′00″	
	B	1	Direct	34°20′00″	34°20′00″
	B	2	Reversed	68°40′30″	34°20′30″

The first angle is obtained by subtracting the initial circle reading on A (0°00′00″ in this example) from the circle reading on B, yielding a first angle of 34°20′00′. The second angle equals the second circle reading minus the first reading on point B, or (68°40′30″)–(34°20′00″) = 34°20′30″. This second angle should agree with the first angle to within ± the least count of the transit, ±30″ in this case. Should this difference be greater, the set should be repeated. This set is acceptable. The final angle is then the average of angle 1 and angle 2: [(34°20′00″) + (34°20′30″)]/2 = 34°20′15″. The final angle can also be calculated by dividing the cumulated angle minus the initial observation by the number of repetitions; in this case, [(68°40′30″) − (0°00′00″)]/2 equals 34°20′15″.

The preceding example dealt with an interior angle. Next, consider the procedure for turning a deflection angle by repetition (Art. 6.10). Assume that the deflection angle at B in Fig. 6.12 is to be observed. The transit is set at point B and with the A vernier set on zero, a backsight is taken on point A with the telescope in the direct position. Next, the telescope is plunged so that the line of sight falls along the extension of line AB and the A vernier is still on zero. The upper clamp is loosened and a sight is taken on point C using the upper clamp and tangent screw. The deflection angle is observed on the clockwise circle using the A vernier. Assume that the first deflection angle at B from A to C is 22°35′00″ to the right. To *double* the angle, the lower clamp is loosened and a second backsight is taken on point A using the lower clamp and lower tangent screw. The telescope is reversed along the backsight and the angle 22°35′00″ is on the horizontal circle at the A vernier. Again the telescope is plunged, the upper clamp loosened, and a sight is taken on point C with the upper clamp and tangent screw. The telescope is now in the direct position and the angle observed on the horizontal circle at the A vernier is the doubled deflection angle. Notes for the example are shown below.

Station	From/to	Rep.	Tel.	Circle reading	Deflection angle
B	A	0	Direct	0°00′00″	
	C	1	Reversed	22°35′00″	22°35′00″ R
	C	2	Direct	45°10′30″	22°35′30″ R

Note: Engineer's transit number 37, least count 30″.

The procedure for calculating the first and second deflection angle is similar to that described for interior angles in the preceding example. Since the first angle agrees with the second to within the least count of the transit, the set is acceptable. The final average angle is 22°35′15″, the average of the two angles. The procedure just outlined is called observing angles by *double deflection*. Note that when deflection angles are observed, the direction of the deflection must be recorded.

Whenever possible, angles should always be turned by repetition the minimum number of times, once direct and once reversed. In this way systematic instrumental errors are compensated and blunders are eliminated, even though a high order of accuracy is not achieved.

When an angle is turned more than twice, the number of repetitions is always some multiple of two. Thus, an angle repeated twice direct and twice reversed involves four repetitions. Similarly, angles can be turned three, four, five, and six times direct and reversed involving six, eight, ten, and twelve repetitions, respectively. There should always be an equal number of direct and reversed observations.

When observing an angle by repetition using an engineer's transit with direct-reading verniers, it is customary to observe both the *A* and *B* verniers. In this way the systematic error between the two verniers is averaged and the accuracies possible by repetition are more nearly achieved.

The note form for turning an angle with six repetitions using a 30″ engineer's transit is shown below:

					Vernier	
Station	From/to	Rep.	Tel.	Circle reading	*A*	*B*
O	*M*	0	Direct	0° 0′	00″	00″
	N	1	Direct	80°20′	00″	00″
	N	6	Reversed	122°01′	00″	00″

Note that the circle reading for the first repetition on point *N* is recorded and then the angles are cumulated without recording until the final repetition is completed. The first angle is the difference between circle observation one on point *N* and circle observation zero on point *M*. For this set, the initial angle, used *only* as a check value, is 80°20′00″. To calculate the average angle, it is necessary to know the number of times the horizontal circle index has passed zero. The total number of degrees passed in six repetitions is the initial approximate angle times the number of repetitions or (6)(80°20′) = 482°. Since zero was passed once, 360° must be added to the cumulated angle recorded for the sixth repetition before calculating the average angle. Thus, the average angle for this set is [(122°01′00″) + 360°]/6 = 80°20′10″. This average angle agrees with the initial check angle to within ± the least count of the transit, so that the set of observations is acceptable. Figure 6.46*a* shows a field note form for angles by repetition.

The circle reading for the first sight does not have to be zero, as illustrated by the following notes for six repetitions of angle *EFG* using a transit having a least count of 20″:

					Vernier	
Station	From/to	Rep.	Tel.	Circle reading	*A*	*B*
F	*E*	0	Direct	136°02′	00″	$\overline{40}''$
	G	1	Direct	216°21′	00″	00″
	G	6	Reversed	257°58′	00″	$\overline{40}''$

The reading on the *B* vernier for repetition zero on point *E* is $\overline{40}''$. The bar means that the next lower minute was read, so the average circle reading is $\frac{1}{2}$[(136°02′00″)

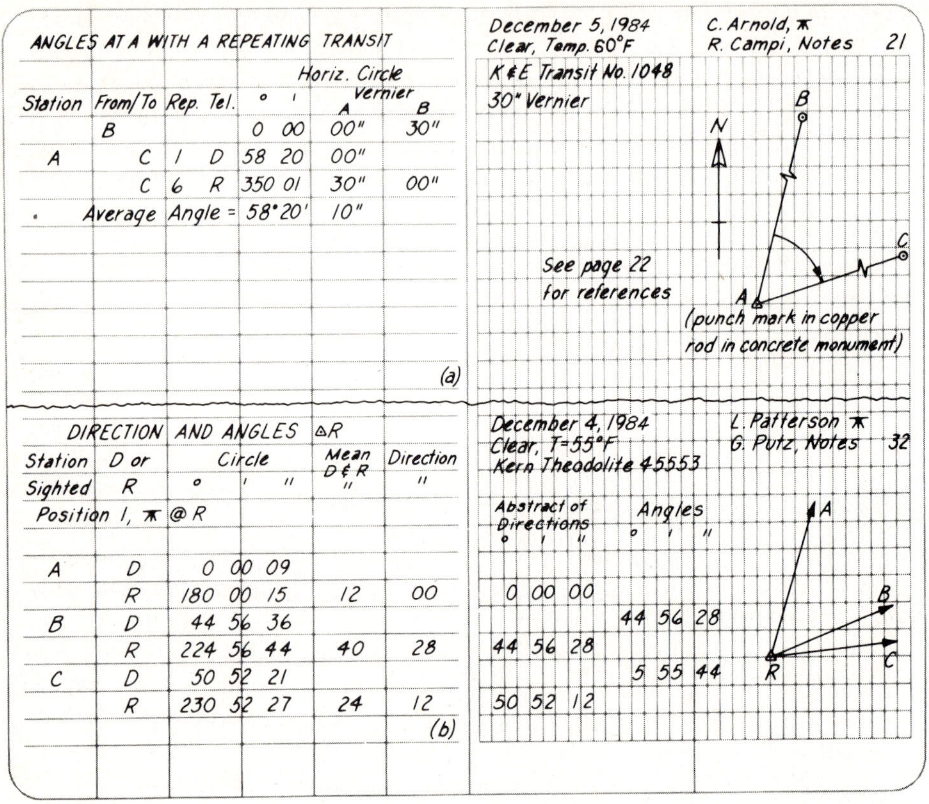

Fig. 6.46 (*a*) Field notes for angles by repetition using a repeating instrument. (*b*) Field notes for directions with a theodolite.

+(136°01′40″)] = 136°01′50″. The approximate angle is therefore (216°21′00″) − (136°01′50″) = 80°19′10″. Six repetitions of this angle or 6(80°19′10″) = 481°55′00″. This value added to 136°01′50″ equals 617°56′50″, indicating that zero was passed once and 360° must be added to the final circle reading to calculate the final angle. Thus, [(257°57′50″ + 360°) − (136°01′50″)]/6 = 80°19′20″, the final average angle.

When angles are repeated with a repeating theodolite, readings 180° apart on the horizontal circle are automatically averaged in the optical reading process. Notes for angle *SRT* observed 12 times with a repeating theodolite having a least count of 10″ given below:

Station	From/to	Rep.	Tel.	Circle reading
R	*S*	0	Direct	70°10′30″
	T	1	Direct	170°30′30″
	T	12	Reversed	194°11′30″

The approximate angle is 100°20′00″. Twelve times this approximate angle plus the first circle reading equals 1274°10′30″. Thus, the circle index passed zero three times. The average angle equals [(194°11′30″) + (1080°) − (70°10′30″)]/12 = 100°20′05″, which is within the least count or ±10″ of the approximate angle.

Angles by repetition should not be attempted unless the instrument operator is experienced both in the instrument setup and in observing angles. Speed and accuracy in observation are essential if the repetition procedure is to be successful. Otherwise, considerable time can be wasted without obtaining the desired results.

6.40. Directions and angles with the direction theodolite As noted in Art. 6.30, the direction theodolite has only a single horizontal motion equipped with one clamp and tangent screw for azimuth. Thus, it is possible to observe *directions only*, and angles are computed by subtracting one direction from another. For example, assume that the angle *AOB* (Fig. 6.45*a*) is to be measured with a direction instrument. The theodolite is set up over point *O*, leveled and centered, and a sight is taken on point *A* using the horizontal clamp and tangent screw for azimuth (Fig. 6.35). The horizontal circle is then viewed through the optical-viewing system and the circle reading is observed and recorded (Art. 6.30). Assume that this reading is 89°34′22″. The horizontal clamp is then released and the telescope is sighted on point *B*. Once again the sight is taken with the horizontal clamp and the tangent screw for azimuth. The horizontal circle is observed through the optical-viewing system and this reading is recorded as being 102°10′52″. These two observations constitute directions which have a common reference direction that is completely arbitrary. The clockwise angle = (102°10′52″) − (89°34′22″) = 12°36′30″. Note that on a direction theodolite, there is only one horizontal circle graduated in a clockwise direction so that only clockwise angles can be derived from observations of directions. In Fig. 6.45*b*, point *B* is to the left of point *A*. Suppose that the theodolite is set up on point *O*, the initial sight is on point *A*, and the horizontal circle observation is 10°15′12″. The horizontal clamp is loosened and the telescope is rotated to sight on *B*, where the circle reading is 351°50′32″. The clockwise angle from *A* to *B* is therefore (351°50′32″) − (10°15′12″) = 341°35′20″. Again, the angle is clockwise from the initial pointing on *A*. Note that in neither case was any attempt made to set the horizontal circle on zero or on an integral value. There is a control which allows the horizontal circle to be advanced to a desired approximate value (Fig. 6.35). This control is a coarse motion only and does not permit fine settings.

6.41. Angles using the direction theodolite Measuring angles with the direction theodolite involves observing directions with the telescope direct and reversed, calculating mean directions to the observed points, and determining mean angles from the mean directions. Figure 6.46*b* shows notes for directions observed from a theodolite set at point *R* to points *A*, *B*, and *C*. Referring to these notes, the procedure for observing the directions is as follows:

1. Make an initial pointing and record the horizontal circle observation. Normally, the telescope is in the direct position and the leftmost point is selected as the initial sighting. The value of the initial circle reading is governed by the number of repetitions to be observed (Art. 9.6).

2. Release the horizontal clamp and sight each point in succession, recording the circle reading for each sight. These observations constitute the circle readings to A, B, and C with the telescope direct.
3. Reverse the telescope, sight point C, record the circle reading, and then sight on points B and A in reverse directions, recording the circle observation with the telescope in the reversed position in each case. These reversed readings are different by exactly 180° from the direct observation plus or minus a few seconds which represent the collimation error of the instrument (Art. 6.51). The observations taken as described constitute *one position*.

The means of the direct and reversed readings are calculated and recorded in column 4 of the notes (Fig. 6.46*b*). Then an *abstract of directions* is calculated and recorded in column 5 of the notes. This abstract is determined by assuming a direction of 0°00′00″ along the initial sight from R to A. Thus, the direction from R to B is (44°56′40″) − (0°00′12″) = 44°56′28″ and the direction from R to C is (50°52′24″) − (0°00′12″) = 50°52′12″. Finally, angles ARB and RBC are computed by subtracting successive directions as given in the abstract.

Depending on the accuracy desired, anywhere from 1 to 16 positions of directions may be observed. In such cases the horizontal circle would be advanced for each successive position so as to use the entire circle. The final abstract of directions and angles would be the mean of all the positions.

6.42. Laying off angles by repetition If it is desired to establish an interior angle with a precision greater than that possible by a single observation, the methods of the preceding section may be employed in the following manner: In Fig. 6.47*a*, OA represents a fixed line and AOB the angle which is to be laid off to establish the line OB. The engineer's transit is set up at O, the vernier is set at 0°, and a sight is taken to A. The vernier is set as closely as possible to the given angle and a trial point B' is established with the line of sight in its new position. The angle AOB' is then measured by repetition, and the line OB' is measured. The angle AOB' must be corrected by an angular amount $B'OB$ to establish the correct angle AOB. The correction, which is too small to be laid off accurately by angular measurement, is applied by offsetting the distance $B'B = OB'$ tan (or sin) $B'OB$, thus establishing the point B beside B'. It is convenient to remember that the tangent or sine of 1′ = 0.0003 and the tan or sine of 1″ = 4.848×10^{-6}. As a check, the angle AOB is measured by repetition.

Example 6.3 Suppose that an angle of 30°00′ correct to the nearest 05″ is to be laid off and that the transit to be employed reads to the nearest 01′. Let the total value of AOB' after six repetitions be 180°02′, correct to the nearest 30″. Then the measured value of AOB' is 180°02′/6 = 30°00′20″ correct to the nearest 05″, and the correction to be applied to AOB' is 20″. Suppose that OB' = 400 ft. Then the length of the offset $B'B$ equals tan 20″ × 400 ft = (0.0001)(400) = 0.04 ft.

In route surveying, highway center lines are frequently set by deflection angles. Referring to Fig. 6.47*b* assume that it is desired to establish the direction of the line from station (Art. 5.42) 16 + 10.450 to station 18 + 84.900 by turning the deflection angle 45°10′00″ to the right from a backsight on station 14 + 55.250. The instrument is set on station 16 + 10.450, the horizontal circle index is set on zero, and with

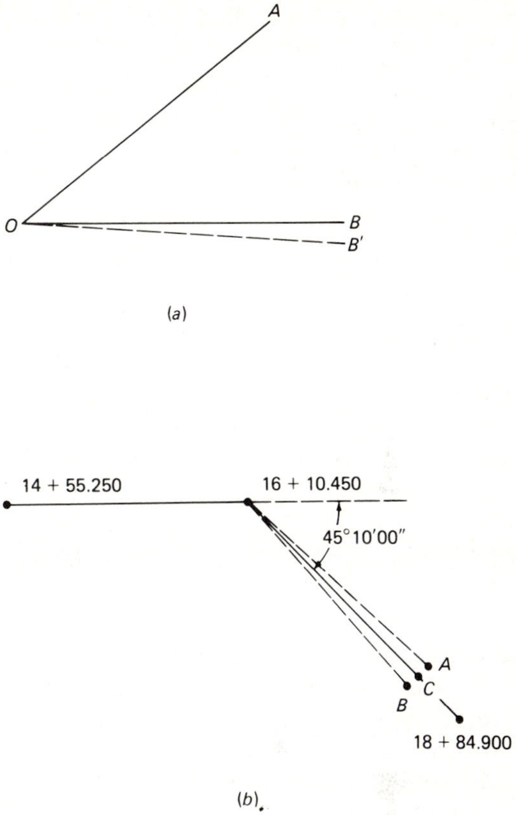

Fig. 6.47 Laying off horizontal angles. (*a*) Interior angle. (*b*) Deflection angle.

the telescope in the direct position a backsight is taken on 14 + 55.250 using the lower clamp and tangent screw. The telescope is plunged, the upper clamp is loosened, the angle 45°10′00″ is set off on the clockwise circle with the horizontal circle index, and point *A* is set on line along the line from 16 + 10.450 to 18 + 84.900. With 45°10′00″ still set on the horizontal circle, release the lower clamp and take a second backsight on 14 + 55.250 using the lower clamp and tangent screw. At this point the telescope is in the reversed position. Next, the telescope is plunged a second time, the upper clamp is loosened, and twice the deflection angle, or 90°20′00″, is set off on the horizontal circle using the upper clamp and upper tangent screw. This procedure is called setting the line by *double deflection angle*. If the instrument is in excellent adjustment, the last direction will coincide with the first point set at *A*. However, this rarely occurs, so that a second point is set at *B* on the top of the same stake set for *A*. Point *C* is set such that $BC = AC$ and the line from station 16 + 10.450 to *C* represents the best estimate of the line that deflects 45°10′00″ to the right.

6.43. Measurement of vertical angles The method for measuring vertical angles varies with the type of instrument used. The engineer's transit has a fixed vertical vernier and spirit level attached to the telescope. Repeating and direction theodolites have vertical circles on which zero corresponds to the zenith when a control spirit level bubble is centered. Theodolites of a more recent design are equipped with some form of vertical circle index compensator actuated by force of gravity, which automatically ensures that zero on the vertical circle coincides with the zenith (Art. 6.30).

Using the engineer's transit, the vertical angle to a point is its angle of elevation (+) or depression (−) from the horizontal. The transit is set up and leveled as when measuring horizontal angles. The plate bubbles should be centered carefully. The telescope is sighted approximately at the point, and the horizontal axis is clamped. The horizontal cross hair is set exactly on the point by turning the telescope tangent screw, and the angle is read by means of the vertical vernier. Note that the vernier is on the outside or below the vertical circle which moves with respect to the fixed vernier when the telescope is rotated about the horizontal axis (Fig. 6.48). For a positive or elevation vertical angle with the telescope in the direct position, the vertical circle is read to the right of zero and the vernier is read on the right side of the index. A negative or depression vertical angle (telescope direct) is read to the left of zero and on the left side of the vernier index. Figure 6.48 illustrates positive and negative vertical angles of $+10°53'$ and $-9°43'$, respectively.

On theodolites, zenith angles from 0° to 90° are above horizontal (+vertical angles, Fig. 6.49*a*), while zenith angles from 90° to 180° are below horizontal (−vertical angles, Fig. 6.49*b*). Before reading the vertical circle, the vertical circle control bubble is carefully centered. This ensures that zero on the vertical circle coincides with the zenith. The vertical circle is then viewed through the reading microscope (Fig. 6.33*a*). Figures 6.32*b* and 6.33*c* provide example vertical circle readings on different styles of theodolites. Zenith angles are observed in a similar manner on theodolites equipped with vertical circle index compensators. On these instruments, there is no need to center a vertical circle bubble, since the compensator automatically makes the vertical circle index coincide with the zenith. Details concerning the compensators can be found in Art. 6.30. Theodolites shown in Figs. 6.32*a*, 6.33*a*, and 6.34*a* are equipped with compensators.

Given the zenith angle, observed on a theodolite, the vertical angle can be computed. For example, a zenith angle of $85°10'45''$ yields a vertical angle of

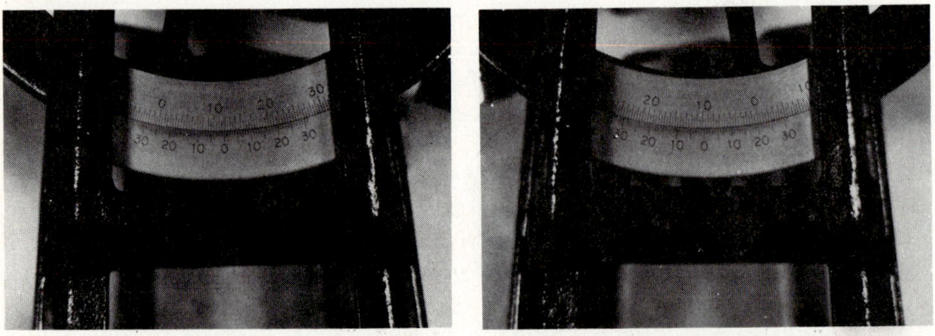

(*a*) (*b*)

Fig. 6.48 Vertical circle readings with engineer's transit. (*a*) Positive vertical angle. (*b*) Negative vertical angle.

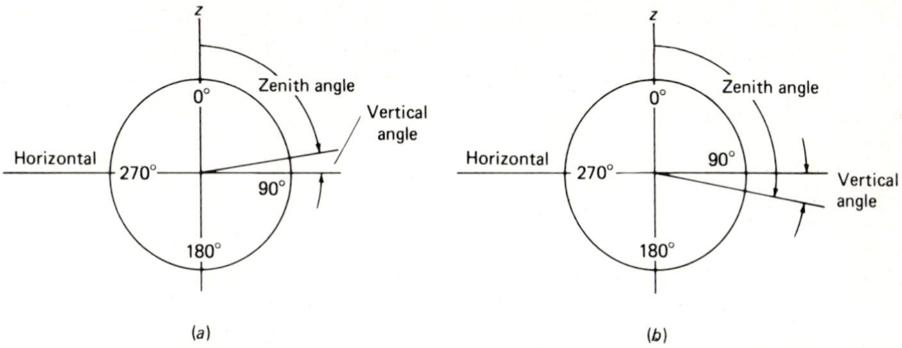

Fig. 6.49 Zenith angles. (*a*) Above horizontal. (*b*) Below horizontal.

(90°00′00″) − (85°10′45″) = +4°49′15″. A zenith angle of 94°12′44″ corresponds to a vertical angle of (90°00′00″) − (94°12′44″) = −4°12′44″. Both of these angles are observed with the telescope in the direct position. With the telescope reversed and assuming zero index error (Art. 6.45), corresponding zenith angles are 360° − (85°10′45″) = 274°49′15″ and 360° − (94°12′44″) = 265°47′16″. These zenith angles then yield respective vertical angles of (274°49′15″) − 270° = +4°49′15″ and (265°47′16″) − 270° = −4°12′44″. Note that the zenith angles are unambiguous, since they are positive in all cases.

6.44. Vertical angles by double centering The method of observing vertical or zenith angles by double centering consists of reading once with the telescope direct and once with it reversed and taking the mean of the two values thus obtained.

Double centering is used in astronomical observations and in similar measurements of vertical angles to distant objects. In traversing, a similar result is obtained by measuring the vertical angle of each traverse line from each end. For each line, the height of instrument above the point (h.i.) must be equal to the target height and the telescope should be in the same position (direct or reversed) for the two observations. The final result is taken as the mean of the two values.

6.45. Index error *Index error* is the error in an observed angle due to (1) lack of parallelism between the line of sight and the axis of the telescope level bubble or between the line of sight and axis of the vertical circle level bubble, (2) displacement (lack of adjustment) of the vertical vernier, and/or (3) for a transit having a fixed vertical vernier, inclination of the vertical axis.

The index correction is equal in amount but opposite in sign to the index error. Thus, if the observed vertical angle is +12°14′ and if the index error is determined to be +02′, the correct value of the angle is

$$+12°14' - 02' = +12°12'$$

Methods of determining the index error (and, therefore, the correction) are given in the following paragraphs.

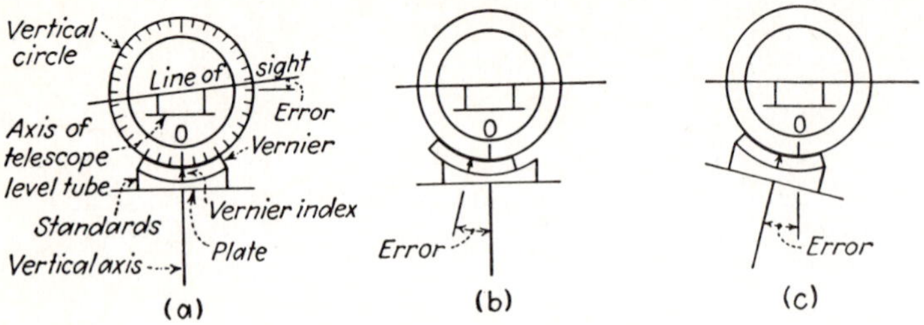

Fig. 6.50 (*a*), (*b*), and (*c*) Sources of error in measurement of vertical angles.

1. Lack of parallelism between the line of sight and axis of telescope level If the axis of the telescope level is not parallel to the line of sight and if the vertical vernier reads zero when the bubble is centered (Fig. 6.50*a*), an error in vertical angle results. This error can be rendered negligible for ordinary work by careful adjustment on the instrument (Art. 6.51, adjustment number 5). The combined error due to this cause and to displacement of the vertical vernier can be eliminated by double centering. The index error due to the two causes can be determined by comparing a single reading on any given point with the mean of the two readings obtained by double centering to the same point. Thus, if the observed vertical angle to a point is $+2°58'30''$ with telescope direct and is $+2°55'30''$ with telescope reversed, the index error for readings with telescope direct is

$$\frac{+2°58'30'' - 2°55'30''}{2} = +1'30''$$

and the correction to angles observed with the telescope in the direct position is $-1°30'$.

Repeating and direction theodolites are not equipped with telescope level bubbles. However, there can be index error in these instruments due to lack of parallelism between the line of sight and the level bubble used to orient the vertical circle. As with the engineer's transit, this type of error can be detected by observing vertical circle readings on a given point with the telescope direct and reversed. Thus, if direct and reversed zenith angles are 94°12′44″ and 265°47′24″, respectively, the index error is

$$\frac{94°12'44'' - 360° + 265°47'24''}{2} = 0°00'04''$$

This type of error can be eliminated by double centering or can be made nearly negligible for most ordinary surveys by adjustment of the vertical circle level bubble (Art. 6.53, adjustment number 4).

2. Displacement of vertical vernier Displacement of the vertical vernier on an engineer's transit (Fig. 6.50*b*) introduces a constant index error. The error can be rendered negligible by careful adjustment (Art. 6.51, adjustments 6 and 7). The combined index error due to this cause and to lack of parallelism between the line of sight and the axis of the telescope level can be eliminated by double centering; or the combined error can be determined as described in the preceding paragraph. For a transit having a fixed vertical vernier, the error due to displacement of the vertical vernier alone can be determined—provided that the line of sight is in adjustment—by leveling the transit carefully, leveling the telescope, and reading the vertical vernier. For a transit having a movable vertical vernier with control level, the error due to displacement of the vertical vernier alone can be determined by leveling both the telescope level and the vernier level, and reading the vertical vernier.

3. Inclination of vertical axis Assuming a fixed vertical vernier, inclination of the vertical axis (Fig. 6.50*c*) introduces an index error varying with the direction of the telescope and equal in amount to the angle the vernier is displaced about the horizontal axis while the instrument is directed toward the point. This error can be compensated by careful leveling of the transit before each observation and keeping the plate level bubbles in position for the specified direction. It is *not eliminated* by double centering. If the line of sight and vertical vernier are in adjustment, this error alone can be determined for each direction of pointing by leveling the telescope and reading the vertical vernier.

6.46. Precise leveling of the transit or theodolite When using the engineer's transit for astronomical observations or for measurement of horizontal angles requiring steeply inclined sights, it is usually desired to level the transit with greater precision than is possible through the use of the plate levels. In such cases the vertical axis is made truly vertical by means of the telescope level as follows. First the transit is leveled with the plate levels in the usual manner. With the telescope over one pair of opposite leveling screws, the bubble of the telescope level is centered using the telescope tangent screw. The telescope is rotated end for end about the vertical axis; then the bubble is brought halfway back to center by means of the leveling screws, the plate levels being disregarded. The process is repeated alternately for both pairs of opposite leveling screws until the bubble of the telescope level remains centered for any direction of pointing.

6.47. Operations with the transit or theodolite: prolonging a straight line If a straight line such as *AB* (Fig. 6.51) is to be prolonged to *P* (not already defined on the ground), which is beyond the limit of sighting distance or is invisible from *A* and *B*, the line is extended by establishing a succession of stations *C*, *D*, etc., each of which is occupied by the transit.

As an example, line *AB* in Fig. 6.51 may be prolonged to *C* by: (1) occupying *A*, sighting on *B*, and extending the line to *C*; (2) occupying *B*, sighting *A*, and plunging the telescope to establish point *C*; (3) occupying *B*, sighting *A* with zero set on the horizontal circle index, and turning 180° to set point *C*; and (4) occupying *B*, sighting *A*, and extending the line to *C* by *double centering*.

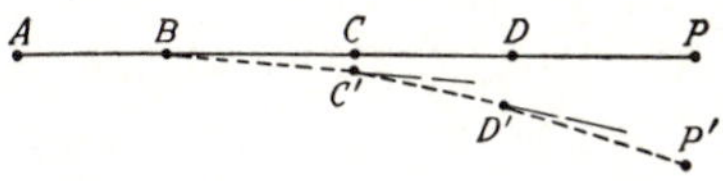

Fig. 6.51 Prolonging a straight line.

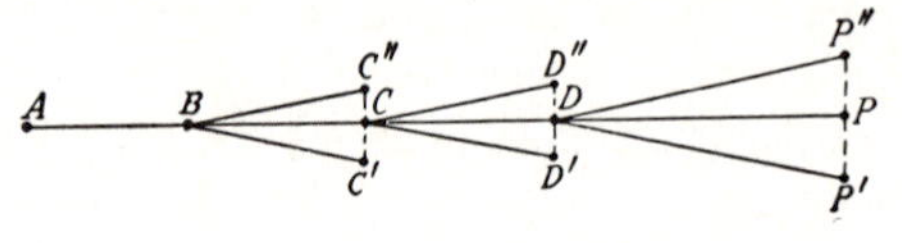

Fig. 6.52 Double sighting to prolong line.

Methods 1 through 3 are susceptible to excessive error propagation due to observational limitations and systematic instrumental errors as the line is extended ultimately to P by successive setups at B, C, and D. These procedures should be used only for surveys where a low order of accuracy is acceptable. The dashed line to C', D', and P' in Fig. 6.51 shows how the errors might propagate using method 2 above with an instrument in which the optical line of sight is not exactly perpendicular to the horizontal axis.

Method 4, double centering, should be employed for all surveys requiring from moderate to high accuracy. If the line AB (Fig. 6.52) is to be prolonged to some point P, the transit is set up at B and a backsight is taken to A with the telescope in its *direct* position. The telescope is plunged, and a point C' is set on line. The transit is then revolved about its vertical axis, and a second backsight is taken to A with telescope *reversed*. The telescope is plunged, and a point C'' is established on line beside C'. It is evident that C' will be as far on one side of the true prolongation of AB as C'' is on the other. Midway between C' and C'', a point C is set defining a point on the correct prolongation of AB. In a similar manner the next point D is established by setting up at C, double centering to B, and setting points at D', D'', and D. The process is repeated until the desired distance is traversed.

6.48. Prolonging a line past an obstacle Figure 6.53 illustrates one method of prolonging a line AB past an obstacle where the offset space is limited. The transit is set up at A, a right angle is turned, and a point C is established at a convenient distance from A. Similarly the point D is established, the distance BD being made equal to AC. The line CD, which is parallel to AB, is prolonged; and points E and F are established in convenient locations beyond the obstacle. From E and F right-angle offsets are made, and G and H are set as were C and D; GH then defines the prolongation of AB. The distance AH is determined by measuring the length of the lines AB, DE, and GH. If measurement of distance is to be carried forward with high accuracy, it is necessary to erect the perpendiculars AC, BD, etc. with greater than ordinary care; and if the line is to be prolonged with high accuracy, it is essential not only that the offset distances be measured carefully but also that AB and EF, the distances between offsets, be as long as practicable.

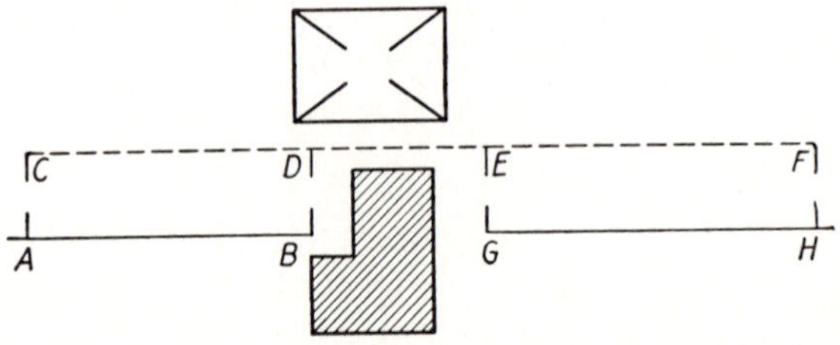

Fig. 6.53 Prolonging line past obstacle by perpendicular offsets.

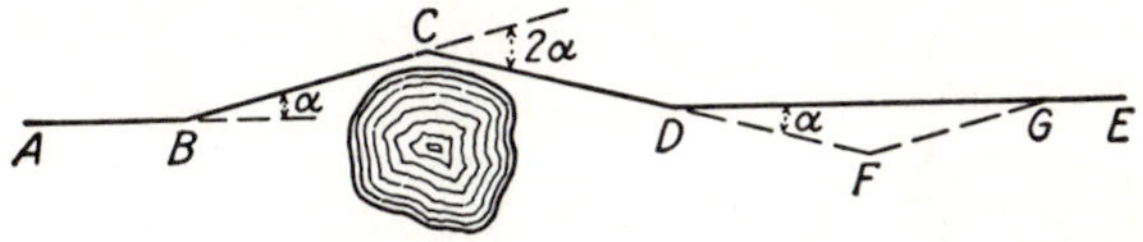

Fig. 6.54 Prolonging line past obstacle by angles.

Another method of prolonging a line AB past an obstacle is illustrated by Fig. 6.54. A small angle α is turned off at B, and the line is prolonged to some convenient point C which will enable the obstacle to be cleared. At C, the angle 2α is turned off in the reverse direction, and the line is prolonged to D, with CD made equal to BC. The point D is then on the prolongation of AB; and DE, the further prolongation of AB, is established by turning off the angle α at D. If there were another obstacle between D and E, as there might often be in wooded country, the line CD is prolonged to some point, as F, from which the obstacle can be cleared; and so a zigzag course is followed until it is possible to resume traversing on the direct prolongation of the main line AB. As compared with the method of perpendicular offsets, this method is more convenient in the field, but it requires computation to determine the length BD.

6.49. Running a straight line between two points If the terminal points A and B of a line are fixed and it is desired to establish intervening points on the straight line joining the terminals, the method to be employed depends upon the length of the line and the character of the terrain. Two common cases are considered below:

Case 1. Terminals intervisible The transit is set up at A, a sight is taken to B, and intervening points are established on line (Fig. 6.55).

When the intervening points are to be set with high precision or when the adjustment of the instrument is uncertain and the vertical angles are large, the intervening stations are set by double-sighting.

Case 2. Terminals not intervisible, but visible from an intervening point on line The location of the line at the intervening point C is determined by trial, as follows: In Fig. 6.55, A and B represent the terminals both of which can be seen from the vicinity of C. The transit is set up on the estimated location of the line near C, a backsight is taken to A, the telescope is plunged, and the location of the line of sight at B is noted. The amount that the transit must be shifted laterally is estimated; and the process is repeated until, when the telescope is plunged, the line of sight falls on the point at B. This process is known as "balancing in." The location of the transit should then be tested by double centering; the test will also disclose whether or not the line of sight and the horizontal axis are in adjustment.

To locate the intermediate point truly on line, trials are made first with the telescope in, say, its direct position for backsights to A, until an intermediate point,

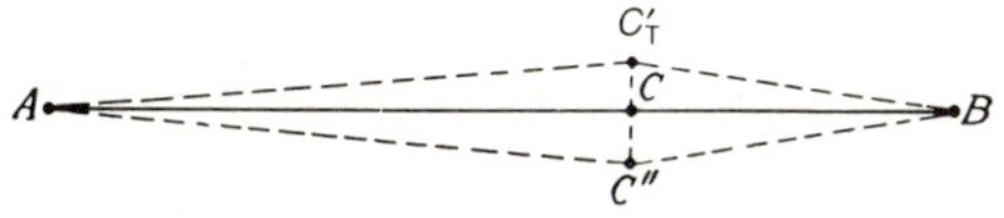

Fig. 6.55 Balancing in.

as C', is determined. Then a second series of trials is made with the telescope reversed for backsights to A, until the corresponding point C'' is located. Then, for reasons previously explained, the true line is at C, halfway between C' and C''. Other intermediate points may then be established by setting up the transit at C and proceeding as in case 1.

6.50. Adjustment of the engineer's transit: desired relations For a transit in perfect adjustment the relations stated in the following paragraphs should exist. The number of each paragraph (1 through 10) is the same as that of the corresponding adjustment described in the following articles. For adjustments 1 to 5, Fig. 6.56 shows the desired relations among the principal lines of the transit.

1. The vertical cross hair should lie in a plane perpendicular to the horizontal axis so that any point on the hair may be employed when measuring horizontal angles or when running lines.
2. The axis of each plate level should lie in a plane perpendicular to the vertical axis so that when the instrument is leveled the vertical axis will be truly vertical; thus horizontal angles will be measured in a horizontal plane, and vertical angles will be measured without index error due to inclination of the vertical axis.
3. The line of sight should be perpendicular to the horizontal axis at its intersection with the vertical axis. Also, the optical axis, the axis of the objective slide, and the line of sight should coincide. If these conditions exist, when the telescope is rotated about the horizontal axis the line of sight will generate a plane when the objective is focused for either a near sight or a far sight, and that plane will pass through the vertical axis.
4. The horizontal axis should be perpendicular to the vertical axis so that when the telescope is plunged the line of sight will generate a vertical plane.
5. The axis of the telescope level should be parallel to the line of sight so that the transit may be employed in direct leveling and so that vertical angles may be measured without index error due to lack of parallelism.

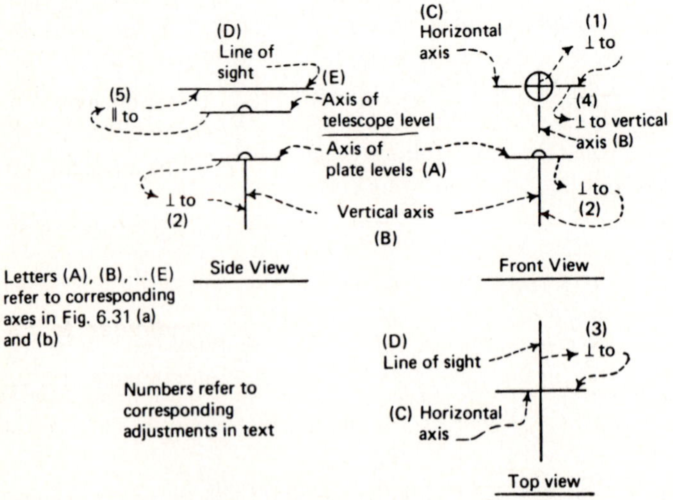

Fig. 6.56 Desired relations among principal lines of transit.

6. If the transit has a fixed vernier for the vertical circle, the vernier should read zero when the plate bubbles and telescope bubble are centered, in order that vertical angles may be measured without index error due to displacement of the vernier.
7. If the vertical vernier is movable and has a control level, the axis of the control level should be parallel to that of the telescope level when the vernier reads zero.
8. The optical axis and the line of sight should coincide (see 3, above).
9. If the transit is equipped with a striding level† for the horizontal axis, the axis of the striding level should be parallel to the horizontal axis. Thus, when the bubble of the striding level is centered and the instrument is plunged, the line of sight (if in adjustment) will generate a vertical plane.
10. On transits and theodolites equipped with an optical plummet, the vertical collimation line should coincide with the plumb line.
11. The axis of the objective slide should be perpendicular to the horizontal axis (see 3, above).

6.51. Adjustments In the description of the following adjustments (except 7 and 9) it is assumed that the objective slide does not permit adjustment, but that it is permanently fixed in the telescope tube so far as lateral motion is concerned; and that the maker has so constructed the instrument that the optical axis and the axis of the objective slide coincide and are perpendicular to the horizontal axis. This ideal construction is never exactly attained; but in most instruments the departure is so slight that it need not be considered in ordinary transit work, and in precise surveying the resulting errors are eliminated by methods of procedure.

The adjustments of the transit are more or less dependent upon one another. For this reason, if the instrument is badly out of adjustment, time will be saved by first making corrections roughly for related adjustments until all the tests have been tried and then repeating the tests and corrections in the same order. The plate levels will not be disturbed by other adjustments and should be corrected exactly before other adjustments are attempted.

Any movement of the screws controlling the cross-hair ring is likely to produce both lateral displacement and rotation of the ring; hence any considerable adjustment of the line of sight is likely to disturb the vertical hair so that it will no longer remain on a point when the telescope is rotated about the horizontal axis. The adjustment of the telescope level depends upon the unaltered position of the horizontal cross hair and hence should not be tested until the line of sight and horizontal axis have been corrected.

The transit adjustments commonly made are 1 to 7. Adjustments 8 to 10 may be required occasionally for some instruments.

1. To make the vertical cross hair lie in a plane perpendicular to the horizontal axis

Test. Sight the vertical cross hair on a well-defined point not less than 200 ft or 60 m away. With both horizontal motions of the instrument clamped, swing the

†A spirit level (sensitivity of 05″ to 15″ per 2-mm division) constructed so it can be placed on the horizontal instrumental axis and then removed when no longer needed.

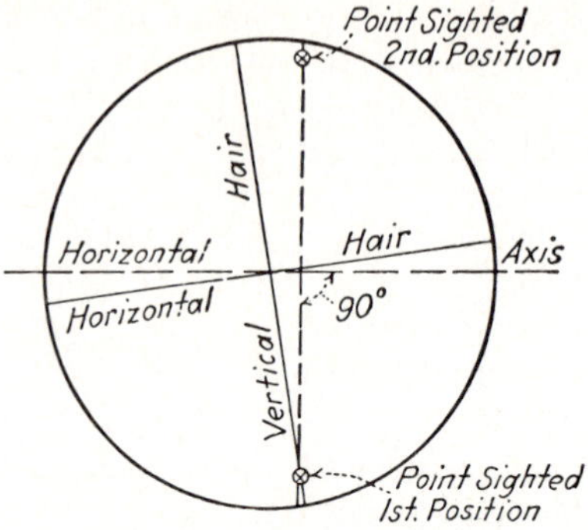

Fig. 6.57 Adjustment of vertical cross hair of transit.

telescope through a small vertical angle, so that the point traverses the length of the vertical cross hair. If the point appears to move continuously on the hair, the cross hair lies in a plane perpendicular to the horizontal axis (see Fig. 6.57).

Correction. If the point appears to depart from the cross hair, loosen two adjacent capstan screws and rotate the cross-hair ring in the telescope tube until the point traverses the entire length of the hair. Tighten the same two screws. This adjustment is similar to adjustment 2 of the dumpy level (Art. 5.47), with the terms *vertical* and *horizontal* interchanged.

2. *To make the axis of each plate level lie in a plane perpendicular to the vertical axis*

Test. Rotate the instrument about the vertical axis until each level tube is parallel to a pair of opposite leveling screws. Center the bubbles by means of the leveling screws. Rotate the instrument end for end about the vertical axis. If the bubbles remain centered, the axis of each level tube is in a plane perpendicular to the vertical axis (see Fig. 6.58).

Correction. If the bubbles become displaced, bring them *halfway* back by means of the adjusting screws. Level the instrument again and repeat the test to verify the results. This is the method of reversion.

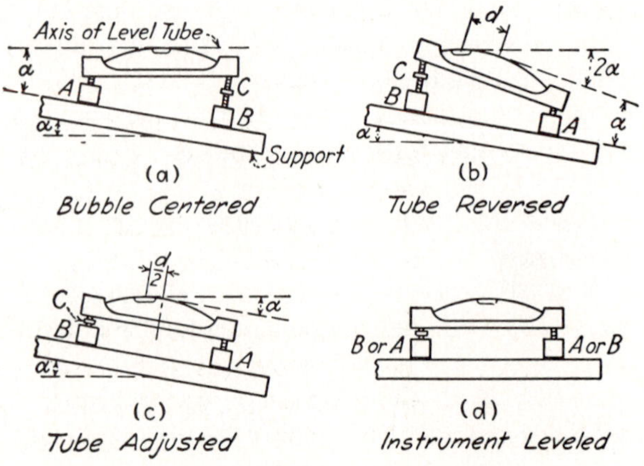

Fig. 6.58 (*a*) through (*d*) Adjustment of level tube by reversion.

The steps involved in adjustment are shown in Fig. 6.58. In view (*a*) is shown the level tube out of adjustment by the amount of the angle α, but with the bubble centered; the support is therefore not level, and the vertical axis is not vertical. In view (*b*) the level tube has been lifted and reversed end for end, which is effected by rotating the instrument 180° about the vertical axis. The axis of the level tube now departs from the horizontal by 2α, or *double* the error of the setting. In view (*c*) the bubble has been brought back *halfway* to the middle of the tube by means of the adjusting screw *C*, without moving the support; the tube is now in adjustment. Finally, in view (*d*) the bubble is again centered by raising the low end (and/or lowering the high end) of the support; the support is now level and the adjustment may be checked by reversing the tube again.

If it were desired to level the support in the direction of the tube without taking time to adjust the tube, this could be accomplished by centering the bubble as in view (*a*), Fig. 6.58; reversing the tube as in view (*b*); and raising the low end (and/or lowering the high end) of the support until the bubble is brought halfway back to the center of the tube. This position of the bubble corresponds to the error of setting of the tube; and whenever the bubble is in this position the support will be level.

3. To make the line of sight perpendicular to the horizontal axis

Test. Level the instrument. Sight on a point *A* (see Fig. 6.59) about 500 ft or 150 m away, with telescope in the direct position. With both horizontal motions of the instrument clamped, plunge the telescope and set another point *B* on the line of sight and about the same distance away on the opposite side of the transit. Unclamp the upper motion, rotate the instrument end for end about the vertical axis, and again sight at *A* (with telescope reversed). Clamp the upper motion. Plunge the telescope as before; if *B* is on the line of sight, the desired relation exists.

Correction. If the line of sight does not fall on *B*, set a point *C* on the line of sight beside *B*. Mark a point *D*, *one-fourth* of the distance from *C* to *B*, and adjust the cross-hair ring (by means of the two opposite horizontal screws) until the line of sight passes through *D*. The points sighted should be at about the same elevation as the transit.

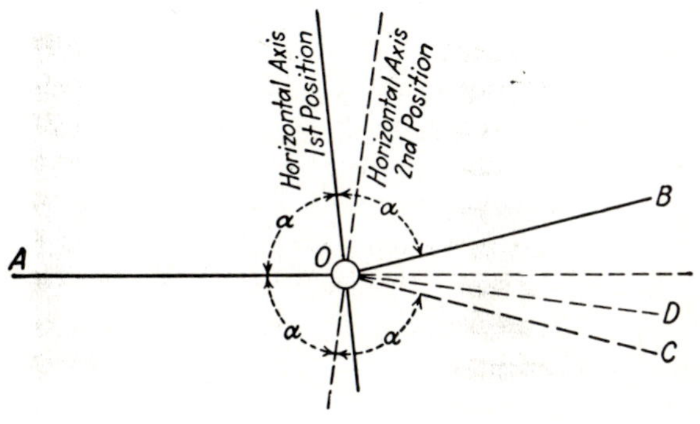

Fig. 6.59 Adjustment of line of sight of transit.

4. To make the horizontal axis perpendicular to the vertical axis

Test. Set up the transit near a building or other object on which is some well-defined point A at a considerable vertical angle. Level the instrument very carefully, thus making the vertical axis truly vertical. Sight at the high point A (see Fig. 6.60) and, with the horizontal motions clamped, depress the telescope and set a point B on or near the ground. If the horizontal axis is perpendicular to the vertical axis, A and B will be in the same vertical plane. Plunge the telescope, rotate the instrument end for end about the vertical axis, and again sight on A. Depress the telescope as before; if the line of sight falls on B, the horizontal axis is perpendicular to the vertical axis.

Correction. If the line of sight does not fall on B, set a point C on the line of sight beside B. A point D, halfway between B and C, will lie in the same vertical plane with A. Sight on D; elevate the telescope until the line of sight is beside A; loosen the screws of the bearing cap, and raise or lower the adjustable end of the horizontal axis until the line of sight is in the same vertical plane with A.

The high end of the horizontal axis is always on the same side of the vertical plane through the high point as the point last set.

In readjusting the bearing cap, care should be taken not to bind the horizontal axis, but it should not be left so loose as to allow the objective end of the telescope to drop of its own weight when not clamped.

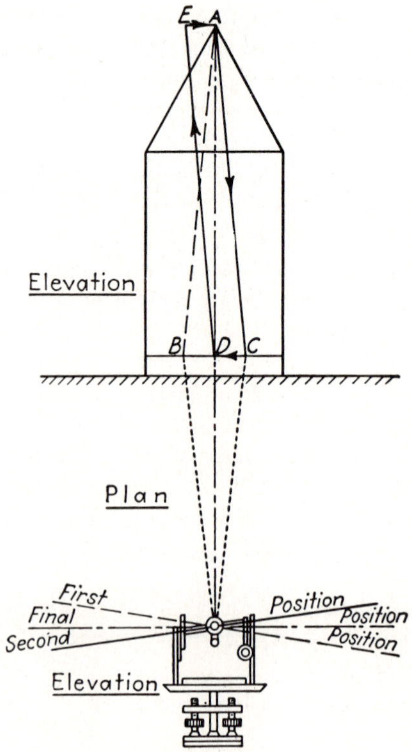

Fig. 6.60 Adjustment of horizontal axis of transit.

5. To make the axis of the telescope level parallel to the line of sight

Test and correction. Proceed the same as for the two-peg adjustment of the dumpy level (Art. 5.47, adjustment 3), except as follows: With the line of sight set on the rod reading established for a horizontal line, the correction is made by raising or lowering one end of the telescope level tube until the bubble is centered.

6. (For Transit Having a Fixed Vertical Vernier) ***To make the vertical circle read zero when the telescope bubble is centered***

Test. With the plate bubbles centered, center the telescope bubble and read the vertical vernier.

Correction. If the vernier does not read zero, loosen it and move it until it reads zero. Care should be taken that the vernier will not bind on the vertical circle as the telescope is rotated about the horizontal axis.

7. (For Transit Having a Movable Vertical Vernier with Control Level) ***To make the axis of the auxiliary level parallel to the axis of the telescope level when the vertical vernier reads zero***

Test. Center the telescope bubble, and by means of the vernier tangent screw move the vertical vernier until it reads zero.

Correction. If the bubble of the control level which is attached to the vertical vernier is not at the center of the tube, bring it to the center by means of the capstan screws at one end of the tube.

6.52. Special adjustments In addition to the usual adjustments just described, the following adjustments may be made as required by the type or the condition of the transit.

8. To make the line of sight, insofar as defined by the horizontal cross hair, coincide with the optical axis

Test. Set two pegs, one about 25 ft or about 8 m and the other 300 to 400 ft or about 90 to 100 m from the transit. With the vertical motion clamped, take a rod reading on the distant point, and without disturbing the vertical motion read the rod on the near point. Plunge the telescope, rotate the instrument about the vertical axis, and set the horizontal cross hair at the last rod reading with the rod held on the near point. Sight to the distant point. If the desired relation exists, the first and last readings on the distant rod will be the same.

Correction. If there is a difference between the rod readings, move the horizontal cross hair by means of the upper and lower adjusting screws until it intercepts the distant rod at the mean of the two readings. Repeat the process, until by successive approximations the error is reduced to zero.

9. (For Transit Equipped with Striding Level) ***To make the axis of the striding level parallel to the horizontal axis***

Test. By means of the level screws center the striding level bubble. Lift the level from its supports and turn it end for end. If it is in adjustment, the bubble will again be centered.

Correction. If the bubble is displaced, bring it halfway back to the center by means of the capstan screw at one end of the level tube (Fig. 6.58). Relevel the instrument by means of the leveling screws and repeat the test until the adjustment is perfected.

10. *Adjustment of the optical plummet*

Field adjustment is performed as follows. The instrument is leveled and centered over a clearly marked point using a plumb bob suspended from the screw that secures the instrument to the foot plate. The plumb line is removed and the cross in the optical plummet is very carefully focused on the point. Any deviations between the point and cross are due to maladjustment of the optical plummet and can be corrected by means of adjusting screws usually located near the optical plummet's eyepiece. For best results, this test should be conducted in an area protected from the wind. A heavy plumb bob should be used.

A more accurate procedure for checking and adjusting an optical plummet is to apply the reversion principle. After centering the instrument over a carefully marked point, it is necessary to rotate the entire assembly of instrument, leveling head, foot plate, and tripod (or trivet) 180° about the vertical axis. This procedure requires auxiliary equipment and is best performed in a laboratory.

6.53. Adjustment of theodolites Adjustment of repeating and direction theodolites requires that the same fundamental relationships must be satisfied in the same sequence as detailed for the engineer's transit. However, there are certain differences which are outlined in the following paragraphs.

1. *Adjustment of the plate level bubble* The theodolite is set up and the circular level bubble is centered by means of the three level screws (Fig. 6.43). Next, the axis of the plate level bubble tube is aligned with a pair of level screws and the bubble is centered. Rotate the theodolite 180° about the vertical axis; if the bubble leaves its central position, correct one-half the error with the level screws and one-half the error using the adjusting screw at one end of the bubble tube. If the error is only one or two graduations, do not make the adjustment. Bring the bubble halfway to center with the level screws and note the position of the bubble for future operations. Next, rotate the instrument 90° about the vertical axis from its original position and center the plate level bubble with the remaining screw. Then rotate the instrument 180° and note the position of the bubble. If the deviation is small (one or two divisions), note the position for future reference and bring the bubble halfway to center with the level screws but do not make an adjustment. If the deviation is large, bring the bubble halfway to center with the level screws and remove the remaining half with the adjusting screw usually located at one end of the level bubble tube.

2. *Adjustment of the circular level bubble* Level the theodolite using the plate level bubble (Art. 6.34). The circular bubble in the tribach of the leveling head should be centered. If the circular bubble is not centered, it should be adjusted by using the adjusting screws located around the bubble case. The adjustment procedure is the same as described for automatic levels in Art. 5.48.

3. Optical plummet Adjustment of the optical plummet in a theodolite is the same as described for the engineer's transit equipped with an optical collimator (Art. 6.52, adjustment number 10).

4. Vertical circle level bubble The instrument is leveled and the vertical circle level bubble is centered using the vertical tangent screw. Using the vertical lock and tangent screw, the horizontal cross hair is then sighted on a well-defined object point about 400 ft or 120 m away and the vertical circle reading is observed. Assume that this observation is 92°01′30.0″. The telescope is reversed, the vertical circle level bubble is centered, and the horizontal cross hair is sighted on the same object point with the vertical tangent motion. If the zenith angle from this second sight is 360° − (267°58′40.0″) = 92°01′20.0″, the index error is 05″ and the true value of the zenith angle is 92°01′25.0″. To make the adjustment, 267°58′35.0″ is set on the vertical circle using the altitude tangent screw and observing the optical micrometer while the horizontal cross hair remains on the object point. This manipulation causes the vertical circle level bubble to move off center. The bubble is centered by loosening one of the adjusting screws at one end of the level bubble and tightening the other. When the correction has been made, the entire procedure should be repeated to test the quality of the adjustment.

For theodolites equipped with self-indexing vertical circles, the test for the error is the same but adjustment is accomplished by moving the cross-hair reticle up or down until the horizontal cross hair is on the target while the horizontal circle and micrometer are on the correct angle. Note that this operation is the same as adjustment 8 (Art. 6.52) to ensure that the line of sight as defined by the horizontal cross hair coincides with the optical axis.

Locations of adjusting screws are not standardized on theodolites. Reference should be made to the manufacturer's instruction manual prior to attempting any adjustment.

6.54. Errors in determining angles and directions: general Except in field astronomy, a measured angle is always closely related to a measured distance; and in general there should be a consistent relation between the precision of measured angles and that of measured distances (see Art. 3.7). From the standpoint of both precision achieved and expediting the work, it is important that the surveyor be able to: (1) visualize the effect of errors in terms of both angle and distance, (2) appreciate what degree of care must be exercised to keep certain errors within specified limits, and (3) know under what conditions various instrumental errors can be eliminated.

Errors in transit work may be instrumental (Art. 6.55), personal (Art. 6.56), or natural (Art. 6.57).

6.55. Instrumental errors Errors due to instrumental imperfections and/or non-adjustment are all systematic, and they can either be eliminated or reduced to a negligible amount by proper methods of procedure.

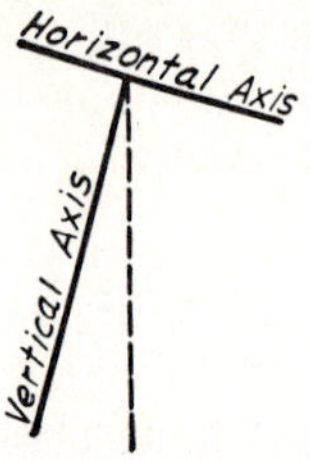

Fig. 6.61 Inclination of vertical axis due to maladjustment of plate level bubbles.

1. Errors in horizontal angles caused by nonadjustment of plate levels When the bubbles of plate levels in nonadjustment are centered, the vertical axis is inclined, and hence measured angles are not truly horizontal angles. Also the horizontal axis is inclined to a varying degree depending upon the direction in which the telescope is sighted. There will be one vertical plane which will include the vertical axis in its inclined position; this is illustrated by Fig. 6.61, in which the horizontal axis and the vertical axis are in the plane of the paper. When the horizontal axis is rotated about the vertical axis until it is normal to the plane of the paper, it becomes truly horizontal and the line of sight will generate a vertical plane when the telescope is plunged; hence no error in direction is introduced regardless of the angle of elevation to the point sighted. As the transit is rotated about its vertical axis, the horizontal axis becomes inclined, making a maximum angle with the horizontal when it reaches the plane of the paper. With the horizontal axis in this position, the line of sight generates a plane making an angle with the vertical equal to the error in the position of the vertical axis; and with the line of sight inclined at a given angle, the maximum error in determining the direction of a line is introduced. The larger the vertical angle, the greater the error in direction. The error cannot be eliminated by double centering.

The diagram of Fig. 6.62 shows for various vertical angles (values of α) the errors introduced in horizontal angles due to an inclination of 01′ in the vertical axis or one space on the plate levels of the ordinary transit. The values of H are the horizontal angles which the line of sight makes with the vertical plane in which lies the vertical axis in its inclined position (i.e., with the plane of the paper, Fig. 6.61). Within reasonable limits the error in horizontal angle varies directly as the inclination of the

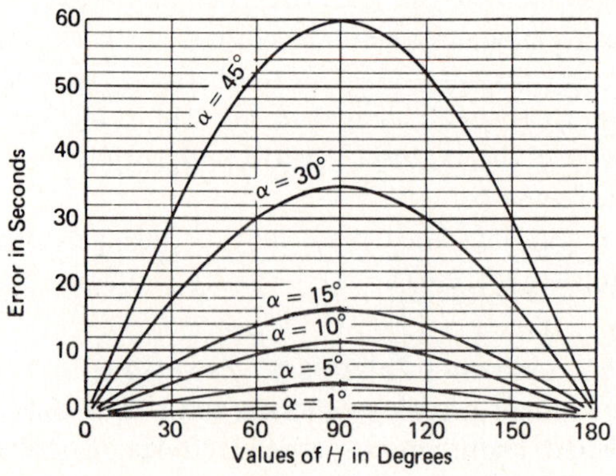

Fig. 6.62 Errors in horizontal angles for a 1′ inclination of the vertical axis (α = vertical angle).

vertical axis, hence a similar diagram for an inclination of 02′ would show ordinates twice as great as those of Fig. 6.62.

Although the diagram may not be of much practical value, it serves to illustrate some noteworthy facts.

(*a*) For observations of ordinary accuracy taken in flat country where the vertical angles are rarely greater than 3° and usually much less, the plate bubble may be out several spaces without appreciably affecting the precision of horizontal angular measurements. For example, if an angle were measured between $H = 0°$ and $H = 90°$, with bubble out two spaces and $\alpha = 3°$, the error would be about 06″; or if in prolonging a straight line the telescope were plunged from the position $H = 90°$ to $H = 270°$ with both backsight and foresight taken at a vertical angle of +3°, the error would be doubled and for the bubbles out two spaces (vertical axis in error 02′), the angular error introduced in the direction of the line would be 12″.

(*b*) For angular measurements of higher precision, such as when measuring an angle by repetition, the plate levels must be in good adjustment and the bubbles must be centered with reasonable care even though the survey is conducted over fairly smooth ground. For example, if a horizontal angle were measured between the positions $H = 0°$ and $H = 90°$, $\alpha = 5°$, and the vertical axis were inclined 30″, the error in horizontal angle would amount to 02.2″.

(*c*) In rough country where the vertical angles are large, even for surveys of ordinary precision the plate levels must be in good adjustment and the bubbles must be carefully centered if errors in horizontal angles or in the prolongation of lines are to be kept within negligible limits. For example, if a line were prolonged by plunging the telescope from the position $H = 90°$ to $H = 270°$, α for both backsight and foresight being +30° and the vertical axis being inclined 01′, the diagram (Fig. 6.62) shows that the error introduced is $2 \times 34.6'' = 01'09.2''$. In other words, the angle at the station at which the instrument was set instead of being a true 180° would be 180°01′09.2″, and beyond the station the established line would depart from the true prolongation about 0.1 ft in each 300 ft or 33 mm in 100 m.

2. Errors in vertical angles due to nonadjustment of plate levels These errors obviously vary with the direction in which the instrument is pointed. With the fixed vertical vernier they are eliminated by observing (for each sighting) the index error of the corresponding observed vertical angle (Art. 6.45).

Certain styles of repeating and direction theodolites have vertical circle level bubbles which are centered prior to each vertical angle eliminating the error caused by nonadjustment of plate level bubbles (Art. 6.51, test for adjustment 2). Theodolites equipped with automatic vertical circle indexing are also immune to this type of error.

It may be noted further that nonadjustment of the plate levels causes an inclination of the plane of the vertical arc. This source of error may be considered negligible.

3. Line of sight not perpendicular to horizontal axis If the telescope is not reversed between backsight and foresight, if the sights are of the same length so that it is not required to refocus the objective, and if both points sighted are at the same angle of inclination of the line of sight, no error is introduced in the measurement of horizontal angles even though the error due to this lack of adjustment is large. If the instrument is plunged between backsight and foresight, the telescope rotating about the horizontal axis generates a cone, the apex of which falls at the intersection of the line of sight and the horizontal axis. The resultant error in the observed angle is double the error of adjustment.

With the line of sight out of adjustment by a given amount, the effect of the error depends on the vertical angle to the point sighted. In Fig. 6.63, *OA* and *DB* are horizontal and are perpendicular to the horizontal axis *OH* of the instrument; *e* is the angle between the nonadjusted line of sight and a vertical plane normal to *OH* (that is, *e* is the error in direction for a horizontal sight *OB*); *E* is the error in direction for an inclined sight *OC*; *h* is the actual vertical angle to *C*, the point sighted; and *OB* is made equal to *OC*. Then

$$\begin{aligned}\sin e &= \frac{AB}{OB}\\ &= \frac{AB}{OB}\frac{OD}{OD} = \frac{AB}{OD}\frac{OD}{OB} = \frac{AB}{OD}\frac{OD}{OC}\\ &= \sin E \cos h\end{aligned}$$

or

$$\sin E = \sin e \sec h \tag{6.1}$$

For all ordinary cases Eq. (6.1) may be taken as

$$E = e \sec h = e \sec \alpha \qquad \text{(approx.)} \tag{6.2}$$

where α is the observed vertical angle.

For example, assume that the optical line of sight makes an angle of 89°58′00″ with the horizontal axis of the instrument. In turning a horizontal angle, if the line of sight on the backsight is horizontal and is inclined with a vertical angle of 40°00′ on the foresight, the error in the observed horizontal angle due to the lack of adjustment is

$$E = (02')\sec 40°00' = (02')(1.30541) = 0°02'37''$$

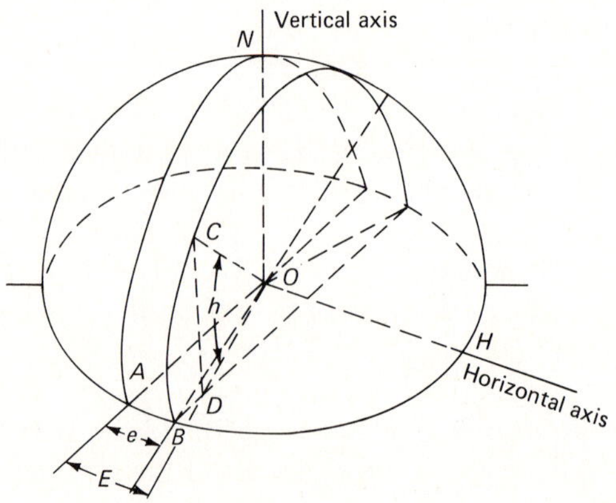

Fig. 6.63

For two direct pointings (the backsight and the foresight) there will be a value of E for each, and the error in the angle is the difference between them. In the measurement of a deflection angle by the method in which the telescope is inverted between backsight and foresight, the error in angle is the *sum* of the two values of E.

The error may be eliminated by taking the mean of two angular observations, one with the telescope in the direct position and the other with the telescope reversed. In prolonging a line, errors are avoided by the method of double centering described in Art. 6.47.

4. Horizontal axis not perpendicular to vertical axis No error is introduced in horizontal angles so long as the points sighted are at the same angle of inclination of the line of sight. The angular error in the observed direction of any line depends both on the angle by which the horizontal axis departs from the perpendicular to the vertical axis and on the vertical angle to the point sighted. In Fig. 6.64, OH is perpendicular to the vertical axis; OH' is the horizontal axis in nonadjustment with the vertical axis by the angular amount e'; OA is horizontal and is perpendicular to OH and OH'; α is the observed vertical angle to C, the point sighted; B is directly beneath C and in the same horizontal plane with OA; angles OAB, OAC, and ABC are right angles; and θ is the angular error in direction. From the figure,

$$\begin{aligned}\tan\theta &= \frac{BA}{OA}\\ &= \frac{BA}{OA}\frac{CA}{CA} = \frac{BA}{CA}\frac{CA}{OA}\\ &= \sin e' \tan\alpha \qquad (6.3)\end{aligned}$$

or with sufficient precision,

$$\theta = e' \tan\alpha \qquad \text{(approx.)} \qquad (6.4)$$

Thus, for example, if a horizontal angle were measured between D, to which the vertical angle is $-30°$, and E, to which the vertical angle is $+15°$, the horizontal axis being inclined 02′, then the error in horizontal angle is

$$\theta = 02'[\tan 15° - (-\tan 30°)] = 32'' + 01'09'' = 01'41''$$

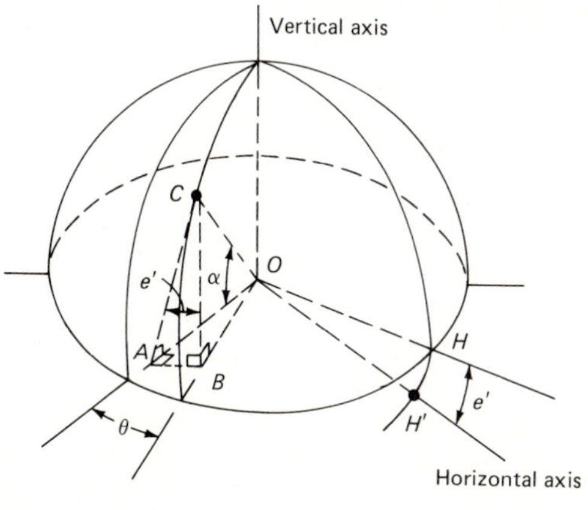

Fig. 6.64

The preceding example is sufficient to show that the error in an observed horizontal angle may become large. Obviously, the sign of the error in a horizontal angle depends upon the direction of displacement of the horizontal axis from its correct position. Hence, if any angle is measured with the telescope first in the direct and then in the reversed position, one value will be too great by the amount of the error and the other will be correspondingly too small; thus, the error is eliminated by taking the mean of the two values.

5. Effect of lack of coincidence between line of sight and optical axis Under these conditions if the line of sight is perpendicular to the horizontal axis for one position of the objective, it will not be perpendicular for other positions, but will swing through an angle as the objective is moved in or out. If an angle is measured without disturbing the position of the objective, no error is introduced. For most instruments the error from this source is not sufficiently large to be of consequence in ordinary transit or theodolite work. It is eliminated by taking the mean of two angles, one observed with the telescope direct and the other with it reversed.

6. Errors due to eccentricity With the transit in good condition, errors due to eccentricity of verniers and/or eccentricity of centers are of no consequence in the ordinary measurement of angles. In any case, such errors are eliminated by taking the mean of readings indicated by opposite verniers.

7. Imperfect graduations Errors from this source are of consequence only in work of high precision. They can be reduced to a negligible amount by taking the mean of several observations for which the readings are distributed over the circle and over the vernier.

8. Lack of parallelism between axis of telescope level and line of sight This introduces an error in leveling which is compensated by equalizing backsight and foresight distances (see Art. 5.40). An error is also introduced into vertical angles where compensation is achieved by observing two vertical angles, one with the telescope direct and one with the telescope reversed (Art. 6.43). The mean of the two observations is the best estimate of the vertical angle.

9. Nonadjustment of the vertical vernier This produces a constant error in the measurement of vertical angles. If the transit is equipped with a full vertical circle, the error can be eliminated by taking the mean of two values, one observed with the telescope direct and the other with the telescope reversed.

Summary Summing up, it is seen that with regard to the instrumental errors:

1. Errors in horizontal angles due to nonadjustment of plate levels or of horizontal axis become large as the inclination of the sights increases.
2. The maximum error in horizontal angles due to nonadjustment of the line of sight is introduced when the telescope is plunged between backsight and foresight readings. If the

telescope is not plunged between backsight and foresight readings when observing horizontal angles, no error is introduced when these distances are equal and the inclination of the line of sight is the same for both backsight and foresight.

3. Errors due to instrumental imperfections and/or nonadjustment are all systematic, and without exception they can be either eliminated or reduced to a negligible amount by proper procedure. In general, this procedure consists in obtaining the mean of two values —one observed before and one after a reversal of the horizontal plate by plunging the telescope and rotating it about the vertical axis. One of these values is as much too large as the other is too small. An exception is the error in either horizontal or vertical angle due to inclination of the vertical axis, which cannot be so eliminated but which can be eliminated, so far as its *systematic* character is concerned, by releveling the plate bubbles in addition to the reversal of the plate. However, for precise work the usual practice would be to make the vertical axis truly vertical by means of the telescope level, and then to proceed in the ordinary manner.

6.56. Personal errors Personal errors arise from the limitations of the human eye in setting up and leveling the transit and in making observations.

1. Effect of not setting up exactly over the station This produces an error in all angles measured at a given station, the magnitude of the error varying with the direction of pointing and inversely with the length of sight. It is convenient to remember that 0.1 ft in approximately 300 ft or 3 cm in about 100 m yields an angle of 01′. Thus, if the transit were offset 0.05 ft or 1.5 cm from the ends of lines 150 ft or 50 m long, respectively, the error in the observed direction would be 01′, but if the lines were 600 ft or 200 m long, the error would be only 15″. In general, the error may be kept within negligible limits by reasonable care. Many instrument operators waste time by exercising needless care in setting up when the sights to be taken are long.

When using a plumb bob, centering to within 0.02 ft or 6 to 9 mm is possible without difficulty. Centering errors of less than 0.003 ft or 1 mm are possible with the optical plummet or telescopic centering rod.

2. Effect of not centering the plate bubbles exactly This produces an error in horizontal angles after the manner described in the preceding section for plate levels out of adjustment. The error from this source is small when the sights are nearly level, but may be large for steeply inclined sights (see Fig. 6.62). The average transit operator does not appreciate the importance of careful leveling for steeply inclined sights; on the other hand, unnecessary care in leveling is often used when sights are nearly horizontal. Since the error in horizontal angle is caused largely by inclination of the horizontal axis, the striding level is a necessity on precise work.

3. Errors in settling and reading the vernier These are functions of the least count of the vernier and of the legibility of scale and vernier lines. For the usual 01′ transit the readings and settings can be made to the nearest 30″; for the 30″ transit these values are to the nearest 15″. The use of a reading glass enables closer reading, particularly for finely graduated circles. Also, in reading the vernier it is helpful to observe the position of the graduations on both sides of the ones that appear to coincide, and to note that the unmatched graduations appear to lack coincidence by the same amount.

Horizontal and vertical circle readings with an optical theodolite having a least count of 1″ can be observed with a standard deviation of from 2″ to 3″.

4. Not sighting exactly on the point This is likely to be a source of rather large error on ordinary surveys where sights are taken on the range pole of which often only the upper portion is visible from the transit. The effect upon a direction is, of course, the same as the effect of not setting up exactly over the station. For short sights greater care should be taken than for long sights, and the plumb line should be employed instead of the range pole.

5. Imperfect focusing (parallax) The error due to imperfect focusing is always present to a greater or less degree, but with reasonable care it can be reduced to a negligible quantity. The manner of detecting parallax is described in Art. 5.10.

Summary All the personal errors are random and hence cannot be eliminated. They form a large part of the resultant error in transit and theodolite work. Of the personal errors, those due to inaccuracies in reading and setting the vernier or reading and setting the optical micrometer and to not sighting exactly on the point are likely to be of greater magnitude.

6.57. Natural errors Sources of natural errors are (1) settlement of the tripod, (2) unequal atmospheric refraction, (3) unequal expansion of parts of the telescope due to temperature changes, and (4) wind, producing vibration of the transit or making it difficult to plumb correctly.

In general, the errors resulting from natural causes are not large enough to affect appreciably the measurements of ordinary precision. However, large errors are likely to arise from settlement of the tripod when the transit or theodolite is set up on boggy or thawing ground; in such cases the instrument may be kept relatively stable through the use of extra-long tripods or of small platforms for the tripod legs. Settlement is usually accompanied by an angular movement about the vertical axis as well as linear movements both vertically and horizontally. When horizontal angles are being measured, usually a larger error is produced by the angular displacement of the circle between backsight and corresponding foresight than by the movement of the transit laterally from the point over which it is set. Errors due to adverse atmospheric conditions can usually be rendered negligible by choosing appropriate times for observing.

For measurements of high precision the methods of observing are such that instrumental and personal errors are kept within very small limits, and natural errors become of relatively great importance. Natural errors are generally random, but under certain conditions systematic errors may arise from natural causes. On surveys of very high precision, special attempt is made to establish a procedure which will as nearly as possible eliminate natural systematic errors. Thus the instrument may be

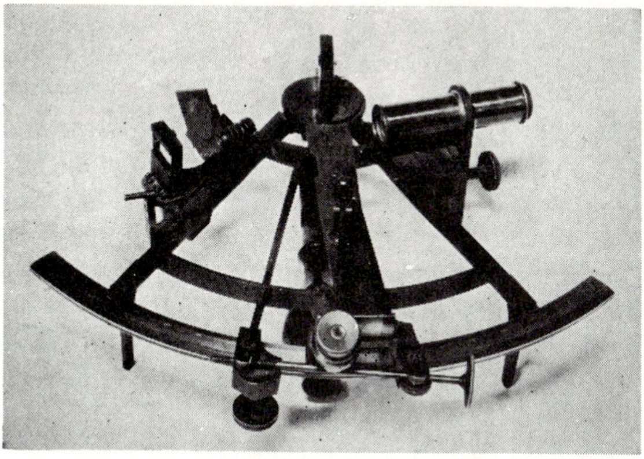

Fig. 6.65 Sextant.

set up on a masonry pier and protected from sun and wind; also certain readings may be made at night when temperature and atmospheric conditions are nearly constant.

6.58. Sextant The sextant (Fig. 6.65) is well suited to hydrographic work and has the added advantage of measuring angles in any plane. It is used principally by navigators and surveyors for measuring angles from a boat, but it is also employed on exploratory, reconnaissance, and preliminary surveys on land.

The essential features of the sextant are illustrated in Fig. 6.66, with the instrument in position for measuring a horizontal angle *FEL*. An index mirror *I* is rigidly attached to a movable arm *ID*, which is fitted with a vernier, clamp, and tangent screw and which moves along the graduated arc *AB*. A second mirror *H*, called the *horizon glass*, having the lower half of the glass silvered and the upper half clear, is rigidly attached to the frame. A telescope *E*, also rigidly attached to the frame, points into the mirror *H*.

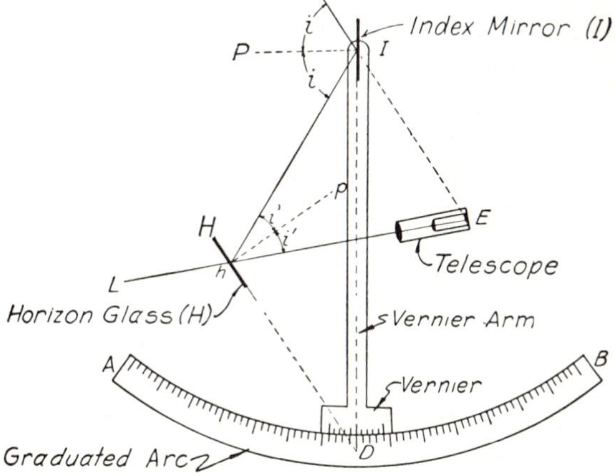

Fig. 6.66 Essential features of a sextant.

With signals at L and F and the eye at E, it is desired to measure the angle FEL. The ray of light from signal L passes through the clear portion of glass H on through the telescope to the eye at E. The ray of light from signal F strikes the index mirror at I and is reflected to h and then through the telescope to E. Each set of rays forms its own image on its respective half of the objective. When the arm ID is moved, these images can be made to move over each other and there will be one position in which they coincide. An observation with the sextant consists in bringing the two images into exact coincidence and reading the angle on the vernier scale on limb AB.

To prove that the angle FEL equals two times angle IDh, that is, that the angle between the signals is equal to twice the angle between the mirrors: Draw IP and hp normal to the two mirrors; then the angles of incidence and reflection of the two mirrors are i and i', respectively. By trigonometry,

$$\begin{aligned} FEL &= FIh - IhE \\ &= 2i - 2i' \\ &= 2(i - i') \end{aligned}$$

Also,

$$\begin{aligned} IDh &= HhI - hID \\ &= (90° - i') - (90° - i) \\ &= i - i' \end{aligned}$$

Therefore, FEL (the angle between the objects) equals twice the angle IDh (the angle between the mirrors).

6.59. Measuring angles with the sextant The handle of the sextant is held in the right hand, and the plane of the arc is made to coincide with the plane of the two objects between which the angle is to be measured. The sextant is turned in the plane of the objects until the left-hand object can be viewed through the telescope and the clear portion of the horizon glass. With the instrument held in this position, the index arm is moved with the left hand until the images of the two objects coincide. The final setting is made with the tangent screw, and a test for coincidence is made by twisting the sextant slightly in the hand to make the reflected image move back and forth across the position of coincidence. When the setting is thus verified, the vernier arm is clamped and the angle is read.

The precision of horizontal angle measurement with the sextant depends upon the size of the angle and upon the length of sight. It is evident from Fig. 6.66 that the angle FEL actually measured has its vertex E not at the eye but at the intersection of the sight rays FE and LE from the points sighted. The distance to this intersection will increase as the angle decreases, and for small angles the vertex may be at a considerable distance back of the observer. The effect of this factor on the horizontal angle is essentially the same as the error caused by improper centering of the transit (Art. 6.56). Hence the sextant is not an instrument of precision for small angles (say, less than 15°) and short distances (say, less than 1000 ft or about 300 m). If objects sighted are at a great distance away, the angular error is usually small (about 1′ of arc).

Vertical angles can be measured with the sextant in a manner similar to that just described for horizontal angles. Most commonly, the altitude of a celestial body is observed, as in navigation. The vertical angle between the body and the sea horizon is measured and is corrected for dip, or angular distance of the observer above the horizon. On land, if the sea horizon is not visible, an artificial horizon is used; this consists of a horizontal reflecting surface, such as that of a small vessel of mercury, near the sextant. The vertical angle between the celestial body and its reflection in the artificial horizon is measured; this angle is twice the altitude of the body.

6.60. Problems

6.1 Describe the methods by which points on or near the surface of the earth may be located.
6.2 Describe the differences among true, grid, and magnetic meridians.
6.3 Given the following bearings: N35°40′E, S46°32′E, N60°30′W, S32°59′W, N89°40′W, N79°59′E, N65°20′W, S78°42′W, N10°13′W, S15°16′W. Express these directions as (*a*) azimuths from the north; (*b*) azimuths from the south.
6.4 Express the following azimuths from the north as bearings: 329°15′, 45°00′, 181°01′, 172°31′, 146°40′, 242°13′, 10°38′, 92°59′, 349°21′, 319°40′.
6.5 Given the following forward azimuths: A_{AB} = 273°30′, A_{AC} = 42°15′, A_{DC} = 195°25′, and A_{DE} = 160°45′. Convert these directions to back azimuths.
6.6 In Prob. 6.5, determine the clockwise horizontal angles between lines *AB* and *AC*, and *DC* and *DE*.
6.7 The following azimuths are reckoned from the north: *AB*, 187°12′; *BC*, 273°47′; *CD*, 318°48′; *DE*, 0°48′; *EF*, 73°00′. What are the corresponding bearings? What are the deflection angles between consecutive lines?
6.8 Given the bearings for the following pairs of lines: *AB* = N24°21′E, *AC* = N31°20′W; *BD* = S89°41′E, *BF* = S5°15′W; *XR* = N14°25′E, *XS* = S21°52′E; and *EQ* = N50°40′W, *EP* = S21°15′E. Compute the clockwise horizontal angles between *AC* and *AB*, *BF* and *BD*, *XR* and *XS*, *EQ* and *EP*.
6.9 The interior angles of a five-sided closed polygon *ABCDE* are as follows: *A*, 117°36′; *B*, 96°32′; *C*, 142°54′; *D*, 132°18′. The angle at *E* is not measured. Compute the angle at *E*, assuming the given values to be correct.
6.10 The magnetic bearing of a line is S47°30′W and the magnetic declination is 12°10′W. What is the true bearing of the line?
6.11 The true bearing of a line is N18°17′W and the magnetic declination is 7°12°E. What is the magnetic bearing on the line?
6.12 In an old survey made when the declination was 2°10′W, the magnetic bearing of a given line was N35°15′E. The declination in the same locality is now 3°15′E. What are the true bearing and the present magnetic bearing that would be used in retracing the line?
6.13 The magnetic azimuth of a line observed in 1948 was 355°30′. At the time of observation, the declination was 15°E. Compute the true azimuth in 1948. It is desired to establish the direction of this line in the field in 1978. Compute the magnetic and true azimuths in 1978. The average annual rate of change in declination has been 14′ East.
6.14 A transit which is to have a least count of 30″ has a horizontal circle with a diameter of 150.0000 mm. What is the width of the smallest division on the horizontal circle? What is the width of the smallest division on the vernier? Calculate these widths to the nearest 0.0001 mm.
6.15 Describe the relationships that should exist among the principal instrumental axes in a perfectly adjusted transit. Illustrate these relationships using a line sketch.
6.16 Thirty spaces on a transit vernier are equal to 29 spaces on the graduated circle, and 1 space on the circle is 15′. What is the least count of the vernier?
6.17 Sixty spaces on a transit vernier are equal to 59 on the graduated circle, and 1 space on the circle is 15′. What is the least count of the vernier?

6.18 A transit for which the circle is graduated 0° to 360° clockwise is used to measure an angle by 12 clockwise "repetitions," 6 with telescope direct and 6 with the telescope reversed. Compute the value of the angle from the following data:

		Vernier	
Telescope	Reading	*A*	*B*
Direct	Initial	48°46′	228°46′
Direct	After first turning	161°09′	
Reversed	After twelfth turning	317°24′	137°25′

6.19 An angle is repeated three times direct and three times reversed with a repeating theodolite having a least count of 0.1′. The circle reading for the initial backsight is 330°28.1′ and after the first repetition is 129°33.0′. The circle reading after the sixth repetition is 204°52.0′. Calculate the average angle.

6.20 A direction theodolite with a least count of 1″ is set over station *D* to measure directions to stations *C*, *B*, and *A*. The observed directions for one position are as follows:

Station	Tel.	Circle
C	Direct	0°10′16″
	Reversed	180°10′26″
B	Direct	48°52′06″
	Reversed	228°52′40″
A	Direct	83°06′48″
	Reversed	263°06′48″

Compile an abstract of average directions and compute the average angles.

6.21 In laying out the lines for a building, a 90° angle was laid off as precisely as possible with a 01′ transit. The angle was then measured by repetition and found to be 89°59′40″. What offset should be made at a distance of 250 ft from the transit to establish the true line?

6.22 The following observations were made to determine an index correction: Vertical angle to point $A = +7°16'$ with telescope direct and $+7°14'$ with telescope inverted. Compute the index correction for observations with telescope direct.

6.23 A vertical angle measured by a single observation is $-12°02'$, and the index error is determined to be $+06'$. What is the correct value of the angle?

6.24 A theodolite equipped with a vertical circle level bubble has zero on the vertical circle oriented toward the zenith. The following direct and reversed zenith angles were observed to determine the index correction: 80°05′20″ and 279°54′16″. Compute the index correction for the instrument. The same instrument is used to measure a zenith angle of 91°16°35″. Calculate the (*a*) corrected zenith angle; (*b*) corrected vertical angle.

6.25 Two points *A* and *B*, 5280 ft apart, are to be connected by a straight line. A random line run from *A* in the general direction of *B* is found by computation to deviate 03′18″ from the true line. On the random line at a distance 1250.6 ft from *A* an intermediate point *C* is established. What must be the offset from *C* to locate a corresponding point *D* on the true line?

6.26 What error would be introduced in the measurement of a horizontal angle, with sights taken to points at the same elevation as the transit if, through nonadjustment, the horizontal axis was inclined (*a*) 03′? (*b*) 3°? (*c*) If the horizontal axis was inclined 03′, what error would be introduced if both sights were inclined at angles of +30°? (*d*) If one sight was inclined at +30° and the other at −30°?

6.27 A line *AB* is prolonged to *F* by setting up the transit at succeeding points *B*, *C*, *D*, and *E*; backsighting to *A*, *B*, *C*, and *D*, respectively; and plunging the telescope. If the line of sight made an angle of 10″ with the normal to the horizontal axis and the procedure were such that each backsight was taken with the telescope normal, what would be the angular error in the segment *EF*? What would be the offset error (approximate) in the position of *F* if the segments *AB*, *BC*, etc., were each 120 m long?

6.28 In prolonging a straight line the transit is set at *B*, a backsight is taken to *A*, and the telescope is plunged to set point *C* which is 300 m in advance of *B*. If the vertical axis were inclined 01′ with the true vertical in a vertical plane making 90° with the direction of the line, what would be the linear offset error in the located position of *C*: (*a*) If *A* and *B* are at the same elevation, but the vertical angle from *B* to *C* is +15°00′. (*b*) If *A*, *B*, and *C* are all at the same elevation? (*c*) If the vertical angle from *B* to *A* and from *B* to *C* is +15°00′?

6.61. Field problems

PROBLEM 1 MEASUREMENT OF ANGLES BY REPETITION

Object To obtain a more precise determination of the horizontal angles between various stations about a point than would be possible by a single measurement (see Art. 6.39).

Procedure (1) Set up the repeating theodolite or engineer's transit very carefully over the point labeled *O*. (2) Set four chaining pins about 150 ft or 50 m from the instrument, forming four angles about the occupied station *O*. (3) When using a repeating theodolite, set the horizontal circle to zero. If a transit with direct-reading verniers is being used, set the *A* vernier to zero and read the *B* vernier and record the readings. (4) Keep notes in a form similar to those employed for repeated angles in Art. 6.39 and illustrated in Fig. 6.46*a*. (5) With the telescope direct, measure one of the angles clockwise. Record the horizontal circle reading to the least count of the circle or both vernier readings to the least count of the verniers. (6) Leaving the upper motion clamped, again set on the first point and again measure the first angle clockwise with the telescope direct (thus doubling the angle). (7) Continue until three repetitions with the telescope direct have been obtained. (8) In like manner, without resetting the horizontal circle, measure the horizontal angle three repetitions with the telescope reversed, always measuring clockwise. (9) Record the total angle observing to the least count of the instrument used. With instruments having direct-reading verniers, observe both the *A* and *B* verniers and take the average. (10) Compute the average angle by dividing the total angle by the number of repetitions or six in this case. Compare this average angle with the first angle turned. (11) If the average angle fails to agree with the first single observation to within the least count of the instrument, repeat the entire set. (12) Go through the same process (steps 5 through 11) for all other angles about the point. For an instrument with a least count of 1′, the error of horizon closure should not exceed $10''\sqrt{\text{number of angles}}$. (13) Adjust the angles so their sum will equal 360° by distributing the error equally among the mean values.

Hints and precaution (1) Level the instrument very carefully before each repetition but do not disturb the leveling screws while a measurement is being made. (2) Be careful not to loosen the wrong clamp-screw. (3) Do not become confused when computing the total angle turned. Observe how the horizontal circle is graduated and do not omit a full turn. (4) The instrument should be handled very carefully. When the lower motion is being turned, the hands should be in contact with the *lower* plate, not the upper motion. When making an exact setting on a point, the last movement of the tangent screw should be clockwise or against the opposing spring. (5) After each repetition the instrument should be turned on its lower motion in the same direction as that of the measurement. (6) Do not walk around the transit to read the second vernier; rotate it to you, always turning the instrument clockwise in order to avoid possible errors due to slackness and twist. (7) The single measurement is taken as a check on the number of repetitions; it should agree closely but not necessarily exactly with the mean value.

PROBLEM 2 LAYING OFF AN ANGLE BY REPETITION

Object To lay off a given horizontal angle more precisely than is possible with a single setting of the horizontal circle (see Art. 6.42).

Procedure (1) Drive and set tacks in two stakes about 500 ft or 150 m apart. (2) Carefully set up the transit over one end of the line. Sight at the point at the other end, and lay off the given angle. (3) Set a stake on the line of sight about 500 ft or 150 m from the instrument (distance by pacing), and carefully set a tack. (4) By repetition measure the angle laid off, as in the previous problem, making three "repetitions" with telescope direct and three with it reversed. (5) Find the difference between the angle laid off and the required angle, and by trigonometry compute the linear distance that the tack must be moved perpendicular to the line of sight. (6) Set the tack accordingly.

PROBLEM 3 OBSERVING DIRECTIONS WITH A DIRECTION THEODOLITE

Object To observe four positions of directions about a station using a direction theodolite (see Art. 6.40).

Procedure (1) Set a hub with a tack in it as the station to be occupied (station *O*). Set three chaining pins about 100 m from the station in an array such that the angles subtended at the station are less than 100° and more than 30°. Existing points, such as lightning rods, flagpoles, and so on, may be used if they are suitably located and provide well-defined targets. (2) Set the theodolite over the central station leveling and centering according to the instructions in Art. 6.34. (3) Designate the points to be sighted as *A*, *B*, and *C* in a clockwise direction, with *A* to be the initial point sighted. (4) Following the instructions in Arts. 6.40 and 6.41, observe directions to points *A*, *B*, and *C* with the telescope direct and then backward from *C* to *B* to *A* with the telescope reversed. Use the note form shown in Fig. 6.46*b*. (5) Take the mean of the direct and reversed readings and reduce to an abstract of directions assuming a direction of 0°00′00″ from station *O* to *A* (Art. 6.41). This set of measurements constitutes *one position*. (6) Repeat steps (4) and (5) three more times, advancing the horizontal circle about 45° for each initial pointing on *A*. (7) When the four positions have been completed, calculate the averages of the directions in the four positions to yield a final abstract of directions to stations *A*, *B*, and *C* from *O*. Compute the angles *AOB* and *BOC* from these directions.

References

1. Allan, A. L., "A Note on Centering the Instrument," *Survey Review*, October 1977.
2. Berthon-Jones, P., "A Note on Run Error and Error of Half-Interval in Mean Reading Theodolites," *Australian Surveyor*, December 1972.
3. Ericson, K. E., "Electronic Surveying Systems," *Proceedings, 37th Annual Meeting, ACSM*, 1977, p. 209.
4. Gort, A. E., "The Hewlett-Packard 3829A Electronic Total Station," *Proceedings, Fall Technical Meeting, ACSM*, October 1977, pp. 308–326.
5. Jackson, F. W., "A Method of Adjusting the Optical Plummet of the Wild T-2," *Survey Review*, July 1971.
6. Kissam, P., *Surveying for Civil Engineers*, 2d ed., McGraw-Hill Book Company, New York, 1981.
7. Moffitt, F. H., and Bouchard, H., *Surveying*, 7th ed., Harper & Row, Publishers, New York, 1982.
8. Navyasky, M., "Adjusting Observed Directions by Angles Using Equitable Weights," *Surveying and Mapping*, December 1974.
9. Webber, J. T., "A Source of Error in Optical Plummets," *Australian Surveyor*, March 1971.

CHAPTER 7
Stadia and tacheometry

7.1. General Tacheometry is the procedure by which horizontal distances and differences in elevation are determined indirectly using subtended intervals and angles observed with a transit or theodolite on a graduated rod or scale. The distances and elevations thus obtained are usually of a lower order of accuracy than is possible by taping, EDM, or differential leveling. However, the results are very adequate for many purposes. Horizontal distances can be determined with a relative accuracy of 1 part in 300 to 400, while differences in elevation can be obtained to within ± 0.1 ft or ± 0.03 m.

Applications of tacheometry include traverse and leveling for topographic surveys, location of detail for topographic surveys, leveling as well as field completion surveys for photogrammetric mapping, and hydrographic surveys.

The most common tacheometric method practiced in the United States is stadia surveying, in which an engineer's transit or telescopic alidade and graduated leveling rod are used. Stadia surveying by this procedure is discussed first. Sections on self-reducing theodolites and subtense bars follow.

7.2. Stadia method Equipment for stadia measurements consists of a telescope with two horizontal cross hairs called *stadia hairs* and a graduated rod called a *stadia rod* or stadia board.

The process of taking stadia measurements consists of observing, through the telescope, the apparent locations of the two stadia hairs on the rod, which is held in a vertical position. The interval thus determined, called the *stadia interval* or *stadia reading*, is a direct function of the distance from the instrument to rod as developed in Art. 7.4. The ratio of distance to stadia interval is 100 for most instruments.

7.3. Stadia hairs and stadia rods The telescopes of transits, theodolites, plane-table alidades, and many levels are furnished with stadia hairs in addition to the regular cross hairs. One stadia hair is above and the other an equal distance below the horizontal cross hair. Stadia hairs are usually mounted on the same reticle and in the same plane as the horizontal and vertical cross hairs. Under these conditions, the stadia hairs are not adjustable and the distance between the hairs remains unchanged. Figure 7.1 shows two styles of stadia cross hairs: the stub stadia hairs (Fig. 7.1*a*), and the stadia hairs which go completely across the cross-hair ring (Fig. 7.1*b*).

A conventional rod such as the Philadelphia rod is adequate for stadia work where the sights do not exceed 200 ft or 60 m. For longer sights stadia rods are advisable and provide more accurate results. Stadia rods are described in Art. 5.23 and illustrated in Fig. 5.22.

7.4. Principle of stadia Figure 7.2*a* illustrates the principle upon which the stadia method is based. In the figure, the line of sight of the telescope is horizontal and the stadia rod is vertical. The stadia hairs are indicated by points a and b. The distance between the stadia hairs is i. The apparent locations of the stadia hairs on the rod are points A and B and the stadia interval is s.

Rays from a passing through the optical center of the lens O and the focal point of the lens F are brought to a focus at A. Similarly, the reverse is true, so that rays from A which pass through F and O are brought to a focus at a.

Since $ab = a'b'$, by similar triangles

$$\frac{f}{i} = \frac{d}{s}$$

Hence, the horizontal distance from the principal focus to the rod is

$$d = \left(\frac{f}{i}\right)s = Ks$$

in which $K = f/i$ is a coefficient called the *stadia interval factor*, which for a particular instrument is a constant as long as conditions remain unchanged. Thus, for a horizontal sight the distance from principal focus to rod is obtained by multiplying the stadia interval factor by the stadia interval. The horizontal distance from center of instrument to rod is then

$$D = Ks + (f + c) = Ks + C \tag{7.1}$$

in which C is the distance from center of instrument to principal focus. This formula is employed in computing horizontal distances from stadia intervals when sights are horizontal.

7.5. Stadia constants The focal distance f is a constant for a given instrument. It can be determined with all necessary accuracy by focusing the objective on a distant point and then measuring the distance from the cross-hair ring to the objective. The

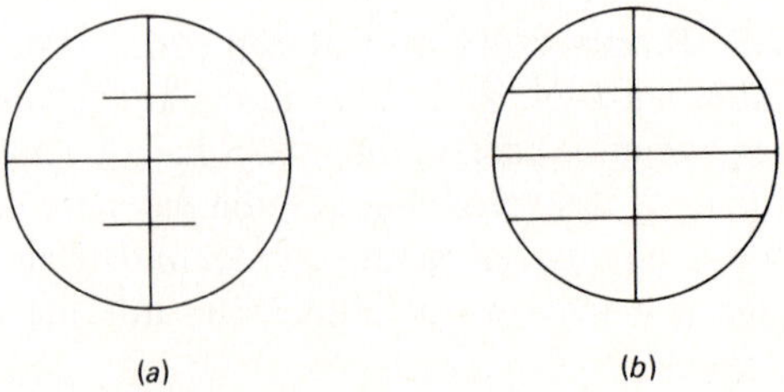

Fig. 7.1 (*a*) and (*b*) Two styles of stadia hairs.

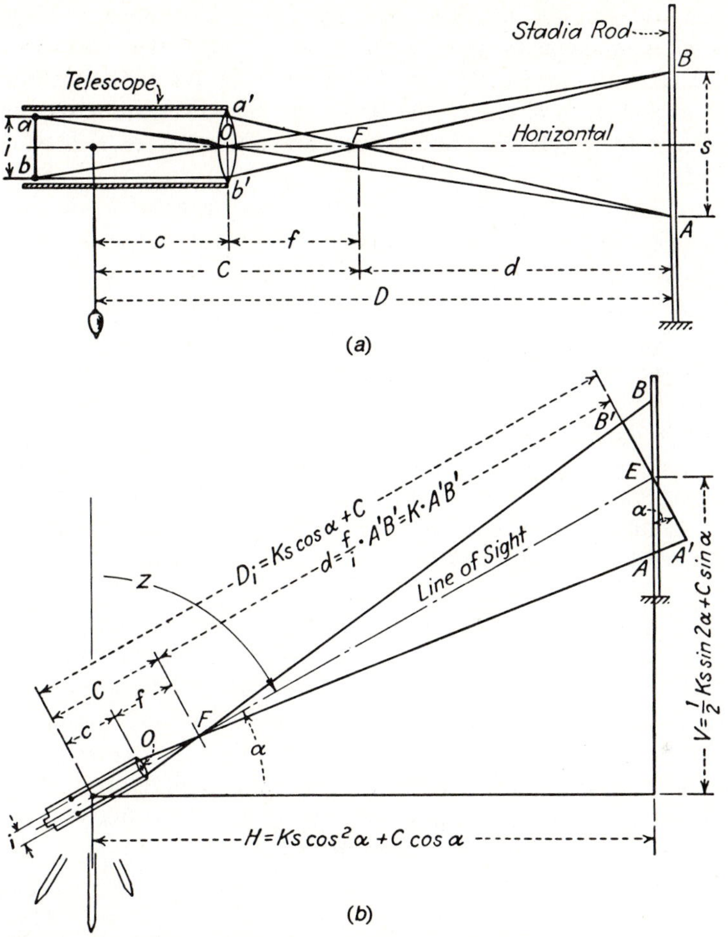

Fig. 7.2 (*a*) Horizontal stadia sight. (*b*) Inclined stadia sight.

distance *c*, though a variable depending upon the position of the objective, may for all practical purposes be considered a constant. Its mean value can be determined by measuring the distance from the vertical axis to the objective when the objective is focused for an average length of sight.

Usually, the value of $C = f + c$ is determined by the manufacturer and is stated on the inside of the instrument box. For external-focusing telescopes, under ordinary conditions C may be considered as 1 ft without error of consequence. Internal-focusing telescopes (Art. 5.8) are so constructed that C is zero or nearly so; this is an important advantage of internal-focusing telescopes for stadia work.

7.6. Stadia interval factor The nominal value of the stadia factor $K = f/i$ is usually 100. The interval factor can be determined by observation. The usual procedure is to set up the instrument in a location where a horizontal sight can be obtained. With a tape lay off, from a point $C = f + c$ in front of the center of the instrument, distances of 100 ft, 200 ft, etc., up to perhaps 1000 ft, and set stakes at the points established. The stadia rod is then held on each of the stakes, and the stadia interval is read. The stadia interval factor is computed for each sight by dividing the distance from the principal focus to the stake by the corresponding stadia interval, and the mean is taken as the best estimate. Owing to errors in observation and perhaps to errors from natural sources, the values of K for the several distances are not likely to agree exactly.

To overcome any prejudicial tendencies on the part of the instrument operator, observations may be made on the rod held on stakes at random distances from the instrument, these distances being measured later with the tape.

For use on long sights, where the full stadia interval would exceed the length of the rod, the stadia interval factor may be determined separately for the upper stadia hair and horizontal cross hair and for the lower stadia hair and horizontal cross hair.

With adjustable stadia hairs, the interval factor is made 100 by moving the hairs until their rod interval is $\frac{1}{100}$ of the distance from the principal focus to the rod, this distance being measured with a tape. The stadia hairs are so adjusted that the horizontal cross hair bisects the space between them, each in its turn being moved vertically until the distance between it and the horizontal hair is the proper half-interval, as indicated by the rod interval.

7.7. Inclined sights In stadia surveying, most sights are inclined, and usually it is desired to find both the horizontal and the vertical distances from instrument to rod. The problem therefore resolves itself into finding the horizontal and vertical projections of an inclined line of sight. For convenience in field operations the rod is always held vertical.

Figure 7.2*b* illustrates an inclined line of sight, AB being the stadia interval on the vertical rod and $A'B'$ being the corresponding projection normal to the line of sight. The length of the inclined line of sight from center of instrument is

$$D_i = \frac{f}{i}(A'B') + C \tag{7.2}$$

For all practical purposes the angles at A' and B' may be assumed to be 90°. Let $AB = s$; then $A'B' = s\cos\alpha$. Making this substitution in Eq. (7.2), and letting $K = f/i$, the inclined distance is

$$D_i = Ks\cos\alpha + C \tag{7.3}$$

The horizontal component of this inclined distance is

$$H = Ks\cos^2\alpha + C\cos\alpha \tag{7.4}$$

which is the general equation for determining the horizontal distance from center of instrument to rod, when the line of sight is inclined.

The vertical component of the inclined distance is

$$V = Ks\cos\alpha\sin\alpha + C\sin\alpha \tag{7.4a}$$

The equivalent of $\cos\alpha\sin\alpha$ is conveniently expressed in terms of double the angle α, or

$$V = \tfrac{1}{2}Ks\sin 2\alpha + C\sin\alpha \tag{7.5}$$

which is the general equation for determining the difference in elevation between the center of the instrument and the point where the line of sight cuts the rod. To determine the difference in ground elevations, the height of instrument and the rod reading of the line of sight must be considered.

If the zenith angle z (Fig. 7.2b) is observed, Eqs. (7.4) and (7.5) become

$$H = Ks \sin^2 z + C \sin z \tag{7.5a}$$

and

$$V = \tfrac{1}{2} Ks \sin 2z + C \cos z \tag{7.5b}$$

Equations (7.4) through (7.5b) are known as the *stadia formulas for inclined sights.*

7.8. Permissible approximations More approximate forms of the stadia formulas are sufficiently precise for most stadia work. Usually, distances are computed only to feet or decimetres and elevations to tenths of feet or centimetres. Under these conditions, for side shots where vertical angles are less than 3°, Eq. (7.4) for horizontal distances may properly be reduced to the form

$$H = Ks + C \tag{7.6}$$

which is the same as for horizontal sights (Art. 7.4). But for traverses of considerable length, owing to the systematic error introduced, this approximation should not be made for vertical angles greater than perhaps 2°.

Owing to unequal refraction and to accidental inclination of the rod, observed stadia intervals are in general slightly too large. To offset the systematic errors from these sources, frequently on surveys of ordinary precision the constant C is neglected. Hence, in any ordinary case Eq. (7.4) may, with sufficient precision, be expressed in the form

$$H = Ks \cos^2 \alpha \qquad \text{(approx.)} \tag{7.7}$$

where α is the vertical angle or

$$H = Ks \sin^2 z \qquad \text{(approx.)} \tag{7.7a}$$

in which z is the zenith angle.

Also, Eq. (7.5) may often be expressed with sufficient precision for ordinary work in the form

$$V = \tfrac{1}{2} Ks \sin 2\alpha \qquad \text{(approx.)} \tag{7.8}$$

or

$$V = \tfrac{1}{2} Ks \sin 2z \qquad \text{(approx.)} \tag{7.8a}$$

where α and z are the vertical angle and zenith angle, respectively. However, the error in elevation introduced through using Eq. (7.8) or (7.8a) may not be negligible, as for large vertical angles it amounts to several tenths of a foot or up to a decimetre.

Equations (7.7) through (7.8a) are simple in form and are most generally employed.

When K is 100, the common practice is to multiply mentally the stadia interval by 100 at the time of observation, and to record this value in the field notebook. This distance Ks is often called the *stadia distance*. Thus, if the stadia interval were 7.37 ft, the stadia distance recorded would be 737 ft.

For external-focusing telescopes, the degree of approximation in using Eqs. (7.7) through (7.8*a*) may be greatly reduced either by adding 0.01 ft or 3 mm to the observed stadia interval s or—when K is 100—by adding 1 ft or 0.3 m to the observed stadia distance Ks. It is convenient to add the correction mentally and to record the corrected value in the field notebook. The notes should state that corrected values are recorded.

7.9. Stadia reductions Horizontal distance and difference in elevation can be computed from stadia observations by using (1) an electronic calculator by the exact Eqs. (7.4) through (7.5*b*) or approximate Eqs. (7.7) through (7.8*a*), (2) stadia tables, and (3) stadia slide rules.

Example 7.1 The following data were obtained by stadia observation: vertical angle = +8°10′, $s = 2.50$ ft. The stadia interval factor is known to be 100 and $C = 0.75$ ft. Calculate H and V.

SOLUTION By exact equations (7.4) and (7.5),

$$\begin{aligned} H &= Ks\cos^2\alpha + C\cos\alpha \\ &= (100)(2.50)(\cos^2 8°10') + (0.75)(\cos 8°10') \\ &= (100)(2.50)(0.9798) + (0.75)(0.9899) \\ &= 245.0 + 0.7 \\ &= 245.7 = 246 \text{ ft} \\ V &= \tfrac{1}{2}Ks\sin 2\alpha + C\sin\alpha \\ &= \tfrac{1}{2}(100)(2.50)(0.2812) + (0.75)(0.1421) \\ &= 35.2 + 0.11 \\ &= 35.3 \text{ ft} \end{aligned}$$

If approximate equations (7.7) and (7.8) are used and $C = 1.0$ is assumed, we may add 0.01 ft to the stadia interval. Thus, $s = 2.50 + 0.01 = 2.51$:

$$\begin{aligned} H &= (100)(2.51)(\cos^2 8°10') = (251)(0.9798) \\ &= 246 \text{ ft} \\ V &= \tfrac{1}{2}(100)(2.51)\sin 2\alpha = (125.5)(0.2812) \\ &= 35.3 \text{ ft} \end{aligned}$$

and the approximate equations are seen to yield adequate accuracy for these data.

Example 7.2 The stadia interval was 1.372 m at a vertical angle of 20°32′. $C = 0.30$ m and the stadia interval factor is 100. Calculate H and V.

SOLUTION By Eq. (7.4),

$$\begin{aligned} H &= (100)(1.372)(\cos^2 20°32') + (0.30)(\cos 20°32') \\ &= 120.6 \text{ m} \end{aligned}$$

By Eq. (7.5),

$$\begin{aligned} V &= \tfrac{1}{2}(100)(1.372)(\sin 41°04) + (0.30)(\sin 20°32') \\ &= 45.17 \text{ m} \end{aligned}$$

If approximate equations (7.7) and (7.8) are used for this problem, add 0.003 to the stadia interval so that

$$s = 1.372 + 0.003 = 1.375$$

and

$$H = 100(1.375)\cos^2 \alpha = 120.6 \text{ m}$$
$$V = \tfrac{1}{2}(100)(1.375)\sin 41°04' = 45.16 \text{ m}$$

Use of the approximate equations for solution with the large vertical angle resulted in an error of 1 cm in the difference in elevation. Had the telescope of the instrument been internally focusing, as is the case with most modern transits and theodolites, C would have been equal to zero and approximate equations (7.7) and (7.8) could be used without danger of errors.

Example 7.3 Using a repeating theodolite, the stadia interval was 2.174 m and the zenith angle was 72°45′. The stadia interval factor is 100 and $C = 0$. Calculate H and V.

SOLUTION Since $C = 0$, approximate equations (7.7*a*) and (7.8*a*) are adequate. By Eq. (7.7*a*)

$$H = (100)(2.174)(\sin^2 72°45')$$
$$= 198.3 \text{ m}$$

By Eq. (7.8*a*)

$$V = \tfrac{1}{2}(100)(2.174)\sin(145°30')$$
$$= 61.57 \text{ m}$$

7.10. Observation of stadia interval On transit or plane-table surveys the stadia interval is usually determined by setting the lower stadia hair on a foot or decimetre mark and then reading the location of the upper stadia hair. Thus, the stadia interval is mentally computed more easily and with less chance of mistake than if the lower hair were allowed to take a random position on the rod. When the vertical angle is taken to a given mark on the rod, the corresponding stadia interval is observed with the lower hair on the foot or decimetre mark that renders a minimum displacement of the horizontal cross hair from the mark to which the vertical angle is referred.

Thus, if a vertical angle were taken with the line of sight cutting the rod at 4.9 ft and the lower stadia hair fell at 2.3 ft, the telescope would be rotated about the horizontal axis until the lower hair was at 2.0 ft. The center horizontal cross hair would then fall at 4.6 ft.

On metric rods it is more convenient to set the bottom cross hair on the nearest even decimetre. For example, if the vertical angle were taken when the middle cross hair is on 1.495 m, and the lower stadia hair was at 1.23, rotate the telescope in a vertical plane about the horizontal axis until the lower stadia hair coincides with 1.200 and read the top stadia hair as 1.730 m. The interval is then 1.730 − 1.200 = 0.530 m.

Whenever the stadia interval is in excess of the length of the rod, the separate half-intervals are observed and their sum is taken.

For precise stadia work, the readings may be made by means of two targets on the rod, one target on, say, the 2-ft or 0.500-m mark and the other set by the rodman as directed by the instrumentman. To avoid excessive effects of atmospheric refraction, the intercept of the lower cross hair with the rod should not fall nearer the ground than necessary.

Note forms that can be used for recording stadia data in stadia traverses (Art. 8.12) and topographic surveys by the stadia method (Art. 12.10) are shown in Figs. 8.16 and 12.3, respectively.

7.11. Beaman stadia arc The Beaman stadia arc, in modified form known also as the *stadia circle*, is a specially graduated arc on the vertical circle of the transit or the plane-table alidade. It is used to determine distances and differences in elevation by stadia without reading vertical angles and without the use of tables, or stadia slide rules. The stadia arc has no vernier, but settings are read by an index mark.

Horizontal distance In the type of stadia arc shown in Fig. 7.3, the graduations for determining distances are at the left, inside the vertical circle. When the telescope is level (vertical vernier reading zero), the reading of the arc is 100, indicating that the horizontal distance is 100 percent of the observed stadia distance. When an inclined sight is taken, the observed stadia interval is multiplied by the reading of the "Hor." stadia arc, to obtain the horizontal distance from principal focus to rod. Thus, the "Hor." scale is a nonuniform logarithmic scale graduated according to $K\cos^2\alpha$, where α is the vertical angle the telescope makes with the horizontal. For example, if the stadia interval is 4.11 ft, $K = 100$, $C = 0.0$, and the "Hor." reading on the stadia arc is 99, the stadia distance is $(100)(4.11) = 411$ and the horizontal distance $= (4.11)(99) = 407$ ft. In Fig. 7.3, the "Hor." reading is 98. If $s = 1.325$ m, the horizontal distance is $(1.325)(98) = 129.9$ m.

Another type of stadia arc (Fig. 7.4) is graduated to give the *correction* in percent to be subtracted from the observed stadia interval times the stadia interval factor. In Fig. 7.4, the "H" reading opposite the index is estimated to be 2.6. If Ks is 208.4 m, the horizontal distance is $208.4 - (208.4)(0.026) = 203.0$ m. Thus, the H scale is graduated according to $K(1 - \cos^2\alpha)$, where α is the vertical angle of the telescope with the horizontal.

Difference in elevation In Fig. 7.3, the graduations for determining differences in elevation are at the right, inside the vertical circle. When the telescope is level (vertical vernier reading zero), the reading of the arc is zero. When an inclined sight is taken, first the stadia distance is observed in the usual manner, that is, with the lower stadia hair on a foot mark on the rod. The telescope is then either elevated or depressed slightly until the nearest graduation on the "Vert." scale of the stadia arc coincides with the index of the arc (in order to avoid interpolation), and a rod reading is taken at the point where the line of sight strikes the rod. The observed stadia interval is multiplied by the reading of the "Vert." stadia arc, to obtain the vertical distance from center of the instrument to point last sighted on the rod. This

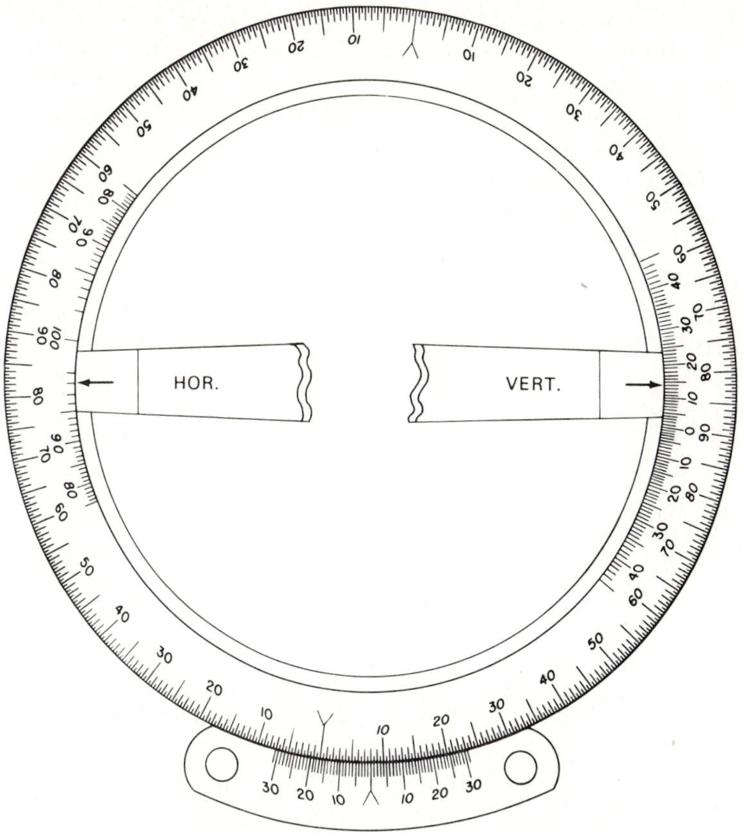

Fig. 7.3 Stadia arc (multiplier type).

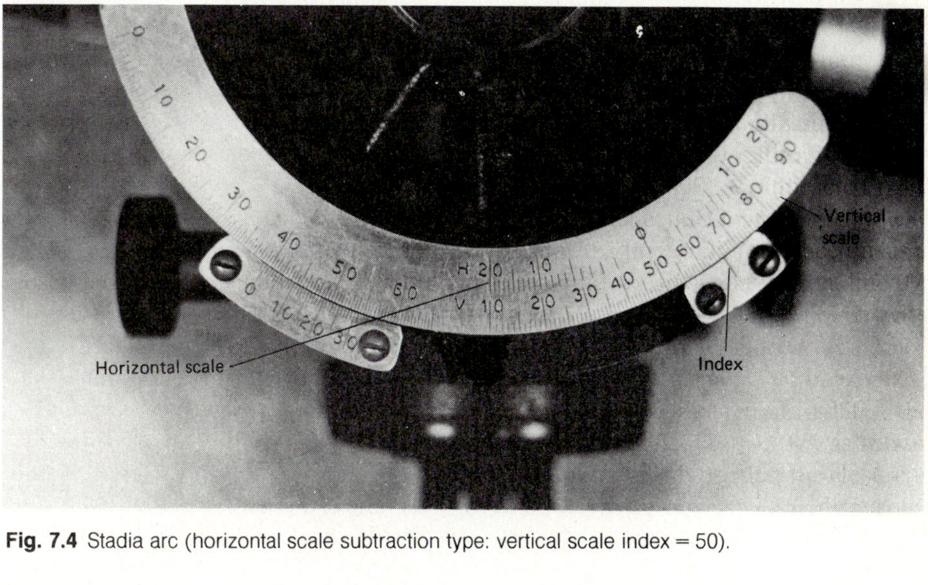

Fig. 7.4 Stadia arc (horizontal scale subtraction type: vertical scale index = 50).

difference in elevation, combined with the height of instrument and the rod reading, gives the difference in elevation between the instrument station and the point on which the rod is held (see Art. 7.12).

In Fig. 7.3, the "Vert." reading is +14 (+ indicates that the line of sight is above the horizon). Assuming that $s = 4.11$, the difference in elevation between the center of the telescope and the point sighted on the rod is (14)(4.11) = 57.5 ft. This value is analogous to V in Eq. (7.5) or (7.8) and the "Vert." scale is graduated according to $\frac{1}{2}K \sin 2\alpha$.

The other type of stadia arc (Fig. 7.4) is so graduated that, when the telescope is level, the reading of the "Vert." stadia arc is 50 instead of 0; when the telescope is elevated, the reading is greater than 50, and when the telescope is depressed, the reading is less than 50. In all cases, 50 is subtracted from the reading of the stadia arc, and the remainder (positive or negative as determined by the subtraction) is multiplied by the observed stadia interval to obtain the vertical distance from center of the telescope to the point sighted on the rod. This arrangement of the scale on the stadia arc avoids mistakes of reading a positive value for a negative value, or vice versa, but it introduces an additional step (subtraction) in the computations. In Fig. 7.4, the V reading has been set to 66. If the stadia interval is 1.431 m, the difference in elevation between the center of the telescope and the point sighted is (1.431)(66 − 50) = 22.9 m.

Observations with the Beaman stadia arc do not include the effect of the instrumental constant $C = f + c$. If, for an external-focusing telescope, more precise results are desired than are yielded by the approximate formulas [Eqs. (7.7) and (7.8)], the observed values should be corrected, particularly if the vertical angles are large. The simplest method is to add 0.01 ft or 0.003 m to the observed stadia interval.

7.12. Difference in elevation In Fig. 7.5 the instrument is at station A with a height of instrument or h.i. above A equal to AB and the rod is held in a vertical position at station D. It is desired to determine the difference in elevation between points A and D given stadia observations and the vertical angle α. First consider the problem when the difference in elevation between points A and D on the ground is desired. This situation occurs in stadia traversing and in topographic surveying from a known point. The h.i. $= AB$, the stadia interval s, the vertical angle α or zenith angle z with the middle cross hair set on E, and the rod reading DE are observed and recorded. The difference in elevation between the telescope and E, which is equal to V, can be calculated by using Eq. (7.5) or (7.8) or if the Beaman arc is used, by the methods described in Art. 7.11. The difference in elevation is Δ el. $= AB + V - DE$ or Δ el. = h.i. + V − rod reading at D. Thus, the elevation at D = the elevation at $A + (AB + V - DE)$. If the middle cross hair is sighted on the rod such that $DE = AB =$ h.i., then Δ el. $= V$ and the elevation at D = the elevation at $A + V$. Obviously, setting the horizontal cross hair on the rod at a value equal to the h.i. prior to reading the vertical angle simplifies calculating the change in elevation between the transit station and the point on which the rod is held.

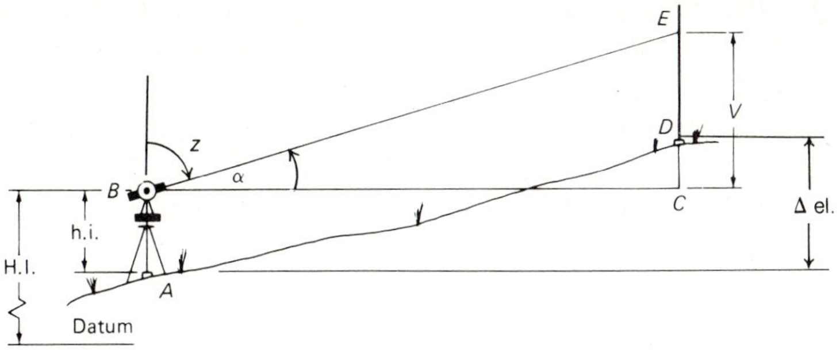

Fig. 7.5 Difference in elevation between instrument and rod stations.

In another situation also common in topographic surveying, it may be desired to determine the difference in elevation between the center of the telescope and the ground point on which the rod is held. Referring to Fig. 7.5, assume that the elevation of station *A* and the height of instrument above *A* or the h.i. are known. Thus, the height of instrument above the datum or H.I. can be calculated and the elevation of *D* can be computed directly from the instrument H.I. Given the stadia interval and α, V can be computed so that the elevation of $D = (\text{H.I. at } B) + V - DE$, or elevation of $D = (\text{H.I. at } B) + V - (\text{rod reading})$. Note that it is important that the rod reading always be recorded.

Figure 7.6 illustrates the situation where the difference in elevation is desired between points *B* and *F* utilizing an intermediate instrument set up at *G*. Assume that a backsight is taken on a rod held at *B*. The stadia interval, rod reading *BC*, and α_B are recorded. Using these data, V_B can be calculated. Next, a foresight is taken on the rod held at *F*. The stadia interval, rod reading *FE*, and vertical angle are observed and recorded so that V_F can be computed. A general expression for

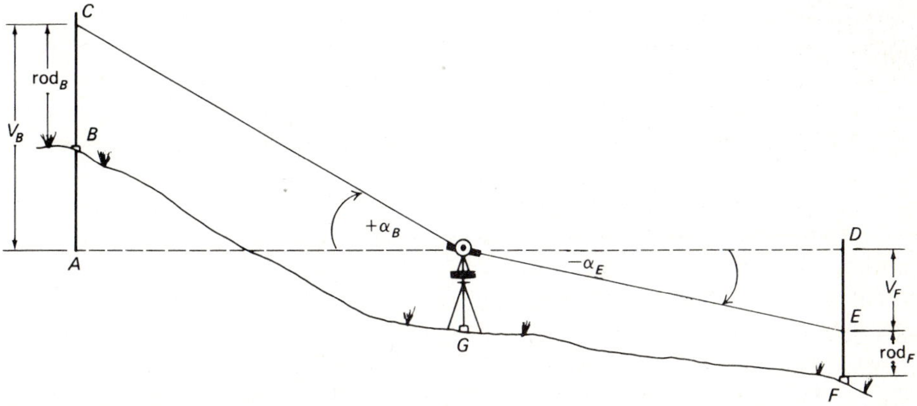

Fig. 7.6 Difference in elevation between two points.

difference in elevation between two points B (the backsight) and F (the foresight) is

$$\Delta\,\text{el.}_{BF} = \text{rod}_B \pm V_B \pm V_F - \text{rod}_F \tag{7.9}$$

in which the sign on V_B is opposite to the sign of α_B and the sign of V_F corresponds to the sign of α_E. Thus, in Fig. 7.6, $\Delta\,\text{el.}_{BF} = BC - V_B - V_F - EF$.

In the reverse direction, if the backsight is on F and the foresight on B, then $\Delta\,\text{el.} = EF + V_F + V_B - BC$.

7.13. Uses of stadia Uses of stadia are as follows:

1. In differential leveling, the backsight and foresight distances are balanced conveniently if the level is equipped with stadia hairs.
2. In profile leveling and cross sectioning, stadia is a convenient means of finding distances from level to points on which rod readings are taken.
3. In rough trigonometric, or indirect, leveling with the transit, the stadia method is more rapid than any other. The line of trigonometric levels is run as described in Art. 5.5, except that stadia intervals are observed and that differences in elevation are computed by the stadia formula. The method is described in Art. 7.14.
4. On traverses of low relative accuracy, where only horizontal angles and distances are required, the stadia method is more rapid than taping. It may be used either in running traverse lines or in locating details from traverse lines. Stadia intervals are observed as each point is sighted. Horizontal angles are measured, but vertical angles are observed only when of sufficient magnitude to make the horizontal distance appreciably different from the stadia distance (say, when greater than 3°) and then are estimated without reading the vernier. The stadia traverse method of running such surveys is described in Art. 8.12.
5. On surveys of low relative accuracy—particularly topographic surveys—where both the relative location of points in a horizontal plane and the elevation of these points are desired, stadia is useful. Both horizontal and vertical angles are measured, and the stadia interval is observed, as each point is sighted; these three observations define the location of the point sighted. The transit-stadia method of making observations when both the horizontal location and the elevation are desired is described in Art. 8.12.
6. Where the plane table is used (Art. 7.18), stadia observations are made with the telescopic alidade (Art. 7.18) in the same manner as with the transit, but horizontal distances and differences in elevation are computed in the field and are plotted immediately instead of being recorded in the form of notes. Methods for plane-table traverse are described in Art. 8.13. Uses of the plane table and alidade in topographic surveying are given in Art. 12.11.
7. Transit or plane-table stadia are useful for field completion surveys required for photogrammetric topographic mapping (Art. 13.15).

7.14. Indirect leveling by stadia Where the required precision is low and the terrain is rolling or rough, the stadia method of indirect leveling is rapid. The transit should preferably be provided with a sensitive control level for the vertical vernier in order that index error may be readily eliminated.

Theodolites and self-reducing tacheometers with automatic vertical indexing are ideal for stadia leveling. With the ordinary transit having a vertical circle reading to single minutes, differences in elevation are usually computed only to the nearest 0.1

ft or to the nearest 0.03 m. The average length of sight in stadia surveying is usually considerably greater than in differential leveling.

In running a line of levels by this method, the transit is set up in a convenient location. A backsight is taken on the rod held at the initial bench mark, first by observing the stadia interval and then by measuring the vertical angle to some arbitrarily chosen mark on the rod. A turning point is then established in advance of the transit, and similar observations are taken, the vertical angle being measured with the middle cross hair set on the same mark as before. The transit is moved to a new location in advance of the turning point, and the process is repeated. The stadia distances and vertical angles are recorded, as is the rod reading, which is used as an index when vertical angles are measured. If it is impractical to sight at this chosen index reading, the vertical angle is measured to some other graduation and this rod reading is recorded in the notes.

Figure 7.7 shows a form of notes for stadia leveling. The bench mark sighted for a backsight is recorded in column 1 and its known elevation is shown on the same line in column 11. The stadia interval for the backsight is recorded in column 2, the vertical angle in column 3, and the rod reading, rod_B, where the middle cross hair cuts the rod is given in column 4. Similar readings for the foresight are recorded one line down in columns 6, 7, and 8. The values V_B and V_F (Fig. 7.6) for backsight and foresight are calculated by Eq. (7.8) and are recorded in columns 5 and 9, respectively. The sign of V_B is opposite the sign of the vertical angle to the backsight, while the sign of V_F corresponds to the sign of the vertical angle to the foresight. The values of rod_B, V_B, V_F, and rod_F are then used in Eq. (7.9) to calculate Δel. between the point on the ground where the rod was held for the backsight and the point on the ground where the rod was held for the foresight. This value is recorded in column 10. Thus, for the notes in Fig. 7.7 for the first instrument setup between B.M.$_{42}$ and T.P.$_1$, $\Delta\text{el.} = 3.2 + 0.7 + 19.6 - 4.4 = 19.1$ ft. This Δel. (recorded in column 10) is applied to the elevation on the ground at the backsight (B.M.$_{42}$) to

STADIA LEVEL FOR INDIAN TRAIL RECONNAISSANCE

July 20, 1984
Fair 70°F
J.L. Black ⊼
V.A. Alden, Rod
15

Station (1)	Backsight (2) Interval	(3) Vert Angle	(4) $(rod)_B$	(5) V_B	(6) Interval	Foresight (7) Vert Angle	(8) $(rod)_F$	(9) V_F	(10) Δ Elev	(11) Elev. ft	Transit
BM 42	4.22	−0°06′	3.2	+0.7						5972.4	No. 132
TP1	3.16	0°00′	9.7	0.0	2.64	+4°16′	4.4	+19.6	+19.1	5991.5	C = 0
TP2	4.05	−2°15′	6.0	+15.9	4.41	+1°02′	7.7	+8.0	+10.0	6001.5	K = 100
TP3					3.94	+2°20′	6.0	+16.0	+31.9	6033.4	

Fig. 7.7 Notes for stadia leveling.

yield the elevation on the ground at the foresight (T.P.$_1$). This method of recording the data is then continued for subsequent setups, as indicated in Fig. 7.7.

7.15. Errors in stadia Many of the errors of stadia are those common to all similar operations of measuring horizontal angles (Arts. 6.54 to 6.57) and differences in elevation (Art. 5.38). Sources of error in horizontal and vertical distances computed from observed stadia intervals are as follows:

1. Stadia interval factor not that assumed This condition produces a systematic error in distances, the error being proportional to that in the stadia interval factor. The case is parallel to that of the tape which is too long or too short. When the value of the interval factor is closely determined by observations as described in Art. 7.6 and the stadia measurements are taken under similar conditions, the error from this source may be made negligible.

2. Rod not standard length If the spaces on the rod are uniformly too long or too short, a systematic error proportional to the stadia interval is produced in each distance. Errors from this source may be kept within narrow limits if the rod is standardized and corrections for erroneous length are applied to observe stadia intervals. Except for stadia surveys of more than ordinary precision, errors from this source are usually of no consequence.

3. Incorrect stadia interval The stadia interval varies randomly owing to the inability of the instrument operator to observe the stadia interval exactly. In a series of connected observations (as a traverse) the error may be expected to vary as the square root of the number of sights. This is the principal error affecting the precision of distances. It can be kept to a minimum by proper focusing to eliminate parallax, by taking observations at favorable times, and by care in observing. Where high precision is required, stadia measurements may be taken by sighting on a rod with two targets, one fixed and the other movable.

4. Rod not plumb This condition produces a small error in the vertical angle. It also produces an appreciable error in the observed stadia interval and hence in computed distances, this error being greater for large vertical angles than for small angles. It can be eliminated by using a rod level.

5. Unequal refraction Unequal refraction of light rays in layers of air close to the Earth's surface affects the sight on the lower stadia hair more than the sight on the upper stadia hair and thus introduces systematic positive errors in stadia measurements. Although errors from this source are of no consequence in ordinary stadia surveying, they may be important on the more precise surveys. The periods most favorable for equal refraction are at times when it is cloudy or, if the sun is shining, during the early morning or late afternoon. On precise stadia surveys where it is necessary to work under a variety of atmospheric conditions, it is appropriate to determine the stadia interval factor for each condition and to apply the factor to all observations taken under that condition. Whenever atmospheric conditions are unfavorable, the sights should not be taken near the bottom of the rod.

6. Errors in vertical angles Errors in vertical angles are relatively unimportant in their effect upon *horizontal distances*. For example, analysis of Eq. (7.7) shows that an uncertainty of 0.1′ in a vertical angle of 5° yields a discrepancy of 0.02 ft in a 300-ft sight and 0.005 m in a 100-m sight. Under the same conditions, an error of 01′ in a 15° vertical angle produces discrepancies of 0.05 ft in the 300-ft sight and 0.02 m in the 100-m sight.

With respect to differences in elevation, analysis of Eq. (7.8) reveals that an uncertainty of 0.1′ in the magnitude of the vertical angle results in a discrepancy of 0.1 ft in a 300-ft sight and 0.03 m in a 100-m sight for a vertical angle of about 5°. Uncertainty in the stadia interval, especially at higher vertical angles, will have a more pronounced effect on elevation differences than will errors in vertical angles.

7.16. Self-reducing tacheometers The Beaman arc provided an example of how an analog device could be incorporated into the engineer's transit so as to simplify the stadia reduction procedure. A similar concept is used for optical theodolites where the major use of the instrument is for tacheometry. In self-reducing, optical-reading tacheometers, the fixed stadia cross hairs are replaced by three variable slightly curved lines as illustrated in Fig. 7.8 for one style of tacheometer. These curves are etched on the vertical circle and are projected onto the plane of the reticle so as to be visible when viewing through the telescope. The bottom curve is a base-line or zero curve, the middle curve is for elevation differences, and the top curve is for the horizontal distances. The interval between the zero curve and middle

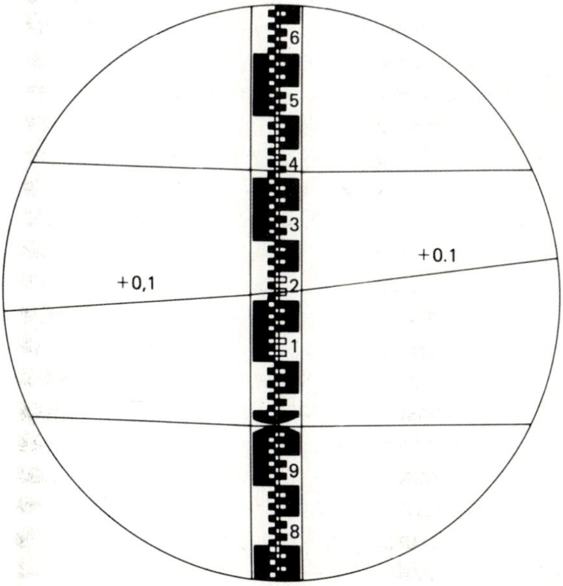

Fig. 7.8 Curved stadia intercepts in a self-reducing tacheometer. (*Courtesy, Wild Heerbrugg Instruments, Inc.*)

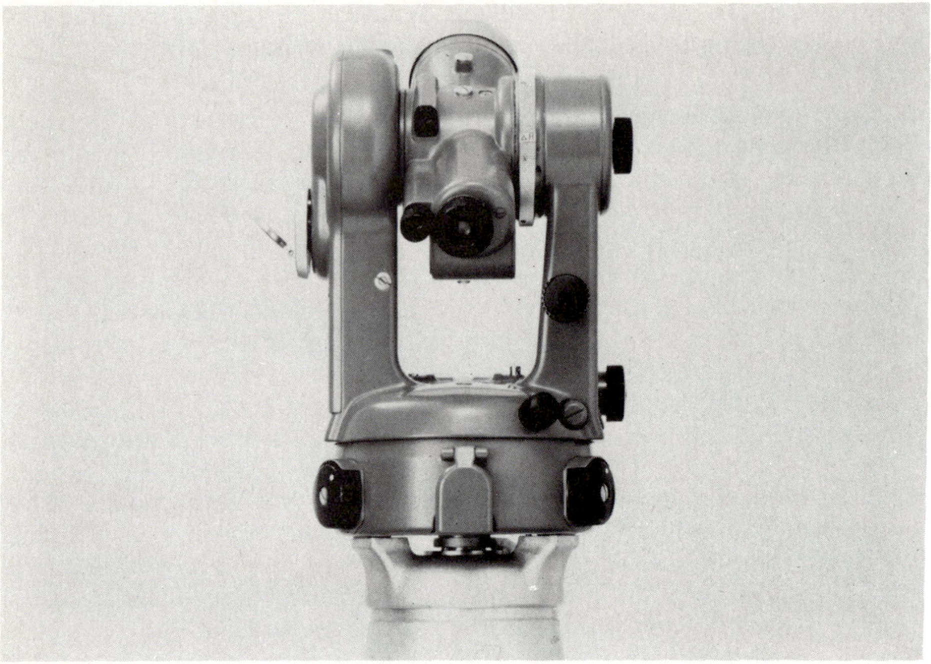

Fig. 7.9 Self-reducing tacheometer, Kern K1-RA. (*Courtesy Kern Instruments, Inc.*)

curve varies according to the $\sin\alpha\cos\alpha$ and when multiplied by the constant above the curve, yields the difference in elevation between the center of the telescope and the point sighted on the rod directly. In Fig. 7.8, the bottom line is on 1.000 m and the middle line is on 1.215 m, yielding an interval of 0.215 m. The factor of 0.1, labeled above the middle line in the field of view, must be multiplied times the interval in centimetres. Thus, the elevation difference between the telescope and the point sighted is (21.5)(0.1) = 2.15 m. The interval between the bottom and top lines varies according to $\cos^2\alpha$ and when multiplied by 100 yields horizontal distance. In Fig. 7.8 this interval is 0.410 m, so that the horizontal distance is (100)(0.410) = 41.0 m. As the inclination of the telescope varies, the shapes of the lines change and the multiplier on the middle curve changes. The constant for the horizontal interval is always 100.

Another style of self-reducing tacheometer equipped with a pendulum compensator for automatic vertical circle index compensation is illustrated in Fig. 7.9. In this instrument, two horizontal lines are projected into the field of view of the telescope (Fig. 7.10*a*). The upper line is a fixed base line. The lower line is on a movable reticle which is raised or lowered mechanically by a cam that changes position as the inclination of the telescope varies. The distance between the two lines changes according to the stadia equation, so that the interval is the distance directly. In Fig. 7.10*a*, the top line is on 1.300 m and the bottom (movable) line is on 1.456 m, yielding an interval of 0.156 m. The horizontal distance is (100)(0.156) = 15.6 m. The

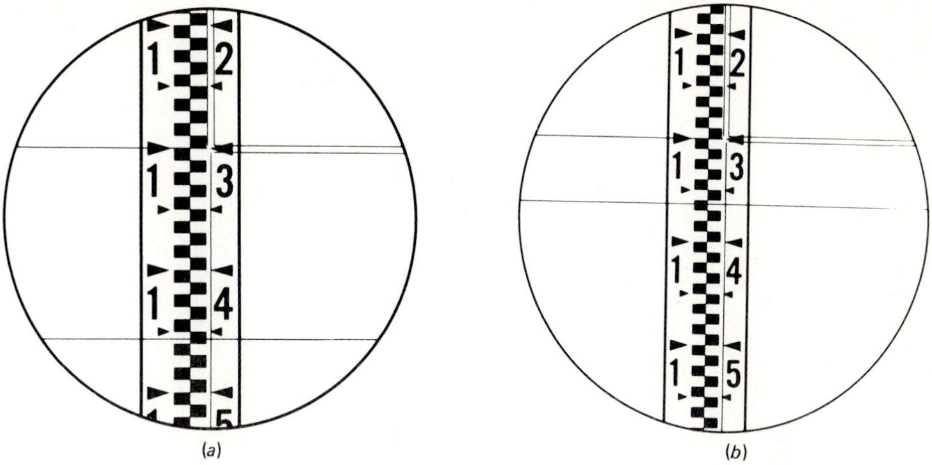

Fig. 7.10 (*a*) and (*b*) Stadia intervals in a self-reducing tacheometer. (*Courtesy Kern Instruments, Inc.*)

same horizontal base line is used to obtain an interval for difference in elevation. A control on the tacheometer permits rotating the cam through 180°. In this position, the cam moves the lower line in such a way that the interval gives the elevation difference between the telescope and the point sighted with the fixed top line. In Fig. 7.10*b*, the upper fixed line is on 1.300 m and the bottom line is on 1.364 m, giving an interval of 0.064 m. Thus, the difference in elevation between the telescope and the point sighted is (100)(0.064) = 6.4 m.

7.17. Subtense bar Another method for indirect distance determination is to observe the angle subtended by the two ends of a horizontal rod of fixed length called a *subtense bar* (Fig. 7.11). The subtense bar has targets at the two ends which are connected by invar wires that are under a slight but constant tension. In this way, effects of temperature changes are minimized. The subtense bar is mounted on a tripod and can be centered over the station to which it is desired to measure. The bar is leveled by centering a circular level bubble and three level screws. The bar can then be oriented perpendicular to the line to be measured, by sighting a small telescope attached to the midpoint of the bar.

Referring to Fig. 7.12, the subtense bar is set over point S and a 1″ direction theodolite is set over point R. Then the *horizontal angle* β between the two targets at the ends of the bar is measured by repetition. The *horizontal* distance RS is

$$RS = \frac{BB'}{2} \cot \frac{\beta}{2}$$

or

$$d = \frac{b}{2} \cot \frac{\beta}{2} \tag{7.10}$$

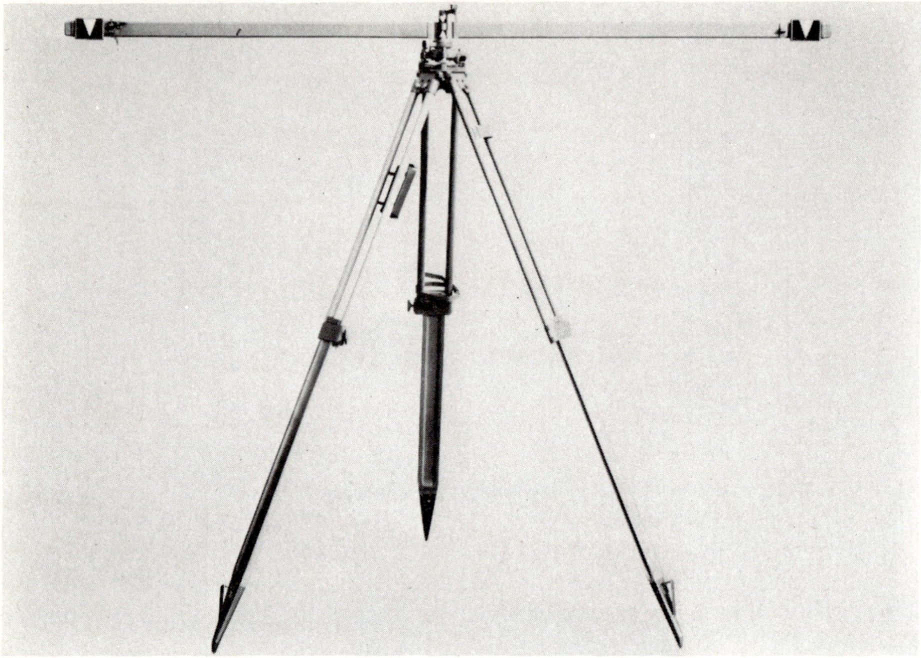

Fig. 7.11 Wild invar subtense bar. (*Courtesy Heerbrugg Instruments, Inc.*)

in which d is the horizontal distance between the two points, b the calibrated length of the subtense bar, and β the observed angle subtended by the two targets at the ends of the subtense bar. Normally, subtense bars are 2 m in length, so that $b = 2$ m.

The angle β must be measured a sufficient number of times to yield accurate values for the distance d. If one assumes standard errors in β and b, it is possible to estimate the standard error in the distance d by applying the rules of error propagation (App. A) to Eq. (7.10). For example, if $\sigma_\beta = 01''$ of arc and $\sigma_b = 0.0002$ m, then it can be shown (beyond the scope of this textbook) that $\sigma_d = 0.008$ m for a distance of 50 m.

Listed in Table 7.1 are estimated standard errors for various distances measured by a subtense bar based on $\sigma_\beta = 01''$ and $\sigma_b = 0.0002$ m. Note that the relative accuracy in the distance d decreases rapidly as the distance increases. With distances of over 400 m or about 1300 ft, the maximum relative accuracy possible is about 1 part in 1000. Obviously a direction theodolite with a least count of 01″ must be used to observe multiple readings in order to achieve these relative accuracies.

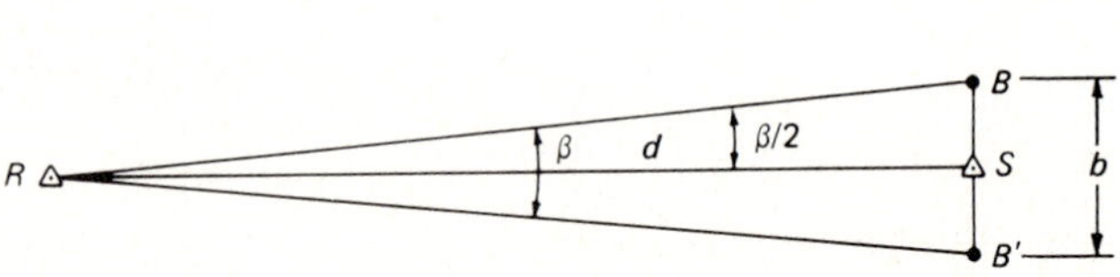

Fig. 7.12 Geometry of subtense bar reduction.

Table 7.1 Estimated Standard Deviations in Distances by Subtense Bar

Distance, m	σ_d, m	Relative accuracy, σ_d/d
50	0.008	1:6250
100	0.026	1:3846
150	0.057	1:2630
200	0.099	1:2020
300	0.220	1:1364
400	0.390	1:1026
500	0.608	1:822
600	0.875	1:686

If 12 repetitions with a first-order theodolite were made to observe β, a value of $\sigma_\beta = 0.6''$ is possible. Propagation of this value through Eq. (7.10) for a distance of 150 m leads to $\sigma_d = 0.036$, or a relative accuracy of 1 part in 4000 can be maintained for distances of moderate length when the proper equipment and procedures are employed.

Should the vertical distance between occupied points also be desired, the vertical angle α can be observed to the subtense bar. The difference in elevation between the center of the telescope and the center of the subtense bar is

$$V = d \tan \alpha \tag{7.11}$$

in which V is the difference in elevation, d the horizontal distance computed from Eq. (7.10), and α the vertical angle.

7.18. Plane table and alidade The surveying system made up of the *plane table* and *alidade* consists of (1) a drawing board mounted on a tripod, and (2) an alidade having the vertical plane of the line of sight fixed parallel to a straightedge that rests upon but is not attached to the board (Fig. 7.13*b*). A sheet of drawing paper, called a *plane-table sheet*, is fastened to the board.

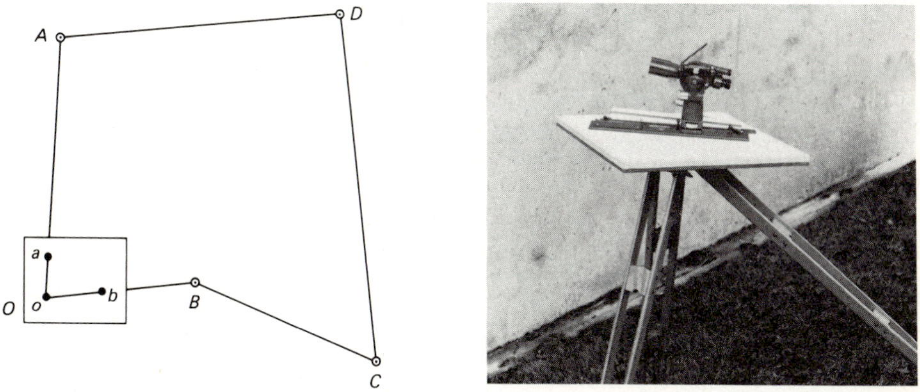

(*a*) (*b*)

Fig. 7.13 (*a*) Plotting positions of points on a plane table. (*b*) Plane table and alidade.

The location of any object is determined as follows (Fig. 7.13*a*). With the straightedge through the plotted point *o* representing the station *O* occupied by the instrument, the line of sight is directed to the object *A*, and a line *oa* of indefinite length is drawn along the straightedge on the plane-table sheet; this line represents the direction from station to object. The measured distance *OA* between station and object is then plotted to scale from *o*, thus locating *A* on the map at *a*.

The term "plane table" is somewhat ambiguous, being used sometimes to designate only the board with its supporting tripod and sometimes (more generally) both the table proper and its accompanying alidade.

By means of the plane table, points on the ground to which observations are made can be plotted immediately in their correct relative positions on the drawing, all angles being plotted graphically. The plane-table method is especially adapted to securing the details of the map. On extensive surveys the primary points of horizontal and vertical control are generally established by other more precise methods. It is a valuable and commonly used means of completing the compilation of maps from aerial photographs (Chap. 13), especially where the ground is hidden by overgrowth.

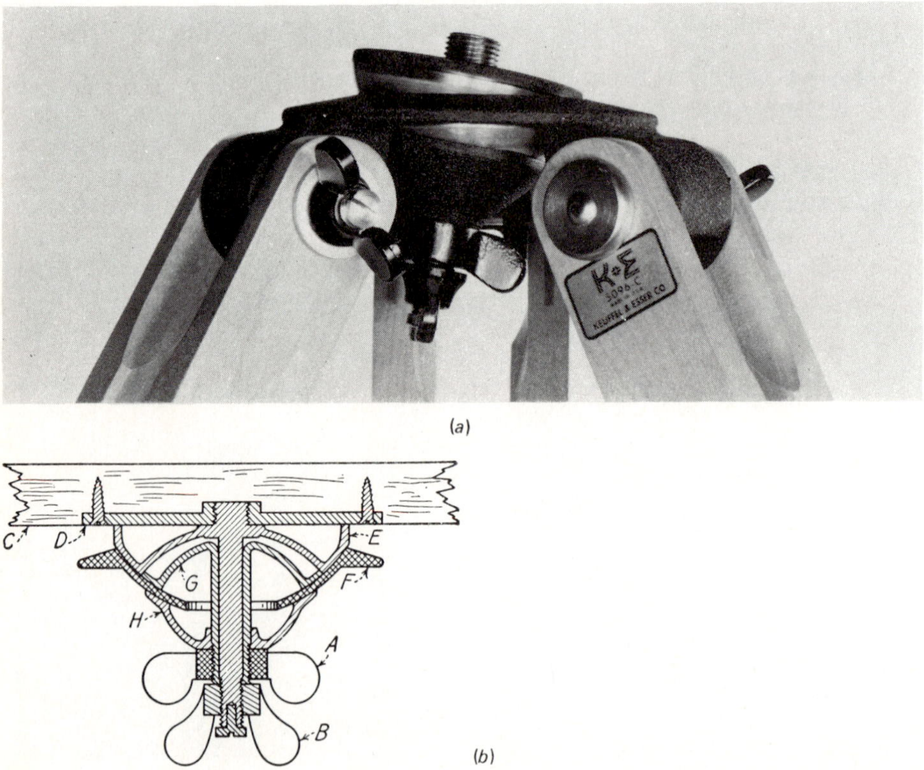

Fig. 7.14 (*a*) Johnson head for the plane table. (*Courtesy Keuffel & Esser Co.*) (*b*) Cross section of the Johnson head.

The *plane table* is mounted on the tripod by means of a Johnson head (Fig. 7.14), which permits the table to be rotated in azimuth and leveled without disturbing the tripod. The drawing board is 18 by 18 in (45.7 × 45.7 cm), 18 by 24 in (45.7 × 61.0 cm), or 24 by 31 in (61.0 × 78.7 cm). Into the underside of the board is fastened a circular brass plate, shown as *D* in Fig. 7.14*b*. By means of the threaded opening in the plate, the board can be fastened to the upper casting *E* of the tripod head. The head comprises a ball-and-socket joint and a vertical spindle; the cup *F* is supported by the tripod (not shown). When clamp *A* is loosened, the grip of parts *G* and *H* on the cup is released, and thus the board can be leveled. The clamp is then tightened to fix the board in a horizontal plane. When clamp *B* is loosened, the board can be rotated about the vertical axis and thus be oriented. The plane-table sheet is held in position by countersunk screws in the top of the board or by drafting tape placed along the edges of the sheet.

There are other styles of plane-table mounts. One is illustrated in Fig. 7.15, where the plane table is mounted on a heavy metal casting, so arranged that the board can be leveled by means of three leveling screws. A clamp and tangent screw permit the board to be rotated to any position in azimuth.

Telescopic alidade The telescopic alidade is designed to provide precision in the control of the table, especially to make possible the stadia method of measuring distances. The base, or plate, of the alidade consists of a brass blade or straightedge beveled on one edge and chrome-plated on the bottom. Upon one end of the plate is mounted either a circular level or a pair of level tubes at right angles to each other.

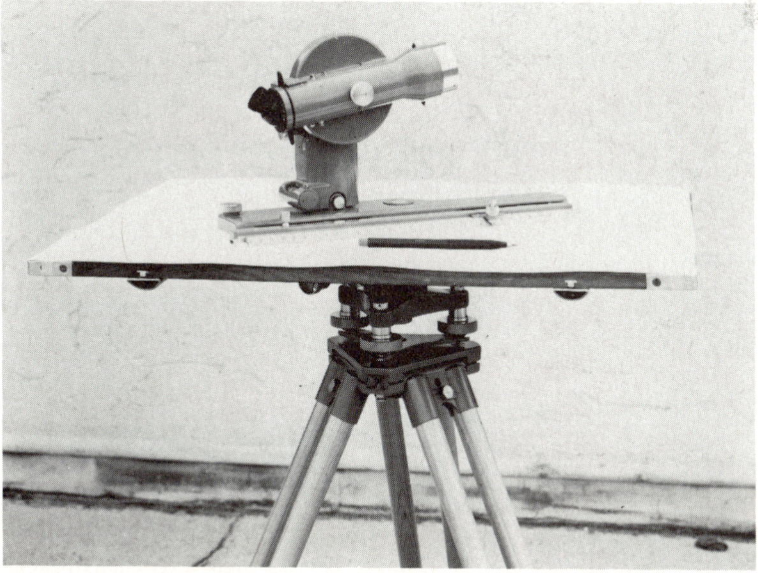

Fig. 7.15 Plane table with European-style Wild RK1 self-reducing alidade. (*Courtesy Wild Heerbrugg Instruments, Inc.*)

Upon the other end of the plate is mounted a trough compass consisting of a magnetic needle mounted in a narrow box with a short graduated arc at the end. In the center of the plate is mounted a column which supports a telescope similar to that of the transit.

Most alidades of modern design are equipped with an enclosed glass vertical circle which is observed by means of a microscope. The glass circle is graduated according to the stadia equations so that H and V readings (similar to the Beaman arc, Art. 7.11) are read directly and can be easily converted to horizontal distance and difference in elevation. A vertical circle that can be read directly to 30″ is also visible. Figure 7.16 shows a *self-indexing* alidade (of the American style) in which a damped pendulum automatically brings the index of the vertical arc to the correct scale reading even though the board is not quite level. The telescope is internal focusing; therefore, the stadia constant $f + c$ is zero. The reticle is fixed in the telescope, and the telescope is fixed to its axis; thus, neither the line of sight nor the axis of the telescope needs adjustment. No striding level or collars are required. The design is simple, and most parts are enclosed. A telltale circle on the bull's-eye level indicates when the blade is tilted beyond the range of the pendulum. A detachable elbow eyepiece is provided. The only field adjustment is to zero the index of the vertical arc when the line of sight is horizontal, as follows. With the line of sight on the rod reading established for a horizontal line by means of the two-peg test (Art. 5.47, adjustment 3), the index is brought to the zero mark (or equivalent scale graduation) by means of a capstan-head screw which controls its setting.

Figure 7.17 shows the field of view visible when looking through the reading microscope of the alidade shown in Fig. 7.16. Three sets of graduations are seen: the

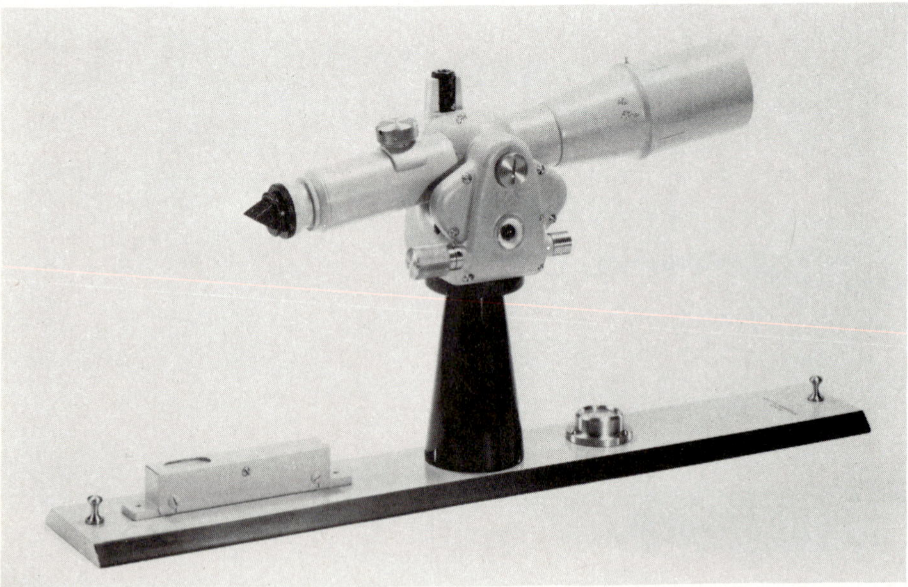

Fig. 7.16 Self-indexing alidade of the American style. (*Courtesy Keuffel & Esser Co.*)

zenith angle of 103°42′; the *V* reading [an optical version of the Beaman stadia arc (Art. 7.11)] which is 27.0, from which 50 (the *V* reading when the telescope is horizontal) must be subtracted to yield −23.0 that must be multiplied by the stadia interval to get the difference in elevation between the center of the telescope and the point sighted on the rod; and the *H*-scale multiplier, which is 94.4, multiplied by the stadia interval to give horizontal distance. If the stadia interval is 1.431 m, $V = (-23.0)(1.431) = -32.9$ m and $H = (94.4)(1.431) = 135.1$ m.

When using this style of self-indexing alidade, the instrument operator sights on the rod; observes *s*, the stadia interval; sets to the nearest even *V*-scale reading; reads *H*; and observes where the center cross hair cuts the rod. The values of *s*, *V*, *H*, and the rod reading are recorded and the horizontal distance and difference in elevation are calculated.

Another style of alidade with optical reading of the *H*, *V* scales and vertical angle but without the automatic self-indexing is shown in Fig. 7.18. This alidade is equipped with a vertical clamp and tangent screw to control movement of the telescope and another tangent screw to control a spirit level bubble attached to the vertical circle. When this bubble is centered, the vertical circle index coincides with the horizontal. The telescope contains stadia hairs with $K = 100$ and is internally focusing so that $f + c$ equals zero. The operation of this instrument is similar to that of the self-indexing instrument with one exception: Prior to reading or setting the *V* scale and observing the rod, the vertical circle index bubble must be centered. Figure 7.19 illustrates the circle readings visible when looking through the microscope of this style of alidade.

Note that for the vertical angle and *V* scale, there is a sign visible in the field of view. These signs appear for angular values of up to ±3° and up to the *V* graduation of ±5. Beyond these graduations, the tilt of the telescope is sufficient so that the operator can easily tell whether the inclination is plus or minus.

For this example, $V = +6$, the vertical circle = +3°27′, and *H* is estimated as 0.4. With a stadia interval of 0.348 m, the difference in elevation between the center of the telescope and the point sighted is $(6)(0.348) = +2.09$ m and the horizontal distance is $[100 - (0.4)](0.348) = 34.7$ m. Assuming the H.I. of the alidade to be 31.22 m, the middle horizontal cross hair on 1.25 m with *V* on +6, and the vertical

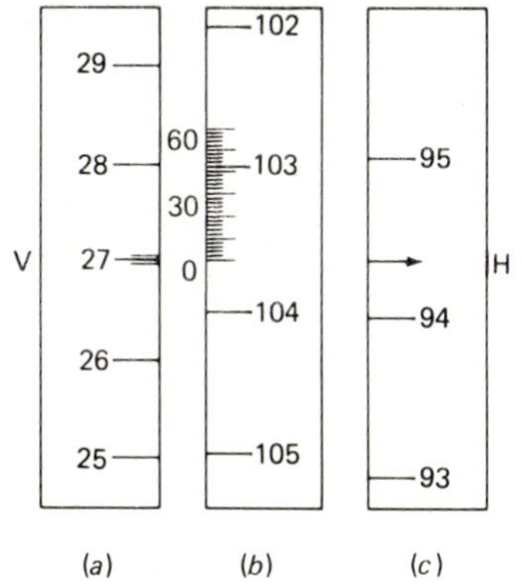

Fig. 7.17 View through scale-reading eyepiece of an American style self-indexing alidade. (*Courtesy Keuffel & Esser Co.*)

Fig. 7.18 Telescopic microptic alidade.

circle index bubble centered, the elevation of the point = 31.22 + 2.09 − 1.25 = 32.06 m.

A European-style self-reducing alidade is shown in Fig. 7.15. This alidade contains cross hairs with intervals which vary with the inclination of the telescope. The stadia intervals, when multiplied by a constant that appears in the field of view, provide horizontal distance and difference in elevation directly. The principle is similar to that described in Art. 7.16 for self-reducing tacheometers.

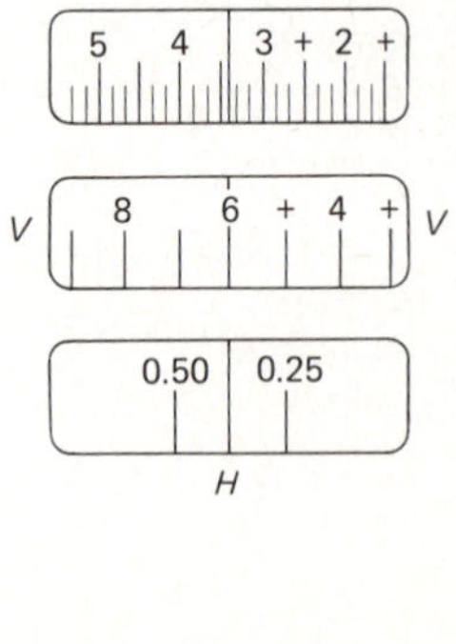

Fig. 7.19 Circle readings on microptic alidade.

Digital alidade A digital alidade is illustrated in Fig. 7.20. This instrument has the same general configuration as a standard optical alidade with the important differences that slope distances are determined by an EDM (Arts. 4.29 through 4.38) and vertical angles are sensed by an encoder. The key components of this device are (see Fig. 7.20): (1) a compact electro-optical EDM; (2) an automatic angle sensor; and (3) a microcomputer contained in the instrument base. The EDM, with a built-in telescope for sighting, is mounted on a bracket so that the coplanar optical axis of telescope and EDM can be rotated in a vertical plane. Control of this rotation is by a vertical clamp and vertical tangent screw. The angle sensor has absolute encoding and registers vertical angles with respect to horizontal as defined by the surface of the plane table on which the alidade rests. The plane table (not shown in Fig. 7.20) is mounted on a leveling head similar to the one shown in Fig. 7.15 and can be leveled by centering two spirit level bubbles (sensitivity 20″) mounted at right angles and placed on the surface of the plane table. Vertical angles can be determined with an estimated standard error of about 1′ with this value totally dependent on the accuracy in leveling the plane table. Sights are taken on a reflector mounted on an adjustable rod held at the point for which a location is

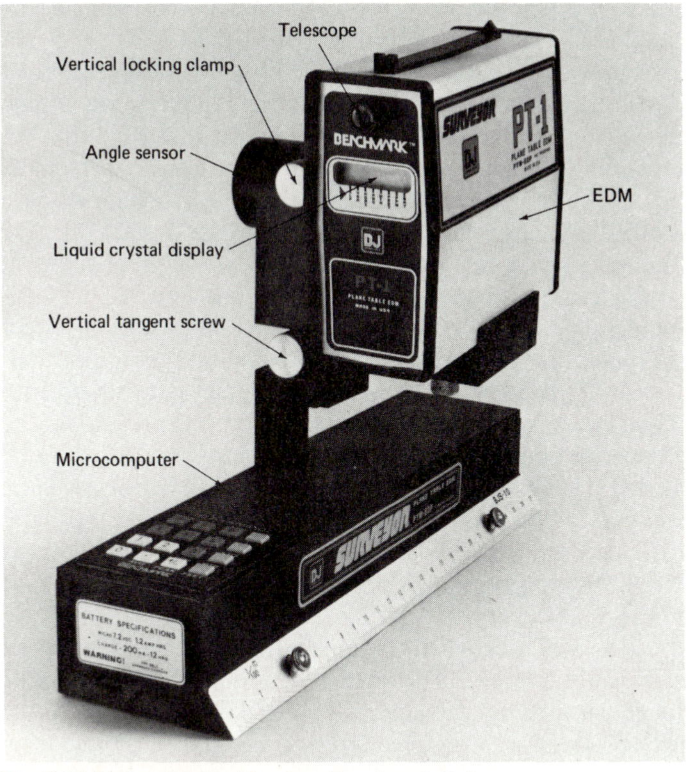

Fig. 7.20 Digital alidade. (*Courtesy Benchmark Co.*)

desired. Slope distance and vertical angle are measured by sighting the reflector and operating the appropriate controls on the keyboard mounted on the base of the instrument. Observed slope distance and vertical angle are then processed in the microcomputer to yield horizontal distance and difference in elevation to the point sighted. These values are registered on a liquid crystal display by pressing the proper key in the keyboard. Removable scales of various ratios can be attached to the base of the alidade. The EDM has a range of 0.5 km (1640 ft) to 1 km (3280 ft) when sighting on one or three reflectors, respectively, and under average atmospheric conditions. The estimated standard error in measurement of distances is 5 mm + 10 ppm.

7.19. Setting up and orienting the plane table The plane table is set up approximately waist-high, so that the topographer can bend over the board without resting against it. The tripod legs are spread well apart and planted firmly in the ground. The board is leveled by centering the circular bubble on the blade of the alidade. When the board is level, the upper locking clamp is tightened. Since few tables are sufficiently rigid to remain level as the alidade is shifted about, no special attempt is made to see that the board is perfectly level each time an observation is made when using an optical alidade. However if the digital alidade described in Art. 7.18 is being used, a stable, rigid board attached to a leveling head similar to the one illustrated in Fig. 7.15 must be employed and every effort should be made to ensure that the plane table is level at all times.

For plotted angles to be theoretically correct, the plotted location of the station at which the plane table is set should be exactly over the corresponding point on the ground. Practically, the degree of care exercised in bringing the plotted point over the ground point depends upon the scale of the map. For map scales smaller than perhaps 1:600, the plane table is set up over the station without any attempt to place the plotted point vertically above the station point. For maps of larger scale, the table is set up roughly and oriented approximately, and then it is shifted bodily until the point on the paper is practically over the station point, as indicated by plumbing. Either a hook-shaped plumbing arm may be used to support the plumb line under the plotted point, or the plumb line and point may be sighted from two directions approximately at right angles to each other. In any case, the aim is to set up with sufficient care so that the plotted position of lines drawn from the station will be shown correctly within the scale of the map.

The table may be oriented (1) by use of the *magnetic compass*, (2) by *backsighting*, or (3) by solving the *three-point problem* (Art. 9.12). As soon as the table is oriented, it is clamped in position and all mapping at the station is carried on without disturbing the board.

1. Orientation by compass For rough mapping at small scale, orientation by the magnetic compass is sufficiently accurate. This method is susceptible to the same errors as those encountered when using the surveyor's compass, but an error in the plotted direction of one line introduces no systematic errors in the lines plotted from succeeding stations.

If the compass is fixed to the drawing board, the board is oriented by rotating it about the vertical axis until the fixed bearing (usually magnetic north) is observed. If the compass is attached to the alidade or to a movable plate, the edge of the blade or the plate is aligned with a meridian previously established on the plane-table sheet, the board is turned until the needle reads north and the lower locking clamp *B* (Fig. 7.14*b*) is tightened.

2. Orientation by backsighting For mapping at intermediate or large scale, the board is oriented by backsighting along an established line, the direction of which has been plotted previously but the length of which need not be known. The method is equivalent to that employed in azimuth traversing with the transit. Greater precision is obtainable than with the compass, but an error in direction of one line is transferred to succeeding lines.

The plane table is set up as at *B* (Fig. 7.21) on the line *AB*, which has previously been plotted as *ab*, the board is leveled, the straightedge of the alidade is placed along the line *ba*, and the board is oriented by rotating it until the line of sight falls at *A*. Finally, the lower clamp *B* (Fig. 7.14*b*) is tightened. The length of *AB* or *ab* need not be known. Preferably the longest line available should be used for precision in orienting.

When the straightedge or blade of the alidade is placed along a previously plotted line passing through the plotted location of the station occupied and that of another station or object (such as line *ab*, Fig. 7.21), and then the board is turned until the line of sight cuts the station or object, the sight is called a *backsight*. When a station or object is sighted and a ray is drawn through the plotted location of the station occupied toward the station or object sighted, the sight is called a *foresight*. Line *bc*, the plotted location of line *BC* in Fig. 7.21, is an example of a foresight.

When a known station is sighted and a line is drawn through the plotted location of that station toward the station occupied, such as line *be* (Fig. 7.21), the sight is called a *resection*. Resection is also a general term applied to the process of determining the location of the station occupied (Art. 9.12).

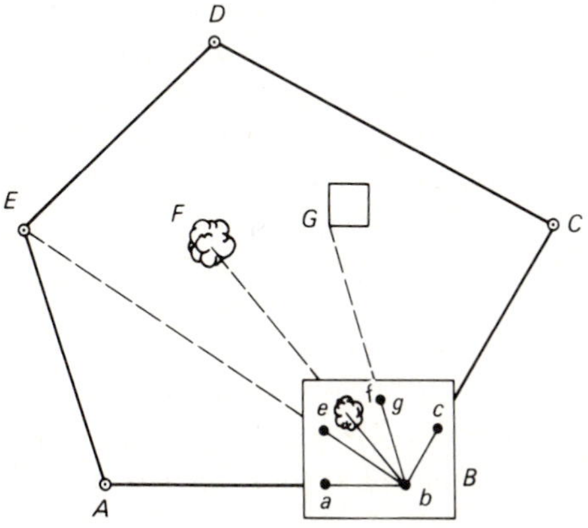

Fig. 7.21 Plane-table setup.

When the table has been oriented, the direction to any object in the landscape may be drawn on the map by pivoting the alidade about the plotted location of the plane-table station, pointing the alidade toward the distant object, and drawing a line along the straightedge.

Thus, in Fig. 7.21 the plane table is shown in position over station *B*. The plotted location of the plane-table station is indicated at *b* on the plane-table sheet. The alidade is pivoted about this point; and as sights are taken to points as *A*, *E*, and *C*, rays are drawn along the edge of the ruler.

Stadia data are recorded for each sight and are reduced to horizontal distance and difference in elevation to each point. As the distance to each point is calculated, the position to the point is plotted to scale along the corresponding ray, thus locating points such as *F* and *G* on the plane-table sheet at *f* and *g*. This procedure is called *radiation*.

7.20. Sources of error The major sources of error in plane-table work are the same as those which affect transit work and plotting, and the discussion relating to those subjects need not be repeated here. However, the following three sources of error should be considered:

1. Setting over a point Because plotted results only are required, it is not necessary to set the plotted location of the plane table over the corresponding ground point with any greater precision than is required by the scale of the map (see Art. 7.19).

2. Drawing rays The accuracy of plane-table mapping depends largely upon the precision with which the rays are drawn; consequently, the rays should be of considerable length. To avoid confusion, however, only enough of each ray is drawn to ensure that the plotted point will fall upon it, with one or two additional dashes drawn near the end of the alidade straightedge to mark its direction. Fine lines are desirable both for precision and for legibility.

3. Instability of the table If it is manipulated with care, the plane table can be oriented with considerable precision; however, a principal source of error in its use arises from the fact that its position is subject to continual disturbance by the topographer while work is progressing. Errors from this source can be kept within reasonable limits (*a*) by planting the tripod firmly in the ground, (*b*) by setting the table approximately waist-high so that the topographer can bend over it without leaning against it, (*c*) by avoiding undue pressure upon or against the table, and (*d*) by testing the orientation of the board occasionally and correcting its position if necessary. This test is always applied before a new instrument station is plotted.

7.21. Applications for plane table and alidade The plane table and alidade are especially adapted for securing details for mapping. The use of the plane table for topographic mapping is discussed in Art. 12.11. It is also a valuable and commonly used method for field completion and checking of photogrammetrically compiled maps (Art. 13.15), especially where the ground is concealed by vegetation. The plane table can also be used for indirect leveling (Art. 7.14), graphical triangulation (Art. 9.1), graphical resection and intersection (Arts. 9.11 and 9.12), and for traversing (Art. 8.13).

7.22. Problems

7.1 To determine the stadia interval factor, a transit is set up at a distance *C* back of the zero end of a level base line 800.0 ft long, the base line being marked by stakes set every 100.0 ft. A rod is then held at successive stations along the base line. The stadia interval and each half-interval observed at each location of the rod are tabulated below. Compute the lower, upper, and full interval factor for each distance, and find the average value for the lower interval, the upper interval, and the full interval.

Distance	Lower interval		Upper interval		Full interval	
− *C*, ft	Feet	Factor	Feet	Factor	Feet	Factor
100	0.49		0.50		0.99	
200	0.98		0.99		1.97	
300	1.47		1.48		2.95	
400	1.97		1.98		3.96	
500	2.46		2.47		4.94	
600	2.95		2.97		5.92	
700	3.45		3.47		6.91	
800	3.94		3.96		7.89	
Average						

7.2 A stadia interval of 2.415 m is observed with a theodolite for which the stadia interval factor is 100.0 and *C* is 0.305 m. The zenith angle is 98°10′30″ with the middle cross hair set on 2.000 m. If the instrument has a height of instrument h.i. of 1.38 m above the point over which it is set and the point has an elevation of 100.05 m, calculate the horizontal distance and elevation of the point sighted by the (*a*) exact stadia equations; (*b*) approximate stadia equations.

7.3 The following observations are taken with a transit for which the interval factor is 100.0 and *C* is 1.00 ft:

Observation	Stadia interval	Vertical angle
	(i)	
(*a*)	10.00 ft	+0°30′
(*b*)	10.00 ft	+10°00′
(*c*)	10.00 ft	+25°00′
	(ii)	
(*d*)	3.000 m	+0°40′
(*e*)	3.000 m	+10°10′
(*f*)	3.000 m	+25°10′

By means of Eqs. (7.4) and (7.5), Art. 7.7, compute the horizontal distance and differences in elevation. By means of the approximate equations (7.7) and (7.8), Art. 7.8, determine the same quantities and note the errors introduced by the approximations.

7.4 What would be the amount and sign of error introduced in each computed horizontal distance and difference in elevation if the observations of Prob. 7.3, part (i) were taken (*a*) with a 12-ft rod that was unknowingly 0.5 ft out of plumb with top leaning toward the transit? (*b*) With the top of the rod leaning 0.5 ft away from the transit? (*c*) What conclusions may be drawn from these results?

7.5 What error will be introduced in each computed horizontal distance and difference in elevation if in the observations of Prob. 7.3, part (i) or (ii): (*a*) Each vertical angle contains an error of 01′? (*b*) Each stadia interval is in error by $\frac{1}{1000}$ of the interval? (*c*) What conclusions may be drawn from these results?

7.6 In determining the elevation of point *B* and the distance between two points *A* and *B*, a transit equipped with a Beamen stadia arc is set up at *A* and the following data are obtained: $V = 38$, $H = 3.0$, stadia interval = 1.311 m, h.i. = 1.28 m, and the line of sight at 2.62 m on rod. The instrument constants are $K = 100.0$ and $C = 0.305$ m. The Beamen stadia arc has index marks of $H = 0$ and $V = 50$ for a horizontal line of sight. The elevation of point *A* is 38.28 m. Compute the distance *AB* and the elevation of point *B*.

7.7 Following are the notes for a line of stadia levels. The elevation of B.M.$_1$ is 637.05 ft. The stadia interval factor is 100.0 and $C = 1.25$ ft. Rod readings are taken at height of instrument. Determine the elevations of remaining points. Record notes and elevations in a proper note form.

	Backsight		Foresight	
Station	Stadia interval, ft	Vertical angle	Stadia interval, ft	Vertical angle
B.M.$_1$	4.26	−3°38′		
T.P.$_1$	2.85	−1°41′	3.18	+2°26′
T.P.$_2$	3.30	+0°56′	2.71	−4°04′
T.P.$_3$	2.26	+2°09′	4.45	−0°38′
B.M.$_2$			3.09	+7°27′

7.8 Notes for a line of stadia levels are shown below.

	Backsight			Foresight		
Station	Stadia interval, m	Vertical angle	Rod, m	Stadia interval, m	Vertical angle	Rod, m
B.M.$_{10}$	1.241	−4°24′	2.41			
T.P.$_1$	2.041	+3°23′	1.05	1.506	+8°10′	1.52
T.P.$_2$	1.721	−6°30′	1.45	1.432	−3°05′	3.05
T.P.$_3$	2.421	−10°05′	2.04	0.901	+10°08′	0.24
B.M.$_{11}$				1.321	+4°20′	1.43

The elevation of B.M.$_{10}$ is 42.852 m. The stadia interval factor is 100.0 and $C = 0.31$ m. Determine elevations for the turning points and B.M.$_{11}$. Record notes and elevations in a proper note form.

7.9 Following are stadia intervals and vertical angles for a transit-stadia traverse. The elevation of station *A* is 418.6 ft. The stadia interval factor is 100.0 and $C = 1.00$ ft. Rod readings are taken at height of instrument. Compute the horizontal lengths of the courses and the elevations of the transit stations.

Station	Object	Stadia interval, ft	Vertical angle
B	*A*	8.50	+0°48′
	C	4.37	+8°13′
C	*B*	4.34	−8°14′
	D	12.45	−2°22′
D	*C*	12.41	+2°21′
	E	7.18	−1°30′

7.10 Following are stadia intervals and vertical angles taken to locate points from a transit station the elevation of which is 415.7 ft. The height of instrument above the transit station is 4.6 ft, and rod readings are taken at 4.6 ft except as noted. The stadia interval factor is 100.0 and $C = 1.00$ ft. Compute the horizontal distances and the elevations.

Object	Stadia interval, ft	Vertical angle
43	7.04	$-0°58'$
44	8.25	$+2°30'$ on 2.1
45	7.56	$-0°44'$ on 9.2
46	7.25	$+1°20'$ on 6.0
47	3.72	$-5°36'$

7.11 A transit equipped with a stadia arc is used in locating points from a transit station the elevation of which is 233.39 m. The stadia arc has index marks of $H = 100$ and $V = 50$ for a horizontal line of sight. The instrument constants are $K = 100.0$ and $C = 0$ (internal-focusing telescope). The height of instrument above the transit station is 1.37 m. Compute the horizontal distances and the elevations.

Object	Stadia interval, m	Rod reading, m	Stadia arc readings V	Stadia arc readings H
114	0.994	1.10	18	88.3
115	2.390	1.77	35	97.7
116	0.664	1.43	39	98.8
117	0.506	1.31	76	92.6
118	2.481	1.95	69	96.2

References

1. Allan, A. L., "An Alidade for Use with Electro-Optical Distance Measurement," *Survey Review*, July 1977.
2. Asplund, E. C., "A Review of Self Reducing, Double Image Tacheometers," *Survey Review*, January 1971.
3. Moffitt, F. H., and Bouchard, H., *Surveying*, 7th ed., Harper & Row, Publishers, New York, 1982.
4. Instruction Manual, *Self-Reducing Engineer's Tachymeter, K1-RA*, Kern & Co. Ltd., Aaru, Switzerland.
5. Wolf, P. R., Wilder B. P., and Mahun, G., "An Evaluation of Accuracies and Applications of Tacheometry," *Surveying and Mapping*, Vol. 38, No. 3, September 1978.

PART III
SURVEY OPERATIONS

CHAPTER 8
Traverse

8.1. General A *traverse* consists of a series of straight lines connecting successive established points along the route of a survey. The points defining the ends of the traverse lines are called *traverse stations* or *traverse points*. Distances along the line between successive traverse points are determined either by direct measurement using a tape or electronic distance measuring (EDM) equipment, or by indirect measurement using tacheometric methods. At each point where the traverse changes direction, an angular measurement is taken using a transit or theodolite. With minor modifications, these practices are common to land, city, topographic, and hydrographic surveys and to location surveys such as those for highways and railways.

8.2. Traverse party The *traverse party* is usually composed of an instrument operator, a head tapeman, and a rear tapeman. The transit operator directs the party, operates and cares for the instrument, and takes notes. The head tapeman performs the duties of that position as described in Arts. 4.10 to 4.12, gives line as directed by the transit operator, is responsible for the accuracy and speed of taping operations, and attends to the setting and marking of stakes. The rear tapeman carries out the duties of that position as described in Arts. 4.10 to 4.12, often carries and drives stakes, and assists in removing obstructions to the vision of the transit operator. In heavily wooded terrain, additional employees are engaged to clear the transit line of trees and brush. The head tapeman keeps in close communication with those clearing the line and assists in their direction. Where sights are long, a flagman may be employed to give sights to the transit operator. Where many observations are taken, the notes may be taken by a recorder, who may also act as the chief of party.

8.3. Equipment for the traverse party The equipment of the party usually consists of a transit or theodolite, steel tape of desired length, or EDM device plus supporting equipment: range poles and plumb bobs, stakes and hubs, tacks, axe or hammer, field notebook, taping pins, and marking crayon. Where *forced centering* (Art. 6.30) is to be employed, three tripods compatible with theodolite, sighting targets, and/or reflectors are required. Also included are devices for marking stations, such as ordinary or special nails, a cold chisel, spray paint, and colored plastic flagging or tape. A light tripod with plumb bob is useful for sighting on traverse stations where vision is obscured near the ground. Portable sending and receiving radios (such as citizens' band radios) may be used to expedite communication.

For stadia or tacheometric traverses (Art. 8.12), distance measuring equipment is omitted and a level or stadia rod is included in addition to the other items listed above.

8.4. Traverse stations Any temporary or permanent point of reference over which the transit is set up is called a *traverse station*. On most surveys the traverse station is a peg, called a *hub*, driven flush with the ground and having a tack driven in its top to mark the exact point of reference for measurements. On pavements the traverse station may be a driven nail, a cross cut in the pavement or curb, or a tack set in a hole drilled with a star drill and filled with lead wool. In land surveying, the stations are often iron pipes, stones, or other more or less permanent monuments set at the corners. In mountainous country, often the station marks are cut in the natural rock.

The location of a hub is usually indicated by a flat *guard stake* extending above the ground and driven sloping so that its top is over the hub. This guard stake carries the number or letter of the traverse station over which it stands. Usually, the number is marked with keel or lumber crayon on the underside of the guard stake and reads down the stake. The hubs are usually square, say 2 by 2 in (5 by 5 cm) and the guard stakes are usually flat, perhaps $\frac{3}{4}$ by 3 in (2 by 8 cm).

To avoid the necessity for using a rear flagman to give backsights, as far as practical each traverse station is marked by a temporary signal, such as a lath or a stick set on line, with a piece of paper, cloth, or tape attached. When forced centering is employed, sighting targets mounted on tripods are used instead.

Lines connecting traverse stations are called *traverse lines*. Both in the field notes and as part of the identification mark left in the field, a traverse station number is preceded by the symbol $\odot$.

8.5. Purpose of the traverse Traversing is a convenient, rapid method for establishing horizontal control. It is particularly useful in densely built up areas and in heavily forested regions where lengths of sight are short so that neither triangulation nor trilateration is suitable. Traverses are made for numerous purposes, including:

1. Property surveys to locate or establish boundaries
2. Supplementary horizontal control for topographic mapping surveys
3. Location and construction layout surveys for highways, railways, and other private and public works
4. Ground control surveys for photogrammetric mapping

Frequently, the traverse is employed to densify the number of coordinated control points in a triangulated or trilaterated horizontal control network.

8.6. Types of traverse The two general classes of traverse are the open traverse and the closed traverse.

The *open traverse* originates at a point of known position and terminates at a point of unknown position. No computational check is possible to detect errors or blunders in distances or directions in this type of traverse. To minimize errors, distances can be measured twice, angles turned by repetition, magnetic bearings observed on all lines, and astronomic observations made periodically. In spite of these precautions, an open traverse is a risky proposition and should not be used for any of the applications discussed in Art. 8.5, since the results are uncertain and will always be subject to question. Open traverses are necessary in certain types of mine surveying. Hence, there are applications of an open traverse when circumstances dictate their use.

Closed traverses originate at a point of known position and close on another point of known horizontal position. A point of known horizontal position is given either by geographic latitude and longitude or by X and Y coordinates on a rectangular coordinate grid system. In Fig. 8.1, traverse $MABCDQ$ originates at known point M with a backsight along line MN of known azimuth and closes on known point Q, with a foresight along line QR also of known azimuth. This type of traverse is preferable to all others since computational checks are possible which allow detection of systematic errors in both distance and direction.

A traverse that originates and terminates on a single point of known horizontal position is called a *closed-loop traverse*. In Fig. 8.2, traverse 1-2-3-4-5-6-1, which originates and closes on point 1, is an example of a closed-loop traverse. This type of traverse permits an internal check on the angles, but there is no way to detect systematic errors in distance or errors in the orientation of the traverse. A closed-loop traverse should not be used for major projects. To minimize the chance of errors in a closed-loop traverse, all distance measuring equipment must be carefully calibrated and astronomic observations must be made periodically.

A closed-loop traverse that originates and terminates on a point of assumed horizontal position provides an internal check on angles but no check on systematic errors in distance. An additional disadvantage is that the points are not located on

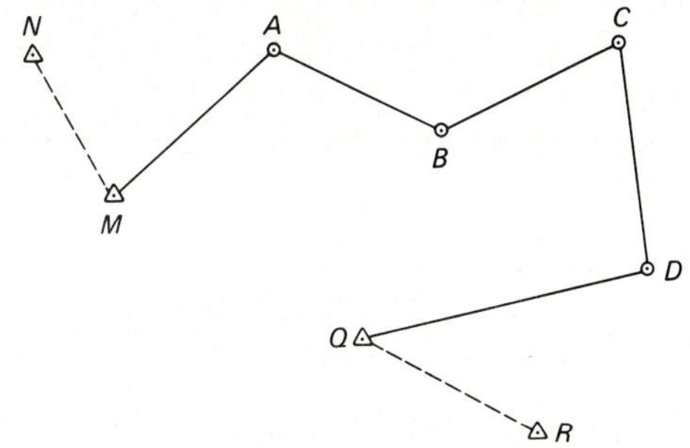

Fig. 8.1 Closed traverse.

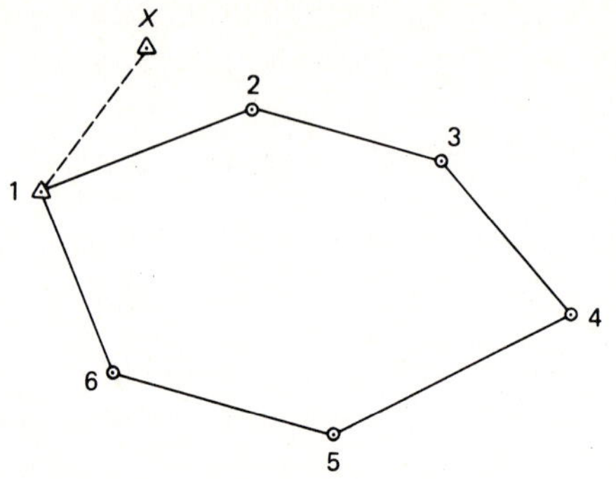

Fig. 8.2 Closed-loop traverse.

any datum and have no relationship with points in any other survey, although the relative positions of the points are determined. The use of such a method of running a traverse is to be discouraged except under extenuating circumstances, where no feasible alternative is possible.

Each of the types of traverse discussed above can also be described according to the method of turning the angles in the traverse. Thus, there are traverses by deflection angles, interior angles, angles to the right, and the azimuth method. Each of these methods for running traverse is discussed in the following articles.

8.7. Deflection-angle traverse This method of running traverses is probably more commonly employed than any other, especially on open traverses where only a few details are located as the traverse is run. It is used almost entirely for the location surveys for roads, railroads, canals, and pipelines. It is employed to a less extent in land surveying and in establishing control traverses for topographic and hydrographic surveys.

Successive transit stations are occupied, and at each station a backsight is taken with the *A* vernier set at zero and the telescope reversed. The telescope is then plunged, the foresight is taken by turning the instrument about the vertical axis on its upper motion, and the deflection angle is observed. The angle is recorded as right R or left L, according to whether the upper motion is turned clockwise or counterclockwise. Usually, it is considered good practice to observe the deflection angle at least twice, once with the telescope direct and once reversed as described in Art. 6.39. This process is called *turning the angle by double deflection*.

Figure 8.3 shows a closed deflection-angle traverse that originates at point *R* at one end of line *RS* having a known azimuth of 170°30′00″ as determined from a previous survey. The traverse closes on point *M* at one end of line *MN* which has an azimuth of 104°42′30″ also determined from a previous survey.

A portion of the field notes for the traverse is illustrated in Fig. 8.4. Angles are recorded on the left with the necessary details sketched on the right. All data are recorded from the bottom to the top of the page, a procedure that is also followed in preliminary route surveys. Angles are turned by double deflection and magnetic bearings are observed forward and backward from each traverse station as a rough

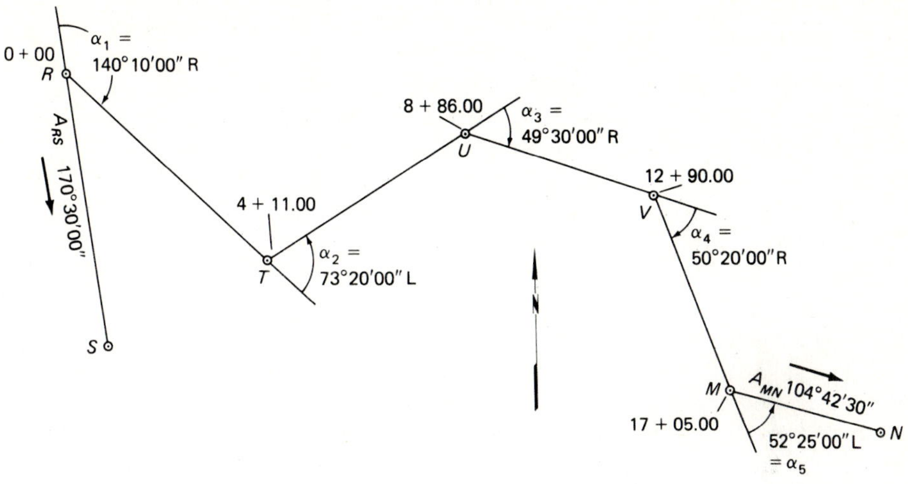

Fig. 8.3 Deflection-angle traverse between lines of known direction.

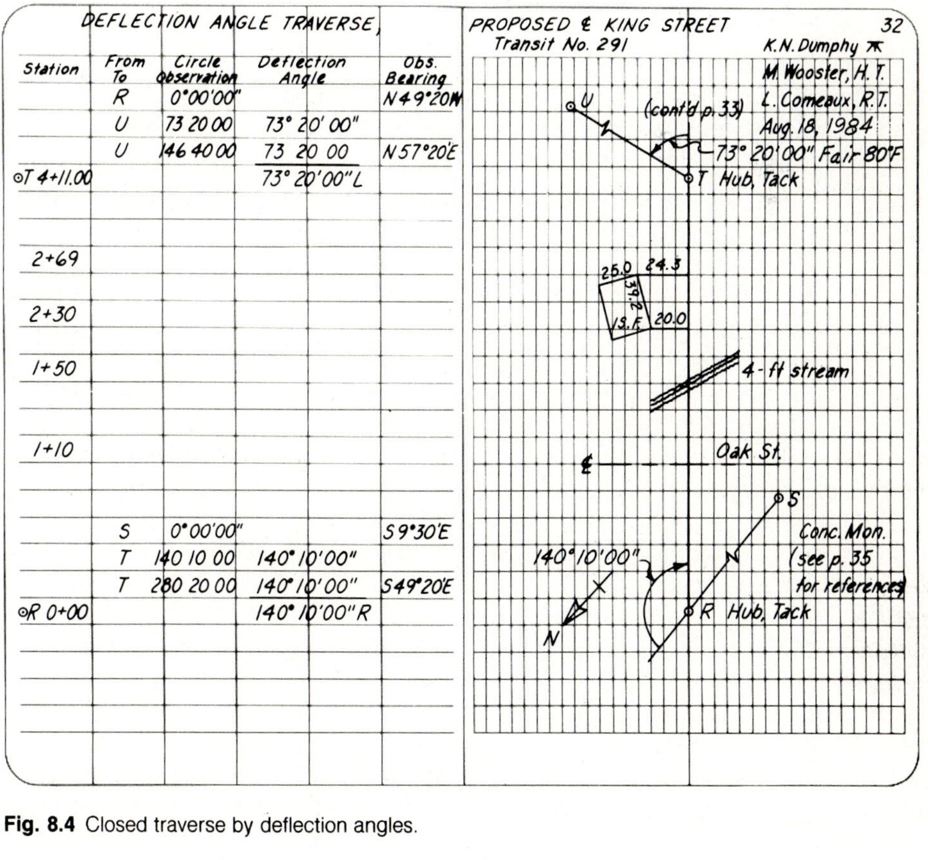

Fig. 8.4 Closed traverse by deflection angles.

check on the angles. To check the angular closure, azimuths are calculated from known line *RS* to known line *MN*.

To compute the azimuth of an unknown line, a deflection angle to the *right* is added to the *forward azimuth* of the previous line. A deflection angle to the *left* is subtracted from the forward azimuth of the previous line. Thus, in Fig. 8.3, the azimuth of line *RT* is 350°30′00″ + 140°10′00″ − 360°00′00″ = 130°40′00″. Similarly, the azimuth of line *TU* is 130°40′00″ − 73°20′00″ = 57°20′00″. Figure 8.5 illustrates computation of azimuths using deflection angles to the left and right, respectively. The results of these computations for the example traverse in Fig. 8.3 are shown in Table 8.1. The process of computing azimuths for unknown lines is continued through lines *UV*, *VM*, to *MN*, where the calculated azimuth fails to agree with the fixed azimuth of 104°42′30″ by 02′30″, which is the angular closure for the traverse. The equation by which this error of closure may be computed for the example traverse is

$$A_{SR} + \alpha_1 - \alpha_2 + \alpha_3 + \alpha_4 - \alpha_5 - 360° = A_{MN} \tag{8.1}$$

The general relationship for a traverse that contains n deflection angles is

$$A_1 + \sum_{i=1}^{n} \alpha_{R_i} - \sum_{i=1}^{n} \alpha_{L_i} = A_2 + 360° \tag{8.2}$$

in which A_1 is the forward azimuth at station 1, the origin of the traverse, and A_2 is the forward azimuth at station 2, the closing point for the traverse.

If adjustment of the observed angles to fit the known directions is desired (assuming equal weights for all angles), the closure error may be distributed equally among the five deflection angles. For this example 30″ will be subtracted from each

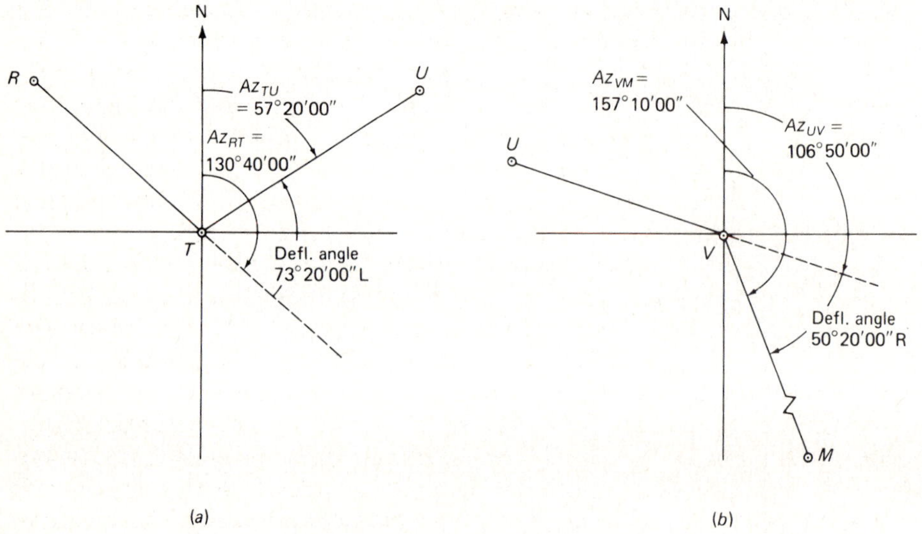

Fig. 8.5 (*a*) Azimuth with deflection angle to the left. (*b*) Azimuth with deflection angle to the right.

Table 8.1 Calculation and Adjustment of Azimuths in a Closed Traverse by Deflection Angles

From	To	Azimuth	Correction	Adjusted azimuth	Adjusted bearing
S	*R*	350°30′00″		350°30′00″	N9°30′00″W
	+∠*R*	140°10′00″R			
		490°40′00″			
	−	360°00′00″			
R	*T*	130°40′00″	−00′30″	130°39′30″	S49°20′30″E
	−∠*T*	73°20′00″L			
T	*U*	57°20′00″	−01′00″	57°19′00″	N57°19′00″E
	+∠*U*	49°30′00″R			
U	*V*	106°50′00″	−01′30″	106°48′30″	S73°11′30″E
	+∠*V*	50°20′00″R			
V	*M*	157°10′00″	−02′00″	157°08′00″	S22°52′00″E
	−∠*M*	52°25′00″L			
M	*N*	104°45′00″	−02′30″	104°42′30″	S75°17′30″E
M	*N*	104°42′30″			

Angular error of closure = +2′30″

deflection angle where angles to the right are positive and angles to the left are negative. When azimuths have already been calculated, as in this case, it is simpler to apply the cumulative correction directly to the unadjusted azimuth. Thus, the correction to the first azimuth from *R* to *T* is −30″, to the second from *T* to *U* is −01′00″, etc., as shown in Table 8.1. The same result is achieved using adjusted deflection angles to compute the azimuths.

Figure 8.6 illustrates a closed-loop traverse with deflection angles. This traverse originates and closes on traverse point *A*. The notes are recorded down the page in the order of occupying the points and a sketch is drawn on the right-hand page. Magnetic bearings have been observed forward and backward from each station. For any closed-loop traverse the initial azimuth A_1 equals the closing azimuth A_2. Then from Eq. (8.2), the sum of the deflection angles should equal 360°. The sum of the average values of the angles is recorded at the bottom of the right-hand page, where a closure error of 02′30″ is indicated. If an adjustment is desired and assuming that angles are of equal weight, this error is distributed equally among the five angles. Since the left deflection angles have a sum that is too large, 30″ is subtracted from each left deflection angle and 30″ is added to the right deflection angle.

The next step is to calculate directions for each of the traverse lines. Since this is a closed-loop traverse and no angle was observed to a line of known direction, line *AB* will be assumed to have the observed magnetic bearing of S30°00′W which with 18°E declination (Art. 6.14) gives a true bearing of S48°00′W. This bearing is then converted to an azimuth of 228°00′00″ from the north for determination of azimuths for the balance of the lines. The procedure is similar to that employed for the previous example except that the closure check is on the original line *AB*, as shown in Table 8.2, which contains the calculation of the directions. Note that the closure is exact, as it should be, since the angles had been balanced previously. The directions, as determined for this example, are not absolute since they are based on a magnetic

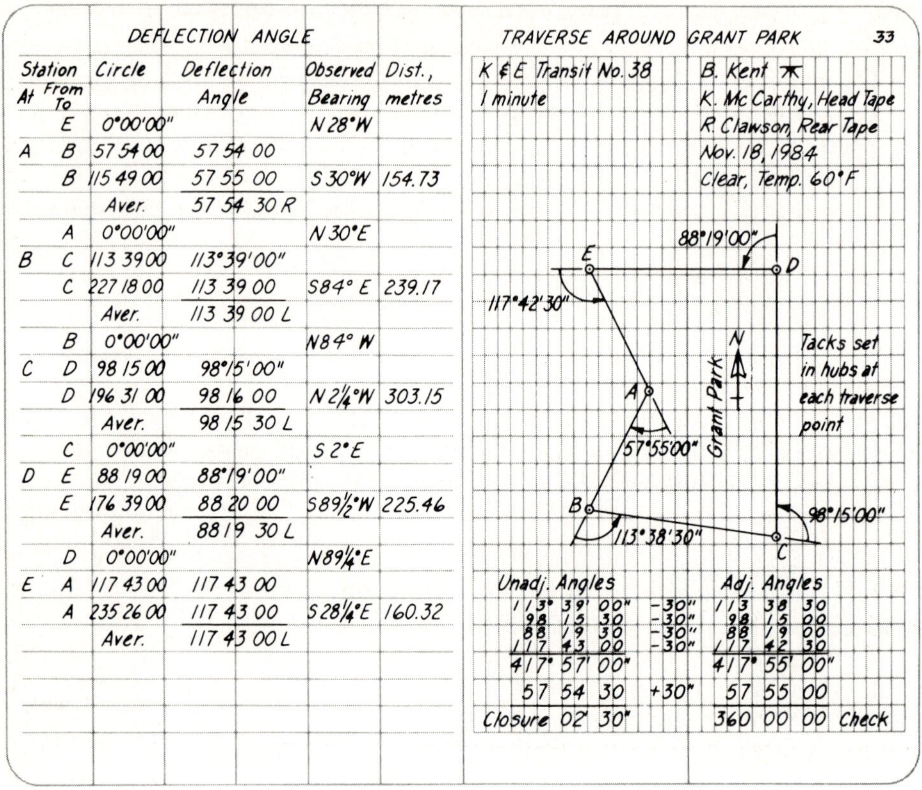

DEFLECTION ANGLE

Station At	From To	Circle	Deflection Angle	Observed Bearing	Dist., metres
	E	0°00'00"		N 28°W	
A	B	57 54 00	57 54 00		
	B	115 49 00	57 55 00	S 30°W	154.73
	Aver.		57 54 30 R		
	A	0°00'00"		N 30°E	
B	C	113 39 00	113°39'00"		
	C	227 18 00	113 39 00	S84° E	239.17
	Aver.		113 39 00 L		
	B	0°00'00"		N84° W	
C	D	98 15 00	98°15'00"		
	D	196 31 00	98 16 00	N 2¼°W	303.15
	Aver.		98 15 30 L		
	C	0°00'00"		S 2°E	
D	E	88 19 00	88°19'00"		
	E	176 39 00	88 20 00	S89½°W	225.46
	Aver.		88 19 30 L		
	D	0°00'00"		N89¼°E	
E	A	117 43 00	117 43 00		
	A	235 26 00	117 43 00	S 28¼°E	160.32
	Aver.		117 43 00 L		

Fig. 8.6 Deflection-angle traverse, closed loop.

Table 8.2 Azimuths and Bearings in a Closed-Loop Traverse

From	To	Azimuth	Bearing
A	*B*	228°00′00″	S48°00′00″W
	−∠*B*	113°38′30″L	
B	*C*	114°21′30″	S65°38′30″E
	−∠*C*	98°15′00″L	
C	*D*	16°06′30″	N16°06′30″E
	−∠*D*	88°19′00″L	
	+	360°00′00″	
D	*E*	287°47′30″	N72°12′30″W
	−∠*E*	117°42′30″L	
E	*A*	170°05′00″	S9°55′00″E
	+∠*B*	57°55′00″R	
A	*B*	228°00′00″ checks	S48°00′00″W

bearing that is only approximate. In practice, an angle should be turned to a line of known azimuth or an astronomic direction should be determined.

The arithmetical check of the sum of the angles should be performed immediately on completing the traverse in order to detect blunders or excessively large errors before leaving the field. The angular error of closure should not exceed the estimated standard deviation for observing an angle from a single setup times the square root of the number of instrument stations. In practice, this estimated standard deviation in angular measurement is usually taken equal to 0.5 to 1.0 times the least count of the instrument used to turn the angles.

8.8. Interior-angle traverse Field operations for this method of traversing are not materially different from those used for the deflection-angle traverse. At each station the vernier is set at zero, and a backsight to the preceding station is taken. The instrument is then turned on its upper motion until the advance station is sighted and the interior angle is observed. Except for surveys of a very low order of accuracy, all interior angles should be turned at least twice, once with the telescope direct and once with the telescope reversed. Notes may be kept in a form similar to that shown for deflection angles in Fig. 8.6.

A closed-loop traverse run by the method of interior angles is illustrated in Fig. 8.7. In this traverse, point 1 was occupied first and an angle was observed between line 1*Q*, which has a known azimuth, and traverse line 12. After the clockwise interior angle at 1 was observed, points 2, 3, 4, 5, and 6 were occupied with a clockwise interior angle being observed by repetition (at least twice) at each traverse station.

Directions for the traverse are reckoned using the angle *Q*12 turned from the line of known azimuth 1*Q* to line 12. Thus, the azimuth of line 12 is 8°30′00″ + 97°50′00″ = 106°20′00″. The azimuth of each succeeding traverse line is then calculated by adding the clockwise interior angle to the *back azimuth* of the preceding line. For example, the forward azimuth of line 23 = 106°20′00″ +

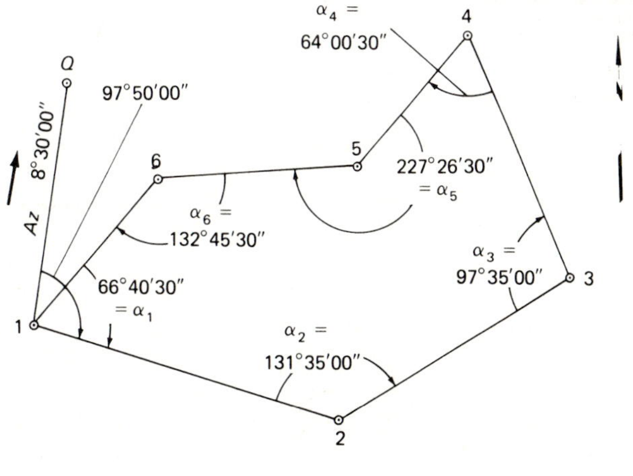

Fig. 8.7 Closed-loop traverse with interior angles observed.

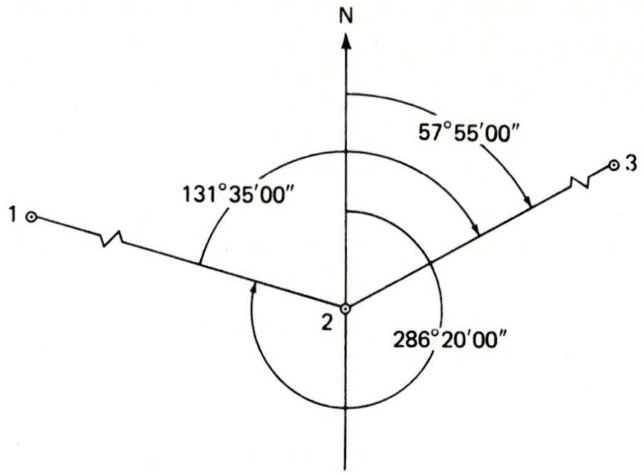

Fig. 8.8 Calculation of an azimuth using an interior angle.

$180°00'00'' + 131°35'00'' - 360°00'00'' = A_{12} + \alpha_2 - 180° = 106°20'00'' + 131°35'00'' - 180° = 57°55'00''$, as illustrated in Fig. 8.8. Similarly, the azimuth of line 34 $= A_{23} + 180° + \alpha_3 = 57°55'00'' + 180° + 97°35'00'' = 335°30'00''$, etc., as illustrated in Table 8.3.

For the example traverse, the angular error of closure can be checked by taking the sum of the angles α_i in Table 8.3 or

$$\alpha_1 + \alpha_2 + \alpha_3 + \alpha_4 + \alpha_5 + \alpha_6 = 4(180°)$$

Table 8.3 Azimuths Using Interior Angles

Line	Angle	Unadjusted azimuth	Correction	Adjusted azimuth
12		106°20′00″		106°20′00″
21		286°20′00″		
	$+\alpha_2$	131°35′00″		
23		57°55′00″ $= A_{12} - 180° + \alpha_2$	−00′30″	57°54′30″
32		237°55′00″		
	$+\alpha_3$	97°35′00″		
34		335°30′00″ $= A_{23} + 180° + \alpha_3$	−01′00″	335°29′00″
43		155°30′00″		
	$+\alpha_4$	64°00′30″		
45		219°30′30″ $= A_{34} - 180° + \alpha_4$	−01′30″	219°29′00″
54		39°30′30″		
	$+\alpha_5$	227°26′30″		
56		266°57′00″ $= A_{45} - 180° + \alpha_5$	−02′00″	266°55′00″
65		86°57′00″		
	$+\alpha_6$	132°45′30″		
61		219°42′30″ $= A_{56} - 180° + \alpha_6$	−02′30″	219°40′00″
16		39°42′30″		
	$+\alpha_1$	66°40′30″		
12		106°23′00″ $= A_{61} - 180° + \alpha_1$	−03′00″	106°20′00″
12		106°20′00″		
	Angular error of closure =	03′00″		

Table 8.4 Checking and Adjustment of Interior Angles

Station	Observed angles	Correction	Adjusted angles	
1	66°40′30″	−30″	66°40′00″	
2	131°35′00″	−30″	131°34′30″	
3	97°35′00″	−30″	97°34′30″	
4	64°00′30″	−30″	64°00′00″	
5	227°26′30″	−30″	227°26′00″	
6	132°45′30″	−30″	132°45′00″	
Sum =	720°03′00″		720°00′00″	checks
$(n-2)180°$ =	720°00′00″			
Closure =	03′00″			

In general, for any polygon with n sides, the sum of the interior angles is given by

$$\alpha_1 + \alpha_2 + \cdots + \alpha_n = (n-2)(180°) \tag{8.3}$$

Thus, the most rapid and simple field check of observed interior angles is to calculate the sum of the observations and compare with $(n-2)180°$.

Table 8.4 shows the sum of the observed interior angles for the example traverse. The angular error of closure is 03′, the same value as determined previously.

Table 8.5 Azimuth Calculation Using Adjusted Interior Angles

Line	Angle	Azimuth		Bearing
12		106°20′00″		S73°40′00″E
21		286°20′00″		
	$+\alpha_2$	131°34′30″		
23		57°54′30″		N57°54′30″E
32		237°54′30″		
	$+\alpha_3$	97°34′30″		
34		335°29′00″		N24°31′00″W
43		155°29′00″		
	$+\alpha_4$	64°00′00″		
45		219°29′00″		S 39°29′00″W
54		39°29′00″		
	$+\alpha_5$	227°26′00″		
56		266°55′00″		S 86°55′00″W
65		86°55′00″		
	$+\alpha_6$	132°45′00″		
61		219°40′00″		S39°40′00″W
16		39°40′00″		
	$+\alpha_1$	66°40′00″		
12		106°20′00″	checks	

If adjusted angles are desired at this point, then assuming that all angles were observed with equal precision, the error is distributed equally among the angles and $180''/6 = 30''$ is subtracted from each observed angle, as shown in Table 8.4.

The simplest way to compute adjusted azimuths is to apply a cumulative correction to the unadjusted azimuths in Table 8.3. Thus, corrections of 30″, 01′00″, 01′30″, 02′00″, 02′30″, and 03′00″ are applied respectively to the computed values of $A_{23}, A_{34}, \ldots, A_{12}$. These corrections and adjusted azimuths are listed in Table 8.3. An alternative procedure is to use adjusted angles to compute the adjusted azimuths and bearings, as shown in Table 8.5.

Note that in this closed-loop traverse, although the calculated azimuths are internally consistent, the absolute orientation is based entirely on one angle observed between lines $1Q$ and 12. This aspect of a closed-loop traverse is a weakness in this procedure which was mentioned in Art. 8.6. To eliminate the weakness, another angle should be observed from some other traverse point to another independent line of known azimuth. If points are not available to permit this second tie, an astronomic observation for azimuth should be made for line 34 or 45 so as to provide a redundancy in orientation of the traverse. Short traverses, such as the one illustrated in Fig. 8.7, may not require such a tie, depending on the nature of the survey. Long closed-loop traverses should always be tied to at least two lines of known direction.

Allowable closures for interior-angle traverses are similar to those for deflection-angle traverses. The sum of the interior angles should not deviate from $(n-2)180°$ by more than the square root of the number of instrument setups times the estimated standard deviation in observing the angles.

8.9. Traverse by angles to the right This method can be used in open, closed, or closed-loop traverses. A repeating transit or theodolite or a direction theodolite may be employed to turn the angles. When a repeating instrument is used, the backsight to the preceding station is taken with the *A* vernier set on zero. Upper clamp loosened, the instrument is turned on the upper motion, a foresight is taken on the next station, and the angle turned to the right is observed on the clockwise circle using the *A* vernier. The characteristics of a direction theodolite are such that angles derived from directions observed between backsight and foresight are always angles to the right (Art. 6.40). As in the other methods for observing traverse angles, it is advisable to observe the angles twice, once with the telescope direct and once with the telescope reversed.

A traverse by angles to the right is shown in Fig. 8.9. This traverse originates at line *YZ* for which the azimuth is 250°00′00″ and closes on line *TU* with a given azimuth of 319°59′45″ from *T* to *U*. Angles for this traverse were derived from directions observed with a direction theodolite. Figure 8.10 shows the notes from which the angles were determined. Note that directions were observed with the telescope direct and reversed and angles are calculated from the average directions.

Azimuths are calculated by adding the angle to the right to the back azimuth of the previous line. Computation of these azimuths is shown in Table 8.6. The angular closure is 25″. This condition of closure can be expressed by the following equation

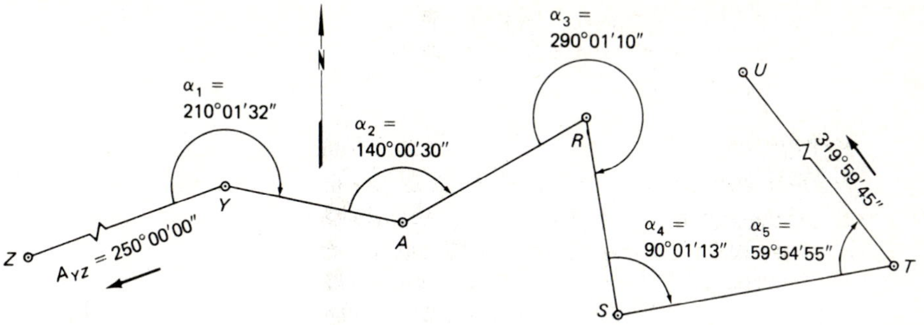

Fig. 8.9 Traverse by angles to the right with a direction instrument.

GRANT FARM TRAVERSE

Station	From/To	Tel.	Circle	Average Direction	Angle
	Z	D	10°01′13″	10°01′15″	
Y		R	190 01 17		210°01′32″
	A	D	220 02 45	220 02 47	
		R	40 02 49		
	Y	D	90 05 22	90 05 25	
A		R	270 05 28		140 00 30
	R	D	230 05 52	230 05 55	
		R	50 05 58		
	A	D	5 10 02	5 10 05	
R		R	185 10 08		290 01 10
	S	D	295 11 13	295 11 15	
		R	115 11 17		
	R	D	170 04 11	170 04 14	
S		R	350 04 17		90 01 13
	T	D	260 05 26	260 05 27	
		R	80 05 28		
	S	D	270 00 00	270 00 01	
T		R	90 00 02		59 54 55
	U	D	329 54 53	329 54 56	
		R	149 54 59		

Dec. 1, 1984
Clear 50°F 11
R. Hoyt ⊼
C. Roup R.T.
K. Krone H.T.

Wild T-2
Direction Theodolite
No. 5238

Note: Traverse alignment and references are on Page 10

Fig. 8.10 Notes for angles to the right traverse.

Table 8.6 Azimuths Calculated by Angles to the Right

Line	Azimuth	Correction	Corrected azimuth
YZ	250°00′00″		
	210°01′32″		
YA + 360°	460°01′32″		
	−180°00′00″		
AY	280°01′32″	05″	280°01′37″
	140°00′30″		
AR + 360°	420°02′02″		
	−180°00′00″		
RA	240°02′02″	10″	240°02′12″
	290°01′10″		
RS + 360°	530°03′12″		
	−180°00′00″		
SR	350°03′12″	15″	350°03′27″
	90°01′13″		
ST + 360°	440°04′25″		
	−180°00′00″		
TS	260°04′25″	20″	260°04′45″
	59°54′55″		
TU	319°59′20″	25″	319°59′45″
TU	319°59′45″ fixed		
	Angular closure = 25″		

for this example:

$$A_{YZ} + \alpha_1 + \alpha_2 + \alpha_3 + \alpha_4 + \alpha_5 - (4)(180°) - A_{TU} = 0 \tag{8.4}$$

which in general becomes

$$A_1 + \alpha_1 + \alpha_2 + \cdots + \alpha_n - (n-1)(180°) - A_2 = 0 \tag{8.5}$$

in which A_1 and A_2 are fixed azimuths from the north at lines of origin and closing, respectively, and there are n traverse stations (not counting fixed stations Z and U).

If adjusted azimuths are desired at this point, assuming observations of equal precision, the error of closure is distributed equally among the five angles, or $-25''/5 = 05''$ per angle. The adjustment is made by applying a cumulative correction to each unadjusted azimuth. The corrections and adjusted azimuths are also listed in Table 8.6.

8.10. Azimuth traverse An advantage of the azimuth method is that the simple statement of one angular value, the azimuth, fixes the direction of the line to which it refers. The method is used extensively on topographic and other surveys where a large number of details are located by angular and linear measurements from the traverse stations. Any angular error of closure of a traverse becomes evident by the difference between initial and final observations taken along the first line. The reference meridian may be either true or assumed.

Successive stations are occupied, beginning with the line of known or assumed azimuth. At each station the transit is "oriented" by setting the *A* vernier or horizontal circle index to read the back azimuth (forward azimuth ±180°) of the preceding line and then backsighting to the preceding traverse station. The instrument is then turned on the upper motion, and a foresight on the following traverse station is secured. The reading indicated by the horizontal circle index on the clockwise circle is the azimuth of the forward line.

Figure 8.11 shows a closed-loop traverse run by the azimuth method. The traverse is begun at station 1 with a backsight along line 15 for which the azimuth is known to be 270°28′ by a solar observation. With the instrument set over station 1, the *A* vernier is set to 270°28′ on the clockwise circle and a backsight is made on station 5 using the lower motion. The instrument is then turned on the upper motion and a foresight is taken on station 2. Since zero is oriented toward true north (Fig. 8.11), the clockwise circle read at the *A* vernier (350°30′) is the azimuth from the north of line 12. Note that any other angle observed on the clockwise circle to another point from this setup is also an azimuth from the north. When the instrument is moved to point 2, the back azimuth of line 12 is computed by subtracting 180° from the

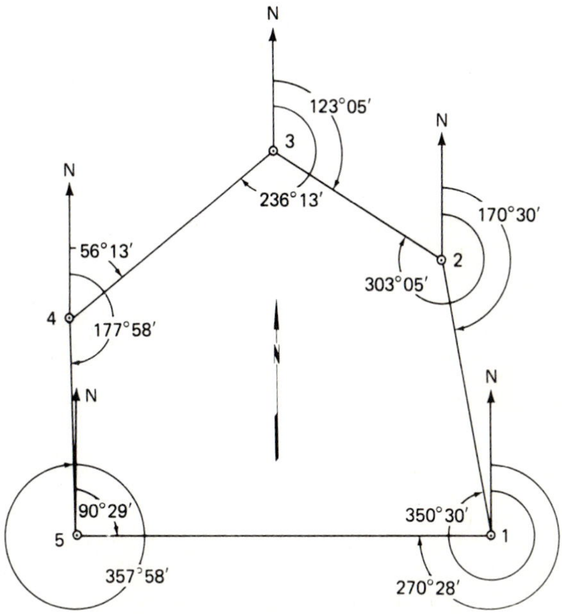

Fig. 8.11 Azimuth traverse.

forward azimuth (350°30′ − 180° = 170°30′), and this value is set on the clockwise circle before a backsight, using the lower motion, is taken on station 1. Once again, the clockwise circle is oriented with zero toward true north and when the instrument is turned on the upper motion for a foresight on station 3, the angle observed on the clockwise circle at the *A* vernier (303°05′) is the azimuth of line 23 from true north. This operation is repeated for each successive traverse station. When the final setup is made over station 5, the azimuth observed on the foresight along line 51 should equal the azimuth of line 15 minus 180°, or in this case 270°28′ − 180″ = 90°28′. Any deviation between this last observation and the fixed direction of the closing line is the angular closure. Figure 8.12 illustrates the notes recorded for the traverse shown in Fig. 8.11. The closure error is 01′. Since angles have been observed only to the nearest minute, this error could be applied to one angle to provide adjusted directions. Magnetic bearings are observed and a check against blunders is made by noting that computed bearings vary from corresponding magnetic bearings by about 9°10′, the amount of the magnetic declination.

AZIMUTH TRAVERSE AT HIGH-WATER LINE — 25

Proposed Mill Pond, El. 741.36
Silver Creek, Penn.
(For Land Damage Est.)

Sta.	Obj.	Dist.	Azimuth	Mag. B.	Cal. Bear.
1	5		270°28′	N80½°W	N89°32′W
	2	689.32	350°30°	N	N9°30′W
2	1		170°30′	S	S9°30′E
	3	509.66	303°05′	N48°W	N56°55′W
3	2		123°05′	S48¼°E	S56°55′E
	4	678.68	236°13′	S65½°W	S56°13′W
4	3		56°13′	N65¼°E	N56°13′E
	5	572.50	177°58′	S 7°W	S2°02′E
5	4		357°58′	N 7¼°E	N2°02′W
	1	1082.71	90°29′	S80°E	S89°31′E
	Angular closure = 01′				

Gurley transit No. 191

J. Stanbois ✱
F. Lowe
June 15, 1984
Cloudy, Warm

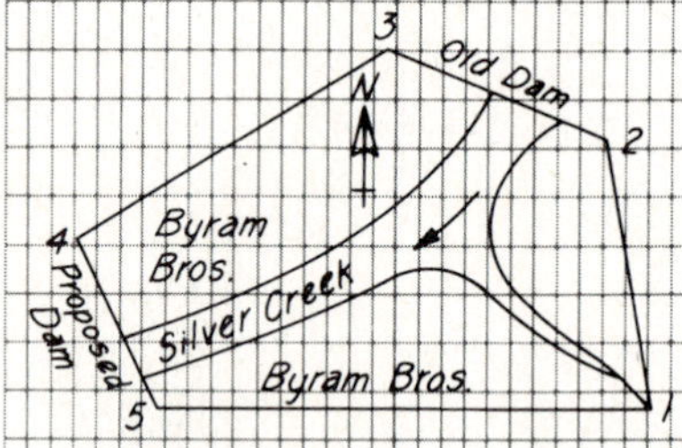

True azimuth of line 1-5 found by solar observation.
Mag. declination = 9° 10′ W

Fig. 8.12 Notes for a short, closed azimuth traverse.

8.11. Compass traverse Use of the compass was described in Art. 6.19. When a compass traverse is performed, forward and back bearings are observed from each traverse station and distances are taped. If local attraction exists at any traverse station, both the forward and back bearings are affected equally. Thus, interior angles computed from forward and back bearings observed at that station are independent of local attraction. Field notes for a closed-loop compass traverse are shown in Fig. 8.13. The declination of 20°15′E was set off on the compass (Fig. 6.20, Art. 6.14) so that bearings are referred to the true meridian.

Checking the compass traverse is accomplished by computing the interior angle from the observed bearings. Since these angles are independent of local attraction, the sum of these interior angles provides a legitimate indication of the angular error in the traverse. The method for computing interior angles from bearings is illustrated in Fig. 8.14. Thus, at station A the unadjusted interior angle is $180° + 28°00' + 30°40' = 238°40'$ and the unadjusted angle at B is $180° - (30°40' + 83°50') = 65°30'$. Computed interior angles are listed in column four of the notes shown in Fig. 8.13. The sum of these angles reveals a closure error of 25′. Assuming that all bearings are

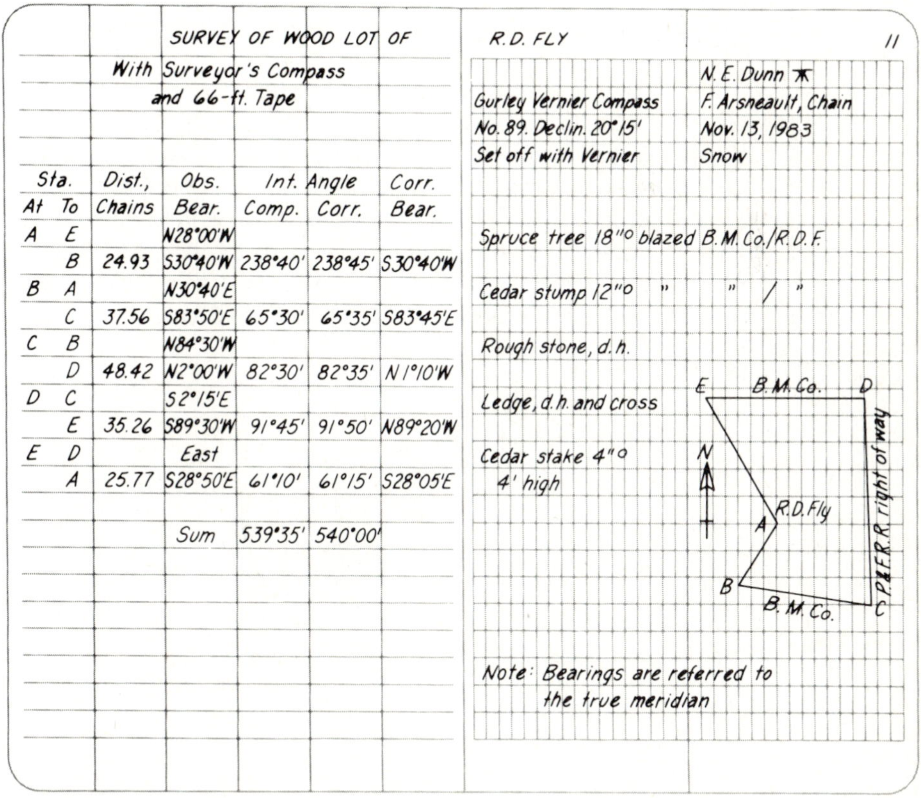

SURVEY OF WOOD LOT OF R. D. FLY 11

With Surveyor's Compass and 66-ft. Tape

Sta. At	To	Dist., Chains	Obs. Bear.	Int. Angle Comp.	Int. Angle Corr.	Corr. Bear.
A	E		N28°00'W			
	B	24.93	S30°40'W	238°40'	238°45'	S30°40'W
B	A		N30°40'E			
	C	37.56	S83°50'E	65°30'	65°35'	S83°45'E
C	B		N84°30'W			
	D	48.42	N2°00'W	82°30'	82°35'	N1°10'W
D	C		S2°15'E			
	E	35.26	S89°30'W	91°45'	91°50'	N89°20'W
E	D		East			
	A	25.77	S28°50'E	61°10'	61°15'	S28°05'E
			Sum	539°35'	540°00'	

Fig. 8.13 Notes for a compass survey.

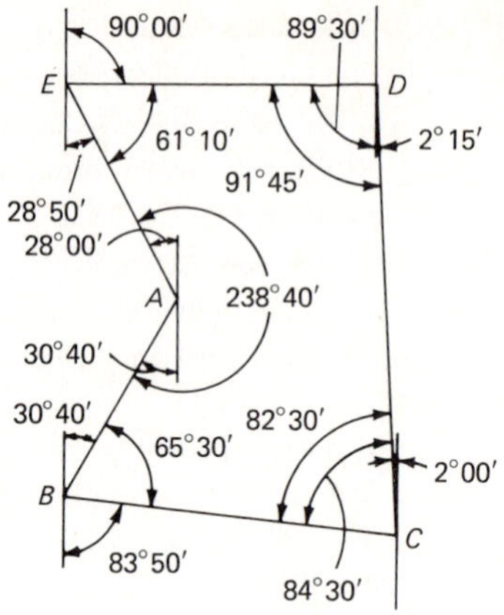

Fig. 8.14 Observed bearings and computed interior angles. Σ Int. Angles = 539°35′

of equal precision and noncorrelated, this error is distributed equally among the five interior angles so that 05′ is added to each unadjusted interior angle to provide an adjusted angle. Adjusted interior angles are listed in column 5 of the notes in Fig. 8.13.

Since none of the traverse lines has an absolute direction that is known to be correct, it is necessary to select a line affected least by local attraction. Examination of column 3 of the notes in Fig. 8.13 indicates that points A and B are most likely to be free of local attraction since the forward and back bearings of AB are numerically equal and opposite in direction. Hence, the correct forward bearing from A to B can be taken as S30°40′W. Holding the direction of AB fixed, the bearing of BC is $180° - (30°40' + 65°35') =$ S83°45′E, where 65°35′ is the adjusted interior angle at B. The remainder of the adjusted bearings are computed similarly, as illustrated graphically in Fig. 8.15. These adjusted bearings are listed in column 6 of the notes shown in Fig. 8.13.

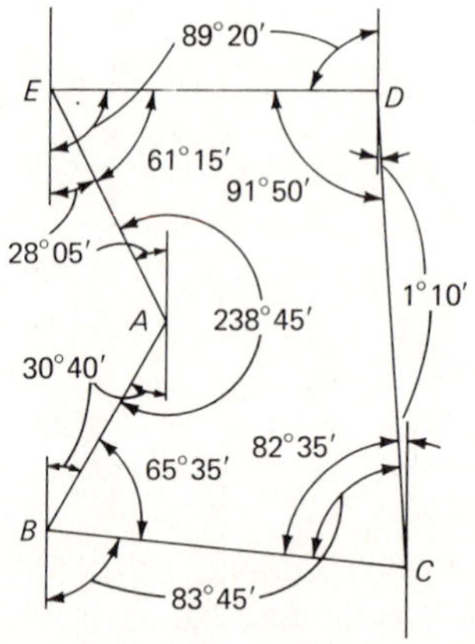

Fig. 8.15 Corrected bearings and corrected interior angles. Σ Int. Angles = 540°00′

If the error in the sum of interior angles exceeds 10′ to 15′ times the square root of the number of angles, it is likely that a blunder in reading the compass has occurred, and the field measurements should be repeated. If the error is within permissible limits but cannot be distributed equally among the angles in amounts of 05′ or 10′, the greater corrections (in multiples of 05′) should be applied arbitrarily to those angles for which the conditions of observing were estimated to be least favorable. The precision of most compass measurements does not justify computations with a precision closer than multiples of 05′.

8.12. Stadia traverse When distances between traverse stations are determined by the stadia method (Chap. 7), the resulting survey is called a stadia traverse. This type of traverse is appropriate for certain types of preliminary and reconnaissance surveys, rough surveys for boundary locations, and topographic mapping surveys. Where only horizontal locations are required, the stadia method provides an excellent means for establishing control points on surveys of low precision.

On topographic and similar surveys (see Chap. 12), the elevation and horizontal position at each traverse station are desired. An advantage of the stadia traverse is that, if desired, elevations may be determined concurrently with horizontal position. The procedure consists of observing directions by the azimuth method and distances by stadia. In addition, differences in elevation are determined by observing vertical or zenith angles or stadia arc readings and stadia intervals. If a self-reducing instrument is being used only the rod interval, horizontal distance (rod interval times a given constant), and difference in elevation (rod interval times a given constant) are recorded. The party consists of an observer, a rodman, and a recorder.

Field notes for a small closed-loop traverse run with a transit or theodolite are shown in Fig. 8.16. The instrument is set up over traverse station 1, the elevation and location of which are known. The index error of the vertical circle is observed and recorded and the height of instrument (h.i.) above the station over which the instrument is set is measured with a rod or tape to the nearest centimetre or nearest tenth of a foot. If the direction of a line in the traverse is not known, a magnetic direction along one line can be observed. In the example traverse, a magnetic bearing is first observed along line 14. This value is converted to an azimuth and corrected for declination. These calculations are shown at the top left of the sample notes in Fig. 8.16. This azimuth is then set off on the clockwise horizontal circle using the *A* vernier and station 4 is resighted with the lower motion. The stadia interval and vertical angle for the sight on station 4 are observed and recorded in columns 2 and 5, respectively, in the notes shown in Fig. 8.16. In this example, the middle cross hair was set on the value of the h.i. = 1.49 m before reading the vertical angle, so that it was not necessary to record a rod reading. Should it not be possible to set the middle cross hair on the h.i., because of some obstruction on the line of sight, the rod reading sighted is recorded directly above the vertical angle in column 5 of the notes (Fig. 8.16 from point 2 to 3). When observations along the backsight to station 4 are completed, the upper clamp is loosened and a foresight is taken on station 2, where the stadia interval, vertical angle, and azimuth are observed and recorded. The azimuth is then observed on the clockwise horizontal circle. Thus, this and all subsequent azimuths in the traverse are observed according to the *azimuth method*, as outlined in Art. 8.10. If a self-reducing instrument is being used for the observations, only the rod interval and rod reading at the middle or fixed cross hair are observed and recorded in columns 2 and 5, respectively. If the middle cross hair is set on a rod reading equal to the value of the h.i., it need not be recorded and the difference in elevation is the rod interval multiplied by the observed constant (Art. 7.16).

When all observations are completed at the first station, the instrument is moved to station 2, where the h.i. is measured (1.64 m) and a backsight is taken on station 1 with the back azimuth of line 12 (145°19′ + 180° = 325°19′) set on the clockwise horizontal circle at the *A* vernier or the clockwise index of the horizontal circle. When the stadia interval and vertical angle from 2 to 1 are observed, it is possible to check these readings against the corresponding readings from 1 to 2. The stadia intervals should agree to within 2 or 3 mm and the vertical angles should have opposite signs and agree numerically to within a few minutes. Note that if index error in the vertical angle is present, there will be a systematic difference between forward and backward vertical angles. A foresight is then taken on station 3, where the stadia interval, vertical angle, and azimuth are observed and recorded.

PROBLEM 7
STADIA TRAVERSE FOR CONTROL

Point	Interval, metres	Horiz. Dist., metres	Azimuth ° ′	Vert. Angle ° ′	Δ Elev., metres	Adj. Elev.
Mag. Bear. 1-4 S 75° W, Internal focussing transit f+c=0, Index Error=0				Mag α = 255°00′ Decl. = 17 30 E True α = 272°30′		
	Inst. @ 1 h.i. = 1.49m					30.00 Fixed
4	0.500	49.5	272°30′	−5°46′	−5.00	
2	0.738	73.8	145°19′	−1°13′	−1.57	28.43
	Inst. @ 2 h.i. = 1.64m					
1	0.738	73.8	325°19′	1°15′	+1.61	
3	0.853	85.2	261°55′	Rod=3.12 −1°48′	−4.16	24.28
	Inst. @ 3 h.i. = 1.43m					
2	0.856	85.4	81°55′	+2°48′	+4.18	
4	0.747	74.7	354°51′	+0°33′	+0.72	25.02
	Inst. @ 4 h.i. = 1.43m					
3	0.750	75.0	174°51′	−0°33′	−0.72	
1	0.497	49.2	92° 34′	+5°44′	+4.94	30.00 (Check)
		Disc. = +04′				

May 19, 1983 — 29
Warm, Clear
Temp. 21° C
Party 41: Brown, Jones, Orini, Svensson

Equip: Transit No. 14, 2 Plumb Bobs, 20-m Tape No. 4, Hand Lens

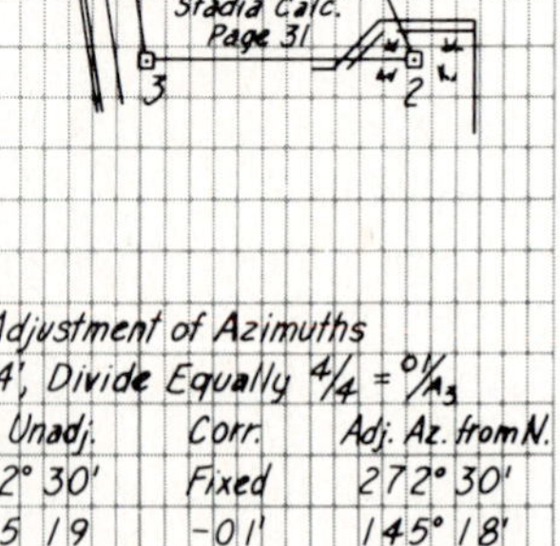

Adjusted Elevations

	Aver. Dist., m	Aver. Δ Elev., m	Corrections	Adjusted Δ Elev.
1-2	73.8	−1.59	0.07× 74/283 = +0.02	−1.57
2-3	85.3	−4.17	0.07× 85/283 = +0.02	−4.15
3-4	74.8	+0.72	0.07× 75/283 = +0.02	+0.74
4-1	49.4	+4.97	0.07× 49/283 = +0.01	+4.98
	283.3	−0.07	+0.06	0.00

Adjustment of Azimuths
Disc. = +04′, Divide Equally 4/4 = 01/A₃

	Az. Unadj.	Corr.	Adj. Az. from N.
1-4	272° 30′	Fixed	272° 30′
1-2	145 19	−01′	145° 18′
2-3	261 55	−02′	261° 53′
3-4	354 51	−03′	354° 48′
4-1	92 34	−04′	92° 30′

Fig. 8.16 Stadia traverse for horizontal and vertical control.

The procedure described in the previous paragraphs is repeated for each station in the traverse. When the final station is occupied, the azimuth along the initial line of known direction (line 41 in this example) should correspond to $180° \pm$ the known azimuth or $272°30' - 180°00' = 92°30'$. The difference is the angular error of closure, which is 04′ in this example.

Horizontal distances and differences in elevation are computed using the stadia intervals and vertical angles. These values are recorded in columns 3 and 6 in the notes. Horizontal distances are averaged and average differences in elevation are used to check the closure error in elevation. These values are tabulated at the bottom of the left page in the notes. The closure error in elevation for this example is -0.07 m, which is distributed among the traverse stations in proportion to the lengths of the traverse sides (Art. 5.41). Since the total length around the traverse is 283.3 m, the correction to the difference in elevation between stations 1 and 2 is $(74/283)(0.07) = +0.02$ m. The remaining corrections are computed in a similar manner and are applied to the average differences in elevation, as shown at the bottom of the left page of notes in Fig. 8.16. These corrected differences in elevation are then used to calculate adjusted elevations for traverse stations listed in column 7 of the notes.

The error in angular closure of 04′ is distributed equally among the four unadjusted azimuths by applying a cumulative correction to each observed direction. Thus, the correction per interior angle is $04'/4 = 01'$ per angle or the corrections are 01′, 02′, 03′, and 04′ per azimuth for lines 12, 23, 34, and 41, respectively. These values are tabulated at the bottom of the right page of the sample notes in Fig. 8.16.

Larger angular closures can be tolerated in a stadia traverse than one in which distances are determined with a tape or EDM equipment. If the estimated standard deviation of a horizontal distance measured by stadia is assumed to be 0.3 m (Art. 7.15), an angular error compatible with this error in a distance of 100 m (Art. 3.7) is $\tan^{-1} 0.3/100 \simeq 10'$ of arc. Consequently, angular closure errors of up to 10′ are reasonable in a stadia traverse where average lengths of lines are $\sim$ 100 m.

8.13. Plane-table traverse Traversing with the plane table involves the same principles as running a traverse with a transit or theodolite. As each successive station is occupied, the table is oriented, sometimes with the compass but usually by taking a backsight on the preceding station. A foresight is then taken to the next station and its location is plotted on the plane-table sheet. Distances and differences in elevations are determined by stadia using a conventional or self-reducing alidade. The plane table and alidade are described in Art. 7.18. Use of the plane table is outlined in Art. 7.19.

The plane-table traverse is used for the control of small surveys and extension of control for larger surveys in order to locate points from which map details can be plotted advantageously.

In Fig. 8.17, a series of traverse stations *A*, *B*, *C*, *D*, and *E* are represented. The initial setup is on station *A*, from which *E* and *B* are visible. The plane table is set up at *A*, leveled, and *a*, representing station *A* on the ground, is plotted on the plane-table sheet in such a location that the other stations will fall within the limits

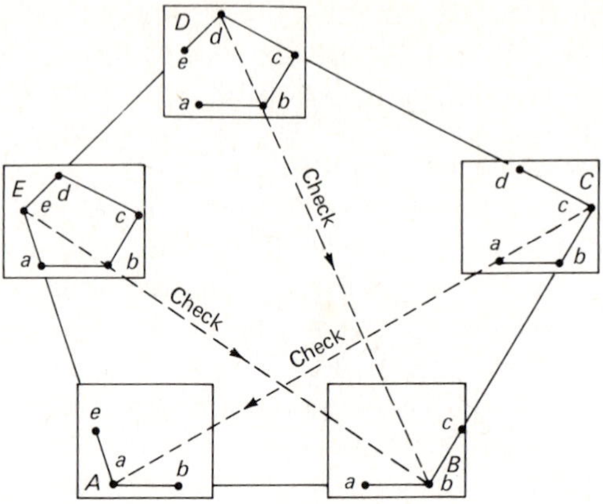

Fig. 8.17 Traverse with a plane table.

of the sheet. Using the plotted position of *a* as a pivot, the blade of the alidade is directed toward station *E*, stadia readings are observed and reduced to a horizontal distance, and the position of *E* is plotted as *e* on the plane-table sheet to the scale of the traverse. Next, the blade of the alidade is rotated about *a* and a foresight is taken on station *B*. Stadia observations are made on station *B* and are reduced to the horizontal distance *AB*. A line in the direction of *B* is drawn from *a* on the plane table and point *b* is plotted to scale.

The plane table is then set over station *B*, is leveled, the blade of the alidade is aligned with *ba*, and the plane table is rotated until the vertical cross hair of the alidade is on line with station *A*. Stadia readings are made on *A* and reduced to a second determination of the distance *BA*. The average of the two distances is plotted from *a* to establish an improved position for *b* on the plane table. The blade of the alidade is rotated about *b* and a foresight is taken on *C* along with the stadia observations necessary to determine distance *BC*. A line in the direction of *C* is drawn from *b* and the position of *c* is plotted.

The procedure outlined is repeated for each station in the traverse. When the last station *E* is occupied, the board is oriented by a backsight on station *D*, where the necessary stadia observations are made to permit a second determination of the

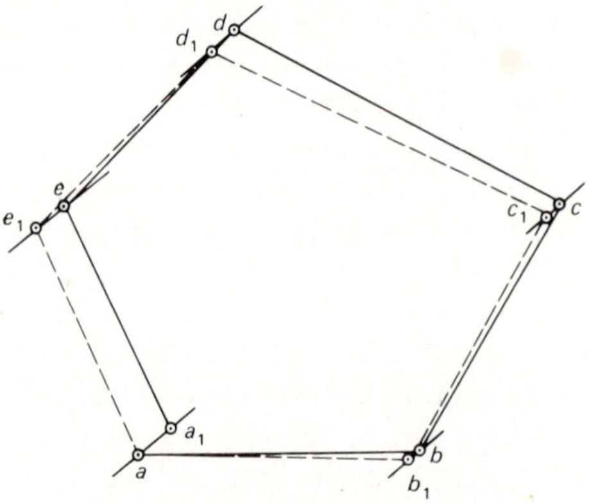

Fig. 8.18 Adjustment of the plane-table traverse.

horizontal distance *DE*, and an improved position for *e* is plotted using the average distance *DE*. Finally, a foresight is made on station *A*, permitting a second determination for the horizontal distance *EA*. The average distance *EA* is plotted to scale along the line from *e* in the direction of *A* so as to establish a final position for *a* which does not necessarily coincide with its original plotted position. The difference between the initial and final positions for *a* represents the *closure* for the stadia traverse. In Fig. 8.18, the closure is represented by distance aa_1, where *a* and a_1 are the initial and final locations of the first station. To adjust the traverse graphically, draw lines parallel to aa_1 through each plotted traverse station. The increments bb_1, cc_1, dd_1, and ee_1 are in proportion to the distance of the station from the fixed initial station *a*. The adjusted traverse lines are shown as dashed lines in Fig. 8.18.

8.14. Referencing traverse stations Many of the hubs marking the location of highways, railroads, and other public and private works are bound to be uprooted or covered during the progress of construction and must be replaced, often more than once, before construction is completed. Such hubs marking traverse stations are usually tied by angular and/or linear measurements to temporary wooden hubs called *reference hubs*, or to other objects that are not likely to be disturbed. A traverse station is said to be *referenced* when it is so tied to nearby objects that it can be replaced readily. The manner of referencing a traverse station is indicated in the notes by an appropriate sketch. The precision of referencing should be comparable to that of the measurements between traverse stations. Particularly on land surveys, the corners should be tied to nearby objects which can be found readily, which are not likely to be moved or obliterated, and which are of a more-or-less permanent character.

Figures 8.19 through 8.21 illustrate several methods of referencing a traverse station. The station shown in Fig. 8.19 is tied by linear measurements to three specific points located on semipermanent structures or objects located around the traverse station. Location or relocation of the traverse station is accomplished by intersecting the reference distances. Where possible, reference points should be chosen so as to be easily identifiable and ought to be located on objects or structures

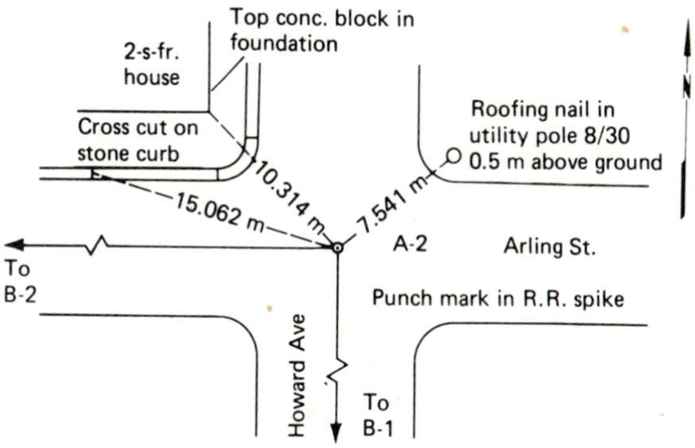

Fig. 8.19 Reference to a traverse station in an urban area.

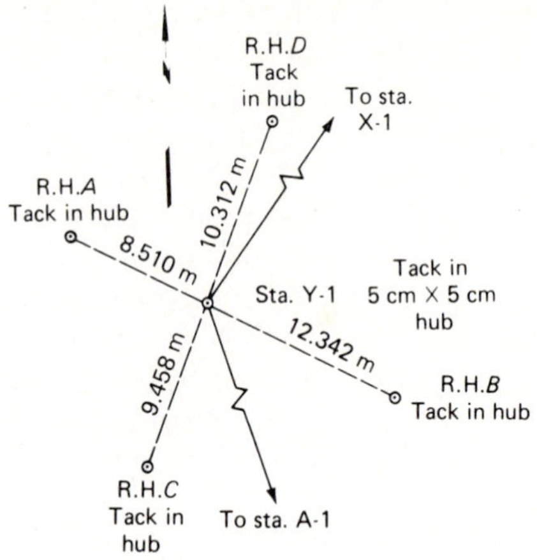

Fig. 8.20 Referencing by reference hubs.

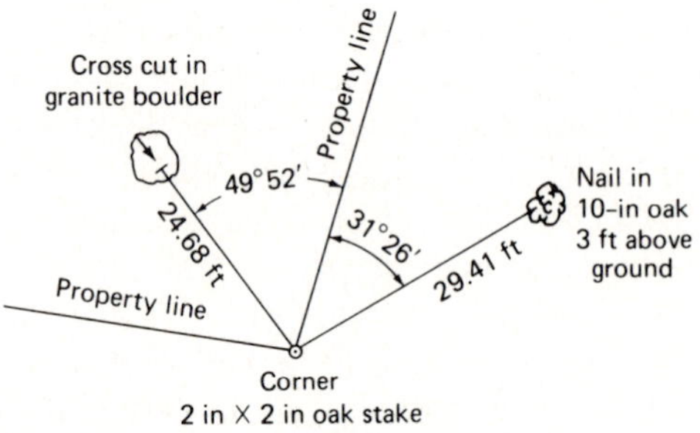

Fig. 8.21 References to a property corner.

of a permanent nature. Tie lines should intersect at favorable angles (greater than 30° and less than 150°), so that the station can be relocated with certainty.

Figure 8.20 shows a station referenced by setting reference hubs on a line passing through the station and taking linear measurements from these hubs to the station. In this example, four reference hubs are set at the ends of lines *AB* and *CD*, which intersect at traverse station *Y*-1. With the measurements as shown, any two reference hubs may be destroyed and the station may still be relocated.

A typical method for referencing property corners in land surveying is illustrated in Fig. 8.21. Both angular and linear ties are employed.

8.15. Traverse computations Traverse operations in the field yield observed angles or directions and measured distances for a set of lines connecting a series of traverse stations. Angles, or directions converted to angles, can be checked to obtain the angular error of closure that may be distributed among the angles to provide preliminary adjusted azimuths or bearings. When the observed distances have been corrected for all systematic errors (Arts. 4.13 to 4.25), the preliminary directions and corrected distances are suitable for use in traverse computations which are performed in a plane rectangular coordinate system.

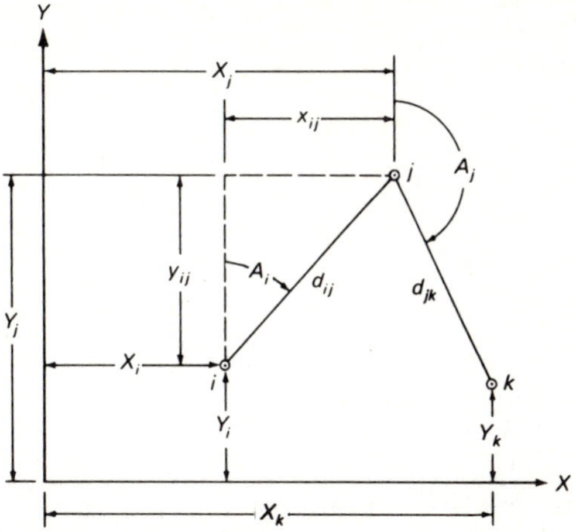

Fig. 8.22 Departure and latitude of lines.

Computations with plane rectangular coordinates are extremely adaptable for use with pocket calculators, desktop calculators, and high-speed electronic computers. Coordinated points have an unambiguous designation that is ideal for storage and retrieval from data banks (see Arts. 1.11 and 3.15) and is compatible with many methods for plotting the locations of points, including automatic plotting techniques.

The system within which the coordinates are computed may have a national, state, regional, local, or completely arbitrary origin (Chap. 11). Plane coordinate computations in the United States are usually performed in one of the state plane coordinate systems established by the National Ocean Survey for each state.

Computations with plane coordinates are illustrated by reference to Fig. 8.22, where traverse line *ij* has a reduced horizontal distance of d_{ij} and azimuth of A_i. Let x_{ij} and y_{ij} be designated as the departure and latitude for line *ij*, so that

$$
\begin{aligned}
x_{ij} &= d_{ij} \sin A_i \\
y_{ij} &= d_{ij} \cos A_i
\end{aligned}
\qquad (8.6)
$$

The algebraic signs of the departure and latitudes for a traverse line depend on the signs of the sine and cosine of the azimuth of that line. Figure 8.23 shows the algebraic signs of the sine and cosine functions for each of the four quadrants. Pocket calculators, desktop calculators, and electronic computers with internal routines for trigonometric functions yield the proper sign automatically given the azimuth. When bearings are used to describe the direction of the lines, the sines and cosines are positive since bearings are always equal to or less than 90° and the algebraic signs of the departures and latitudes are taken from the quadrant. Consequently, the azimuth is the logical means of defining direction in traverse calculations when modern calculators are used.

Please observe that the terms "departure" and "latitude," as well as *X* and *Y*, are specified in that order throughout this text. Note that several hand-held calculators

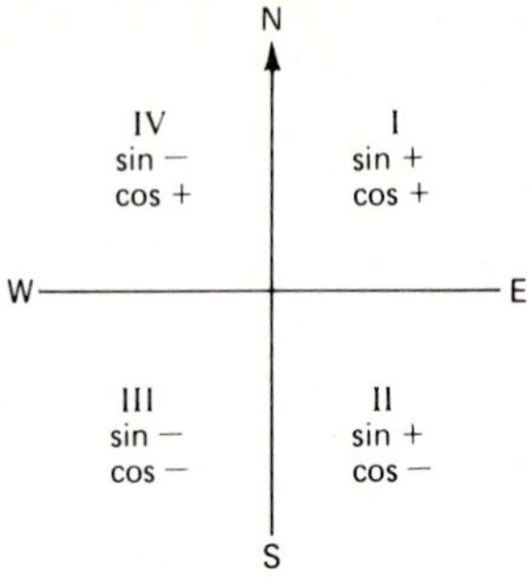

Fig. 8.23 Signs of azimuth functions.

with direct rectangular to polar or polar to rectangular keys use the reverse notation of latitude and departure, Y and X.

For example, in Fig. 8.22, let $A_i = 43°30'00''$ and $d_{ij} = 432.182$ m. Applying Eqs. (8.6), the departure $= x_{ij} = 432.182 \sin 43°30'00'' = 432.182(0.6883546) = 297.494$ m and the latitude $= y_{ij} = 432.182 \cos 43°30'00'' = 432.182(0.7253744) = 313.494$ m. Continuing, if $A_j = 153°25'40''$, $d_{jk} = 385.151$ m, then the departure $= x_{jk} = 385.151(0.4473255) = 172.288$ m and the latitude $= y_{jk} = 385.151(-0.8943712) = -344.468$ m.

Given that the coordinates of station i are X_i and Y_i (Fig. 8.22), the coordinates of station j are

$$\begin{aligned} X_j &= X_i + x_{ij} \\ Y_j &= Y_i + y_{ij} \end{aligned} \tag{8.7}$$

If the coordinates of station i in the example are $X_i = 142{,}482.352$ m, $Y_i = 43{,}560.805$ m, then the coordinates of station j are

$$\begin{aligned} X_j &= 142{,}482.352 + 297.494 = 142{,}779.846 \text{ m} \\ Y_j &= 43{,}560.805 + 313.494 = 43{,}874.299 \text{ m} \end{aligned}$$

and the coordinates of station k are

$$\begin{aligned} X_k &= 142{,}779.846 + 172.288 = 142{,}952.134 \text{ m} \\ Y_k &= 43{,}874.299 - 344.468 = 43{,}529.831 \text{ m} \end{aligned}$$

Consequently, a major portion of traverse computations consists of calculating departures and latitudes for successive traverse lines and cumulating the values to determine the coordinates for consecutive traverse stations.

The inverse solution is also possible, given the coordinates for the two ends of a traverse line. Thus, the distance between stations i and j in Fig. 8.22 is

$$d_{ij} = \left[(X_j - X_i)^2 + (Y_j - Y_i)^2 \right]^{1/2} \tag{8.8}$$

and the azimuth of line ij from the north is

$$A_{Nij} = \arctan \frac{X_j - X_i}{Y_j - Y_i} \tag{8.9}$$

The azimuth of line ij from the south is

$$A_{S_{ij}} = \arctan \frac{X_i - X_j}{Y_i - Y_j} \tag{8.10}$$

In Eqs. (8.6) through (8.10), due regard should be paid to algebraic signs of the trigonometric functions (Fig. 8.23) and the differences between coordinates at the two ends of a line.

When coordinates for all the traverse points (or all of the departures and latitudes) for all lines have been computed, a check is necessary on the accuracy of the observations and the validity of the calculations. In a closed traverse, the algebraic sum of the departures should equal the difference between the X coordinates at the beginning and ending stations of the traverse. Similarly, the algebraic sum of the latitudes should equal the difference between the Y coordinates at the beginning and ending stations. If Eqs. (8.7) are used to calculate coordinates directly, the calculated coordinates should agree with the given values for the final or closing station. It follows that in a closed-loop traverse, the algebraic sum of the departures and the algebraic sum of the latitudes each must equal zero. For a traverse containing $i = 1, 2, \ldots, n$ stations, starting at station $i = 1$ and terminating at station $i = n$, the foregoing conditions can be expressed as follows:

$$\begin{aligned} X_n - X_1 &= \sum_{i=1}^{n-1} x_{i,i+1} = \sum_{i=1}^{n-1} \text{departures} \\ Y_n - Y_1 &= \sum_{i=1}^{n-1} y_{i,i+1} = \sum_{i=1}^{n-1} \text{latitudes} \end{aligned} \tag{8.10a}$$

Equations (8.10a) are rarely satisfied exactly in practice, owing to random observational errors and uncorrected systematic errors in observations. The amounts by which Eqs. (8.10a) fail to be satisfied are called the *errors in closure in position* or simply the *closures* for a traverse. The closure *corrections* dX and dY, which are of opposite signs to errors, for a traverse defined as for Eqs. (8.10a) are

$$\begin{aligned} dX &= (X_n - X_1) - \sum_{i=1}^{n-1} x_{i,i+1} \\ dY &= (Y_n - Y_1) - \sum_{i=1}^{n-1} y_{i,i+1} \end{aligned} \tag{8.10b}$$

When preliminary traverse computations are completed, closure corrections are evaluated and an adjustment is performed. The subject of traverse adjustment is covered in Art. 8.18. First consider the complete traverse computations for a closed traverse.

8.16. Computations for a closed traverse Figure 8.24 shows a traverse that originates at station 1, and closes on station 5. Observed distances, corrected for systematic errors, and angles to the right observed once direct and once reversed with a 30″ repeating theodolite, are given in Table 8.7.

Table 8.7 Data for Closed Traverse

Line	Distance, m	Angle to the right	
12	305.41	$\alpha_1 = 104°00'30''$	$A_{N_1} = 43°31'30''$
23	359.61	$\alpha_2 = 63°20'00''$	$A_{N_5} = 106°17'30''$
34	612.15	$\alpha_3 = 258°11'00''$	
45	485.12	$\alpha_4 = 120°45'00''$	$\sigma_d = 0.01$ m
		$\alpha_5 = 236°30'30''$	$\sigma_\alpha = 15''$

The angular error of closure is 01′ [Eq. (8.5) and Table 8.8]. Adjusted azimuths are calculated according to the procedure in Art. 8.9. Distribution of the angular error equally among the five angles leads to a correction of $-12''$ per angle. Observation of the angles twice with a 30″ instrument does not warrant this precision. Thus, the corrections to angles are rounded arbitrarily to $-10''$ each to angles 1, 2, and 3 and $-15''$ to angles 4 and 5. Table 8.8 shows azimuth calculations and adjusted azimuths.

Distances, adjusted azimuths, fixed coordinates for the beginning and ending stations, calculated azimuths, and calculated coordinates are listed in Table 8.9. Note that coordinates can be calculated directly using Eqs. (8.7) and tabulation of departures and latitudes is not necessary unless these values are to be used for some specific purpose. They are included here for illustrative purposes. The closure corrections in X and Y coordinates at station 5 are 0.297 m and 0.196 m, respectively. These values may also be calculated by applying Eqs. (8.10*b*) to yield $dX = (X_5 - X_1) - \Sigma\,\text{dep.} = 1297.839 - 1297.542 = 0.297$ and $dY = (Y_5 - Y_1) - \Sigma\,\text{lat.} = 164.710 - 164.514 = 0.196$ m. The values for closure are here shown graphically in Fig. 8.25; the fixed and calculated positions for station 5 are shown to a larger scale. Using Eq. (8.8), the resultant closure, $d_c = (dX^2 + dY^2)^{1/2}$, is the distance from the calculated position at 5′ to the fixed location 5. In a closed traverse of this type, the resultant closure is caused by random variations in the observations

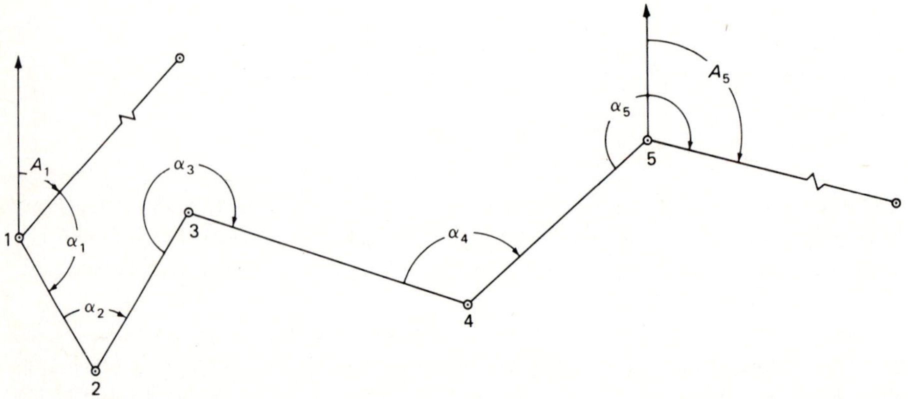

Fig. 8.24 Closed traverse.

Table 8.8 Azimuth Calculations

Line	Unadjusted azimuth	Correction	Adjusted azimuth
A_1	43°31′30″		
	α_1 104°00′30″		
12	147°32′00″	−10″	147°31′50″
	α_2 63°20′00″		
32	210°52′00″		
23	30°52′00″	−20″	30°51′40″
	α_3 258°11′00″		
43	289°03′00″		
34	109°03′00″	−30″	109°02′30″
	α_4 120°45′00″		
54	229°48′00″		
45	49°48′00″	−45″	49°47′15″
	α_5 236°30′30″		
	286°18′30″		
	−180°00′00″		
A_5	106°18′30″	−01′00″	106°17′30″
	106°17′30″ fixed		
	Angular error = 01′00″		

and uncorrected systematic errors in distance and direction. For this example, $d_c = [(0.297)^2 + (0.196)^2]^{1/2} = 0.36$ m.

The ratio of the closure d_c to the distance traversed provides an indication of the goodness of the survey and is often referred to as the relative accuracy ratio for the traverse. Surveys are frequently classified according to relative accuracy ratios such as 1/5000, 1/10,000, etc. When blunders or uncorrected systematic errors in distance or directions are present, the closure and the consequent relative accuracy ratio will be too large. The relative accuracy ratio provides a measure of the relative merits of various traverses, but should not be misconstrued as an indication of the absolute accuracy in position for each station in the traverse. For the example traverse, the relative accuracy ratio is $0.36/1762 = 1/4894$.

Specifications for various classifications of traverse and corresponding relative accuracies are discussed in Art. 8.22.

8.17. Computations for a closed-loop traverse When a traverse originates and closes on the same station, the algebraic sum of the departures and the algebraic sum of the latitudes each must equal zero. As an example, consider the traverse illustrated in Fig. 8.7 for which the adjusted azimuths are listed in Table 8.5, and station 1 is held fixed with given coordinates.

Distances, azimuths, calculated departures and latitudes, given coordinates for station 1, and calculated coordinates for the rest of the traverse stations are listed in Table 8.10. The sum of the departures and the sum of the latitudes are 0.22 and −0.17 ft, respectively. These algebraic sums represent the errors of closures for the

Table 8.9 Traverse Computations for Closed Traverse of Fig. 8.24

Station	Distance, m	Azimuth	Departures x	Latitudes y	Coordinates, m X	Y
1					4321.404	6240.562
	305.41	147°31′50″	163.959	−257.668		
2					4485.363	5982.894
	359.61	30°51′40″	184.465	308.694		
3					4669.828	6291.588
	612.15	109°02′30″	578.654	−199.717		
4					5248.482	6091.871
	485.12	49°47′15″	370.464	313.205		
5					5618.946	6405.076
	1762.29		$\Sigma = 1297.542$	$\Sigma = 164.514$	[5619.243]	[6405.272]
			$X_5 - X_1 = 1297.839$	$Y_5 - Y_1 = 164.710$		
Closure corrections:			$dX = +0.297$	$dY = +0.196$	$dX = 0.297$	$dY = 0.196$

$$d_c = (dX^2 + dY^2)^{1/2} = [(0.297)^2 + (0.196)^2]^{1/2} = 0.36 \text{ m}$$

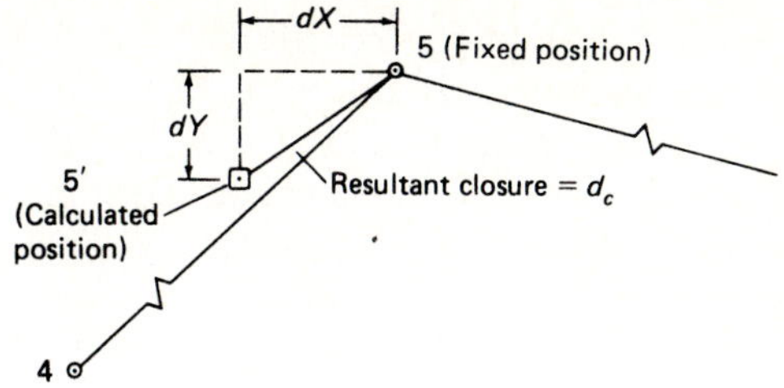

Fig. 8.25 Closure in a closed traverse.

Table 8.10 Computations for a Closed-Loop Traverse of Fig. 8.7

Station	Distance, ft	Azimuth	Departures x	Latitudes y	Coordinates, ft X	Coordinates, ft Y
1					4382.09	6150.82
	405.24	106°20′00″	+388.89	−113.96		
2					4770.98	6036.86
	336.60	57°54′30″	+285.17	+178.83		
3					5056.15	6215.69
	325.13	335°29′00″	−134.92	+295.82		
4					4921.23	6511.51
	212.91	219°29′00″	−135.38	−164.33		
5					4785.85	6347.18
	252.19	266°55′00″	−251.82	−13.56		
6					4534.03	6333.62
	237.69	219°40′00″	−151.72	−182.97		
1					4382.31	6150.65
	1769.76	Closure errors	$\Sigma = 0.22$	$\Sigma = -0.17$	(4382.09)	(6150.82)
					$dX = -0.22$	$dY = +0.17$
	$d_c = [(0.22)^2 + (0.17)^2]^{1/2} = 0.28$ ft				Closure corrections	

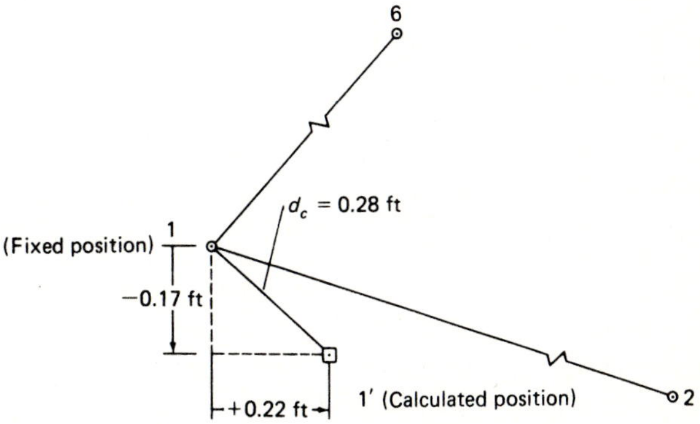

Fig. 8.26 Closure in a closed-loop traverse.

loop traverse. According to Eqs. (8.10*b*), the closure corrections are $dX = 0.0 - 0.22 = -0.22$ ft, $dY = 0.0 - (-0.17) = +0.17$ ft. Figure 8.26 shows that the resultant closure d_c, calculated by Eqs. (8.8), is 0.28 ft. For a closed-loop traverse of this type, the resultant closure is a function of random variations in the observations and uncorrected systematic errors in angles or directions. Systematic errors in distance measuring equipment will cancel and are not revealed by the mathematical closure of the traverse. Note also that although the directions are consistent within the traverse, the entire polygon could be rotated with respect to the grid system with no indication of this rotation in the results from the traverse computations. A second tie to a line of known direction is necessary to reveal this type of error.

8.18. Traverse adjustment Traverse adjustment is for the purpose of providing a mathematically closed figure and at the same time yielding the best estimates for horizontal positions of all the traverse stations. Methods of adjustment may be classified as approximate and rigorous. Consider the approximate adjustments first.

Traditional methods of approximate traverse adjustment have been developed to accommodate prevailing conditions in certain combinations of angular and linear precision in the observations. In this respect there are three combinations which are still common.

1. The precision in angles or directions exceeds its equivalent in linear distance observation (Art. 3.7). This combination exists in the stadia survey, where azimuths can be determined to within $\pm 30''$ with little difficulty, while the relative precision in distance measurements will never exceed 1 part in 300 to 400.
2. Precision in angles or directions is essentially equal to its equivalent in the precision of distances. This combination is typified by a traverse with an engineer's transit and a steel tape, where the terrain is fairly regular and the distances are corrected for all possible systematic errors.
3. Precision in distances exceeds that in angles and directions. This combination is prevalent in traverses containing long lines measured with calibrated EDM equipment.

Approximate adjustments include the *transit rule*, the *compass rule*, and the *Crandall method*. The transit rule was developed for the first combination of precisions in observations. Unfortunately, the transit rule is valid only when the traverse lines are parallel with the grid system used for the traverse computations and therefore is not considered further in this text. The compass rule was developed for the second combination and can be shown to be rigorous when the condition that the angular precision equals the precision in linear distances is rigidly enforced. The compass rule is developed further in Art. 8.19. The Crandall method is a rather complicated procedure which is more rigorous than either the compass or transit rules but requires substantially more computations. This method is not developed further in this book. The interested reader is referred to Refs. 3, 6, and 13 at the end of the chapter for detailed developments of these and other approximate methods.

The method of least squares provides the most rigorous adjustment, which allows for variation in precision in the observations, minimizes random variations in the observations, provides the best estimates for positions of all traverse stations, and yields statistics relative to the accuracies of adjusted observations and positions. This method does require more of a computational effort than the approximate adjustments. However, it is well within the capabilities of desktop and even some of the more sophisticated hand-held calculators. Theoretical development and examples of traverse adjustment by the method of least squares can be found in Refs. 8, 10, 11, and 13. The student of surveying should study these references and become familiar with adjustment of traverse by the method of least squares. The results are well worth the effort.

8.19. Adjustment of a traverse by the compass rule For any traverse station i, let

δX_i = correction to X_i
δY_i = correction to Y_i
dX_t = total closure correction of the traverse in the X coordinates [by Eqs. (8.10b)]
dY_t = total closure correction of the traverse in the Y coordinates [by Eqs. (8.10b)]
L_i = distance from the initial station to station i measured along the lines of the traverse
L = total length of traverse

then the corrections are

$$\delta X_i = \left(\frac{L_i}{L}\right) dX_t \quad \text{and} \quad \delta Y_i = \left(\frac{L_i}{L}\right) dY_t \tag{8.11a}$$

In an alternative procedure, corrections may be applied to the departures and latitudes prior to calculating coordinates. In this case,

$$\delta x_{ij} = \left(\frac{d_{ij}}{L}\right) dX_t \quad \text{and} \quad \delta y_{ij} = \left(\frac{d_{ij}}{L}\right) dY_t \tag{8.11b}$$

where δx_{ij} and δy_{ij} are respective corrections to the departure and latitude of line ij, which has a length of d_{ij}, and dX_t, dY_t, and L are as defined previously.

As an example, consider adjustment, by the compass rule, of the coordinates for the traverse shown in Fig. 8.24 and calculated in Table 8.9. The closure corrections in departures and latitudes are 0.297 and 0.196 m, respectively. According to Eqs. (8.11a), $\delta X_2 = (305.41/1762.29)(0.297) = 0.051$ m, so that X_2 adjusted = 4485.363 + 0.051 = 4485.41 m rounded to two decimal places. Similarly, $\delta Y_2 = (305.41/1762.29)(0.196) = 0.034$ m and Y_2 adjusted = 5982.894 + 0.034 = 5982.93 m. To calculate corrections to X_3 and Y_3, the cumulated distance to station 3 is used in Eqs. (8.11a). Thus, $\delta X_3 = (665.02/1762.29)(0.297) = 0.112$ m and $\delta Y_3 = (665.02/1762.29)(0.196) = 0.074$ m, yielding X_3 adjusted = 4669.828 + 0.112 = 4669.94 m and Y_3 adjusted = 6291.588 + 0.074 = 6291.66 m. Distances, corrections, and adjusted coordinates are listed in Table 8.11.

Traverse adjustment by the compass rule can also be accomplished by balancing (adjusting) the departures and latitudes using Eqs. (8.11b) prior to computing the coordinates.

Table 8.11 Adjustment of Coordinates by the Compass Rule (Traverse of Fig. 8.24, Calculated in Table 8.9)

Station	Distance, m	Corrections δX_i	Corrections δY_i	Adjusted coordinates, m X	Adjusted coordinates, m Y
1				4321.404	6240.562
2	305.41	0.051	0.034	4485.41	5982.93
3	665.02	0.112	0.074	4669.94	6291.66
4	1277.17	0.215	0.142	5248.70	6092.01
5	1762.29	0.297	0.196	5619.243	6405.272

Table 8.12 Corrections to Latitudes and Departures (Traverse of Fig. 8.7, Calculated in Table 8.10)

Station	Distance, ft	Departures, ft	Latitudes, ft	Corrections Dep.	Corrections Lat.	Adjusted Departures	Adjusted Latitudes
1							
	405.24	+388.89	−113.96	−0.05	0.04	+388.84	−113.92
2							
	336.60	+285.17	+178.83	−0.04	0.03	+285.13	+178.86
3							
	325.13	−134.92	+295.82	−0.04	0.03	−134.96	+295.85
4							
	212.91	−135.38	−164.33	−0.03	0.02	−135.41	−164.31
5							
	252.19	−251.82	−13.56	−0.03	0.03	−251.85	−13.53
6							
	237.69	−151.72	−182.97	−0.03	0.02	−151.75	−182.95
1							
	1769.76	0.22 ft	−0.17 ft	−0.22	0.17	$\Sigma = 0.00$	$\Sigma = 0.00$

Table 8.13 Adjusted Coordinates (Traverse of Fig. 8.7, Calculated in Table 8.10)

Station	Adjusted Departures, ft	Adjusted Latitudes, ft	Adjusted Coordinates, ft X	Adjusted Coordinates, ft Y
1			4382.09	6150.82
	+388.84	−113.92		
2			4770.93	6036.90
	+285.13	+178.86		
3			5056.06	6215.76
	−134.96	+295.85		
4			4921.10	6511.61
	−135.41	−164.31		
5			4785.69	6347.30
	−251.85	−13.53		
6			4533.84	6333.77
	−151.75	−182.95		
1			4382.09	6150.82

To illustrate this procedure of applying the compass rule adjustment, consider the traverse shown in Fig. 8.7 for which calculations are given in Table 8.10. For this traverse, dX and dY, calculated by substituting the sums of departures and latitudes into Eqs. (8.10*b*), are -0.22 ft and $+0.17$ ft, respectively. Thus, applying Eqs. (8.11*b*), the correction to the departure of line 1-2, $\delta x_{12} = (405.24/1769.76)(-0.22) = -0.05$ ft and the correction to the latitude of line 1-2, $\delta y_{12} = (405.24/1769.76)(+0.17) = +0.04$.

Corrections and balanced (adjusted) departures and latitudes are listed in Table 8.12. The sums of corrections to latitudes and departures should be computed and checked against their respective closures. Note that the sum of the latitudes was different from the closure in latitudes by 0.01 ft due to roundoff error. This discrepancy was corrected by examining the third place of the corrections and noting that the correction to the latitude of line 5-6 lay closest to the next highest hundredth, so that 0.01 ft was added to this correction. The sums of the adjusted departures, and latitudes should also be checked to ensure against any error in applying the corrections.

Adjusted coordinates computed using the balanced departures and latitudes are given in Table 8.13.

When calculating coordinates with balanced departures and latitudes, the computations should always be checked by cumulating the balanced departures and latitudes of the final line so as to provide a check on the coordinates of the closing station. Regardless of whether coordinates are adjusted directly or departures and latitudes are balanced first, the final adjusted coordinates must be identical.

8.20. Adjusted distances and azimuths When the traverse has been adjusted, the final step in traverse computations is to calculate adjusted (corrected) distances and directions using the adjusted coordinates or adjusted departures and latitudes in Eqs. (8.8) and (8.9).

Table 8.14 Adjusted Distances and Azimuths (Closed traverse in Fig. 8.7)

	Adjusted	
Station	Distance, ft	Azimuth
1		
	405.18	106°19′45″
2		
	336.59	57°54′01″
3		
	325.18	335°28′43″
4		
	212.92	219°29′33″
5		
	252.21	266°55′30″
6		
	237.69	219°40′28″
1		

For the example traverse (Fig. 8.7) calculated in Table 8.10, and using adjusted coordinates from Table 8.13, the corrected distance from 2 to 3 using Eq. (8.8) is

$$d_{23} = \left[(5056.06 - 4770.93)^2 + (6215.76 - 6036.90)^2\right]^{1/2}$$
$$= 336.59 \text{ m}$$

and the direction of line 2-3 (from the north) by Eq. (8.9) is

$$A_{N_{23}} = \arctan \frac{X_3 - X_2}{Y_3 - Y_2}$$
$$= \arctan \frac{285.13}{178.86} = 57°54'01''$$

Corrected distances and azimuths are listed for all the lines in this example traverse in Table 8.14. These are the values that would be used on the plan or map for which the survey was intended.

8.21. Coordinate transformations When surveyed points are coordinated, a number of survey problems can be solved conveniently by *coordinate transformations*. Coordinates in a local system may be required in the state plane coordinate system (Chap. 11); or it may be desired to convert coordinates in several local systems into one regional framework. Any time that it becomes necessary to convert from one coordinate system to another, a transformation is needed.

Coordinate transformations in two- and three-dimensional space can be found in Ref. 10.

8.22. Traverse accuracy and error propagation In general, the accuracy of traverses is judged on the basis of the resultant closure (Art. 8.16) of the traverse. This resultant closure is a function of the accuracies in the measurement of lengths and directions and hence varies with the length of the traverse. Consequently, a *relative accuracy* based on the resultant closure and length of traverse is computed. Thus, for the example traverse computed in Table 8.9, the resultant closure is 0.36 m for a traverse having a total length of 1762 m. The relative accuracy for this traverse is 0.36/1762.29, or 1 part in 4894.

To achieve a desired relative accuracy for a given traverse, specifications are written to govern the traverse field operations and the types of equipment that should be employed. Table 8.15 shows specifications recommended by the U.S. Federal Geodetic Control Committee for first-, second-, and third-order traverses. Note that second- and third-order traverses are each divided into class I and class II surveys. The relative accuracy or *positional closure* for each order and class of survey is expressed as a ratio (e.g., 1:10,000 for a third-order class I traverse) or as a constant times the square root of the distance traversed, which is designated by K [e.g., $0.4(K)^{1/2}$ for a third-order class I traverse]. The latter criterion is to be used for longer traverses. It is recommended that the relationship yielding the smallest possible positional closure be utilized.

Although the positional closure is an indication of the overall quality of the traverse and is used for traverse classification, it does not yield information on the precision of point locations determined in a traverse. The techniques of error propagation (see App. A, Art. A.1) must be employed to determine statistical measures of the accuracy for each point in the traverse. The interested reader should consult Refs. 8 and 10 for further details concerning this type error propagation.

8.23. Calculation of areas of land: general One of the primary objects of most land surveys is to determine the area of the tract. A closed traverse is run, in which the lines of the traverse are made to coincide with property lines where possible. Where the boundaries are irregular or curved or where they are occupied by objects which make direct measurement impossible, they are located with respect to the traverse line by appropriate angular and linear measurements. The lengths and bearings of all straight boundary lines are determined either directly or by computation, the irregular boundaries are located with respect to traverse lines by perpendicular offsets taken at appropriate intervals, and the radii and central angles of circular boundaries are obtained. The following sections explain the several common methods by means of which these data are employed in calculating areas.

In ordinary land surveying, as discussed herein, the area of a tract of land is taken as its projection upon a horizontal plane, and it is not the actual area of the surface of the land. For precise determinations of the area of a large tract, such as state or nation, the area is taken as the projection of the tract upon the earth's spheroidal surface at mean sea level.

8.24. Methods of determining area The area of a tract may be determined by any of the following methods:

1. By plotting the boundaries to scale as described in Chap. 10; the area of the tract may then be found by use of the planimeter or it may be calculated by dividing the tract into triangles and rectangles, scaling the dimensions of these figures, and computing their areas mathematically. This method is useful in roughly determining areas or in checking those that have been calculated by more exact methods. Its advantage lies in the rapidity with which calculations can be made.
2. By mathematically computing the areas of individual triangles into which the tract may be divided (Art. 8.25). This method is employed when it is not expedient to compute the departures and latitudes of the sides.
3. By calculating the area from the coordinates of the *corners* of the tract (Art. 8.26).
4. By calculating the area from the double meridian distances and the latitudes of the *sides* of the tract (Art. 8.27).
5. For tracts having irregular or curved boundaries, the methods of Arts. 8.28 to 8.31 are employed.

For computation of the areas of cross sections, see Art. 14.19.

8.25. Area by triangles When the lengths of two sides and the included angle of any triangle are known, its area is given by the expression

$$\text{area} = \tfrac{1}{2}ab\sin C \tag{8.12}$$

Table 8.15 Classification, Standards of Accuracy, and General Specifications for Horizontal Control

Classification	First-order	Second-order Class I	Second-order Class II	Third-order Class I	Third-order Class II
		Traverse			
Recommended spacing of principal stations	Network stations 10–15 km; other surveys seldom less than 3 km	Principal stations seldom less than 4 km except in metropolitan area surveys where the limitation is 0.3 km	Principal stations seldom less than 2 km except in metropolitan area surveys where the limitation is 0.2 km	Seldom less than 0.1 km in tertiary surveys in metropolitan area surveys; as required for other surveys	
Horizontal directions or angles					
Instrument	0.″2	0.″2 or 1.″0 [a]	0.″2 or 1.″0 [a]	1.″0	1.″0
Number of observations	16	8 or 12 [a]	6 or 8 [a]	4	2
Rejection limit from mean	4″	4″ or 5″ [a]	4″ or 5″ [a]	5″	5″
Length measurements					
Standard error[b]	1 part in 600,000	1 part in 300,000	1 part in 120,000	1 part in 60,000	1 part in 30,000
Reciprocal vertical angle observations					
Number of and spread between observations	3 D/R—10″	3 D/R—10″	2 D/R—10″	2 D/R—10″	2 D/R—20″
Number of stations between known elevations	4–6	6–8	8–10	10–15	15–20

Astro azimuths					
Number of courses between azimuth checks	5–6	10–12	15–20	20–25	30–40
Number of observations/night	16	16	12	8	4
Number of nights	2	2	1	1	1
Standard error	0.″45	0.″45	1.″5	3.″0	8.″0
Azimuth closure at azimuth checkpoint not to exceed[c]	1.″0 per station or $2''\sqrt{N}$	1.″5 per station or $3''\sqrt{N}$; metropolitan area surveys seldom to exceed 2.″0 per station or $3''\sqrt{N}$	2.″0 per station or $6''\sqrt{N}$; metropolitan area surveys seldom to exceed 4.″0 per station or $8''\sqrt{N}$	3.″0 per station or $10''\sqrt{N}$; metropolitan area surveys seldom to exceed 6.″0 per station or $15''\sqrt{N}$	8″ per station or $30''\sqrt{N}$
Position closure[d]					
After azimuth adjustment	0.04 m $\sqrt{K}$ or 1 : 100,000	0.08 m $\sqrt{K}$ or 1 : 50,000	0.2 m $\sqrt{K}$ or 1 : 20,000	0.4 m $\sqrt{K}$ or 1 : 10,000	0.8 m $\sqrt{K}$ or 1 : 5,000

[a]May be reduced to 8 and 4, respectively, in metropolitan areas.

[b]Use Eq. (2.19), Art. 2.15.

[c]N is the number of stations for carrying an azimuth.

[d]K is the distance in kilometres.

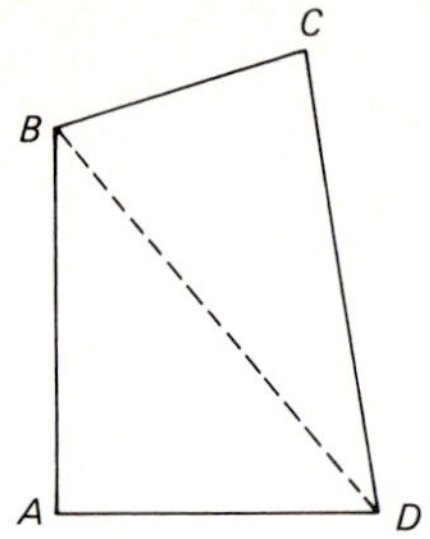

Fig. 8.27

When the lengths of the three sides of any triangle are given, its area is determined by the equation

$$\text{area} = \sqrt{s(s-a)(s-b)(s-c)} \tag{8.13}$$

in which $s = \frac{1}{2}(a+b+c)$ and a, b, c are the lengths of the sides (App. B).

In surveying small lots as for a city subdivision, it is common practice to omit the determination of the error of closure of each lot (a practice not condoned in this book) and hence the computation of departures and latitudes is unnecessary. Under such circumstances the area may be calculated by dividing the lot, usually quadrangular in shape, into triangles, as illustrated by Fig. 8.27, for each of which two sides and the included angle have been measured. By Eq. (8.12), the areas of *ABD* and *BCD* are computed; the sum of these two areas is the area of the lot. The area thus found can be checked independently by computing the areas of the two triangles *ABC* and *CDA* formed by dividing the quadrilateral by a line from *A* to *C*.

The accuracy of the field work may be investigated by determining the lengths of the diagonals. *BD* can be determined by solving either triangle *ABD* or triangle *BCD*. The field measurements are satisfactory if the length of *BD*, computed by solving one of the triangles, is the same as that computed by solving the other.

Figure 8.28 illustrates a survey made by a single setup of the transit at *O*, such as might be the case for a small lot where the property lines *ABCD* are obstructed or where the transit cannot be set up at the corners. Under these circumstances the angles about *O* and the distances *OA*, *OB*, *OC*, and *OD* are measured in the field. Since in each triangle two sides and the included angle are known, the area can be determined by use of Eq. (8.12). If in addition to the foregoing measurements, the lengths of the sides *AB*, *BC*, etc., are measured, the area of the lot can be checked independently by solving each triangle by Eq. (8.13). In general, the area of a triangle computed from two different sets of measurements will not be identical, owing to random measuring errors. Therefore, an alternative would be to use the method of least squares (see Ref. 8) to compute the area using all redundant measurements.

8.26. Area by coordinates When the points defining the corners of a tract of land are coordinated with respect to some arbitrarily chosen coordinate axes or are

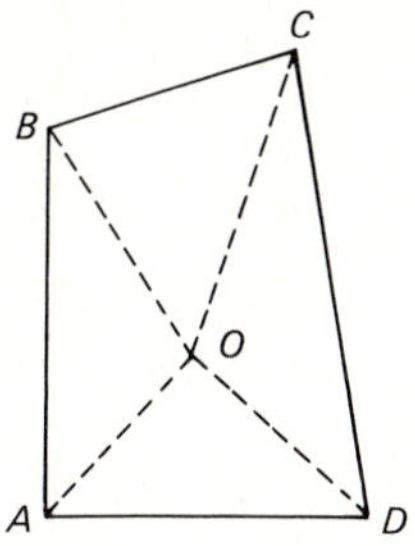

Fig. 8.28

given in a regional system (such as a state plane coordinate system), these coordinates are useful not only in finding the lengths and bearings of the boundaries but also in calculating the area of the tract. Essentially the calculation involves finding the areas of trapezoids formed by projecting the lines upon one of a pair of coordinate axes, usually a true meridian and a parallel at right angles thereto.

In Fig. 8.29, 123451 represents a tract the area of which is to be determined, where each point 1, 2, ..., 5 has coordinates $X_1, Y_1, \ldots, X_5, Y_5$, as shown in the figure.

The area of the tract can be computed by summing algebraically the areas of the trapezoids formed by projecting the lines upon the reference meridian; thus,

$$\text{area } 123451 = \text{area } 23cb + \text{area } 34dc - \text{area } 45fd$$
$$- \text{area } 15fa - \text{area } 21ab$$

or

$$\begin{aligned}\text{area} = {} & \tfrac{1}{2}(X_2 + X_3)(Y_2 - Y_3) + \tfrac{1}{2}(X_3 + X_4)(Y_3 - Y_4) \\ & - \tfrac{1}{2}(X_4 + X_5)(Y_5 - Y_4) - \tfrac{1}{2}(X_5 + X_1)(Y_1 - Y_5) \\ & - \tfrac{1}{2}(X_1 + X_2)(Y_2 - Y_1) \end{aligned} \tag{8.14}$$

By multiplication and a rearrangement of terms in Eq. (8.14), there is obtained

$$\begin{aligned} 2 \times \text{area} = {} & -[Y_1(X_5 - X_2) + Y_2(X_1 - X_3) + Y_3(X_2 - X_4) \\ & + Y_4(X_3 - X_5) + Y_5(X_4 - X_1)] \end{aligned} \tag{8.15}$$

and in general for any polygon having n stations

$$\begin{aligned} 2 \times \text{area} = {} & Y_1(X_2 - X_n) + Y_2(X_3 - X_1) + \cdots + Y_{n-1}(X_n - X_{n-2}) \\ & + Y_n(X_1 - X_{n-1}) \end{aligned} \tag{8.16}$$

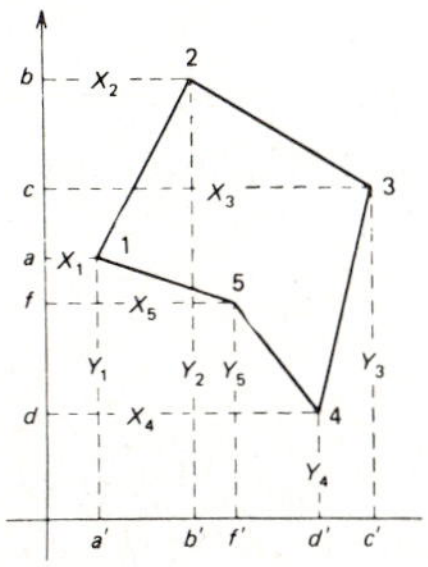

Fig. 8.29 Area by coordinates.

or, equivalently,

$$2 \times \text{area} = X_1(Y_2 - Y_n) + X_2(Y_3 - Y_1) + \cdots$$
$$+ X_{n-1}(Y_n - Y_{n-2}) + X_n(Y_1 - Y_{n-1}) \tag{8.17}$$

Note that a negative sign for the final value is the result of the direction that calculations progress around the figure and is of no concern. The total area will always be positive.

Example 8.1 Given the following data, find the required area by applying Eq. (8.16).

Corner	1	2	3	4	5
X coordinate, m	300	400	600	1000	1200
Y coordinate, m	300	800	1200	1000	400

SOLUTION

$$\begin{aligned} 2 \times \text{area} &= 300(400 - 1200) + 800(600 - 300) + 1200(1000 - 400) + 1000(1200 - 600) \\ &\quad + 400(300 - 1000) \\ &= -240{,}000 + 240{,}000 + 720{,}000 + 600{,}000 - 280{,}000 \\ &= 1{,}040{,}000\ \text{m}^2 \\ &= \frac{1{,}040{,}000}{2} = 520{,}000\ \text{m}^2 \end{aligned}$$

Equation (8.16) can also be expressed in the form

$$2 \times \text{area} = X_2Y_1 + X_3Y_2 + X_4Y_3 + \cdots + X_nY_{n-1} + X_1Y_n$$
$$- X_1Y_2 - X_2Y_3 - \cdots - X_{n-1}Y_n - X_nY_1 \tag{8.18}$$

When this form is employed, computations can be made conveniently by tabulating each X coordinate below the corresponding Y coordinate as follows:

$$\frac{Y_1}{X_1} \times \frac{Y_2}{X_2} \times \frac{Y_3}{X_3} \times \cdots \frac{Y_n}{X_n} \times \frac{Y_1}{X_1} \tag{8.19}$$

Then in expression (8.19), the difference between the sum of the products of the coordinates joined by full lines and the sum of the products of the coordinates joined by dotted lines is equal to twice the area of the tract. The expression is equally valid if Y_i and X_i are inverted.

Example 8.2 Compute the area enclosed by the polygon formed by the closed-loop traverse calculated in Art. 8.17 and adjusted in Art. 8.19. The adjusted coordinates are listed in Table 8.13. Use Eq. (8.17).

SOLUTION

$$\begin{aligned} 2 \times \text{area} &= 4382.09(6036.90 - 6333.77) + 4770.93(6215.76 - 6150.82) \\ &\quad + 5056.06(6511.61 - 6036.90) + 4921.10(6347.30 - 6215.76) \\ &\quad + 4785.69(6333.77 - 6511.61) + 4533.84(6150.82 - 6347.30) \end{aligned}$$

Observing the rules of significant figures the double area is

$2 \times \text{area} = 314{,}501 \text{ ft}^2$

and

$\text{area} = 157{,}250 \text{ ft}^2 = 3.6100 \text{ acres}$

8.27. Area by double meridian distances and latitudes Certain closed-loop traverses, such as those performed to locate property corners, are not coordinated. The closure for this type of survey should always be checked by calculating departures and latitudes for the traverse. When the area enclosed by the traverse is also desired, it can be calculated with the adjusted departures and latitudes using the *double-meridian-distance* method as described in this section.

The departures and latitudes of all the lines or courses are determined as described in Art. 8.15, and the survey is adjusted. A reference meridian is then assumed to pass through some corner of the tract, usually for convenience the most westerly point of the survey; the double meridian distances of the lines are computed as described herein; and double the areas of the trapezoids or triangles formed by orthographically projecting the several traverse lines upon the meridian are computed. The algebraic sum of these double areas is double the area within the traverse.

The meridian distance of a point is the total departure or perpendicular distance from the reference meridian; thus in Fig. 8.30 the meridian distance of B is Bb and is positive. The meridian distance of a straight line is the meridian distance of its midpoint. The *double meridian distance* of a straight line is the sum of the meridian distances of the two extremities; thus, the double meridian distance of BC is $Bb + Cc$. It is clear that if the meridian passes through the most westerly corner of the traverse, the double meridian distance of all lines will be positive, which is a convenience (although not a necessity) in computing.

The length of the orthographic projection of a line upon the meridian is the latitude of the line; thus, in Fig. 8.30 the latitude of BC is bc and is negative, and that of DF is df and is positive.

From the figure it is seen that each projection trapezoid or triangle, for which a course in the traverse forms one side, is bounded on the north and south by meridian distances and on the west by the latitude of that course. Thus, the projection trapezoid for BC is $BCcb$. Therefore, the double area of any triangle or

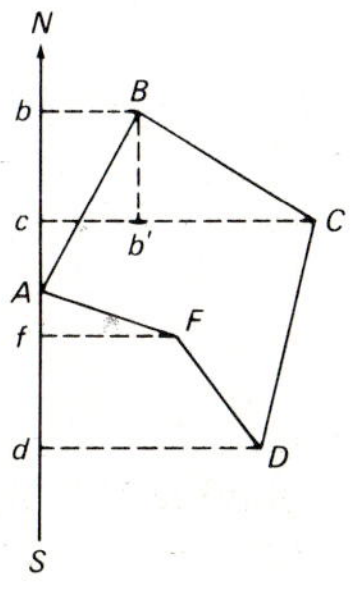

Fig. 8.30 Area by double meridian distances.

trapezoid formed by projecting a given course upon the meridian is the product of the double meridian distance (D.M.D.) of the course and the latitude of the course, or

$$\text{double area} = \text{D.M.D.} \times \text{latitude} \tag{8.20}$$

In computing double areas, account is taken of signs. If the meridian extends through the most westerly point, all double meridian distances are positive; hence, the sign of a double area is the same as that of the corresponding latitude. Thus, in the figure the double areas of *AbB*, *DdfF*, and *FfA* are positive, the latitudes *Ab*, *df*, and *fA* being positive; while the double areas of *CcbB* and *DdcC* are negative, the latitudes *bc* and *cd* being negative. Since the projected areas *outside* the traverse are considered once as positive and once as negative, the algebraic sum of their double areas is zero. Therefore, the algebraic sum of all double areas is equal to twice the area of the tract *within* the traverse.

Whether this algebraic sum of the double areas is a positive or negative quantity is determined solely by the order in which the lines of the traverse are considered. If the reference meridian passes through the most westerly corner, then a clockwise order of lines, as in the figure, results in a negative double area, and a counterclockwise order results in a positive double area. The sign of the area is not significant.

Computation of D.M.D. When the reference meridian passes through a traverse point, there is an intimate relation between the departures and double meridian distances. In Fig. 8.30 it is seen that the D.M.D. of *Ab* is *bB*, which is equal to the departure of the course in both magnitude and sign. The D.M.D. of *BC* is equal to $bB + cb' + b'C$, which is equal to the D.M.D. of *Ab*, plus the departure of *AB*, plus the departure of *BC*. Similar quantities make up the D.M.D.'s of *CD* and *DF*. For the last line *FA* the D.M.D. is *fF*, which is equal in magnitude but opposite in sign to the departure of *FA*.

Following are three convenient rules for determining D.M.D.'s, which are deduced from the relations just illustrated:

1. The D.M.D. of the first course (reckoned from the point through which the reference meridian passes) is equal to the departure of that course.
2. The D.M.D. of any other course is equal to the D.M.D. of the preceding course, plus the departure of the preceding course, plus the departure of the course itself.
3. The D.M.D. of the last course is numerically equal to the departure of the course but with opposite sign.

The first two rules are employed in computing values. The third rule is useful as a check on the correctness of the computations.

Area within a closed polygon by the D.M.D. method The steps employed in calculating the area within a closed-loop traverse by the D.M.D. method are as follows:

1. Compute and adjust departures and latitudes for all lines in the traverse (Arts. 8.15 to 8.19).

2. Assume that the reference meridian passes through the most westerly point in the survey and compute the D.M.D.'s by the rules of the preceding section, using the corrected departures.
3. Compute the double areas by multiplying each D.M.D. by the corresponding corrected latitude.
4. Find the algebraic sum of the double areas, and determine the area by dividing this sum by 2.

Example 8.3 Determine the area enclosed by the closed-loop traverse of Fig. 8.7 computed in Art. 8.17. Use the adjusted departures and latitudes listed in Table 8.13. Establish the reference meridian through station 1 and compute D.M.D.'s in a counterclockwise direction. The D.M.D. of the first line equals the departure of $12 = 388.84$ ft. The D.M.D. of the second line 23 equals the departure of the previous course, 388.84 + the D.M.D. of the previous course, 388.84 + the departure of the course itself, $285.13 = 1062.81$. The double area for line $12 = (388.84)(-113.92) = -44{,}296.7\ \text{ft}^2$. There are five significant figures in the data, so that six digits are carried in the calculations. D.M.D.'s for each line, latitudes, double areas, and the final area are listed in Table 8.16.

8.28. Area of tract with irregular or curved boundaries

If the boundary of a tract of land follows some irregular or curved line, such as a stream or road, it is customary to run a traverse in some convenient location near the boundary and to locate the boundary by offsets from the traverse line. Figure 8.31 represents a typical

Table 8.16 Computations for Area by Double Meridian Distance

Line			D.M.D., ft	Latitude, ft	Double area +	Double area −
12			388.84	−113.92		44,296.7
	Dep. 12		388.84			
	Dep. 23		285.13			
23	D.M.D.	=	1062.81	+178.86	190,094	
	Dep. 23		285.13			
	Dep. 34		−134.96			
34	D.M.D.	=	1212.98	+295.85	358.860	
	Dep. 34		−134.96			
	Dep. 45		−135.41			
45	D.M.D.	=	942.61	−164.31		154,880
	Dep. 45		−135.41			
	Dep. 56		−251.85			
56	D.M.D.	=	555.35	−13.53		7,513.89
	Dep. 56		−251.85			
	Dep. 61		−151.75			
61	D.M.D.	=	151.75	−182.95		27,762.7
					548,954	−234,453
					−234,453	
					2|314,501 = 2 × area	
					area = 157,250 ft² = 3.6100 acres	

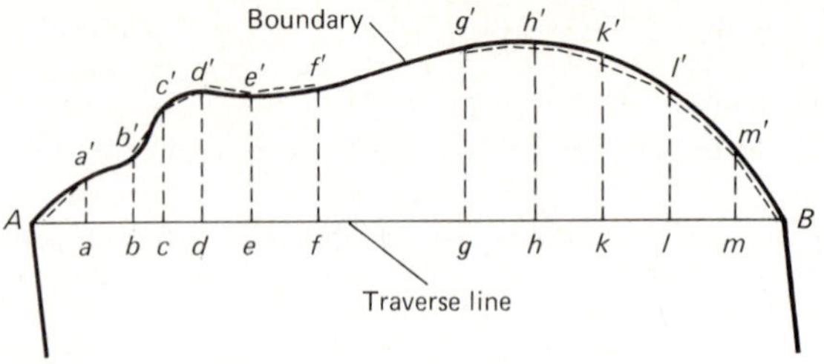

Fig. 8.31 Irregular boundary.

case, AB being one of the traverse lines. The determination of area of the entire tract involves computing the area within the closed traverse, by methods which have already been described, and adding to this the area of the irregular figure between the traverse line AB and the curved boundary. The offset distances are aa', bb', etc., and the corresponding distances along the traverse line are Aa, Ab, etc. Where the boundary is irregular, as from a' to f', it is necessary to take offsets at points of change and hence generally at irregular intervals. Where a segment of the boundary is straight, as from f' to g', offsets are taken only at the ends. Where the boundary is a gradual curve, as from g' to m', ordinarily the offsets are taken at regular intervals.

If the offsets are taken sufficiently close together, the error involved in considering the boundary as straight between offsets is small as compared with the inaccuracies of the measured offsets. When this assumption is made, the assumed boundary takes some such form as that illustrated by the slashed lines $g'h'$, $h'k'$, etc., in Fig. 8.31, and the areas between offsets are of trapezoidal shape. Under such an assumption, irregular areas are said to be calculated by the *trapezoidal rule* (Art. 8.29).

Where the curved boundaries are of such definite character as to make it justifiable, the area may be calculated somewhat more accurately by assuming that the boundary is made up of segments of parabolas as first suggested by Simpson. Under this assumption, irregular areas are said to be computed by *Simpson's one-third rule* (Art. 8.30).

8.29. Offsets at regular intervals: trapezoidal rule Let Fig. 8.32 represent a portion of a tract lying between a traverse line AB and an irregular boundary CD, offsets $h_1, h_2, \ldots, h_n$ having been taken at the regular intervals d. The summation of the areas of the trapezoids comprising the total area is

$$\text{area} = \frac{h_1 + h_2}{2}d + \frac{h_2 + h_3}{2}d + \cdots + \frac{h_{n-1} + h_n}{2}d \tag{8.21}$$

$$= d\left(\frac{h_1 + h_n}{2} + h_2 + h_3 + \cdots + h_{n-1}\right) \tag{8.22}$$

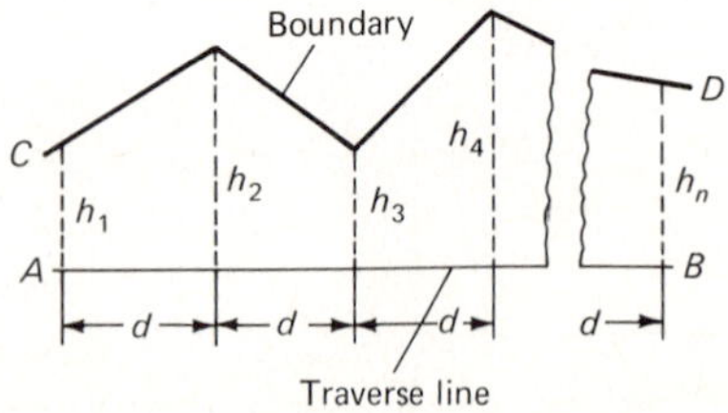

Fig. 8.32 Area by trapezoidal rule.

Equation (8.22) may be expressed conveniently in the form of the following rule:

Trapezoidal rule *Add the average of the end offsets to the sum of the intermediate offsets. The product of the quantity thus determined and the common interval between offsets is the required area.*

Example 8.4 By the trapezoidal rule find the area between a traverse line and a curved boundary, rectangular offsets being taken at intervals of 5 m, and the values of the offsets in metres being $h_1 = 3.2$, $h_2 = 10.4$, $h_3 = 12.8$, $h_4 = 11.2$, and $h_5 = 4.4$. By the foregoing rule,

$$\text{area} = 5\left(\frac{3.2 + 4.4}{2} + 10.4 + 12.8 + 11.2\right) = 191 \text{ m}^2$$

8.30. Offsets at regular intervals: Simpson's one-third rule In Fig. 8.33 let AB be a portion of a traverse line, DFC a portion of the curved boundary assumed to be the arc of a parabola, and h_1, h_2, and h_3 any three consecutive rectangular offsets from traverse line to boundary taken at the regular interval d.

The area between traverse line and curve may be considered as composed of the trapezoid $ABCD$ plus the area of the segment between the parabolic arc DFC and the corresponding chord DC. One property of a parabola is that the area of a segment (as DFC) is equal to two-thirds the area of the enclosing parallelogram (as $CDEFG$). Then the area between the traverse line and curved boundary within the length of $2d$ is

$$\text{area}_{1,2} = \frac{h_1 + h_3}{2} 2d + \left(h_2 - \frac{h_1 + h_3}{2}\right) 2d\left(\frac{2}{3}\right)$$
$$= \frac{d}{3}(h_1 + 4h_2 + h_3)$$

Similarly for the next two intervals,

$$\text{area}_{3,4} = \frac{d}{3}(h_3 + 4h_4 + h_5)$$

The summation of these partial areas for $(n - 1)$ intervals, n being an odd number and representing the number of offsets, is

$$\text{area} = \frac{d}{3}\left[h_1 + h_n + 2(h_3 + h_5 + \cdots + h_{(n-2)}) + 4(h_2 + h_4 + \cdots + h_{(n-1)})\right] \tag{8.23}$$

Equation (8.23) may be expressed conveniently in the form of the following rule, which is applicable if the number of offsets is odd.

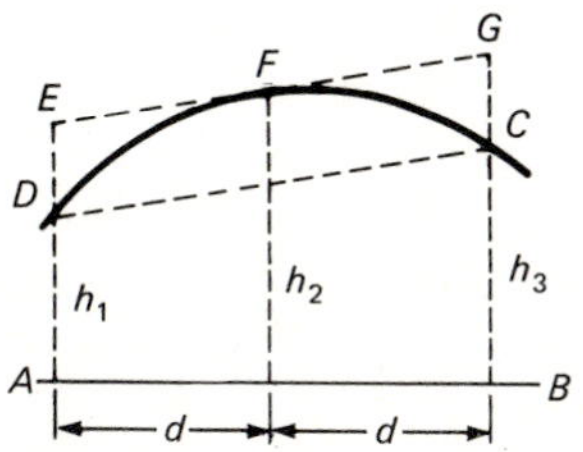

Fig. 8.33 Area by Simpson's rule.

Simpson's one-third rule *Find the sum of the end offsets, plus twice the sum of the odd intermediate offsets, plus four times the sum of the even intermediate offsets. Multiply the quantity thus determined by one-third of the common interval between offsets, and the result is the required area.*

Example 8.5 By Simpson's one-third rule find the area between the traverse line and the curved boundary of Example 8.4.

SOLUTION

$$\text{area} = \tfrac{5}{3}[3.2 + 4.4 + 2(12.8) + 4(10.4 + 11.2)] = 199 \text{ m}^2$$

If the total number of offsets is *even*, the partial area at either end of the series of offsets is computed separately, in order to make n for the remaining area an odd number and thus make Simpson's rule applicable.

Simpson's rule is also useful in other applications. The prismoidal formula (Art. 14.22) embodies a three-dimensional form of Simpson's rule for computing volumes of earthwork.

Results obtained by using Simpson's rule are greater or smaller than those obtained by using the trapezoidal rule, depending upon whether the boundary curve is concave or convex toward the traverse line. Some appreciation of the variations between the two methods will be gained by studying the foregoing examples. It will be seen that the two results differ by more than 4 percent. Under average conditions the difference will be much less than this, but in an extreme case it may be much larger.

In general, Simpson's rule is more accurate than the trapezoidal rule. The latter approaches the former in accuracy to the extent that the irregular boundary has curves of contrary flexure thereby producing the compensative effects mentioned above.

When the interval along the traverse line is irregular, the trapezoidal rule may be applied by using Eq. (8.22) and using the appropriate interval $d_1, d_2, \ldots, d_n$. Simpson's one-third rule is based on an equal interval for each pair of adjacent segments. Thus, if it is desired to utilize Simpson's one-third rule, a regular interval should be used in the field measurements.

8.31. Area of segments of circles A problem of frequent occurrence in the surveying of city lots and of rural lands adjacent to the curves of highways and railways is that of finding the area where one or more of the lines of the boundary is the arc of a circle.

In Fig. 8.34, *ABCDEQF* may be taken as a boundary of this character, for which it is convenient to run a traverse along the straight portions of the boundary and to make the chord *EF* the closing side of the traverse, the length of the chord $EF = c$ and the middle ordinate $PQ = M$ being measured in the field.

In calculating the area, it is convenient to divide the tract into two parts: (1) that within the polygon formed by the traverse *ABCDEFA*, for which the area is found by the coordinate method or the double-meridian-distance method, and (2) that between the chord *EPF* and the arc *EQF*, which is the segment of a circle. The area

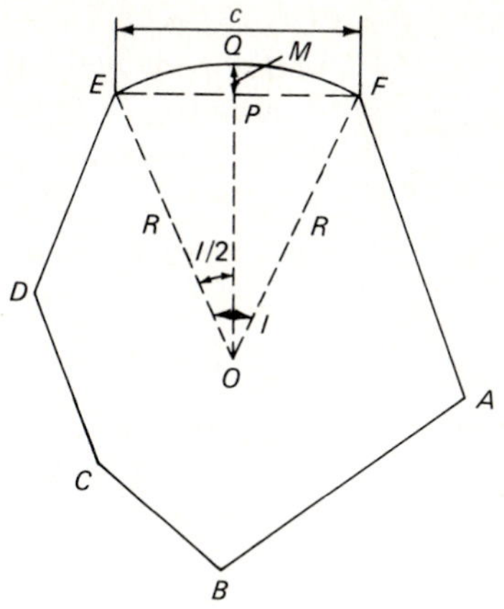

Fig. 8.34

of this segment is found exactly by subtracting the area of the triangle *OEPF* from the area of the circular sector *OEQF*. If I is the angle and R is the radius whose arc is *EQF*, then by Eq. (14.8), Art. 14.4,

$$\tan\frac{1}{4}I = \frac{2M}{c} \tag{8.24}$$

and

$$R = \frac{c}{2\sin\frac{1}{2}I} \tag{8.25}$$

The area of the circular sector *OEQF* is $A_s = \pi R^2 I°/360$, in which $I°$ is expressed in degrees. The area of triangle *OEF* is

$$A_t = \frac{cR\cos(I/2)}{2} = \frac{R^2}{2}\sin I$$

The area of the segment is exactly

$$\text{area} = A_s - A_t = R^2\left(\frac{\pi I°}{360} - \frac{\sin I}{2}\right) \tag{8.26}$$

Example 8.6 Find the area of a circular segment when the chord length is 80.00 m and the middle ordinate is 10.00 m.

SOLUTION By Eq. (8.24),

$\tan\frac{1}{4}I = 0.2500$

$\frac{1}{4}I = 14.036°$ $\quad I = 56.145°$

By Eq. (8.25),

$$R = \frac{c/2}{\sin(I/2)} = \frac{80}{(2)(0.47059)} = 85.00 \text{ m}$$

By Eq. (8.26),

$$\text{area} = (85.000)^2\left[\frac{(56.145°)\pi}{360} - \frac{\sin 56.145°}{2}\right]$$

$$= 539.9 \text{ m}^2$$

An alternative method of finding the area of the tract *ABCDEQFA* (Fig. 8.34) is to divide the area into a rectilinear polygon *ABCDEOFA* and the circular sector *OEQF*, and to add the two areas. The polygon has one more side than the one used above, but there is no need to compute the area of a circular segment.

Approximation by parabolic segment The area of a parabolic segment is

$$(\text{area})_p = \tfrac{2}{3}cM \tag{8.27}$$

in which the letters have the same significance as before. This expression may be employed for finding the approximate areas of circular segments, the precision decreasing as the size of the central angle I increases. The following example illustrates the error involved in applying this expression to the conditions of Example 8.6.

Example 8.7 By Eq. (8.27) find the approximate area of the circular segment of Example 8.6, and determine the percentage of error introduced through using the approximate expression Eq. (8.27).

SOLUTION The area by Eq. (8.27) is

$$(\text{area})_p = \tfrac{2}{3} \times 80 \times 10 = 533.3 \text{ m}^2$$

This value is

$$\frac{539.9 - 533.3}{539.9} 100 = 1.2 \text{ percent too low}$$

When the central angle is small, the error involved in using Eq. (8.27) for circular arcs is often negligible; thus, when $I = 30°$, the error is less than 0.2 percent. But for large values of I, the error introduced is so great as to render the approximate expression of little use; thus, when $I = 90°$, the error is about 3 percent, and when $I = 180°$, the error is about 15 percent.

Alternative method When tangents to the curve are property lines, it is sometimes more convenient to establish the traverse, as illustrated by Fig. 8.35. Here *KA* and *FC*, which are tangent to the curve *ABC*, are run to an intersection at *D*, and the distances *AD* and *CD* and the angle *I* are measured. Also, *E* is usually measured as a check.

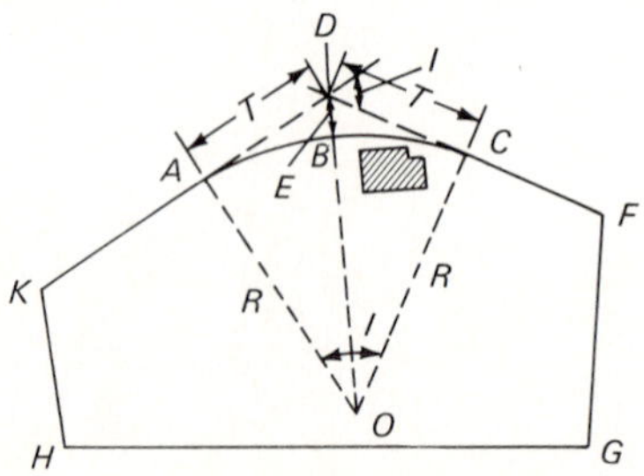

Fig. 8.35

The work of finding the area is conveniently divided into two parts: (1) that of calculating the area within the polygon *ADCFGHKA* by the coordinate method or the double-meridian-distance method, and (2) that of calculating the external area between the arc *ABC* and the tangents *AD* and *CD*. The latter area subtracted from the former is the required area.

The external area may be found by subtracting the area of the circular sector $OABC = A_s = \pi R^2 I°/360$ from the area $OADC = TR$, in which T is the tangent distance $AD = CD$, and R is the radius of the curve. If R is unknown, it may be found by the relation

$$R = \frac{T}{\tan \frac{1}{2} I} \qquad \text{[see Eq. (14.3), Art. 14.5]} \qquad (8.28)$$

8.32. Partition of land Some of the simpler cases frequently encountered in the subdivision of irregular tracts of land will be described in the succeeding articles. Methods of subdividing the U.S. public lands are given in Chap. 15.

Where a given tract is to be divided into two or more parts, a resurvey is run, the departures and latitudes are computed, the survey is adjusted, and the area of the entire tract is determined. The corrected departures and latitudes are further employed in the computations of subdivision.

Area cut off by a line between two points In Fig. 8.36 let *ABCDEFG* represent a tract of land to be divided into two parts by a line extending from *A* to *D*. A survey of the tract has been made, the adjusted coordinates have been computed for the corners, and the area has been computed.

It is desired to determine the length and direction of the cutoff line *AD* without additional field measurements, and to calculate the area of each of the two parts into which the tract is divided.

When coordinates are available for *A* and *D*, the simplest approach is to use Eqs. (8.8) and (8.9) developed in Art. 8.15 to calculate the distance and azimuth of *AD*. Then the areas of the traverse *ABCD* and *ADEFG* can be computed by the coordinate method (Art. 8.26).

8.33. Area cut off by a line running in a given direction In Fig. 8.37, *ABCDEFG* represents a tract of known dimensions, for which the corrected departures, latitudes, and adjusted coordinates are given; and *DH* represents a line

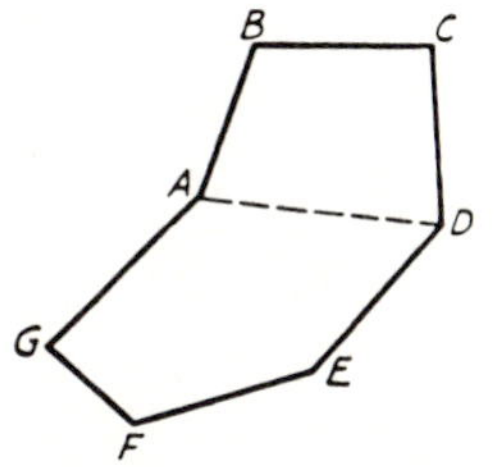

Fig. 8.36

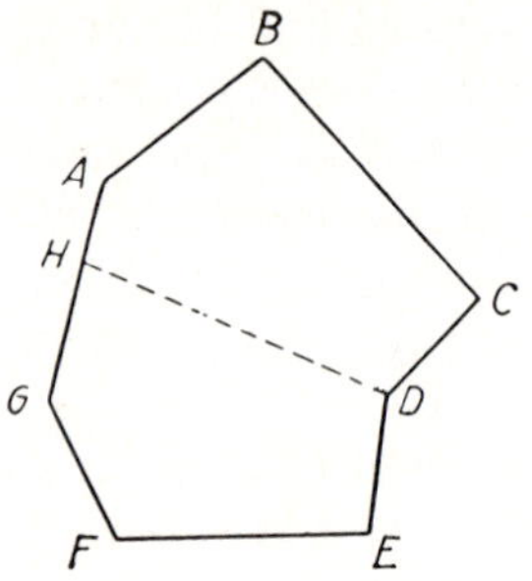

Fig. 8.37

running in a given direction which passes through the point *D* and divides the tract into two parts.

It is desired to calculate from the given data the lengths *DH* and *HA* and the area of each of the two parts into which the tract is divided.

The direction of *DH* is given and the coordinates of *D* are known, so that the point slope Eq. (B.46) can be formed. Since coordinates are known for *G* and *A*, Eq. (B.48) can be written for line *AG*. These two equations, one for line *DH* and one for line *GA*, can be solved simultaneously to obtain the coordinates of *H*. The distance *HA* can be calculated with Eq. (8.8), using the coordinates of *H* just computed and the known coordinates for *A*. Areas for *ABCDHA* and *HDEFGH* can be computed by the coordinate method [Eq. (8.16) or (8.17)].

In the field the length and direction of the side *DH* are laid off from *D*, and a check on field work and computations is obtained if the point *H* thus established lies on the line *GA* and if the computed distance *HA* agrees with the observed distance. The area computations may be checked by observing that the sum of the areas of the two parts, each computed independently, is equal to the area of the entire tract.

8.34. To cut off a required area by a line through a given point In Fig. 8.38, *ABCDEF* represents a tract of land of known dimensions, for which the corrected departures, latitudes, and adjusted coordinates for each point are given; and *G* represents a point on the boundary through which a line is to pass cutting off a required area from the tract. The area for the total tract can be calculated by the coordinate method and a sketch of the tract can be prepared.

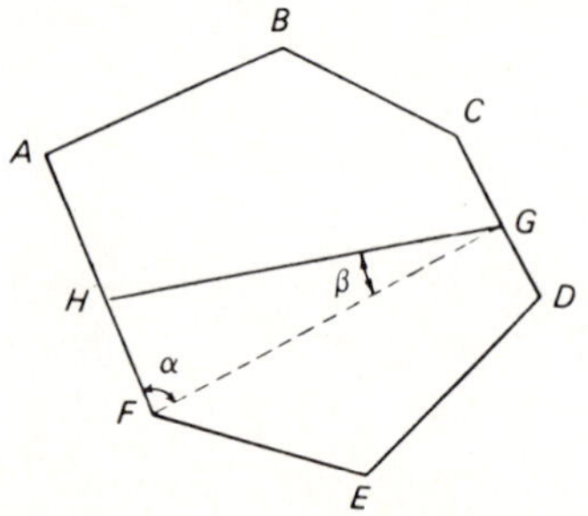

Fig. 8.38

To find the length and direction of the dividing line, the procedure is as follows. A line *GF* is drawn to that corner of the traverse which, from inspection of the sketch, will come nearest to being on the required line of division. Since distance *CG* is specified, the coordinates of *G* can be calculated. Then the distance and azimuth of *GF* are calculated using Eqs. (8.8) and (8.9). The area enclosed by the traverse and the cutoff line *GF*, *ABCGFA*, is computed by the coordinate method. The difference between this area and the amount specified is found.

In the figure it is assumed that *FABCG* has an area greater than the desired amount, *GH* being the correct position of the dividing line. The triangle *GFH* represents this excess area. The angle α may be computed from known directions. Then one side *FG*, one angle α, and the area of the triangle are known. The length *HF* is computed from the equation for area, area $= \frac{1}{2}(ab \sin C)$, so that from Fig. 8.38,

$$HF = \frac{2 \times \text{area } GFH}{FG \sin \alpha} \tag{8.29}$$

The triangle is then solved for angle β and length *GH*. From the known direction of *GF* and the angle β, the azimuth of *GH* is computed. The departures and latitudes of the lines *FH*, *GH*, and *HA* are computed and coordinates are calculated for point *H*.

In the field the length *GH* is laid off in the required direction, and a check on field work and computations is obtained if the point *H* thus established falls on the line *FA* and if the computed distance *HF* or *HA* agrees with the measured distance.

8.35. To cut off a required area by a line running in a given direction In Fig. 8.39, *ABCDEF* represents a tract of land of known dimensions and area, which is to be divided into two parts, each of a required area, by a line running in a given direction. The figure is assumed to be drawn at least roughly to scale, and the corrected departures and latitudes, and adjusted coordinates for the corners are known.

Through the corner that seems likely to be nearest the line cutting off the required area, a trial line *DG* is drawn in the given direction. Then in the closed traverse *GBCDG* the departures and latitudes of *BC* and *CD* and the directions of *DG* and *GB* are known, and the lengths of two sides *DG* and *GB* are unknown.

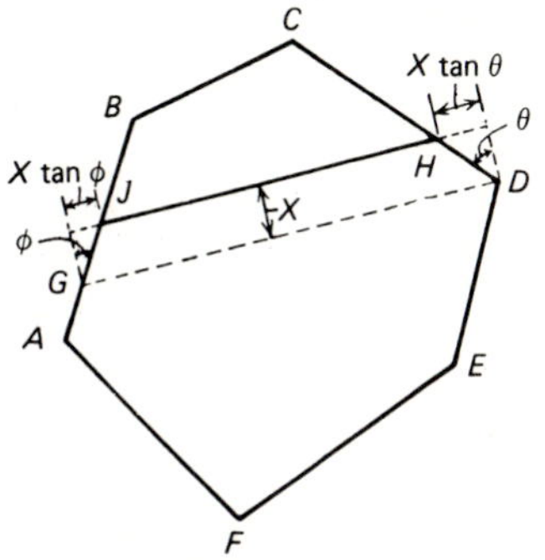

Fig. 8.39

Using these data, the coordinates of G can be found by forming the point-slope equation, Eq. (B.46), for line DG and Eq. (B.48) for line AB and solving these two equations simultaneously for X_G and Y_G.

Next, the area cut off by trial line DG is calculated. The difference between this area and that required is represented in the figure by the trapezoid $DGJH$ in which the side DG is known. The angles at D and G can be computed from the known directions of adjacent sides, and in this way θ and ϕ are determined. Tan θ and tan ϕ are positive if they fall inside the trapezoid and negative if they fall outside. With this convention, the

$$\text{area of trapezoid} = (DG)x + \frac{x^2}{2}(\tan\theta + \tan\phi) \tag{8.30}$$

in which x is the altitude of the trapezoid. (In the figure both angles lie outside the trapezoid so that both tangents are negative.) When the known values for DG, θ, and ϕ are substituted into Eq. (8.30), it is a quadratic equation in the form $ax^2 + bx + c = 0$ which may be solved for x using $x = [-b \pm (b^2 - 4ac)^{1/2}]/2a$.

In the field points H and J are established on the lines CD and AB, at the calculated distances from the adjacent corners. The side JH is then measured. If this measured value agrees with the computed value, the field work and portions of the computations are verified. A further check on the computations is introduced by calculating the area $BCHJ$ and comparing it with the required area of this figure.

Example 8.8 Determine the length and direction of the line passing through point 5, which divides the area of the tract enclosed by the traverse of Fig. 8.7 into two equal areas. Adjusted coordinates are listed in Table 8.13 and the total area within the tract is 157,250 ft^2 (see Example 8.2).

SOLUTION Use of coordinates and analytical geometry provides the most direct solution. To simplify the computations, all coordinates in Table 8.13 are translated so that the Y' axis passes through station 1 and the X' axis through station 2 (Fig. 8.40). The X' and Y' translated values are given in Table 8.17.

Assume line 5-2 as a first approximation to the division line. The distance and direction of 5-2 by Eqs. (8.8) and (8.9) are

$$d_{5\text{-}2} = [(388.84 - 403.60)^2 + (0.0 - 310.40)^2]^{1/2} = 310.751 \text{ ft}$$

$$\tan A_{52} = \frac{388.84 - 403.60}{0 - 310.40}$$

$$A_{52} = 182°43'21''$$

$$\text{angle } 521 = A_{52} - A_{12} = 76°23'36''$$

The total area of the tract, calculated in Example 8.2 by the coordinate method, is 157,250 ft^2, so that area$/2 = 78{,}625$ ft^2.

The area enclosed by stations 1256, using expression (8.19), is

$$2(\text{area})_{1256} = \frac{113.92}{0.00} \times \frac{0.00}{388.84} \times \frac{310.40}{403.60} \times \frac{296.87}{151.75} \times \frac{113.92}{0.00}$$

$$(\text{area})_{1256} = 83{,}200 \text{ ft}^2$$

$$-\frac{\text{total area}}{2} = 78{,}625$$

$$\text{excess area} = 4{,}575 \text{ ft}^2$$

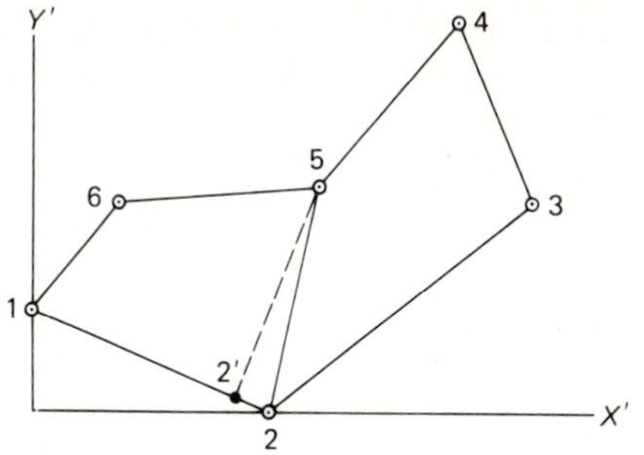

Fig. 8.40

so that the area enclosed by 1256 is too large by 4575 ft². By Eq. (8.29) and referring to Fig. 8.40, the distance 22′ from station 2 to the corrected division line 52′ is

$$22' = \frac{(2)(4575)}{(310.751)\sin 76°23'36''} = 30.2951 \text{ ft}$$

The coordinates of 2′ by Eqs. (8.6) and (8.7) are

$$X'_{2'} = 388.84 + (30.2951)(\sin 286°19'45'') = 388.84 + (30.2951)(-0.959662) = 359.767 \text{ ft}$$
$$Y'_{2'} = 0.00 + (30.2951)(\cos 286°19'45'') = 0.00 + (30.2951)(0.281155) = 8.5178 \text{ ft}$$

The area enclosed by 12′561 is

$$2(\text{area})_{12'561} = \frac{0.00}{113.92} \times \frac{359.77}{8.52} \times \frac{403.60}{310.40} \times \frac{151.75}{296.87} \times \frac{0.00}{113.92}$$

$$(\text{area})_{12'561} = 78{,}625 \text{ ft}^2$$

that is equal to total area / 2. The area enclosed by 2′23452′ is computed as a check

$$2(\text{area})_{2'23452'} = \frac{359.77}{8.52} \times \frac{388.84}{0.00} \times \frac{673.97}{178.86} \times \frac{539.01}{474.71} \times \frac{403.60}{310.40} \times \frac{359.77}{8.52}$$

$$(\text{area})_{2'23452'} = 78{,}625 \text{ ft}^2$$

By Eqs. (8.8) and (8.9), $d_{52'} = 305.05$ ft and $A_{52'} = 188°15'40''$.

Table 8.17 Data for Computing Parting-Off Land from Traverse of Fig. 8.7 (See Tables 8.13 and 8.14)

	Corrected		Translated coordinates	
Station	Distances, ft	Azimuths	X', ft	Y', ft
1			0.00	113.92
	405.18	106°19′45″		
2			388.84	0.00
	336.59	57°54′01″		
3			673.97	178.86
	325.18	335°28′43″		
4			539.01	474.71
	212.92	219°29′33″		
5			403.60	310.40
	252.21	266°55′30″		
6			151.75	296.87
	237.69	219°40′28″		
1				

Example 8.9 The tract enclosed by the traverse of Fig. 8.7 and also shown in Fig. 8.41 is to be divided so that one-fourth of the enclosed area lies south of a line parallel to the X and X' coordinate axes. Compute coordinates for the intersection of this dividing line with the traverse sides. Data listed in Table 8.17 may be used for the computations.

SOLUTION The initial position for the dividing line is taken passing through station 1, as illustrated in Fig. 8.41. An approximate area for triangle 127 made using distances scaled from the sketch reveals that this line results in an area that is ~ 20 percent too small, but it does provide a suitable starting line. As in Example 8.8, analytical geometry and coordinates provide the best method of solving the problem. Since the line 17 is parallel to the X' axis,

$$Y'_7 = Y'_1 = 113.92 \tag{1}$$

Equation (B.48) for line 23 is

$$\frac{Y'_7 - Y'_2}{Y'_3 - Y'_2} = \frac{X'_7 - X'_2}{X'_3 - X'_2}$$

$$\frac{Y'_7 - 0.00}{178.86 - 0.00} = \frac{X'_7 - 388.84}{673.97 - 388.84}$$

which yields

$$285.13Y'_7 - 178.86X'_7 = -69{,}547.92 \tag{2}$$

Substitution of Y'_7 from (1) into (2) gives

$$-178.86X'_7 = -102{,}029.93$$

$$X'_7 = 570.446 \text{ ft} = d_{17}$$

The area of 127 by area $= \frac{1}{2}(ab)$, where $a = 570.446$ and $b = 113.92$, is

$$\text{area} = \tfrac{1}{2}(570.446)(113.92) = 32{,}493 \text{ ft}^2 \tag{3}$$

The total area, calculated in Example 8.2 is 157,250 ft^2, so that (area)/4 = 39,313 ft^2. Thus, the area enclosed by 127 is too small by 39,313 − 32,493 = 6820 ft^2. Accordingly, the dividing line is moved parallel to itself to position 89, a distance x from the original line. Using Eq. (8.30), the trapezoidal area enclosed by 1897 is

$$\text{area}_{1897} = \frac{x}{2}(d_{17} + d_{17} - x\tan\theta + x\tan\phi) = xd_{17} + x^2\left(\frac{\tan\phi - \tan\theta}{2}\right) \tag{4}$$

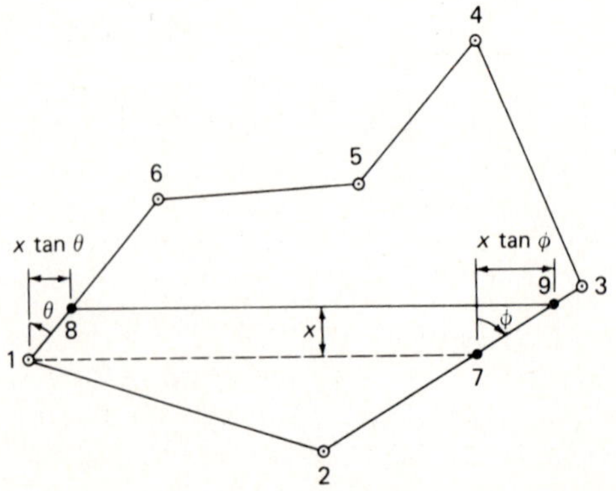

Fig. 8.41

in which $\theta = 39°40'28''$ and $\phi = 57°54'01''$ are the azimuths of lines 16 and 23 as found in Table 8.17 and shown on Fig. 8.41. Substitution of known values into Eq. (4) gives

$$6820 = 570.466x + \left(\frac{1.59415 - 0.829463}{2}\right)x^2 = 570.446x + 0.382345x^2$$

which is of the form $ax^2 + bx + c = 0$ and can be solved by

$$x = \frac{-b \pm (b^2 - 4ac)^{1/2}}{2a} = \frac{-570.446 \pm [(570.446)^2 - (4)(0.382345)(-6820)]^{1/2}}{(2)(0.382345)}$$

$$= \frac{-570.446 \pm 579.516}{0.764691} = \frac{9.070}{0.764691} = 11.8613 \text{ ft}$$

Distances d_{18} and d_{79} are calculated using x and the cosines of angles θ and ϕ:

$$d_{18} = \frac{x}{\cos\theta} = \frac{11.8613}{0.769684} = 15.411 \text{ ft}$$

$$d_{79} = \frac{x}{\cos\phi} = \frac{11.8613}{0.531394} = 22.321 \text{ ft}$$

Coordinates for 8 and 9 are computed by Eqs. (8.6) and (8.7) as follows:

$$X'_8 = X'_1 + d_{18}\sin 39°40'28'' = 0.00 + (15.411)(0.638425) = 9.84 \text{ ft}$$

$$Y'_8 = Y'_1 + d_{18}\cos 39°40'28'' = 113.92 + 11.86 = 125.78 \text{ ft}$$

$$X'_9 = X'_7 + d_{79}\sin 57°54'01'' = 570.446 + (22.321)(0.847124) = 589.35 \text{ ft}$$

$$Y'_9 = Y'_7 + d_{79}\cos 57°54'01'' = 113.92 + 11.86 = 125.78 \text{ ft}$$

As a final check, the area of the segment enclosed by points 12981 is calculated using the computed coordinates and expression (8.19).

$$2(\text{area}) = \frac{0.00}{113.92} \times \frac{388.84}{0.00} \times \frac{589.35}{125.78} \times \frac{9.84}{125.78} \times \frac{0.00}{113.92}$$

$$= 78{,}625 \text{ ft}^2$$

$$\text{area} = 39{,}313 \text{ ft}^2$$

which is equal to one-fourth of the total area. Final coordinates for points 8 and 9 translated into the original system (see Example 8.8) are:

$X_8 = 4391.93$ ft $\quad Y_8 = 6162.68$ ft $\quad X_9 = 4971.44$ ft $\quad Y_9 = 6162.68$ ft

8.36. Radial surveys A traverse performed from a single station is called a *radial traverse*. The station utilized should be a coordinated point in the primary control network for the project. A second station is necessary for a backsight and a third known point is useful for check purposes. A theodolite and EDM device or an electronic tacheometer (Chap. 9) are required since the procedure involves observing directions and distances.

The method is best used where the terrain is open with few restrictions to the line of sight. It may be employed to determine locations for traverse stations or to set off the positions of precalculated points.

Figure 8.42 shows a radial traverse executed from known control point *A* with a backsight on known point *H*. It is desired to locate the positions of traverse points, *B*, *C*, *D*, *E*, and *F*. If a theodolite and separate EDM device are being used, the theodolite is set over station *A*, a backsight taken on station *H*, and directions are

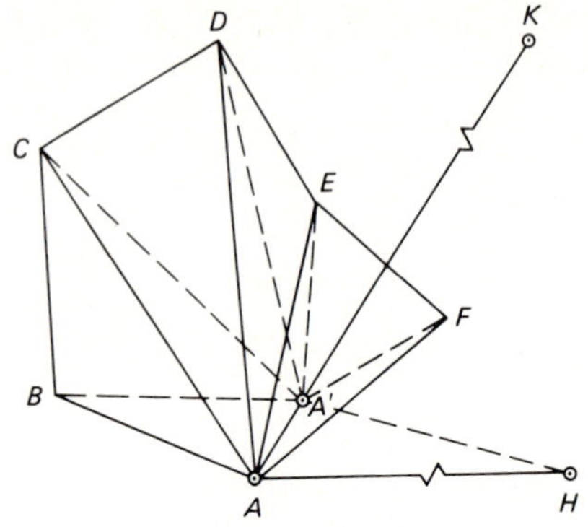

Fig. 8.42 Radial traverse.

observed to all traverse stations *B* through *F*. In addition, a sight is taken on known control point *K* and auxiliary point *A′* is set along the line *AK*. When the directions to these points have been observed a sufficient number of times to ensure the desired accuracy, the theodolite is removed from *A* and is replaced by the EDM device. Distances are observed from *A* to stations *B*, *C*, *D*, *E*, *A′*, and *F*. If an electronic surveying system or electronic tacheometer is employed (see Art. 9.14), directions and distances can be obtained simultaneously. Since lines *AH* and *AK* are of known directions, the azimuths of lines *AB*, *AC*, *AD*, *AE*, *AA′*, and *AF* can be determined. Using EDM distances (corrected for all systematic errors) and azimuths, departures and latitudes are calculated [Eqs. (8.6)] for lines *AB*, *AC*, *AD*, *AE*, *AA′*, and *AF*, permitting computation of coordinates for *B*, *C*, *D*, *E*, *A′*, and *F* [Eqs. (8.7)].

To provide a check on the work, the entire operation is repeated from auxiliary station *A′* with a backsight on station *K* yielding a second set of azimuths and horizontal distances from *A′* to all of the traverse stations *A* through *F*. These extra measures provide a second set of coordinates for all traverse points. In this way, blunders and undetected systematic errors can be found and corrected. If the values for the two sets of coordinates for each point are in reasonable agreement, the mean of each pair of coordinates is used as the final position for the station. Alternatively, the data from both points *A* and *A′* can be combined in a simultaneous least-squares solution (see Refs. 8 and 10). Distances and directions between these coordinated stations can then be calculated for lines *AB*, *BC*, *CD*, *DE*, *EF*, and *FA* using Eqs. (8.8) and (8.9). If distance *A′H* is observable, the triangle *AA′H* should also be calculated as a check.

Note that the entire procedure could be performed with a backsight on a single line such as *AH* or *AK*. However, the extra known point provides an additional check on direction and if the distance can be observed from *A′* to *K* a redundancy on the position of *A′* is furnished. In either case, an auxiliary station such as *A′* is absolutely necessary to provide a check on each point located.

In order to improve the final positions adopted for the stations, rigorous adjustment methods can be performed to reconcile the redundant data. The linearized form of the distance and direction Eqs. (8.8) and (8.9) can be formed for each ray from *A* and *A′* to the traverse points. The systems of equations so formed could then be solved by the method of least squares. Adjustment of combined triangulation and trilateration may also be applicable. Details for these adjustments can be found in Refs. 8, 10, and 13.

Procedures for setting off the precalculated locations of points such as in construction layout are developed in Chap. 14.

It should be emphasized that there is no standard solution for the application of radial methods to surveying problems. Each problem is unique and the basic rules as outlined must be used to suit the situation in the field.

8.37. Traverse in three dimensions When elevations are also determined in the process of running a stadia or plane-table traverse, or a traverse to establish control points in a mine where *X*, *Y*, and *Z* coordinates are required, these methods constitute traverses in three dimensions. Such three-dimensional surveys are suitable when horizontal and vertical positional accuracy requirements are low (horizontal, 1 part in 300 to 400 and vertical position to tenths of feet or to a few centimetres). When EDM equipment and direction theodolites or electronic tacheometers or surveying systems (Chap. 9) are employed for establishing horizontal and vertical control, relatively high accuracies in both horizontal and vertical positions are possible. In other words, trigonometric leveling as described in Art. 5.5 is performed in conjunction with any one of the several methods of running a traverse described in Arts. 8.7 to 8.10. Since the direction theodolite with its 01″ nominal least count in the vertical circle is necessary for trigonometric leveling, the angles-to-the-right traverse (Art. 8.9) is the most appropriate.

The minimum equipment (also the least expensive) required for a three-dimensional traverse consists of a direction theodolite with a nominal least count in horizontal and vertical circles of 01″ (Art. 6.30); an EDM instrument and reflectors or remote unit; two sighting targets adaptable to a tripod (Fig. 8.43); and three tripods compatible with theodolite, EDM instrument, reflectors, and target so that forced centering can be employed (Art. 6.30).

Assume that a light-wave EDM instrument and 01″ theodolite are to be used to run a three-dimensional traverse from station *A* to station *E* as shown in Fig. 8.44. Line *AK* is of known direction and stations *A* and *E* have known *XY* coordinates and elevations. Line *EH* is also of known direction. The procedure for running the traverse is as follows. The theodolite is set over station *A* and tripods are centered over stations *K* and *B*. Sighting targets are placed on the tripods at *K* and *B*. All heights of instrument and targets above the traverse stations (h.i.'s) are observed and recorded. The angle to the right from *K* to *B* and the zenith angle from *A* to *B* are observed. The theodolite is removed from the tripod at *A* and is replaced by the EDM device. A reflector replaces the sighting target on the tripod at *B*. The h.i. for the EDM at *A* and reflector at *B* are recorded. While the slope distance from *A* to *B* is being observed and recorded, the tripod at *K* is moved ahead and centered over station *C* where the sighting target is set in place and the h.i. observed and recorded. The theodolite is moved to the tripod at *B*, the EDM is moved toward station *B*, and a sighting target is placed on the tripod at *A*. After the necessary h.i.'s are recorded, the zenith angle from *B* to *A*, the angle to the right from *A* to *C*, and the zenith angle

Fig. 8.43 Target adaptable to tripod. (*Courtesy the Lietz Co.*)

from B to C are observed. The sighting targets at A and C are replaced by reflectors and the theodolite at B is replaced by the EDM instrument. All h.i.'s must be measured and recorded. The slope distances BA and BC are observed and recorded. Next, the tripod at A is moved ahead to D with a sighting target, the theodolite is set on the tripod at C, the EDM device is moved toward C, and a sighting target is placed on the tripod at B. The necessary h.i.'s are recorded, and the zenith angles from C to B and from C to D and the horizontal angle to the right from B to D are measured and recorded. The EDM instrument replaces the theodolite at C, reflectors are set on the tripods at B and D, and slope distances are measured from C to B and

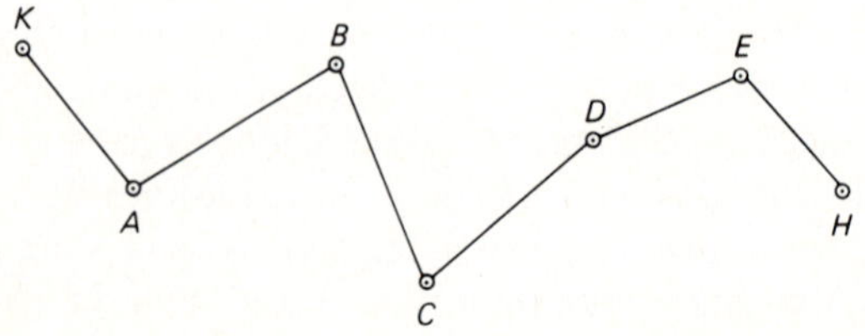

Fig. 8.44

C to *D*. This sequence of moving theodolite, EDM device, sighting targets, and reflectors ahead is continued until station *E* is occupied by the theodolite and the EDM instrument in that order, yielding the zenith angle from *E* to *D*, the closing horizontal angle from *D* to known line *EH*, and the slope distance *ED*. Care must be observed throughout to measure and record all h.i.'s to reflectors, targets, theodolite, and EDM instrument. It is also necessary to observe and record the necessary meteorological data to permit correcting the distances measured by EDM for systematic errors caused by atmospheric conditions (Art. 4.33).

Zenith angles should be measured a minimum of once with the telescope direct and once with the telescope reversed (Art. 6.44). Directions ought to be measured with a sufficient number of repetitions (Art. 6.39) to provide the accuracies specified for horizontal position (see Table 8.15). Azimuths are then computed from the angles determined from the averaged repeated directions. Measured slope distances should be corrected for all EDM systematic errors (Arts. 4.32, 4.34, and 4.35) and reduced to horizontal distance using the average zenith angles (Art. 4.37). Distances and directions so determined are then used to perform traverse computations as described in Art. 8.16.

Differences in elevations can also be calculated using average zenith angles and corrected slope or horizontal distances according to the methods outlined in Art. 5.5.

Electronic tacheometers or electronic surveying systems (Art. 9.14) used with combined sighting targets and reflectors (Fig. 9.10*a*, *b*) expedite the running of three-dimensional traverse considerably. In this case, distance and directions are measured from one instrument setup and there is no need to replace sighting targets with reflectors, and vice versa. Also, the data-reduction process is simplified. The simplest of the electronic tacheometers (Fig. 9.9*a b*) provides direct output of horizontal distance and difference in elevation between points sighted. The more sophisticated systems (Figs. 9.10 and 9.11) have data-storage and -processing capacity using a microcomputer built into the instrument. Additional details concerning these systems can be found in Art. 9.14.

8.38. Problems

8.1 Describe the two general classes of traverses, pointing out advantages, disadvantages, and applications for each class.

8.2 A traverse is initiated and closed on station *George*, which has known *X* and *Y* coordinates. What type of traverse is this? Should a traverse of this type be employed for a major project? Explain.

8.3 The interior angles of a five-sided closed traverse are as follows: *A*, 114°34′; *B*, 94°30′; *C*, 140°50′; *D*, 130°20′. The angle *E* is not measured. Compute the angle *E*, assuming the given values to be correct.

8.4 (*a*) What are the deflection angles of the traverse of Prob. 8.3? (*b*) What are the computed bearings if the bearing of *AB* is due north?

8.5 Following are the deflection angles of a closed traverse: *A*, 85°20′L; *B*, 10°11′R; *C*, 83°32′L; *D*, 63°27′L; *E*, 34°18′L; *F*, 72°56′L; *G*, 30°45′L. Compute the error of closure. Adjust the angular values on the assumption that the error is the same for each angle.

8.6 In Prob. 8.5, the azimuth of line *AB* from the north is 72°30′. Compute azimuths for the rest of the lines in the traverse and show a check on the calculations.

8.7 The clockwise interior angles observed with a 01′ transit in a traverse that starts and closes on point J are as follows: $J = 127°30′$; $A = 83°30′$; $B = 251°50′$; $C = 101°20′$; $D = 85°50′$; $E = 91°40′$; $F = 251°00′$; and $G = 87°28′$. Compute the error of closure and distribute this error assuming that all angles are of equal precision. Calculate azimuths for the lines in the traverse assuming that line AJ has a known azimuth from the north of $A_{AJ} = 206°30′$.

8.8 Comment on the relative and absolute angular accuracy of the data in Prob. 8.7.

8.9 The following deflection angle traverse was run with a 30″ repeating theodolite from a line of known direction LM to another independent line of known direction TU. Compute the angular error of closure and adjusted azimuths for each line. Comment on the relative and absolute angular accuracy of this traverse.

Station	Deflection angle	Azimuth from north
L		212°45′
M	39°47′L	
N	17°28′L	
O	14°08′L	
P	3°11′L	
Q	49°59′L	
R	32°18′R	
S	18°44′R	
T	7°31′L	
U		131°43′

8.10 The following are bearings taken for an open compass traverse. Correct for local attraction.

Line	Forward bearing	Back bearing
AB	N37°15′E	S36°40′W
BC	S65°30′E	N66°15′W
CD	S31°00′E	N31°00′W
DE	S89°15′W	N89°45′E
EF	N46°30′W	S46°45′E
FG	N15°00′W	S14°45′E

8.11 The following are bearings taken on a closed compass traverse. Compute the interior angles and correct them for observational errors. Assuming the observed bearing of the line AB to be correct, adjust the bearings of the remaining sides.

Line	Forward bearing	Back bearing
AB	S37°30′E	N37°30′W
BC	S43°15′W	N44°15′E
CD	N73°00′W	S72°15′E
DE	N12°45′E	S13°15′W
EA	N60°00′E	S59°00′W

8.12 Following are stadia intervals and vertical angles for a transit-stadia traverse. The elevation of station A is 418.6 ft; the stadia interval factor, 100.0; $C = 1.00$ ft. Rod readings are taken at height of instrument. Compute the horizontal lengths of the courses and the elevations of the traverse stations.

Station	Object	Stadia interval, ft	Vertical angle
B	*A*	8.50	+0°48′
	C	4.37	+8°13′
C	*B*	4.34	−8°14′
	D	12.45	−2°22′
D	*C*	12.41	+2°21′
	E	7.18	−1°30′

8.13 Following are stadia intervals and vertical angles for a transit-stadia traverse. The elevation of station *A* is 160.434 m. The stadia interval factor is 100 and $C = 0$. Rod readings are taken at the height of instrument.

Station	Azimuth	Interval, m	Vertical angle	Horizontal distance	Differences in elevation	Elevation, m
A						160.434
	85°06′	0.997	+4°32′			
B						
	10°18′	1.896	+3°51′			
C						
	265°00′	1.551	−5°04′			
D						
	173°04′	1.859	−2°10′			
A						

(*a*) Calculate the horizontal distance for each line to the nearest decimetre. Determine differences in elevation between stations to the nearest centimetre. (*b*) Determine elevations of transit stations and distribute the error of closure in proportion to the distance.

8.14 Given the following notes for a closed-loop traverse. The coordinates for station *A* are $X_A =$ 457,200.054 m, $Y_A = 167{,}640.842$ m. Compute the error of closure and coordinates for each traverse station adjusted according to the compass rule.

Course	Azimuth	Distance, m
AB	0°42′	372.222
BC	94°03′	164.988
CD	183°04′	242.438
DA	232°51′	197.145

8.15 Given the following adjusted azimuths and distances for a closed traverse that starts at station $B(X_B = 450{,}237.821$ m and $Y_B = 152{,}321.065$ m) and closes on station $J(X_J = 449{,}032.330$ m and $Y_J = 150{,}140.510$ m). Compute the error in closure for the traverse and coordinates for each traverse station adjusted by the compass rule. Determine the relative or positional accuracy for this traverse. Discuss the value of this measure of accuracy in terms of absolute position. Outline a better way to describe the absolute positional accuracy of a traverse station for which coordinates are available.

Course	Azimuth	Distance, m
AB	142°08′	
BC	181°37′	349.301
CD	296°13′	158.740
DE	323°46′	248.869
EF	249°51′	221.407
FG	214°03′	567.538
GH	195°45′	852.068
HJ	191°28′	750.936
JK	138°42′	

8.16 Given the following plane rectangular coordinates for traverse stations 3 and 4. Compute the distance and azimuth of the line from 3 to 4.

Station	*X*, m	*Y*, m
3	525,341.821	148,241.305
4	524,210.613	148,000.006

8.17 Compute adjusted distances and azimuths between all traverse points using adjusted coordinates from Prob. 8.14.

8.18 A square field contains 40 acres. What are its dimensions in chains, in rods, in feet, and in metres?

8.19 How many acres are there in a rectangular tract 50 × 100 ft? In a tract 400 × 400 ft? In a tract 2640 × 2640 ft? Express these areas in square metres and in hectares (Art. 1.8).

8.20 A triangle has sides 150.30, 300.85, and 245.62 m in length. Calculate the area of the triangle expressed in square metres, hectares, and acres.

8.21 What is the area of a triangle having sides of length 219.0, 317.2, and 301.6 ft? Of a triangle having two sides of length 1167.1 and 392.7 ft and an included angle of 39°46′?

8.22 The mutually bisecting diagonals of a four-sided field are 480 and 360 ft. The angle of intersection between the diagonals is 100°. Find the interior angles and the lengths of the sides.

8.23 In the following tabulation are given *X* and *Y* coordinates (in metres) for stations in a closed traverse. Calculate the area enclosed by the traverse using the coordinate method.

	Station			
Coordinate	*A*	*B*	*C*	*D*
X	5000.0	5102.5	5202.8	5102.5
Y	5000.0	5153.4	4900.0	4874.5

8.24 In the following tabulation are given the departures and latitudes of an adjusted closed traverse. Calculate the area (*a*) by the D.M.D. method; (*b*) by the coordinate method.

Course	Departure, ft	Latitude, ft
AB	W213.6	S198.7
BC	W174.4	N181.1
CD	E 89.2	N334.1
DE	E110.7	N224.9
EA	E188.1	S541.4

8.25 (*a*) Find the error of closure of the following traverse. Adjust the survey by the compass rule, and calculate the area in square metres by the D.M.D. method.

Course	Bearing	Length, m
AB	S45°45′E	89.733
BC	N65°30′E	80.284
CD	N35°15′E	95.585
DE	N64°15′W	119.482
EF	S59°00′W	60.107
FA	S25°30′W	77.261

(*b*) The coordinates for *A* are $X_A = 520{,}484.183$ m and $Y_A = 424{,}323.640$ m. Calculate coordinates for the traverse stations and determine the area in hectares by the coordinate method.

8.26 A traverse *ABCD* is established inside a four-sided field, and the corners of the field are located by angular and linear measurements from the traverse stations, all as indicated by the following data. Compute the departures and latitudes, and adjust the traverse by the compass rule. Compute the coordinates of each traverse point and of each property corner, using *D* as an origin of coordinates. Compute the length and bearing of each side of the field *EFGH*, and tabulate results. Calculate the area of the field by the coordinate method.

Course	Bearing	Length, ft
AB	S89°38′E	296.4
AE	N20°00′W	34.2
BC	S43°20′W	333.9
BF	N35°20′E	16.9
CD	S80°21′W	215.6
CG	S73°00′E	27.6
DA	N27°24′E	314.2
DH	S36°30′W	15.7

8.27 Given the following offsets from traverse line to irregular boundary, measured at points 10 m apart. By the trapezoidal rule (Art. 8.29) calculate the area between traverse line and boundary.

Distance, m	Offset, m	Distance, m	Offset, m
0	0.00	50	8.60
10	5.06	60	3.63
20	10.70	70	9.36
30	11.98	80	13.23
40	12.80	90	6.86

8.28 Given the data of Prob. 8.27, calculate the required area by Simpson's one-third rule. Note that the number of offsets is even.

8.29 Following are offsets from a traverse line to an irregular boundary, taken at irregular intervals. Calculate the area between traverse line and boundary by means of the trapezoidal rule.

Distance, ft	Offset, ft	Distance, ft	Offset, ft
0	18.5	100	44.1
25	37.7	170	53.9
60	58.2	200	46.0
70	40.5	220	34.2

8.30 Following are offsets from a traverse line to an irregular boundary taken at irregular intervals. Calculate the area between the traverse line and the boundary using the trapezoidal rule.

Distance, m	Offset, m	Distance, m	Offset, m
0	5.15	30	13.40
8	10.49	55	17.54
20	16.74	60	14.05
22	13.54	70	9.52

8.31 In Fig. 8.34, what is the area of the circular segment *EQF* if the length of the chord *c* is 249.080 m and the middle ordinate *M* is 27.179 m?

8.32 In Fig. 8.34, what is the area of the circular segment *EQF* if the chord length *c* is 600.00 ft and the middle ordinate *M* is 7.85 ft?

8.33 Solve Probs. 8.31 and 8.32 using the approximate expression [Eq. (8.27)] of Art. 8.31. Compare the results with those of Probs. 8.31 and 8.32, and for each case compute the percentage of error introduced through use of the approximate expression.

8.34 A curved corner lot is similar in shape to that shown in Fig. 8.35. The tangent distances *T* are each 50.0 ft and the intersection angle *I* is 40°. What is the area between the circular curve *ABC* and the tangents *AD* and *CD*? What is the external distance *E*?

8.35 Given the data of Table 8.13, Art. 8.19, and illustrated in Fig. 8.7. Find the area of each of the two parts into which the tract is divided by a meridian line through the point 5.

8.36 Given the data of Prob. 8.35. Find the length and direction of a line that runs through 3 and divides the tract into two equal parts.

8.37 Given the data of Prob. 8.35. The tract is to be divided into two equal parts by a north-south line. Compute the length of the dividing line, and compute the distances from the ends of the line to adjacent traverse stations.

8.39. Field problems

PROBLEM 1 DEFLECTION-ANGLE TRAVERSE

Object To measure deflection angles at each station of a traverse having *n* sides using the method of double deflection angles as described in Art. 6.39. A repeating transit or theodolite having a least count of 01′ or 30″ should be used. It is desirable that the instrument be equipped with a magnetic compass.

Procedure

1. Set *n* traverse stations each consisting of a hub with a tack at locations which provide maximum length of sight and optimum geometric strength for the area designated. If the traverse is to be used again, reference each station (Art. 8.14).
2. Occupy each station with the repeating transit or theodolite moving either direction around the traverse.
3. Level and center the instrument over the first station and loosen the compass needle.
4. Set the horizontal circle index or the *A* vernier to zero and take a backsight on the previous station with the telescope direct. Observe and record the magnetic bearing along that line.

5. Measure the deflection angle twice, once with the telescope direct and once with it reversed according to the procedure outlined in Art. 6.39. Be sure to observe and record the magnetic bearing along the line of the foresight when the telescope is in the direct position. The second angle should agree with the first angle to within the least count of the instrument. If it does not, repeat the measurement.
6. Repeat steps 3 to 5 for each station, recording the angles in a form similar to that illustrated in Fig. 8.6. Be sure to indicate the direction of the deflection angle as right (R) or left (L).
7. Calculate the sum of the average deflection angles. This sum should check the theoretical value [by Eq. (8.2)] to within $\sqrt{n}$ × (least count of the instrument).
8. Calculate adjusted angles by distributing the angular error of closure equally among the angles.

Hints and precautions

1. When lengths of sights are short, best results can be obtained by sighting directly on a plumb bob or pencil held on the traverse station being sighted.
2. When sighting into the sun, the notekeeper should shade the objective lens and eyes of the instrument operator.
3. Be sure to tighten the clamp before using the tangent screw when using either upper or lower motions.
4. Concentrate on each sight to be sure that the lower clamp and tangent screw are used on the backsight and upper clamp and tangent screw are used on the foresight.

PROBLEM 2 INTERIOR-ANGLE TRAVERSE

Object To measure interior angles at each station of a closed traverse having *n* sides, using the method for measuring angle by repetition described in Art. 6.39. Use a repeating transit or theodolite having a least count of 30″ or 01′.

Procedure

1. Set *n* traverse stations as described for Field Problem 1.
2. Occupy each station, moving around the traverse in either a clockwise or a counterclockwise direction.
3. Level and center the instrument over the first station and set the horizontal circle index or *A* vernier to zero.
4. Take a backsight on the previous station with the telescope direct and using the lower clamp and tangent screw. Record 0°00′00″ for this sight in a form similar to that used for deflection angles (Fig. 8.6).
5. Loosen the upper clamp and turn the alidade toward the next traverse station with the telescope direct, taking a foresight with the upper clamp and upper tangent screw.
6. Read and record the interior angle.
7. Loosen the lower clamp and resight the first station with the telescope reversed using the lower clamp and tangent screw.
8. Repeat step 5, doubling the angle with the telescope reversed. The difference between the first and second angles should not exceed the least count of the instrument. If it does, repeat the set.
9. Repeat steps 3 to 8 for each traverse station. When all angles have been measured, check the sum of the average angles. This sum should not deviate from $(n-2)(180°)$ by more than $\sqrt{n}$ × (least count of the instrument). The difference is the angular error of closure.
10. Distribute the error of closure equally among the *n* angles.

Hints and precautions

1. Turn all interior angles in the same direction.
2. Be sure to observe the correct horizontal circle; that is, when turning an interior angle to the right, read the clockwise circle; when turning an angle to the left, read the counterclockwise circle. Observe all precautions given for Field Problem 1.

PROBLEM 3 STADIA TRAVERSE WITH TRANSIT

Object To establish horizontal and vertical control for a topographic survey. The azimuth method (Art. 8.10) is used to establish directions and stadia (Chap. 7 and Art. 8.12) is employed to get distances and differences in elevations.

Procedure

1. Set stakes around the perimeter of the area to be surveyed, locating stations so as to provide maximum length of sight and strong geometry.
2. Establish the elevation of one station by differential leveling from the nearest bench mark (Arts. 5.31 to 5.36).
3. Occupy the point of known elevation (the initial station), measure the height of instrument above the station (h.i.) with a rod, and take a sight on the station that is to be the backsight.
4. Release the magnetic compass and observe the magnetic bearing along the line of the backsight. Convert this value to a magnetic azimuth and apply the declination to give a true azimuth. Release both motions of the repeating transit, set the true azimuth of the line on the clockwise horizontal circle, and resight the backsight station using the lower motion.
5. Have a rod placed on the station used for a backsight. Sight the rod observing the stadia interval and vertical angle according to the procedure described in Art. 8.12.
6. Record stadia interval, vertical angle, and true azimuth using the note form of Fig. 8.16.
7. Release the upper motion and take a foresight on the next station using the upper clamp and tangent screw.
8. Observe and record the stadia interval, vertical angle, and azimuth (on the clockwise horizontal circle) to this station.
9. Compute and record the back azimuth of this line by adding 180° to the forward azimuth.
10. Set up the instrument on the station just sighted and with the back azimuth just computed set on the clockwise horizontal circle, sight the station previously occupied using the lower clamp and tangent screw. The instrument is now oriented.
11. Release the upper motion and sight the next station forward, observing and recording the stadia interval, vertical angle, and azimuth. Proceed in this manner until all stations have been occupied.
12. The forward azimuth from the last station to the initial station should coincide with 180° $\pm$ the azimuth previously recorded for this line (step 4). Any discrepancy is the angular error of closure which should not exceed $\sqrt{n} \times$ (least count of the instrument) where $n =$ number of instrument setups. Distribute the discrepancy to each forward azimuth in proportion to the number of angles involved in determining that azimuth, leaving the first observed side fixed.

References

1. Angus-Leppan, P., "Practical Application of Accuracy Standards in Traversing," *Australian Surveyor*, March 1973.
2. Adler, R. K., et al., "Precise Traverse in Major Geodetic Networks," *Canadian Surveyor*, September 1971.
3. Baarda, W., and Albercia, J., "The Connection of Geodetic Adjustment Procedures with Method of Mathematical Statistics," *Bulletin Géodésique*, Vol. 66, 1962.
4. Berton-Jones, P., "Simplified Traverse Design," *Australian Surveyor*, September 1971.
5. Berton-Jones, P., "The Notion of Permissible Misclosure in Traversing," *Australian Surveyor*, September 1970.
6. Crandall, C. L., "The Adjustment of a Transit Survey as Compared with that of a Compass Survey," *Transactions of the American Society of Civil Engineers*, Vol. 45, 1901, pp. 453–464.
7. Chrzanowski, A., *Design and Error Analysis of Surveying Projects*, Lecture Notes No. 47, Department of Surveying Engineering, University of New Brunswick, Fredericton, New Brunswick, Canada, 1977.

8. Davis, R. E., Foote, F. S., Anderson, J. M., and Mikhail, E. M., *Surveying: Theory and Practice*, 6th ed., McGraw-Hill Book Company, New York, 1981.

9. Matson, D. F., "Angular Reference Ties," *Surveying and Mapping*, Vol. 33, No. 1, March 1973, p. 33.

10. Mikhail, E. M., *Observations and Least Squares*, Harper & Row, Publishers, New York, 1976.

11. Moffitt, F. H., and Bouchard, H., *Surveying*, 7th ed., Harper & Row, Publishers, New York, 1982.

12. Obenson, G., "Absolute Traverse Distortions of Some Adjustment Methods," *Survey Review*, July 1975.

13. Richardus, P., *Project Surveying*, John Wiley & Sons, Inc., New York, 1966, p. 277.

14. Tãrczy-Hornoch, A., "Remarks on Computations for Missing Elements of Closed Traverses," *Surveying and Mapping*, Vol. 32, No. 4, p. 523.

CHAPTER 9
Other survey operations

9.1. Triangulation and trilateration Triangulation and trilateration are employed extensively to establish horizontal control for topographic mapping; charting lakes, rivers, and ocean coast lines; and for the surveys required for the design and construction of public and private works of large extent.

A *triangulation system* consists of a series of joined or overlapping triangles in which an occasional line is measured and the balance of the sides are calculated from angles measured at the vertices of the triangles. The lines of a triangulation system form a network that ties together all the *triangulation stations* at the vertexes of the triangles.

A *trilateration system* also consists of a series of joined or overlapping triangles. However, for trilateration all of the lengths of the triangle's sides are measured and the few directions or angles observed are only those required to establish azimuth. Trilateration has become feasible with the development of EDM equipment (Chap. 4) which makes practical the measurement of all lengths with a high order of accuracy under almost all field conditions. Field procedures for the establishment of trilateration stations are similar to those discussed in this chapter for triangulation.

A combined triangulation and trilateration system consists of a network of triangles in which all the angles and all the distances are observed. Such a combined system represents the strongest network for creating horizontal control that can be established by conventional terrestrial methods.

Triangulation and/or trilateration may be used for a simple topographic survey covering a few acres or hectares; for design and construction of bridges and pipeline crossings; for topographic maps of counties or regions; or for extending first-, second-, and third-order horizontal control throughout an entire continent.

The most notable example of control establishment by triangulation and trilateration in the United States is the transcontinental horizontal system established by the U.S. National Ocean Survey. This system was designed to provide horizontal control for the entire continent. A permanent reference point for datum, called the North American Datum of 1927, has been established at Meades Ranch in Osborne County, Kansas. All of the horizontal control surveys of the United States, Canada, and Mexico are referred to this point. Originally, this control network was extended purely by triangulation. With the advent of EDM devices, trilateration and traverse were used to supplement and strengthen the original network. Recently, data from all of the modern positioning systems such as analytical photogrammetric triangulation (Art. 9.13), inertial systems (Art. 9.15), and satellite doppler measurements (Art. 9.16) are being employed with traditional triangulation and trilateration data in a readjustment of the entire system to a new datum that is expected to be available in 1985 (see Ref. 19 at the end of the chapter).

9.2. Accuracy of horizontal control systems Accuracies required for horizontal control depend on the type of survey and the ultimate use of the control points. In the United States, standards of accuracy for geodetic control surveys are prepared by the Federal Geodetic Control Committee (FGCC) and have been reviewed by the American Society of Civil Engineers, the American Congress on Surveying and Mapping, and the American Geophysical Union.

These standards provide for three orders of accuracy: first, second, and third; the latter two of which are subdivided into classes I and II. First-order or primary horizontal control provides the principal framework for the national control network. It is used for earth crustal movement studies in areas of seismic and tectonic activity, for testing defense and scientific equipment, for studying the performance of space vehicles, and for engineering projects of high precision and extending over long distances.

Second-order class I or secondary horizontal control consists of the networks between first-order arcs and detailed surveys in areas where the land values are high. Surveys of this class include the basic framework for densification of control. Secondary horizontal control strengthens the entire network and is adjusted as part of the National Network.

Second-order class II surveys are utilized to establish control for inland waterways, the interstate highway system, and for extensive land subdivision and construction. This class of control contributes to and is published as part of the National Network.

Third-order class I and class II or supplementary surveys are used to establish control for local improvements and developments, topographic and hydrographic surveys, or for other such projects for which they provide sufficient accuracy. Third-order control may or may not be adjusted to the National Network. The surveying engineer should know that third-order class I surveys constitute the lowest order of accuracy permissible for specifying points in the state plane coordinate systems (Chap. 11).

Table 9.1 contains a tabulation of standards and principal uses of geodetic control taken from the U.S. National Ocean Survey (NOS) publication *Classification, Standards of Accuracy and General Specifications of Geodetic Control Surveys*. Standards of accuracy and general specifications for horizontal control established by triangulation and trilateration are shown in Table 9.2.

9.3. Triangulation figures In a narrow triangulation system a chain of figures is employed, consisting of *single triangles*, *polygons*, *quadrilaterals*, or combinations of

Table 9.1 Standards for the Classification of Geodetic Control and Principal Recommended Uses

Horizontal control					
		Second-order		Third-order	
Classification	First-order	Class I	Class II	Class I	Class II
Relative accuracy between directly connected adjacent points (at least)	1 part in 100,000	1 part in 50,000	1 part in 20,000	1 part in 10,000	1 part in 5000
Recommended uses	Primary National Network, metropolitan area surveys, scientific studies	Area control that strengthens the National Network; subsidiary metropolitan control	Area control that contributes to, but is supplemental to, the National Network	General control surveys referenced to the National Network; local control surveys	

Vertical control					
	First-order		Second-order		
Classification	Class I	Class II	Class I	Class II	Third-order
Relative accuracy between directly connected points or bench marks (standard error)	0.5 mm$\sqrt{K}$	0.7 mm$\sqrt{K}$	1.0 mm$\sqrt{K}$	1.3 mm$\sqrt{K}$	2.0 mm$\sqrt{K}$
Recommended uses	Basic framework of the National Network and metropolitan area control; regional crustal movement studies; extensive engineering projects; support for subsidiary surveys		Secondary framework of the National Network and metropolitan area control; local crustal movement studies; large engineering projects; tidal boundary reference; support for lower-order surveys	Densification within the National Network; rapid subsidence studies; local engineering projects; topographic mapping	Small-scale topographic mapping; establishing gradients in mountainous areas; small engineering projects; may or may not be adjusted to the National Network

(K is the distance in kilometres between points)

Table 9.2 Classification, Standards of Accuracy, and General Specifications for Horizontal Control

Classification	First-order	Second-order Class I	Second-order Class II	Third-order Class I	Third-order Class II
		Triangulation			
Recommended spacing of principal stations	Network stations seldom less than 15 km; metropolitan surveys 3 to 8 km and others as required	Principal stations seldom less than 10 km; other surveys 1 to 3 km or as required	Principal stations seldom less than 5 km or as required	As required	As required
Strength of figure					
R_1 between bases					
Desirable limit	20	60	80	100	125
Maximum limit	25	80	120	130	175
Single figure					
Desirable limit					
R_1	5	10	15	25	25
R_2	10	30	70	80	120
Maximum limit					
R_1	10	25	25	40	50
R_2	15	60	100	120	170
Base measurement					
Standard error[a]	1 part in 1,000,000	1 part in 900,000	1 part in 800,000	1 part in 500,000	1 part in 250,000
Horizontal directions					
Instrument	0.″2	0.″2	0.″2 or 1.″0	1.″0	1.″0
Number of positions	16	16	8 or 12	4	2
Rejection limit from mean	4″	4″	5″ or 5″	5″	5″
Triangle closure					
Average not to exceed	1.″0	1.″2	2.″0	3.″0	5.″0
Maximum seldom to exceed	3.″0	3.″0	5″.0	5″.0	10.″0

[a]See Art. 2.15.

Table 9.2 (Continued)

Classification	First-order	Second-order		Third-order	
		Class I	Class II	Class I	Class II
		Triangulation			
Side checks					
In side equation test, average correction to direction not to exceed	0.″3	0.″4	0.″6	0.″8	2″
Astro azimuths					
Spacing figures	6–8	6–10	8–10	10–12	12–15
Number of observations/night	16	16	16	8	4
Number of nights	2	2	1	1	1
Standard error	0.″45	0.″45	0.″6	0.″8	3.″0
Vertical angle observations					
Number of and spread between observations	3 D/R—10″	3 D/R—10″	2 D/R—10″	2 D/R—10″	2 D/R—20″
Number of figures between known elevations	4–6	6–8	8–10	10–15	15–20
Closure in length					
(also position when applicable) after angle and side conditions have been satisfied should not exceed	1 part in 100,000	1 part in 50,000	1 part in 20,000	1 part in 10,000	1 part in 5000

	Trilateration				
Recommended spacing of principal stations	Network stations seldom less than 10 km; other surveys seldom less than 3 km	Principal stations seldom less than 10 km; other surveys seldom less than 1 km	Principal stations seldom less than 5 km; for some surveys a spacing of 0.5 km between stations may be satisfactory	Principal stations seldom less than 0.5 km	Principal stations seldom less than 0.25 km
Geometric configuration					
Minimum angle contained within, not less than		25°	25°	20°	15°
Length measurement					
Standard error[a]	1 part in 1,000,000	1 part in 750,000	1 part in 450,000	1 part in 250,000	1 part in 150,000
Vertical angle observations					
Number of and spread between observations	3 D/R—10″	3 D/R—10″	2 D/R—10″	2 D/R—10″	2 D/R—20″
Number of figures between known elevations	4–6	6–8	8–10	10–15	15–20
Astro azimuths					
Spacing figures	6–8	6–10	8–10	10–12	12–15
Number of observations/night	16	16	16	8	4
Number of nights	2	2	1	1	1
Standard error	0.″45	0.″45	0.″6	0.″8	3.″0
Closure in position					
after geometric conditions have been satisfied should not exceed	1 part in 100,000	1 part in 50,000	1 part in 20,000	1 part in 10,000	1 part in 5000

[a]See Art. 2.15.

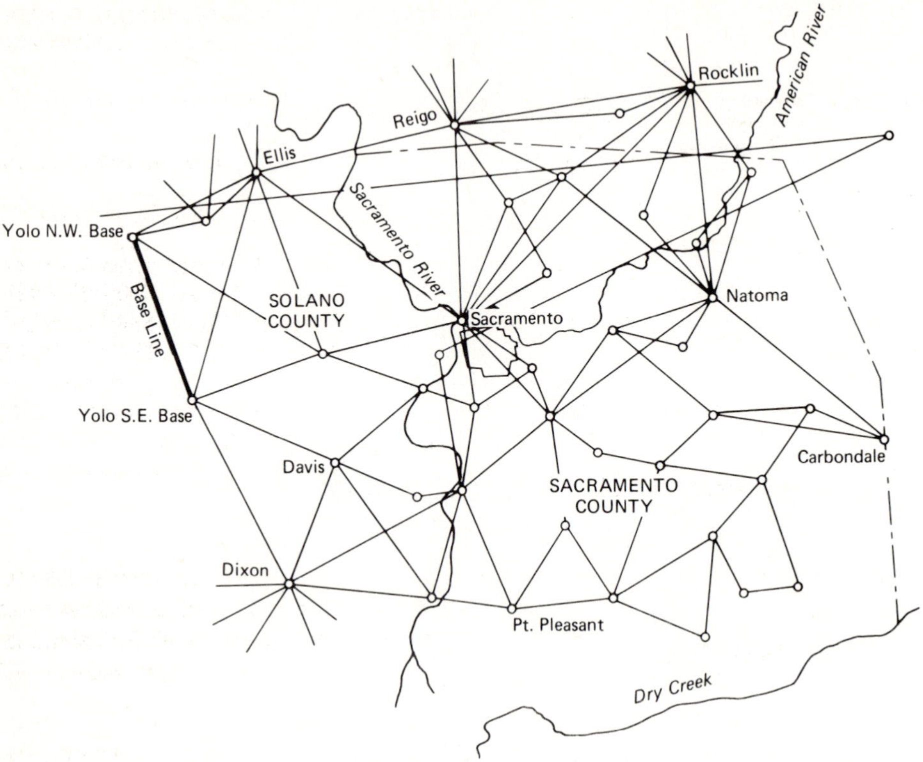

Fig. 9.1 Area triangulation.

these figures. A triangulation system extending over a wide area is likewise divided into figures irregularly overlapping and intermingling, as illustrated by the system of Fig. 9.1. A base line is measured (see Art. 9.6) to give the triangulation figure scale. Then, most or all of the angles in the figure are measured. Data reduction includes computing the sides of the triangles and ultimately the coordinates for all the stations in the network. The computations for such a system can be arranged to afford checks on the computed values of most of the sides. As many sides as possible are included in the routes through which the computations are carried from one base line to the next.

1. Chain of triangles In the chain of single triangles (Fig. 9.2) there is but one route by which distances can be computed through the chain. If AC is the base line

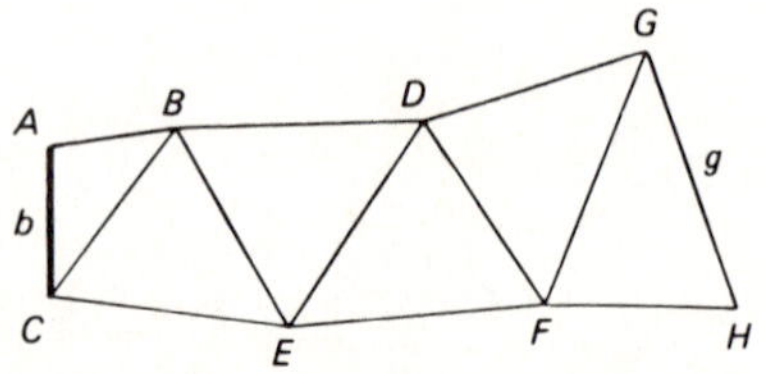

Fig. 9.2 Chain of single triangles.

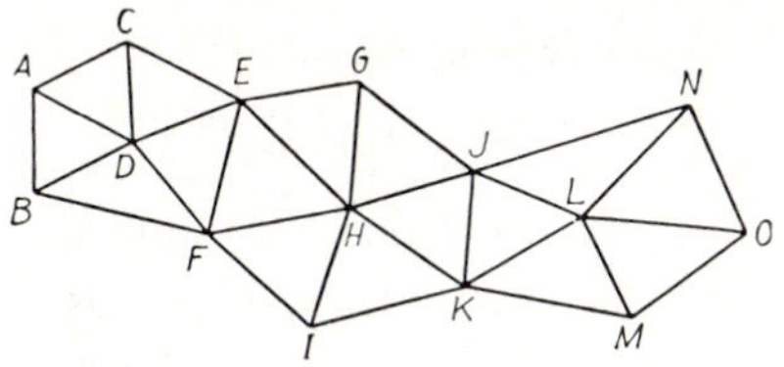

Fig. 9.3 Chain of polygons.

whose length b is measured, and if all the angles of the triangles are observed, the length of the triangle sides in the chain (as AB, BC, CE, etc.) may be calculated progressively along the chain from the measured base line to the triangle side farthest removed from the base line. If two lines are measured as base lines, one at each end of the system, the calculations may be carried from each toward the other to a triangle side somewhere between them.

2. Chain of polygons In triangulation, a polygon, or "central-point figure," is composed of a group of triangles, the figure being bounded by three or more sides and having within it a station which is at a vertex common to all the triangles. A chain of such composite figures is illustrated in Fig. 9.3, in which $BACEF$ is a five-sided polygon with D as the central point, and $FEGJKI$ is a six-sided polygon with H as the central point.

3. Chain of quadrilaterals Figure 9.4 illustrates another type of triangulation figure, in which the individual triangles overlap one another. This type usually occurs in the form of the quadrilateral, of which the figures $ABDC$, $DCEF$, etc., are examples. In the individual quadrilateral there is no triangulation station at the intersection of the diagonals.

9.4. Choice of figure Of the three forms of chains of triangulation figures, the chain of single triangles is the simplest, requiring the measurement of fewer angles than either of the other two. This type of system, however, has the obvious weakness that the only check (except between bases) is in the sum of the angles of each triangle considered by itself. To reach the same precision in the determination of lengths, base lines would need to be placed closer together. This type of chain is not employed in work of high precision, but it is satisfactory where less precise results are required.

For more precise work, quadrilaterals or polygons are used; quadrilaterals are best adapted to long, narrow systems and polygons to wide systems.

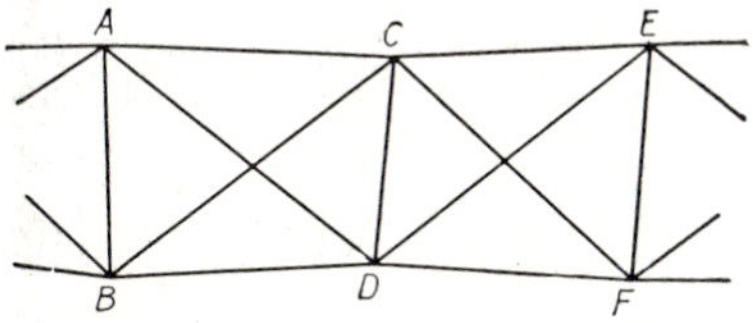

Fig. 9.4 Chain of quadrilaterals.

9.5. Reconnaissance The initial and one of the most important stages of a triangulation project is the preliminary reconnaissance. Reconnaissance consists of selection of stations, determination of the size and shape of the resulting triangles, the number of stations to be occupied, and the number of angles or directions to be observed. The intervisibility and accessibility of stations, the usefulness of stations in later work, the cost of the necessary signals, and the convenience of base-line measurements are considered.

After a preliminary study of all available maps, survey information, and aerial photographs of the area, the person in charge makes an on-site inspection, choosing the most favorable locations for stations. If the information required cannot be obtained from existing maps, photographs, etc., angles and distances to other stations are estimated or measured roughly en route, so that the suitability of the system as a whole can be evaluated before detailed work is begun.

Measurement of base lines is generally performed using EDM equipment (Arts. 4.30 through 4.38), which provides optimum accuracy with minimum expenditure of money and effort. For this case, the base line should be located so as to yield good geometric strength in the triangulation network. If the base line is to be taped, a further requirement is that it be located on uniform terrain suitable for high-precision measurement. If towers are needed for the observation of directions, the necessary heights and number of towers should be determined.

Information acquired during reconnaissance is then used for determination of strength of figure and/or error propagation.

Strength of figure is evaluated in order to ensure uniform accuracy throughout the network. The strength of figure is a function of:

1. The geometric strength of the triangles that make up the network. Ideally, the triangles should be equilateral.
2. The number of stations occupied for angle or direction measurements. Lines occupied at only one end should be avoided whenever possible.
3. The number of conditions used in adjusting the network. This number should be large in proportion to the number of observations since they correspond to the number of redundant measurements.

Error propagation can be performed by using the trigonometric law of sines and propagating the error through the network rigorously (see App. A, Art. A.1).

Evaluation of strength of figure and error propagation in triangulation networks is beyond the scope of this textbook. For further details consult Ref. 16 at the end of this chapter.

9.6. Measurements for triangulation As noted in Art. 9.1, the basic measurements for triangulation consist of angles with measured base lines at specified intervals.

Directions from which angles are determined for triangulation are usually observed with a direction theodolite (Art. 6.30). First-order triangulation requires a first-order optical-reading direction theodolite with a least count of less than 1 second (Fig. 9.5). A second-order direction theodolite (Figs. 6.33 and 6.34) which has

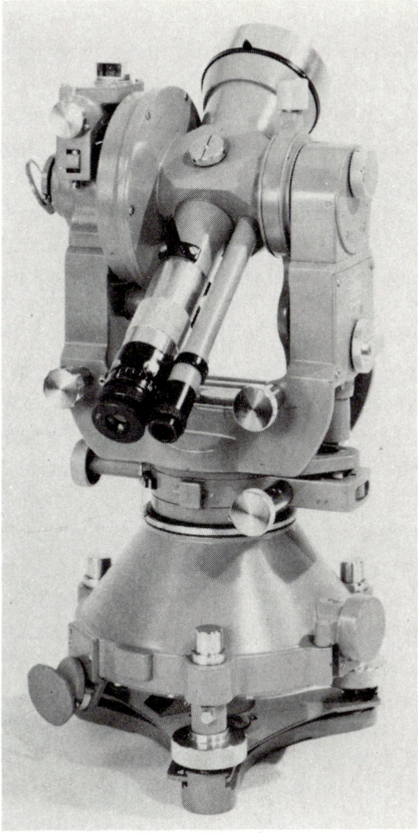

Fig. 9.5 Wild T-3, first-order theodolite. (*Courtesy Wild Heerbrugg Instruments, Inc.*)

a least count of 1 second is adequate for second- and third-order triangulation. The use of optical-reading theodolites is developed in Arts. 6.30, 6.40, and 6.41. Repeating theodolites and transits of good quality can be used to observe angles directly in third-order triangulation but are not recommended because of the excessive effort required to obtain results of sufficient accuracy.

To satisfy first-, second-, and third-order specifications as given in Table 9.2 extra care must be exercised to (1) center the instrument and targets very carefully, (2) eliminate phase† in the targets, (3) protect the instrument from the heating effects of the sun and vibrations caused by wind, and (4) provide an instrument support of adequate stability. This last requirement is extremely important and frequently leads to use of a concrete pedestal for support of the station and instrument. Scheduling of observations when adverse horizontal refraction is at a minimum is also advisable. Operations at night using illuminated signals will help reduce bad atmospheric conditions and assist in achieving optimum results.

†Signals should be of such shape and orientation that they will be as nearly free as possible from phase, or asymmetry of sighting. If the signal is illuminated from the side, the observer tends to sight to one side of its center. This phenomenon is called *phase*.

Directions should be observed using procedures that minimize collimation, circle, and micrometer errors in the instrument. The procedure for observing one position of directions from a single setup with the telescope direct and reversed is given in Art. 6.39. Depending on the accuracy desired, from 2 to 16 positions may be required (Table 9.2). In order to minimize errors in the horizontal circle and in the micrometer, each position is advanced to a new initial circle reading and micrometer readings should be spread over the range of the micrometer. In general, the horizontal circle is advanced by an interval I, where

$$I = \frac{360°}{mn} + \frac{\text{micrometer drum range (in minutes)}}{n} \tag{9.1}$$

in which n is the number of positions to be observed and m the number of verniers on the instrument.

Mean directions and mean angles calculated from these directions are determined and the results are then checked in the field so that bad measurements can be rejected and repeated while still on the site. These angles or directions are then used in subsequent adjustment of the triangulation.

Utilization of EDM equipment vastly simplifies the measurement of base lines in triangulation operations. The primary requirement for measurement of a base line by an EDM instrument is that an unobstructed measuring path exist between the two ends of the line.

Care should be exercised to correct for meteorological conditions, systematic instrumental errors, and slope of the line of sight as described in Arts. 4.32 to 4.37. With respect to instrumental errors, the EDM system should be calibrated periodically on a base line of known length (Arts. 4.34 and 4.35). Vertical angles or elevations at the two ends of the line must be obtained in order to allow reduction of slope distances to horizontal distances (Art. 4.37). At least one elevation is required to permit reduction to sea level.

When taping is employed for base-line measurements of ordinary precision either the steel tape or the invar tape may be employed, but for measurements of higher precision the invar tape is always used. Invar is a nickel-steel alloy for which the coefficient of thermal expansion may be as low as 0.0000002 ft/ft per 1°F or 0.00000036 m/m per °C (about one-thirtieth that of steel). The invar tapes commonly used have a coefficient of expansion one-eighth to one-tenth that of steel. Often a "long tape" (length 50 m or 200 to 500 ft) is used.

The length of the tape should be precisely determined by comparison with a standard of known length. The National Bureau of Standards at Washington, D.C., for a nominal fee, will compare the tape with a length that has been precisely determined and will issue a certificate showing the actual length of the tape under stated conditions as regards tension, temperature, and supports (see Art. 4.16). It is desirable to have the tape compared with the standard under the conditions of tension and support that will be employed in the field work, so that no corrections for sag or tension will be necessary.

A detailed description of precise base-line taping is beyond the scope of this textbook. The interested reader should consult Ref. 16.

9.7. Triangulation computations and adjustment The primary objective of triangulation computations is to determine the best possible planimetric positions, usually defined by X and Y coordinates, for the triangulated stations. To achieve this objective it is necessary to (1) correct all base-line measurements for systematic errors and (2) reduce observed directions or angles to average values and check triangle closures. Depending on the characteristics of the triangulation network, there are several other routine reductions which may be required. These are: (1) reduction of horizontal distances to sea level where base-line distances are reduced to a common sea-level datum; (2) *reduction to center* when eccentric triangulation stations have been occupied; and (3) computation of and correction for *spherical excess* where lines and triangles are of sufficient length and size so that the assumption of plane triangles is no longer valid.

Reduced directions or angles and corrected and reduced distances are then used to compute azimuths and lengths for each unknown side in the triangulation network. Next, these values are employed to determine the X and Y plane coordinates (preferably in the state plane system; see Chap. 11) for each triangulation station. In order to do this, coordinates for at least two stations must be known beforehand. This overall procedure is usually called *position computation*. When position computations are performed manually, care should be exercised always to calculate the coordinates of an unknown station with two sets of independent angles in order to guard against mistakes in the computations.

The X and Y plane coordinates, so determined, and the reduced measured data are then used as approximations in a rigorous, simultaneous least-squares adjustment of the triangulation. Depending on the technique of least-squares adjustment employed, the output will consist of: (1) adjusted angles or directions; or (2) adjusted angles or directions and adjusted X and Y coordinates for all unknown triangulation stations. If the adjustment is by the first technique, azimuths are recalculated with adjusted angles or directions and then the position computations are repeated to give adjusted X and Y coordinates for each triangulation station. When the second technique is used, adjusted angles or directions and X and Y coordinates for unknown stations are obtained directly from the least-squares adjustment.

Both techniques yield statistics useful in evaluating the quality of the measurements and adjusted positions of the stations and both will produce identical results.

The theory and details involved in the reduction to center, reduction to sea level, spherical excess, and the least-squares adjustment of triangulation are beyond the scope of this textbook. The readers interested in pursuing these aspects should consult Ref. 16 at the end of this chapter.

Approximate methods for adjustment of triangulation have been developed and are fully documented. Even for small problems, these methods require a substantial effort to formulate and to solve. Since these methods are not rigorous, the answers are not the best possible values, and no rigorous statistics for evaluation of the quality of observations and adjusted positions are available. Consequently, the use of these approximate methods is not recommended for the adjustment of triangulation networks.

9.8. Trilateration A trilateration network consists of a system of joined or overlapping triangles in which all lengths are measured and only enough angles or directions are observed to establish azimuth. The continuing development and refinement of EDM equipment has made trilateration feasible and, given suitable geometry, this procedure is now competitive with triangulation for establishing horizontal control.

Theoretically, a pentagon or hexagon would be the ideal figure for trilateration. However, from a practical standpoint, establishing even a single pentagon or hexagon in the field would be an almost impossible task.

The U.S. National Ocean Survey recommends use of the traditional braced quadrilateral, a center-point quadrilateral with an additional measurement (Fig. 9.6*a*), or a center-point pentagon with an additional measurement (Fig. 9.6*b*). They further recommend that angles be measured wherever possible to supplement the trilateration, thus increasing the number of conditions.

The procedure for establishing a trilateration network is similar to that required for triangulation. All available data, such as existing maps, references to existing control, and aerial photographs, are assembled for the area in question and preliminary locations for trilateration stations are chosen in the office. A field reconnaissance follows during which the intervisibility of points, the accessibility of stations, and strength of the figure chosen are verified.

The specifications in Table 9.2 under "Trilateration" should be followed closely. EDM equipment should be chosen that will yield required standard deviations in length and no triangle within the figure should contain an angle less than the specified minimum angles (15 to 25°), particularly for higher-order surveys. Because few, if any, angles will be observed, few signals must be maintained. The construction of towers or stands for instruments is generally unnecessary. Reflectors or antenna units for EDM equipment can be extended up to 13 m above the station without serious effects on the precision of the measurements. In this way distances of up to 20 km can be measured without the need for towers. This represents a real advantage for trilateration, since towers are costly.

Distances measured by EDM equipment must be corrected for all systematic instrumental errors and for the effects of atmospheric conditions. Slope distances must be reduced to horizontal sea-level distances (Arts. 4.37 and 9.7). Vertical angles from both ends of the lines or elevations at the ends of the lines must be obtained.

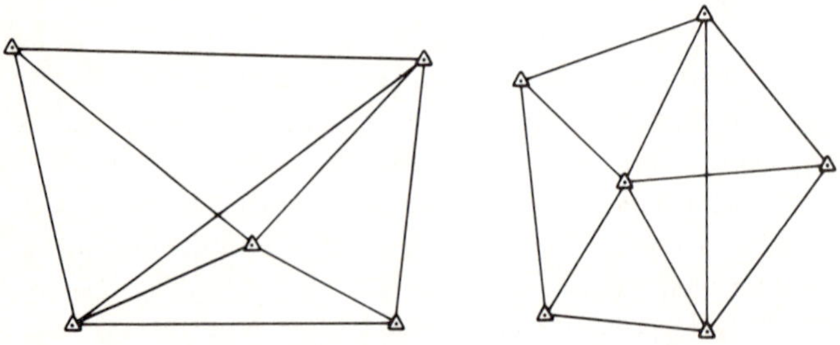

(*a*) (*b*)

Fig. 9.6 (*a*) Center-point quadrilateral with brace. (*b*) Center-point pentagon with additional distance.

Where the lines are long for an extensive network, elevations can be obtained with sufficient accuracy for this purpose by interpolating from the contours on a topographic map or by use of barometric leveling. Where a relatively small network containing lines of only a few kilometres in length is being surveyed, vertical angles and trigonometric levels (Art. 5.5) are usually employed for slope corrections and reduction to sea level.

9.9. Adjustment of trilateration The observed values in a pure trilateration network are the measured distances. There must be enough measured distances to provide redundant observations. With redundant observed distances, the best procedure for adjustment is the method of least squares. Output from the adjustment consists of adjusted *X* and *Y* coordinates for each station of unknown position in the network, adjusted distances, and statistics useful in evaluating accuracies of the coordinates for each adjusted station in the network. For further information, the reader should consult Ref. 16 at the end of this chapter.

9.10. Combined triangulation and trilateration Under circumstances that prohibit achieving the necessary strength of figure by pure trilateration or using triangulation alone, it is quite feasible to perform a combination of triangulation and trilateration. This procedure can extend from the addition of a few angles or distances to a trilateration or triangulation network, respectively, to the observation of all angles and all distances in the net. Such an operation can be called a combined control survey and the subsequent adjustment is designated as a combined adjustment.

From an operational standpoint in the field, combined control surveys are certainly practical. This is particularly true for third-order control surveys of smaller extent where electronic tacheometers or surveying systems (Art. 9.14) permit observation of distance and direction from a station using a single instrument.

Adjustment of this combined system is accomplished by combining the triangulation with the trilateration adjustment and then solving the entire system in a single rigorous least-squares adjustment.

Details and worked examples concerning the formulation and calculations involved in trilateration and combined triangulation and trilateration and their respective adjustments can be found in Ref. 16.

9.11. Location of points by intersection When coordinates of a point are given and the azimuth and distance to a second point are also known, it is possible to compute the coordinates of the second point (Art. 8.15). Similarly, if the coordinates are given for the two ends of a line and directions are observed from each end of this line to a third point not on the line, then coordinates of that third point can be calculated. This procedure is called *location by intersection*.

In Fig. 9.7, *B* and *D* are points of known coordinates from which angles α and β have been observed so as to locate point *C* by intersection. The distance *c* and

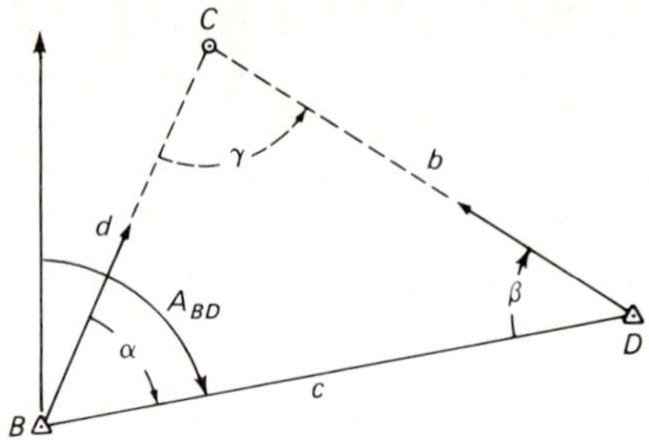

Fig. 9.7 Location of a point by intersection.

azimuth A_{BD} from B to D can be found using Eqs. (8.8) and (8.9):

$$c = \left[(X_D - X_B)^2 + (Y_D - Y_B)^2\right]^{1/2}$$

$$\tan A_{BD} = \frac{X_D - X_B}{Y_D - Y_B}$$

Also in triangle DBC, $\gamma = 180° - (\alpha + \beta)$, so that

$$d = \frac{c \sin \beta}{\sin \gamma} \qquad \text{and} \qquad b = \frac{c \sin \alpha}{\sin \gamma} \tag{9.2}$$

where d is the distance B to C and b the distance D to C. The azimuths of lines BC and DC can be calculated and the coordinates for C may be computed by Eqs. (8.6) and (8.7) using data from line BC as a check on calculations with data from line DC.

Several other methods for solving the intersection problem can be found in the references at the end of this chapter and in other surveying textbooks.

9.12. Resection When angles between lines to three points of known position are observed from a point of unknown position, the coordinates of the unknown point can be calculated. This procedure is called location by *resection*.

The resection problem is illustrated graphically in Fig. 9.8. In the figure, B and A are of known position and O is of unknown position. If the angle α is observed between OB and OA, the position of O is indeterminate because O can be anywhere on the circle circumscribing the triangle OAB. Additional information is needed to make the problem determinate. When the direction of OA or OB is known, the

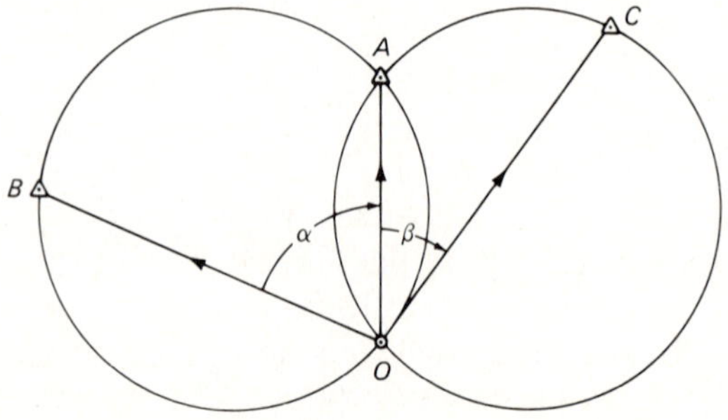

Fig. 9.8 Three-point resection.

problem is determinate. However, this information is usually not available. It is more convenient to observe angle β, subtended by AC, where C is a third known point. Then O, A, and C lie on the circle that circumscribes triangle OAC. Since O is on the circles that circumscribe both of the triangles OAB and OAC, it must lie at one of the two intersection points of the two circles. As A is one of these two intersections, O is uniquely determined. This solution presumes that the three known points A, B, C and O do not fall on the circumference of one circle.

Should the two circumscribing circles tend to merge into one circle, the problem will be less stable and finally becomes indeterminate again when the two circles coincide. Points should be selected in the field so as to avoid this situation. It can be shown that this condition is present when $\angle BAC + \alpha + \beta = 180°$.

Development of the calculations required to solve the resection problem is beyond the scope of this textbook. Numerous methods for solving the resection problem can be found in the references at the end of this chapter.

9.13. Modern positioning systems Position determinations on the surface of the earth have long been accomplished by a combination of directional and distance measurements as well as by astronomic observations. These procedures are presently being challenged by more advanced techniques that are based on modern systems and equipment. In this chapter, only an introductory coverage of the following four classes of systems will be given:

1. Electronic positioning systems
2. Inertial positioning systems
3. Doppler positioning systems
4. Global positioning system (GPS)

Each of the classes will be discussed in a separate section. The electronic positioning systems are an outgrowth of developments in regular surveying equipment where both directional and distance measuring capabilities are combined in one compact instrument. Inertial positioning systems draw on the extensive experience with navigational components used on board aircraft. Such systems are used either on helicopters or on ground vehicles. Finally, doppler and global positioning systems rely on signals received from artificial satellites.

9.14. Electronic positioning systems Electronic positioning systems consist of specially designed short- to medium-range electronic distance measuring (EDM) devices mounted on a theodolite framework which permits determination of horizontal and vertical angles. The net result is a single instrument that can be used to determine distances and directions from a single instrument setup. There are many such systems now available, some of which have minor differences and some that are different in fundamental ways. Generally speaking, electronic positioning systems may be classified into three groups:

1. Optical-reading repeating or direction theodolites with an attached but removable EDM device. These systems are called *combined theodolite and EDM*.

2. Optical- or digital-electronic-reading theodolites with an attached EDM device and built-in microprocessor. These systems are designated as *computerized theodolite and EDM*. Optional equipment includes a solid-state memory device for storing data.
3. *Integrated digitized electronic systems* consisting of a digitized theodolite, microprocessor, and EDM device incorporated into a single system. These systems are designated *electronic tacheometers* and can be equipped with a solid-state memory device to store data.

In addition to the systems listed above, there are also various devices equipped with lasers. Instruments of this type do not permit direct determination of distances or angles but are used primarily for alignment problems.

Figure 9.9 shows a combined theodolite and EDM system in which a specially designed light-weight EDM device is attached to an optical-reading theodolite of the type described in Art. 6.30. The system illustrated is also equipped with an interface which permits attaching a pocket-sized electronic calculator used to reduce and process data. The EDM device is a light-wave instrument that utilizes infrared light rays modulated by two or three frequencies (EDM instruments are described in Arts. 4.29 to 4.36). Slope distance between instrument and reflector is determined by phase comparison of the transmitted with the reflected light beam. In the system shown in Fig. 9.9, this measurement is made simply by turning the instrument on and then pressing a button. The slope distance is then registered on a liquid crystal

Fig. 9.9 Combined theodolite and EDM, Kern DKM 2-AT, DM 502, and interface DIF41 for HP 41C attached. (*Courtesy Wild Heerbrugg Instruments, Inc.*)

display eight seconds after the measurement was initiated. The slope distance is also automatically transmitted to the calculator where it is used with observed angles to determine horizontal distance and difference in elevation and to perform other routine survey computations. The EDM and interface for the calculator are removable so that the theodolite can be used independently if desired. Normally, the entire system is employed, with distance and direction being measured from a single setup.

Combined reflector and targets for use with these systems are shown in Fig. 9.10*a* and *b*. The operator views the reflector through the telescope and centers the cross hairs on the target. This operation also centers the EDM transmitter optics on the reflector for maximum return.

The computerized theodolite and EDM systems are essentially similar to the combined theodolite and EDM except for one important difference. A compact solid-state computer or microprocessor is built into the system. This microprocessor is used for data reduction. Figure 9.11*a* illustrates a system with a specially designed EDM unit attached to a theodolite having digitized horizontal and vertical circles. The EDM unit can be detached, permitting the digitized theodolite to be used independently. Another such system shown in Fig. 9.11*b* has an optical-reading horizontal circle, digitized vertical circle, and an integrated EDM unit with optics which are coaxial with the theodolite telescope. In both systems, measured slope distances and zenith angles are reduced automatically by the microprocessor to provide output of horizontal distances and differences in elevation which are shown on the liquid crystal display registers.

Data collectors or terminals can be connected to each system. These devices, which are small, hand-held units equipped with a keyboard for control, contain

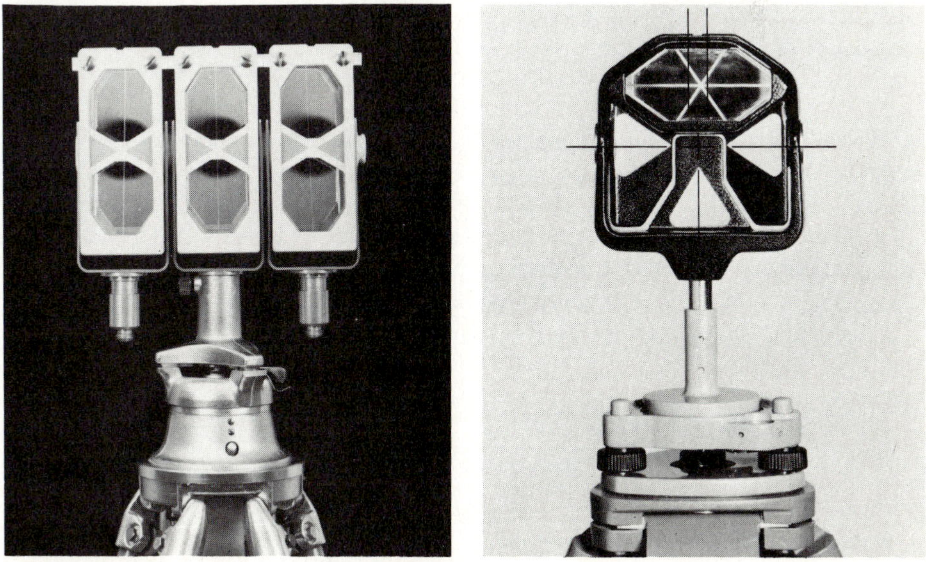

(*a*) (*b*)

Fig. 9.10 Reflectors/targets for combined theodolite/EDM systems. (*a*) Three reflectors coupled. (*Courtesy Kern Instruments, Inc.*) (*b*) Single reflector. (*Courtesy Wild Heerbrugg Instruments, Inc.*)

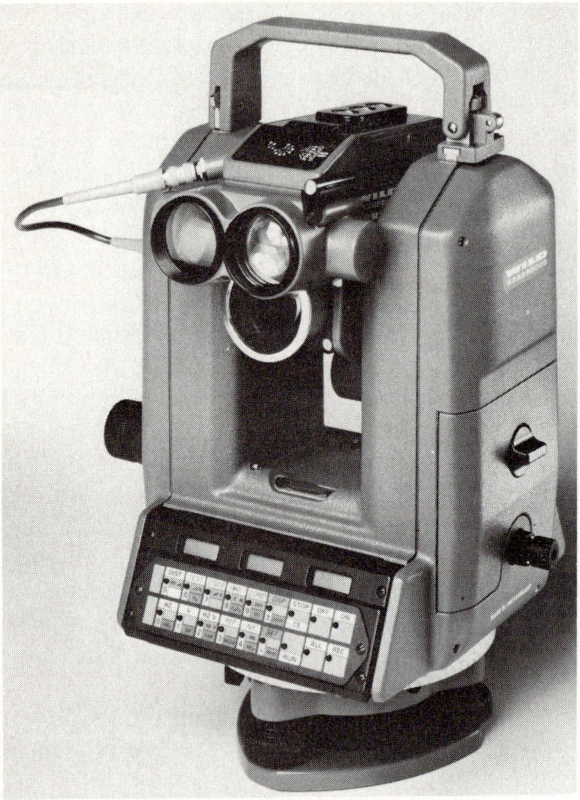

Fig. 9.11 Computerized theodolite and EDM systems. (*a*) Wild T 2000 with D-14. (*Courtesy Wild Heerbrugg Instruments, Inc.*)

solid-state or magnetic-bubble memories with from 400 to 4000 lines of storage where a line could consist of up to 10 words. Use of these data collectors permits storage in the field of substantial amounts of data that are processed later in the office. Thus, the data collector constitutes a form of electronic field book. In the office, an interface allows transferring data from the collector to an electronic computer for subsequent calculation, adjustment, and plotting.

Integrated digitized electronic surveying systems (electronic tacheometers) consist of the following components:

1. An electronic digitized theodolite
2. An EDM unit or distance meter
3. A microprocessor
4. Keyboard or control panel and display register
5. Data storage unit and/or field computer

The major differences between these systems and the computerized theodolite and EDM are that the electronic tacheometers have: digitized horizontal and vertical circles; a completely integrated EDM with an optical path that is coaxial with the

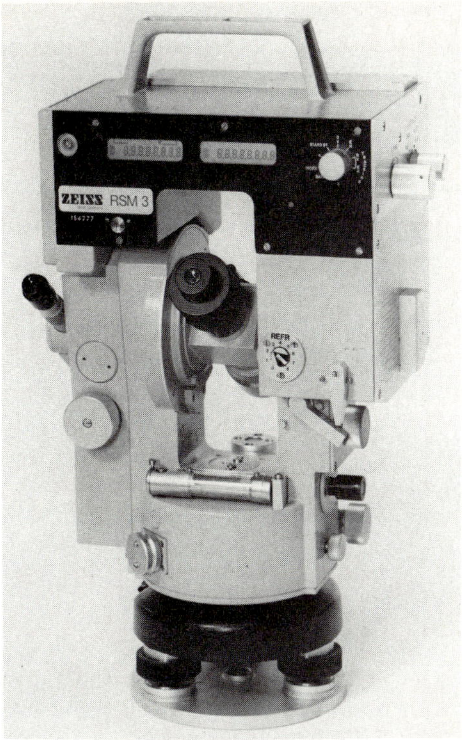

Fig. 9.11 (*b*) Zeiss RMS3. (*Courtesy Carl Zeiss, Inc.*)

theodolite telescope; and a more powerful built-in microprocessor which permits more extensive reduction of the measured data. Figure 9.12*b* illustrates such a system which contains a microcomputer with sufficient power and capacity to not only reduce angle and slope distances to horizontal distances and differences in elevation, but also to perform certain routine survey calculations such as coordinate computations, determination of position by resection, calculations for setting off distances, and collimation and index corrections. Also available for this system are a data collector of the type described previously and a data processing unit (Fig. 9.12*a*). This unit is used for additional calculations with computed or stored data and to transfer information from the data collector to a larger compatible electronic computer.

Under good atmospheric conditions, ranges in distances for all the systems fall within common boundaries and are from 1 to 1.5 km (3281 to 4921 ft) with one reflector and from 1.6 to 2.5 km (5249 to 8200 ft) with three reflectors. Accuracies in distances vary from 3 to 10 mm (0.010 to 0.033 ft) + 2 to 5 ppm. Angular accuracies in horizontal and vertical angles vary from 1″ to 3″ of arc. All systems contain a *tracking mode* in which an updated distance is displayed in real time at intervals of from 1 to 4 seconds. This last feature is especially useful for construction layout operations.

As mentioned previously, the data collector unit available for computerized theodolites and electronic tacheometers constitutes an electronic field notebook with

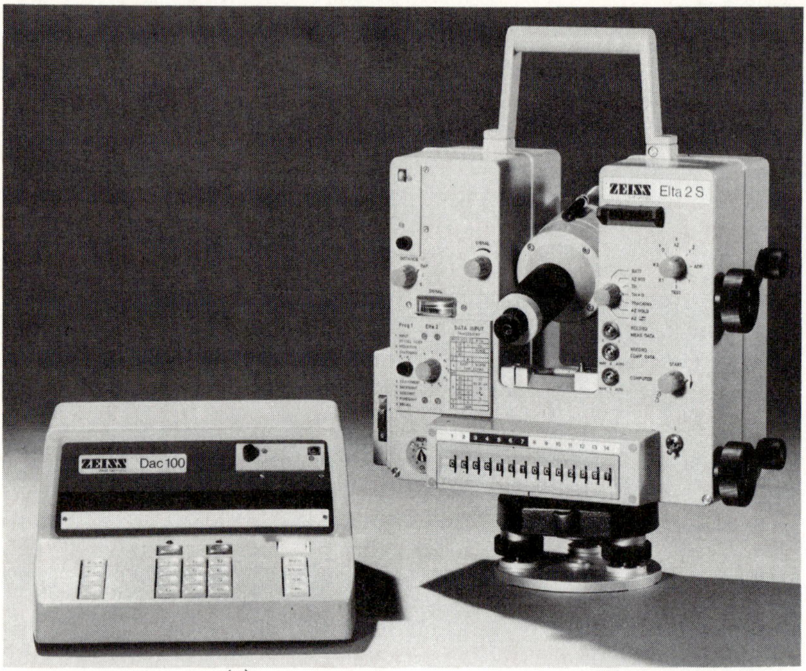

(*a*) (*b*)

Fig. 9.12 Integrated electronic surveying system. (*a*) Data processing unit. (*b*) Electronic tacheometer. (*Courtesy Carl Zeiss, Inc.*)

numbered lines. Each line contains an identification number (assigned by the computer) and then several items of measured or calculated data. Each brand of storage unit has unique formatting potential well beyond this brief description. However as an example, after the computer-assigned number a line might contain: (1) slope distance, azimuth, and zenith angle; or (2) horizontal distance, azimuth, difference in elevation, and point number; or (3) the *X* coordinate, *Y* coordinate, elevation, and point number; and so forth. The final record of these data is a hard-copy listing made either in the field with a portable printing calculator designed for the purpose or in the office after the data have been read onto magnetic tape or into an electronic computer, which, in turn, is connected to a printer. The *data storage unit should not be considered a substitute* for the standard field book, only as a supplement. Conventional field notes still serve to provide a record of descriptive information such as sketches, references for traverse stations, and other identification pertinent to the job being performed.

The systems described in this section can be applied to practically any surveying task discussed in this text. Naturally, due regard must be given to the precision of the system versus the specified accuracies required for the application. Also, the procedures as detailed for standard surveying equipment may need modifications so as to exploit the features built into these systems. In general, there are applications for all of these systems in control surveys by traverse (Chap. 8) and by combined triangulation/trilateration (Art. 9.10); topographic surveying (Chap. 12), and construction layout (Chap. 14). It should be noted that all of these systems provide data that are very adaptable to forming digital terrain models (DTM). In particular, the computerized theodolites and electronic tacheometers automatically provide data that are directly convertible to a DTM by connecting the data storage unit to a compatible electronic computing system.

9.15. Inertial positioning systems The inertial positioning system is described by its promoters as fulfilling the surveyor's dream of having a "black box" that will instantaneously provide position when set down on any point of the earth's surface. It requires no line-of-sight clearance, no towers, and is unaffected by refraction and other atmospheric conditions. In fact, it can be operated day or night during hot or cold weather, rain or shine, and with the speed of a motor vehicle or a helicopter. All of these apparent advantages should not mislead the student, as it must be realized that an inertial system is a product of the very highly sophisticated advanced aerospace technology. It requires well-trained personnel to operate it, and more important, to service it. Furthermore, to obtain results with the potential high precision of the system, very careful planning and execution of the survey project according to well-defined guidelines is necessary.

In simple terms this type of surveying system is basically an inertial platform with precise *gyroscopes* that keep its three mutually orthogonal axes properly oriented in space. Each axis is controlled by a gyroscope.

The essence of an inertial survey system is depicted in Fig. 9.13, where three gyroscopes are used to orient the platform in the east-west, north-south, and up-down directions. Each axis also contains a sensitive accelerometer which measures the acceleration component along that axis. An on-line computer monitors these three components of acceleration by sampling the acceleration along each axis every 17 ms (0.017 s). Since acceleration is the second derivative of distance traveled with respect to time, then by integrating the measured acceleration values twice with respect to time, the three components of the distance traveled are obtained. When these acceleration components (corrected to compensate for earth curvature) are

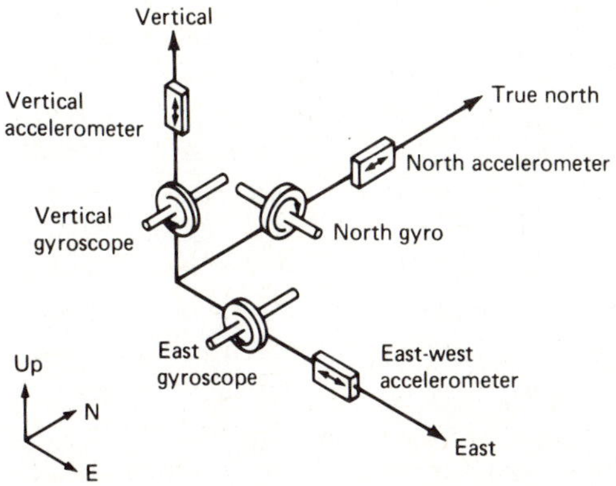

North Alignment

Horizontal accelerometers level the platform
Gyroscopes sense earth rate
Electronics rotates axis to north sensed by gyroscopes

Survey

Accelerometers measure acceleration in each coordinate axis
Electronics and computer integrates acceleration into distance traveled
Distance traveled in each axis is resolved by computer into latitude, longitude, and elevation

Fig. 9.13 Inertial survey system. (*Courtesy Litton Guidance and Control Systems.*)

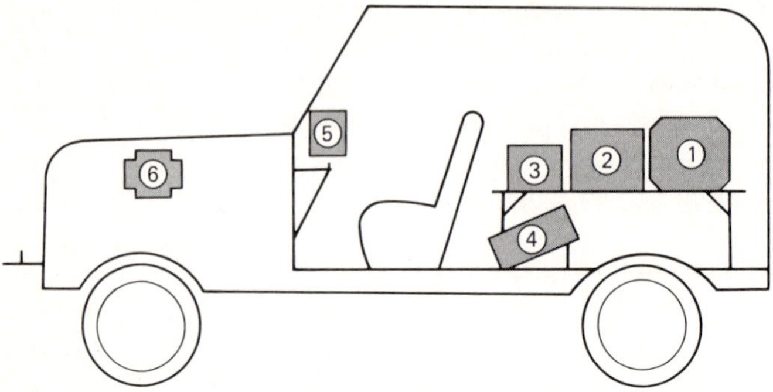

Fig. 9.14 Schematic of inertial surveying system: (1) inertial measuring unit; (2) power supply; (3) computer; (4) cassette recorder; (5) display and control; (6) power source. (*Courtesy L. F. Gregerson, Ref. 8.*)

processed in the on-line computer, determination of spatial positions is possible as the inertial platform is moved from point to point.

The inertial system may be mounted in either a motor vehicle or a helicopter. At the time of this writing (1983), three systems, Honeywell, Ferranti, and the earliest Litton system, are already in use. The Litton system's vehicle-mounted version is schematically depicted in Fig. 9.14; a photograph from the back of the vehicle is shown in Fig. 9.15.

In operation, the platform is first aligned over an initial point of known position and values for latitude, longitude, and elevation of the point are input into the

Fig. 9.15 Inertial system mounted in the back of a truck. (*Courtesy SPAN International, Inc.*)

computer. Calibration and alignment at this initial point usually require about an hour. Then the vehicle containing the platform is moved at a fairly constant speed (30 to 50 km/h or about 18 to 30 mi/h) along the survey route, making periodic stops at 3- to 5-min intervals for calibration purposes (called *zero-velocity-update* or *ZUPT stops*). At each unknown point where a position is to be found, a ZUPT is performed and the system determined latitude, longitude, and elevation are stored in the computer memory for subsequent processing. When the final control point of known spatial position is reached, its previously known coordinates are entered into the on-line computer and an appropriate smoothing routine may be made to eliminate large errors. A subsequent final adjustment can then be performed off-line with a minicomputer. In an alternative approach, the raw data are also adjusted or smoothed off-line so as to free the on-line computer for other more essential tasks.

Inertial survey systems have been introduced only quite recently and hence will continue to undergo modifications and improvements for several years to come. Therefore, the following accuracy data, adopted from Ref. 22, are applicable only at the time of that reference (1977).

Mean errors for any intermediate point in a traverse which runs in both directions between two control points are as follows:

In a vehicle, 60 km (37 mi) maximum length, flat gravity field:

Latitude and longitude	20 cm
Elevation	15 cm

In mountainous regions, or on longer distances, some deterioration is to be expected.

In a helicopter, 160 km (100 mi) maximum, where the terrain is flat:

Latitude and longitude	60 cm
Elevation	50 cm

It should be emphasized that these numbers are the result of experience gained up to 1977. If the student is interested in a more thorough analysis of errors in inertial survey systems, Refs. 23 and 38 should be consulted.

Other factors that are important if maximum accuracy of inertial systems is to be achieved include the following:

1. *Accuracy of basic control* The positional accuracy can be no better than that of the basic control to which the inertial survey is tied. The highest-order horizontal and vertical control should be used.
2. *Rate of change of latitude and longitude* Best results are obtained when the rate of change of latitude and longitude are linear with respect to time. Thus, there should be as little meandering as possible between control points, and a uniform rate of travel is necessary. Careful advance planning of the survey is needed to satisfy this requirement, which also results in minimum travel time between points.
3. *Double-run traverses* Control traverses should be run in both directions between known terminal control points. The results from the two determinations can be averaged.
4. *Gravity anomalies* Avoid large gravity anomalies when possible. If large gravity gradients are anticipated, zero-velocity-update periods should be taken at shorter intervals—$1\frac{1}{2}$ to 2 min.

5. *Smooth travel* Travel should be performed in as smooth a manner as possible. Avoid holes in pavements, rough pavement, sudden stops, needless changes in direction, and abrupt accelerations.

Applications for inertial positioning systems consist of those projects where it is most economical. These may be categorized as (1) large projects either in total area or number of points; (2) surveys in remote areas where support is an important factor; (3) surveys where weather, terrain, traffic, environmental restrictions, or the client's desire to remain unnoticed may place restraints on survey personnel and operations; and (4) projects with limited time for completion due to weather, economics, or contract restrictions.

9.16. Doppler positioning systems Both doppler and global positioning systems utilize high-frequency radio signals broadcast from orbiting satellites and detected by specially designed terrestrial receivers.

The concept underlying the doppler system is the *doppler phenomenon*. The classical example usually cited for this phenomenon is the continuously changing pitch of a train whistle as it approaches, travels past, then travels away from a stationary observer. The same principle applies to a continuous tone or frequency signal transmitted from a satellite as it passes over an observer on the earth's surface. As the satellite gets closer to the observer, the received frequency continuously changes. When these frequency changes are properly monitored by the observer, they provide an indirect measure of the changes in distance between the satellite and the observer. If the position of the satellite is known with sufficient accuracy, and if a sufficiently large number of satellite passes is observed, the location of the observer can be determined with a high level of accuracy.

To determine position accurately, several passes of different satellites are needed. Prior to 1967, the U.S. Navy Navigational Satellite System was used only for military applications by providing accurate position determination for marine vessels. In 1967, the system was made available to nonmilitary users. At the present time there are six satellites in the system, with circular orbits having an altitude of 600 nautical miles (about 1075 km) above the earth. The period of revolution for each satellite is about 107 min. With this configuration, a location near the equator can observe only from 8 to 12 passes, while 40 to 50 passes can be observed from the arctic region. Each satellite continuously transmits the following:

1. Two stable frequencies, one 150 mHz and the other 400 mHz
2. Precise timing signals
3. Orbital parameters, which are called the *Broadcast Ephemeris*, every 2 min predicting the location of the satellite

An observer needs the following to make use of the foregoing transmitted data:

1. An antenna capable of receiving the signals transmitted by the satellite
2. A receiver to detect, amplify, and decode the signal
3. A medium suitable for recording the observational data for subsequent processing and analysis
4. Recording of temperature, humidity, and pressure at the observer's location during each pass in order to apply corrections for systematic errors due to these factors

Although navigational receivers can be used for doppler surveying, several commercial firms have designed special receivers for the purpose. Such a system is shown in Fig. 9.16. This system is designed to accept both transmitted frequencies, thus allowing for the correction for ionospheric refraction, which is a significant systematic error. The system is essentially composed of an antenna, usually collapsible, receiver, recording medium such as paper or magnetic tape, and a rugged carrying case. The receiver derives its time signal from a very precise internal clock instead of using the satellite data, as in the case of navigational receivers. This allows a considerable reduction in the number of observable passes for a given accuracy requirement.

Accuracies in position of 2 to 3 m (root mean square errors, see Chap. 2 and App. A) can be obtained using one doppler receiver at an individual station and the Broadcast Ephemeris. In this procedure, called *point positioning*, about 100 satellite passes must be measured. Improved internal relative accuracy is possible by employing two or more receivers (one in a fixed position) in a method referred to as *translocation*. Root mean square errors in position of less than 1 m are achieved

Fig. 9.16 Doppler receiving system. (*Courtesy JMR Instruments, Inc.*)

when the so-called *short-arc geodetic adjustment* is utilized. In this last procedure, an optimum of four receivers are used and an off-line least-squares adjustment of all parameters (including orbital elements) is necessary.

The results obtained by all these methods can be improved by using the *Precise Ephemeris* derived by the U.S. Defense Mapping Agency (DMA) from tracking data. However, the Precise Ephemeris is available for only two satellites and must be acquired from DMA, a process which takes about 6 months. Consequently, in most practical applications, the Broadcast Ephemeris is employed.

In view of the excellent results obtainable when using doppler systems, some of the complications that can arise should also be mentioned. Since many of the most obvious applications are located in remote inaccessible areas, helicopters and camps may be required. The doppler systems, which are relatively expensive, should *not* be left unattended. Thus, personnel, transport, and good communications are necessary. If only one set is used, the observation time necessary to obtain a good position may extend to a week or 10 days. When a single doppler set is employed with the Precise Ephemeris (to ensure better results), data must be sent to the U.S. Defense Mapping Agency for computation. This process takes 6 months. The other option is to use translocation or short-arc techniques for which a minimum of two doppler sets is needed and four sets are recommended for optimum accuracy and efficiency. Thus, the initial investment is more than doubled or quadrupled, since computer programs to reduce the data must be purchased or developed in house.

Assuming that the finances and work force can be justified for a project, doppler positioning can be used to establish control for (1) surveys to establish right of way, (2) property surveys, and (3) primary and secondary control networks.

Frequently, doppler positioning may best be utilized as a supplement rather than as a substitute for traditional and photogrammetric surveying procedures. For example, in a topographic survey of a large area containing difficult terrain, doppler positioning can be used to establish control at widely separated points. Photogrammetric methods supplemented by ground surveys could then be utilized to complete the control network and perform topographic mapping. Similar examples can be cited for virtually every type of survey.

9.17. Global positioning system In an extension of satellite geodesy, the U.S. Department of Defense (DoD) is currently developing a *global positioning system* that will eventually include a minimum of eighteen satellites placed in three orbital planes each containing six satellites. Five of these satellites are already in orbit. The entire system is supposed to be operational in 1987.

The satellites in this system will transmit two frequencies including a *clear acquisition* signal and a *precise coded* signal. In order to exploit the full potential of the system, a receiver with the capability of detecting and decoding from three to four satellite signals sequentially or simultaneously will be required. For security reasons, DoD will probably deny use of the precise coded signal to the civil community, resulting in substantial degradation of the system's positioning capability (i.e., doppler satellite techniques would be better). However, it is expected that relative accuracies of ± 2 *cm* will be obtainable in 1 to 2 h of observation time using receivers now being developed and tested. These receivers are designed to exploit *radio interferometry* and can operate using the clear acquisition signal. They are light (less than 45 kg), portable, and fairly simple to operate.

Figure 9.17 illustrates the antenna and electronics unit for such a system under operating conditions. The antenna, which contains the radio receiving electronics in the thin table-shaped unit (see Fig. 9.17), is mounted on a tripod and can be precisely centered over a survey point. In Fig. 9.17, the operator is measuring the height of instrument above the survey point over which the antenna is centered. The electronics unit containing power supplies, quartz crystal oscillator, clock, and microcomputer is mounted in the back of the jeep (Fig. 9.17). A minimum of two such systems is required to determine the relative position of an unknown point with respect to a survey point of known location. The antenna of one system is centered over the survey point of known position while that of the second system is placed over the point for which a position is desired. Satellite orbital data and approximate coordinates for the point(s) to be located must then be loaded into the computer memory of each system. Both systems are programmed to become active when the GPS satellites are within range so that signals emitted by the satellites are received simultaneously at both points. The system works by measuring phase difference between microwave signals at the two survey points. The multiple signals received simultaneously from satellites at different positions are used to resolve ambiguities and enables determining the relative three-dimensional position of the unknown point with respect to the known point. This measuring process takes from $1\frac{1}{2}$ to 2 h to complete. The system is operational under all kinds of weather conditions and the

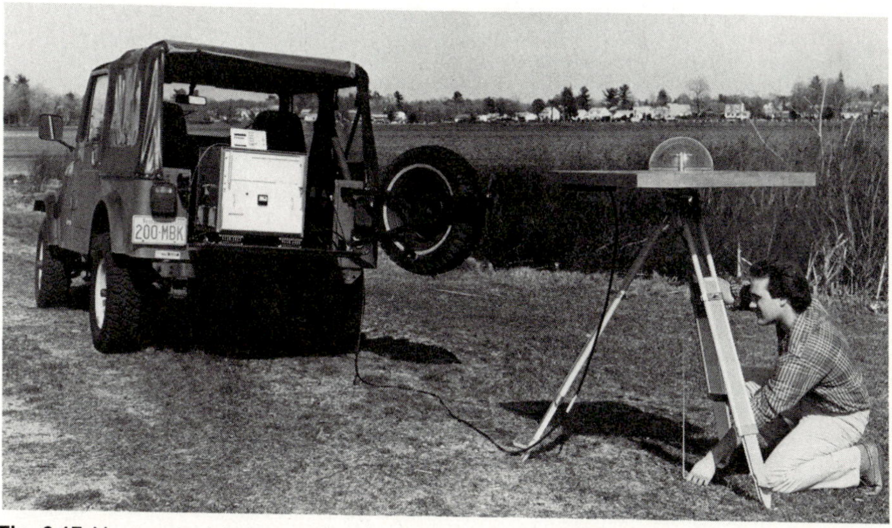

Fig. 9.17 Macrometer interferometric unit and antenna. (*Courtesy Macrometrics, Inc., Woburn, Massachusetts.*) (*Note:* Macrometer is a trademark, registration pending, of Macrometrics, Inc.)

two points occupied need not be intervisible. All measurement data are stored in the solid-state memory of the microcomputer contained in the electronics unit. Processing the data to yield three-dimensional positions is performed later at a field or office station using a desktop microcomputer with associated hardware and necessary programs.

The equipment shown in Fig. 9.17 has been tested on lines of from 124 m to about 60 km in length. Advertised relative accuracies with the available units are said to be from 1 part in 100,000 to 1 part in 200,000.

Technological developments in the field of satellite geodesy are extremely rapid and dynamic. At this time, the effects of these systems on the field of surveying are not clear. It is obvious that the impact of the GPS system on geodesy and control surveying will be substantial. Some operational receivers compatible with the GPS system will be operational in 1983 or 1984. Readers interested in further details concerning these systems should consult Refs. 6, 7, 13, 14, and 15 at the end of this chapter.

9.18. Problems

9.1 Describe: (*a*) triangulation system, (*b*) trilateration system and cite examples of the application of these types of surveys to practical problems.

9.2 Using Table 9.1, give relative accuracies between bench marks used for: (*a*) construction of a 50-mi section of the Interstate highway system; (*b*) a small subdivision consisting of fifty $\frac{1}{4}$-acre lots; (*c*) the vertical control network to be established for detecting subsidence along a fault line; and (*d*) supplementing the National Network of Vertical Control.

9.3 A county ordinance states that subdivision monuments must have coordinates in the state plane system (see Art. 11.16). What order and class of survey should govern the control survey used to establish these monuments?

9.4 Describe several geometric figures used to extend triangulation, drawing sketches of each type. What figures are best adapted to precise work?

9.5 Describe the process of reconnaissance for triangulation.

9.6 What are the basic measurements in triangulation and how are they obtained?

9.7 Eight positions of directions in a quadrilateral are to be observed with a theodolite having two verniers and a micrometer drum range of 10′. Compute the nominal circle readings for each of the eight positions.

9.8 A base line is to be measured by EDM in a network done to first-order class I specifications. What steps should be taken to ensure satisfaction of these specifications?

9.9 What initial reductions should be performed with the basic measurements of triangulation before positions are computed?

9.10 Describe the procedure for position computation in triangulation.

9.11 What is the major objective of triangulation computations?

9.12 Describe the differences between trilateration and triangulation. What is the best type of geometric figure for trilateration?

9.13 What are the basic measurements and what types of equipment are used to obtain these measurements for trilateration?

9.14 Most triangulation and trilateration networks contain extra or redundant measurements. What is the best type of adjustment for these networks? Justify your answer.

9.15 Under what conditions would combined triangulation and trilateration be necessary? What types of equipment are best suited for this type of survey?

9.16 The coordinates for stations *C* and *D* are as follows:

Station	*X*, m	*Y*, m
C	323,484.123	124,231.305
D	323,888.059	124,832.170

Horizontal angles measured with a 01″ theodolite from stations *C* and *D* to *B* are angle $BCD = 65°37'24.8''$ and angle $CDB = 80°49'41.9''$. Station *B* lies north of *C* and *D*. Compute the coordinates for station *B*.

9.17 Station *C* is located by intersection from triangulation points *B* and *A* for which the coordinates are $X_A = 1{,}421{,}231.304$ ft and $Y_A = 521{,}304.009$ ft. The distance and azimuth from the north of the line from *A* to *B* are 897.859 ft and 235°20′32″, respectively. The measured horizontal angles are: angle $BAC = 80°27'35.8''$ and angle $CBA = 54°14'37.8''$. Compute the coordinates of point *C* by intersection. Point *C* is north of *A* and *B*.

9.18 Describe the procedure for determining distance and direction using a combined theodolite and EDM.

9.19 "An electronic tacheometer may be considered a complete surveying system." Discuss this statement, describing the various components of such a system.

9.20 Explain the various applications of electronic positioning systems in which special features of such systems are exploited.

9.21 Describe the basic idea of an inertial positioning system and its principal components.

9.22 "With the invention of inertial survey systems, classical survey methods will soon be obsolete." Discuss whether this statement is likely to be true or false, expressing your feelings for the role of such systems and their effects on the field of surveying.

9.23 Describe in detail the operational steps involved in a survey with a vehicle-mounted inertial survey system.

9.24 Discuss concisely four areas of application for inertial survey systems.

9.25 What is the doppler phenomenon and how is it exploited in satellite doppler position systems?

9.26 Enumerate the factors that affect the accuracy of satellite doppler surveying.

9.27 Describe the two methods for determining location by doppler surveying: point positioning and translocation.

9.28 Discuss how satellite doppler positioning, inertial survey systems, and electronic survey systems could be utilized to perform the control and topographic surveys required for the design of a dam in a remote region where the terrain is obscured by rain forest.

References

1. Allan, A. L., "A Proof of the Barycentric Resection Formulae," *Survey Review*, July 1975.
2. Allman, J. S., and Hoar, G. J., "Optimization of Geodetic Networks," *Survey Review*, January 1973.
3. Atia, K. A., "The Adjustment of Trilateration Networks Observed in Paris," *Survey Review*, July 1976.
4. Ball, W. E., "Testing an Airborne Inertial Survey System for B.L.M. Cadastral Survey Application in Alaska," *Proceedings, Annual Meeting, ACSM*, 1975.
5. Ball, W. E., "Adjustment of Inertial Survey System Errors," *Proceedings, Annual Meeting, ACSM*, 1978.
6. Bossler, J. D., "A Note on Global Positioning System Activities," *Bulletin American Congress on Surveying and Mapping*, No. 74, August 1981.
7. Bossler, J. D., Goad, C. C., and Bender, P. L., "Using the Global Positioning System for Geodetic Positioning," *Bulletin Géodésique*, No. 54, 1980, pp. 553–563.
8. Brown, Duane C., *A Primer on Satellite Doppler Surveying*, DBA Systems Technical Note 75-002, April 1975.

9. Brown, D. C., *A Test of Short Arc versus Independent Point Positioning Using Broadcast Ephemeris of Navy Navigational Satellites*, DBA Systems Technical Note 75-001, 1976.

10. Brown, D. C., "Doppler Positioning by the Short Arc Method," Presented to the International Geodetic Symposium, Physical Science Laboratory, New Mexico State University, Las Cruces, N.M., October 12–14, 1976.

11. Burke, K. F., "Why Compare Triangulation and Trilateration?" *Proceedings, Annual Meeting, ACSM*, Washington, D.C., March 1971, p. 244.

12. Chrzanowski, A., and Steeves, P., "Control Networks with Wall Monumentation, A Basis for Integrated Survey Systems in Urban Areas," *The Canadian Surveyor*, Vol. 31, No. 3, September 1977, p. 211.

13. Counselman III, C. C., and Shapiro, I. I., "Miniature Interferometric Terminals for Earth Surveying," *Bulletin Gèodèsique*, No. 53, 1979, pp. 139–163.

14. Counselman III, C. C., "The MacrometerTH Interferometric Surveyor," *Proceedings International Symposium on Land Information at the Local Level*, University of Maine at Orono, 1982, pp. 233–241.

15. Counselman III, C. C., "The First Genuinely Mites: The Macrometer Interferometric Surveyor," Report to the *GTCS Bulletin*, February 11, 1983.

16. Davis, R. E., Foote, F. S., Anderson, J. M., and Mikhail, E. M., *Surveying: Theory and Practice*, 6th ed., McGraw-Hill Book Company, New York, 1980.

17. Dracup, J. F., "Tests for Evaluating Trilateration Surveys," *Proceedings, Fall Convention, ACSM*, Seattle, Wash., 1976, p. 96.

18. Erickson, Kent E., "Electronic Surveying Systems," *Proceedings ACSM, ASP / ACSM Fall Convention*, Little Rock, Ark., October 1977.

19. Gergen, J. G., "The New Adjustment of the North American Datum," *Bulletin American Congress on Surveying and Mapping*, No. 79, February 1982.

20. Gort, Alfred F., "The Hewlett Packard 3829A Electronic Total Station," *Proceedings ACSM, ASP / ACSM Fall Convention*, Little Rock, Ark., October 1977.

21. Gregerson, L. F., and Carriere, R. J., "Inertial Surveying System Experiments in Canada," *Proceedings, IUGG*, Grenoble, France, August 1975.

22. Gregerson, L. F., "A Description of Inertial Technology Applied for Geodesy," *70th Annual Convention of CIS*, May 1977.

23. Hadfield, M. J., "Field Test Results on the GEOSPIN ESG Inertial System," *A.S.C.E. Specialty Conference on Engineering Applications of Space-Age Technology*, Nashville, Tenn., June 16–19, 1982.

24. Hothem, L. D., Strange, W. E., and White, M., "Doppler Satellite Surveying System," *Journal of the Surveying and Mapping Division, ASCE*, No. SU1, Proceedings Paper 14132, November 1978.

25. Klinkenberger, H., "Coordinates Systems and the Three-Point Problem," *The Canadian Surveyor*, July 1953.

26. Krakiwsky, E. J., Wells, D. E., and Thompson, D. B., "Geodetic Control from Doppler Satellite Observations for Lines Under 200 Km," *The Canadian Surveyor*, Vol. 32, No. 2, June 1972.

27. Laurila, S., *Electronic Surveying and Navigation*, John Wiley & Sons, Inc., New York, 1976.

28. Mancini, A., and Huddle, J., "Gravimetric and Position Determination Using Land Based Inertial System," *Proceedings, Annual Convention, ACSM*, 1975.

29. Mattson, D. F., "Determination of Position with Intersecting Circles," *Surveying and Mapping*, September 1973.

30. O'Brien, L. S., "Investigations of Inertial Survey Systems by the Geodetic Survey of Canada," *The Canadian Surveyor*, Vol. 30, No. 5, December 1976, p. 385.

31. Pinch, M. C., "An Evaluation of the Accuracy of an Autotape Survey," *The Canadian Surveyor*, March 1971.

32. *Proceedings of the International Symposium on Problems Related to the Redefinition of the North American Geodetic Networks*, University of New Brunswick, Fredericton, New Brunswick, Canada, *The Canadian Surveyor*, Vol. 28, No. 5, December 1974.

33. Rainsford, H. F., *Survey Adjustments and Least Squares*, Constable and Company Ltd., London, 1957.

34. Rawlinson, C., "Automatic Angle Measurement in the AGA 700 Geodimeter," *Survey Review*, April 1976.

35. Reynolds, W. F., *Manual of Triangulation Computation and Adjustment*, Spec. Publ. No. 138, U.S. Department of Commerce, Coast and Geodetic Survey (now National Ocean Survey, NOAA), Washington, D.C., 1965.

36. Richardus, P., *Project Surveying*, John Wiley & Sons, Inc., New York, 1966.

37. Schellens, Dieter F., "Surveying with the Zeiss REG ELTA 14 in Densely Populated Areas," *Proceedings, Annual Meeting, ACSM*, 1973, p. 451.

38. Todd, M. S., "Rapid Geodetic Survey System (RGSS) White Sands Tests for Position, Height and Anomalous Gravity Vector Components," U.S.A. ETL, Ft. Belvoir, Va. 22060, 1980.

39. Vincenty, T., "Desk Top Computers: Resection," *Survey Review*, April 1973.

40. Vincenty, T., "Length Ratios and Scale Unknowns in Trilateration," *Surveying and Mapping*, September 1975.

41. Wolf, P. R., and Johnson, S. D., "Trilateration with Short Range EDM Equipment and Comparison with Triangulation," *Surveying and Mapping*, Vol. 34, No. 4, December 1974, pp. 337–346.

PART IV
MAPPING

CHAPTER 10
Mapping and map drafting

10.1. General The various classes, types, and general categories of maps are defined and described in Art. 1.11. It was noted there that the two main types of maps are the line map prepared by a draftsman using traditional drafting procedures from field data and/or a photogrammetically prepared manuscript, and orthophoto maps produced from a stereo-pair of aerial photographs by photogrammetric methods. An increasing proportion of the line drawn maps are being compiled on a flat-bed plotter under the control of an electronic computer. Thus, compilation and production methods for all kinds of maps are undergoing substantial modifications due to the rapidly changing technologies now adaptable to mapping procedures.

Also defined in Art 1.11 are the two general categories of maps: planimetric maps, which show the plane positions of natural and constructed features; and topographic maps, which show positions of all features as well as the spatial configuration of the terrain called *relief*.

In this chapter emphasis is placed on characteristics of topographic maps; methods of illustrating relief on topographic maps; contours and how contours are plotted; map scales and contour intervals; topographic map construction; types of information that should appear on maps; methods of map reproduction; a brief review of automatic computer-aided map drafting; and tests and standards for the accuracy of maps.

10.2. Datums for mapping In surveying and mapping, datums are necessary to correlate measurements used to determine elevations and horizontal positions for points at different locations. A horizontal datum and a vertical datum are used in surveying.

A *horizontal datum* is defined by the latitude and longitude of an initial point, the azimuth of a line from that point, and the two radii needed to define the geometric reference surface (the spheroid) which best approximates the surface of the earth in the region of the surveys. In the conterminous United States and Alaska, the horizontal datum now being used is the North American Datum of 1927 and the reference surface is the Clarke Spheroid of 1866. This datum is scheduled to be revised in 1985 when a new adjustment based on all data gathered to date and satellite observations will be completed.

A *vertical datum* is the surface to which elevations or depths are referred. All surveys and mapping in the conterminous United States are based on the National Geodetic Vertical Datum (NGVD) of 1929, formerly called the Sea Level Datum of 1929. The datum is based on a best fit to 26 mean sea-level tide stations in the United States and Canada and is the result of a general adjustment of level networks in the United States and Canada in 1929. This vertical datum will also undergo small changes in the 1985 readjustment. The elevations of the National Vertical Control Network are tied to the National Geodetic Vertical Datum. Thus, the elevations that are normally used in topographic mapping, geodetic surveys, engineering studies, and engineering construction surveys are referred to the NGVD.

10.3. Topographic maps A topographic map shows by the use of suitable symbols (1) the spatial configuration of the earth's surface, which includes such features as hills and valleys; (2) other natural features such as forests, lakes, and streams; and (3) the physical changes wrought upon the earth's surface by the works of man, such as buildings, roads, canals, and cultivation. The distinguishing characteristic of a topographic map, as compared with other maps, is the representation of the terrestrial relief.

Topographic maps are used in many ways. They are a necessary aid in the design of any engineering project that requires a consideration of land forms, elevations, or gradients, and they are used to supply the general information necessary to the studies of geologists, economists, and others interested in the broader aspects of the development of natural resources.

The preparation of general topographic maps is largely in the hands of governmental organizations. The principal examples are the topographic maps of areas of the United States compiled and published by the U.S. Geological Survey. These maps are published in quadrangle sheets, which usually include territory ranging from $7\frac{1}{2}$ by $7\frac{1}{2}$ minutes at a scale of 1 : 24,000 (1 : 25,000 in the metric system) to 4° by 12° at a scale of 1 : 1,000,000. A portion of a typical map, the scale of which is 1 : 24,000, is shown in Fig. 10.1. Note that the figure has been reduced so the scale is no longer 1 : 24,000. Altogether there are many federal agencies engaged in surveying and mapping. The central source of information regarding all federal maps and aerial photographs is the National Cartographic Information Center, Reston, Virginia 22092. Likewise, many maps are available from state, county, and city agencies.

10.4. Representation of relief Relief may be presented by *relief models*, *shading*, *color gradients*, *hachures*, *form lines*, or *contour lines*. Of the symbols used on maps, only contour lines indicate elevations directly and quantitatively; they have by far the widest use. They are the principal topic of the next several sections. Form lines are similar to contour lines but are not true to scale and are therefore only qualitative.

10.5. Contours and contour lines A *contour* is an imaginary line of constant elevation on the ground surface. It may be thought of as the trace formed by the

Fig. 10.1 Portion of a typical topographic map of the U.S. Geological Survey. Scale is 1: 24,000 (2000 ft / in). Contour interval 20 ft. (*Courtesy U.S. Geological Survey.*)

intersection of a level surface with the ground surface, for example, the shoreline of a still body of water.

If the locations of several ground points of equal elevation are plotted on a drawing, a line joining these points is called a *contour line*. Thus, contours on the ground are represented by contour lines on the map. Loosely, however, the terms "contour" and "contour line" are often used interchangeably. On a given map, successive contour lines represent elevations differing by a fixed vertical distance called the *contour interval*.

The use of contour lines has the great advantage that it permits the representation of relief with much greater facility, and with far greater definiteness and accuracy, than do other symbols. It has the disadvantage that the map is not as legible to the layman.

10.6. Characteristics of contour lines The principal characteristics of contour lines can be illustrated by reference to Fig. 10.2. For the purpose of this discussion the slope of the river surface is disregarded. The stage of the river at the time of the field survey was at an elevation of 510 ft; hence, the shoreline on the map marks the position of the 510-ft contour line. For this map, the contour interval is 5 ft. If the river were to rise through a 5-ft stage, the shoreline would be represented by the 515-ft contour line; similarly, the successive contour lines at 520 ft, 525 ft, etc., represent shorelines that the river would have if it should rise farther by 5-ft stages.

The principal characteristics of contour lines are as follows:

1. The horizontal distance between contour lines is inversely proportional to the slope. Hence, on steep slopes (as at the railroad and at the river banks in Fig. 10.2) the contour lines are spaced closely.
2. On uniform slopes the contour lines are spaced equally.

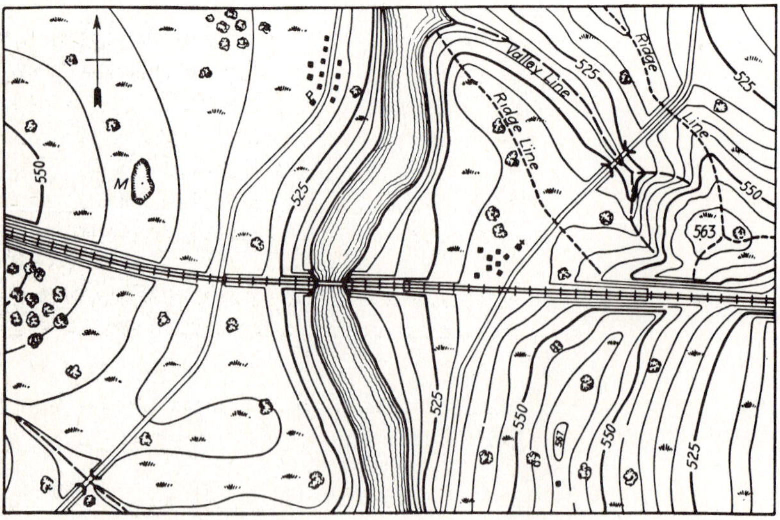

Fig. 10.2 Contour lines.

3. Along plane surfaces (such as those of the railroad cuts and fills in Fig. 10.2) the contour lines are straight and parallel to one another.
4. As contour lines represent level lines, they are perpendicular to the lines of steepest slope. They are perpendicular to ridge and valley lines where they cross such lines.
5. As all land areas may be regarded as summits or islands above sea level, evidently all contour lines must close upon themselves either within or without the borders of the map. It follows that a closed contour line on a map always indicates either a summit or a depression. If water lines or the elevations of adjacent contour lines do not indicate which condition is represented, a depression is shown by a hachured contour line, called a *depression contour*, as shown at *M* in Fig. 10.2.
6. As contour lines represent contours of different elevation on the ground, they cannot merge or cross one another on the map, except in the rare cases of vertical surfaces (see bridge abutments of Fig. 10.2) or overhanging ground surfaces as at a cliff or a cave.
7. A single contour line cannot lie between two contour lines of higher or lower elevation.

Other special features of contours are shown in Figs. 10.3 and 10.4. Figure 10.3 shows contours with a 20-m interval for mountainous terrain. On the right is a

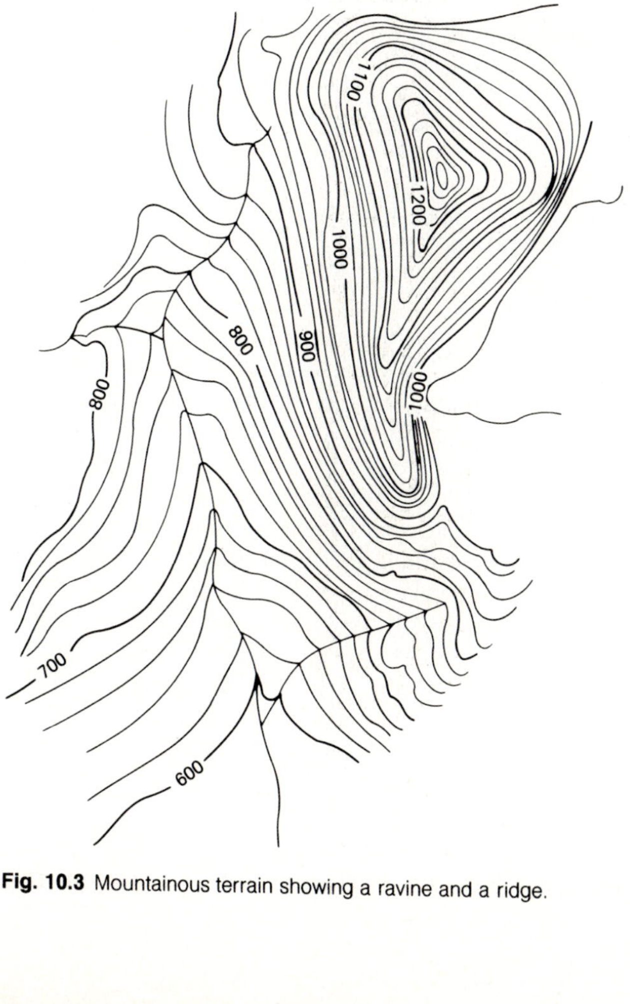

Fig. 10.3 Mountainous terrain showing a ravine and a ridge.

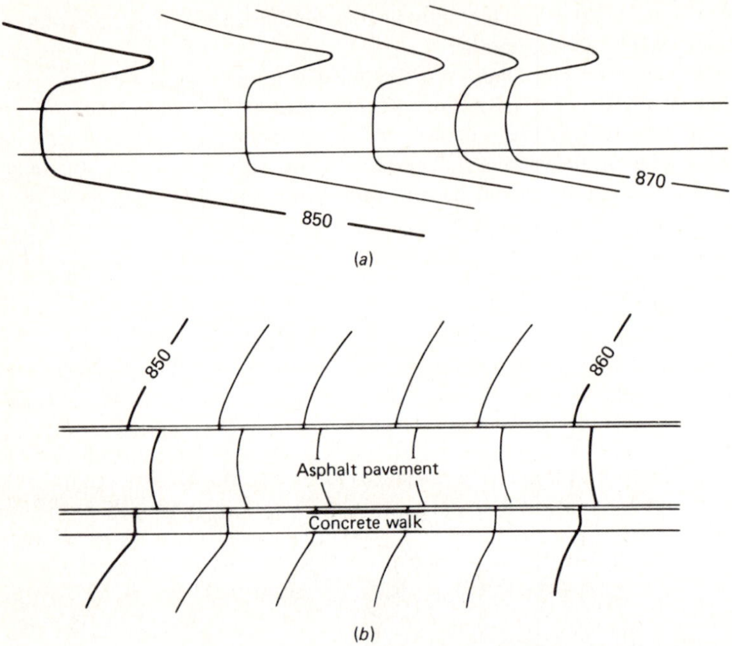

Fig. 10.4 (*a*) Contours crossing a paved road. (*b*) Contours crossing a paved road with a curb.

well-defined ridge. Note that the U shape of the contours points down the slope on a ridge. On the left is a ravine with several small streams leading into it. In contrast to the ridge, the V-shaped contours point up the slope to indicate a ravine, flow line of a stream, or a ditch. Figure 10.4 shows how contours cross artificial features. In Fig. 10.4*a*, the contours cross the paved road built in cut on the upper side and on fill on the lower side. On the upper side there is a ditch, indicated by the V-shaped contours pointing up the slope. The contours cross the pavement as a smooth line (the pavement is relatively smooth) which is slightly convex down the slope since most paved roads are crowned and are slightly higher in the center than on the two edges. In Fig. 10.4*b*, contours cross a paved road with curbs. Most curbs are 0.5 ft or 0.15 m high, so the contour crosses the curb at right angles and runs along the face of the curb until it reaches the contour elevation in the gutter line. Thus, if the grade of the road is 5 percent, the distance from where the contour crosses the top of the curb to the point where it meets the contour elevation in the gutter is $0.5/0.05 = 10$ ft, or about 3 m. Then the contour crosses the pavement slightly convex down the slope to the gutter line on the other side, follows the face of the curb until the contour elevation at the top of the curb is reached, and crosses the curb and sidewalk at right angles (assuming that the sidewalk is horizontal). The contour interval in Fig. 10.4*a* is 5 ft and that in Fig. 10.4*b* is 2 ft.

10.7. Choice of map scale and contour interval It is possible to choose a map scale consistent with the purpose of the survey if the approximate size of plotting error is known. For example, if it is known that (with reasonable care in plotting) the standard deviation in measuring a distance between two definite points on the map is $\frac{1}{50}$ in (0.5 mm), and if it is known that the purpose of the survey will be met if the standard deviation in scaled distances is 10 ft (3 m), then a map scale of 1 : 6000 (500 ft/in) satisfies the conditions.

Other factors that influence choice of map scale alone are (1) the clarity with which features can be shown, (2) the cost (the larger the scale, the higher the cost), (3) the correlation of map data with related maps, (4) the desired size of the map sheet, and (5) physical factors such as the number and character of features to be shown, the nature of the terrain, and the necessary contour interval.

The contour interval may be thought of as the scale by which the vertical distances or elevations are measured on a map. The choice of a proper contour interval for a topographic survey and map is based upon four principal considerations: (1) the desired accuracy of elevations read from the map, (2) the characteristic features of the terrain, (3) the legibility of the map, and (4) the cost.

1. *Accuracy* Let it be assumed that two maps are equally accurate, so that the standard deviation in elevations, read from the map, of points chosen at random is one-half of a contour interval. Assume one map to have a contour interval of 5 ft (1 m) and the other 2 ft (0.5 m). It is evident that the standard deviation in elevations of points chosen at random on one map is $2\frac{1}{2}$ ft (0.5 m) and on the other 1 ft (0.25 m). Therefore, the more refined the scale (i.e, the smaller the interval), the more refined should be the measurements of the elevations of chosen points.
2. *Features* Often field conditions exist where characteristic features require the use of a contour interval that would otherwise be inappropriate. Thus, if the shape of the terrain is such as to show much variation within a small area, or in other words, if the topography is of *fine texture*, then a smaller contour interval is required to show the greater complexity of configuration. On the other hand, if the landscape is composed of large regular forms or is of *coarse texture*, then a larger interval may be used. Where the texture varies considerably within an area to be mapped, or where it is desired to define certain parts of the area more clearly than the general area (as for construction purposes), a smaller contour interval may be used for parts of the area.
3. *Legibility* A map otherwise excellent may be rendered useless and its appearance disfigured by a mass of contour lines that obscure other essential features. In general, contour lines should not be spaced on the map more closely than 20 or 30 to the inch (8 to 12 per cm), although the legibility of the map depends largely upon the fineness and precision with which the lines are drawn. The lithographed maps of the U.S. Geological Survey and of the U.S. National Ocean Survey yield good results with much closer spacing.
4. *Cost* The smaller the contour interval, the higher the cost, especially if the usual accuracy of one-half contour interval is maintained.

Obviously, the contour interval and map scale are interrelated. In general, the smaller the scale, the larger the contour interval. For the most part, this interrelationship depends upon the purpose and scale of the map, the character of the terrain represented, and the degree of land development in the area to be mapped.

Traditionally, maps have been classified according to scale as large, intermediate, and small scale with accompanying contour intervals as follows (comparable metric scales shown in parentheses):

Large scale	10 to 200 ft/in (1 : 100 to 1 : 2000)
Contour interval	0.2 to 10 ft (0.1 to 2 m)
Intermediate scale	200 to 1000 ft/in (1 : 2000 to 1 : 10,000)
Contour interval	1 to 20 ft (0.2 to 5 m)
Small scale	1000 to 5,000,000 ft/in (1 : 10,000 to 1 : 100,000,000)
Contour interval	50 to 1000 ft (5 to 200 m)

The American Society of Civil Engineers (ASCE) Surveying and Mapping Division has a more detailed classification for map scales and contour intervals.

1. *Design maps* These maps are used in the design and construction of specific engineering work of all kinds. Scales vary from 10 to 200 ft/in (1 : 100 to 1 : 2000) with contour intervals between 0.2 and 10 ft (0.1 and 1 m), depending on the type of project, land use, and terrain characteristics. Two subcategories are given with this group:
   a. *Critical design maps* These are maps used on projects having critical space, orientation, position, or elevation restrictions. An example is a highway interchange in an urban area.
   b. *General design maps* These are maps prepared for projects that do not have such rigid restrictions with respect to location. An example is a map prepared for a rural water distribution system.
2. *Planning maps* These include a large group of maps used in planning engineering work or in overall planning at the urban, regional, national, and international levels. These maps may be used for geological studies, land use, agricultural production, and population studies; for public service planning; and for atlases. Map scales range from 100 to 5,000,000 ft (1000 mi)/in (1 : 1000 to 1 : 100,000,000) and contour intervals from 1 to 1000 ft (0.2 to 200 m).

10.8. Topographic map construction Normally, the construction of a topographic map consists of three operations: (1) the plotting of the horizontal control, or skeleton upon which the details of the map are placed; (2) the plotting of details, including the map location of points of known ground elevation, called *ground points*, by means of which the relief is to be indicated; and (3) the construction of contour lines at a given contour interval, the ground points being employed as guides in the proper location of the contour lines. A ground point on a contour is called a *contour point*.

Current trends in the production of topographic maps, for use in engineering design and planning and for maps used by the public, are toward photogrammetric compilation of maps (Art. 13.15) and automatic or semiautomatic construction of the map. General details relevant to automated mapping systems can be found in Art. 10.18. Emphasis in the sections that follow is on the conventionally drawn line map. Where appropriate, references are given to the comparable automatic procedure.

10.9. Methods for plotting horizontal control Horizontal control may be plotted by (1) the coordinate method, (2) the tangent method, (3) the chord method, or (4) by use of a protractor and scale. The coordinate method, the most general and accurate procedure for plotting positions of points, is adaptable to plotting by hand or by automatic methods. The latter three methods are of substantially lower accuracy than the coordinate method and are suitable primarily for rough preliminary plotting of control for maps covering small areas. In any case, distances are measured with the engineer's scale; for precise work the points are pricked with a needle and a reading glass is employed. Traverse and triangulation stations are indicated by appropriate symbols, and control lines are drawn carefully with a hard pencil having a fine point. Points are numbered or lettered to conform to the system employed in the field work. Usually, the control is not shown on the finished map.

The precision of plotting the control depends upon the instrument used and upon the care with which the work is done. With a well-sharpened hard pencil and with the use of reasonable care, points can be plotted within perhaps $\frac{1}{100}$ in (0.3 mm). With a fine needle and a reading glass and with the use of great care, with the eye directly above the needle, points can be plotted within perhaps $\frac{1}{200}$ in (0.13 mm).

10.10. Plotting by the coordinate method The control point positions in horizontal control surveys for topographic maps (Chap. 12) are usually calculated and adjusted in the state plane coordinate system (Chap. 11). Thus, data from field operations that have been calculated and adjusted in the office, known to be correct to within specified limits, are available for direct plotting by the coordinate method. This plotting can be performed with high precision, since all measurements are linear distances measured from orthogonal axes.

In order to plot by the coordinate method, a series of grid lines are drawn on the base sheet for the map. These grid lines are sets of X and Y axes of orthogonal coordinate systems. In the state plane coordinate system (and in most other coordinate systems), the Y grid axis is aligned toward grid north. These grid lines are spaced at some regular, uniform interval (50, 100, 200, 500, 1000 ft or m suitable for the scale of the map being compiled. When drawing the grid lines, extreme care must be exercised to ensure that the lines are straight and of uniform weight and that the X grid lines are perpendicular to the Y grid lines. Each X and Y grid line is labeled at the edges of the map sheet with its coordinate value. Grid lines should be drawn as quickly as possible on stable-base material under uniform temperature conditions so that all measurements are consistent. The positions of the control points are then plotted by laying off the differences between the coordinates for the point and the coordinates of a pair of intersecting grid lines close to the control point.

Figure 10.5 shows a set of coordinate grid lines for plotting the stations of the traverse calculated in Art. 8.16 and adjusted in Art. 8.19. For a small traverse such as this one, it is desirable to locate the grid lines so as to center the traverse approximately on the map sheet. This is done by examining the adjusted coordinates (Table 8.11), noting the maximum and minimum values for X and Y coordinates, and then calculating the maximum changes in X and Y coordinates. These distances reduced to map scale, subtracted from the dimensions of the map sheet and divided by 2, yield the distances that should be measured from the edges of the map sheet to the control points having maximum and minimum X and Y coordinates.

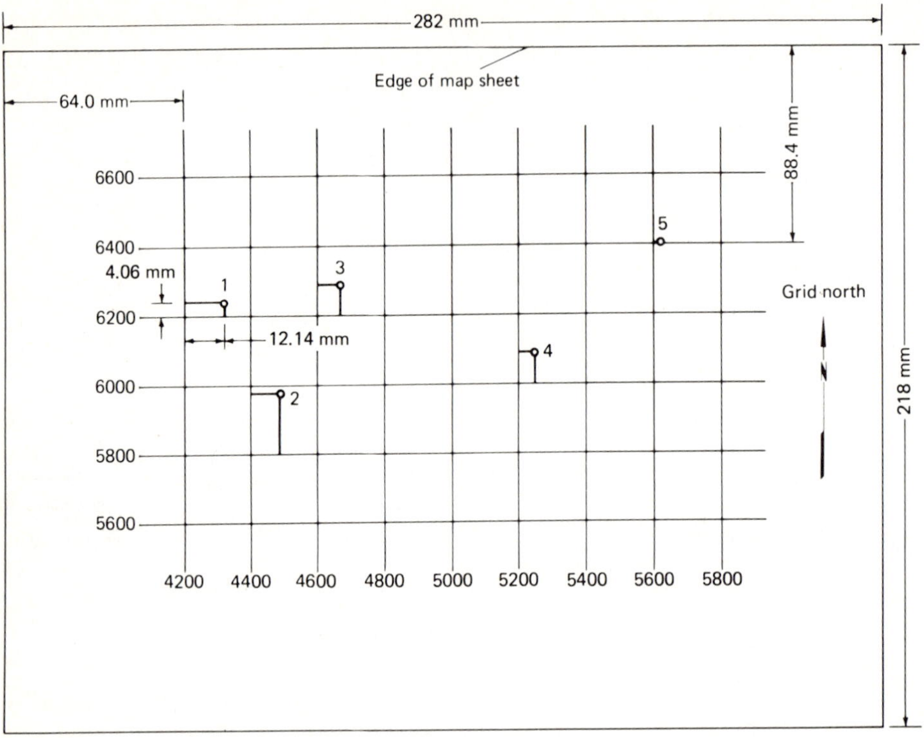

Fig. 10.5 Traverse stations of Art. 8.16, Table 8.11, plotted by rectangular coordinates.

For the example in Fig. 10.5 and using coordinates from Table 8.11, the following data are available:

Map scale = 1 : 10,000
Map sheet dimensions are 282 by 218 mm (11 by $8\frac{1}{2}$ in)
Grid interval = 200 m on the ground or 20 mm at map scale
Maximum change in X = 1297.84 m on the ground or 129.8 mm at map scale
Maximum change in Y = 422.34 m on the ground or 42.2 mm at map scale
Westernmost station is 1, for which X = 4321.40 m
Northernmost station is 5, for which Y = 6405.27 m

With a grid interval of 200 m and in view of the values of coordinates for the westernmost and northernmost stations, it is necessary to determine the locations for the westernmost Y grid line at X = 4200 m and for the northernmost X grid line at Y = 6400 m. It is possible to work either at map scale or in metres on the ground. Assume that the best metric scale available for the work is graduated only in millimetres, so that working at map scale is more convenient. Since the change in X exceeds that in Y, the X axis is aligned with the long dimension of the map sheet. The location of the Y grid line for X = 4200 m is determined from the left or western edge of the sheet. Since the maximum change in X coordinates is 129.8 mm at map scale, the distance from the left edge of the sheet to the grid line is (282 − 129.8)/2

$-(4321.4-4200)/10 = 64.0$ mm at map scale. This distance is measured from the left edge of the map sheet and the Y grid line for $X = 4200$ m is drawn parallel to the edge of the sheet. The location of the X grid line for $Y = 6400$ m is determined from the top of the map sheet next. The maximum change in Y is 42.2 mm at map scale, so that the distance from the top of the sheet to the X grid line for $Y = 6400$ m is $(218 - 42.2)/2 - (6405.3 - 6400)/10 = 88.4$ mm at map scale. This distance is then scaled from the top of the map sheet (Fig. 10.5) and the X grid line for $Y = 6400$ m is drawn through this point and perpendicular to the Y grid line. The balance of the grid lines are then drawn parallel to these two controlling lines at intervals of 20 mm at map scale or 200 m on the ground. After the grid lines are labeled, the position of station 1 is plotted by subtracting $X = 4200$ and $Y = 6200$ from the X and Y coordinates for 1 to yield increments in the X and Y directions of 121.4 m and 40.6 m, respectively. These distances are scaled along the grid lines and perpendiculars are erected, the intersection of which defines the control point (see Fig. 10.5). The remaining stations are plotted in similar fashion. When all points have been plotted, the distances between points should be scaled as a check against the adjusted distances between the stations of the traverse. In this way the accuracy of the plotting is verified.

The grid lines are retained on the finished drawing or map or at the very least, the grid tick intersections are left. These grid lines or grid intersection points are invaluable for scaling the coordinates of points on the map. They can also be useful to evaluate dimensional changes in the map-base material and allow more realistic scaling of distances and locations from a given map sheet.

Rectangular coordinates can be plotted more rapidly and accurately by mechanical methods using a coordinatograph or flat-bed plotter. These devices consist of a flat, smooth drawing surface mounted on a stand or legs. A beam is then mounted along one edge of the drawing surface in a fixed position. Another beam, at right angles to the first beam, is mounted on wheels or a slide such that it can move back and forth across the entire drawing surface of the table. A chuck for holding a pencil, inking pen, scribing device, needle, or magnifier is mounted on the second beam in such a way that the chuck can be moved the entire length of the beam. In the most elementary form of coordinatograph, the movable beam and pencil chuck are driven by threaded screws turned by hand wheels at the end of each beam. Scales are mounted parallel to the beams, and the hand wheels used to turn the screw are also graduated, so that it is possible to plot grid lines at a specified interval or the positions of points referred to a given origin. The least count on most flat-bed plotters is from 0.001 to 0.004 in (0.025 to 0.10 mm). Figure 10.6 shows a modern flat-bed plotter.

Flat-bed plotters of the most recent design contain servomotors to drive the position of the movable beam and chuck. These plotters are under the control of an electronic computer. The coordinates of the points to be plotted are read into the computer from magnetic tape or punched cards or may be introduced manually using a keyboard. These systems are very adaptable to the automatic plotting of coordinate grid lines and map details when the data are on a suitable input form, such as magnetic tape or punched cards, and a program for processing the data is

Fig. 10.6 Flat-bed plotter. (*Courtesy Wang Laboratories, Inc.*)

available. The data would consist of the X and Y coordinates of points to be plotted plus the necessary identifying information for each point. The program is necessary to read the data to the computer and to convert the X and Y coordinates for each point into impulses that drive the plotter to the specified locations. A plotter, which can be tilted for convenience and to save space, connected to a minicomputer with a cathode ray tube for visual display of program instructions and input is shown in Fig. 10.7.

The advantages of plotting horizontal control by rectangular coordinates are as follows: (1) adjusted coordinates are used to plot the locations so it is known that the data are correct to within the closure of the control survey network; (2) each point is plotted independently so that there is no cumulation of error; (3) high precision and uniform accuracy can be maintained on large map sheets and when multiple map sheets are required; and (4) the method is very adaptable to automatic plotting routines and is compatible with modern data-storage procedures.

10.11. Other methods for plotting control When distances and directions are given between consecutive control points (e.g., in traverse), it is possible to plot the relative positions of these points by setting off the direction of the line or the angle between two lines and then scaling the distance along the line from one point to the next. Thus, the basic procedure in these methods is that of plotting angles.

The protractor and scale method The protractor is a device for laying off and measuring angles on drawings. The usual form for mapping consists of a full circle or semicircular arc of metal, plastic, or paper graduated in degrees or fractions of a degree. Some have radial scales by means of which a distance and an angle may be plotted at one operation. Others have one or more radial arms with scales.

Fig. 10.7 Coordinate plotter, computer, and interactive terminal. (*Courtesy California Computer Products, Inc.*)

The procedure is illustrated by Fig. 10.8*a* where the direction of the initial line *CD* is fixed and the length *CD* is scaled to establish point *D*. Orienting the protractor at the forward point *D*, set off the deflection angle α to the next point and draw a light line in the direction of *E*. Scale the distance *DE* to fix point *E*. Continue the process, orienting each succeeding line with respect to the previous one until the traverse closes on the initial line or another plotted line of known position.

The tangent method Here the angle is set off by linear measurement from the previous line extended, as shown in Fig. 10.8*b* where line *CD* is drawn 10 units to establish point *d*. A perpendicular offset is erected at *d* and the distance $dd' = Dd \tan \alpha$ is scaled to set point *d'*. The line *Dd'* is drawn and the distance *DE* is scaled to fix point *E*.

The chord method This method is much like that of plotting by tangents as just described. Referring to Fig. 10.8*c*, an arc *Dd* of 10 or 20 units radius is struck. The

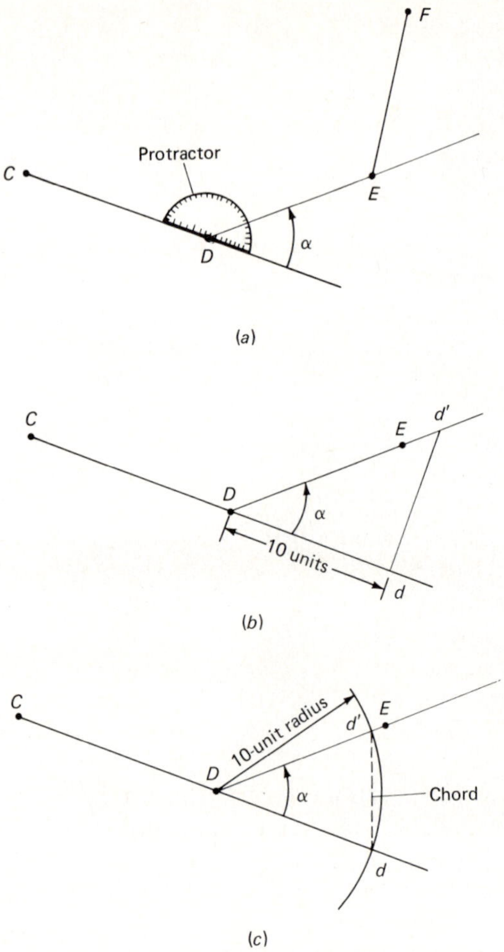

Fig. 10.8 Methods for plotting control. (*a*) Protractor and scale method. (*b*) Tangent method. (*c*) Chord method.

chord distance $dd' = 2Dd \sin \alpha/2$ is then scaled from point d to set point d' on the arc. The distance DE is scaled along line Dd' to fix point E.

10.12. Plotting the details The work of plotting the details is less refined than that of plotting horizontal control, but the aim is still to plot objects of a definite size within the allowable error.

The method of plotting details is a function of the field procedure used for acquiring topographic details. If details were located by horizontal distance and direction (angle, bearing, or azimuth) from a given control point with a backsight on another control station, the directions can be set off with a protractor and the distances with a scale to position each individual point. Data of this type would be available when details are located in the field by stadia (Art. 12.10) or with one of the electronic positioning systems illustrated in Figs. 9.9 and 9.11.

When data are located in the field by the grid or checkerboard method (Art. 12.13) or cross sections (Art. 12.12), the grid lines or profile center line and cross section lines must be plotted on the map sheet in the correct relationship to the horizontal control points. Data are then plotted as recorded at the appropriate grid intersections or at the proper offset along the cross-section lines.

If data are located by radial surveys from a control point using one of the electronic surveying systems (Figs. 9.9, 9.11, and 9.12, Art. 9.13), the solid-state memory of the data storage unit represents the field record of all measurements to the points so located. When the job is completed, the storage unit is taken to the office and the data can be converted to coordinates by connecting the unit to a compatible electronic calculator. The output, consisting of *X*, *Y*, *Z* coordinates for each point located, is stored on magnetic tape, or disk storage. The output in this form can then be used as input to another electronic computer utilized to drive a drum or flat-bed plotter for automatic plotting by the coordinate method of all the details on the map sheet (Fig. 10.7). Naturally, such an approach requires compatible field and office equipement with the proper programs to process the field data and drive the plotter.

10.13. Plotting the contours When the horizontal control points and details have been plotted, the map sheet contains the planimetric details and an array of points for which elevations are known and are plotted on the map sheet. These elevations may be randomly located control points on the terrain or may be positioned on a regular grid or cross-section pattern. The next task in compiling the topographic map is to use this array of points in order to plot the contours.

Regardless of the number of ground points whose plotted locations are known, it is evident that any contour line must be drawn, to some degree, by estimation. This condition requires that the draftsman use considerable skill and judgment so that the contour lines may best represent the actual configuration of the ground surface.

Contour lines are shown for elevations that are multiples of the contour interval. They are drawn as fine smooth freehand lines of uniform width. Each fifth contour line is made heavier than the rest, and sometimes these lines are drawn first to facilitate the location of intermediate contour lines.

Elevations of contours are indicated by numbers placed at appropriate intervals; usually, only the fifth or heavier contour lines are labeled. However, in areas of low relief where contours are spaced widely apart, each line may be labeled with its elevation. The line is broken to leave a space for the number. As far as possible, the numbers are faced so as to be read from one or two sides of the map; but on some maps the numbers are faced so that the top of the number is uphill. *Spot elevations* are shown by numbers at significant points, such as road intersections, bridges, water surfaces, summits and depressions.

Since contours ordinarily change direction most sharply where they cross ridge and valley lines and since the gradients of ridge and valley lines are generally fairly uniform, these lines are important aids to the correct drawing of the contour lines. Special care is taken in the field to locate the ridge and valley lines, and usually these lines are drawn first on the map. Examples of such locations are shown plotted as the ridge line from *a* to *b* and the valley line from *c* to *d* in Fig. 10.9. The stream lines are drawn through those points which represent valleys, and the contour crossings are spaced along them before any attempt is made to interpolate or to draw the contour lines. This procedure aids in the interpretation of the data. For example, in the square bounded by the points *D*-5, *D*-6, *E*-6, and *E*-5, the contour lines are made to show the head of the valley, the existence of which is indicated only by the valley line previously drawn in the square below. In the figure shown, the ridge line is rather indefinite, but sketching the location at the ridge and spacing the contour lines across the ridge does help in drawing the rest of the contours.

10.14. Drawing contours by interpolation The process of spacing the contour lines proportionally between plotted points is called *interpolation*. For example, in Fig. 10.9 the elevations of points *A*-2 and *B*-2 are 848.0 and 852.0 m, respectively. The contour interval for this map has been taken as 2 m, and the 848- and 852-m contour lines pass through the corresponding points. Under the assumption that the slope is uniform, the 850-m contour line passes through a point midway between *A*-2 and *B*-2. If a 1-m interval were used, the additional 849- and 851-m contour lines would be drawn through the quarter points of the line from *A*-2 to *B*-2.

The procedure is usually not so simple as in the case just cited. The elevations at the corners of the other squares in the figure are mostly of such values that the contour lines do not pass through them; further, on many maps the points of known elevation are spaced irregularly. Under such conditions, the interpolations may be made by estimation on the map, by arithmetical computations, or by graphical means, as follows:

1. Estimation Since each contour map is the result of more or less interpretation by the draftsman, in many cases it is not inconsistent with the other methods of map

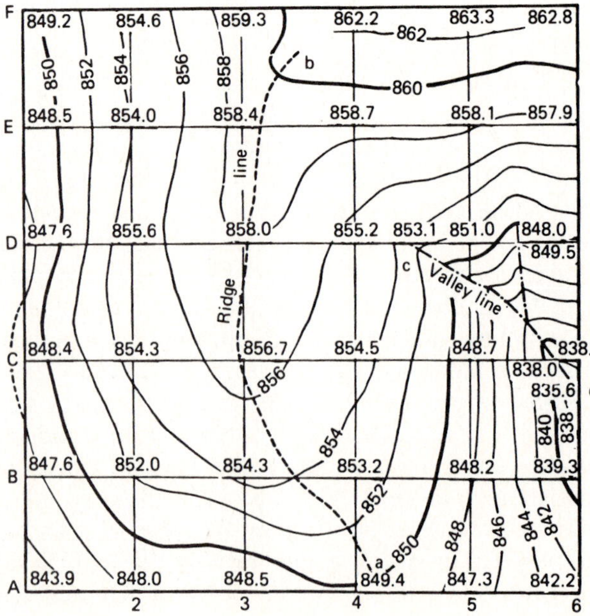

Fig. 10.9 Contour lines by the grid method.

construction if the interpolation is made by careful estimation supplemented by approximate mental computations. This method is most commonly used on intermediate-scale and small-scale maps.

2. Computation Where considerable precision is desired in the map, the errors of estimation may be eliminated by simple arithmetical computations made with the aid of a hand-held calculator. For example, the elevations of points *E*-6 and *F*-6 (Fig. 10.9) are 857.9 and 862.8 m, respectively. The contour interval is 2 m; hence, the difference in elevation between *E*-6 and the 858-m contour is 0.1 m. Then, since the total difference in elevation is 4.9 m, the proportional part of the distance from *E*-6 to *F*-6 to locate the 858-m contour line is 0.1/4.9 of the map distance between these points. Similarly, the proportional parts for the 860- and 862-m contour lines are, respectively, 2.1/4.9 and 4.1/4.9 of the distance from *E*-6 to *F*-6. The computed map distances are plotted to scale.

3. Graphical means The computations indicated in the previous paragraph become laborious if many interpolations are to be made, and accordingly various means of graphical interpolation are in use. One method is illustrated in Fig. 10.10. Parallel lines are drawn at equal intervals on drafting film, each fifth or tenth line being made heavier than, or of a different color from, the rest and being numbered as shown. If it is desired to interpolate the position of, say, the 52- and 54-ft or -m contours between *a* with elevation of 50.7 and *b* with elevation of 55.1, the line on the drafting film corresponding to 0.7 ft or m (scale at left end) is placed over point *a* on the tracing, and the film is turned about *a* as a center until the line corresponding to 5.1 ft or m (scale at same end) covers point *b* on the tracing. The interpolated points are at the intersections of lines 2.0 and 4.0 (representing elevations 52 and 54) and the line *ab*, and may be pricked through the drafting film. Had the known points been much closer together, the scale at the right end of the film would have been used, and thus the value of each space would have been doubled; or if the scale were small, the contour interval large, and the topography rugged, each space might represent 1 ft or 0.2 m. Thus by assigning different values to the spaces, a single piece of drafting film so prepared can be made to suit a variety of conditions.

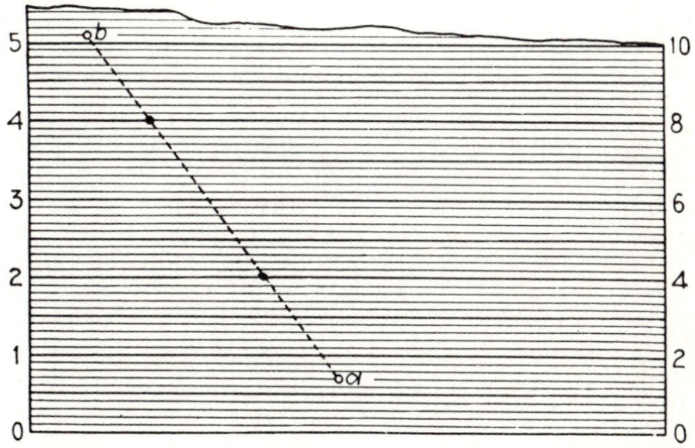

Fig. 10.10 Graphical interpolation of contour lines.

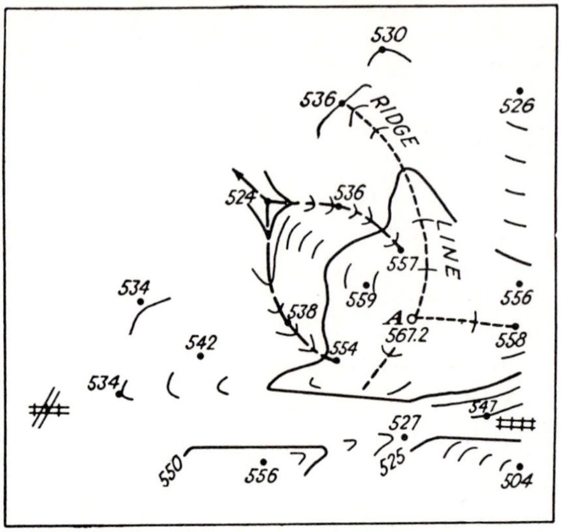

Fig. 10.11 Interpolation of contours using controlling points.

Another convenient graphical means of interpolation is the use of a rubber band graduated at equal intervals with lines that form a scale similar to that just described for the drafting film. The band is stretched between two plotted points so that these points fall at scale divisions corresponding to their elevations. The intermediate contour points are then marked on the map.

Contours partially plotted by interpolation using data obtained by the controlling point method (Art. 12.9) are illustrated in Fig. 10.11. Figure 10.12 shows contours plotted by interpolation with data from cross sections (Art. 12.12).

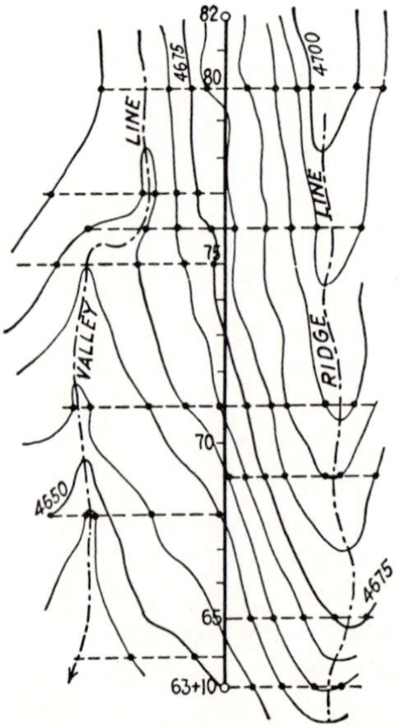

Fig. 10.12 Interpolation of contours from cross-section data.

10.15. Automatic plotting of contours from a digital terrain model Modern surveying and photogrammetric equipment enables rapid acquisition of planimetric positions and elevations, all referred to a specified datum, for large numbers of points. This dense network of points which defines the configuration of the terrain is called a *digital terrain model* (DTM) or *digital elevation model* (DEM). If the points located in the field are sufficiently close and of large enough numbers, then any one of the methods for acquiring topographic details (Chap.12) yields a DTM. However, in order to be useful in a practical sense, the DTM data must be recorded in "real time" on a medium that is compatible with an electronic computer. Consequently, the DTM data that are usable for mapping are provided by the integrated electronic surveying systems (Art. 9.13) and digitized photogrammetric stereoplotters (Art. 13.14), all of which allow recording the data in real time on magnetic tape, core memory, or disk storage.

The most efficient procedure to utilize this enormous quantity of data (10,000 to 1,000,000 points per photogrammetric stereomodel) is to use an electronic computer to process the information into a form that permits automatic plotting of contours on a flat-bed plotter (Art. 10.10). Some rather complex computer programs are required for this purpose. There are many such programs now available for reducing the DTM to contours. Reference 10 at the end of this chapter contains a number of papers on the subject.

Figure 10.13 illustrates two contour maps of the same area plotted at a scale of 1:2500 and with a contour interval of 2.5 m (adapted from Ref. 1). Figure 10.13*a* was plotted by manual interpolation using about 6100 points located by an electronic tacheometric survey with an instrument similar to the one illustrated in Fig. 9.12. The contour map shown in Fig. 10.13*b* is of the same area and was plotted automatically from a DTM consisting of the same 6100 points located by the electronic tacheometric survey. The maximum distance between points in the field survey was about 15 m. The interval for the superimposed grid for the DTM was also 15 m on the ground. The data were processed on a large electronic computer. Output from these computations was then used to drive an automatic plotter on which the contours were drawn. Obviously, the two maps agree quite well.

10.16. Information shown on maps Maps are classified variously according to their specific use or type, but in general either they become a part of public records of land division or they form the basis of a study for the works of man. In general, the following information should appear on any map:

1. The direction of the meridian and basis for directions (astronomic, grid, magnetic, etc)
2. A graphical scale of the map with a corresponding note stating the scale at which the map was drawn
3. A legend or key to symbols other than the common conventional signs
4. An appropriate title
5. On topographic maps, a statement of the contour interval (Art. 10.5)
6. A statement giving the datum to which horizontal and vertical control are referenced
7. A statement giving the mapping projection
8. A statement giving the coordinate systems for which grid lines or grid ticks are published on the map

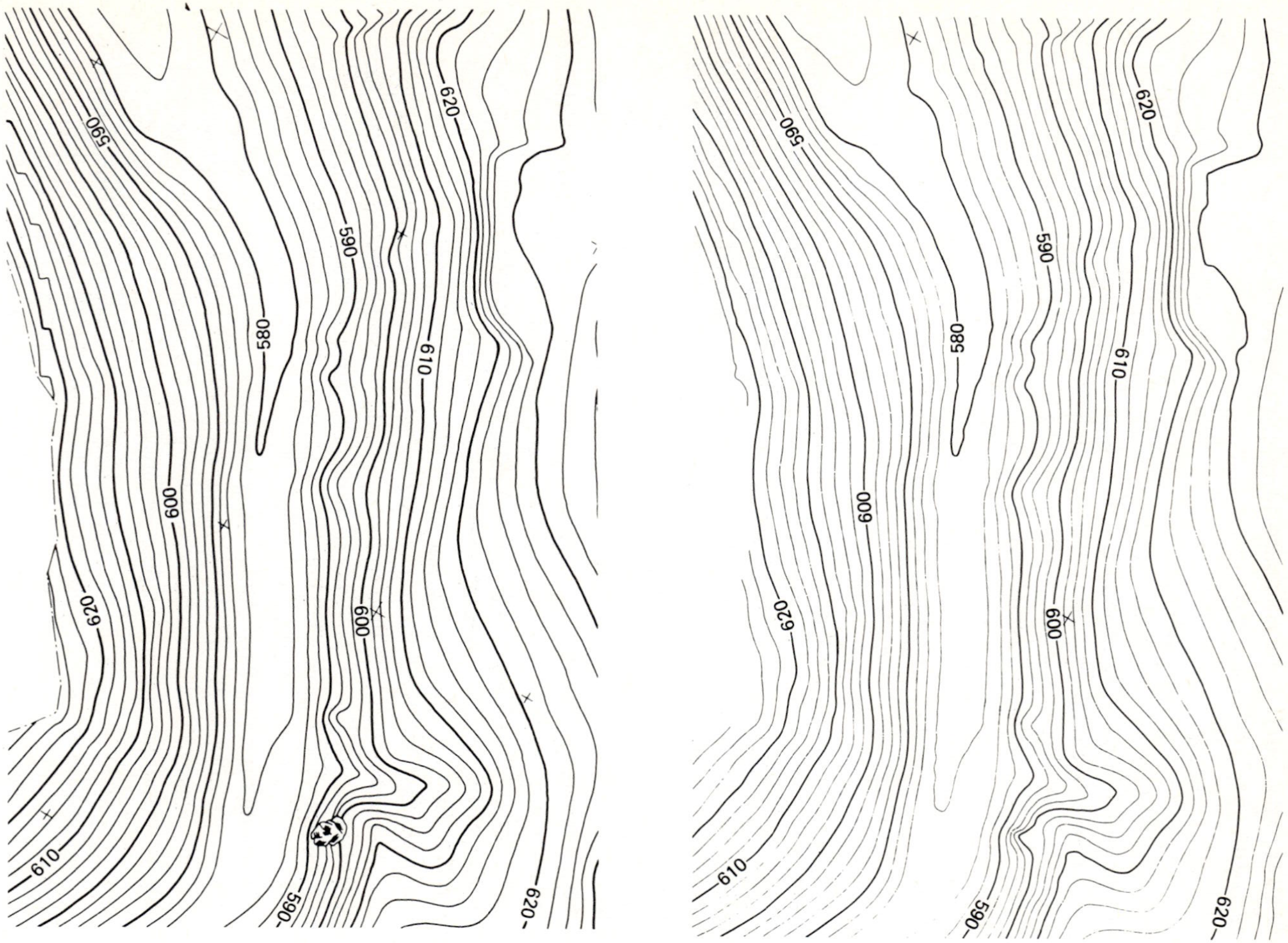

(a) (b)

Fig. 10.13 Contours plotted manually and automatically (2.5-m intervals, scale 1:2500). (*a*) Contours drawn by manual interpolation. (*b*) Contours drawn automatically using a 15-m grid interval. (*Courtesy Technical Institute for Photogrammetry, University of Stuttgart.*)

In addition, a map that is to become a part of a public record of land division should contain the following information:

1. The length of each line
2. The bearing or azimuth of each line or the angle between intersecting lines
3. The location of the tract with reference to established coordinate axes
4. The number of each formal subdivision, such as section, block, or lot
5. The location and kind of each monument set, wtih distances to reference marks
6. The location and name of each road, stream, landmark, etc.
7. The names of all property owners, including owners of property adjacent to the tract mapped
8. A full and continuous description of the boundaries of the tract by bearing and length of sides; and the area of the tract
9. The witnessed signatures of those possessing title to the tract mapped; and if the tract is to be an addition to a town or city, a dedication of all streets and alleys to the use of the public
10. A certification by the surveyor that the map is correct to the best of his or her knowledge

Explanatory notes may be employed to give such information as the sources of data for the map and the precision of the survey.

The map scale should be indicated on the map sheet numerically and graphically. Numerical designations include the representative fraction, where one unit on the drawing represents a stated number of units on the ground (e.g., 1/25,000 or 1 : 25,000); or a statement that 1 in on the map represents some whole number of feet on the ground such as 1 in per 200 ft. The representative fraction is independent of units and is the method of designating scale that is utilized internationally.

The graphical scale is a line subdivided into map distances corresponding to convenient units of length on the ground. Various forms of graphical scales are shown in Fig. 10.14.

If distances are to be determined accurately from a map, a *graphical scale should always be shown*. The material on which the map is drawn changes dimensions over time, and often the scale of a map is distorted in the reproduction process. Since the graphical scale will change in the same ratio as the map, it provides a means of obtaining accurate distances by scaling from the map.

The direction of the meridian is indicated by a needle or feathered arrow pointing north, of sufficient length to be transferred with reasonable accuracy to any part of the map. The astronomic or true meridian is usually represented by an arrow with *full* head; the magnetic meridian by an arrow with *half* head. When both are shown,

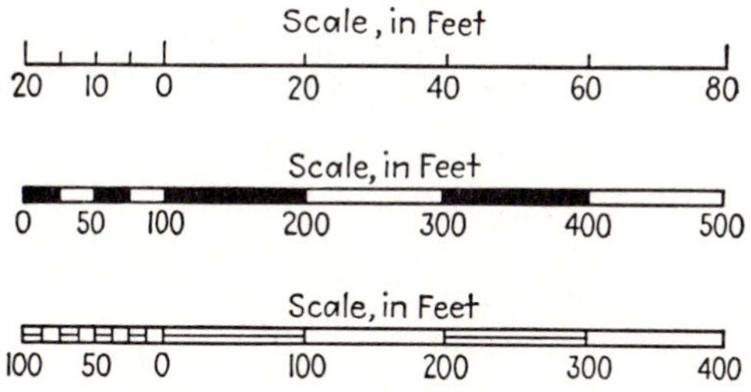

Fig. 10.14 Graphical scales.

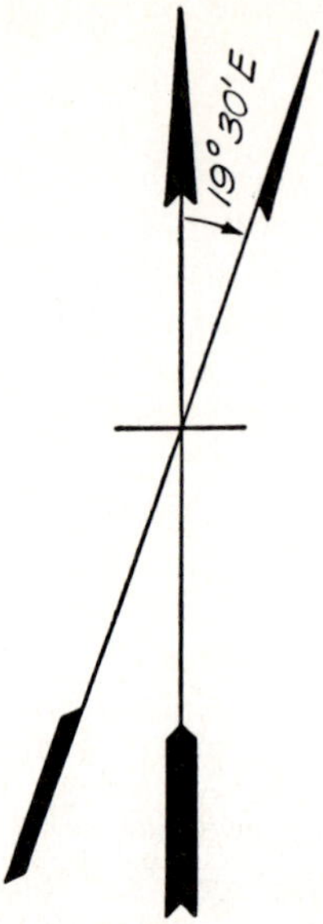

Fig. 10.15 Meridian arrows.

the angle between them should be indicated. The general tendency is to make needles and arrows too large, blunt, and heavy. A simple design is shown in Fig. 10.15.

Preferably, the top of a map should represent north, although the shape of the area covered or the direction of some principal feature of a project may make another orientation preferable.

Map titles should be so constructed that they will readily catch the eye. The best position for the title is the lower right-hand corner of the sheet, except where the shape of the map makes it advantageous to locate the title elsewhere. The space occupied by the title should be in proportion to the size of the map; the general

Fig. 10.16 Title for a map.

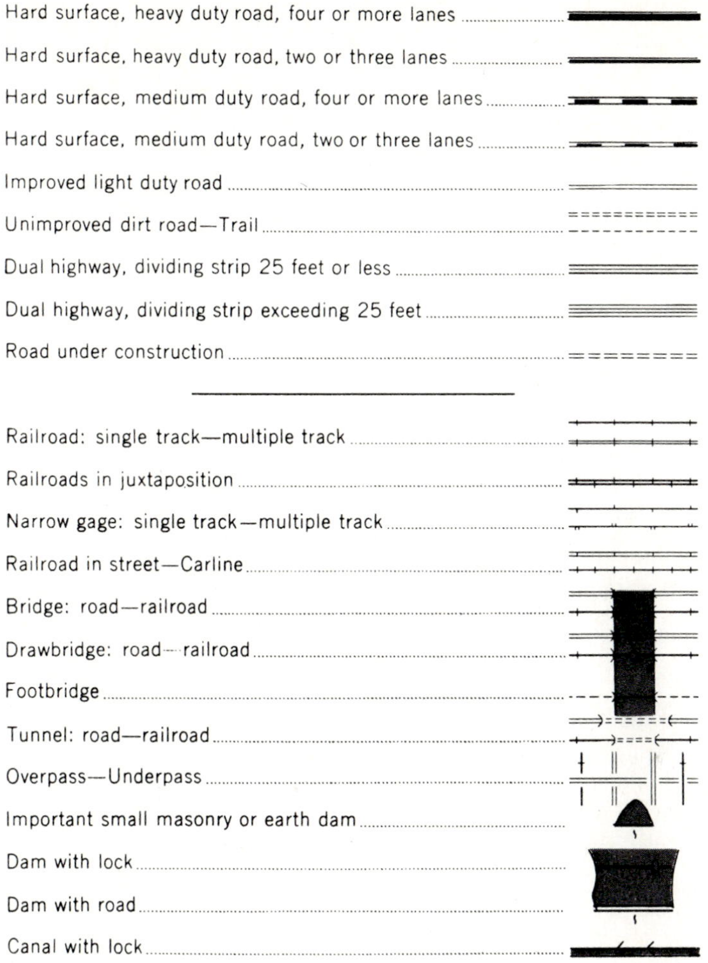

Fig. 10.17*a* Symbols for roads, railroads, and dams (*Courtesy U.S. Geological Survey.*)

tendency is to make the title too large. The title should include the type or purpose of the map, the name of the tract, location of the tract, the scale and contour interval of the map (unless shown elsewhere), name of the engineer and/or draftsman, and the date. A simple form of title is shown in Fig. 10.16.

Explanatory notes or legends are often of assistance in interpreting a drawing. They should be as brief as circumstances will allow, but at the same time should include sufficient information as to leave no doubt in the mind of the person using the drawing. A key to the symbols representing various details ought to be shown unless the symbols are conventional. The nature and source of data upon which the drawing is based ought sometimes to be made known. For example, the data for a map may be obtained from several sources, perhaps partly from old maps, partly from old survey notes, and partly from new surveys; the surveys have been made

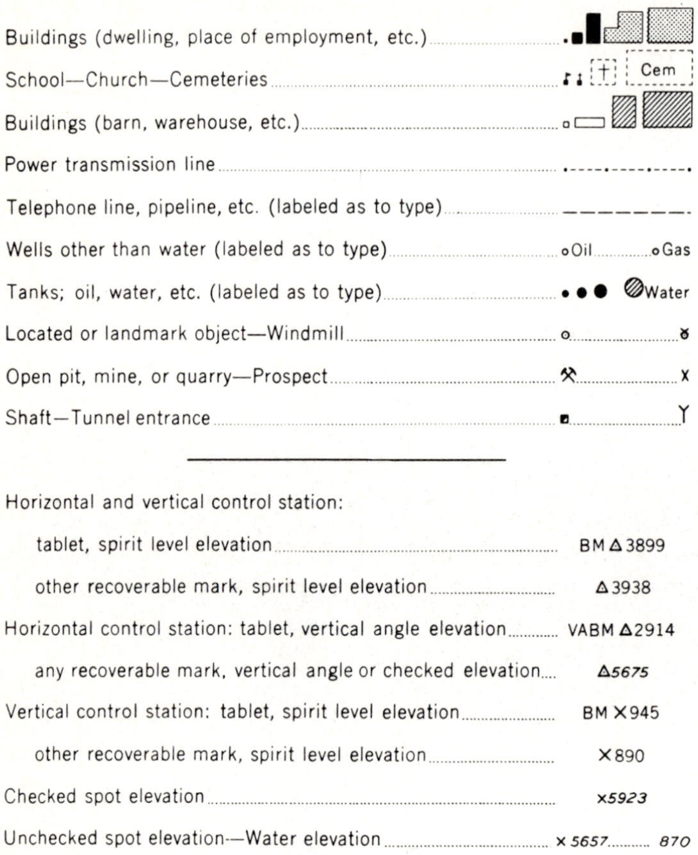

Fig. 10.17*b* Symbols for structures and stations. (*Courtesy U.S. Geological Survey.*)

with a certain precision; the direction of the meridian has been determined by astronomical observation; and elevations are referred to a certain datum as indicated by a certain bench mark of a previous survey.

10.17. Symbols Objects are represented on a map by *symbols*, many of which are conventional. Some topographic map symbols published by the U.S. Geological Survey are shown in Fig. 10.17. A chart showing these symbols is available from the Survey, without charge.

Where feasible, the information shown on topographic maps is distinguished by colors in which the symbols are printed. Black is used for engineered or cultural features such as roads, buildings, names, and boundaries. Blue is used for water or hydrographic features, such as lakes, rivers, canals, and glaciers. Brown is used to show the relief or configuration of the ground surface as portrayed by contours or hachures. Green is used for wooded or other vegetative cover, with typical patterns to show such features as scrub, vineyards, or orchards. Red emphasizes important

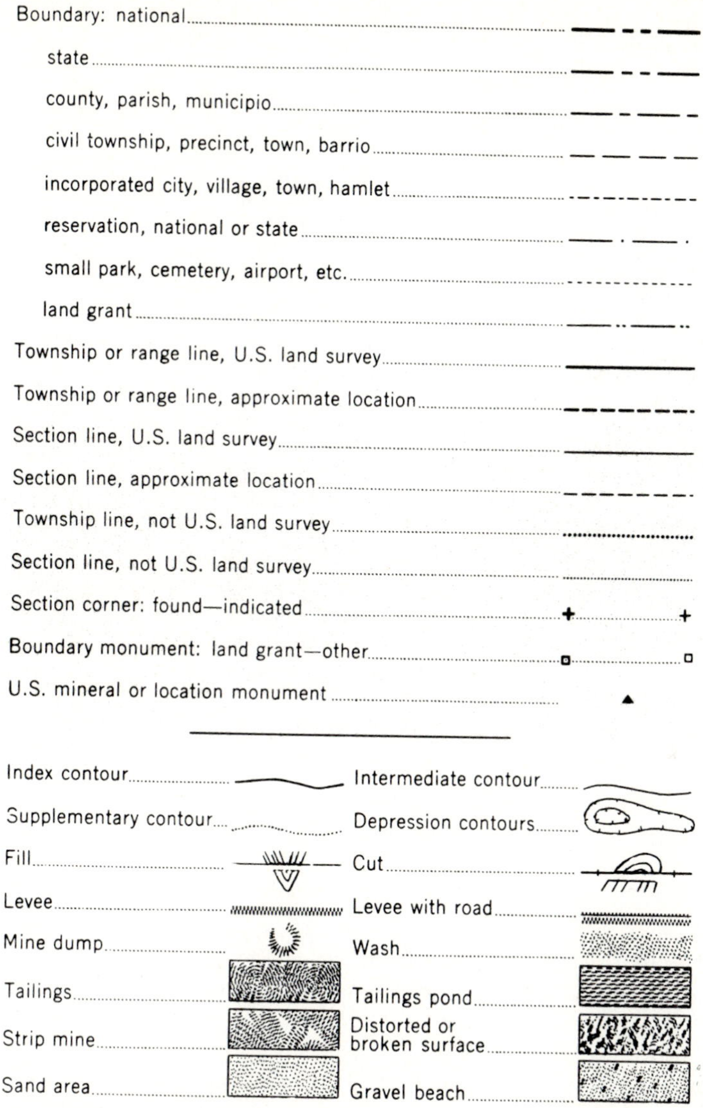

Fig. 10.17c Symbols for boundaries, contours, and excavations. (*Courtesy U.S. Geological Survey.*)

roads and public-land subdivision lines and shows built-up urban areas. Where colors are not available or where the map is to be reproduced by contact or photographic printing and will therefore be in one color, black is used throughout; the conventional shapes of the symbols are the same as where color is used.

The methods used in drawing symbols are illustrated in Fig. 10.18.

Grass The symbol for grass consists of a series of lines drawn radially toward a point.

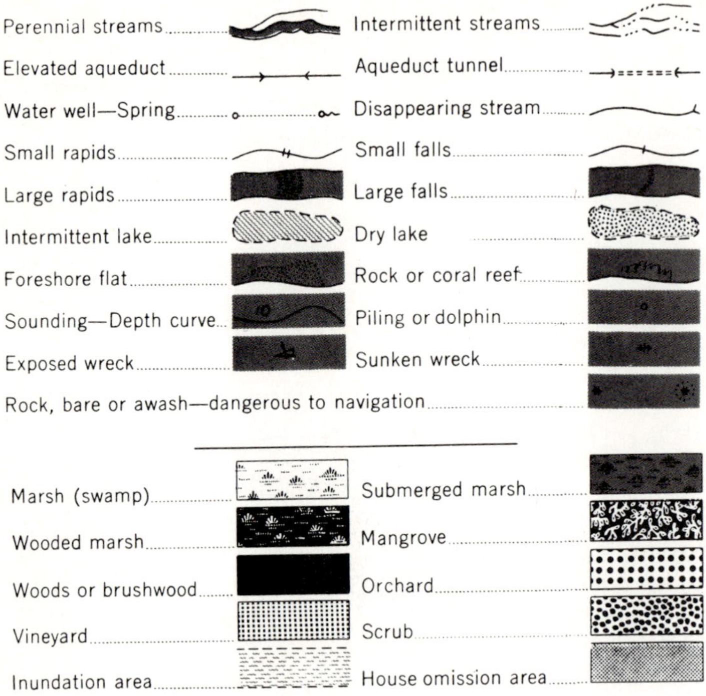

Fig. 10.17*d* Symbols for hydrography and land classification. (*Courtesy U.S. Geological Survey.*)

Marsh The symbol for marsh consists of the grass symbol beneath which the water surface is shown by either a single or a double line drawn slightly longer than the base of the grass tuft.

Trees Tree symbols may be drawn either in plan or in elevation. The latter practice is better adapted to reconnaissance sketches or elevation drawings of a terrain, whereas on most topographic drawings the symbol shown in plan is more suitable.

Water lines To draw the water-line symbol for lakes and ponds, the draftsman begins by sketching the shoreline as a heavy line, then drawing a fine line as close as possible to the shoreline. Fine lines are then drawn successively, each one in close conformity with the preceding line, and the spacing between the lines is increased uniformly outward from the shore.

Contour lines Details concerning the plotting of contours can be found in Art. 10.13.

10.18. Automated mapping The possibility of automated mapping is created when data from modern ground surveys, existing survey records and maps, and

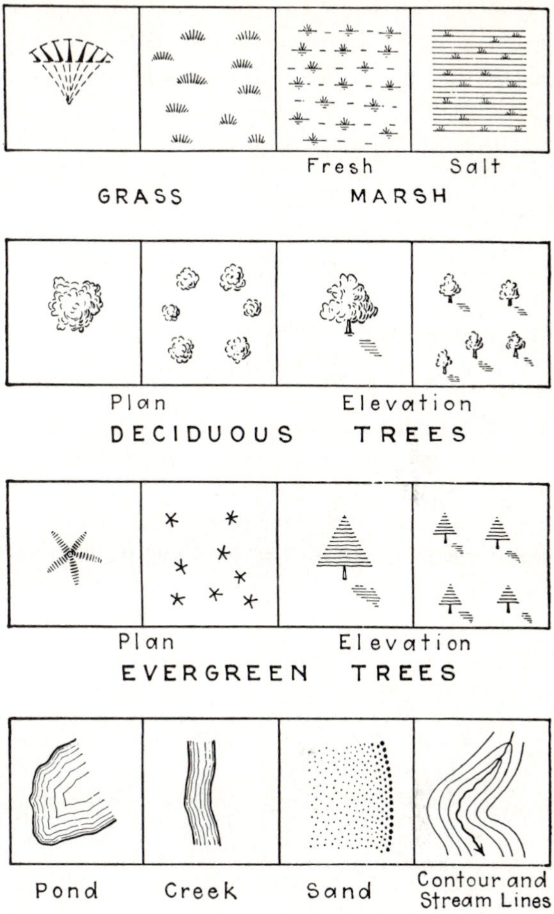

Fig. 10.18 Details of map symbols.

photogrammetric mapping procedures are pooled in a data storage bank under the control of an electronic computer that drives an automated flat-bed plotter.

In Art. 9.14, an electronic surveying system is described which stores the X, Y coordinates and elevations of located points in a solid-state device that can later be connected to a computer and read onto magnetic tape, disk storage, or into core memory. Existing maps, photogrammetric stereomodels (Art. 13.14), and the scanned elements of orthophotos (Art. 13.18) can be digitized and stored on tape, disk, or in core memory. These data constitute a digital terrain model (DTM) (Art. 10.15).

Given a system consisting of an electronic computer interfaced with a flat-bed plotter, the DTM stored in peripheral storage or core memory (called a data bank) can be automatically converted into a topographic map. The automatic plotting of contours is discussed in Art. 10.15. Planimetric details can also be plotted from stored X, Y coordinates of specified objects. Naturally, an extensive library of computer programs or *software* compatible with the electronic computer system is required for processing and plotting the data. A sufficient number of known ground-control points must also be available. Locations for these control points can be established by ground-control surveys, analytical photogrammetric triangulation (Art. 13.16), and/or by one of the electronic positioning systems (Chap. 9).

If only a few copies of the map are needed for design and planning purposes, the final product can be a traditional line map with contours, or an orthophoto map with an overlay of contours. All legends, grid lines, and title blocks can be automatically drawn. When the map is for wide distribution, color separation sheets can be individually compiled automatically. These sheets are then used to prepare plates for map printing.

Map data can be stored in the data bank such that they can be retrieved selectively. Thus, in the foregoing example or preparation of color separation sheets, one could select to plot planimetry, hypsography (contours), forest cover, and hydrography, each separately. On most mapping systems, the capability of selective compilation is controlled by an interactive terminal consisting of a cathode ray tube and keyboard, which gives the operator the option of intervening and controlling the mapping procedure (see Fig. 10.7). This feature not only permits selective plotting but also means that data can be input manually or stored data may be reviewed, edited, and re-stored for compilation. Given the proper programs, calculations with the stored data are also possible. Thus, map revision can be a major function of the system.

Fig. 10.19 Automatically compiled map (orthophoto map with contours). (*Courtesy Riverside County, California, Flood Control District.*)

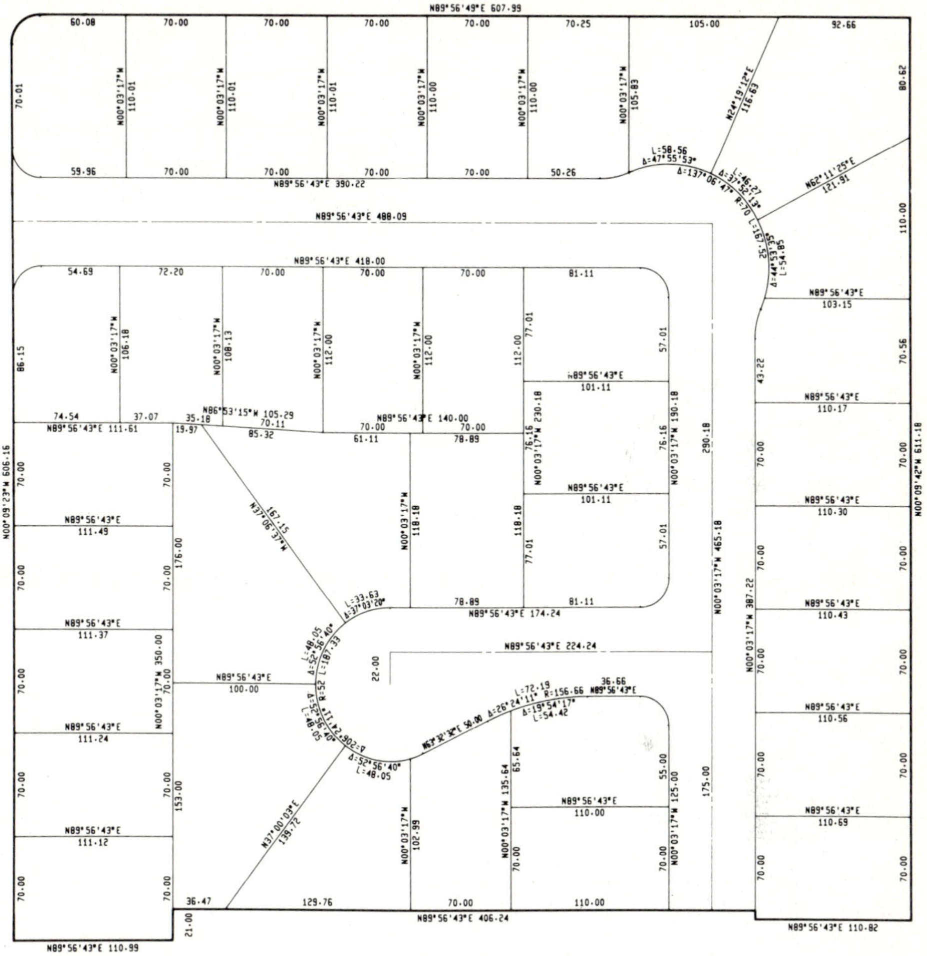

Fig. 10.20 Subdivision map drawn automatically. (*Courtesy CONCAP Computing Systems, Oakland, California.*)

An automatically compiled general planning map consisting of an orthophoto map with an overlay of contours is shown in Fig. 10.19. Figure 10.20 illustrates a portion of a subdivision development map that was calculated and plotted using an interactive terminal interfaced with a minicomputer and a flat-bed plotter.

A useful byproduct of the DTM which can be produced by an automatic mapping system is the capability of compiling three-dimensional perspective terrain models from a specified vantage point in space. Figure 10.21 illustrates a perspective terrain model automatically plotted from a DTM. Design plans can be incorporated in these models thus providing views from as many points as are necessary of the finished project. Planning group members generally find such a display more comprehensible and useful than topographic maps with design plans included.

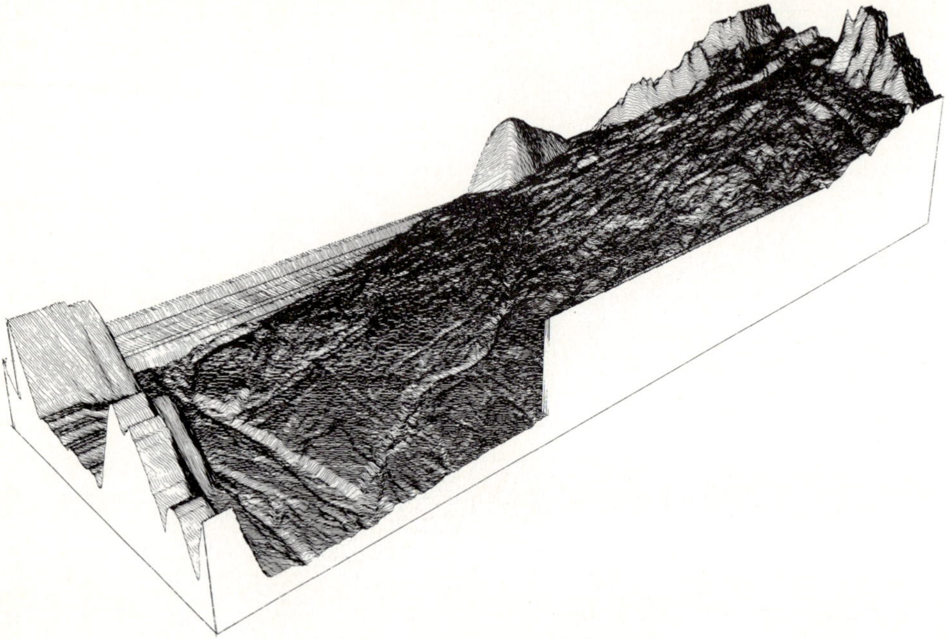

3-D PERSPECTIVE PALM CANYON, PALM SPRINGS, CALIFORNIA - MODELS 1, 2, 3 - X-Y ROTATION=135°

HORIZON ANGLE=20.7° - VIEW ELEVATION=3000' - DISTANCE FROM VIEW POINT TO FOCAL POINT=6000'

Fig. 10.21 Three-dimensional perspective terrain model, Palm Canyon, Palm Springs, California. (*Courtesy Riverside County, California, Flood Control District.*)

10.19. Problems

10.1 Describe the various classes, types, and categories of maps.

10.2 Discuss the difference between planimetric and topographic maps.

10.3 Define horizontal datum and vertical datum as used in the United States.

10.4 What are the different ways in which relief can be represented on a flat sheet of paper?

10.5 Discuss the principal characteristics of contour lines.

10.6 On a map at a scale of 1 : 5000 and with a contour interval of 1 m, the distance scaled on the map between two adjacent contours is 15.0 mm. What is the slope of the ground in percent?

10.7 The following tabulation gives elevations of points over the area of a 60- by 100-ft city lot. The elevations were obtained by the checkerboard method, using 20-ft squares. Point *A*-1 is in the northwest corner of the lot, and point *F*-1 is at the southwest corner. Plot the contours, using a horizontal scale of 1 : 120 (1 in / 10 ft) and a contour interval of 2 ft.

	Elevation, ft			
Point	1	2	3	4
A	322.9	327.0	327.5	328.4
B	326.6	331.0	333.3	332.2
C	327.4	333.3	335.7	333.5
D	326.6	334.6	337.0	334.2
E	327.5	333.0	337.4	337.7
F	328.2	333.6	338.3	341.2

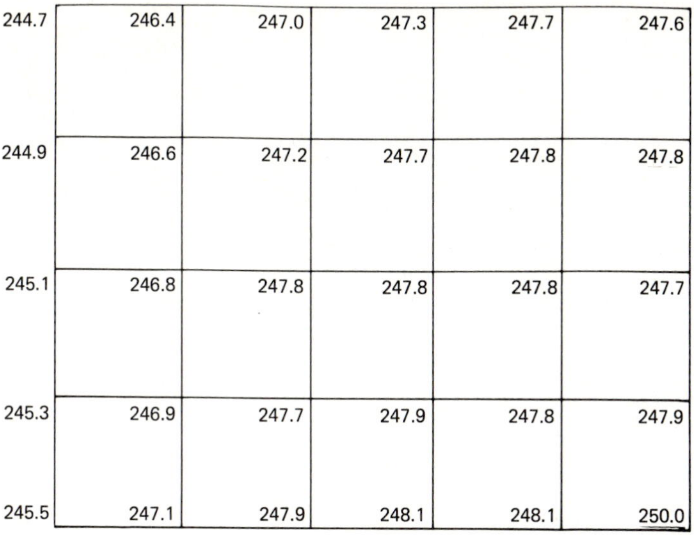

Fig. 10.22

10.8 The grid interval is 50 m for the layout illustrated in Fig. 10.22, and elevations shown at grid intersections are in metres. Draw the layout of grid intersections to a map scale of 1:1500 and plot contours with an interval of 0.2 m.

10.9 It is possible to plot and scale map details to within 0.5 mm. What scale of map is required to permit scaling distances between well-defined points to within 1 m?

10.10 In Prob. 10.9 plotting accuracy was used to select a map scale. What other factors affect the choice of map scale?

10.11 Discuss the criteria that govern the selection of the contour interval for a topographic map.

10.12 Discuss the relationship between contour interval and map scale. What are the principal factors that affect this relationship?

10.13 Unadjusted azimuths, distances, and coordinates for traverse *ABCDEF* are as follows:

Point	Azimuth	Distance, ft	*X*, ft	*Y*, ft
A			62,480.5	49,487.7
	0°00′	338		
B			62,480.5	49,825.7
	351°00′	307		
C			62,432.5	50,128.9
	67°45′	792		
D			63,165.5	50,428.8
	142°30′	822		
E			63,665.9	49,776.7
	244°30′	624		
F			63,102.7	49,508.0
	268°00′	620		
A				

Plot the positions of the traverse stations by the method of coordinates to a scale of 1:2400. Use a coordinate-grid-line interval of 500 ft. Scale the distances between the plotted points as a check on the plotting accuracy. Since the coordinates are unadjusted, there will be an error of closure which is reflected in the distance and direction of the line *FA*. Scale and record these errors in distance and direction.

10.14 What is the most desirable form for data that are to be plotted automatically by a computer controlled flat-bed plotter?

10.15 Discuss the concepts involved in the automatic plotting of contours.

10.16 What information should appear on any map?

10.17 What information should appear on a map of a subdivision that is recorded and becomes a part of the public record?

10.18 What are the major sources of data for automatic mapping procedures?

10.19 What are the major components of an automatic mapping system?

10.20 Discuss the overall concepts involved in the automatic plotting of planimetric details and contours.

10.20. Office problem

TOPOGRAPHIC MAP CONSTRUCTION

Object To construct a complete topoographic map using data collected from field problems in Chap. 12. Relief will be represented by contours.

Procedure

1. Calculate adjusted coordinates and elevations for all horizontal and vertical control points.
2. Select a map scale and contour interval compatible with the size and terrain configuration of the area to be mapped and intended purpose of the map (Art. 10.7). For a student project such as this, it is wise to select an area and scale that will allow plotting the map on a single sheet.
3. Plot the control points on the base map sheet by the coordinate method (Art. 10.10). Label each point, mark the elevation next to each vertical point, and mark the distance and azimuth on each line for which these data are available. The base map should be drawn on a good-quality, heavy detail paper or polyester drafting film.
4. The method for plotting details is a function of the procedure employed to collect data in the field. Use one of the methods for plotting details suggested in Art. 10.12.
5. Plot all details, draw the planimetric features, and draw the contours by interpolation (Art. 10.14). Use the standard symbols and include a title on the base map. When the base map has been smooth-drafted (all line work is in pencil at this point), make a tracing of the base map on a good-quality tracing paper or polyester drafting film. The tracing may be in pencil, ink, or scribed on coated drafting film. Include all the information that is required on a map (Art. 10.16). This final tracing constitutes the map manuscript from which reproductions would be made.

References

1. Ackermann, F., "Experimental Investigation into the Accuracy of Contouring from DTM," *Photogrammetric Engineering and Remote Sensing*, Vol. 44, No. 12, December 1978.
2. American Society of Civil Engineers, Manual on "Map Uses, Scales, and Accuracies for Engineering and Associated Purposes," Unpublished committee report, 1978.
3. American Society of Civil Engineers, "Selection of Maps for Engineering and Planning," Task Committee for Preparation of a Manual on Selection of Map Types, Scales and Accuracies for Engineering and Planning, *Journal of the Surveying and Mapping Division, ASCE*, No. SU1, July 1972.
4. Collins, S. H., "Terrain Parameters Directly from Digital Terrain Model," *The Canadian Surveyor*, Vol. 29, No. 5, December 1975, p. 507.
5. Keates, J. S., *Cartographic Design and Production*, John Wiley & Sons, Inc., New York, 1973.

6. Kloosterman, B., Norgren, R. R., and Sharp, W. R., "Computer Assisted Color Separation for the Production of Thematic Maps," *The Canadian Surveyor*, Vol. 28, No. 1, March 1974.

7. Lee, K. S., et al., "CRT Light Head for Automatic Drafting Systems," *The Canadian Surveyor*, September 1977.

8. Mikhail, Edward, M., *Observations and Least Squares*, Harper & Row, Publishers, New York, 1976, pp. 32, 33.

9. National Oceanic and Atmospheric Administration and U.S. Geological Survey, Stang, Paul R., and Baxter, Franklin S., eds., *Coastal Zone Mapping Handbook*, March 1976.

10. *Proceedings of the Digitial Terrain Models Symposium, American Society of Photogrammetry in cooperation with ASCM*, St. Louis, Mo. May 9–11, 1978, ASP, Falls Church, Va.

11. Robinson, A. H., Sale, R., and Morrison, J., *Elements of Cartography*, 4th ed., John Wiley & Sons, Inc., New York, 1978.

12. Schut, G. H., "Review of Interpolation Methods for Digital Terrain Models," *The Canadian Surveyor*, Vol. 30, No. 5, December 1976, p. 389.

13. Stine, G. E., et al., "Automation in Cartography," *Surveying and Mapping*, December 1971.

14. Thompson, M. M., et al., "On Map Accuracy Specifications," *Surveying and Mapping*, March 1971.

15. Topographic Instructions of the U.S. Geological Survey, *Color-Separation Scribing*, Book 4, Chaps. 4B1–4B3, Government Printing Office, Washington, D.C., 1961.

CHAPTER 11
Map projections and state plane coordinates

11.1. Introduction In plotting a map of a small and limited area, the curvature of the earth need not be considered. A level surface on the earth is assumed to be a plane, and points are plotted on the map in terms of rectangular coordinates from two orthogonal axes representing the east-west and north-south directions.

For maps of larger areas this simple method is not satisfactory because the curvature of the earth can no longer be ignored. The geometric shape of the earth is a *spheroid* with a polar diameter about one-third of 1 percent shorter than the equatorial diameter. Therefore, a plane passing through the equator would cut it in a circle, while a plane through the poles (*meridional* plane) intersects it in an ellipse. However, with that slight difference between polar and equatorial dimensions, the ellipse is very nearly a circle. Consequently, for our purposes here, the earth is assumed to be a *sphere*. This will make it easier for the beginning student to visualize the various map projections to be discussed.

Regardless of whether the earth is considered a sphere or spheroid, it is not possible to develop its *surface exactly* onto a plane, just as it is impossible to flatten a section of orange peel without tearing it. It follows, then, that whatever procedure is used to represent a large area on a map, there will *always* be some distortion. To minimize the distortion as well as to develop several possibilities, points on the map are represented in terms of parallels of latitude and meridians of longitude. Position on the earth in terms of latitude and longitude (e.g., New York City is at 40°45′N latitude and 74°00′W longitude) is transformed into scaled linear dimensions on the map. This is accomplished by using the dimensions of the earth and a selected set of criteria for representing the curved earth on the flat map. Such a transformation from latitude ϕ and longitude λ to a map's X, Y coordinates is the function of *map projections*.

On an *ideal* map *without distortion*, the following conditions must be satisfied: (1) All distances and areas on the map would have correct relative magnitude, (2) all azimuths and angles would be correctly shown on the map, (3) all great circles on the earth would appear as straight lines on the map, and (4) geodetic latitudes and longitudes of all points would be correctly shown on the map.

As noted earlier, because of the shape of the earth, it is impossible to satisfy *all* of these conditions in the same map. It is possible, however, to satisfy one or more of the four conditions. Thus, there result several classes of map projection:

1. *Conformal* projection results in a map showing the correct angle between any pair of short intersecting lines, thus making small areas appear in correct *shape*. As the scale varies from point to point, the shapes of larger areas are incorrect.
2. An *equal-area* projection results in a map showing all areas in proper relative *size*, although these areas may be much out of shape and the map may have other defects.
3. In an *equidistant* projection, distances are correctly represented from one central point to other points on the map.
4. In an *azimuthal* projection, the map shows the correct *direction* or azimuth of any point relative to one central point.

11.2. Types of map projections Although all map projections are carried out by computing X, Y for each pair of ϕ, λ, there are two methods of projection: geometric and mathematical. In *geometric* projection a surface that can be developed into a plane (such as a plane, a cone, or a cylinder) is selected such that it either cuts or is tangent to the earth. A point is then selected as the projection center from which straight lines are connected to points on the earth and extended until they intersect the selected mapping surface. In *mathematical* projection, there is no one particular projection point; instead, an equation is used to compute the location X, Y of the point on the map from its position ϕ, λ on the earth. In the following sections the various projections are introduced in groups, beginning with those on a plane tangent to the earth sphere.

11.3. Map projection to a plane The easiest geometric map projection to visualize is that in which the projection surface is a plane tangent to the sphere at any point. The point of tangency becomes the central point in the map. Then there are three possibilities for the point used as projection center: (1) when the center of the sphere is used (Fig. 11.1), the projection is called *gnomonic*; (2) when the end of the

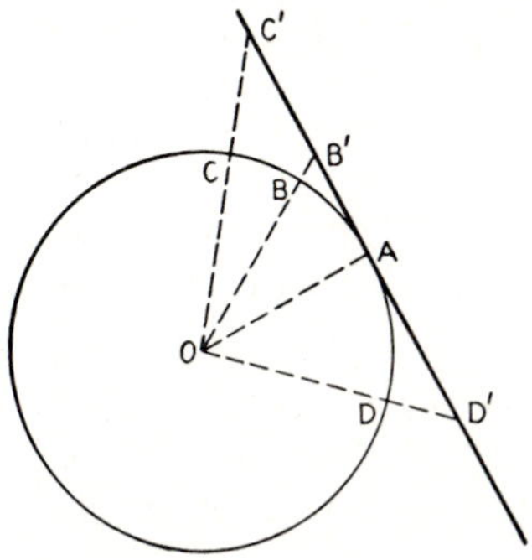

Fig. 11.1 Gnomonic projection.

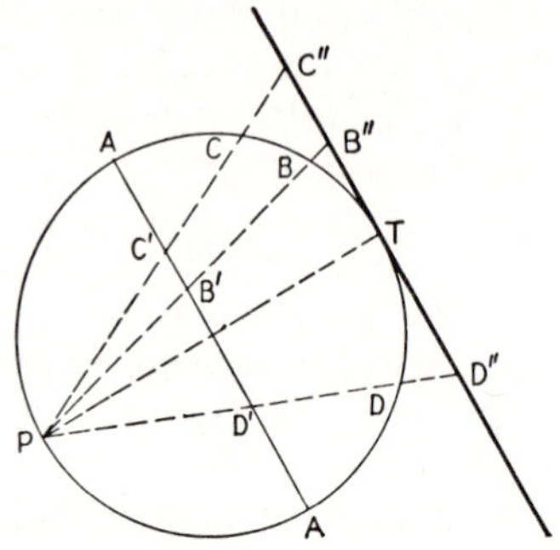

Fig. 11.2 Stereographic projection.

diameter opposite to the point of tangency is used (Fig. 11.2), the projection is called *stereographic*; and (3) when the projection center is at infinity, in which case the projection lines become parallel and perpendicular to the tangent plane, the projection is called *orthographic* (Fig. 11.3).

11.4. Gnomonic projection This, considered to be the oldest true map projection, is assumed to have been devised by the great abstract geometrist Thales in the sixth century B.C. It is a geometric projection to a plane tangent to the sphere at any point such as A in Fig. 11.1, with the projection center at the center of the sphere. Lines from the earth's center 0 through all points to be mapped, A, B, etc., are extended until they intersect the tangent plane in the map points A, B', C', etc. The new points A, B', C', etc., in the plane represent the gnomonic projection of the original points. This projection has the following properties:

1. Great circles show as straight lines and therefore meridians are straight lines.
2. Shapes and sizes undergo extreme distortion as we go farther from the central tangent point.
3. Azimuths of lines drawn from the tangent point to other map points are correct, because such lines are great circles. Therefore, this is an *azimuthal* projection.
4. Since the point of projection is at the center of the sphere, it is not possible to map an entire hemisphere.

This projection is used mostly for navigational purposes. For other purposes only a limited area around the central point may be used.

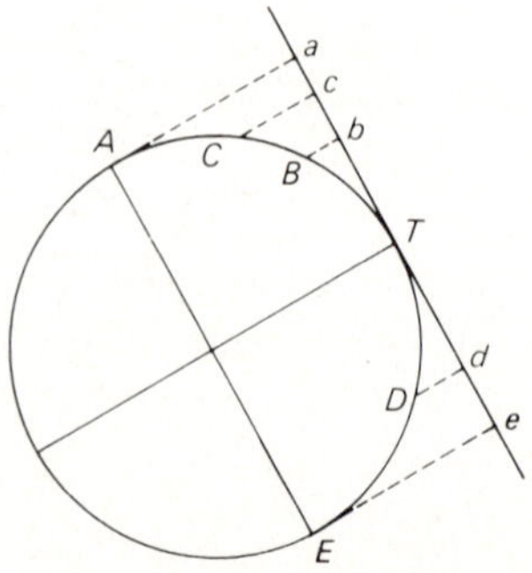

Fig. 11.3 Orthographic projection.

11.5. Stereographic projection This projection is credited to the Greek astronomer Hipparchus dating back to the second century B.C. It is a geometric projection to a tangent plane with the point of projection *P* diametrically opposite to the point of tangency *T* (Fig. 11.2). Rays from *P* to surface points *B*, *C*, etc., are extended until they intersect the tangent plane. The points of intersection B'', C'', etc., are the map points on the stereographic projection on the tangent plane. On any other intersecting plane, such as *A*-*A* in Fig. 11.2, an equivalent projection is possible, with only a reduction scale factor from the projection on the tangent plane.

This projection is both azimuthal and conformal. It has the following properties:

1. It is the only projection on which circles on the earth still appear as circles on the map.
2. Like the gnomonic projection, azimuths of lines from the central point are correct because such lines are great circles.
3. The scale increases as we move away from the tangent point and therefore its main defect is that areas are not correctly shown.

This is an excellent projection for general maps showing a hemisphere. It is also used for plotting ranges from central objects and for navigational purposes in higher latitudes, above 80°, when the plane is tangent at the pole.

11.6. Orthographic projection This is an ancient projection, which was used during the Renaissance for artistic representation of the globe. Recent space applications have revived its use. It is again a geometric projection on a tangent plane, with the projection lines parallel to each other and perpendicular to the tangent plane (Fig. 11.3). If the tangent point is one of the poles, each parallel of latitude will be shown correctly to scale as a circle, but the distance between successive parallels becomes rapidly smaller as we move farther from the center of the map. This means that unlike the gnomonic and stereographic projections, the scale in the orthographic projection *decreases* away from the central tangent point. On the other hand, this is an azimuthal projection just like the gnomonic and stereographic, so that the azimuths from the tangent point are correct since the lines through it are great circles. The representation of the globe on this projection appears as it would in a photograph taken from deep space. Therefore, the orthographic projection is also used for map representations of the moon and planets.

11.7. Conical projections Unlike a sphere, both a cone and a cylinder can be developed into a plane without distortion, and therefore both are used for map projection. A cone fits over a sphere, touching it in a small circle of latitude which is called the *standard parallel*, when the cone apex lies on the polar axis (Fig. 11.4*a*). As the height of the cone increases, the standard parallel gets closer to the equator (Fig. 11.4*b*). Finally, when the standard parallel reaches the equator, the elements of the cone become parallel, and the cone becomes a cylinder (Fig. 11.4*c*). When the cone height decreases, the standard parallel moves to higher latitudes (Fig. 11.4*d*), and finally the cone becomes a plane touching the sphere at one point, the pole (Fig.

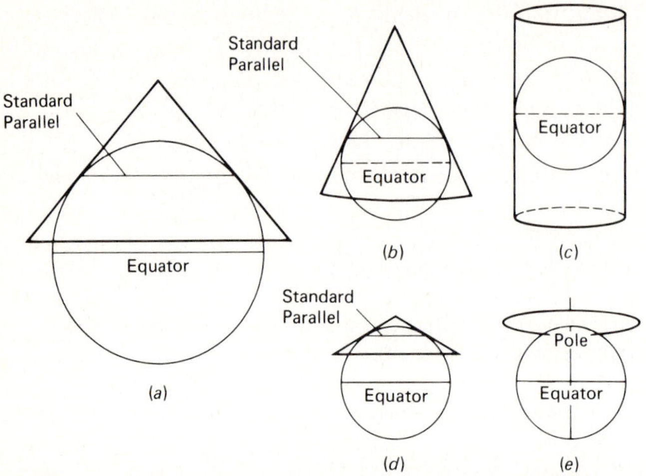

Fig. 11.4 (*a*) Cone tangent to a sphere. (*b*) Cone tangent to a sphere with a standard tangent closer to the equator. (*c*) Cylinder tangent to a sphere at the equator. (*d*) Cone with standard parallel at higher latitudes. (*e*) Plane tangent to a sphere at the pole. (*Courtesy National Geographic Society.*)

11.4*e*). Thus, the projection of a sphere to a plane and a cylinder are actually limiting cases (at least geometrically) to conical projection.

Since a tangent cone has one parallel of latitude common to the sphere, its representation on the map will be at true scale, while scale distortions increase as the distance is extended to the north and the south of the parallel. For this reason, many conical projections use a cone that intersects the sphere in *two standard parallels* to minimize scale distortions. Although *geometric* projection on the cone is possible, the more important and useful conical projections are based on *mathematical* projection.

11.8. Albers equal-area projection This is a mathematical projection on a cone with two standard parallels, developed by H. C. Albers in 1805. Its main characteristic is that small areas on the map are equal, at the scale of the map, to corresponding small areas on the earth. Other properties of this projection include:

1. Scale is true along the standard parallels, is small between them, and large beyond them along *other parallels*. *Meridional* scale is large between and small beyond standard parallels.
2. Distances and directions measured on the map are reasonably accurate.
3. Meridians are straight lines meeting at a point that is the center of the concentric circular arcs representing the parallels. Therefore, meridians and parallels meet at right angles.
4. Parallel spacing increases away from the standard parallels.

This projection is useful for maps requiring equal-area representation and is used for areas at midlatitudes with extensive east-west dimensions.

11.9. Polyconic projection Instead of a single cone, a series of conical surfaces may be used, points on the surface of the earth being considered as projected to a

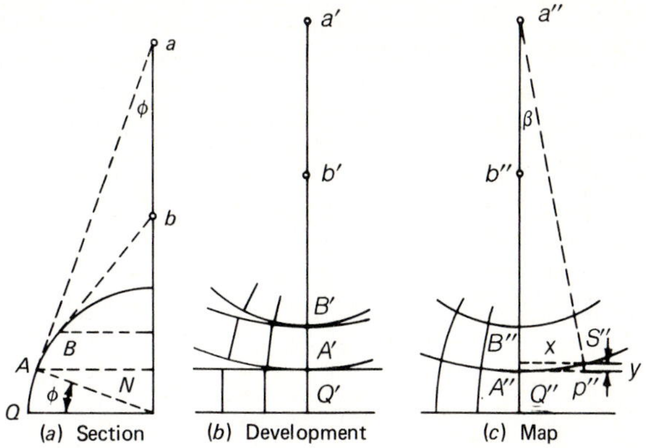

Fig. 11.5 Polyconic projection.

series of frustums of cones that are fitted together. These conical surfaces are then developed each way from a central meridian. Owing to differences in radii, the resulting strips would not exactly fit together when laid flat, but spaces would appear between them, such spaces increasing in width as the distance from the central meridian increases (Fig. 11.5*b*). To avoid such spaces, the north-south scale must be modified along the various meridians. Upon such a system of lines, points such as point S'', Fig. 11.5*c*, are plotted by latitude and longitude.

In Fig. 11.5 it is seen that each parallel of latitude appears on the map as the arc of a circle having as radius the corresponding tangent distance; the parallel through A has a radius Aa, that through B has a radius Bb, and so on. The centers of these circles all lie on the central meridian of the map. The length of each tangent distance Aa, etc., is $N \cot \phi$, in which N is the length of the normal or vertical at latitude ϕ extended to its intersection with the earth's axis. For the assumption that the earth is a sphere, N is equal to the radius of the sphere. More precisely, since the earth is actually an ellipsoid in shape,

$$N = \frac{a}{\sqrt{1 - \varepsilon^2 \sin^2 \phi}} \tag{11.1}$$

in which a is the earth's equatorial radius and ε is the eccentricity of the ellipse in the meridian section ($\varepsilon^2 = 0.0067686580$). The length of the tangent distance $N \cot \phi$ varies with the latitude.

The distances $Q'A'$, $A'B'$, etc., along the central meridian on the map (Fig. 11.5*b*) are true scale representations of the corresponding arc distances QA, AB, etc., in the meridian section (Fig. 11.5*a*). The parallels drawn on the map may be selected with as small a difference in latitude as may be desired, and each one is drawn with its own particular radius as shown, with the center of the arc on the central meridian at the proper tangent distance (to scale) above the point where the parallel cuts the central meridian.

It should be observed that the method of drawing these arcs of the parallels of latitude on the map is such that each parallel is separately developed as the circumference of the base of its own distinct cone, and that the spacing between them increases with increasing differences of longitude from the central meridian, thereby changing the north-south scale of the map from place to place as the longitude difference increases.

These meridians are curved, concave toward the central meridian, but if the parallels are drawn close enough together, each meridian may be drawn as a series of straight lines from parallel to parallel. On the network of parallels and meridians so prepared, points are plotted by latitude and longitude. Near the central meridian there is little error in such a map, but the error increases in proportion to the square of the difference in longitude along any one parallel. The variation with difference in latitude is not in direct proportion.

It is to be noted that along the central meridian and along every parallel the map is true to scale; that along the other meridians the scale is somewhat changed; that near the central meridian the parallels and meridians intersect nearly at right angles; and that areas of great extent north and south may be mapped with a very small distortion.

Although better adapted to mapping an area of great extent in latitude than for an area of great extent east and west, the polyconic projection is sufficiently accurate for maps of considerable areas, and it is widely used by the U.S. Geological Survey and the U.S. National Ocean Survey.

11.10. Conformal mapping Conformal projection was defined in the introduction as that which preserves the angle between any pair of intersecting short line segments. The cartesian coordinates X, Y on the map are calculated from the geodetic coordinates ϕ, λ using complicated functions which may be written symbolically as

$$\begin{aligned} X &= f_1(\lambda, \phi) \\ Y &= f_2(\lambda, \phi) \end{aligned} \tag{11.2}$$

In order for the map projection to be conformal, additional mathematical conditions are imposed on Eqs. (11.2). The details of these relations are beyond the scope of this discussion and for them the reader can consult Ref. 2 at the end of this chapter.

11.11. Lambert conformal conic projection This projection was developed by Johann Heinrich Lambert in 1772, the same year in which he developed the transverse Mercator projection. The Lambert conformal conic projection is one of the most widely used projections in the United States and even worldwide. It is used for the state plane coordinate systems of states (or zones thereof) of greater east-west than north-south extent. It is a conic projection with two standard parallels, as shown in Fig. 11.6. Therefore, meridians are straight lines meeting in a point outside the map limits; parallels are arcs of concentric circles, and both sets of lines meet at right angles. Scale is true only along the two standard parallels; it is compressed between and expanded beyond them.

For example, along the two standard parallels, a distance calculated from grid coordinates (i.e., coordinates on the map projection) will be the same as the corresponding geodetic distance at sea level. Anywhere else within the limits of the

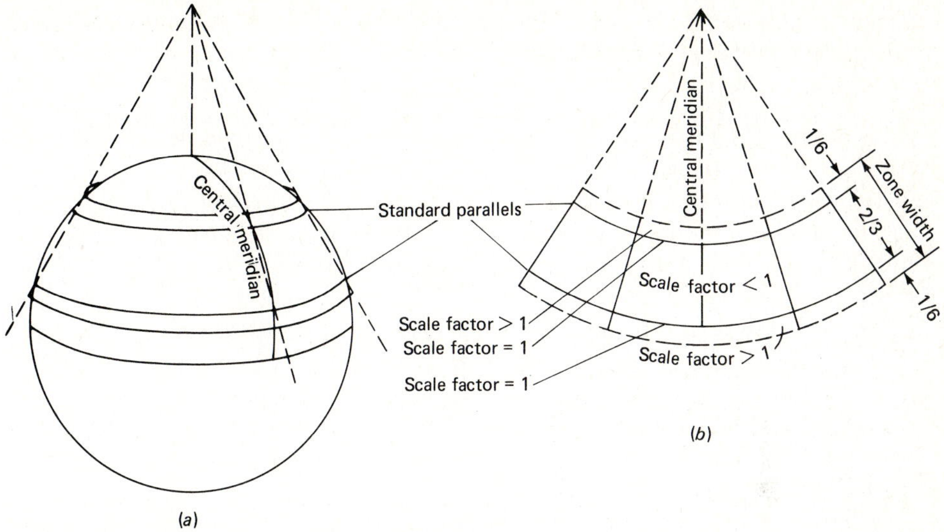

Fig. 11.6 (*a*) and (*b*) Lambert conformal conic projection.

projection, these two distances will *not* be the same because of scale distortion arising from the map projection. The *scale factor k* is defined as that factor by which the sea-level geodetic distance is multiplied to yield the grid (or map projection) distance. As shown in Fig. 11.6*b*, k is less than unity inside the two standard parallels; that is, sea-level distances are compressed to smaller grid distances. Conversely, k is larger than unity outside the two standard parallels so that the sea-level distances are expanded to larger grid distances.

The equations for this projection, which must satisfy the conformality conditions, are given in Art. 11.17 [Eqs. (11.4) through (11.6)]. The apex of the cone is the center of the concentric circles representing the parallels and the point in which all the meridians intersect. Unlike the Albers projection, the spacing of parallels increases away from the standard parallels. For a map of the United States on this projection, scale errors at any point need not exceed 2 percent. Computation of state plane coordinates in the Lambert conformal conic projection is explained in detail in Art. 11.17.

11.12. Mercator projection This is a projection on a cylinder tangent to the earth at the equator (Fig. 11.4*c*). It is a *mathematical* projection, however, and not geometric in order to enforce the conformality conditions. It was created by Mercator in 1569 as a result of his efforts to have *rhumb* line, or *loxodrome*, or the line of constant bearing on the globe, appear as a straight line on the map. Thus, true scale is only at the equator. The meridians are equally spaced vertical straight lines, and parallels are unequally spaced horizontal lines. The parallel spacing increases toward the poles, which in this projection are at infinity on the map. Thus, while the scale at any one point of intersection of a meridian and a parallel is *equal in all directions around the intersection* (which is the basis of conformality), the scale expands rapidly at high latitudes. Thus, if s is the scale number at the equator and s_ϕ is the scale number at any latitude ϕ, then

$$s_\phi = s \cos \phi \tag{11.3}$$

As an example, if the scale at the equator is 1:20,000, it would be 1:10,000 at a latitude $\phi = 60°$ because $s_\phi = s \cos 60° = (20{,}000)(0.5) = 10{,}000$.

Aside from conformality, the particular feature of the Mercator projection is that the rhumb line on the earth is plotted as a straight line on the map, a property that renders it invaluable for purposes of navigation. The shortest course between two points is determined by drawing a great circle on a gnomonic chart where it will appear as a straight line. Selected points, at convenient distances apart, of this great circle are then plotted on the Mercator chart, after making any necessary corrections on account of shoals, wind, currents, etc. The rhumb line connecting any two adjacent points indicates the true bearing of the course, which is read by means of a protractor. This true bearing, corrected for magnetic declination, gives the compass bearing to be used in steering.

Owing to the rapid variation of scale, maps constructed on the Mercator projection give very inaccurate information as to relative sizes of areas in widely different latitudes. For example, on the map Greenland appears larger than South America, whereas in fact South America is nine times as large as Greenland. Consequently, such a map is not suited to general use, although because of its many other advantages it is widely published.

11.13. Transverse Mercator projection This projection was developed by Lambert in 1772, analytically derived by Gauss 50 years later, and then formulas more suitable for calculations were devised by Kruger in 1912. This is perhaps the reason that it is one of the most widely used conformal map projections.

A transverse Mercator projection is the ordinary Mercator projection turned through a 90° angle so that it is related to a central meridian in the same way that

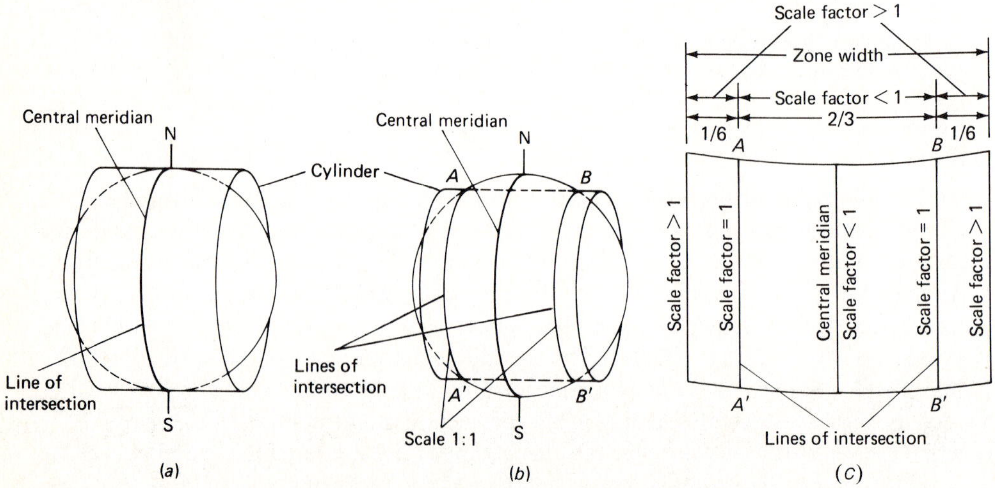

Fig. 11.7 Transverse Mercator projection. (*a*) Cylinder with one standard line. (*b*) Cylinder with two standard lines. (*c*) Schematic of a zone of the transverse Mercator projection.

the ordinary Mercator is related to the equator (Fig. 11.7*a*). Because the cylinder is tangent to the globe at a meridian, the scale is true along that meridian, which is called the *central meridian*. Although, like the regular Mercator, this projection is conformal, it does not retain the straight-rhumb-line property of the Mercator. Other properties of this projection include: (1) both the central meridian and a line perpendicular to it are represented by straight lines; (2) other meridians are complex curves that are concave toward the central meridian; (3) parallels are concave curves toward the pole; and (4) the scale is true only along the central meridian.

This projection is used for the UTM system of coordinates (see Art. 11.14) and for the state plane coordinate systems of states (or zones thereof) of greater north-south than east-west extent. For these systems the transverse Mercator projection cylinder is made to cut the surface of the sphere along two standard lines parallel to the central meridian instead of being tangent to the sphere as in the ordinary Mercator projection (Fig. 11.7*b*).

Figure 11.7*c* is a schematic of a zone of the transverse Mercator projection. Two-thirds of the zone is inside the two intersection meridians, and one-sixth of the zone on either side. The scale is true along the intersection meridians, compressed between, and expanded beyond them. Thus, as shown in Fig. 11.7*c*, the scale factor k (defined in Art. 11.11) is unity along the intersection meridians, less than unity between them, and greater than unity beyond these meridians.

For the state plane coordinate systems, zones are limited in width to about 158 mi (about 254 km) so that the scale distortion from the projection does not exceed 1 part in 10,000 (i.e., scale factor of 0.9999). For the UTM system (see Art. 11.14), a wider zone is used and the allowable scale distortion is one part in 2500 (a scale factor of 0.9996).

11.14. Universal transverse Mercator (UTM) This projection is based entirely on the transverse Mercator projection discussed in the preceding section. Its specifications are (UTM 1958):

1. Transverse Mercator projection is in zones that are 6° wide.
2. The reference ellipsoid is Clarke 1866 in North America.

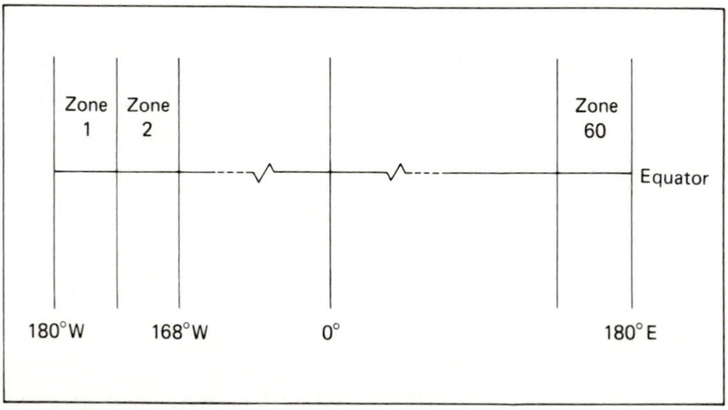

Fig. 11.8 Universal transverse Mercator zones.

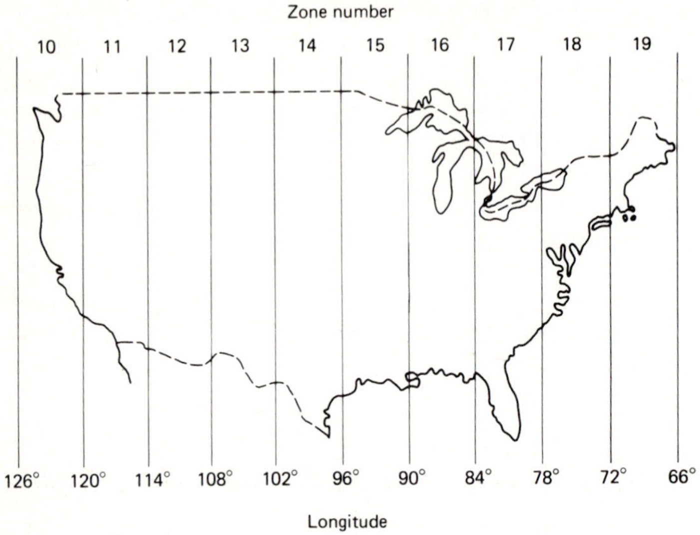

Fig. 11.9 UTM zones in the United States.

3. The origin of longitude is at the central meridian.
4. The origin of latitude is at the equator.
5. The unit of measure is the metre.
6. For the southern hemisphere, a false northing of 10,000,000 m is used (to correspond to the equator).
7. A false easting of 500,000 m is used for the central meridian of each zone.
8. The scale factor at the central meridian is 0.9996.
9. The zones are numbered beginning with 1 for the zones between the meridians of 180° W and 174° W and increasing to 60 for the zone between meridians 174° E and 180° E (Fig. 11.8).
10. The latitude for the system varies from 80° N to 80° S.

The UTM zones in the United States are shown in Fig. 11.9. An individual zone is illustrated in Fig. 11.10. Interested readers should consult Ref. 2 at the end of this chapter for more details on the UTM projection.

11.15. Transverse Mercator in 3° zones The UTM projection system is the transverse Mercator projection in 6° zones. It is also possible to have a transverse Mercator projection in 3° zones. Since the zone width is only 3° and the scale factor of 0.9999 is assigned to the central meridian, a smaller scale error can be expected throughout the zone. This scale error is 1/10,000 on the central meridian as compared to 1/2500 for the UTM.

11.16. State plane coordinate systems Each state in the United States has a plane coordinate system which is based on one or more zones of either the Lambert conic conformal projection (Art. 11.11) or the transverse Mercator projection (Art.

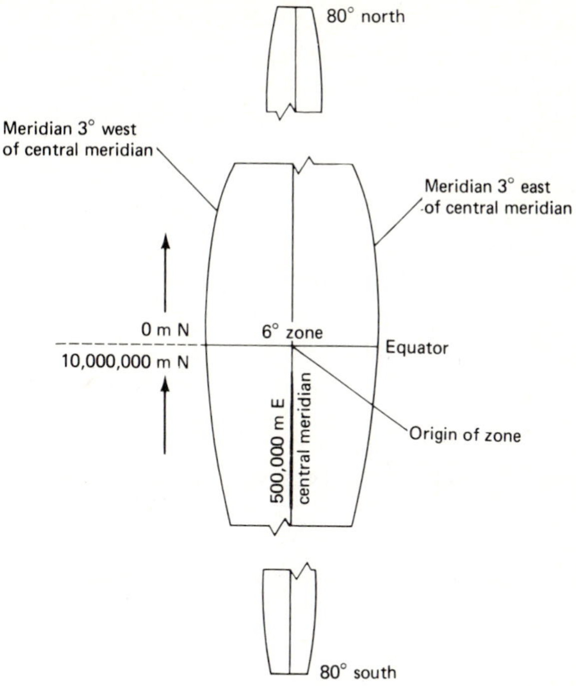

Fig. 11.10 *X* and *Y* coordinates of the origin of a UTM grid zone. *(From Ref. 17.)*

11.13). The Lambert projection is used in states having a large east-west dimension; the transverse Mercator projection is employed for states with a longer north-south dimension. If the north-south dimension of the Lambert projection and the east-west dimension of the transverse Mercator projection are held to 158 mi or 254 km, the maximum distortion in the projection is 1 part in 10,000. In states that exceed these dimensions, more than one zone of the same projection or perhaps two different projections are used. For example, New York has three transverse Mercator zones and one Lambert zone which covers Long Island; two transverse Mercator zones cover the peninsula of Florida, while a Lambert zone covers the western portion of the state; California has seven Lambert zones; and Vermont is covered by one transverse Mercator zone. There are a total of 126 zones (including Alaska and Hawaii) in the state plane coordinate systems in the United States. These systems are ratified by legislation in 35 states.

The present state plane coordinate systems, introduced in the early 1930s and tied to the 1927 North American Datum, utilize the foot as the basic unit.

When the North American Datum is redefined (scheduled to occur in 1985) two plane coordinate systems are to be published: (1) the universal transverse Mercator (UTM 1958, as described in Art. 11.14), and (2) the state plane coordinate (SPC) system. The constants used to define both of these systems will be in metric units and both systems will be referred to the North American Datum. Thus, the present state plane coordinate systems are to be retained with only a change in units and a slight shift in the respective origins.

Since new parameters and projection tables are not yet available, all state plane coordinate computations demonstrated in the following sections are in feet, using constants defined according to the 1927 North American Datum. The basic theory related to computing X and Y coordinates from projection parameters will remain the same.

11.17. State plane coordinates using the Lambert conformal conic projection

The coordinate system is illustrated in Fig. 11.11 which shows that the origin of coordinates is shifted from the cone apex A by a Y value of R_b and the central meridian is given an X value designated by C. It follows from the figure that the state plane coordinates are given by

$$\begin{aligned} X &= C + R\sin \ell(\lambda_0 - \lambda) = C + R\sin\theta \\ Y &= R_b - R\cos \ell(\lambda_0 - \lambda) = R_b - R\cos\theta \end{aligned} \tag{11.4}$$

Values of C, ℓ, and R_b are given for each state plane coordinate system or zone thereof. In addition, values for R are tabulated in terms of the latitude ϕ, and θ is tabulated as a function of the longitude λ (see Table 11.1).

If the latitude ϕ and the longitude λ are required from given coordinates X, Y (called the *inverse transformation*), they may be obtained by rearranging Eqs. (11.4) or

$$R\sin\theta = X - C$$
$$R\cos\theta = R_b - Y$$

Division of the first by the second equation yields

$$\tan\theta = \frac{X - C}{R_b - Y}$$

or

$$\theta = \arctan\left(\frac{X - C}{R_b - Y}\right) \tag{11.5a}$$

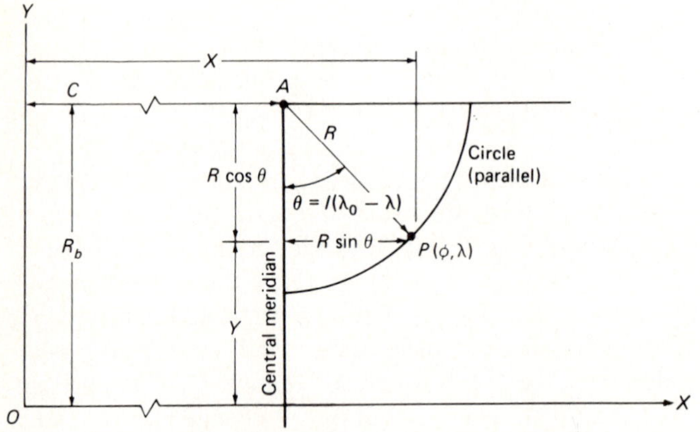

Fig. 11.11 State plane coordinates on the Lambert projection.

Table 11.1 Lambert Projection for California Zone I
Table I

Latitude	R, ft	Y', y value on central meridian feet	Tabular difference for 1″ of latitude	Scale expressed as a ratio
41°16′	24,088,135.79	704,300.44	101.20383	0.9999230
17	24,082,063.56	710,372.67	101.20433	0.9999252
18	24,075,991.30	716,444.93	101.20483	0.9999275
19	24,069,919.01	722,517.22	101.20533	0.9999299
20	24,063,846.69	728,589.54	101.20583	0.9999324
41°21′	24,057,774.34	734,661.89	101.20650	0.9999350
22	24,051,701.95	740,734.28	101.20717	0.9999376
23	24,045,629.52	746,806.71	101.20767	0.9999404
24	24,039,557.06	752,879.17	101.20817	0.9999432
25	24,033,484.57	758,951.66	101.20883	0.9999461
41°26′	24,027,412.04	765,024.19	101.20950	0.9999491
27	24,021,339.47	771,096.76	101.21000	0.9999522
28	24,015,266.87	777,169.36	101.21067	0.9999554
29	24,009,194.23	783,242.00	101.21133	0.9999586
30	24,003,121.55	789,314.68	101.21183	0.9999619

Table II

Longitude		θ	
120°56′	+0°	41′	50.9158″
57	+0°	41′	11.6827″
58	+0°	40′	32.4497″
59	+0°	39′	53.2166″
121°00′	+0°	39′	13.9836″
121°01′	+0°	38′	34.7505″
02	+0°	37′	55.5174″
03	+0°	37′	16.2844″
04	+0°	36′	37.0513″
05	+0°	35′	57.8183″
121°06′	+0°	35′	18.5852″
07	+0°	34′	39.3521″
08	+0°	34′	00.1191″
09	+0°	33′	20.8860″
10	+0°	32′	41.6530″

$C = 2{,}000{,}000$ $\qquad Y_0 = 547{,}078.17$

Central meridian = 122°00′ $\qquad \ell = 0.6538843192$

$R_b = 24{,}792{,}436.23$ ft $\qquad \dfrac{1}{2\rho_0^2 \sin 1''} = (2.38)(10^{-10})$

Then

$$\lambda_0 - \lambda = \frac{\theta}{\ell} \tag{11.5b}$$

where λ_0 is the longitude of the central meridian of the zone. Also,

$$R = \frac{R_b - Y}{\cos\theta} = \frac{X - C}{\sin\phi} \tag{11.6}$$

Then from the state plane coordinate tables available, the value of ϕ is obtained directly as a function of R except for linearly interpolating for the seconds portion of the angle.

The U.S. National Ocean Survey publishes tables for each state plane coordinate projection (Ref. 4). These tables can be obtained from the U.S. Government Printing Office or from NOAA, National Geodetic Information Center OA/C184, 6001 Executive Blvd., Rockville, Maryland 20852. A portion of the *Plane Coordinate Projection Tables* for California, Spec. Publ. No. 253 (1952), is illustrated in Table 11.1. Table I in this publication gives the value for R in feet for every minute of latitude within the zone. The tabular difference for 01″ of latitude is provided for the purpose of obtaining the values sought by interpolation. Thus, given ϕ to some decimal part of a second, R can be determined to the nearest hundredth of a foot. Conversely, given R, the latitude can be found. The scale of the projection is given as a ratio for each minute of latitude. Table II contains the mapping angle θ [it can be derived from Eq. (11.5*b*)] for each minute of longitude. The constants listed at the bottom include ℓ, C, R_b discussed previously and $1/2\rho_0^2 \sin 1''$ used in converting from grid to geodetic or geodetic to grid azimuth for long lines.

More recently the U.S. National Ocean Survey has published a different set of tables with constants for use in computations with automatic data-processing equipment. Like the rigorous procedure, this topic is considered beyond the scope of this text. Interested readers should consult Ref. 2 at the end of this chapter for further details.

Example 11.1 The latitude and longitude for a control point 8 in northern California zone I are

$\phi = 41°25'01.100''$N $\lambda = 121°02'20.500''$W

Compute X and Y plane coordinates for the point on the California Lambert projection zone I, using data in Table 11.1.

SOLUTION Obtain R by interpolation from Table I of Table 11.1:

R for 41°25′	=	24,033,484.57
(101.20883)(1.1″)	= −	111.33
R for 41°25′01.100″	=	24,033, 373.24

Next, find θ by interpolation from Table II of Table 11.1:

θ for 121°02′ $= +0°37′55.5174″$

$\Delta\theta = \ell\Delta\lambda = (0.6538843192)(20.500″) = - 13.4046″$

θ for 121°02′20.500″ $= 0°37′42.1128″$

$\theta = 0.6283646667°$ $\quad \sin\theta = 0.0109668124$
$\cos\theta = 0.9999398627$
Then, by Eq. (11.4),

$$R \sin\theta = 263{,}569.50$$
$$C = 2{,}000{,}000.00$$
$$X_8 = 2{,}263{,}569.50 \text{ ft}$$

$$R_b = 24{,}792{,}436.23$$
$$-R\cos\theta = -24{,}031{,}927.94$$
$$Y_8 = 760{,}508.29 \text{ ft}$$

Example 11.2 The X and Y California state plane coordinates, zone 1, for control point 10 are $X = 2{,}269{,}500.94$ ft and $Y = 778{,}680.42$ ft. Compute the latitude and longitude for the point.

SOLUTION First, calculate the longitude. From Table 11.1, $\ell = 0.6538843192$, $R_b = 24{,}792{,}436.23$ ft, $C = 2{,}000{,}000$ ft. From Fig. 11.11 and using Eq. (11.5*a*),

$$\theta = \arctan \frac{X - C}{R_b - Y}$$

Thus,

$$\theta = \arctan\left[\frac{(2{,}269{,}500.94) - (2{,}000{,}000)}{(24{,}792{,}436.23) - (778{,}680.42)}\right] = \arctan 0.01122277340 = 0.°6429905562$$

or

$$\theta = 0°38′34.7660″$$

With this value of θ and ℓ from the tables, compute $\lambda_0 - \lambda$ with Eq. (11.5*b*), in which θ is expressed in degrees:

$$\lambda_0 - \lambda = \frac{\theta}{\ell} = \frac{0.°6429905562}{0.6538843192}$$
$$\lambda = 122° - 0.°9833399232 = 121.°0166601$$
$$= 121°00′59.9764″\text{W}$$

Next compute the latitude. From the computation above,

$$R_b - Y = 24{,}013{,}755.81 \text{ ft}$$
$$\cos\theta = 0.9999370306$$

Solve for R using Eq. (11.6):

$$R = \frac{R_b - Y}{\cos\theta} = 24{,}015{,}268.04 \text{ ft}$$

Using the extract from Special Publication 253 in Table 11.1, determine the latitude ϕ, by interpolation from Table 1 as follows:

R for 41°27′	= 24,021,339.47
R for ϕ	= 24,015,268.04
Difference	6,071.43

Also from Table I, column 4, the tabular difference for 1″ is 101.21000, so that

$$\text{difference in seconds} = \frac{6071.43}{101.21000} = 59.9884''$$

and the latitude is

$$\phi = 41°27'59.9884''\text{N}$$

The method presented here, and demonstrated by Examples 11.1 and 11.2, represents the simplest approach. Practitioners interested in using the electronic computer for state plane coordinate computation should use formulations more suitable for programming (see Ref. 3 at the end of this chapter).

11.18. Grid and geodetic azimuths and distances on the Lambert conformal projection In Fig. 11.12 the line AP from the apex of the cone to point P is a geodetic meridian and all lines parallel to the central meridian are grid-north meridians. The azimuth of a line between two points can be referred to a grid-north meridian by *grid azimuth* or to a geodetic meridian by *geodetic azimuth*. For lines of up to 5 mi or 8 km in length, grid and geodetic azimuths differ by the mapping angle θ, as shown in Fig. 11.12. As demonstrated in Art. 11.17, θ can be obtained as a function of λ from Table II of Table 11.1 in the state plane coordinate tables (Example 11.1), or can be computed by Eq. (11.5a) when the X and Y coordinates for the point are given (Example 11.2). The relationship between geodetic and grid azimuth expressed as an equation is

$$\text{grid azimuth} = \text{geodetic azimuth} - \theta \tag{11.7}$$

in which the sign of θ is determined from Eq. (11.5b).

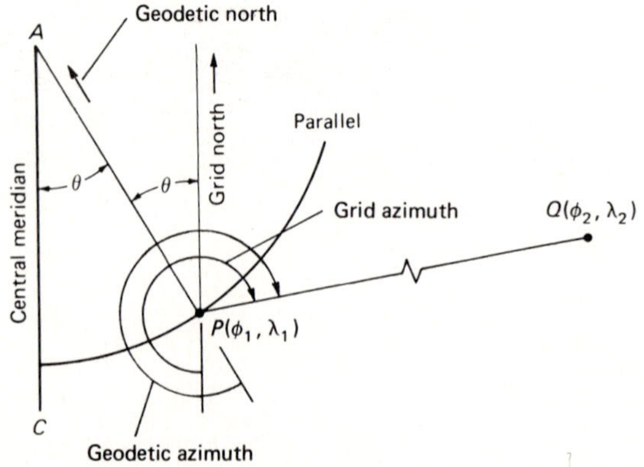

Fig. 11.12 Geodetic and grid azimuths on the Lambert projection.

When lines exceed 5 mi or 8 km in length a second term must be included in Eq. (11.7) so as to achieve the same accuracy as would be possible in geodetic computations. The modified equation is

$$\text{grid azimuth} = \text{geodetic azimuth} - \theta + \frac{X_2 - X_1}{2\rho_0^2 \sin 1''}\left(Y_1 - Y_0 + \frac{Y_2 - Y_1}{3}\right) \tag{11.8}$$

for azimuth from point 1 to point 2, with map coordinates (X_1, Y_1) and (X_2, Y_2), respectively. When a line is close to or in a grid north or south direction, $X_2 - X_1$ in Eq. (11.8) is either very small or zero, so that the correction term is negligible or vanishes.

Grid distances or distances computed on the map projection are calculated using the given X and Y coordinates in Eq. (8.8) in Art. 8.15. In order to obtain the geodetic distance at sea level, the scale factors k are taken from the projection tables (Table 11.1, Table I). The relationship between grid and geodetic distance at sea level is then

$$\text{grid distance} = (\text{geodetic distance at sea level})(\text{scale factor})$$

As a rule, the average of the two scale factors at the two ends of the line, k_{av}, is used in this computation.

If it is necessary to convert grid distances to geodetic distances on the terrain or reduce measured distances to grid distances, the elevation of the line above sea level must be known and a sea-level correction is made (Art. 9.7). Table 11.2 lists sea-level factors, used for this correction, for elevations between 500 and 10,000 ft.

The sea-level correction factor is calculated from:

$$\text{sea-level correction factor} = \frac{R}{R + h}$$

in which R is the average radius of the earth (20,906,000 ft or 6,372,000 m) and h is the average elevation of the line or area above sea level.

A combined scale and sea-level correction factor may also be calculated where:

$$\text{combined factor} = (\text{scale factor})\ (\text{sea-level correction factor})$$

Table 11.2 Sea-Level Correction Factors

Elevation *h*, ft	Sea-level factor	Elevation *h*, ft	Sea-level factor
500	0.999976	5,500	0.999737
1,000	0.999952	6,000	0.999713
1,500	0.999928	6,500	0.999689
2,000	0.999904	7,000	0.999665
2,500	0.999880	7,500	0.999641
3,000	0.999857	8,000	0.999617
3,500	0.999833	8,500	0.999593
4,000	0.999809	9,000	0.999570
4,500	0.999785	9,500	0.999546
5,000	0.999761	10,000	0.999522

Then,

$$\text{grid distance} = (\text{geodetic distance at elevation } h)\,(\text{combined factor})$$

Example 11.3 The coordinates for points 8 and 10 as calculated in Example 11.1 and given in Example 11.2 are as follows:

Point	X, ft	Y, ft
8	2,263,569.50	760,508.29
10	2,269,500.94	778,680.42

Determine the grid and geodetic azimuths from both ends of the line. Also, compute the grid distance and geodetic sea-level distance between points 8 and 10. Assuming that the average elevation of the line is 2500 ft above sea level, calculate the geodetic distance on the ground at this elevation.

SOLUTION Since these are plane coordinates, the grid azimuth is given by Eqs. (8.10). Thus, the grid azimuth from the south of the line from point 8 to point 10 is

$$\tan A_{8-10} = \frac{X_8 - X_{10}}{Y_8 - Y_{10}} = \frac{-5931.44}{-18{,}172.13} = 0.3264031239$$

so that the grid azimuth from south is

$$A_{8-10} = 198°04'36.64''$$

The grid azimuth from south of the line from point 10 to point 8 is exactly 180° different from the value above. Thus,

$$A_{10-8} = 18°04'36.64''$$

Next, compute the geodetic azimuth of the line from point 8 to point 10, from the south, by Eq. (11.8). From Example 11.1

$$\theta = 0°37'42.1128''$$

and from Table 11.1

$$\frac{1}{2\rho_0^2 \sin 1''} = (2.38)(10^{-10}) \qquad \text{and} \qquad Y_0 = 547{,}078.17$$

Thus, the last (or correction) term in Eq. (11.8) is

$$(5{,}931.44)(2.38)(10^{-10})\left(760{,}508.29 - 547{,}078.17 + \frac{18{,}172.13}{3}\right) = 0.31''$$

and by Eq. (11.8)

geodetic azimuth = grid azimuth + θ − correction term

or

$$\begin{aligned}\text{geodetic azimuth} &= 198°04'\ 36.64''\\ &\quad +0°37'\ 42.11''\\ &\quad -0°00'\ 0.31''\\ &= 198°42'\ 18.44''\end{aligned}$$

Note that the distance between the two points is less than 5 mi. Therefore, the correction term in Eq. (11.8) could have been neglected. It was included in this example to demonstrate to the student both the way to calculate it, and the fact that the resulting value (0.31″) is indeed small and can be neglected for the level of precision involved.

Similarly, the geodetic azimuth from south of the line from point 10 to point 8 is obtained by first calculating the correction term, or

$$(-5{,}931.44)(2.38)(10^{-10})\left(778{,}680.42 - 547{,}078.17 - \frac{18{,}172.13}{3}\right) = -0.32''$$

Then,

$$\begin{aligned}\text{geodetic azimuth 10–8} &= 18°04'\ 36.64''\\ &\quad +0°38'\ 34.77''\\ &\quad +0°00'\ 00.32''\\ &= 18°43'\ 11.73''\end{aligned}$$

Note that the geodetic azimuth from 8 to 10 does not differ from the geodetic azimuth 10-8 by exactly 180° as do corresponding grid azimuths. Figure 11.13 illustrates the grid and geodetic azimuths from 8 to 10 for Example 11.3. The student can make a similar sketch for the geodetic and grid azimuths from 10 to 8 as an exercise to verify, graphically, the signs of θ and the correction term in Eq. (11.8) for this part of Example 11.3. The grid distance from 8 to 10 calculated by Eq. (8.8) is

$$\begin{aligned}d_{8-10} &= [(X_{10} - X_8)^2 + (Y_{10} - Y_8)^2]^{1/2}\\ &= [(5931.44)^2 + (18{,}172.13)^2]^{1/2} = 19{,}115.656 \text{ ft}\end{aligned}$$

The scale factors from Table I of Table 11.1 for points 8 ($\phi_8 = 41°25'01.100''$N) and 10 ($\phi_{10} = 41°27'\ 59.988''$) are

$$k_8 = 0.9999461 + \left(\frac{1.1''}{60''}\right)(0.0000030) \qquad = 0.9999462$$

$$k_{10} = 0.9999522 + \left(\frac{59.988''}{60.''}\right)(0.0000032) \quad = 0.9999554$$

$$\text{Average scale factor } k_{av} \qquad = 1.9999016/2 = 0.9999508$$

Then, the geodetic distance between points 8 and 10 at sea level is

$$\text{geodetic distance 8–10 at sea-level} = \frac{19{,}115.656}{0.9999508} = 19{,}116.597 \text{ ft}$$

The sea-level correction factor may be calculated from $R/(R+h)$, or from Table 11.2, for $h = 2500$, is 0.999880. Therefore, the geodetic distance on the ground at elevation 2500 ft is geodetic distance 8–10 at elevation 2500 ft:

$$\frac{19{,}116.597}{0.999880} = 19{,}118.891 \text{ ft}$$

A combined factor could have been calculated as

$$\text{combined factor} = (0.9999508)(0.999880) = 0.9998308$$

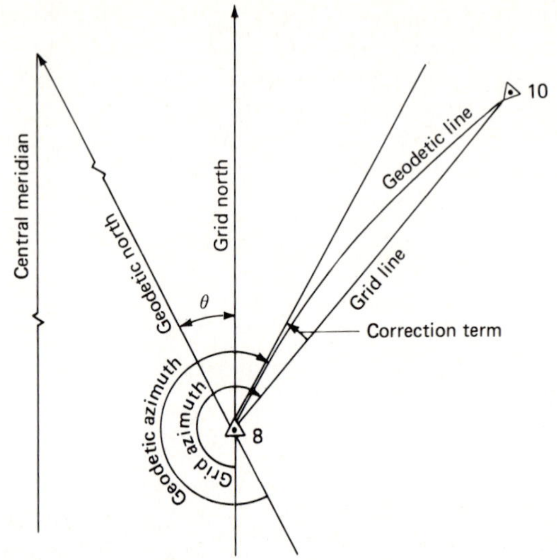

Fig. 11.13 Grid and geodetic azimuth with correction term, Lambert projection.

and the terrain distance at the 2500-ft elevation computed directly from the grid distance as

$$d_{8-10}\,(\text{terrain at } h = 2500 \text{ ft}) = \frac{19{,}115.656}{0.9998308} = 19{,}118.891 \text{ ft}$$

11.19. Transverse Mercator state plane coordinate projection For the state systems, the transverse Mercator projection cylinder cuts the surface of the sphere at two standard lines, parallel to and equally spaced from the central meridian as shown in Fig. 11.7*b*. As mentioned in Art. 11.13, and depicted in Fig. 11.7*b* and *c*, the projection distance between points along these standard lines equals the distance between corresponding points on the sphere. Between the two standard lines, the projection distance is less than the corresponding distance on the sphere and the scale of the projection is too small. Beyond the two standard lines, the projection distance exceeds the corresponding distance on the sphere and the scale of the projection is too large. Thus, the scale on an east-west line varies from point to point while the scale on a grid north-south line is constant throughout even though it may not be 1 : 1. If the two standard circles are set so the distance between them is about 96 km (60 mi), and the total width of the projection is about 254 km (158 mi), then errors due to scale change throughout the projection will not exceed 1 : 10,000. Since the scale is constant along the grid meridians, this type of projection is used in states that have a long north-south dimension.

For the transverse Mercator state plane projections, the central meridian is parallel to the Y axis, which has an arbitrarily assigned X coordinate of $X_0 = 500{,}000$ ft (current projection tables are in feet so all calculations must be in feet). In this manner there will be no negative X coordinates within the system. The X coordinate axis is along a line perpendicular to the central meridian at a latitude selected such that there will be no negative Y coordinates within the system. Still assuming a

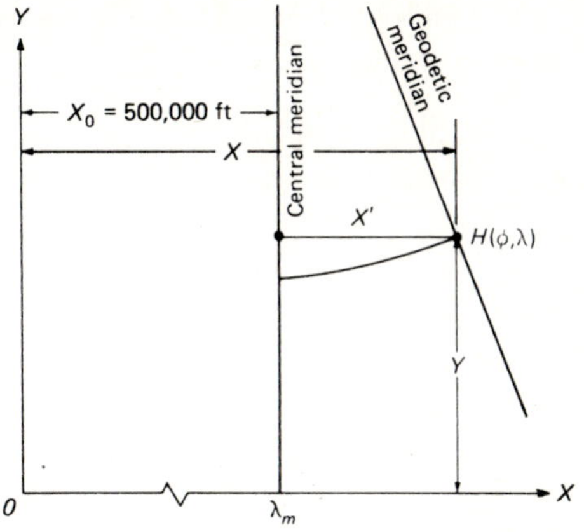

Fig. 11.14 Coordinates on the transverse Mercator projection.

sphere, the Y coordinate for a point H (refer to Fig. 11.14) would be the length of meridian from the origin to a perpendicular from the central meridian passing through H. The X coordinate of point H would be the length of this perpendicular from the central meridian to the point plus X_0.

Tables have been published in order to make it possible to calculate the transverse Mercator projection coordinates. For example, Table 11.3 shows extracts from Spec. Publ. No. 257 of the U.S. Coast and Geodetic Survey (presently known as the National Geodetic Survey) *Plane Coordinate Projection Tables for Arizona.* In conjunction with these tables, the equations used to calculate the state plane coordinates X, Y from the latitude ϕ and longitude λ are:

$$\Delta\lambda = \lambda_m - \lambda \tag{11.9}$$

where λ_m designates the longitude of the central meridian of the zone, and

$$X' = H\Delta\lambda'' \pm ab \tag{11.10}$$

where the tabulated quantities H and a are functions of latitude and b is a function of $\Delta\lambda$. When the sign of the product ab is positive, the *numerical* value of $H\Delta\lambda''$ is increased regardless of its sign, and when the sign of ab is negative the numerical or absolute value of $H\Delta\lambda''$ is decreased. The $\Delta\lambda''$ is calculated from Eq. (11.9) and expressed in seconds of arc. Next,

$$X = X' + X_0 = X' + 500{,}000 \text{ ft} \tag{11.11}$$

$$Y = Y_0 + V\left(\frac{\Delta\lambda''}{100}\right)^2 \pm c \tag{11.12}$$

where the tabulated values Y_0 and V are functions of latitude and c is based on $\Delta\lambda''$. The $\pm c$ means that whatever sign the tabulated value of c has is used directly, i.e., it is algebraically added.

The convergence of the meridian through the station with respect to the central meridian of the zone is given by (see Fig. 11.15)

$$\Delta\alpha'' = \Delta\lambda'' \sin\phi + g \tag{11.13}$$

where both $\Delta\alpha''$ and $\Delta\lambda''$ are expressed in seconds of arc, and the tabulated value of

Table 11.3 Transverse Mercator Projection for Arizona, East and Central Zones

Table I

Lat.	Y_0 ft	ΔY_0 per second	H	ΔH per second	V	ΔV per second	a
31°55′	333,411.52	101.041 33	86.190 082	259.12	1.104 780	5.28	−0.930
31°56′	339,474.00	101.041 50	86.174 535	259.26	1.105 097	5.27	−0.929
31°57′	345,536.49	101.041 83	86.158 980	259.36	1.105 413	5.27	−0.927
31°58′	351,599.00	101.042 17	86.143 419	259.51	1.105 729	5.25	−0.926
31°59′	357,661.53	101.042 33	86.127 849	259.61	1.106 044	5.25	−0.925
32°10′	424,350.36	101.045 33	85.956 107	260.96	1.109 488	5.18	−0.911
32°11′	430,413.08	101.045 50	85.940 450	261.06	1.109 799	5.17	−0.909
32°12′	436,475.81	101.045 83	85.924 787	261.21	1.110 109	5.17	−0.908
32°13′	442,538.56	101.046 17	85.909 115	261.31	1.110 419	5.17	−0.907
32°14′	448,601.33	101.046 33	85.893 437	261.44	1.110 729	5.17	−0.906

Table II

$\Delta\lambda''$	b	Δb	c
1700	+5.498	+0.223	−0.061
1800	+5.721	+0.206	−0.067
1900	+5.927	+0.188	−0.073
2000	+6.115	+0.169	−0.079
2100	+6.284	+0.149	−0.085
2200	+6.433	+0.128	−0.091
2300	+6.561	+0.105	−0.096

Table III for g

Latitude	$\Delta\lambda''$: 0″	1000″	2000″	3000″	4000″	5000″	6000″
31°	0.00	0.00	0.02	0.08	0.19	0.37	0.64
32°	0	0	0.02	0.08	0.19	0.38	0.65
33°	0	0	0.02	0.08	0.19	0.38	0.65
34°	0	0	0.02	0.08	0.19	0.38	0.65
35°	0	0	0.02	0.08	0.19	0.38	0.65

Table IV

Y	P	ΔP	X'	d
0	1.43434	1562	0	0.00
100,000	1.44996	1570	50,000	+0.01
200,000	1.46566	1580	100,000	+0.02
300,000	1.48146	1589	150,000	+0.04
400,000	1.49735	1598	200,000	+0.07

Table V for M

	East and central zones	
Y	M	ΔM
0	0.005 9177	644
100,000	0.005 9821	648
200,000	0.006 0469	652
300,000	0.006 1121	656
400,000	0.006 1777	659
500,000	0.006 2436	664

Table VI for *e*

	X'			
Y	200,000	300,000	400,000	500,000
0	0.0	0.1	0.3	0.6
1,000,000	0.0	0.1	0.4	0.8
2,000,000	0.0	0.2	0.5	1.0

Table VII for scale ratios

X', ft	Scale expressed as a ratio
150,000	0.9999258
155,000	0.9999275
160,000	0.9999293
165,000	0.9999312
170,000	0.9999331
175,000	0.9999351
180,000	0.9999371
185,000	0.9999392
190,000	0.9999413
195,000	0.9999435

$\lambda_m = 110°10'00.000''$ $\quad 1/(6\rho_0^2 \sin 1'')_g = 0.7872 \times 10^{-10}$

$X_0 = 500{,}000$

g is a function of $\Delta\lambda''$. The quantity g *always* increases the value of $\Delta\lambda'' \sin\phi$. Then, conversion from geodetic to grid azimuth is given by an equation similar to Eq. (11.8), or

$$\text{grid azimuth} = \text{geodetic azimuth} - \Delta\alpha - \text{correction term} \tag{11.14a}$$

$$\text{correction term} = \frac{(Y_2 - Y_1)(2X_1' + X_2')}{(6\rho_0^2 \sin 1'')_g} \tag{11.14b}$$

noting that

$$\text{grid azimuth 1–2, from south} = \frac{X_1 - X_2}{Y_1 - Y_2}$$

where X_1' and X_2' are X coordinates of the initial and terminal points of the line referred to the central meridian, and $1/(6\rho_0^2 \sin 1'')_g$ is a constant given in the projection tables.

The "correction term" in Eqs. (11.14) may be neglected for most work. However, for lines 5 mi long or more, this term should be included.

The inverse solution for calculating ϕ and λ from given X and Y coordinates makes use of the following equations:

$$Y_0 = Y - P\left(\frac{X'}{10{,}000}\right)^2 - d \tag{11.15}$$

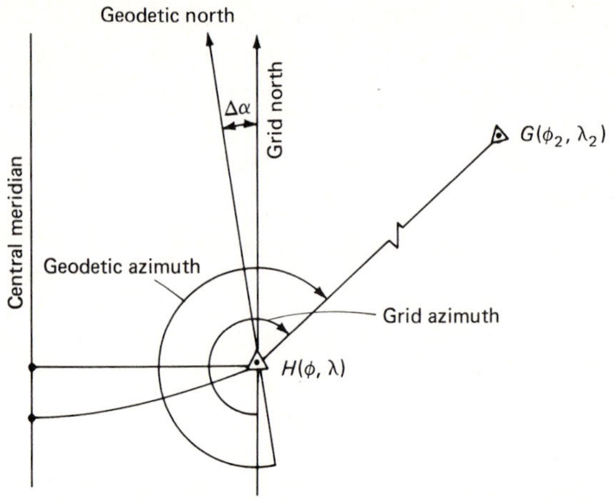

Fig. 11.15 Geodetic and grid azimuths on the transverse Mercator projection.

where X' is calculated from the given X using Eq. (11.11); the tabulated value of d is a function of X' and that for P is a function of Y, as shown in Table IV of Table 11.3. With Y_0 obtained from Eq. (11.15) as an entry, find the latitude ϕ in Table I of Table 11.3.

Next, given ϕ, extract H from Table I of Table 11.3 and calculate an approximate $\Delta\lambda''$ using

$$\text{approximate } \Delta\lambda'' = \frac{X'}{H} \tag{11.16}$$

Now, use the latitude ϕ to determine a value for a and the approximate $\Delta\lambda''$ to obtain b from Tables I and II, respectively, of Table 11.3. Then the correct $\Delta\lambda$ is

$$\Delta\lambda'' = \frac{X' \mp ab}{H} \tag{11.17}$$

When the sign of the product ab is negative, *increase* the numerical or absolute value of X' and vice versa. Then,

$$\lambda = \lambda_m - \Delta\lambda$$

Finally, the convergence of the meridian, $\Delta\alpha$, is given by

$$\Delta\alpha'' = MX' - e \tag{11.18}$$

where e decreases the numerical, or absolute, value of MX', and M is based on Y and e on X' and Y, both obtained from the Table 11.3. The $\Delta\alpha$ is taken positive (+) for points east, and negative (−) for points west, of the central meridian.

The tables for the transverse Mercator projection also contain scale factors for converting sea-level geodetic distances to grid distances. These factors, expressed as ratios, are listed versus X' in increments of 5000 ft (Table VII of Table 11.3).

The following numerical examples demonstrate the various aspects of calculations regarding the transverse Mercator projection using Table 11.3. Example 11.4 covers the direct problem or the transformation from geographic to state plane coordinates.

Example 11.5 considers the inverse problem, from state plane to geographic coordinates. Example 11.6 demonstrates the calculation of grid and geodetic azimuths, while Example 11.7 illustrates the use of the scale factor.

Example 11.4 The geographic coordinates of triangulation station P in the Arizona East transverse Mercator projection zone are

$\phi = 31°55'11.200''\text{N}$

$\lambda = 109°34'09.500''\text{W}$

Calculate X and Y coordinates for the station using the tabulated values in Table 11.3.

SOLUTION By Eq. (11.9), with $\lambda_m = 110°10'00.000''$,

$$\Delta\lambda = (110°10'00.000'') - (109°34'09.500'') = 0°35'50.500''$$

$$\Delta\lambda'' = 2150.500'' \qquad \left(\frac{\Delta\lambda''}{100}\right)^2 = 462.465$$

Using ϕ in projection Table I of Table 11.3, $H = 86.187180$, $V = 1.104839$, and $a = -0.930$. Using $\Delta\lambda''$ in projection Table II of Table 11.3, $b = +6.359$. Then $ab = -5.91387$ and by Eq. (11.10)

$$\begin{aligned} X' &= H\Delta\lambda'' \pm ab = (86.187180)(2150.500) - 5.914 \\ &= 185{,}339.6166 \end{aligned}$$

By Eq. (11.11)

$$X = X' + 500{,}000 = 685{,}339.62 \text{ ft}$$

Using $\Delta\lambda''$ in projection Table II of Table 11.3, $c = -0.088$ and using ϕ in projection Table I, $V = 1.104839$ and $Y_0 = 334{,}543.1829$. By Eq. (11.12)

$$\begin{aligned} Y &= Y_0 + V\left(\frac{\Delta\lambda''}{100}\right)^2 \pm c \\ &= 334{,}543.1829 + (1.104839)(462.465) - 0.088 \\ &= 335{,}054.04 \text{ ft} \end{aligned}$$

Example 11.5 The coordinates for station Q on the transverse Mercator projection for the Arizona East zone are:

$X = 692{,}435.81$ ft
$Y = 353{,}400.22$ ft

Compute the latitude and longitude for the station.

SOLUTION

$X = 692{,}435.81$ ft
$X_0 = 500{,}000.00$ ft

$X' = 192{,}435.81$ ft [by Eq. (11.11)]
$P = 1.48994$ (projection Table IV, using Y, Table 11.3)
$d = 0.065461486$ (projection Table IV, using X', Table 11.3)

By Eq. (11.15)

$$Y_0 = Y - P\left(\frac{X'}{10{,}000}\right)^2 - d = 353{,}400.22 - 1.48994\left(\frac{192{,}435.81}{10{,}000}\right)^2 - 0.07$$

$= 352{,}848.405$ ft

$= 351{,}599.000$ ft for $31°58'$ (projection Table I, Table 11.3)

$1249.405 \text{ ft}/101.04217 \text{ ft}/'' = 12.365''$

$\phi = 31°58'12.365''$

$H = 86.140209$ (projection Table I, using ϕ, Table 11.3)

By Eq. (11.16)

$$\Delta\lambda'' \text{ (approximate)} = \frac{X'}{H} = \frac{192{,}435.81}{86.140209} = 2233.984''$$

$a = -0.926$ (projection Table I, using ϕ, Table 11.3)

$b = +6.476$ (projection Table II, using approximate $\Delta\lambda''$, Table 11.3)

By Eq. (11.17)

$$\Delta\lambda'' = \frac{X' \mp ab}{H} = \frac{192{,}435.81 + 5.997}{86.140209} = 2234.053''$$

$\Delta\lambda = 0°37'14.053''$

$\lambda = \lambda_m - \Delta\lambda = (110°10'00.000) - (0°37'14.053'')$

$= 109°32'45.947''$W

Example 11.6 Determine grid and geodetic azimuths (from the south) of the line from station P (Example 11.4) to station Q (Example 11.5) which have state plane coordinates on the traverse Mercator projection of Arizona East zone as follows:

Station	X, ft	Y, ft
P	685,339.62	335,054.04
Q	692,435.81	353,400.22

SOLUTION P is the origin (station 1) and Q the terminus of the line (station 2). Thus, the grid azimuth from the south of the line PQ is

$$A_{12}(\text{south}) = \arctan\frac{X_1 - X_2}{Y_1 - Y_2} = \frac{685{,}339.62 - 692{,}435.81}{335{,}054.04 - 353{,}400.22}$$

$$= \arctan(0.3867938721) = 201°08'46.194''$$

The convergence of the meridian is given by Eq. (11.13), if ϕ, λ are given, or

$$\Delta\alpha'' = \Delta\lambda'' \sin\phi + g$$

With $\Delta\lambda'' = 2150.500''$ in Example 11.4, and g obtained by interpolation from projection Table III of Table 11.3 as 0.029 (using the given ϕ and $\Delta\lambda''$ in Example 11.4) then by Eq. (11.13)

$$\Delta\alpha'' = (2150.500'')\sin(31°55'11.200'') + 0.029 = 1137.0658''$$

or

$$\Delta\alpha = 0°18'57.066''$$

Another alternative is to calculate $\Delta\alpha$ using the plane coordinates applying Eq. (11.18). Thus, for station 1 (point P),

$$
\begin{aligned}
X' &= X - 500{,}000 \\
&= 685{,}339.62 - 500{,}000 \\
&= 185{,}339.62 \text{ ft}
\end{aligned}
$$

By interpolation from projection Table V of Table 11.3, the value of M is 0.0061351 and from projection Table VI $e = 0$. Thus

$$
\begin{aligned}
\Delta\alpha'' &= MX' - e \\
&= (0.0061351)(185{,}339.62) \\
&= 1137.077103''
\end{aligned}
$$

or

$$\Delta\alpha = 0°18'57.077''$$

The difference in $\Delta\alpha$ from both methods is not significant for the precision of this calculation. Since the line PQ is less than 5 mi, it is not necessary to evaluate the second, or correction, term. Thus, by Eq. (11.14*a*)

$$
\begin{aligned}
\text{geodetic azimuth } PQ &= \text{grid azimuth } PQ + \Delta\alpha \\
&= 201°08'46.19'' \\
&\quad +0°18'57.07'' \\
&= 201°27'43.26''
\end{aligned}
$$

Figure 11.16 illustrates the relationship between grid and geodetic azimuths for the line from station P to station Q. The grid line on the projection is represented by the straight line from P to Q, with the grid azimuth measured clockwise from grid south to the grid line. The geodetic line is a curve concave toward the central meridian. The geodetic azimuth is

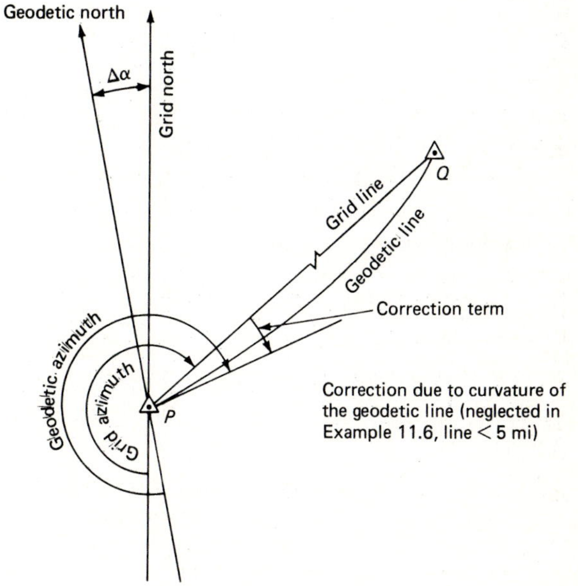

Fig. 11.16 Grid and geodetic azimuth on the transverse Mercator projection.

measured from the geodetic meridian (south) to a tangent to the curving geodetic line at station P.

Example 11.7 Calculate the grid and geodetic distances between the stations P and Q used in Example 11.6.

SOLUTION The grid distance is

$$d_{PQ} = [(X_1 - X_2)^2 + (Y_1 - Y_2)^2]^{1/2}$$
$$= [(685{,}339.62 - 692{,}435.81)^2 + (335{,}054.04 - 353{,}400.22)^2]^{1/2}$$
$$= 19{,}670.7456 \text{ ft} = 19{,}670.75 \text{ ft}$$

In order to obtain the geodetic distance, the grid distance must be modified by the average of the scale factors k_1, k_2 for stations P and Q, respectively. In order to get k_1 and k_2, we need X_1' and X_2'. For $X_1' = 185{,}339.62$, $k_1 = 0.9999393$, and for $X_2' = 192{,}435.81$, $k_2 = 0.9999424$, both scale factors extracted by interpolation from Table VII of Table 11.3. Then, the average scale factor is $k_{av} = (k_1 + k_2)/2 = 0.99994085$.

The geodetic distance PQ at sea level $= d_{PQ}/k_{av} = 19{,}671.91$ ft.

11.20. Applications for state plane coordinates Designation of the location for a survey control point by a pair of X and Y coordinates is of great value in any survey. When these coordinates are in one of the state plane coordinate systems, such a location assumes even more significance since the position so described is unique within the system and is related to the National Network for horizontal control.

Control surveys by traditional triangulation, trilateration (Arts. 9.1 through 9.10), and traverse (Chap. 8), as well as by any of the electronic positioning systems (Art. 9.13), are compatible with the use of state plane coordinates. Applications are feasible for photogrammetric ground control, topographic mapping projects of all types, route surveys of all kinds, construction layout, property corner location, mine surveys, plus many other surveying tasks. A primary requirement for any of these surveys, if performed on the state plane system, is that it be initiated at one horizontal control monument and close on another monument both with known X and Y state plane coordinates. With due regard to reduction to sea level and compensation for scale factors in measured and calculated distances (Art. 11.18), all subsequent computations and adjustments are then performed in the state plane coordinate system. All horizontal positions so determined are ideally suited for subsequent storage and retrieval of the information in and from a data bank (Arts. 1.11 and 3.15). The use of the state plane coordinate systems has unsurpassed utility for practically all kinds of surveys and should be encouraged whenever possible. Readers interested in further details for applications of these systems should consult Refs. 2, 6, and 10 at the end of this chapter.

11.21. Problems

11.1 "Any map projection has some distortions." Is this statement true or false, and why?

11.2 What are the four conditions that an ideal distortion-free map must satisfy?

11.3 Describe briefly four classes of map projections.

11.4 Explain the basic differences between geometric map projection and mathematical map projection.

11.5 The Lambert conformal conic projection and the transverse Mercator projection are two of the most commonly used map projections; discuss the principal differences between these two projections.
11.6 Enumerate the specifications of the universal transverse Mercator projection.
11.7 "The shape of a state, or its north-south extent relative to its east-west extent, influences the type of map projection used for its state plane coordinate system." Discuss this statement fully.
11.8 The latitude and longitude for station 1 are 41°20′45.618″N and 121°05′50.623″W. Compute on the Lambert projection, California zone 1: (*a*) the mapping angle θ for station 1; (*b*) X and Y state plane coordinates for the station; (*c*) the scale factor at station 1. Use the projection tables (Table 11.1).
11.9 The state plane coordinates for point 2 on the Lambert projection for California zone 1 are $X_2 = 2{,}300{,}000.10$ ft and $Y_2 = 770{,}421.89$ ft. The average elevation of the terrain between point 2 and point 10 of Example 11.2 is 5000 ft above mean sea level. Calculate (*a*) the grid azimuth from station 2 to station 10; (*b*) the geodetic azimuth from station 2 to station 10 and from station 10 to station 2; (*c*) the grid distance, geodetic distance at mean sea level, and geodetic distance at average terrain elevation between stations 2 and 10. Assume an average scale factor between 2 and 10 of 0.99995315.
11.10 In Prob. 11.9, compute the geographic coordinates for station 2, using the projection tables in Table 11.1.
11.11 The state plane coordinates for station R on the transverse Mercator projection, Arizona East zone, are $X = 691{,}500.41$ ft and $Y = 353{,}200.05$ ft. (*a*) Compute the grid distance and grid azimuth from station P (Example 11.4) to R; (*b*) determine the geodetic azimuth from the south of the line from P to R.
11.12 Compute geographic coordinates for station R in Prob. 11.11.
11.13 Calculate the convergence of the meridians and the scale factor for station R in Prob. 11.11. Also compute the sea-level geodetic distance and the geodetic azimuth from the south of the line from R to P. (Refer to Examples 11.4, 11.6, and 11.7.)
11.14 The distance between two stations calculated from their state plane coordinates is 10,500.825 m. The scale factor and average terrain elevation are 1.0004 and 1000 m above mean sea level, respectively. What is the horizontal distance between these two stations as it would be measured on the ground?
11.15 Calculate the ϕ and λ for point P in Example 11.4, given $X = 685{,}339.62$ ft and $Y = 335{,}054.04$ ft, and check your answer against the values for ϕ and λ given in the example.
11.16 In the area described for Prob. 11.14, the state plane coordinates for two control points A and B are $X_A = 2{,}140{,}982.13$ ft, $Y_A = 640{,}283.85$ ft; and $X_B = 2{,}141{,}116.82$ ft, $Y_B = 640{,}496.27$ ft. The coordinates for a construction control point C, which is to be established from A, are $X_C = 2{,}139{,}885.06$ ft and $Y_C = 639{,}480.09$ ft. Compute (*a*) the clockwise angle to be turned from an instrument set up at A with a backsight on B in order to direct the line of sight toward point C, which is to be established; (*b*) the distance that must be measured over level terrain from A to set point C.
11.17 Control point 6 has geographic coordinates of $\phi = 41°29′30.205″$N and $\lambda = 121°08′45.638″$W. Compute on the Lambert projection, California zone 1: (*a*) the mapping angle θ for station 6; (*b*) state plane coordinates for the station; and (*c*) the scale factor at station 6. Use Table 11.1.
11.18 The X and Y coordinates for station 5 on the Lambert state plane projection for California zone 1 are $X_5 = 2{,}247{,}830.75$ ft and $Y_5 = 734{,}485.79$ ft. Compute the latitude and longitude for station 5. Use Table 11.1.
11.19 The average elevation of the terrain between station 5 in Prob. 11.18 and station 6 in Prob. 11.17 is 3000 ft above mean sea level. Compute the (*a*) grid azimuth from station 5 to station 6; (*b*) geodetic azimuth from station 5 to station 6 and from station 6 to station 5; and (*c*) grid distance and geodetic distance between stations 5 and 6 at mean sea level.

References

1. Adams, O. S., and Claire, C. N., *Manual of Plane Coordinate Computation*, Spec. Publ. No. 193, U.S. Department of Commerce, Coast and Geodetic Survey (now National Ocean Survey, NOAA), Rockville, Md., 1935; reprinted 1971.
2. Davis, Raymond E., Foote, Francis S., Anderson, James M., and Mikhail, Edward M., *Surveying: Theory and Practice*, 6th ed., McGraw-Hill Book Company, New York 1981, Chap. 14.

3. Claire, C. N., *State Plane Coordinates by Automatic Data Processing*, Publ. 62-4, U.S. Department of Commerce, Environmental Sciences Services Administration (now National Ocean Survey, NOAA), Rockville, Md., 1968; reprinted with corrections 1973/1976.
4. Coast and Geodetic Survey, *Plane Coordinate Projection Tables*, Spec. Publ. series, U.S. Department of Commerce, National Ocean Survey, NOAA, Rockville, Md., also Government Printing Office, Washington, D.C., for some states.
5. Dracup, J. F., *Fundamentals of the State Plane Coordinate Systems*, National Geodetic Survey Information Center, C18, NOAA, Rockville, Md., October 1974.
6. Dracup, J. F., "The New Adjustment of the North American Datum." *ACSM Bulletin*, November 1977.
7. Gergen, J. G., "New Adjustment of the North American Horizontal Datum, Article 26: Interim Status Report," *ACSM Bulletin*, No. 26, February 1982.
8. Krakiwsky, E. J., "Conformal Map Projections in Geodesy," University of New Brunswick Lecture Notes No. 37, September 1973.
9. Meade, B. K., "Coordinate Systems for Surveying and Mapping," *Surveying and Mapping*, Vol. 33, No. 3, September 1973.
10. Moffitt, F. H., and Bouchard, H., *Surveying*, 7th ed., Harper & Row, Publishers, New York, 1982.
11. National Geographic Society, *The Round Earth on Flat Paper*, National Geographic Society, Monograph, 1947.
12. Raisz, Erwin, *Principles of Cartography*, McGraw-Hill Book Company, New York, 1962.
13. U.S. Department of the Army Technical Manual, *Field Manual, Map Reading*, FM 21-6, 1973.

PART V
TYPES OF SURVEYS

CHAPTER 12
Control and topographic surveying

12.1. General The distinguishing feature of a topographic survey is the determination of the location, both in plan and elevation, of selected ground points which are necessary for plotting contour lines and the planimetric location of features on the topographic map. A topographic survey consists of (1) establishing over the area to be mapped a system of horizontal and vertical *control*, which consists of key stations connected by measurements of high precision; and (2) locating the details, including selected ground points, by measurements of lower precision from the control stations.

Topographic surveys fall roughly into three classes, according to the map scale employed, as follows:

Large scale 1 : 1200 (1 inch to 100 ft) or larger
Intermediate scale 1 : 1200 to 1 : 12,000 (1 in to 100 ft to 1 in to 1000 ft)
Small scale 1 : 12,000 (1 in to 1000 ft) or smaller

Because of the range in uses of topographic maps and variations in the nature of the areas mapped, topographic surveys vary widely in character.

Topographic surveys can be performed by aerial photogrammetric methods, by ground survey methods, or by some combination of these two procedures. The largest portion of almost all of the small and intermediate scale as well as some large-scale topographic mapping is now performed by photogrammetric methods (Art. 13.15). This photogrammetric operation includes establishing portions of the horizontal control in addition to compilation of the topographic map. However, ground survey methods are still applicable for large-scale topographic mapping of small areas and for field completion surveys which are usually needed for photogrammetrically compiled topographic maps. The discussions in this chapter are directed primarily toward the various procedures for topographic surveys by ground survey methods.

12.2. Planning the survey The choice of field methods for topographic surveying is governed by (1) the intended use of the map, (2) the area of the tract, (3) the map scale, and (4) the contour interval.

1. Intended use of map Surveys for detailed maps should be made by more refined methods than surveys for maps of a general character. For example, the earthwork estimates to be made from a topographic map by a landscape architect must be determined from a map which represents the ground surface much more accurately in both horizontal and vertical dimensions than one to be used in estimating the storage capacity of a reservoir. Also, a survey for a bridge site should be more detailed and more accurate in the immediate vicinity of the river crossing than in areas remote therefrom.

2. Area of tract It is more difficult to maintain a desired precision in the relative location of points over a large area than over a small area. Control measurements for a large area should be more precise than those for a small area.

3. Scale of map It is sometimes considered that, if the errors in the field measurements are not greater than the errors in plotting, the former are unimportant. But since these errors may not compensate each other, the errors in the field measurements should be considerably less than the errors in plotting at the given scale. The ratio between field errors and plotting errors should be perhaps one to three.

The ease with which precision may be increased in plotting, as compared with a corresponding increase in the precision of the field measurements, points to the desirability of reducing the total cost of a survey by giving proper attention to the excellence of the work of plotting points, of interpolation, and of interpretation in drawing the map.

The choice of a suitable map scale is discussed in Art. 10.7.

4. Contour interval The smaller the contour interval, the more refined should be the field methods. The choice of a suitable contour interval is discussed in Art. 10.7.

12.3. Establishment of control Control consists of two parts: (1) *horizontal control*, in which the planimetric positions of specified control points are located by trilateration, triangulation, intersection, resection, or traverse; and (2) *vertical control*, in which elevations are established on specified bench marks located throughout the area to be mapped. This control provides the skeleton, which is later clothed with the *details*, or locations of such objects as roads, houses, trees, streams, ground points of known elevations, and contours.

On surveys of wide extent, a relatively few stations distributed over the tract are connected by more precise measurements, forming the *primary control*; within this system, other control stations are located by less precise measurements, forming the *secondary control*. For small areas only one control system is necessary, corresponding in precision to the secondary control used for larger areas.

12.4. Horizontal control Horizontal control can be established by triangulation, trilateration, traverse, aerial photogrammetric methods, inertial and doppler positioning systems, and the global positioning system.

Triangulation and trilateration (Chap. 9) can be used to establish primary and secondary control for relatively large topographic surveys. These methods are also utilized in areas of lesser extent when field conditions are appropriate (hilly, urban, or rugged mountainous regions). Specifications for triangulation and trilateration are given in Table 9.2, Art. 9.2.

Traverse (Chap. 8) with a theodolite and electronic distance measurement (EDM) instrument can also be used for primary and secondary control surveys. Specifications for various orders and classes of traverse are given in Table 8.15, Art. 8.22.

Horizontal control determination by aerial photogrammetric methods is feasible and particularly applicable to small-scale mapping of large areas. Note that this procedure requires a basic framework of horizontal control points established by traditional methods (trilateration and/or triangulation) or by inertial, doppler, or global positioning systems.

Inertial, doppler, and global positioning systems are most appropriate for establishing primary horizontal control when the topographic survey is of very large extent, covers an inaccessible region, or where the conduct of the survey is governed by special conditions. A description of these modern positioning systems is given in Arts. 9.13 through 9.17.

12.5. Vertical control The purpose of vertical control is to establish bench marks at convenient intervals over the area to serve (1) as points of departure and closure for leveling operations of the topographic parties when locating details, and (2) as reference marks during subsequent construction work.

Vertical control is usually accomplished by direct differential leveling (Art. 5.31), but for small areas or in rough country the vertical control is frequently established by trigonometric leveling (Art. 5.5).

All elevations for topographic mapping should be tied to bench marks which are referred to the sea-level datum (Art. 10.2).

Specifications for first-, second-, and third-order differential levels are given in Table 5.2, Art. 5.39. These specifications may be relaxed somewhat depending on map scale, character of the terrain to be mapped, the contour interval desired, and ultimate use of the survey. Table 12.1 gives ranges of approximate closures applicable to intermediate- and large-scale topographic mapping surveys. The smaller error of closure for a given map and type control is used for very flat regions where a contour interval of 1 ft (0.5 m) or less is required and on surveys which are to be used for determination of gradients of streams or to establish the grades of proposed

Table 12.1 Topographic Survey Vertical Control Specifications

Scale of map	Type of control	Length of circuit		Maximum error of closure	
		mi	km	ft	mm
Intermediate	Primary	1–20	2–30	0.05–0.3 $\times\sqrt{\text{mi}}$	12–72 $\times\sqrt{\text{km}}$
	Secondary	1–5	2–8	0.1–0.5 $\times\sqrt{\text{mi}}$	24–120 $\times\sqrt{\text{km}}$
Large	Primary	1–5	2–8	0.05–0.1 $\times\sqrt{\text{mi}}$	12–24 $\times\sqrt{\text{km}}$
	Secondary	$\frac{1}{2}$–3	1–5	0.05–0.1 $\times\sqrt{\text{mi}}$	12–24 $\times\sqrt{\text{km}}$

drainage or irrigation systems. The higher errors of closure apply to surveys in which no more exact use is made of the results other than to determine the elevations of ground points for contours having 2-, 5-, and 10-ft or 0.5-, 2-, and 3-m intervals. Where the largest error of closure is specified (0.5 ft $\sqrt{M}$ and 240 mm $\sqrt{K}$)† stadia leveling is adequate.

12.6. Horizontal and vertical control by three-dimensional traverse A three-dimensional stadia survey may be adequate for establishing control of a small area where the contour interval exceeds 1 m or 5 ft and a relative accuracy (Art. 8.22) of less than 1 part in 500 or 600 is satisfactory. The method for performing a stadia traverse with elevations required is described in Art. 8.12.

When higher accuracy is required, a three-dimensional traverse using a theodolite and EDM device or an electronic tacheometer (Chap. 9) can be performed. The procedures involved for a traverse in three dimensions are outlined in Art. 8.37.

12.7. Location of details: general methods It is assumed in the following sections that horizontal and vertical control has been established and that the field party is concerned only with the location of details. If the plane-table method is to be used, horizontal control is plotted on the plane-table sheet (Art. 7.19).

The adequacy with which the resultant map sheet meets the purpose of the survey depends largely upon the task of locating details. Thus, the topographer should be completely informed as to the uses of the map so that the proper emphasis is placed on each part of the work.

The principal instruments used are the engineer's transit, self-reducing tacheometer, the engineer's level (preferably self-leveling), plane table and alidade, hand level, and clinometer. The transit and self-reducing tacheometer have advantages where there are many definite points to be located or where ground cover limits visibility and requires many setups. Conditions favorable to the plane table are open country and many irregular lines to be mapped; the plane table is also advantageous for

†M and K are the lengths of the circuits in miles and kilometres, respectively.

small-scale mapping. Another class of instruments, electronic surveying systems (Art. 9.13), can also be employed to advantage on topographic detail surveys. These systems provide direction, horizontal distance, and difference in elevation between stations. When available, these devices can be used for almost any phase of a topographic survey.

The four principal methods for acquiring topographic detail in the field are the *controlling-point*, *cross-section*, *checkerboard*, and *trace-contour* methods. These methods are analogous to the systems of points used to plot contours as described in Chap. 10. Each of the methods is described in subsequent sections in this chapter. For a given survey, a combination of the methods may be used; for example, the checkerboard method can be used to obtain detail in open, slightly rolling terrain, while the controlling-point method is employed to get data in irregular, wooded portions of the region being mapped. The aim is to locate details with minimum time and effort.

12.8. Precision The precision required in locating such definite objects as buildings, bridges, and boundary lines should be consistent with the precision of plotting, which may be assumed to be a map distance of about 0.5 mm or $\frac{1}{50}$ in. Such less definite objects as shorelines, streams, and edges of woods are located with a precision corresponding to a map distance of perhaps 0.9 to 1.3 mm or $\frac{1}{30}$ to $\frac{1}{20}$ in. For use in maps of the same relative precision, more located points are required for a given area on large-scale surveys than on intermediate-scale surveys; hence, the location of details is relatively more important on large-scale surveys.

Contours The accuracy with which contour lines represent the terrain depends upon (1) the accuracy and precision of the observations, (2) the number of observations, and (3) the distribution of the points located. Although ground points are definite, contour lines must necessarily be generalized to some extent. The error of field measurement in plan should be consistent with the error in elevation, which in general should not exceed one-fifth of the horizontal distance between contours. The error in elevation should not exceed one-fifth of the vertical distance between contours. The purpose of a topographic survey will be better served by locating a greater number of points with less precision, within reasonable limits, than by locating fewer points with greater precision. Thus, if for a given survey the contour interval is 5 ft, a better map will be secured by locating with respect to each instrument station perhaps 50 points whose standard deviation in elevation is 1 ft than by locating 25 points whose standard deviation is only 0.5 ft. Similarly, if the contour interval is 2 m, it is better to have 50 points with a standard deviation of 0.4 m than 25 points with a standard deviation of 0.2 m.

A general principle which should serve as a guide in the selection of ground points may be noted. As an example, let it be supposed that a given survey is to provide a map which shall be accurate to the extent that if a number of well-defined points is chosen at random on the map, the average difference between the map elevations and ground elevations of identical points shall not exceed one-half of a contour interval. Under this requirement, the attempt is made in the field to choose ground

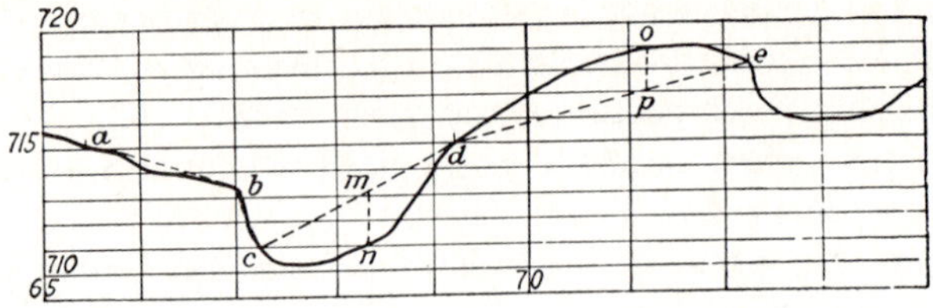

Fig. 12.1 Effect of omission of significant ground points.

points such that a straight line between any two adjacent points will in no case pass above or below the ground by more than one contour interval. Thus, in Fig. 12.1, if the ground points were taken only at *a*, *b*, *c*, *d*, and *e* as shown, the resulting map would indicate the straight slopes *cd* and *de*; the consequent errors in elevation of *mn* and *op* on the profile amount to two contour intervals and show that additional readings should have been taken at the points *n* and *o*. The corresponding displacement of the contours on the map is shown by dotted and full lines in Fig. 12.2.

Angles The precision needed in the field measurements of angles to details may be readily determined by relating it to the required precision of corresponding vertical and horizontal distances. Thus, for a sight at a distance of 1000 ft (300 m), a permissible error of 0.3 ft (0.09 m) in elevation corresponds to a permissible error of 01′ in the vertical angle; likewise, a permissible error of 0.3 ft (0.09 m) in azimuth (measured along the arc from the point sighted) corresponds to a permissible error of 01′ in the horizontal angle. Values for other lengths of sight or degrees of precision are obtained in a similar manner; thus, if it is desired to locate a point to the nearest 2 ft in azimuth (or elevation) and if the length of the sight is 500 ft, the corresponding permissible error in the angle is $\frac{2}{500} = 0.004$ rad-15′.

12.9. Details by the controlling-point method Details may be located by the controlling-point method employing the transit or self-reducing tacheometer and stadia, the plane table and alidade, or electronic systems (Art. 9.14). This method is applicable to practically every type of terrain and condition encountered in topographic mapping. Since only the controlling points that govern the configuration of the land and location of details are located, a measure of economy is achieved that is not possible in the other techniques. However, the topographer must be experienced, since the success of the method depends on selection of the key controlling points. Procedures for use of this method are outlined in subsequent articles.

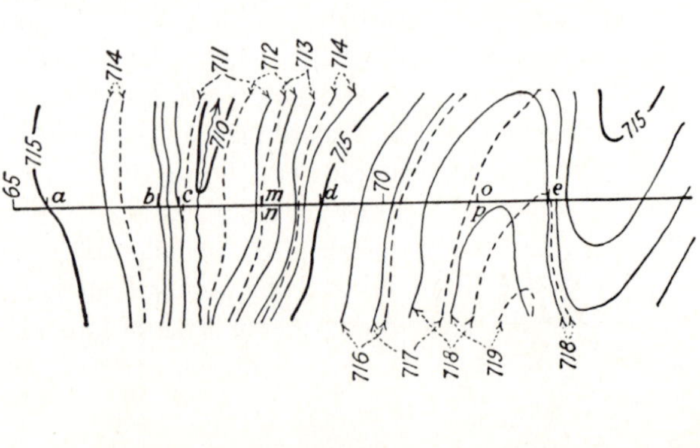

Fig. 12.2 Errors in contour lines due to insufficiency of ground points.

12.10. Transit stadia The personnel of the topography party using the transit or self-reducing tacheometer usually consists of an instrument operator, a notekeeper, and one or two people operating the rods.

In locating ground points, the vertical angles are measured more precisely than the horizontal angles. Accordingly, the vertical circle of the instrument assumes greater importance than the horizontal circle; but because the vertical angles are measured with respect to a horizontal plane, it is important that the horizontal plate be truly horizontal and remain so without need for constant releveling. In this respect, the modern optical repeating theodolites with automatic self-indexing of the vertical circle are a definite asset in this type of work.

Procedure The transit is set up on either a primary or secondary control station, the elevation and position of which are known. The height of instrument (h.i.) of the horizontal axis of the instrument above the station over which it is set is measured with a rod or tape or observed on the plumbing rod if the tripod is so equipped. The vertical circle is checked for index error and the transit is oriented by backsighting along a line the azimuth of which is known, this azimuth having been set off on the clockwise horizontal circle. The upper motion is unclamped and sights to the desired points are taken.

The notekeeper records all stadia intervals, vertical or zenith angles, and azimuths, and describes all points by remarks or sketches so that the draftsman can interpret the data correctly and draw all features properly on the map. Figure 12.3 shows notes of observations taken from station C of a traverse, the elevation of the station

TOPOGRAPHIC DETAILS, BLACK ESTATE 32

Inst. at C; El. 419.5; h.i. = 4.4

Lietz Transit No. 1532 — G. Burke, ⊼
(f+c) = 0 ; K = 100 — M.D. Rand, Notes
F.J. & K.D., Rods
Apr. 4, 1984

Obj.	Az.	Rod Int.	Vert. Ang.	Hor. Dist.	Diff. El.	Elev.	
B	176°14'						
1	10°21'	7.23	-3°11'	721	-40.1	379.4	Water's Edge-Corner . Cloudy, Cold
2	3°14'	7.02	-3°17'	700	-40.1	379.4	" " -On Line
3	352°45'	5.64	-4°11'	561	-41.0	378.5	" "
4	7°18'	5.76	-4°04'	573	-40.7	378.8	" "
5	349°10'	7.17	-2°33'	716	-31.9	387.6	Line (Intervals)
6	16°55'	5.50	-2°50'	549	-27.2	392.3	"
7	315°20'	7.86	-2°41'	784	-36.8	382.7	" (")
8	349°15'	4.13	-5°46'	409	-41.3	378.2	Water's Edge
9	339°30'	5.40	-4°22'	537	-41.0	378.5	Bank Brook 6' wide
10	0°05'	3.71	-4°12'	369	-27.1	392.4	
11	344°40'	4.85	-4°54'	482	-41.3	378.2	" " 15' "
12	25°00'	2.86	0° on 3.2	286	+1.2	420.7	Direct Levels
13	307°45'	4.88	-4°56'	484	-41.8	377.7	Water's Edge
14	319°10'	4.02	-5°56'	398	-41.3	378.2	" "
15	309°45'	5.80	-3°00'	578	-30.3	389.2	
16	318°25'	3.27	-4°36'	325	-26.1	393.4	
B	176°15'	ck.					
17	340°00'	6.34	-3°08'	632	-34.6	384.9	
18	278°35'	2.51	-5°43'	249	-24.9	394.6	
19	276°20'	3.07	-7°56'	301	-42.0	377.5	Water's Edge
20	277°40'	4.24	-5°40'	420	-41.7	377.8	" "
21	264°10'	4.21	-4°47 on 6.7	418	-37.3	382.2	Water's Edge on Bank

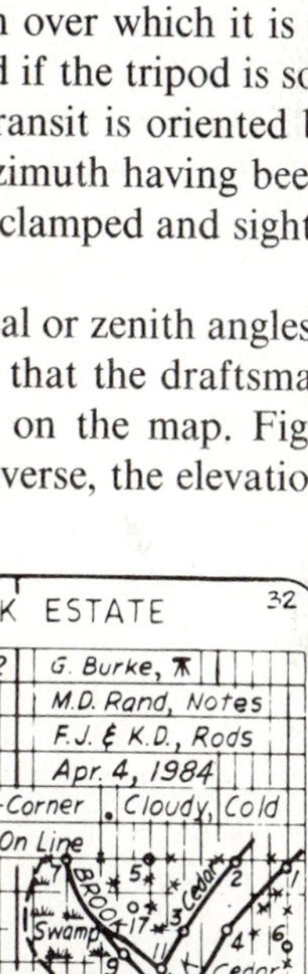

Fig. 12.3 Stadia notes for location of details, with elevations (units are in feet).

having been previously determined as 419.5 ft above the datum. The h.i. of the instrument is 4.4 ft. The transit is oriented by sighting to B, the azimuth of the line CB being set off on the horizontal circle prior to taking the sight. In the first column are given the numbers of the side shots, their locations being shown on the sketch. In the second column are the azimuths of the several points sighted; in the third column are the rod intervals. The measured vertical angles are recorded in the fourth column, and in the following columns are shown, respectively, the computed horizontal distances, differences in elevations, and elevations. Methods for reducing stadia data to horizontal distances and differences in elevation are outlined in Art. 7.9.

In measuring vertical angles it is customary, when practical, to sight at a rod reading equal to the height of instrument above the station over which the transit is set. In this way, the difference in elevation between the center of instrument and the h.i. on the rod is the same as the difference in elevation between the station over which the transit is set and the point on which the rod is held. (see Art. 7.12). This condition is illustrated for the negative change in elevation between C and point 1 (Fig. 12.3) by Fig. 12.4a, where the change in elevation = h.i. − V − rod reading = +4.4 − 40.1 − 4.4 = −40.1 ft. The relationship of h.i., V, and the rod reading for a positive vertical angle is shown in Fig. 7.5, Art. 7.12. When the line of vision is obstructed so that the h.i. cannot be sighted, the sight is taken on some other graduation of the rod, usually a whole number of feet or decimetres above or below the h.i., and this difference in rod readings is recorded in the notes. The notes in Fig. 12.3 show that point 21 was observed in this manner. As shown in Fig. 12.4b, the center cross hair was sighted on 6.7 rather than the h.i. (4.4). Thus, the change in elevation from C to point 21 = h.i. − V − rod reading = 4.4 − 35.0 − 6.7 = −37.3 ft for this example.

When the detail to be observed is at nearly the same elevation as the point over which the transit is set, there is a marked advantage in determining difference in elevation by direct leveling. The notes of Fig. 12.3 show that object 12 was observed in this manner. This case is illustrated in Fig. 12.4c, where the elevation of 12 = elevation of C + h.i. − rod reading = 419.5 + 4.4 − 3.2 = 420.7.

Where measurements are made solely for plotting a map, usually horizontal angles or directions are estimated to 05′ without reading the vernier. Vertical angles are usually measured to minutes, and differences in elevation are computed to tenths of feet or to centimetres. Where elevations to the nearest foot or nearest 3 dm are sufficiently precise, the vertical angles (except for long shots) may properly be read by estimation without use of the vernier.

The usual procedure of observing is to sight on the rod, the lower stadia hair being set on an even foot or decimetre mark such that the horizontal cross hair falls somewhere near the h.i., and to read the stadia interval. The horizontal cross hair is then set on the h.i., the horizontal angle is read and the vertical angle is observed.

When numerous observations are to be taken from a single station, sights to some object the azimuth of which is known are taken at intervals in order to make sure that there is no undetected movement of the lower motion of the transit. The notes of Fig. 12.3 show a check measurement of this kind taken to station B after observations to object 16 had been completed.

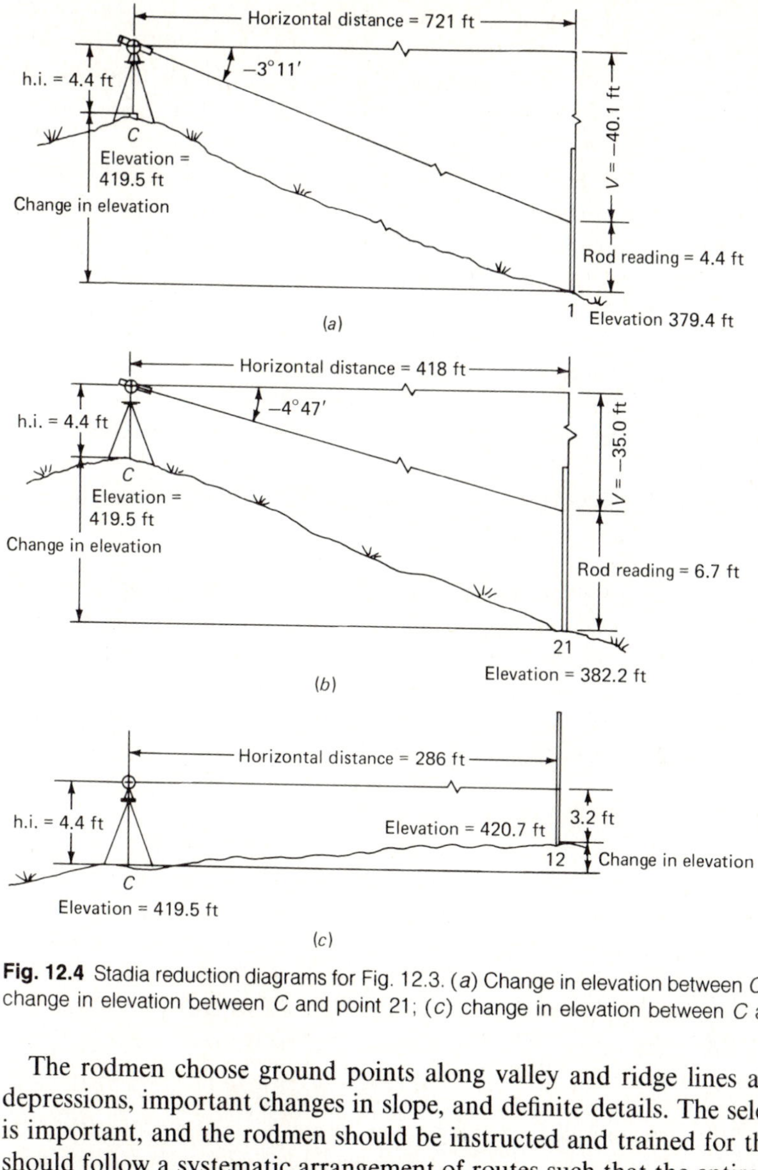

Fig. 12.4 Stadia reduction diagrams for Fig. 12.3. (*a*) Change in elevation between *C* and point 1; (*b*) change in elevation between *C* and point 21; (*c*) change in elevation between *C* and point 12.

The rodmen choose ground points along valley and ridge lines and at summits, depressions, important changes in slope, and definite details. The selection of points is important, and the rodmen should be instructed and trained for their work. They should follow a systematic arrangement of routes such that the entire area is covered and that no important objects are overlooked. They should observe the terrain carefully (often with the aid of a hand level) and report any important features which cannot be seen from the transit station.

When an optical repeating theodolite is used to locate controlling points by stadia, the zenith angle (Fig. 6.49, Art. 6.43) is measured instead of a vertical angle. For example, if such an instrument were used for the measurements in Fig. 12.3, the observations on point 1 would have a zenith angle of 93°11′, on point 2 it would be 93°17′,...,etc. Thus, the vertical angle $= 90° - 93°11' = -3°11'$ and the stadia reductions are performed as indicated by Example 7.1, Art. 7.9. Note that the zenith angle can be used directly to calculate horizontal distance and difference in elevation by using Eqs. (7.5*a*) and (7.5*b*).

If an instrument equipped with a Beaman stadia arc (Art. 7.11) or a self-reducing tacheometer is employed for locating details by stadia, the note form and observational procedure is modified to accommodate the differences in the instrument. Figure 12.5 shows notes recorded for observations using an instrument with a Beaman stadia arc. The point number is recorded in column 1 and the interval is observed and recorded in column 2. When reading the stadia interval there is no need to have the middle cross hair near the value of the h.i. Next, the index on the V scale is set to the nearest integral V graduation using the vertical tangent motion. The values of H and V are observed and recorded in columns 3 and 4 and the rod reading or rod correction is observed and recorded in column 8. The azimuth is observed and recorded in column 6. Then the horizontal distance is calculated and recorded in column 5. As an example, the horizontal distance between B and $21 = (K)(s) - (H)(s) = (100)(2.90) - (3)(2.90) = 281$ ft. Next, calculate the difference in elevation between the instrument and point sighted on the rod, and record in column 7. Again, for point B to point 21, this value is $(V - 50)(s) = (32 - 50)(2.90) = -52.2$. Continuing for the line $B = 21$, the change in elevation between the horizontal axis of the instrument and the ground point = Δel. = $(V - 50)(s)$ − rod reading = $-52.2 - 3.2 = -55.4$ and is recorded in column 9. Note that the height of instrument above point C(4.4 ft) has been added to the elevation of C and recorded as the height of instrument above the datum or H.I. in column 10. The value for change in elevation or Δel. can be applied directly to this H.I. to yield the elevation of ground point 21 recorded in column 11. Thus, elevation of $21 = 428.3 - 55.4 = 372.9$ ft.

When using one of the self-reducing tacheometers such as the one illustrated in Fig. 7.9, a note form similar to that shown in Fig. 12.5 but modified to accommodate the special features of the instrument can be employed. These instruments yield horizontal distance and difference in elevation simply by multiplication of the stadia interval times a constant of 100. Thus, when an experienced operator is available, there would be no need to record the cross-hair readings. The intervals could be observed and mentally multiplied by 100 so that horizontal distance and the change in elevation could be recorded directly. Instruments such as these frequently have an

Pt.	Int. (S)	H	V ⚻	Horiz. Dist. @ C	Az.	(V-50)(I) h.i. =	Rod Corr. (-) 4.4 ft.	ΔElev.	H.I.	Elev. ft	
									428.3	423.9	K = 100; ftc = 0
B					176°14'						
21	2.90	3	32	281	276°40'	−52.2	3.2	−55.4		372.9	Flow Line Stream
22	1.50	5	72	142	175°20'	+33.0	11.0	+22.0		450.3	Top of hill

Fig. 12.5 Note form for details about a point using an instrument equipped with a Beamen stadia arc (units are in feet).

automatic vertical circle index compensator, and there is no need to center a level bubble prior to observing the intervals. Consequently, use of such instruments can yield substantial savings in time when locating details by stadia.

12.11. Details by plane table The personnel of the plane-table party for mapping details by the controlling-point method usually consists of the plane-table operator, computer, and one or more rodmen. The equipment normally includes a plane table, telescopic alidade (preferably of the optical-reading, self-reducing type), scale, small triangles, 6H or 8H pencil, and electronic pocket calculator.

Before the party goes into the field, the horizontal control, the coordinate system, and the outline of the map sheet are adjusted and plotted on the plane-table sheet (Art. 7.18) according to the method developed in Art. 10.10. The elevations of all bench marks either are recorded on the sheet or are in the hands of the computer. It is also advisable to indicate the distance and direction for each course of the control survey on the plane-table sheet.

The instrument operator sets up the table at a convenient station, orients it usually by backsighting on an adjacent station and measures the h.i. of the alidade above point *B*. The rodmen are directed to the controlling points of the terrain, as just described for the transit stadia method. When a rodman holds the rod on a ground point, the instrument operator sights on the rod and observes the data necessary in order to determine horizontal distance and difference in elevation from the instrument to the ground point sighted. These data are called out to the computer, who enters the information into the pocket calculator and makes the necessary reductions to provide horizontal distance to the point and its elevation. The plane-table operator motions to the rodman to move ahead, draws a short portion of the ray on the plane-table sheet from the station to the point sighted at the approximate location of the point, and then plots the point, to the scale of the plane-table sheet, using the distance provided by the computer. Finally, the point is marked with a small circle and the elevation is recorded on the map next to the point. As rapidly as sufficient data are secured, the plane-table operator sketches the contour lines. Other objects of the terrain are located and are drawn either in finished form or with sufficient detail so that they may be completed in the office.

The specific procedure used in observing the rod for each sighting varies somewhat depending on the type of alidade employed. There are three principal types of alidades currently in use: (1) those with a Beaman stadia arc equipped with a vertical circle control level bubble (Art. 7.11); (2) instruments with optically observed scale readings analogous to the Beaman arc, some of which have vertical circle control level bubbles and others that are equipped with automatic vertical circle indexing (Art. 7.18); and (3) alidades with optically observed self-reducing scales with variable stadia intervals which lead directly to the horizontal distance and difference in elevation when multiplied by a constant (Art. 7.16). These last instruments have either a vertical circle control bubble or automatic vertical circle indexing. The electronic digital alidade (Art. 7.18), a recent development, is not yet in common use but appears to have considerable potential for plane-table surveys. The methods for using these instruments and the computations to reduce the measurements to distance and elevation are detailed in the sections indicated above in Chap. 7 and need not be repeated here.

Generally in practice, no record is made of the data observed and used to plot points on the plane table. Since all plotting is done in the field, mistakes or omissions are apparent at once and can be corrected immediately. However, for instructional purposes a record of the data observed can be useful for future reference. When notes of plane-table operations are recorded, a form similar to the ones indicated in Figs. 12.3 and 12.5 for transit stadia can be employed. Naturally, the column for azimuth is deleted and column headings are modified to suit the type of alidade being used.

Figure 12.6 illustrates such a note form designed for observations taken with the alidade shown in Fig. 7.18, an optical-reading instrument with a vertical circle control level bubble. The procedure for obtaining these data for point 1 is as follows. The alidade is sighted on the rod and the stadia interval s is read by setting the bottom stadia cross hair on an integral decimetre or foot graduation. This interval is recorded in column 2. Next, the vertical circle index bubble is centered, the V scale is set to an integral value, -8.0 in this case using the vertical tangent motion, and the rod is read at the middle cross hair (1.25). The values of H, V, and the rod readings are recorded in columns 3, 4, and 7, respectively. The computer calculates the horizontal distance, $(V)(s)$, difference in elevation, and elevation of the point which are recorded in columns 5, 6, 8, and 10, respectively. Note that the difference in elevation is with respect to the horizontal axis of the alidade in this case and is applied directly to the height of instrument above the datum (H.I.) to yield elevation. It must be emphasized that these notes are *not* to be used for plotting the data at a later time. The notes provide a training device which helps systematize party operation during the initial stages of learning to use the plane table. The plotting and drawing of all contours and planimetric detail is done in the field on the plane-table sheet.

The utmost skill of the topographer is used in judging the features of the terrain and representing them on the map with the required precision and with the least expenditure of time.

Many objects are located by the method of intersection (Art. 9.11), the elevations being determined by trigonometric leveling (Art. 5.5).

PLANE	TABLE	SURVEY					ARMSTRONG PROPERTY			May 6, 1984 (24)
Inst. @ A	h.i. = 1.22m									Sunny 70°F
Point	S	H	V	Horiz. Dist.	V×S	Rod Corr. (-)	Δ Elev	H.I.	Elev., m	⊼ - Jones
B								31.22	30.000	Notes - Arnold
										Rod - Adams
1	0.348	0.6	-8	34.6	-2.78	1.25	-4.03		27.19	Top of inlet
2	0.375	0	+2	37.5	0.75	1.12	-0.37		30.85	Edge of sidewalk

Fig. 12.6 Notes for details with a plane table.

Because the plane table permits a ready solution of the three-point problem (Art. 9.12), use is made of this technique to enable the topographer to utilize advantageous instrument stations which have not been included in the control surveys, especially where control has been initially established by triangulation. The elevation of such a station is determined either by stadia leveling or by trigonometric leveling.

Transit and plane table For large-scale maps and where many details are to be sighted, sometimes it is advantageous to use both the transit or self-reducing tacheometer and the plane table. This method saves time in the field, but it may not reduce the total cost, as a larger party is required than for the plane table alone.

The transit or tacheometer is set up and oriented at the control station, the location of which is plotted on the plane-table sheet. The plane table is set up and oriented nearby, and its location is plotted on the map in its correct relation to the transit station. (In some cases the map distance between the transit station and the plane-table station is negligible, and the two points are regarded as identical.) When a rodman has selected a ground point, the transit operator observes the stadia distance and vertical angle to it; the plane-table operator sights in the direction of the point, draws a ray toward it from the plotted location of the plane-table station, plots the point at the correct distance scaled from the plotted location of the transit station, and records on the map the elevation (computed by the transit operator or the computer) of the plotted point.

12.12. Details by the cross-section method In the cross-section method of locating details, the ground points are on relatively short lines perpendicular to the main traverse or base line. The control points are stations on the traverse. These stations are set along the traverse lines at an interval which is a function of the irregularity of the terrain and the desired contour interval. When the unit is the foot, a full station is 100 ft and stations are usually set at 100- or 50-ft intervals. In the metric system, a full station is 100 m, so that stations are generally set at 10- or 20-m intervals. This horizontal control can be established by electronic tacheometer, theodolite, and EDM equipment, transit and tape, or transit stadia (between control stations established by taping or EDM equipment). The elevations of these stations are determined by differential profile leveling and are available to the topography party. Cross sections are particularly appropriate when the area to be mapped is a long, narrow band such as a route survey (Chap. 14).

The cross sections are taken using a level rod plus a transit and tape, level and tape, transit stadia, hand level and tape, or plane table and stadia. The party consists of a topographer and two people to work with the rod and tape. Sometimes a Jacob's staff or other rod about 5 ft long is used as support for the hand level while sights are being taken.

When the transit, rod, and tape are used for cross sections, the transit is set over the specified station and the height of instrument (h.i.) is determined by holding the rod alongside the transit. A sight is taken on another station on the same line and an angle of 90° is turned, establishing the line of the cross section. The rodman moves out along this line to the first break in the terrain. The distance is measured from the center line with the tape and a rod reading is taken with the telescope leveled. Distance and elevation are recorded and the procedure is repeated at the next and all subsequent breaks in terrain along the cross section to the limit of the area to be mapped. If the terrain slope is too steep to allow a rod reading with the telescope level, a vertical angle is turned to a point on the rod equal to the height of the instrument above the center-line station and a slope distance is measured to the point. The difference in elevation equals the slope distance times the sine of the vertical angle. The horizontal distance equals the slope distance times the cosine of the vertical angle. Use of the transit and tape is feasible for cross sections if the strip to be mapped is not too wide and the terrain is not excessively steep and irregular. If it becomes necessary to move the instrument ahead to complete the cross section, consider an alternative method.

Figure 12.7 shows one form of field notes for cross-section data taken from station 23 + 00 to 24 + 00 with 20-m stations. The stations are recorded from the bottom of the page to the top and cross-section data are on the left and right of the page, corresponding to the left and right sides of the center line when facing toward the next higher station. As an example, consider station 24 + 00, where the h.i. at the center line measured from the ground point to the instrumental horizontal axis is

Station		Left					Center Line	Right			15
	H.I.						[428.25]				
	Offset		30	20	17	10	0	10	15	20	30
24+00			2.65	3.45	4.05	3.15	1.65	0.40	+0.75	+2.75	+3.25
	Elev.		425.60	424.80	424.20	425.10	426.60	427.80	429.00	431.00	431.50
							[428.05]				
			30	F.L. Stream 20		10	0	10	20	Ridge 25	30
+80			3.55	3.85		2.95	1.55	0.25	+1.55	+3.05	+2.15
			424.50	424.20		425.10	426.50	427.80	429.60	431.10	430.20
							[427.50]				
			30	F.L. Stream 24	20	10	0	10	20	24	30
+60			4.50	4.30	3.60	2.40	1.60	0.30	+1.40	+1.80	+1.10
			423.00	423.20	423.90	425.10	425.90	427.20	428.90	429.30	428.60
							[427.10]				
		35	F.L. Stream 32	30	20	10	0	10	Ridge 18		30
+40		4.90	5.30	4.20	3.60	2.50	1.50	0.10	+1.00		+0.10
		422.20	421.80	422.90	423.50	424.60	425.60	427.00	428.10		427.20
							[426.45]				
		F.L. Stream 40	36	30	20	10	0	10	Ridge 20		30
+20		5.35	4.00	3.70	2.00	1.20	1.65	0.25	+0.65		0.55
		421.10	422.45	422.75	424.45	425.25	424.80	426.20	427.10		425.90
							[425.20]				
				30	20	10	0	10	Ridge 17	20	30
23+00				4.40	2.30	2.30	1.30	+0.30	+0.90	+0.80	0.30
				420.80	422.90	422.90	423.90	425.50	426.10	426.00	424.90

Fig. 12.7 Notes for the cross-section method (unit are in metres).

1.65 m and the center-line elevation is 426.60 m, taken from the profile notes. Thus, the instrument H.I. is 426.50 + 1.65 = 428.25 m, recorded above the center-line reading and enclosed in brackets. The offset distance to the first break point on the right is 10 m where the rod reading, taken with the telescope level, is 0.40 m and is recorded below the 10-m offset distance. The elevation of this point = 428.25 − 0.40 = 427.8 m, which is calculated immediately and recorded beneath the rod reading. At the next break point to the right of the center line, the ground is higher than the instrumental H.I., so that a vertical angle of +2°52′ was observed to 1.65 m on the rod at a slope distance of 15.0 m, resulting in a change in elevation of +0.75 m and a horizontal distance of 15 m from the center line. This procedure is then repeated to obtain readings 20 and 30 m to the right. Similarly, with the telescope level, rod readings were taken at 10, 17, 20, and 30 m to the left of the center line.

Where specific planimetric details fall within the area being mapped, they are located by plus and offset. For example, the flow line of the stream is located 20 m to the left of station 23 + 80, 24 m left of station of 23 + 60, etc. Similarly, any type of planimetric features, such as buildings, utilities, trees, fences, etc., can be located in the cross-section notes by means of plus and offset, as described at the end of this section.

One of the most common methods for taking cross sections is by using a hand level, rod, and tape. For this operation, the right angle is established previously by the transit, by estimation, or by using a *pentaprism*, a hand-held device for setting off a right angle. The center-line profile is run first using a rod and level. Consequently, elevations for each center-line station are available. First, the direction of the cross section is established and the rod is held on the center line. The hand-level operator is stationed along this cross section in a position which allows comfortable viewing of the rod. Since the hand level has no magnification, lengths of sights should be no more than 15 m or about 50 ft. If the ground is rising, the center-line reading should be as high as possible. If the terrain slopes away from the center line, the reading on the rod at center line should be as low as possible. For example in Fig. 12.8, the original rod reading in the profile notes for the center line was 1.50 m, assuming that the profile was run independently. In order to run the cross section up the slope the hand-level operator is positioned along the line of the cross section so that a backsight of 3.50 m can be taken on the rod held at the center-line station. Thus, the hand level is "up 2 m" and it is the responsibility of the notekeeper to subtract 2 m from all subsequent rod readings taken from that position. In the example, a reading of 2.40 m is taken 10 m right and is recorded as (2.40 − 2) = 0.40 at 10 m right. The next break point is 20 m right, where the rod reading is 1.00 m, which is recorded as +1.00 m, 20 m right. The rod reading is positive since the bottom of the rod is now 1 m above the original H.I. of the instrument used to obtain the center-line reading of 1.50 m, to which all rod readings on the cross section are referred. At this point, the hand-level operator moves ahead to a position which permits a backsight of 3.00 m on the rod still held 20 m right. Now the hand level is "up 4 m" and 4 m must be subtracted from all subsequent rod readings so as to make them consistent with the rest of the cross section. The next reading at 30 m right is 1.60 m, recorded as +2.40 m, 30 m right; and the final reading to the right of center line is 1.40 m, recorded as

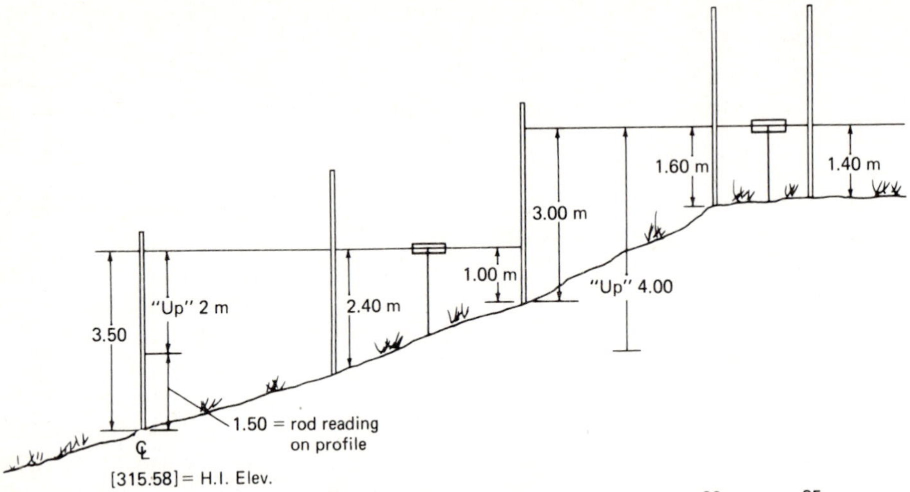

Offset	0	10	20	30	35
Rod reading	1.50	0.40	+1.00	+2.40	+2.60
Elev.	314.08	315.18	316.58	317.98	318.18

Fig. 12.8 Cross section with hand level and tape.

+2.60 m, 35 m right. The note form shown below the sketch is similar to that employed in Fig. 12.7. When running a cross section with a hand level, a stake is placed at the position of the last reading on each section. Then each section is closed on the previous line to that side of the center line and mistakes are avoided. The closures should be within 3 dm or 1 ft. Since all rod readings have been referred to the profile H.I. (recorded above the center-line data and enclosed in brackets [315.58]), the readings are subtracted from or added to this H.I. to yield ground-point elevations as shown in the sample cross-section notes of Fig. 12.8. The same procedure can be applied to terrain that slopes downward from the center line. In this case the height of the hand level will be "down" a given number of metres or feet, which will be added to all readings taken along that cross section.

In regions of slight relief or in urban areas where it is necessary to obtain elevations and locations of pavement, curb lines, sidewalks, utilities, etc., the engineer's level, tape, and rod are frequently used for cross sections. When this procedure is followed, the center-line profile can be run concurrently with the cross sections or as a separate operation. In either case, center-line elevations are available. The procedure is as follows. The level is set up at a location convenient for viewing to the left and right of the center line and the rod is held on a center-line station where a reading is taken. The right angle to set the direction of the cross section may have been previously established by a transit or is now estimated by the rodman. The rodman then moves out along the cross section. Rod readings are taken and distances are measured to significant features and break points along the cross section to the left and right of the center line.

Figure 12.9 illustrates a note form for cross sections using a level, tape, and rod in an urban area. English units are employed in these notes. A plan view of the area

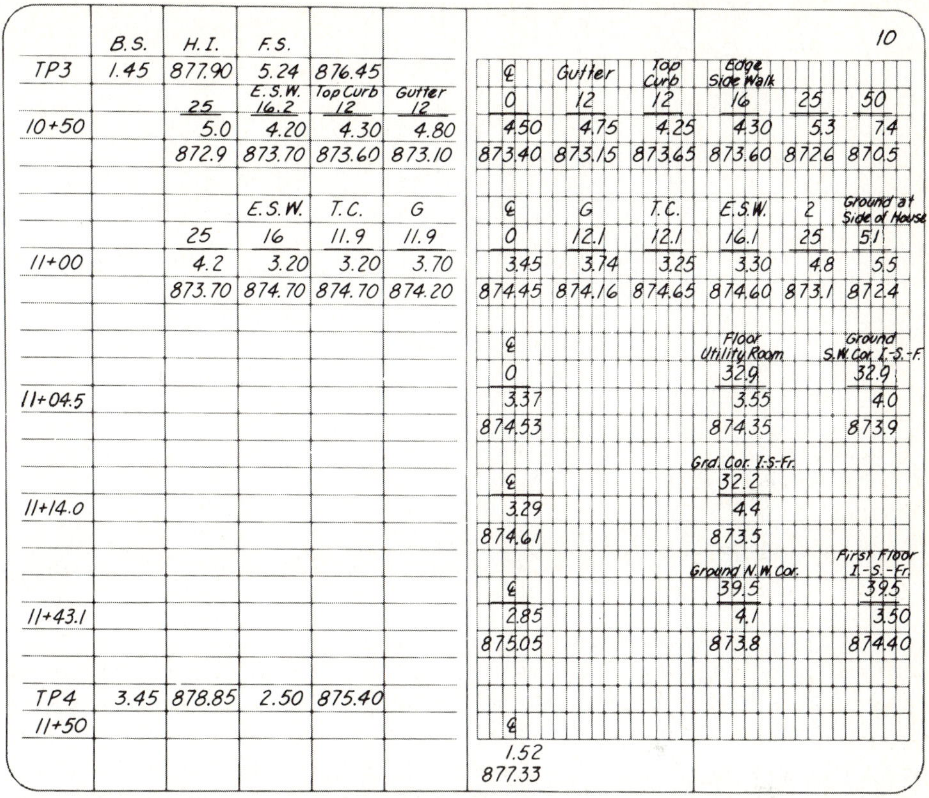

Fig. 12.9 Cross-section notes in an urban area using an engineer's level and rod (units are in feet).

involved is shown in Fig. 12.10. The plan positions for the two instrument stations and two turning points utilized in the notes of Fig. 12.9 are also shown on Fig. 12.10*a*. Locations of instrument setups are usually chosen so as to allow a maximum number of readings from a single setup and also to balance backsight and foresight distances between turning points. For this example, stations are set along the center line of the existing pavement and the center-line profile and cross sections are being run concurrently. Thus, the notes are recorded from the top of the page down, as is generally done for regular profile levels, in contrast to the example in Fig. 12.7, where the cross sections are run independently of the center-line profile. Note that all objects are described and rod readings on semipermanent features such as pavement, curbs, sidewalks, etc., are read to the nearest hundredth of a foot while readings on the ground are to the nearest tenth of a foot. A sectional view of the cross section at station 11 + 00 is shown in Fig. 12.10*b* oriented as one would see it when standing at station 11 + 00 and looking toward station 11 + 50. For convenience, the level is shown in this sectional view even though it does not fall on the line of the cross section as can be seen in Fig. 12.10*a*. Since the center-line profile is also being run, all rod readings (foresights) are subtracted from the H.I. established at T.P.$_3$ (turning point 3) in the profile levels. A new turning point, T.P.$_4$, is taken in

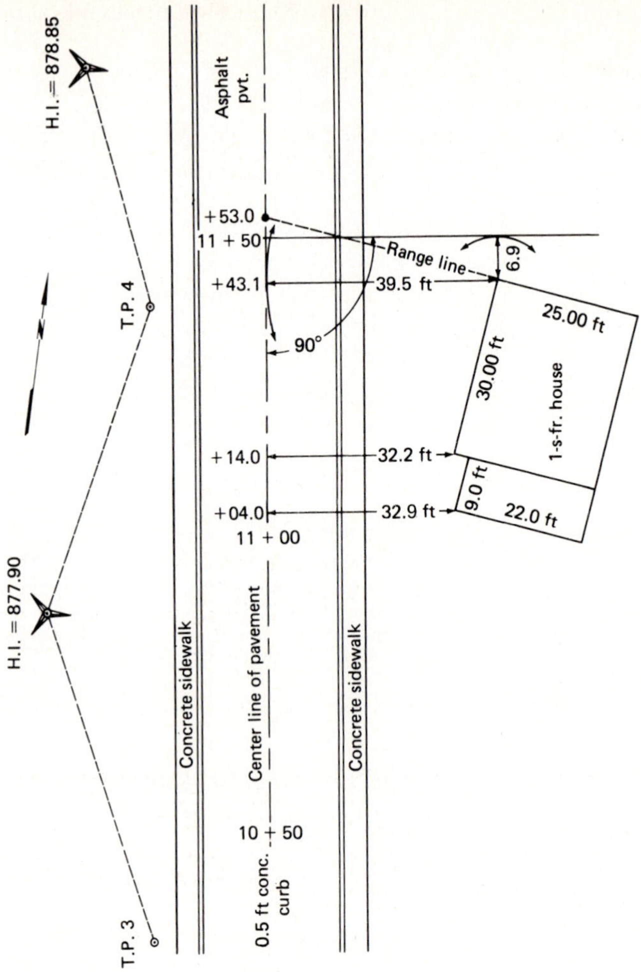

(a)

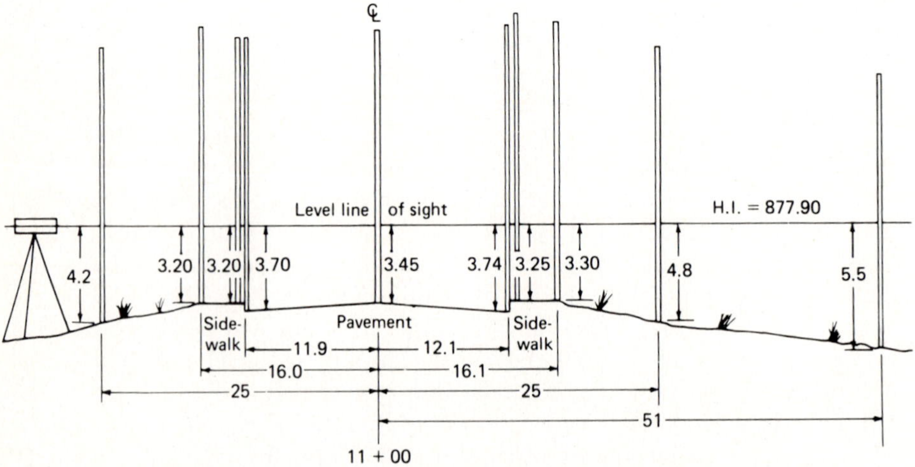

(b)

Fig. 12.10 (*a*) Plan view for cross sections in an urban area and location of details by plus and offset. (*b*) Section through cross section of 11 + 00 from the notes of Fig. 12.9 (units are in feet).

order that the level can be positioned so as to run the cross section at the next station, station 11 + 50. Frequently, when lack of visibility or irregular terrain prevents completing a cross section, a hand level can be used to finish the cross section, assuming that rod readings to the nearest tenth of a foot or to the nearest 0.03 m are sufficiently accurate.

Cross sections with transit stadia are essentially the same as those run using the transit, tape, and rod, except that distances and differences in elevations to break points and other features are determined by the stadia method. This method is particularly advantageous when the terrain is exceptionally steep and irregular.

When cross sections are run using a plane table and alidade, the center line is plotted on the plane-table sheet. The plane table is then set over a station and the plotted center line is aligned with the center line in the field. The alidade is aligned with a perpendicular drawn through the station and the cross section is run along this line. Distances and differences in elevation to break points and features are determined by stadia with the point and elevation being plotted directly on the sheet. Thus, there is no need to keep a separate set of notes. One advantage of utilizing a plane table for cross sections is that all additional planimetric detail can be easily located by the controlling-point method and plotted directly on the plane-table sheet.

The completed topographic map should also show the planimetric position of all significant natural and constructed features in the area. Some of these features may not be obtained by the cross sections where the station interval is 20 m or 50 ft or larger, so that a supplementary operation is required to locate the additional detail. When cross sections are taken by transit or plane-table stadia, these methods are also employed to locate details between the cross sections by distance and direction.

If cross sections are run by engineer's level or a hand level and tape, additional details are usually located by *plus* and *offset* from center or base line. A plus is the numerical value for the center-line station of a line perpendicular to the center line and also passing through the object to be located. The distance from the station to the object is the offset. Figure 12.10*a* illustrates the location of a building by plus and offset. Thus, the northwest corner of the building is located by an offset of 39.5 ft to the right of station 11 + 43.1. A station on the intersection of a *range* line from the north end of the building with the center line is also indicated as being 11 + 53.0. This provides a check on the plus and offset location. For intermediate-scale and many large-scale maps, the plus is determined by estimation or using a pentaprism. When higher precision is required, such as in a property survey where it is desired to show the location of a structure with respect to property lines, a transit and tape are used.

The data from the cross sections are plotted on a master sheet to the scale desired, and the contours are plotted by interpolation as described in Chap. 10. The cross-section data from Fig. 12.7 are plotted in Fig. 12.11. The stations are indicated along the plotted center line, perpendiculars are drawn through these stations, and the points located are plotted on these perpendiculars. The elevations of these points would be recorded on the sheet lightly and then the contours are drawn by interpolation. This portion of the job could be done directly in the field when the plane table is used to collect cross-section data.

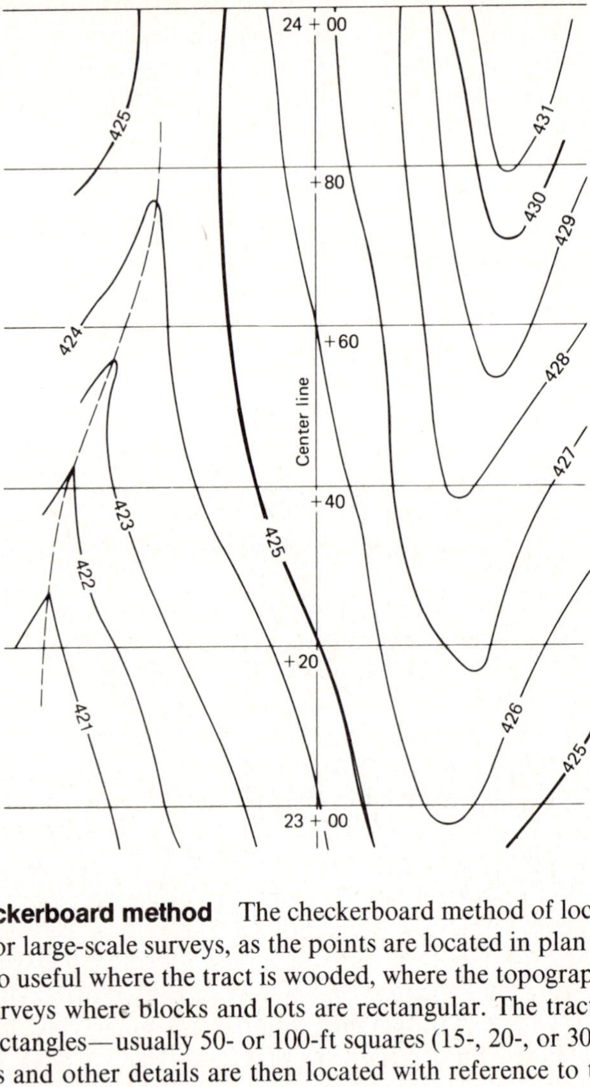

Fig. 12.11 Contour lines (1-m intervals) interpolated from cross-section data given in Fig. 12.7.

12.13. Details by the checkerboard method The checkerboard method of locating details is well adapted for large-scale surveys, as the points are located in plan by tape measurements. It is also useful where the tract is wooded, where the topography is smooth, and on urban surveys where blocks and lots are rectangular. The tract is staked off into squares or rectangles—usually 50- or 100-ft squares (15-, 20-, or 30-m squares). The ground points and other details are then located with reference to the stakes and connecting lines.

The usual procedure is first to run a traverse near the perimeter of the tract such as traverse *A*-6-6′-*K* shown in Fig. 12.12. Next, one side of the traverse is chosen as a base line and is marked off in stations of the desired interval. In Fig. 12.12, if line *AK* is selected as the base line and the interval chosen is 20 m, then $A = 0 + 00$, $B = 0 + 20$, $C = 0 + 40$, etc. Stakes are set at each 20-m station along the base line. Next each station along the base line is occupied with the transit, an angle of 90° is turned from the base line *AK*, and stakes are set at 20-m intervals along lines *BB*′, *CC*′,…, *GG*′. Points *B*′, *C*′, *D*′, *E*′, and *F*′ are set by intersecting the respective grid line with line 66′, usually by estimation. Points 1 through 6 and 1′ through 6′ can be set by ranging in along the appropriate interior grid line and intersecting the

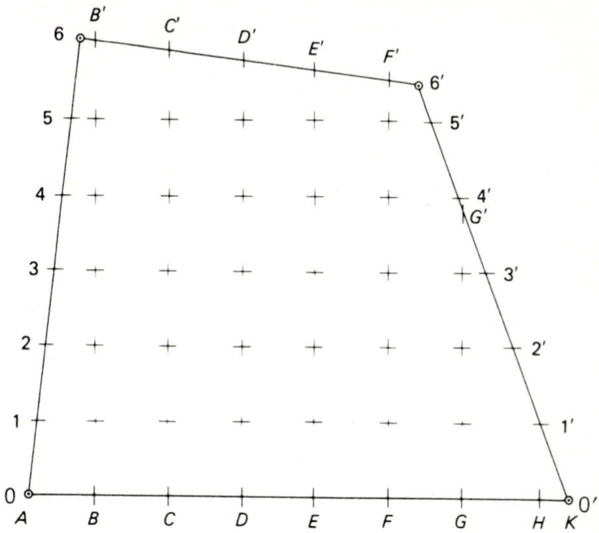

Fig. 12.12 Checkerboard system for topographic survey.

grid line with the traverse line. All points are labeled by number and letter which designate row and column, respectively. Thus, along the grid line from 1 to 1′ there are stations $1A, 1B, \ldots, 1H$. Several different methods for setting the grid intersections are possible, all of which would be satisfactory if the accuracy required for the survey is satisfied. The method outlined would provide fairly precise results for horizontal position of the points.

When all stations have been set and marked, levels are run and an elevation is determined at each grid-line intersection and the intersections of the grid lines with the perimeter traverse lines. An engineer's level or an engineer's transit can be used for the leveling. In addition, all irregularities and significant planimetric features are located from the nearest pair of intersecting grid lines. When all elevations have been determined and details located, the data are plotted on a map sheet on which the traverse and grid lines have been plotted to the desired scale of the map. Contours are drawn by interpolation and all planimetric features are plotted to complete the topographic map of the area.

12.14. Details by the trace-contour method The trace-contour method of locating contour points on the ground is commonly used on large-scale surveys, and sometimes on intermediate-scale surveys where the ground is irregular. Under these conditions, if visibility is good, the trace-contour method is more accurate than the checkerboard method.

Although the transit may be used in this work, either alone or with the engineer's level, the plane table is commonly used because it requires fewer points to be observed and because of the saving time in plotting.

A control survey is run and plotted on the plane-table sheet. The plane table is set up over a control point of known position and elevation, such as point A (plotted as a on the plane-table sheet) in Fig. 12.13, and the plane table is oriented by sighting control point *B*. The height of the instrument above point *A* is measured using the rod and found to be 1.25 m. If the elevation of *A* is known to be 352.40 m, the H.I. = 352.40 + 1.25 = 353.65 m. With the H.I. known, rod readings for specific contours can be calculated. For example, to locate the 351-m contour, the rod reading with the telescope of the alidade leveled should be 353.65 − 351 = 2.65 m. The plane-table operator informs the rodman what reading is required to locate the desired contour, and the rodman estimates where the rod should be held to obtain this reading. The plane-table operator focuses the telescope on the rod, centers the vertical circle level bubble, sets the vertical circle index to zero, and then directs the

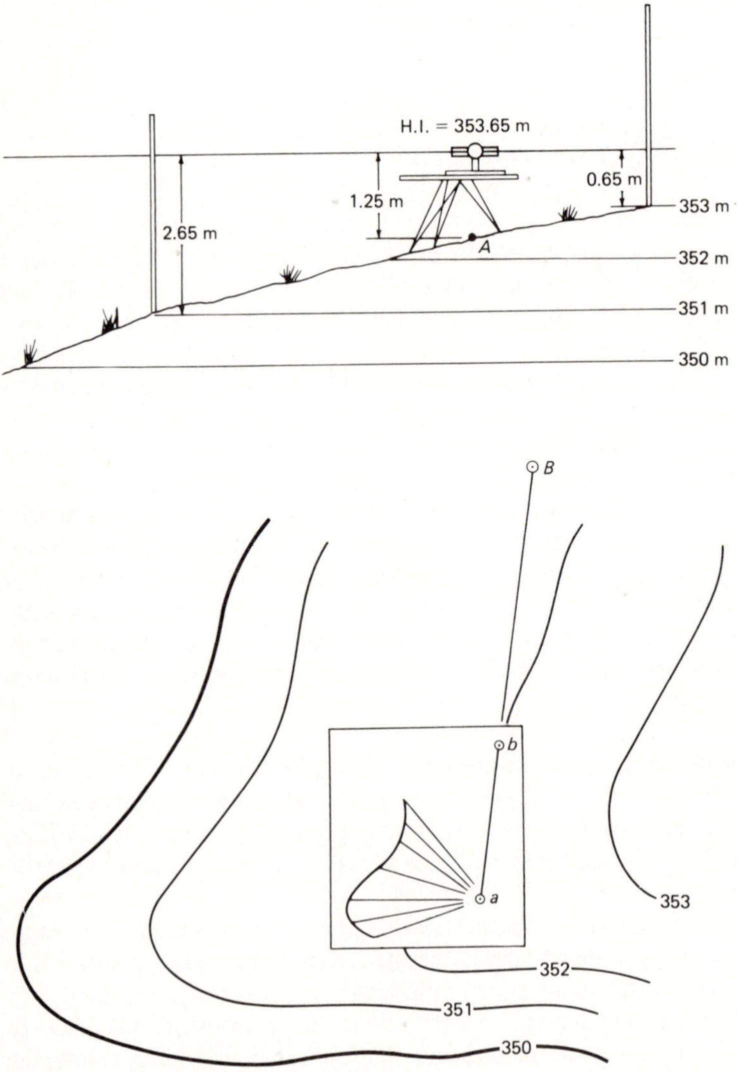

Fig. 12.13 Trace contour method with a plane table.

rodman up or down the slope until the line of sight cuts the rod at 3.65 m. The plane-table operator checks the vertical circle bubble and index reading, observes the stadia interval, draws a ray along the alidade blade, and plots the calculated horizontal distance to the location of the contour on the plane-table sheet. The rodman then moves along the slope to another estimated location of the contour and the procedure is repeated until the contour has been traced throughout the range of the control on the plane-table sheet. As soon as at least three points have been plotted, the plane-table operator should draw and label the contour line interpreting all the irregularities that can be observed as the contour is traced. Readings should not be taken too close together in fairly regular terrain. Where the terrain is highly irregular, rod readings should be taken on all break points in the terrain. When the 351-m contour has been located, the rod reading for the 352-m contour, calculated to be $353.65 - 352 = 1.65$ m, is traced. From the setup shown in Fig. 12.13, the 353-m contour could also be traced, and then a new plane-table setup would have to be made. The trace-contour method is quite accurate but tends to be slow and costly. Thus, this procedure should not be used unless it is absolutely necessary in order to meet rigid accuracy requirements.

When contours are traced using transit stadia or a self-reducing tacheometer, the procedure is the same except that the H.I., stadia interval, rod-reading horizontal distance, and azimuth to the contour locations are recorded in a field notebook. The instrument is set over a known control point and a backsight is taken along a line of known azimuth using the lower motion. The azimuth of the known line is set off on the clockwise horizontal circle before taking this sight. All sights to locate the contour being traced are taken with the upper horizontal motion and with the telescope in a horizontal position.

The engineer's level can also be used to trace contours provided that the telescope has stadia hairs and the level has a horizontal circle. The self-leveling instrument shown in Fig. 5.17 is so equipped.

When contours are traced using transit stadia, self-reducing tacheometer, or engineer's level, the data are plotted in the office. First, a base map must be prepared on which all control points are plotted. Then the contour points are plotted by distance and azimuth from the given control points using a protractor and scale. Finally, the contour points are connected to form contour lines.

12.15. Electronic positioning systems for topographic surveys Reference to Art. 9.14 shows that the combined theodolite and EDM systems and electronic tacheometers automatically provide horizontal distance and difference in elevation between the station occupied and the ground point sighted. Since these are the basic quantities involved in all of the outlined methods for topographic surveying, electronic positioning systems are ideally suited for gathering data for topographic surveys, particularly by the controlling-point method. The procedures for use of such a system for a topographic survey by the controlling-point method are similar to those suggested for acquiring topographic details by transit or self-reducing tacheometer in Art. 12.10. The major differences occur in the operation of the instruments and methods for recording data.

When the electronic tacheometer is used for a controlling-point topographic survey, the instrument is set up over a known control point and oriented by a backsight on another known station. Then the target or signal is placed at a desired ground point and is sighted by the theodolite. On the simplest of these systems, proper manipulation of the controls on theodolite and EDM (Art. 9.14) yields visual output consisting of horizontal distance, direction, and difference in elevation to the point. These data are recorded in the field notebook for subsequent plotting in the office. The more sophisticated electronic tacheometers not only have a visual display of the positional data but also allow for storage of the data in solid-state core memory or on magnetic tape. Voice data entry, described in Art. 3.15, provides a method potentially useful for recording verbal descriptions automatically. Thus, the need for complete field notes is reduced and the bulk of the data are automatically stored for processing in the office. Consequently, the time needed for the operation is limited to the time required to move the signal from one point to another and that necessary to operate the instrument.

Under certain conditions, use of the electronic positioning systems with a plane table is appropriate. The procedure would be similar to the method described in Art. 12.11 for the transit and plane table. Acquisition of data would be substantially faster in this operation so that the limiting time factor would be the ability of the plane-table operator to plot and draw topographic detail. The advantage of this method is that a more reliable map is produced because of the proximity of the map compiler to the terrain.

The electronic digital alidade (described in Art. 7.18) and plane table can provide a powerful system for topographic surveys by the controlling-point method (Art. 12.9). With this equipment, simplified acquisition of data by electronic methods is combined with reliable map compilation by traditional methods on the site.

Electronic positioning systems could also be utilized for cross-section topographic surveys, especially where the cross sections are being run for route alignment in rugged, steep terrain. The procedure would be similar to the method outlined for cross sections using the transit, tape, and rod in Art. 12.12. The transit and tape are replaced by the electronic positioning instrument and the rod is replaced by the reflector or signal. Data are then recorded or stored as previously described in this article for controlling-point topography.

12.16. Topographic surveys and digital terrain models The digital terrain model (DTM) or digital elevation model (DEM) is described in Art. 10.15 along with some of its uses and applications in topographic mapping. Since the DTM is a numerical (X and Y coordinates and elevations) representation of the configuration of the terrain, it can be stored in the memory of a computer and may be processed by a computer program which permits plotting planimetric detail and contours. Thus, the DTM is an extremely flexible method for storing data which describes terrain configuration.

The reason for inserting these comments at this time is to emphasize that the data from each of the methods for obtaining topographic detail constitute a DTM in the most basic form.

The methods for acquiring topographic detail described in this chapter are designed primarily for the small or special purpose topographic survey or map completion survey. The traditional methods for plotting and compiling maps of this type will probably continue to be used well into the future. However, the surveyor or engineer involved with these special mapping problems needs to be aware of the potential of the data being collected. In this way field methods and data recording and storage procedures can be designed so as to be compatible with possible automatic data processing in the future.

12.17. Problems

12.1 Describe the major steps involved in performing a topographic survey.

12.2 Differentiate among large-, intermediate-, and small-scale topographic surveys.

12.3 What factors govern the choice of surveying method used for topographic surveys?

12.4 Discuss the various surveying methods appropriate for establishing horizontal and vertical control for topographic mapping and the conditions that favor each method.

12.5 What sort of equipment would be required for running a three-dimensional traverse to establish horizontal and vertical control for a topographic survey?

12.6 List the four principal methods for acquiring topographic details.

12.7 The controlling-point method is appropriate for locating details in what sort of terrain? What are the advantages of this method?

12.8 The cross-section method of locating details is best adapted to what type of topographic survey?

12.9 Describe how a cross section may be run using a transit and a tape. Assume that the terrain is steep but of constant slope on both sides of the center line. Use a sketch to illustrate your explanation.

12.10 Use a sketch to show how details can be located from a base or center line by plus and offset.

12.11 Describe the principal features of the checkerboard method for locating topographic details.

12.12 What are the advantages and disadvantages of locating topographic details by the trace-contour method?

12.13 Describe how electronic positioning systems can be used for locating topographic details.

12.14 Discuss the relationship between surveys to obtain topographic details and digital elevation models.

12.18. Office problems

Object To obtain a sufficient amount of data for making a topographic map of a small area having a moderate amount of relief by using the controlling-point method (Art. 12.9). Data from this problem can be used in Prob. 1 (Art. 10.20).

Procedure

1. Make rapid reconnaissance of the tract, selecting the most advantageous points for instrument stations from which areas comprising the entire area can be observed.
2. Run a closed traverse through the selected points, observing the stadia intervals and vertical angles. Follow the procedures suggested in Field Problem 3, Chap. 8, using a transit or tacheometer and stadia. If an electronic surveying system (Art. 9.13) is available, this may be used to run the traverse.
3. If details are to be located by transit stadia (Art. 12.10), occupy each of the traverse stations and with the instrument correctly oriented observe the azimuth, stadia distance, and vertical angle to all changes in ground slope and to other natural and artificial features which are within range of the instrument.
4. Include in the notes a sketch drawn approximately to scale. Use a note form similar to Fig. 12.3.
5. Reduce the stadia readings to horizontal distances and differences in elevation. Plot control points and details as directed in the office problem of Chap. 10.

6. If topographic details are to be located by a plane table, adjusted coordinates (Arts. 8.18 and 8.19) and elevations (Art. 5.41) are calculated.
7. Plot the positions of horizontal control points (see the instructions in Art. 10.10) on a plane-table sheet (Art. 7.18) of heavy detail paper or drafting film.
8. Occupy each station with the plane table, locating detail according to the instructions given in Art. 12.11.
9. Draw planimetric features and contours *in the field.* On student projects it is also advisable to keep notes using a form similar to Fig. 12.6.

Hints and precautions

1. If the elevation of a point is not required for mapping, often the point can be located advantageously by the method of intersection, the azimuth being observed from two or more traverse stations.
2. When using transit stadia, it is sometimes advantageous, particularly if there are a large number of details, to plot the map in the field as the work progresses. This is most easily accomplished by using a plane table and a transit as suggested at the end of Art. 12.11.
3. When details are close to the horizontal control station, it is often faster to get the horizontal distance by measuring with a tape rather than reading the stadia interval.

References

1. American Society of Civil Engineers, *Definitions of Surveying and Associated Terms*, Manual of Engineering Practice No. 34, 1978 (rev.).
2. American Society of Civil Engineers, *Technical Procedures for City Surveys*, Manual of Engineering Practice, No. 10, 1963.
3. Bere, C. G. T., "Direct Field Plotting Using Short Range EDM," *Survey Review*, January 1976.
4. Chrzanowski, A., et al., "Control Networks with Wall Monumentation: A Basis for Integrated Survey Systems in Urban Areas," *The Canadian Surveyor*, September 1977.
5. Dearden, J. D., et al., "The BOMB Project, A Control Survey in Southern Ontario," *The Canadian Surveyor*, March 1973.
6. Emerson, J. R., and Schultz, R. J., "A County and University Cooperative Control Survey," *Surveying and Mapping*, June 1972.
7. Wolf, P. R., Wilder, B. P., and Mahun, G., "An Evaluation of Accuracies and Applications of Tacheometry," *Surveying and Mapping*, Vol. 38, No. 3, 1978.

CHAPTER 13
Photogrammetric surveying and mapping

13.1. Introduction The discipline of photogrammetry involves obtaining information about an object *indirectly* by measuring photographs taken of the object. Therefore, unlike surveying procedures, where measurements are usually made *directly* on the object in the field, in photogrammetry the object is first recorded on an intermediate medium, usually photographs, and the measurements are carried out later in the office. Thus, photogrammetry requires the following operations: (1) planning and taking the photography, (2) processing the photographs, and (3) measuring the photographs and reducing the measurements to produce the end results, such as point coordinates or maps.

Two broad categories are involved in photogrammetry: *metric* or *quantitative* activities and *interpretive* or *qualitative* work. Metric photogrammetry involves all quantitative work, such as the determination of ground positions, distances, differences in elevations, areas, volumes, and various types of maps. In the second category, classically called *photo interpretation*, photographs are analyzed qualitatively for purposes of identifying objects and assessing their significance. Photo interpretation relies on human ability to assimilate and correlate such photographic elements as sizes, shapes, patterns, tones, textures, colors, contrasts, and relative location. Accurate and reliable photo interpretation requires extensive training and experience. The photo interpreter is called upon in many fields of application, such as ecology, environmental analysis, forestry, geology, engineering site selection, resource inventory, planning, and of course military intelligence. In recent years, records from other imaging systems such as infrared and radar have been used for interpretive purposes, and therefore the more general name *remote sensing* is used. Civilian applications of remote sensing, however, still employ aerial photography more extensively than other sensor data, although the latter are gaining in acceptance.

In photo interpretation, some measurements are necessary, and in metric photogrammetry some interpretation is required. Therefore, photogrammetry combines both activities and can be broadly defined to include acquisition, measurement, interpretation, and evaluation of photographs, imageries, and other remotely sensed data. Because of this broad definition, photogrammetry has found applications in a large number of fields, ranging from biomedicine and dentistry to aerospace engineering and astronomy, from cloud-chamber measurements in physics to all types of engineering, from tailoring to transportation and urban planning, from accident analysis to forestry, and so on. But by far the most common application is in surveying and mapping. Almost all of the topographic maps produced by federal, state, and private organizations are compiled from aerial photographs. Photogrammetric procedures are also used in deriving supplemental ground control, both horizontal and vertical. Therefore the surveying and engineering students must be aware of the various facets of this field and the ways photogrammetry can assist them in their future professional activities.

Metric photogrammetry has been classically divided into terrestrial and aerial types. In *terrestrial photogrammetry*, the photographs are taken from fixed, often known, points on or near the ground. In *aerial photogrammetry*, a high-precision camera is mounted in an aircraft and photographs are taken in an organized manner as the aircraft flies over the terrain. A more recent branch is *space photogrammetry*, which deals with extraterrestrial photography and imagery where the camera may be fixed on earth, mounted on board an artificial satellite, or located on a moon or a planet. *Close-range photogrammetry* involves applications where the camera is relatively close to the object photographed. Such applications exist in many fields where direct measurement of the object is either impractical, uneconomical, or simply impossible. Examples are found in archaeology, architecture, medicine, and many aspects of industrial applications and experimental engineering laboratory investigations.

This chapter will be devoted to the principles of aerial photogrammetry and its applications in surveying and mapping. An important fact should be established at the outset that an aerial photograph is *not* equivalent to a map except in very unusual circumstances. If the terrain is flat and level, and if the aerial photograph is exposed with the camera pointing perfectly downward (i.e., its optical axis is truly vertical), and assuming no image aberrations, the resulting photograph will be a map with a constant scale. When any of the foregoing restrictions are not met, the photograph will then only *approximate* a map, the degree of approximation depending upon the amount of distortion in the photograph. For example, when the terrain contains relief, the photographic scale at high points will be larger than that at lower points because the higher points are closer to the camera. In order to derive a proper map of the terrain, it is necessary that the ground is imaged in at least two successive aerial photographs. Photogrammetric techniques may then be used to reconstruct, to a scale, a faithful replication of the ground in all three dimensions from the overlapping photographs. Once this is accomplished, horizontal and vertical ground-point positions, planimetric maps, topographic maps, cross sections, and a variety of other products can be extracted. The objective of this chapter is to introduce the student to the various operations and products of photogrammetry so that he or she may determine when to consult a photogrammetrist.

13.2. Cameras and accessories The camera is obviously the first component in the total photogrammetric system since it is used to obtain the photography from which the data can be extracted. The camera is similar in function to the surveying instruments since it is used to gather information about the object in the field. It is, however, quite different with regard to the amount of data it gathers. One sighting by a transit or theodolite yields *one* direction, while one photograph makes possible the determination of an essentially limitless number of directions or as many directions as the number of points identified in the photograph. Consequently, the design of the photogrammetric camera is rather important to guarantee that the photographs obtained are of good metric quality.

Cameras used for terrestrial photogrammetry differ substantially from those used for aerial photography. Since in terrestrial photogrammetry the camera is usually stationary, its design is much simpler than the fast-moving aerial camera. In fact, a camera is combined with a theodolite, thus called a *phototheodolite* (Fig. 13.1), and is used to obtain photographs from known control points. The camera is usually

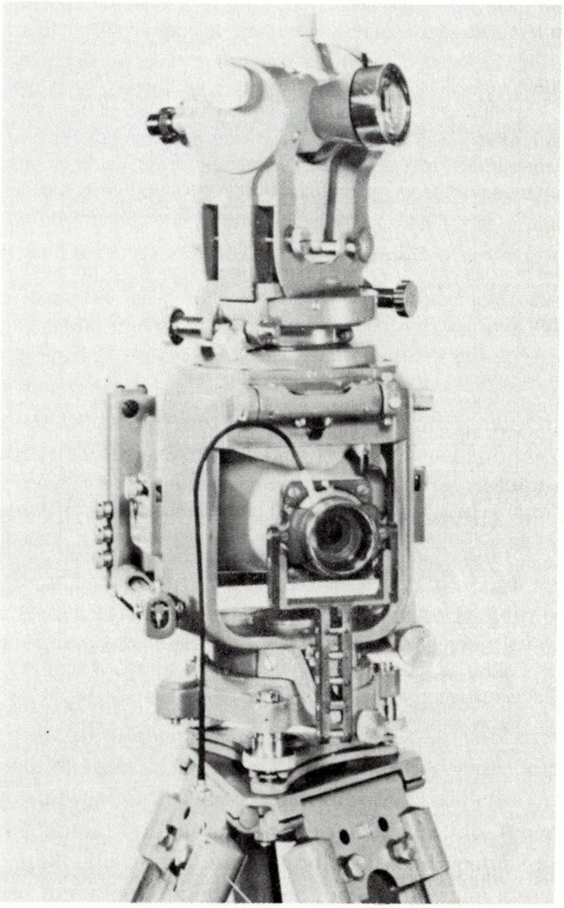

Fig. 13.1 Wild P-30 phototheodolite.

Fig. 13.2 Terrestrial photograph. (*Courtesy Wild Heerbrugg Instruments, Inc.*)

pointed horizontally during exposure. Figure 13.2 is an example of a terrestrial photograph. Note that there are four marks, called *fiducial marks*, one in the middle of each side of the picture. The lines joining each pair of opposite fiducials intersect in the point representing the geometric center of the photograph, called the *principal point*. This point is important in the proper reconstruction of the geometry of the photograph, as will be explained later.

In order to obtain overlapping photographs, the phototheodolite must be moved from one exposure station to another. Another alternative is to use two cameras and expose the two photographs simultaneously. Consequently, manufacturers have produced what is called a *stereometric camera*, where two metric cameras are rigidly mounted at the ends of a base tube with known separation and are synchronized for simultaneous exposure.

Aerial cameras, being in motion during exposure, require a fast lens, a reliable shutter system for a very short exposure time, and a high-speed emulsion on the film. All of these requirements are necessary to guarantee the quality of the aerial photograph, particularly by minimizing the blur due to the motion of the aircraft. Cameras used to gather photographs for metric work, as opposed to those for interpretation, are called *cartographic* cameras and are characterized by geometric

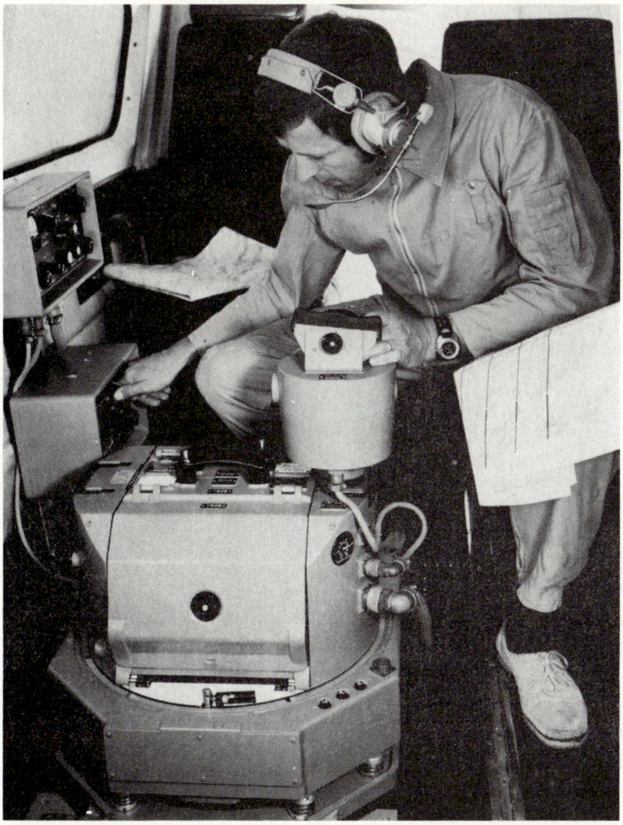

Fig. 13.3 Aerial cartographic camera. (*Courtesy Wild Heerbrugg Instruments, Inc.*)

stability and lenses highly corrected for geometric distortion. Figure 13.3 is a photograph of a modern cartographic camera while Fig. 13.4 is a schematic of the component parts. It is seen that the camera is composed of three main parts: lens cone, camera body, and film magazine. The *lens cone* contains the *lens assembly*, which is composed of a multielement lens, shutter, and diaphragm, and supports the frame that defines the focal plane. The lens is made of several elements, provides for a large angular coverage (about 90° for wide-angle and 120° for super-wide-angle), and is highly corrected for aberrations and distortions. Because the light from the terrain travels through the atmosphere before reaching the lens, it ends up containing a disproportionate amount of blue light. Consequently, aerial cameras are usually provided with filters which prevent some of that blue light from reaching the film. The *shutter* controls the time interval during which the film is exposed, while the *diaphragm* determines the size of the bundle of light allowed to pass through the lens. The *camera body* houses the camera drive motor and mechanism and a recording chamber. The motor and mechanism operate the shutter, film flattening

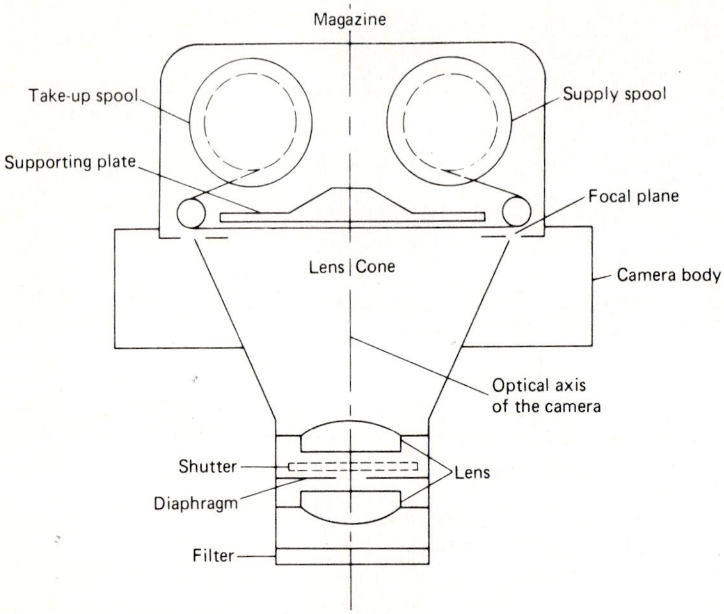

Fig. 13.4 Components of an aerial camera.

device, and advance the film between exposures. The recording chamber allows for printing on each frame several pertinent data such as level bubble, altimeter reading, clock, date, photo and mission numbers, camera number, etc. As shown in Fig. 13.4, the cone and camera body upper surface define the *focal plane*, where a sharp image of the terrain is formed on the emulsion of the film. In the focal plane, a number of so-called *fiducial marks*, usually four, are registered on each frame, as shown in Fig. 13.5*a*, which illustrates *side* fiducial marks. These marks are most important since they define the coordinate system of the photograph. Furthermore, in well-adjusted cameras, the lines connecting opposite pairs of fiducials intersect in a point that falls on the optical axis of the lens. In this way, the photo system can be tied to the terrain system once the camera position and orientation (i.e., the direction of its optical axis) are determined. Although there are other sizes and shapes (e.g., rectangular), the most common type of aerial photographic *format* is that which is square and measures 9 in (23 cm) on the side.

The light entering an aerial camera lens is considered, for all practical purposes, as coming from infinity, owing to the distance between the camera and terrain. Consequently, the image of the terrain will be formed at the focal plane of the camera lens. This is why the emulsion plane of the film is located at a distance from the lens equal to the focal length. Focal lengths of aerial camera lenses are nominally 3.5 in (88 mm), 6 in (153 mm), 8.25 in (210 mm), and 12 in (305 mm), although the 6-in lens is by far the most common.

The *magazine* is light-tight and holds the unexposed and exposed film spools. Magazines vary in capacity from 180 ft (55 m) to 500 ft (150 m) of $9\frac{1}{2}$ in (24 cm)

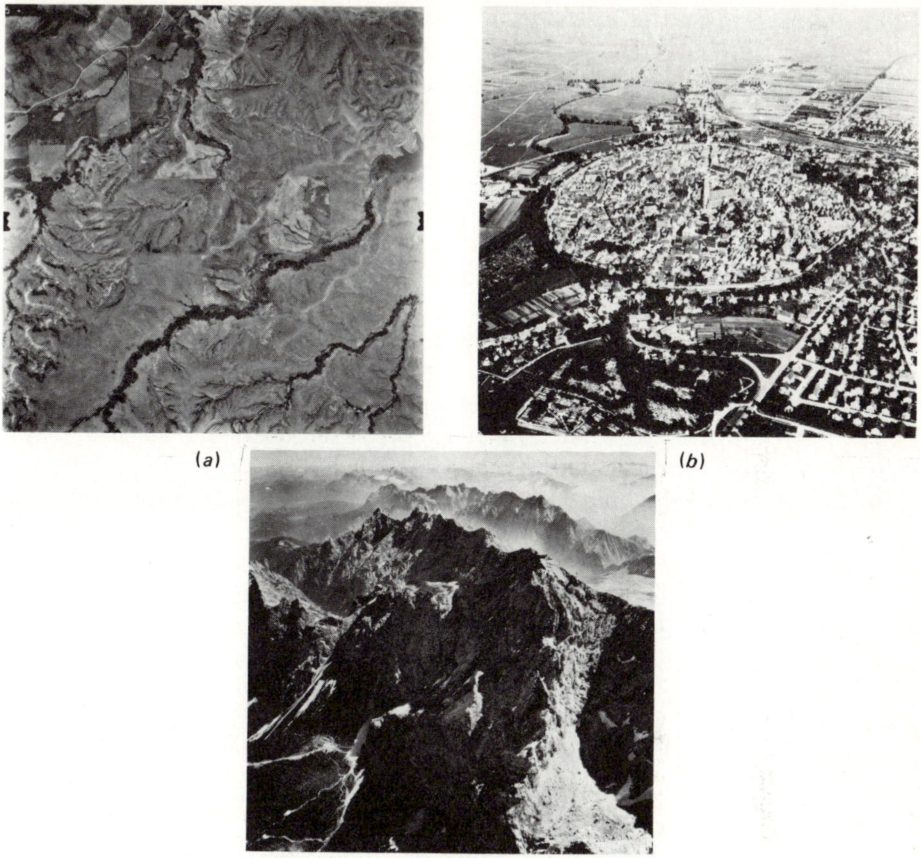

(a) (b) (c)

Fig. 13.5 (*a*) Vertical photograph with four side fiducial marks. (*Courtesy Pacific Aerial Surveys, Oakland, California.*) (*b*) Low-oblique aerial photograph. (*Courtesy Wild Heerbrugg Instruments, Inc.*) (*c*) High-oblique aerial photograph. (*Courtesy Carl Zeiss, Inc.*)

wide thin-base film. Vertical aerial photography is exposed with the intention that the camera optical axis be as truly vertical as possible during exposure. Therefore, it is necessary that the camera be supported in a *stabilized mount* such that it is isolated from aircraft vibrations and perturbations. These mounts employ torquer motors controlled by gyroscopes. Other auxiliary systems used with aerial cameras include viewfinder, intervalometer, and V/H computer. The *viewfinder* provides the photographer with a clear unobstructed view of the terrain below and ahead of the airplane. The *intervalometer* is a timing device that can be set to trigger the camera shutter at a specified time interval between exposures. The V/H (velocity/height) computer develops a voltage that is directly proportional to the aircraft velocity V, and inversely proportional to its altitude H. It may be used to determine the exposure interval as well as the film shift in the direction of flight during exposure to minimize image blur.

13.3. Acquisition and processing of aerial photographs Once a project area has been specified, the first step in the photogrammetric project is to plan the acquisition of the photography in a manner suitable for the desired purpose. For topographic mapping and the production of mosaics or orthophotos (Arts. 13.8 and 13.18), vertical aerial photography is most common. A *vertical* photograph is that taken with the camera axis intended to be vertical, although unavoidable aircraft motion may cause it to tilt a few degrees from the vertical (usually a maximum of 5°, although the average is often 1° or less). An example of a vertical aerial photograph is shown in Fig. 13.5*a*. In small-scale mapping, the camera axis may be intentionally tilted in order to increase the area covered by one photograph. In this case, if the tilt angle, is say, 20°, the photograph is called a *low oblique*, an example of which is shown in Fig. 13.5*b*. When the tilt angle is large enough (50° or 60°) such that the horizon appears in the photograph, it is called a *high oblique*, as shown in Fig. 13.5*c*. The existence of the horizon line in the photograph can be used to advantage when deriving metric information from the high-oblique photograph.

In the case of vertical aerial photography with square format [usually 9 × 9 in (228.6 × 228.6 mm)] the ground-area coverage of a single photograph is accordingly a square. As the airplane flies over the ground, successive photographs are exposed in such a way that each two adjacent photographs cover a common area which is more than half the single photo coverage. This common area is called *forward overlap* or simply *overlap* and is usually 60 percent, as shown in Fig. 13.6. Thus, if photographs along one pass from one end of the project area to the other are laid down on a table, each two successive photographs would overlap by about 60 percent and each three successive photographs by about 20 percent. This type of coverage is necessary to ensure that each area on the ground is covered at least twice so that its three-dimensional geometry may be recovered. The 20 percent *triple overlap* (see Fig. 13.6) is provided to assure the proper connection from one pair of photographs to the next. In some special situations, such as mosaic of rugged terrain, overlap may be increased to say 80 percent. The nominal line passing through the middle of successive photographs is called a *flight line*, and the set of photographs in one line is often referred to as a *strip*.

Once the airplane has traversed the length of the project area, it turns around and returns in the opposite direction as shown in Fig. 13.6. The second strip of

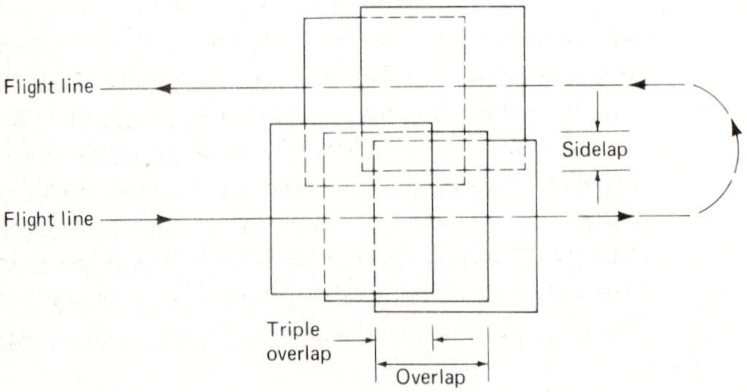

Fig. 13.6 Forward overlap and sidelap.

photography is exposed in such a way that there is at least 20 percent *sidelap* with the first. This again assures no gaps and allows for the tie-in between strips so that the required maps or mosaics can be continuously produced.

After an aerial photography mission, the film is brought to the photo lab for processing. The degree of complexity in the processing of film depends on its type. There are in general four types of emulsions used in aerial photography: (1) *panchromatic black and white* (B & W), which is the most widely used type in metric photogrammetry and interpretation; (2) *color*, which is composed of three emulsion layers to render the different color hues originally in the object, and is used more for interpretation than in mapping; (3) *infrared black and white* (B & W IR) is sensitive to near-infrared light (longer wavelength) and thus penetrates haze and is used mainly for interpretation and intelligence work as it permits detection of camouflage; and (4) *infrared color* (*false color IR*), in which different colors are arbitrarily used to code images from different portions of the light spectrum (thus the name false color) and is employed in various facets of interpretation, such as crop disease detection, pollution monitoring, etc., as well as intelligence.

After processing in the photo lab, the result is either a film negative or a film positive, depending upon the processing procedure used. On a *film negative*, light portions of the original scene appear dark, while dark scene portions appear light, so the scene illumination is reversed on a film negative. On a *film positive*, the original scene light and dark distribution is preserved. It is much more common for B & W films (both panchromatic and IR) to be processed to negatives, while both color and false color films are processed either to positives or negatives. When the result of film processing is negative, positive-tone prints are usually made both for initial inspection, as well as for planning and execution of the various photogrammetric operations. It is a common and good practice to lay all the photographs in a project together in order to determine whether there are any gaps. Also, the tilts of the different photographs are checked in case some are unacceptable. In either case, a re-flight may be required to obtain more suitable photography. Once the photographs are judged as meeting the original specified requirements, the photogrammetric project may then proceed.

13.4. Uses and products of aerial photogrammetry Aerial photographs are used in one of two primary ways, metric and interpretive. However, many of the applications of photogrammetry employ both interpretive and metric techniques. But by far the most extensive use of aerial photographs is in the compilation of topographic maps and orthophotographs, as will be discussed in later sections of this chapter. There are many other uses of aerial photographs. Practically all phases of modern highway design, location, construction, and maintenance make use of aerial photographs. Earthwork quantities, for example, are often computed from a photogrammetric model composed of two overlapping aerial photographs. Maintenance operations, such as those relating to pavement condition and possible erosion of highway banks are determined from the interpretation of photographs. Photographs are also used in a multitude of engineering applications, such as dam settlement and other structural deformation, wave action, channel sedimentation, and traffic studies. In hydrology, analysis of slopes, ground coverage, watershed areas, and snow depth determined from photographs assist in the determination of quantities of runoff water.

In geology, interpretation of photographs for structural form are a valuable supplement to field methods. In agriculture, crop inventory and crop disease analysis are effectively performed using aerial photographs. Aerial photographs are also used for archaeological purposes to explore old sites for new and buried finds.

Aerial photographs are used in land surveying for identification and location of boundary lines and corners (Art. 15.4), and the determination of soil and vegetation cover types. Analytical photogrammetry can be used to determine the location of land boundary corners if they have been well targeted before the aerial photographs are flown. Alternatively, the positions of arbitrarily targeted points can be determined by photogrammetry and other points located by ground surveys using the targeted points as control. Land subdivision can be readily staked out using the photogrammetrically determined points as references.

The uses enumerated above are just a few examples of the extensive range of applications in photogrammetry. In the course of these applications, a variety of products are derived. Some of the more common products include:

1. *Paper-print copies* of the aerial photographs themselves, which may be used for general planning and for field operations.
2. *Mosaics*, which are composed of segments of photographs assembled to give the appearance of a continuous picture of the terrain. Since each is a segment of a perspective view, the mosaic contains distortions and should not be considered as an accurate map (Art. 13.8).
3. *Orthophotos*, specially prepared photographic representations of the terrain without distortions (relief and tilt) which can be used as planimetric maps. In some cases contour lines are superimposed on an orthophoto, resulting in an *orthophotomap* which is then the equivalent of a topographic map (Art. 13.18).
4. *Planimetric maps* (Art. 10.1), photogrammetrically prepared, contain only horizontal position information of terrain features. They are usually compiled from a stereomodel reconstructed in a plotter, as explained in a later section.
5. *Topographic maps* (Chap. 10), which show both planimetric detail and contour lines, compiled photogrammetrically on a stereoplotter.
6. *Thematic maps*, which are portrayals of different "themes" of interest, such as soil type, drainage patterns, transportation network, vegetation, etc. Modern scanning imagery systems are quite suitable for this product, particularly using computer classification techniques.
7. *Digital data*. The photogrammetric model can be used to produce either a listing of ground coordinates of a limited number of required control points or a very dense network of points to represent the total terrain surface. The latter is called the *digital terrain model* (DTM) (Art. 13.17) and is usually in the form of a regular *XY* grid with the elevation of each grid intersection given.

13.5. Analysis of the single photograph As noted earlier, an aerial photograph will be the same as a map only in the very special situation when the terrain is flat and level and the photograph is truly vertical. In general, the photograph is neither taken precisely vertical nor is it common that the terrain be flat and level. Figure

13.7 shows schematically the analysis of both factors. In Fig. 13.7*a* the terrain is assumed to be flat and level between points A and B and between points D and E, and of variable relief in between. Thus, $h_A = h_B$ and $h_D = h_E$. The photograph is *assumed* to be truly vertical. The scale at any point in the photograph such as a is the ratio of the distance from the camera location, C, to the image point, divided by the distance from C to the object point, A, or $S_a = Ca/CA$. From the two similar

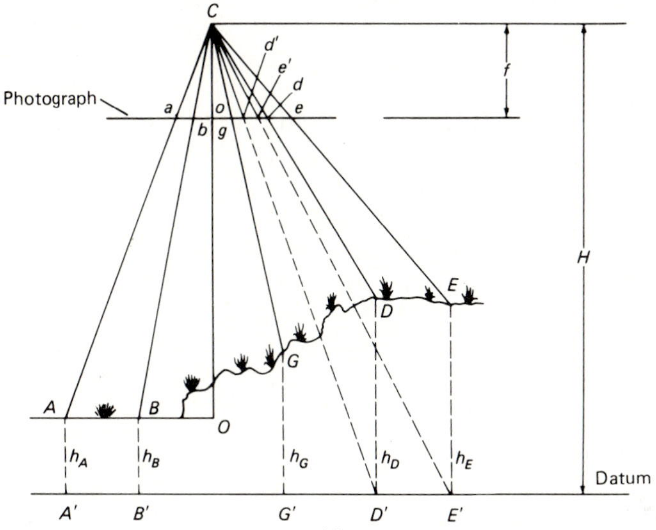

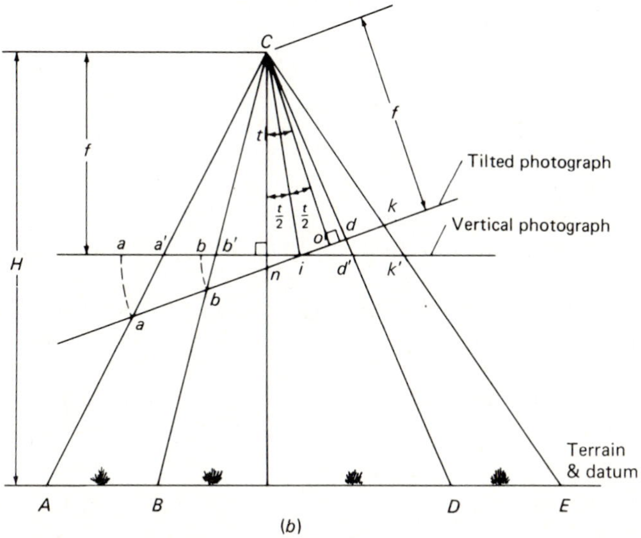

Fig. 13.7 (*a*) Geometry of a single aerial photograph. (*b*) Geometry of a tilted photograph over level terrain.

triangles *Cao* and *CAO*, $S_a = Ca/CA = Co/CO$. Since $Co = f =$ focal length and $CO = H - h_A =$ the flying height above the object point of interest, the *scale at any point* on a truly vertical photograph is

$$S = \frac{f}{H - h} \tag{13.1}$$

in which H, the flying height, and h, the elevation of the point, are both referred to the same datum. It can be seen from Fig. 13.7*a* that the scale is the same for the area between points a and b and is equal to $f/(H - h_A)$. Similarly, the scale is constant between points d and e and is equal to $f/(H - h_D)$. In a region with a *constant scale* on a vertical photograph, the scale can be derived by dividing an image distance by the corresponding ground distance. Hence, $S_a = ab/AB$ and $S_d = de/DE$. Since h_D is larger than h_A, DE is closer to the camera than AB and the scale along *de* is larger than along *ab*. In fact, the two ground distances AB and DE are equal, and the student can see that the image distance *de* is longer than *ab*. It is important, then, to note that unlike a map whose scale *must* by definition be constant, the scale on an aerial photograph is in general not. Therefore, it is better to speak of a photographic scale at a *point* unless a nominal, an average, or an overall scale is specified. These latter scales are approximate and do not apply necessarily to a particular point or area on the photograph. They are usually obtained from a relation similar to Eq. (13.1) using an average value h_{av} for h, or

$$S_{av} = \frac{f}{H - h_{av}} \tag{13.2}$$

The variability of the scale on an aerial photograph, even when truly vertical, is due to the existence of relief. It is clear from Fig. 13.7*a* that if the object points D and E were on the datum at D' and E', they would have been imaged at point d' and e'. The ratio between $d'e'$ and $D'E'$ is equal to f/H and is often called the *datum scale of the photograph*. It is usually the smallest scale unless there are points below the datum. It can be seen that the distance $d'e'$ is, in fact, smaller than the image distance *de*. The effect of relief on the location of image points is called *relief displacement* and is discussed in Art. 13.6.

Example 13.1 Referring to Fig. 13.7*a*, the elevation of point *A* is $h_A = 123.2$ m and the scale at that point on the aerial photograph is 1:32,000. The elevation of point *E* is 275.6 m. Compute the scale at the image for point *E* and the average scale of the photograph if the focal length is $f = 152.4$ mm.

SOLUTION The scale at point *A*, according to Eq. (13.1), is

$$S_A = \frac{f}{H - h_A} \qquad \text{or} \qquad \frac{1}{32{,}000} = \frac{152.4}{(H - 123.2)1000}$$

or

$$H - 123.2 = 152.4 \times 32 = 4876.8 \text{ m}$$

Then

$$H = 4876.8 + 123.2 = 5000 \text{ m}$$

The scale at point E is

$$S_E = \frac{f}{H - h_E}$$

or

$$S_E = \frac{152.4}{(5000 - 275.6)1000} = \frac{1}{31{,}000} \text{ or } 1:31{,}000$$

Because of the shape of the terrain in Fig. 13.7*a*, the mean terrain will be taken as the average of the elevations of points A and E. Thus, $h_{av} = \frac{1}{2}(123.2 + 275.6) = 199.4$ m and the average scale, according to Eq. (13.2), is

$$S_{av} = \frac{152.4}{(5000 - 199.4)1000} = \frac{1}{31{,}500} \text{ or } 1:31{,}500$$

Figure 13.7*b* shows the effect of the second factor, the tilt in the photograph. To isolate and dramatize this effect, the terrain is assumed to be flat and level, with the distances AB and DE being equal. If a truly vertical photograph is exposed from C, it will have a constant scale equal to f/H. This is borne out by observing that the image distance $a'b'$ and $d'k'$ are equal. An aerial photograph exposed from the same point C, but with its optical axis tilted an angle t from the vertical, contains images of the four object points at a, b, d, and k. It is clear that the scale is variable along the tilted photograph. The image distance ab is larger than dk, even though both represent the same distance on the ground. The effect of tilt on image location on a tilted photograph is called *tilt displacement* and is discussed in Art. 13.7.

13.6. Relief displacement on a vertical photograph Figure 13.8*a* shows a section of a vertical plane passing through the camera station C of a truly vertical photograph and an object point A. The datum position of point A is A', and their

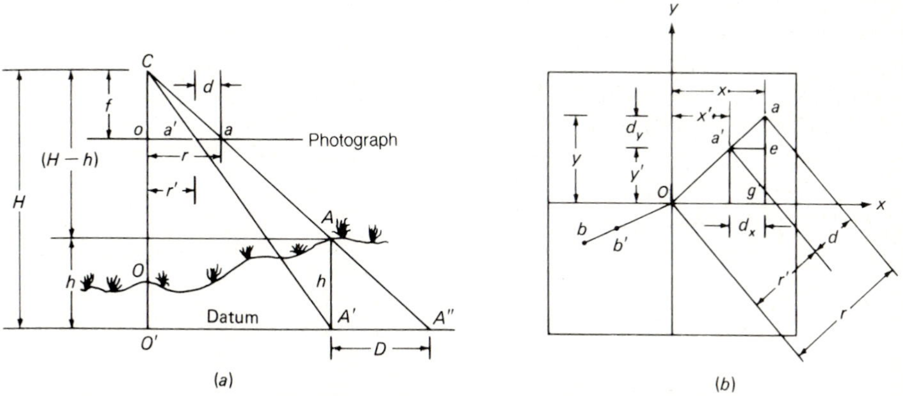

Fig. 13.8 (*a*) and (*b*) Relief displacement in a vertical photograph.

image points are a and a', respectively. Of course, point a is what actually appears on the photograph, while a' is an imaginary point representing the image of A' if it were possible for A' to be seen by the camera. The distance d between a' and a is the *relief displacement on the photograph*. The extension of line CaA intersects the datum in point A'', which represents the datum point that would have been imaged at a. The distance D between A' and A'' is the *relief displacement in the datum plane*. The ratio between d and D is the datum scale of the photograph, or f/H; thus,

$$d = \frac{Df}{H} \tag{13.3}$$

The *principal point* o of a truly vertical photograph is the only point that undergoes no relief displacment. It is seen from Fig. 13.8a that no matter how high the ground point O is above the datum, both O and its datum position O' are imaged at o, since both lie on the vertical line through C. Consequently, all other points are displaced because of relief radially from o as depicted in Fig. 13.8b. The amount of relief displacement depends on the radial distance r between the image point and the principal point o. The farther the image point from o, the larger is the relief displacement, as shown from the relation to be derived.

The two triangles $AA'A''$ and Coa (Fig. 13.8a) are similar and therefore

$$\frac{D}{r} = \frac{h}{f} \qquad \text{or} \qquad D = \frac{rh}{f}$$

which when substituting from Eq. (13.3) leads to

$$d = \frac{rhf}{fH}$$

or

$$d = \frac{rh}{H} \tag{13.4}$$

From Eq. (13.4), the relief displacement is seen to be directly proportional to r and the elevation of the point h, and inversely proportional to camera altitude H; both h and H must be measured above the same datum. One must watch for units in applying Eq. (13.4). In general, d and r may be in the same units, say millimetres while h and H are in metres or feet.

Equation (13.4) gives the total relief displacement for a given image point computed when a scale is used to measure directly the distance r from o to the image. Sometimes, instead of r, the coordinates x, y of the image are measured as shown in Fig. 13.8b. Here, opposite fiducial marks are connected by straight lines to intersect in the principal point o. One line, usually that in the general direction of the flight line, is taken as the x axis, and the other as the y axis, with o as the origin of the coordinate system. One way is to compute r from $r = \sqrt{x^2 + y^2}$ and apply Eq. (13.4) to compute d and scale it either away from or toward o to get a'. Another way is to compute instead the two components d_x, d_y of the relief displacement. In

Fig. 13.8*b* the two triangles *aa′e* and *aog* are similar; hence,

$$\frac{aa'}{ao} = \frac{a'e}{og} = \frac{ae}{ag} \qquad \text{or} \qquad \frac{d}{r} = \frac{d_x}{x} = \frac{d_y}{y}$$

But from Eq. (13.4), $d/r = h/H$, so that

$$d_x = \frac{xh}{H}$$
$$d_y = \frac{yh}{H} \qquad (13.5)$$

Thus, if x, y of the image point are given, d_x and d_y can be evaluated directly from Eq. (13.5). Then the coordinates of a' are simply $x' = x - d_x$ and $y' = y - d_y$. If one is interested in the total relief displacement, then $d = \sqrt{d_x^2 + d_y^2}$. These relations are demonstrated in the following numerical example.

Example 13.2 A vertical photograph is taken with a camera having $f = 151.52$ mm. Two images a and b appear on the photograph, such that a falls on the x axis with $x_a = +78.70$ mm, and b falls on the y axis, with $y_b = -91.30$ mm. The two points A, B on the ground have the same elevation 98 m above datum. Compute the image coordinates of the datum point images a', b' if the datum scale is 1 : 16,500.

SOLUTION The datum scale is $f/H = 1/16{,}500$ or $H = 16{,}500\,f$. Then

$$H = \frac{(16{,}500)(151.52)}{1000} = 2500 \text{ m}$$

Since point a lies on the photo x axis, point a' will also be on the x axis, and the radial distance is equal to x_a. Thus, the relief displacement is [Eq. (13.5)]

$$d_a = d_{x_a} = \frac{x_a h}{H} = \frac{(78.70)(98)}{2500} = 3.09 \text{ mm}$$

and

$$x'_a = x_a - d_{x_a} = 78.70 - 3.09 = 75.61 \text{ mm}$$

Similarly, since b lies on the y axis, then [Eq. (13.5)]

$$d_b = d_{y_b} = \frac{y_b h}{H} = \frac{(-91.30)(98)}{2500} = -3.58 \text{ mm}$$

and the coordinate of b' is

$$y'_b = y_b - d_{y_b} = -91.30 - (-3.58) = -87.72 \text{ mm}$$

13.7. Tilt displacement Figure 13.7*b* shows the relation between a truly vertical photograph and a tilted photograph both taken from the same camera station. The plane of the figure, in which the tilt angle is measured, is called the *principal plane* and is the vertical plane through the optical axis when the tilted photograph is exposed. The trace of that plane on the tilted photograph is a line that passes through the principal point and is called the *principal line*. Only one line is common to both the vertical and tilted photographs. This line is represented by point i in Fig. 13.7*b*. It is called the *isoline* or sometimes the *axis of tilt*, and its intersection with the principal line at i is called the *isocenter*. As can be seen from Fig. 13.7*b* point i lies on the bisector of the tilt angle t. Therefore, the distance along the principal line from principal point o to the isocenter i is

$$oi = f \tan \frac{t}{2}$$

The vertical or plumb line through C pierces the tilted photograph at point n which is called the *nadir point* and also lies on the principal line. The distance on from the principal point to the nadir point is given by (from Fig. 13.7*b*)

$$on = f \tan t$$

Imagine rotating the tilted photograph about its common line with the truly vertical photograph until they coincide. When this is done, images of the same object point do not match, the difference being the *tilt displacement*. Points a and b, being on the downside of the isoline, are displaced radially *away* from the isocenter i, while points d and k are shifted radially *toward* i because they are above the isoline. The amount of tilt displacement is therefore a function of the distance from the image point to the isocenter. Relations exist which may be used to calculate tilt displacement. They are rather involved and are beyond the scope of this textbook.

13.8. Mosaics A mosaic is an assembly of adjacent aerial photographs such that a continuous pictorial display of the terrain is obtained (Fig. 13.9). There are usually three types of mosaics: *uncontrolled*, *controlled*, and *semicontrolled* mosaics, depending upon whether and how much horizontal control has been used in constructing the mosaic. When little or no control is used, the mosaic is uncontrolled. The simplest form of an uncontrolled mosaic is the *index* mosaic (sometimes called index "map"), where overlapping photographs are placed down consecutively and stapled to a board. It is used as an index for the mapping project, usually to monitor its progress.

Controlled mosaics are made both for pictorial display and the extraction of some metric data, particularly for planning. The accuracy of the mosaic depends on minimizing the effects of relief and tilt displacements. Considering relief, the smaller the area around the center of each photograph to be used in the mosaic, the less is the relief displacement, which depends on the radial distance from the center, r [see Eq. (13.4)]. Therefore, long focal length, higher-altitude flying, and increased overlap and sidelap are preferable for photography to be used in a mosaic. With regard to tilt displacement, the regular aerial photographs are usually replaced by *rectified* prints produced by instruments called *rectifiers*. Rectification produces an equivalent truly vertical photograph from the same exposure station. Since the attitude of the camera also varies somewhat from one photograph to another, the rectification process also allows for bringing all rectified prints to the same scale.

Although small central segments of the rectified prints are used, each is still a perspective projection of the terrain. Therefore, horizontal control points are utilized

Fig. 13.9 Aerial mosaic. (*Courtesy Pacific Aerial Surveys, Oakland, California.*)

to *control* the mosaic. Such control is obtained either by ground survey methods or (more frequently) by radial line triangulation. The control is plotted at the required scale on the board on which the mosaic is to be assembled. Each control point must also be identified on the photographic segment to be used. Only the central portion of a photograph is used with its edges being brought to a feather edge using sandpaper. Some small overlap is allowed between adjacent segments. Each segment is then pasted on the board, matching control, if it exists, and other details on adjacent segments as much as possible. Once all segments are pasted and the assembly is finished, the mosaic is then photographed by a large copy camera.

13.9. Stereoscopy *Stereoscopy* refers to the ability of the individual to perceive the object space in three dimensions through using *both* eyes. Each human eye represents a single camera, and therefore *monocular viewing* or viewing with one eye results in *flat perspective* and the person's ability to perceive depth is hampered. *Binocular viewing*, on the other hand, allows the person to view an object from two different locations due to the separation between the eyes. The student can see the similarity between the function of the human eyes and the pair of camera stations

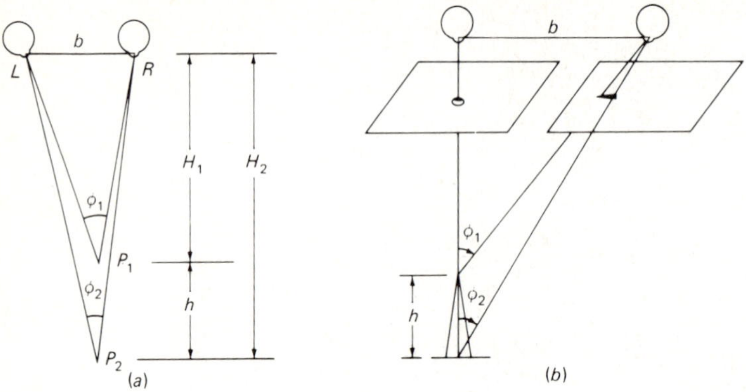

Fig. 13.10 (*a*) Convergence angle in stereoscopic vision. (*b*) Stereoscopic viewing with a pair of overlapping photographs.

when taking overlapping photographs. This is the reason for studying the principles and techniques of stereoscopy in support of photogrammetry.

Viewing with one eye fixes only *one* direction from the eye to the object, which is insufficient for fixing the object's distance from the viewer. When the other eye is utilized, a second direction is fixed and its intersection with the first locates the point. The closer the point to the eyes, the larger is the convergence angle between the two directions, as depicted in Fig. 13.10*a*. Since the angle ϕ_1 is larger than ϕ_2, the observer will perceive point P_1 as closer than point P_2. In fact, the difference in distance from the observer, $H_2 - H_1 = h$, is a function of the difference in convergence angles, $\phi_1 - \phi_2$. The closest distance for distinct comfortable vision (without aids) is 25 cm (10 in), for which the convergence angle is about 15°. This can be computed on the basis of an average eye separation of $b = 65$ mm. The lower limit of the convergence angle for unaided vision is 10 to 20 seconds of arc.

Figure 13.10*b* as compared to Fig. 13.10*a* shows the correspondence between natural binocular vision and stereoscopic viewing of a pair of overlapping photographs. An idealized tower is assumed to have been photographed with the camera first directly overhead and second after having traveled a distance past the tower. If it were possible to view the first photograph with the left eye only and the second photograph with the right eye only, the observer would perceive the tower in three dimensions, as shown schematically in Fig. 13.10*b*. The lines joining the eyes and the two images of the top of the tower will intersect at an angle ϕ_1, while those passing through the images of the tower base make an angle ϕ_2. Since ϕ_1 is larger than ϕ_2 the top of the tower will appear closer to the viewer than its base, and the tower will then be perceived in three dimensions. Stereoscopic viewing of objects and terrain from overlapping photographs is very important to most all operations of both interpretation and metric photogrammetry, as will be shown in subsequent sections of this chapter.

13.10. Topographic information from aerial photographs As we mentioned in Art. 13.9, a three-dimensional view may be perceived from a pair of overlapping

Fig. 13.11 Mirror stereoscope. (*Courtesy Wild Heerbrugg Instruments, Inc.*)

photographs of the object (see Fig. 13.10*b*). For this stereo perception, it is necessary that each photograph be viewed by one eye, a function that is fulfilled through the use of a stereoscope (Fig. 13.11). Although high and low points in the terrain can be ascertained when viewing stereoscopically through a stereoscope, a quantitative measure is not possible until we define *parallax* and show how it can be measured and then converted into differences in elevation. In simple terms parallax is a linear measure on the photographs that corresponds to the parallactic angle ϕ (see Fig. 13.10). The larger the parallactic angle, the larger is the parallax, and the closer is the object to the exposure stations (i.e., the higher the elevation of the terrain point).

The quantitative definition of parallax depends on the simplifying assumptions that the photographs have *zero tilts* and that both photographs are exposed from the same altitude. Any rotation of the camera about its (vertical) axis from one photograph to the other is accommodated by replacing the fiducial system of axes by another. The new system (Fig. 13.12) has its abscissa along the projection of the air base or flight line on the planes of the photographs. Then the x parallax (or simply parallax) is defined by

$$p = x - x' \tag{13.6}$$

where x and x' are the coordinates of any point of interest on the left and right photographs, respectively, with respect to the new axis-of-flight, or base-line, system, and p is the parallax. The algebraic values (i.e., with the proper sign) of both x and x' must be used in Eq. (13.6). The value of p can be used to determine the elevation of terrain points appearing in the overlap area of two aerial photographs. Before this is done, first consider how parallax is measured directly from the photographs.

The fiducial center of each photograph is first determined and assumed to be its principal point. Under the stereoscope, the two photographs are manipulated until

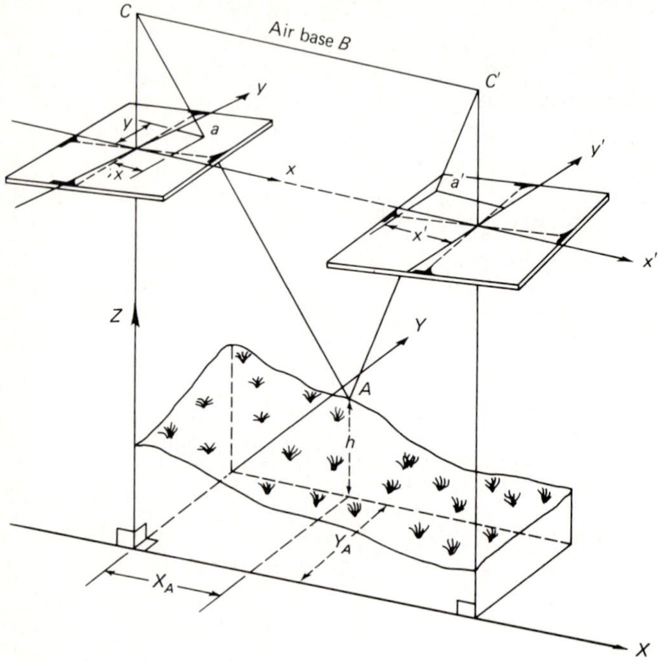

Fig. 13.12 Parallax in stereophotography.

stereoviewing is achieved. Conjugate principal points are then transferred. Lines connecting the pair of principal points on each photograph represent the base and therefore the lines parallel to which parallax is measured. Thus, as shown in Fig. 13.13, the two photographs are laid down such that the base or flight lines are coincident. A simple instrument, called a *parallax bar* and schematically depicted in Fig. 13.13, is used to yield the values of p at various points. The bar has two identical dots, called *half-marks* and identified in the figure, which fuse under the stereoscope and appear as a *floating* dot in the space occupied by the terrain model. When the micrometer m is rotated, the distance between the two half-marks is changed and the dot appears to go up and down. When it appears to be just in

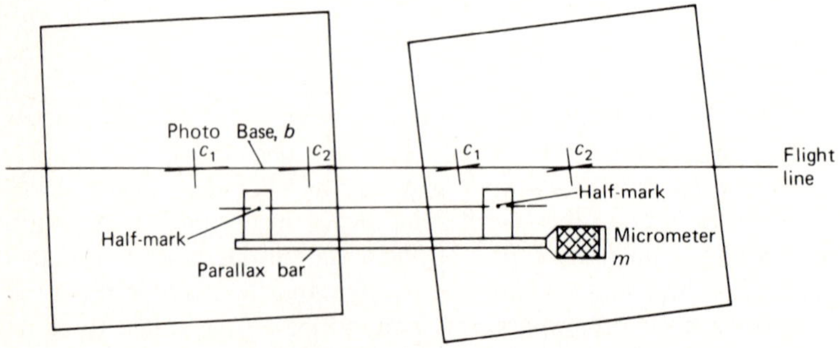

Fig. 13.13 Measurement of parallax with a parallax bar.

contact with the terrain at the point of interest, the reading of the parallax bar would lead to the required value of parallax p, for that point.

Figure 13.14 shows a schematic of two photographs and a terrain point, drawn in a plane through the flight line. The altitude of each photograph is H above datum, while the elevation of the point, A, is h above the same datum. The two coordinates x and x' are shown for the images a and a' on the left and right photographs, respectively. A line Eb is drawn parallel to $E'A$ by construction; thus, c_1b is equal to x'. Since from Eq. (13.6) the parallax is $(x - x')$, the line segment ab is the parallax p. From the fact that the two triangles abE and $EE'A$ are similar,

$$\frac{H-h}{f} = \frac{B}{p} \qquad \text{or} \qquad H - h = \frac{Bf}{p}$$

so that

$$h = H - \frac{Bf}{p} \tag{13.7}$$

Thus, in order to determine the elevation of a point h from a pair of aerial photographs, it is necessary to have the flying height above datum H, the value of the air base B, and the focal length f. Usually, f is known from camera calibration data, B is determined from a radial line plot (Art. 13.11), and H is either given from prior computation or determined on the basis of a known *vertical* control point in

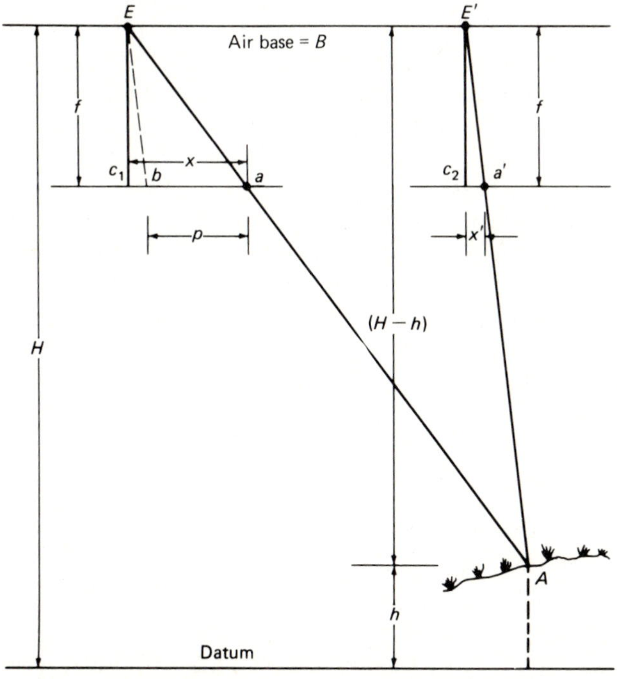

Fig. 13.14 Relation between parallax and point elevation.

the overlap area. In the latter case, h would be known and Eq. (13.7) is rearranged to give H, or

$$H = h + \frac{Bf}{p} \tag{13.7a}$$

Example 13.3 A pair of photographs is taken with a camera the focal length of which is $f = 152.70$ mm. A radial line plot prepared at a scale of 1 : 10,000 was used to determine the distance between the two principal points as 5.65 cm. The measured parallax for a vertical control point having an elevation of 153.2 m is 91.12 mm. Compute the elevation of another point A whose measured parallax is 92.38 mm.

SOLUTION The air base is

$$B = (5.65)(10{,}000) = 56{,}500 \text{ cm} = 565 \text{ m}$$

Using Eq. (13.7) the flying height above datum, H, is

$$H = 153.2 + \frac{(565)(152.7)}{91.12} = 1100.0 \text{ m}$$

Using Eq. (13.7) the elevation of the new point may be determined:

$$h = 1100.0 - \frac{(565)(152.7)}{92.38} = 166.1 \text{ m}$$

The accuracy with which elevations of points are determined from the parallax equations depends on how closely the assumptions made are fulfilled. Thus, the smaller the amount of tilt and difference in altitude between the two photographs, the better is the final accuracy. In general, the obtainable accuracy is consistent with the methodology and material used, particularly if paper prints subject to significant shrinkage are used.

In addition to determining vertical information on a point to point basis, it is possible to plot contour lines, *in perspective*, using a combination mirror stereoscope and parallax bar. The basis for this is Eq. (13.7) which can be rearranged to the form

$$p = \frac{Bf}{H - h} \tag{13.8}$$

Thus, given B, f, and H, the parallax value p for each contour h (i.e., the plane at elevation h in the terrain) can be calculated. The corresponding reading on the parallax bar is then set, fixing the floating dot in the space where the overlap is perceived in stereo. If the stereoscope–parallax bar assembly is moved about such that the floating dot remains in contact with the perceived terrain surface, this motion would describe a contour line. A drawing pencil attached to the assembly would draw such a contour. It is important to note these contours are not orthographic because they are drawn in relation to one of the photographs, which is of course a perspective. There are several simple plotters designed on this principle, such as the *stereocomparagraph*, the *contour finder*, and the K.E.K. plotter, more details on which are given in the references at the end of this chapter.

13.11. Elementary operations for extracting planimetric information from aerial photographs As mentioned previously, a photograph is a perspective projection of the terrain. Therefore, to obtain accurate metric information about the terrain from photographs, rigorous principles must be applied. However, many elementary photogrammetric operations can be based on some simplifying assumptions. Such assumptions include (1) accepting the fiducial center as the principal point of the photograph; (2) lens distortion, earth curvature, and atmospheric refraction are relatively of no significance and can be ignored; (3) only linear corrections for film and paper deformations are considered, since elementary operations occur almost totally with paper prints; and (4) the calibrated focal length is used as the focal length, f.

When overlapping photographs are used, three-dimensional information which may be contained in the photographed object can be recovered. The same applies to terrain photographed by an aerial camera. However, when elementary operations are considered, it is much easier to extract horizontal or planimetric information in a separate operation from that concerned with the height information. Planimetric data can be obtained from overlapping aerial photographs either in the form of horizontal point coordinates, or planimetric line maps, or mosaics. The photogrammetric determination of horizontal control points is accomplished by an operation called *radial line triangulation*, which is fundamentally similar to survey triangulation (see Chap. 9). Such horizontal control is necessary before either plotting planimetric detail or producing a mosaic. Details of radial line triangulation, which are beyond the scope of this text, may be found in the references at the end of this chapter.

13.12. Plotting of planimetry After radial line triangulation is used to produce the horizontal position (within accuracies limited by the assumptions made) of only a finite number of points, it is possible to plot planimetric detail and produce a line map. The simplest procedure is to trace directly on transparent paper (representing the map base) placed over the photographic print. To minimize the effect of tilt and relief displacements, the map sheet should contain as many control points as can be determined by radial triangulation (at least nine well-distributed points per photograph). One control point on the map sheet is made to coincide with its image on the photograph and the map sheet properly oriented. Planimetric detail is then traced around that one point. The sheet is shifted such that another control point coincides with its image and details around it traced, and the process is repeated. Mismatch will occur but can be eliminated by the draftsman. Obviously, the lower the amount of relief (and tilt, in general) the better is the accuracy of this method. Of course, the scale of the planimetric map must be the same as the average scale of the aerial photograph.

Direct tracing is the simplest form of data transfer from photograph to map. To accommodate some change in scale and compensate to some extent for photo tilts, the *camera lucida* principle shown in Fig. 13.15*a* is used. Light reflected off the photo is deflected by the mirror M to the beam splitter (or half-silvered mirror) BS, which directs it to a single eye. Light from the map passes through BS to the eye, which then perceives both the photo and map as if coincident on each other. Thus, the

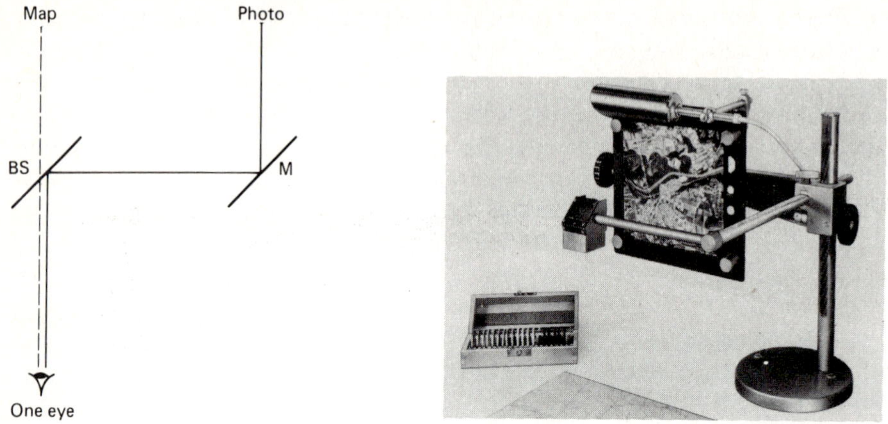

(a) (b)

Fig. 13.15 (*a*) Camera lucida principle. (*b*) Sketch master. (*Courtesy Carl Zeiss, Inc.*)

operator can trace on the map the superimposed photo detail. The ratio between the lengths of the optical paths between photo and eye, and map and eye, determines the scale. The sketch master (Fig. 13.15*b*) is a direct implementation of the camera lucida principle. Figure 13.16*a* is a diagram of a more sophisticated arrangement of the principle used in the *zoom transfer scope* (ZTS), illustrated in Fig. 13.16*b*. The ZTS allows for a continuous change in magnification from 1 to 14 to accommodate a large range of photo and map scale ratios. As shown in the figures, the viewing is binocular, which minimizes eye fatigue. It also allows for *anamorphic* correction, which affects a scale change in only one direction, thus compensating to some extent for geometric distortions in the photograph due to tilt and relief.

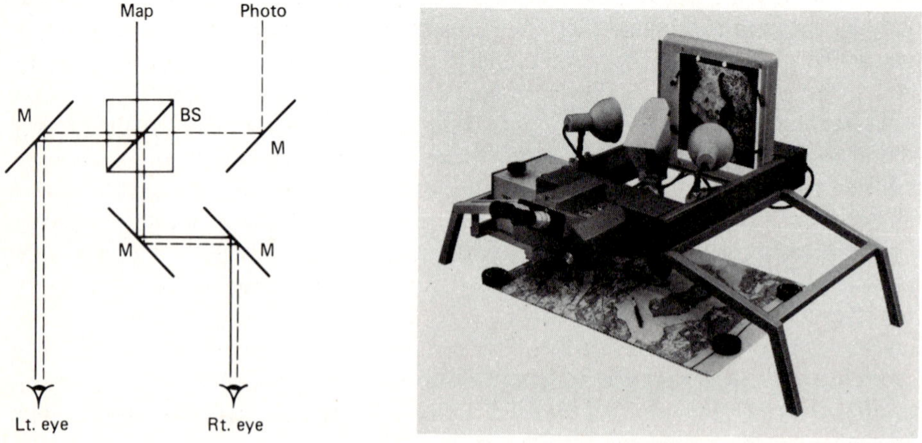

(a) (b)

Fig. 13.16 (*a*) Optical train of zoom transfer scope (ZTS). (*b*) Zoom transfer scope. (*Courtesy Bausch & Lomb, Inc.*)

13.13. Rigorous operations in photogrammetry: general As indicated in Art. 13.11, several assumptions regarding the geometry of aerial photographs can be made which will then allow the deriving of products using relatively simple and elementary procedures. On the other hand, if high-accuracy products are required, the previous simplifications are unsatisfactory and the geometry of the photographs must be rigorously recovered. At the instant a photograph is exposed from an airplane, the aerial camera location may be expressed by three coordinates in the ground space. Also, the attitude of the camera at the time of exposure may be expressed by three angles, usually called *orientation* angles. Thus, attached to each aerial photograph are six elements, generally unknown, designated by the term *exterior orientation*. These elements, when determined, specify the location and attitude of the bundle of light rays that enter the aerial camera lens at the time of exposure. The shape of the bundle, on the other hand, is determined by what is called *interior orientation* elements, which are *three* in number. They express the geometric relationship between the exposure station and the photograph plane. In aerial photography, the interior orientation elements are normally known from camera calibration and consist of the principal point coordinates x_0, y_0 and camera calibrated focal length f.

Given a pair of photographs, their 12 exterior orientation elements (assuming interior orientation is known) must be recovered before the geometry of the photographed terrain can be accurately determined. There are a number of different techniques of accomplishing this, the most common being through the use of analog instruments called *stereoplotters*. They employ optical-mechanical components to simulate and help recover rigorously the geometry of the photographs. There are also digital procedures for this operation (Art. 13.20).

13.14. Photogrammetric stereoplotters A photogrammetric stereoplotter is an instrument of high precision which rigorously reconstructs the geometry of overlapping photographs such that accurate three-dimensional information about the photographed object may be obtained. In the case of aerial photography, planimetric and topographic maps may be compiled, or coordinate listings, both horizontal and vertical, may be derived.

Figure 13.17 is a schematic of a direct-projection stereoplotter. The aerial camera is replaced by a *projector* similar in function to a slide projector. Two such projectors are needed so that the pair of overlapping photographs may be projected simultaneously. Each bundle of rays emerging from the projector lens should duplicate, as nearly as possible, the bundle that entered the aerial camera during the photographic exposure. This duplication is accomplished by properly recovering the three elements of interior orientation. If the projector accepts photographs printed on glass rather than on paper, usually called *diapositives*, which are the same size as the original format used in the aerial camera, the distance between the photograph and the projector lens (i.e., its principal distance) is set equal to the aerial-camera-calibrated focal length. If, on the other hand, the plotter's projector accepts only a reduced-size photograph, the principal distance in the plotter should be the focal length *reduced by the same factor*. For example, if the plates used in the plotter are *one-third* the size

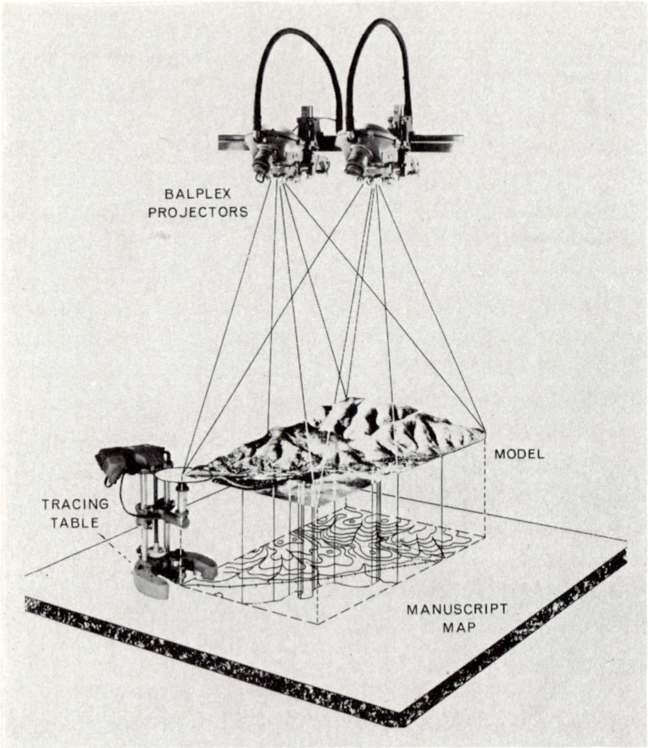

Fig. 13.17 Model representation.

of the original format, the corresponding principal distance would be $f/3$, where f is the calibrated focal length of the camera. The remaining elements of interior orientation are recovered by placing the plate such that its center (principal point) is directly over the optical axis of the projector. This is usually accomplished by matching the plate's fiducial marks to reference marks in the diapositive holder of each projector.

Once interior orientation is completed and the plates are illuminated from above, the ray bundles emerging from the two projectors will each be a replica of the bundle that originally entered the camera. However, the relative position and orientation will not duplicate the original situation during photography because the 12 exterior orientation elements have not yet been recovered. Each projector has six degrees of freedom, including three translations along X, Y, and Z, and three rotations about the X, Y, and Z axes. These are used to recover the exterior orientation elements, as explained in Art. 13.15.

When exterior orientation is properly recovered, a stereoscopic model will result, as shown in Fig. 13.17. In order for the human operator to perceive the model in three dimensions, the image from each projector should be viewed with one eye. One common procedure, called the *anaglyph method*, is to use two glass filters, one *blue-green* and the other *red*, placed in the path of the projecting light rays. The operator then views the model through a pair of spectacles with corresponding blue-green and red lenses. The resulting stereoscopic model is measured by using a *tracing table* (Fig. 13.17), which moves through the model space in all three directions X, Y, Z. It slides freely over the map table (the X and Y motions) and its top surface, called the *platen*, is moved up and down by a knob (the Z motion). All three motions can be monitored and recorded either manually or through the use of automatic devices. The reference measuring point is a small hole in the center of the platen illuminated by a small bulb directly underneath it. Thus, when viewing the model stereoscopically the operator sees within the model a sharp circular white point which can be moved about and brought in apparent contact with the model surface at any desired spot. Directly underneath the measuring mark is a pencil which can be lowered so as to contact the map sheet when plotting.

In another class of stereoplotters called *indirect-viewing* instruments, the operator looks at the photographic plates and consequent stereomodel through optical trains. Thus, these instruments, as far as viewing is concerned, are essentially very sophisticated stereoscopes. However, since there is no direct projection of light rays, the relative geometry of the photographs is completed through optical-mechanical simulation. Instead of a tracing table, a *base carriage* is used. As the carriage is moved in model space, the operator views a different location in the model. Viewing is accomplished through a pair of binoculars which pick up corresponding images via optical trains. The objectives of the optical trains are close to the photographic plates. Within each objective is inserted a half-mark similar to those used in the parallax bar (Fig. 13.13) so that a floating dot appears in the stereomodel viewed through the binoculars.

Plotting in this type of instrument is generally not done in the model space as is the case in direct projection instruments. Instead, the movements of the base carriage are transferred through either pantographs or gear boxes to a plotting table, called a *coordinatograph*. This added feature allows for some additional scale change between the model and the map.

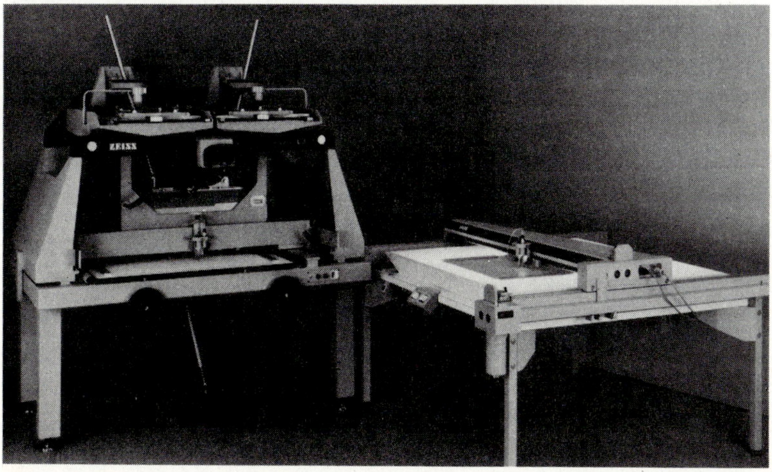

Fig. 13.18 Indirect-viewing mechanical projection plotter, Zeiss Planicart E-3. (*Courtesy Carl Zeiss, Inc.*)

The Zeiss Planicart E3 (Fig. 13.18) is an example of the indirect-viewing mechanical type of projection plotter. It uses full-size diapositives (the same size as the original negative) from both wide-angle and super-wide-angle cameras. Each diapositive can be rotated about each of the three axes, but the projectors can be displaced relative to each other only in one direction, representing the total air base. The model is usually at a scale which is about twice that of the original photography. A linear pantograph can effect an additional scale change of 2.5 times.

There exist a multitude of photogrammetric plotters on the market. Interested students should consult textbooks on photogrammetry (see the references at the end of the chapter) for more extensive treatment of such instruments.

13.15. Orientation and map compilation In the preceding section it was shown that in a photogrammetric stereoplotter both interior and exterior orientation must be properly recovered for the pair of photographs before plotting of a map from the stereomodel can begin. While the procedure for interior orientation was explained, the method for recovering the 12 elements of exterior orientation was not discussed. Referring to Fig. 13.17, note that the plane of the table top where the map manuscript rests represents the XY plane of the plotter; the line normal to it (i.e., the vertical direction of travel of the platen) represents the Z axis. It is possible to bring the two bundles of rays in proper relative registration such that each pair of conjugate rays intersects in a point and one can perceive a model in stereo. The resulting stereomodel, however, may be at an arbitrarily unknown scale and may also be tipped and tilted relative to the plotter's coordinate system. Since, at the time of map plotting the plotter's system represents the object coordinate system, the model must be rotated, translated, and its scale changed until it sits properly into the instrument system just as the terrain was relative to the object coordinate system and the model is at the scale of the map. This operation of manipulating an established (or *restituted*) stereomodel is called *absolute orientation* and relies on having an adequate number of survey control points. Since a model at an unknown scale can be translated along each of the three axes, and rotated about each of the three axes, and its scale changed, absolute orientation involves the recovery of *seven* parameters. This leaves *five* parameters to make the total of 12 exterior orientation elements that must be recovered before one can compile a map from a stereomodel.

The five-parameter operation by which a model is established at an arbitrary scale and arbitrary absolute orientation is called *relative orientation*. It in effect causes each pair of corresponding rays to intersect at a point by recovering the relative tilts of two projectors with respect to each other. There are a number of specific procedures of relative orientation depending upon the type of plotter available and to some extent on the nature of the model (as, for example, the existence of water bodies covering parts of the model). Interested students should consult the references at the end of the chapter for more details.

Absolute orientation is usually accomplished in two separate operations: *scaling*, and *leveling* the model. In scaling, two horizontal control points are plotted on the map manuscript at the scale at which the map is to be compiled. These control points are then used to make the horizontal distance between the corresponding model points equal to the map distance by changing the model base in the instrument. For leveling, an absolute minimum of three vertical control points is needed, since three points determine a plane. In practice, however, the elevations are of four points at the four corners of the stereomodels used, and frequently a fifth point in the center is considered desirable. Usually, two rotations, one front to back, and the other side to side, are effected to make the stereomodel level. Because leveling affects scaling, and vice versa, these two operations are usually repeated alternately to refine the results sufficiently.

Once the model is absolutely oriented relative to a map manuscript that is firmly fastened to the plotting table, map compilation can begin. The plotting of planimetry is performed separately from, and usually before, the drawing of contour lines. The planimetry is drawn totally before contours are added to make sure that the locations of the contour lines fall in the proper relationship to the horizontal features (such as streams and roads). A planimetric feature, for example a road, is plotted by keeping the floating dot always in contact with the terrain. Thus, if the road is ascending or descending, the elevation of the floating dot should be continuously changed to keep it in contact with the apparent terrain surface in the model. It is advisable that the operator compile like features at the same time: all roads first, then all streams, then structures, etc. Whenever the operator encounters difficulty, the manuscript should be annotated with notes to be used later in the completion stage.

Contour plotting is somewhat more difficult than planimetry, particularly for the inexperienced. The floating dot is *fixed* at the elevation representing the contour value and the tracing table, or base carriage, is moved in the *XY* plane *only*, keeping the dot in contact with the terrain surface in the model. It is here that the difficulty is encountered and often a contour line drawn by an operator with little experience will be in the form of a zigzagging line. It is therefore advisable that the operator scan the entire model first to become familiar with its various areas, and then begin with those details that are simpler to compile. Once these areas are finished and as more familiarity with the model terrain is gained, the operator may proceed with the remaining more difficult areas, such as those which are relatively flat and those with ground cover. In fact, it is in such areas that photogrammetric mapping is at a disadvantage compared to field mapping. Thus, when the plotting is finished on the instrument, the compiled manuscript must be checked. Wherever difficulties are encountered, the respective areas are field-completed. Then the manuscript is checked for mistakes, names are added, and its accuracy is determined by comparing photogrammetric measurements to those correspondingly determined in the field. At this point the manuscript is ready for the cartographic and reproduction operations required to produce a finished topographic map.

13.16. Photogrammetric triangulation It was explained in the preceding two sections that a stereomodel can be established such that it becomes a faithful reproduction of a segment of the earth's surface at a desired scale. Consequently, the horizontal and vertical coordinates of any terrain point can be extracted from the stereoplotter. This means that given two horizontal and three (but usually four) vertical control points to absolutely orient a stereomodel, the ground coordinates of any desired number of points in the model can be determined. This analysis can be extended by visualizing a stereoplotter with *three* projectors. Three overlapping photographs can be used and a stereomodel composed of the total overlap (which is equal to the area of the middle photograph) can be relatively oriented. To absolutely orient this large model (composed of two adjacent stereopairs) still involves seven parameters and requires two horizontal and three (preferably four) vertical control points. Once this is accomplished, supplemental control, both horizontal and vertical, can be obtained. When map compilation is required, each stereopair is used separately on a plotter. For each stereopair, enough control will be required to perform absolute orientation. Part of this control would have been field-determined, but the other part is determined photogrammetrically through restituting the three photographs together. This concept can obviously be extended to more photographs than just three and is called *photogrammetric triangulation*. It may be defined as the procedure of establishing the geometric relationships among overlapping and sidelapping photographs for the purpose of determining the positions for supplemental horizontal and vertical control points. In general, the supplemental control thus obtained is used for map compilation. Photogrammetric triangulation is a very powerful tool, since it reduces substantially the control required by field methods, thus markedly improving the economic aspects of photogrammetric mapping. It is a very important aspect of photogrammetry, the study of which is well beyond the scope of this introductory chapter on the subject. Interested students should consult Refs. 1, 4, and 11 at the end of this chapter for further details.

13.17. Digital terrain models As indicated in the preceding two sections, the ground coordinates of any point in a model or strip can be determined once the exterior parameters of the constituent photographs are recovered. The density of points of supplemental control has been relatively sparse, on the order of 10 or at most 20 points per model. These points are usually determined in order to be used for other operations, most notably producing a graphic representation of the terrain in the form of a topographic map. In recent years, however, a totally different product is being photogrammetrically derived, in which the terrain surface is numerically represented by a very dense network of points of known X, Y, Z coordinates. This product is now given the standard name digital terrain model (DTM) or digital elevation model (DEM). The DTM may be produced in a photogrammetric plotter by supplementing it with special processing components that make model digitization possible. The X, Y, Z coordinates are usually stored on magnetic tapes. The data may be in profile form with equal spatial interval or time interval during digitization, or selectively at terrain break points as designated. The data may also be obtained as Z values for an X, Y grid with equal X and Y spacing. Another possibility is to digitize along contour lines, where a large set of X, Y values is developed for each given Z value. These various forms of DTM are produced for a variety of purposes. One of the areas where DTM has been successfully applied is in the automation of road location. Highway engineers have created a DTM, stored it into a computer, and used it over and over to try out different alignments and grades in order to arrive at optimum design factors (Art. 14.25). The increased significance of terrain data banks makes the use of DTM all the more important. During the next decade or two a considerable increase in DTM production, processing, and applications will occur, particularly in relation to automated cartography. In this regard photogrammetric instrumentation and methodology will play an important role.

13.18. Orthophotography One of the more recent products of photogrammetry that is being used increasingly is the orthophoto. It can be defined as a pictorial depiction of the terrain derived from aerial photography in such a way that there are no relief or tilt displacements. Therefore, an orthophoto is equivalent to a planimetric map except that instead of lines and symbols, image tonal variations convey the information. If the photographed terrain is flat and level, a *rectified* aerial photograph would be the same as an orthophoto.

The principle of orthophoto generation is depicted schematically in Fig. 13.19. First a stereomodel is relatively and absolutely oriented as if for plotting a topographic map. Visualizing the plotting a point of detail A, the tracing table is elevated to point A in the model and its map location A_1 is plotted *orthographically* directly underneath on the map plane. When another model point B is to be plotted, the platen is lowered until the floating mark is at B and the point dropped at B_1, on the map. Similarly, point C_1 is plotted when the plane of the tracing-table surface is at C. It is clear then that the relationship between the *horizontal* positions of model points A, B, C is identical to that on the map sheet. If it were convenient, the entire map-sheet plane could have been raised to the level of the point under consideration and the map position recorded. The result in this case would be identical. This is exactly what is done when generating an orthophoto optically. The tracing table is replaced by a large enough sheet of film to cover the total model area. The film is covered by a light-tight screen and rests on a table that can be moved up and down by the operator just as the tracing table. Similar to the floating mark, a small rectangular slit is open in the screen and can be moved in the X and Y directions.

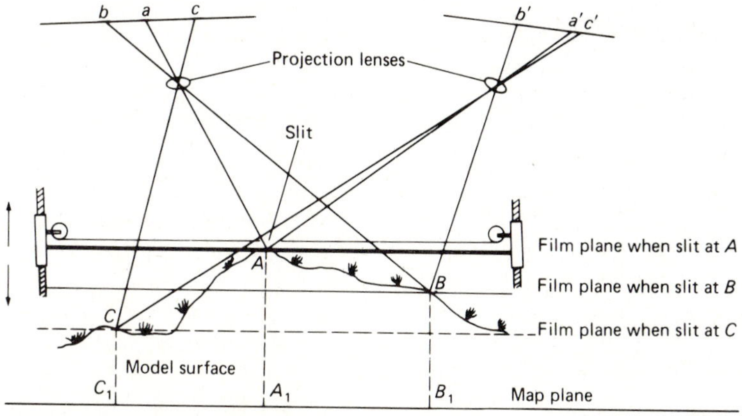

Fig. 13.19 Orthophoto generation.

When the slit is at point A in the model (Fig. 13.19) the light from one projector, usually the blue-green, exposes the film to the point detail within the slit. The film is normally not sensitive to red light and therefore is not affected by the image projected from the second projector. When at another point, B, the total film plane is moved down to eliminate the effect of relief, and the slit area is exposed. It can be seen that the exposed film would be equivalent to the planimetric map.

In actual orthophoto systems, the slit travels automatically either in the X or Y direction of the model at a selected constant speed, thus continuously exposing a film strip as wide as the longer dimension of the slit. This feature frees the operator to concentrate on raising or lowering the film plane to keep the slit always at the model surface. When one strip is finished, the slit is stepped over and an exactly adjacent film strip is exposed until the entire model area is covered. The film is then developed into a negative from which orthophoto prints can be made.

The on-line optical rectification explained above is but one of many methods for the production of an orthophoto. Another method that is widely used employs electronic techniques. In such an approach, the two conjugate images are electronically correlated and the output signal from the correlation circuits are sent to an orthoprinter. In the printer, a cathode ray tube (CRT) is used to display an image which is then orthogonally printed as an orthophoto. Interested students should consult recent textbooks on photogrammetry (such as Ref. 11) for more detail.

Compared to a planimetric map, an orthophoto offers several advantages. Since the details are reproduced in pictorial form, there is a wealth of information on the orthophoto, whereas in a line map the operator has already performed the classification task and selected only those features deemed important. Being an accurate map and rendering abundance of detail makes an orthophoto useful for more applications than either a planimetric map or a regular aerial photograph. Then, if contour lines are added to the orthophoto, it becomes even more useful than a topographic map. Figure 13.20a shows an example of an orthophoto with added contours. For comparison purposes, Fig. 13.20b illustrates a conventional topographic map of a portion of the same area at a larger scale.

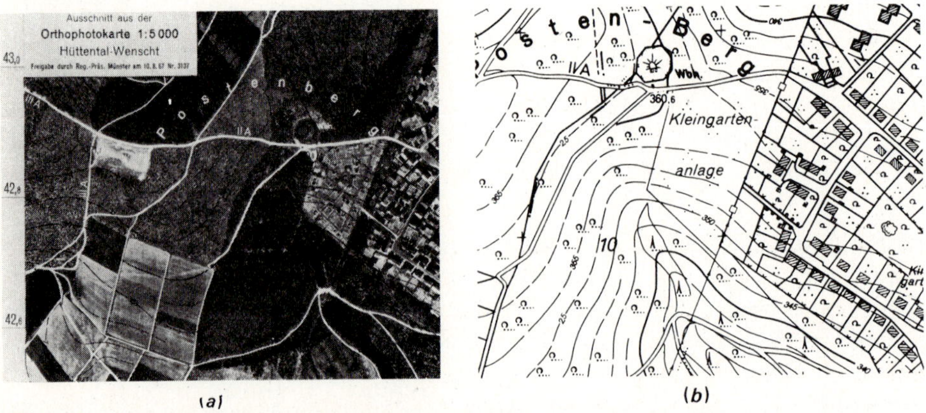

(a) (b)

Fig. 13.20 (*a*) Orthophoto with contours added (scale 1:5000). (*b*) Topographic map of a portion of the same area shown in part (*a*) at a larger scale. (*Courtesy Carl Zeiss, Inc.*)

13.19. Planning of photogrammetric project There are three interrelated phases involved in the planning for a photogrammetric survey: (1) the design of a flight plan which must be followed by the pilot of the airplane while taking the aerial photography; (2) deciding on the amount and location of necessary horizontal ground control and performing the field surveys to obtain such control to the designated accuracy measures; and (3) estimation of costs where the relative costs of various operations are determined so that the most economical combination may be selected without compromising the accuracy requirements. Analysis of all three phases must be made in relation to the particular end product of the project, which may include topographic maps of a wide range of scales, orthophotomaps, mosaics, and lists of supplemental control. Usually, the factors affecting the acquisition of photography for one type of end product may be substantially different from those affecting photography for another product. In general, the most important factors to be considered in planning aerial photography include the following:

1. The intended use of the photography (qualitative or interpretive)
2. The desired product (map, orthophoto, mosaic, numerical data, etc.)
3. Accuracy specifications
4. Size and shape of the area to be covered by photography
5. The amount and disposition of relief in the area to be photographed
6. The scale of photography, which in turn depends on the scale of the final product, the camera to be used, the compilation instruments available, etc.
7. Tilt, crab, and drift as they influence the effective overlap and sidelap between adjacent photographs

When these factors are carefully analyzed for a given project, the basic elements of the flight plan can be resolved. These elements include the flying height above datum (usually mean sea level), the ground distance between successive photo exposures, and the ground spacing between flight lines. Except for applications requiring only one flight line (such as for highway design), the resulting flight lines are usually laid out on the best available map of the area to be photographed. The general scheme for photographic coverage has been discussed and is shown in Fig. 13.6. Although overlap is most commonly taken at 60 percent, it may be increased for some applications (e.g., mosaic). Also, sidelap can be as high as 60 percent, particularly for triangulation of blocks of photographs to produce supplemental control. The actual details of the steps involved in designing a flight plan and the computations associated with it are outside the scope of this chapter.

13.20. Automated photogrammetric systems This chapter has covered the various aspects of photogrammetry involved in the regular production of maps, orthophotos, mosaics, etc. Recently, there has been a significant emphasis in photogrammetry toward automating various functions which are normally performed by a human being. Obviously, details of such aspects are well beyond the scope of this chapter. However, the student should be aware of these developments and their significance in order to be able to call on specialists whenever the use of such systems becomes feasible in future professional work.

Although there are many more possible classes, here we consider automated systems as falling in three categories:

1. Systems where a *correlator* is used to automatically match corresponding imagery on overlapping photographs (thus replacing stereoperception by the human operator)
2. Photogrammetric plotters which operate with the assistance of a digital computer, called *analytical plotters*
3. Totally digital systems in which photographs are scanned and digitized and all photogrammetric operations performed on the digital data

The earliest attempts at automation were in the first category, where the function of the eye-brain combination was being synthesized. Electronic, optical, and digital cross-correlation techniques are used with some success, although they still remain less capable than the far superior ability of the human observer to accomplish stereoviewing. In the third category, very extensive computer capabilities are required, and activities are still in the research stage. In the second category, however, many analytical plotters have already been commercially produced and sold. In an analytical plotter, an on-line digital computer is used to implement on a real-time basis the geometric conditions that exist between the images on the plates and model points. The stereomodel is never fully developed as a projection model as in regular plotters; instead it is incrementally perceived by the operator. The computer continuously computes the locations of conjugate images and transmits X and Y photo shifts to the plates, which are then translated by servomotors to bring the image over the viewing optics. This is done so fast that the operator is never aware of the incremental nature of model formation.

The most distinct advantage of automated systems is flexibility. Analytical plotters usually accommodate a variety of photographic types and can develop several products rapidly. For example, orthophotos could be produced either on-line or off-line, usually automatically without any human intervention. Also, the system can be programmed to automatically produce high-density digital terrain model (DTM) data. These data can themselves be used for producing other useful information (areas, volumes, line of sight, etc.).

13.21. Problems

13.1 Describe concisely the following: (*a*) photo interpretation; (*b*) remote sensing; (*c*) aerial photogrammetry; (*d*) terrestrial photogrammetry; (*e*) close-range photogrammetry.

13.2 Explain the basic differences between a phototheodolite, a stereometric camera, and an aerial camera.

13.3 Describe the following elements of an aerial camera: shutter, diaphragm, camera body, fiducial marks, and magazine.

13.4 "Overlap and sidelap are necessary for proper aerial photographic coverage." Discuss this statement using a sketch to illustrate the two cases.

13.5 Explain the correspondence between stereo perception by the pair of human eyes and the geometry of overlapping vertical photographs.

13.6 Describe fully, with the aid of sketches, the two types of displacements that make an aerial photograph of the terrain different from a map of such terrain.

13.7 Enumerate five different photogrammetric products. Explain each briefly.

13.8 What is the main objective of radial line triangulation? Discuss the assumptions underlying the method.

13.9 Explain the following operations for a stereopair of diapositives to be used in a photogrammetric plotter: (*a*) interior orientation; (*b*) exterior orientation; (*c*) relative orientation; (*d*) absolute orientation.

13.10 How is a topographic map compiled on a stereoplotter?

13.11 Why is photogrammetric triangulation an important operation?

13.12 Explain the principal differences between two photogrammetric products: the digital terrain model and the orthophotograph.

13.13 List the most important factors to be considered in planning vertical aerial photography.

13.14 What is the height of a tower above terrain which appears on a truly vertical photograph with the following data:

flying height above base of tower = 3200 m
distance between principal point and the image of the tower base = 75.11 mm
distance between principal point and the image of the top of the tower = 82.54 mm

13.15 The distance between two well-defined points on a vertical aerial photograph is 72.05 mm. The corresponding distance measured on a U.S. quadrangle map is 55.17 mm. If the map scale is 1 : 24,000, calculate the approximate scale of the photograph. Calculate the flying height above terrain if the focal length of the camera used is 150.00 mm.

13.16 A vertical photograph is taken with a lens having a focal length of 152.40 mm from a flying height of 2670 m above an airport where the elevation is 380 m above sea level. Determine the representative fraction, expressing the scale of the photograph at a point at which the elevation is 221 m above sea level.

13.17 How high above sea level must an aircraft fly in order that photographs at a scale of 1 : 9600 may be obtained, if the focal length is 8.25 in and the average elevation of the terrain is 800 ft?

13.18 A distance measured on a vertical photograph between two points, both lying at ground elevation of 366 m, scales 82.677 mm. The focal length of the camera is 152.908 mm. The distance between the same pair of points measures 25.451 mm on a 1 : 24,000 quadrangle sheet. Compute the flying height above sea level at which the photograph was taken. Calculate the datum scale of the photograph.

13.19 Photographs are to be taken for preparing a highway design map. The lowest elevation in the area to be photographed is 144 m and the highest elevation is 282 m. The minimum photographic scale is to be 1 : 6000. What must be the flying height above sea level if the camera to be used contains a lens with a focal length of 134.63 mm? What will be the maximum scale?

13.20 The distance between two section corners is assumed to be exactly 1 mi. Both corners lie at 450 ft above sea level. If the distance between the images of these two corners scales 2.342 in on a vertical photograph taken with a lens having a 8.262-in focal length, what is the flying height at which the photograph was taken?

13.21 The datum scale of a vertical photograph taken with a lens having a focal length of 152.400 mm is 1 : 8000. A hilltop lies at an elevation of 800 m above sea level and the image of the hilltop is 71.679 mm from the principal point of the photograph. Compute the relief displacement of the hilltop. If this photo should happen to be tilted 2° such that the hilltop is now on the "up side" of the tilted photo, explain the effects of tilt on the displacement caused by relief.

13.22 Two overlapping vertical photographs are taken with a camera having a focal length of 152.00 mm. It is assumed that both photographs are exposed at a flying height of 3500 m above mean terrain, and that the overlap is 60 percent. A control point with elevation of 100.22 m is used as a basis for calculating the elevations of two other points. If the differences in parallax between the control point and the two points are 4.11 mm and −1.73 mm, respectively, calculate the elevations of the two points. Assume the format to be 230 × 230 m.

References

1. American Society of Photogrammetry, *Manual of Photogrammetry,* 4th ed., George Banta, 1980.
2. Anderson, J. M., et al., "Analytic Block Adjustment," *Photogrammetric Engineering,* Vol. 39, No. 10, 1973.

3. Danko, J. O., "Color, the Kelsh, and the PPV," *Photogrammetric Engineering,* Vol. 38, No. 1, 1972.
4. Hallert, B., *Photogrammetry,* McGraw-Hill Book Company, New York, 1960.
5. Hughes, T., et al., "U.S.G.S. Automatic Orthophoto System," *Photogrammetric Engineering,* Vol. 37, No. 10, 1971.
6. Jensen, N., *Optical and Photographic Reconnaissance Systems,* John Wiley & Sons, Inc., New York, 1968.
7. LaPrade, G. L., "Stereoscopy—A More General Theory," *Photogrammetric Engineering,* Vol. 38, No. 12, 1972.
8. Meyer, D., "Mosaics You Can Make," *Photogrammetric Engineering,* Vol. 28, No. 1, 1962.
9. Mikhail, E. M., "A Study in Numerical Radial Triangulation," *Photogrammetric Engineering,* Vol. 34, No. 4, 1968.
10. Moffitt, F. H., and Bouchard, H., *Surveying,* 7th ed., Harper & Row, Publishers, New York, 1982, Chapter 16.
11. Moffitt, F. H., and Mikhail, E. M., *Photogrammetry,* 3rd ed., Harper & Row, Publishers, New York, 1980.
12. Nash, A. J., "Use a Mirror Stereoscope Correctly," *Photogrammetric Engineering,* Vol. 38, No. 12, 1972.
13. Radlinski, W., "Orthophotomaps versus Conventional Maps," *The Canadian Surveyor,* Vol. 22, No. 1, 1967.
14. Stanton, B. T., "Education in Photogrammetry," *Photogrammetric Engineering,* Vol. 32, No. 3, 1971.
15. Thompson, E. H., "Aerial Triangulation by Independent Models," *Photogrammetria,* Vol. 19, No. 7, 1964.
16. U.S. Department of Transportation, Federal Highway Administration, *Reference Guide Outline Specifications for Aerial Surveys and Mapping by Photogrammetric Methods for Highway,* Washington, D.C., 1968.
17. Whiteside, A. E., et al., "Recent Analytical Stereoplotter Developments," *Photogrammetric Engineering,* Vol. 38, No. 4, 1972.
18. Wolf, P. R., *Elements of Photogrammetry,* 2d ed., McGraw-Hill Book Company, New York, 1983.
19. Woodward, L. A., "Survey Project Planning," *Photogrammetric Engineering,* Vol. 36, No. 6, 1970.
20. Yacoumelos, N., "The Geometry of the Stereomodel," *Photogrammetric Engineering,* Vol. 38, No. 8, 1972.

Periodicals

21. *Photogrammetria,* published by the International Society of Photogrammetry, Delft, Holland.
22. *Photogrammetric Engineering and Remote Sensing,* published by the American Society of Photogrammetry, Washington, D.C.
23. *Photogrammetric Record*, published by the British Society of Photogrammetry, London.
24. *Surveying and Mapping*, published by the American Congress on Surveying and Mapping, Washington, D.C.
25. *The Canadian Surveyor,* published by the Canadian Institute of Surveying and Photogrammetry, Ottawa, Canada.

CHAPTER 14
Route and construction surveys

14.1. General The expression "route surveying" used in a very general sense can be applied to the surveys required to establish the horizontal and vertical alignment for transportation facilities. In the most general case, the transportation facilities are assumed to comprise a network that includes the transport of people or goods on or by way of highways, railways, rapid transit guideways, canals, pipelines, and transmission lines. For the past three decades in the United States, highways have been the most highly developed form of transportation facility in the overall network. As a result, route surveys for highways are well defined and widely practiced. Most of the methods that have been developed for highway surveys are equally applicable to the other specified means of transport. Consequently, the emphasis in this chapter is on route surveys for highway alignment, but other topics such as construction surveys for railroads, pipelines, sewers, bridges, buildings, and dams will also be covered briefly.

Surveys of some type are required for practically all phases of route alignment planning, design, and construction work. During the planning phase, a topographic survey (Chap. 12) of the site is performed and maps are prepared to be used in the development of plans for the project. As soon as approval of the project is assured, property line surveys are initiated to be used for acquisition of lands or rights of way. The control network established for these topographic and property surveys contains many of the horizontal and vertical control points which will eventually form the basis for subsequent construction surveys. Consequently, the survey engineer in charge of planning the surveys has the responsibility to organize the initial basic control surveys so as to provide the maximum number of horizontal and vertical control points which are useful not only for topographic and property surveys but also for the construction surveys that follow.

Surveys of some type are required for practically all phases of route alignment planning, design, and construction work. For small projects involving widening or minor improvement of an existing facility, the survey may be relatively simple and may include only the obtaining of sufficient information for the design engineer to prepare plans and specifications defining the work to be done. For more complex projects involving multilane highways on new locations, the survey may require a myriad of details, including data from specialists in related fields to determine the best location; to prepare plans, specifications, and estimates for construction; and to prepare deed descriptions and maps for appraisal and acquisition of the necessary rights of way.

In order to plan and perform the surveys needed to acquire these types of data, the survey engineer must be familiar with (1) the geometry of horizontal and vertical curves and how they are used in the route alignment procedure, (2) the methods of acquiring terrain data utilized in the route design procedure, (3) the procedures followed in processing terrain data to obtain earthwork volumes, (4) establishing on the ground a system of stakes or other markers, both in plan and in elevation, from which measurement of earthwork and structures can be taken conveniently by the construction force, (5) giving line and grade as needed either to replace stakes disturbed by construction or to reach additional points on the structure itself, and (6) making measurements necessary to verify the location of completed parts of the structure (the as-built survey) and to determine the volume of work actually performed up to a given date (usually each month), as a basis of payment to the contractor.

New developments in equipment, such as electronic surveying systems and laser-equipped alignment instruments (Art. 14.31) provide powerful tools which have many applications in these areas.

The detailed methods employed on these surveys vary greatly with the type, location, and size of structure and with the preference of the engineering and construction organizations. Much depends on the ingenuity of the surveyor so that the correct information is given without confusion or needless effort. Consequently, the discussions that follow are of a general nature with a minimum number of specific examples and procedures set forth.

14.2. Route curves for horizontal and vertical alignment: general In highway, railway, canal, and pipeline location, the horizontal curves employed at points of change in direction are arcs of circles. The straight lines connecting these *circular curves* are tangent to them and are therefore called *tangents*. For the completed line, the transition from tangent to circular curve and from circular curve to tangent may be accomplished gradually by means of a segment in the form of a *spiral*. Spiral curves are beyond the scope of this book and interested readers should consult Refs. 9, 10, and 16 at the end of this chapter.

Vertical curves (Art. 14.9) are usually arcs of parabolas. Horizontal parabolic curves are occasionally employed in route surveying and landscaping but they will not be discussed in this textbook.

The subject of route curves is extensive, as indicated by the list of references at the end of the chapter. Herein are discussed only some of the simpler relationships.

14.3. Circular curves: general Once the final alignment of a route is selected, the proposed center line is marked in the field by placing stakes on the ground at 100-ft (English units) or 100-m (metric units) intervals called *full stations* (Art. 5.42). In practice it is also customary to set stakes at one-half station points (every 50 ft with English units) and one-fifth (20-m) or one-tenth (10-m) station points when

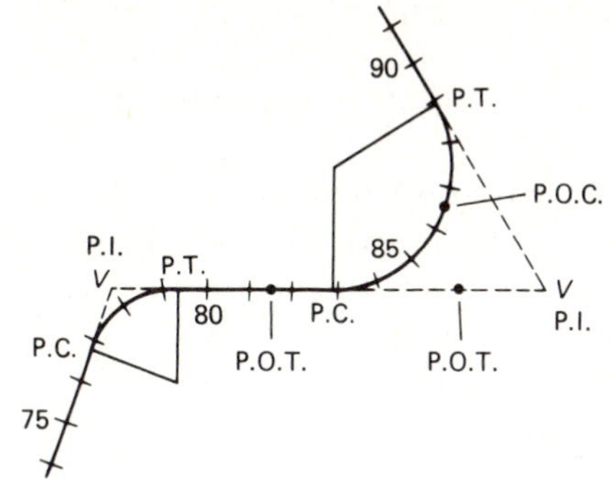

Fig. 14.1 Route stationing.

using metric units. Recall (Art. 5.42) that all stations are reckoned from the beginning of the center line which is designated as 0 + 00. Thus, any point along the center line may be identified by its plus distance from the last full station prior to it. A point that is 13.570 ft or m beyond station 91 is designated 91 + 13.570 and is 9113.570 ft or m from the beginning of the center-line survey. The stationing of a route progresses around a curve in the same manner as along a tangent, as indicated in Fig. 14.1. The point where a circular curve begins is commonly called *the point of curvature*, written P.C.; that where the curve ends is called the *point of tangency*, written P.T.; and that where two tangents produced intersect is called the *point of intersection* or the *vertex*, written P.I. or *V*. A point on the curve is written P.O.C. Other notations are also used; for example, the point of curve may be written T.C., signifying that the route changes from tangent to circular curve, whereas the point of tangency is written C.T. Or the beginning of the curve may be written B.C. and the end of curve E.C. A point on the tangent is written P.O.T. All of these points P.C., P.T., P.O.C., etc, are designated by the appropriate abbreviation and the corresponding stationing.

In the field the distances from station to station on a curve are necessarily measured in straight lines, so that essentially the curve consists of a succession of chords. In this text, a full station is a function of the system of measurement units employed. Thus, for the metric system a full station is 100 m. For the foot system a full station is 100 ft. Where the curve is of long radius, as in railroad practice, the distances along the arc of the curve are considered to be the same as along the chords. In highway practice and along curved property boundaries, the distances are usually considered to be along the *arcs*, and the corresponding chord lengths are computed for measurement in the field.

The sharpness of curvature may be expressed in any of three ways:

1. Radius The curvature is defined by stating the length of radius. This method is often employed in subdivision surveys and sometimes in highway work. The radius is usually taken as a multiple of 100 ft or 100 m.

2. Degree of curve, arc definition Here the curvature is expressed by stating the "degree of curve" D_a, which has traditionally been defined as the angle subtended at the center of the curve by an arc 100 ft long. In the metric system, D_a is defined

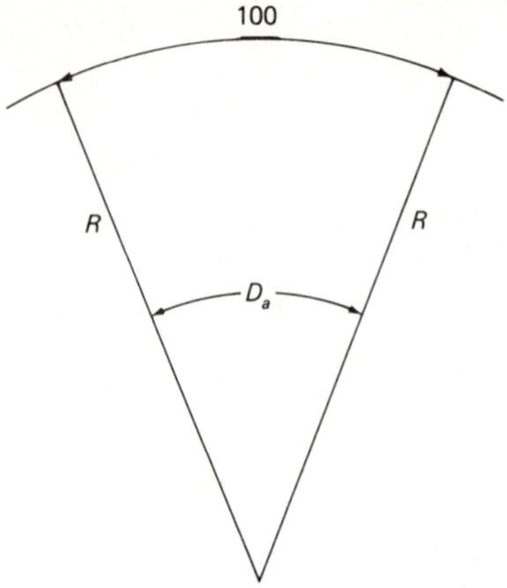

Fig. 14.2 Degree of curve D_a, arc definition.

as the angle subtended by a 100-m arc and, in general, D_a is the angle subtended by an arc equal in length to one full station. The arc definition for degree of curve is the method most frequently followed in highway practice. In Fig. 14.2, D_a is the degree of curve by the arc definition and R is the radius, so that

$$\frac{D_a}{100} = \frac{360°}{2\pi R}$$

or

$$D_a = \frac{36{,}000}{2\pi R} = \frac{5729.578}{R} \tag{14.1a}$$

Equation (14.1*a*) is applicable in both systems of measurement (English and metric).

Since 1 m equals 3.28084 ft, degrees of curve in the two systems differ by the same proportion. Thus, a 1° curve in the metric system is the same curve as a 0°.3048 curve in the foot system. Conversely, a 1° curve in the foot system is equivalent to a 3°.28084 curve in the metric system. Rearranging Eq. (14.1*a*) yields

$$R = \frac{5729.578}{D_a} \tag{14.1b}$$

In design practice, D_a is usually selected on the basis of design speed, allowable superelevation, and friction factor and is rounded *down* to the nearest integral number. The radius is then calculated using Eq. (14.1*b*) for subsequent curve calculations. For example, the radius of a 1° curve is 5729.578 m in the metric system and 5729.578 ft in the foot system, and a 5° curve would have a radius of 1145.916 units in the respective systems.

3. Degree of curve, chord definition Degree of curve according to the chord definition, D_c, is defined as the angle subtended by a *chord* having a length of one full station or 100 ft in the foot system. In the metric system, D_c would be the angle subtended by a chord of 100 m. This definition in the foot system has been followed

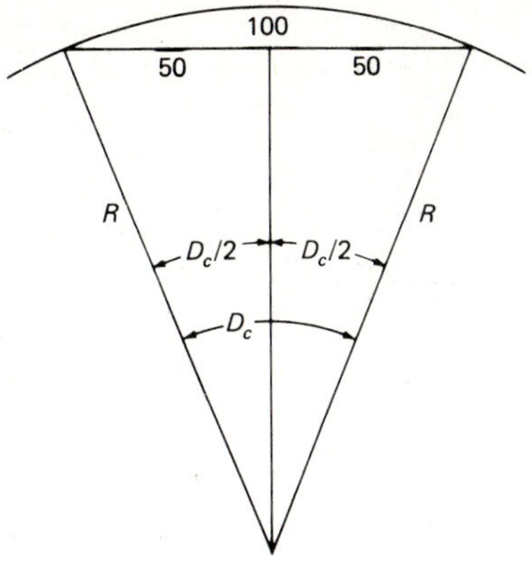

Fig. 14.3 Degree of curve D_c, chord definition.

almost invariably in railroad practice. From Fig. 14.3,

$$\sin\frac{D_c}{2} = \frac{50}{R} \tag{14.2a}$$

Rearranging Eq. (14.2*a*) yields

$$R = \frac{50}{\sin(D_c/2)} \tag{14.2b}$$

Note that the radius of curvature varies inversely as the degree of curve; for example, according to the arc definition, the radius of a 1° curve is 5729.578 ft or m and the radius of a 10° curve is 572.958 ft or m. According to the chord definition, the radius of a 1° curve is 5729.651 ft or m and the radius of a 10° curve is 573.686 ft or m.

Field measurements of the curve with the tape must, of course, be made along the chords and not along the arc. When the arc basis is used, either a correction is applied for the difference between arc length and chord length or the chords are made so short as to reduce the error to a negligible amount. In the latter case, usually 100-ft chords are used for curves up to about 3°, 50-ft chords from 3° to 8°, 25-ft chords from 8° to 25°, and 10-ft chords for curves sharper than 25°. In the metric system, 100-m chords can be used for curves up to 1°, 50-m chords from 1° to 3°, 25-m chords from 3° to 5°, and 10-m chords for curves sharper than 5°. Note that curve layout using 100- or 50-m chords could prove impractical under field conditions where terrain configuration and vegetation present obstacles.

The railroads are the only organizations currently involved in route alignment which use the chord definition for their curves.

14.4. Geometry of the circular curve In discussing circular curves, the following geometrical facts are employed:

1. An inscribed angle (identified by $\delta/2$ in Fig. 14.4) is measured by one-half of the angle at the center of the circle (identified by δ in Fig. 14.4) that subtends the same arc along the circumference of the circle. Inscribed angles that subtend the same or equal length arcs are

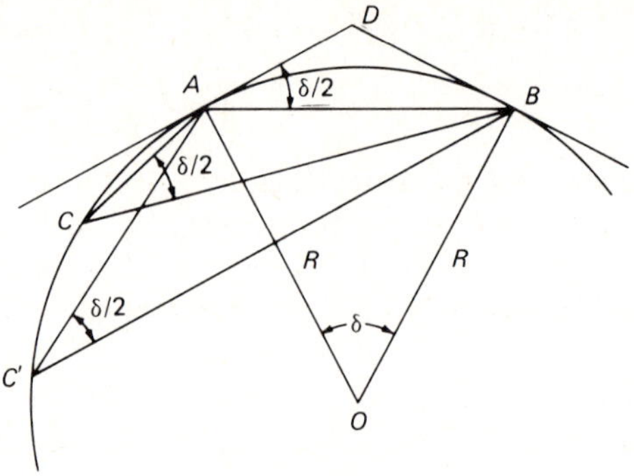

Fig. 14.4 Geometry of a circular curve.

equal. Thus, in Fig. 14.4, the angle ACB (at any point C on the circumference) subtending an arc AB is one-half the central angle AOB subtending the same arc AB; and the angles at the points C and C' are equal.

2. An angle formed by a tangent and a chord is measured by one-half of the angle subtended by its intercepted arc. Thus, in Fig. 14.4, the angle at the point A between AD, the tangent to the curve at that point, and the chord AB, is one-half the central angle AOB subtending the same arc AB. This is a special case of the proposition above, when the point C moves to A.
3. The two tangent distances to a circular curve, from the point of intersection of the tangents to the points of tangency, are equal. In Fig. 14.4, lines AD and DB are tangent to the curve at points A and B, respectively. Thus, distance AD = distance DB.

14.5. Curve formulas Figure 14.5 represents a circular curve joining two tangents. In the field the intersection angle I between the two tangents is measured. The radius of the curve is calculated from the value of D_a or D_c chosen according to the proposed operating conditions for the route and at the same time due regard must be given to fitting the curve to the topography. The line OV bisects the angles at V and at O, bisects the chord AB and the arc ADB, and is perpendicular to the chord AB at

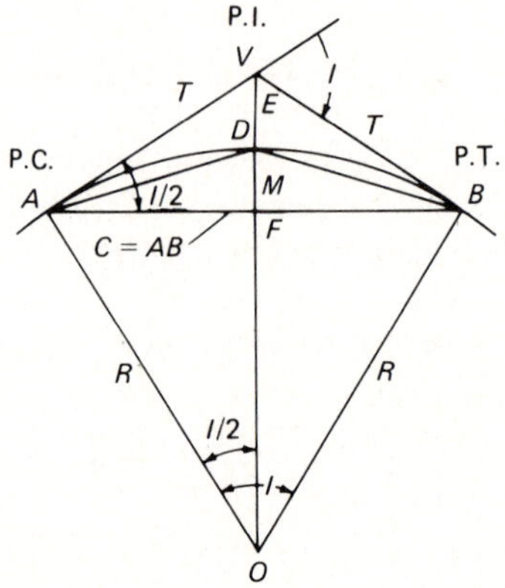

Fig. 14.5 Basis for curve formulas.

F. From the figure, $\angle AOB = I$ and

$$\angle AOV = \angle VOB = \angle VAB = \angle VBA = \frac{I}{2}$$

The chord $AB = C$ from beginning to end of curve is called the *long chord*. The distance $AV = BV = T$ from vertex to P.C. or P.T. is called the *tangent distance*. The distance $DF = M$ from midpoint of arc to midpoint of chord is called the *middle ordinate*. The distance $DV = E$ from midpoint of arc to vertex is called the *external distance*.

Given the radius of the curve $OA = OB = R$ and the intersection angle I, and knowing that $\angle VAO = \angle VBO = 90°$ (since the tangent to a circle is perpendicular to the radius at the point of tangency), then in triangle OAV

$$\frac{T}{R} = \tan\frac{I}{2}$$

$$T = R\tan\frac{I}{2} = \text{tangent distance} \tag{14.3}$$

$$\frac{R}{OV} = \cos\frac{I}{2} = \frac{R}{R+E}$$

$$E = R\sec\frac{I}{2} - R = R\left[\frac{1}{\cos(I/2)} - 1\right] \tag{14.4}$$

$$E = R\,\text{exsec}\,\frac{I}{2} = \text{external distance}$$

From the triangle AOF, in which $AF = C/2$,

$$\frac{AF}{R} = \sin\frac{I}{2} = \frac{C/2}{R}$$

$$C = 2R\sin\frac{I}{2} = \text{long chord} \tag{14.5}$$

$$\frac{OF}{R} = \cos\frac{I}{2} = \frac{R-M}{R}$$

$$M = R - R\cos\frac{I}{2} = R\left(1 - \cos\frac{I}{2}\right)$$

$$M = R\,\text{vers}\,\frac{I}{2} = \text{middle ordinate} \tag{14.6}$$

From the triangle AVF, in which $\angle VAF = I/2$ and $AF = C/2$,

$$\frac{C}{2} = T\cos\frac{I}{2}$$

$$C = 2T\cos\frac{I}{2} = \text{long chord} \tag{14.7}$$

From triangle AVF, in which $\angle DAF = I/4$, since $\angle VAD = I/4$ from Fig. 14.4,

$$\tan\frac{I}{4} = \frac{M}{C/2}$$

$$M = \frac{C}{2}\tan\frac{I}{4} = \text{middle ordinate} \tag{14.8}$$

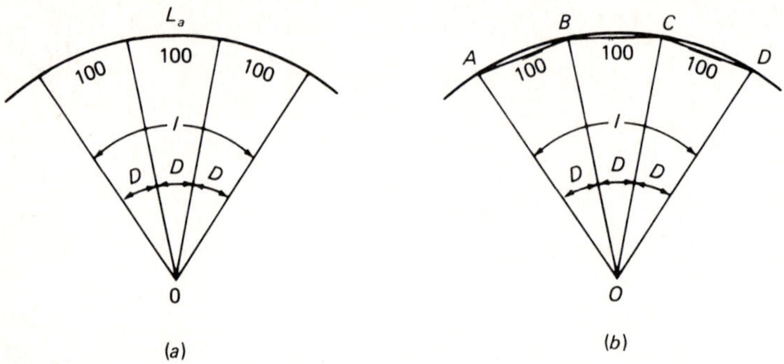

Fig. 14.6 Length of curve. (*a*) Arc definition. (*b*) Chord definition.

14.6. Length of curve The length of the circumference of a circle is $2\pi R$; this is the arc length for a full circle or 360°. As the arc length corresponding to a given radius varies in direct proportion to the central angle subtended by the arc, the length of arc for any central angle I is

$$\text{arc} = \left(\frac{I°}{360°}\right)2\pi R \tag{14.9}$$

in which angle $I°$ is expressed in degrees. This solution is simplified by the use of a table of arc lengths for various angles and for unit radius.

If the degree of curvature is expressed on the *arc* basis, the length of curve L_a is found by referring to Fig. 14.6*a*. The arc L_a subtends the angle I and an arc of 100 units of length subtends angle D_a. By simple proportion

$$\frac{L_a}{I} = \frac{100}{D_a}$$

or

$$L_a = 100\frac{I}{D_a} = \text{length of curve (arc definition)} \tag{14.10}$$

If the degree of curvature is expressed on the *chord* basis, the length of curve is considered to be the sum of the lengths of the chords, normally each 100 ft long. In the metric system, the chord lengths would be 100 m. For these cases, the length of curve (on the chords) is

$$L_c = 100\frac{I}{D_c} = \text{length of curve (chord definition)} \tag{14.11}$$

which is somewhat less than the actual arc length. Thus, if the central angle I of the curve AD (Fig. 14.6*b*) is equal to three times the degree of curve D, as shown, then there are three 100-ft chords between A and D, and the length of "curve" on this basis is 300 ft. Similarly, in the metric system, there would be three 100-m chords between A and D, and L_c is 300 m.

14.7. Laying out a curve by deflection angles Curves are staked out usually by the use of deflection angles turned at the P.C. from the tangent to stations along the

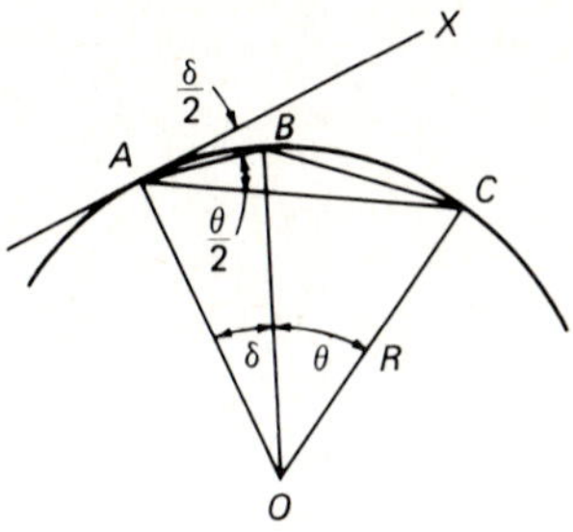

Fig. 14.7 Curve layout by deflection angles.

curve together with the use of chords measured from station to station along the curve. The method is illustrated in Fig. 14.7, in which ABC represents the curve, AX the tangent to the curve at A, and angles XAB and XAC the deflection angles from the tangent to the chords AB and AC.

Assume the transit to be set up at A. Given R, δ, and θ it is required to locate points B and C. Considering point B,

$$\angle XAB = \frac{\delta}{2} \tag{14.12}$$

and by Eq. (14.5)

$$AB = 2R \sin \frac{\delta}{2} \tag{14.13}$$

In the field, point B is located as follows. The deflection angle $XAB = \delta/2$ is set off from the tangent, the distance AB is measured from A, and the forward end of the tape at B is lined in with the transit.

Considering point C,

$$\angle BAC = \frac{\theta}{2} \tag{14.14}$$

$$BC = 2R \sin \frac{\theta}{2} \tag{14.15}$$

$$\angle XAC = \frac{\delta + \theta}{2} \tag{14.16}$$

In Fig. 14.8, let points a, b, c, d represent station points on a simple curve. Point a is an odd distance from the P.C. and the distance dB is also an odd increment. The deflection angles are

$$VAa = \frac{d_1}{2}$$

$$VAb = \frac{d_1}{2} + \frac{D}{2}$$

$$VAc = \frac{d_1}{2} + \frac{D}{2} + \frac{D}{2} = \frac{d_1}{2} + D$$

$$VAd = \frac{d_1}{2} + D + \frac{D}{2} = \frac{d_1}{2} + 3\frac{D}{2}$$

$$VAB = \frac{d_1}{2} + \frac{3D}{2} + \frac{d_2}{2} = \frac{I}{2}$$

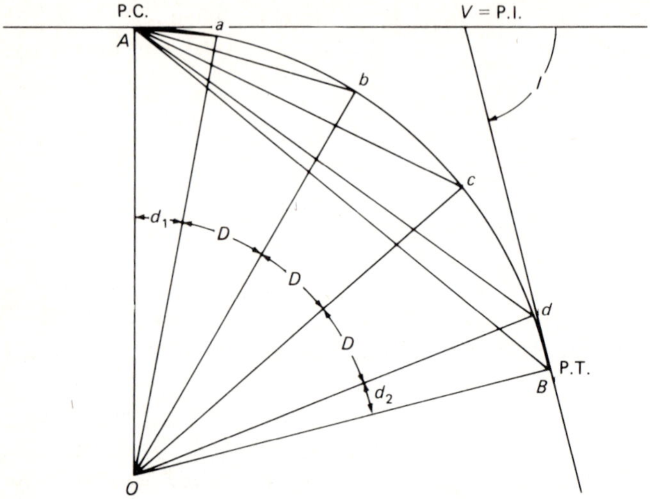

Fig. 14.8 Deflection angles on a simple curve.

Note that the sum of all deflection angles must equal $I/2$, a check on the calculations.

For a curve having degree of curve D (arc or chord) deflection angles per station, half-station, quarter-station, and one-tenth station are $D/2$, $D/4$, $D/8$, and $D/20$, respectively.

For odd lengths of arc or odd stations, it is convenient to calculate the deflection of the arc in minutes per unit of arc. If $D = 9°30'$, then $D/2 = 4°45' = 285'$ per station. Thus, the deflection equals $285'/100$ units $= 2.85'$/unit of arc (ft or m). If the radius is given, then

$$\frac{d}{L} = \frac{360}{2\pi R}$$

so that for $L = 1$ unit of arc (ft or m), $d = 360°/(2\pi R)$ and

$$\frac{d}{2} = \frac{(360°)(60'/°)}{(2)(2\pi R)}$$

or

$$\frac{d}{2}\ \text{(deflection in minutes/unit of arc)} = \frac{1718.873}{R} \tag{14.17}$$

For $D = 9°30'$, $R = 5729.578/9.5 = 603.113$ units and

$$\frac{d}{2} = \frac{1718.873}{R} = 2.85'/\text{unit of arc}$$

measured in feet or metres depending on the system of measurement.

To find $\angle VAa$ in Fig. 14.8 if $D_a = 9°30'$, multiply the arc length Aa by $2.85'$ where the arc length Aa is the difference in stationing between points A and a.

As an alternative method to find $\angle VAa$, label the arc length Aa (Fig. 14.8) as ℓ (the difference in stationing from A to a). Then the following ratio can be written:

$$\frac{\ell}{d_1} = \frac{100}{D_a}$$

from which

$$d_1 = \frac{\ell D}{100} \tag{14.18}$$

From geometrical fact 2 of Art. 14.4, it follows that $\angle VAa$ in Fig. 14.8 is one-half of $\angle d_1$ in Fig. 14.8. Thus, $\angle VAa = \ell D/200$.

A curve is normally located in the field as follows. The P.C. and P.T. are marked on the ground. The deflection angle from the P.C. is computed for each full station on the curve and for any intermediate stations that are to be located. The transit is set up at the P.C., a sight is taken along the tangent, and each point on the curve is located by the deflection angle and by distance measured from the preceding station. The following examples illustrate the procedure and give the usual form of field notes.

Example 14.1 In Fig. 14.9, assume that stations have been set as far as 88 + 00 at *C*. The directions of tangents *CV* and *MV* have been fixed by stakes, but the tangent distances

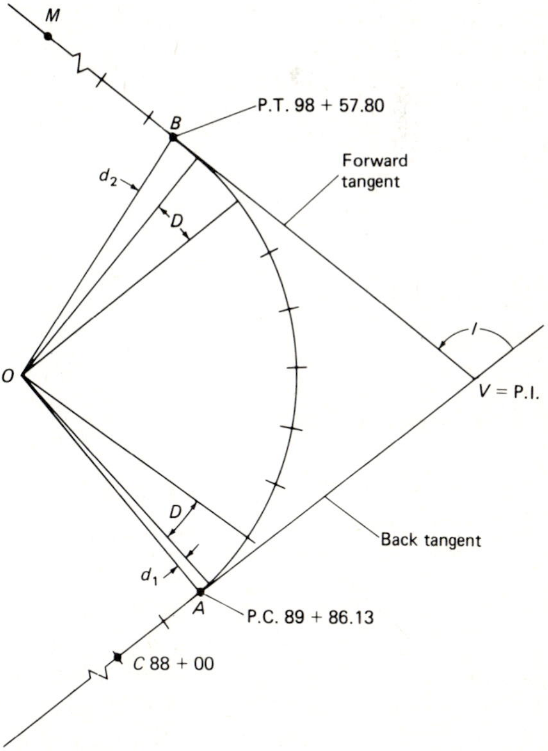

Fig. 14.9 Laying out a simple curve.

have not been measured. The degree of curve D_a is to be 12° according to the arc definition. It is desired to stake full stations on the curve using feet as the unit of measurement.

SOLUTION The tangents *CA* and *MB* are produced to an intersection at *V* and the distance from *C* to *V* is found to be 803.90 ft. Then the transit is set at *V* and the angle *I* is observed by double deflection to be 104°36′ to the left. Since D_a is given, the radius of the curve, tangent distance, and length of curve using Eqs. (14.1*a*), (14.3), and (14.10) can be computed as follows:

$$R = \frac{5729.578}{12} = 477.46 \text{ ft}$$

$$T = R \tan \frac{I}{2} = 617.77 \text{ ft}$$

$$L = \frac{100I}{D} = 871.67 \text{ ft}$$

Thus, the stations at the P.C. and P.T. are:

Station at P.I. = (88 + 00) + 803.90	=	96 + 03.90
Tangent distance	= −	6 17.77
Station at the P.C.	=	89 + 86.13
Length of curve	= +	8 71.67
Station at the P.T.	=	98 + 57.80

The deflection angle for one full 100-ft station is $D/2 = 6°00$. The distance from the P.C. to the first full station is (90 + 00) − (89 + 86.13) = 13.87 ft, and the distance from the last full station on the curve to the P.T. is (98 + 57.80) − (98 + 00) = 57.80 ft. The deflection in minutes per unit of arc equals $(12°)(60'/°)/200 = 3.6'/\text{ft}$, or by the alternative method using Eq. (14.17), the deflection equals $1718.873/477.46 = 3.6'/\text{ft}$ as a check. Thus deflection angles for the odd increments at the beginning and end of the curve are:

$$\frac{d_1}{2} = (13.87)(3.6) = 0°49.93' = 0°49'56''$$

or

$$\frac{d_1}{2} = \frac{(13.87)(12)}{200} = 0.8322° = 0°49'56''$$

as a check.

$$\frac{d_2}{2} = (57.80)(3.6) = 208.08' = 3°28.08' = 3°28'05''$$

or

$$\frac{d_2}{2} = \frac{(57.80)(12)}{200} = 3.4680° = 3°28'05''$$

by the alternative method.

Chord distances for the initial and final odd increments of arc and the full stations are

$$c_1 = 2R \sin \frac{d_1}{2} = (2)(477.46)\sin 0°49'56'' = 13.87 \text{ ft}$$

$$c_2 = (2)(477.46)\sin 3°28'05'' = 57.76 \text{ ft}$$

$$c_{100} = (2)(477.46)\sin 6°00'00'' = 99.82 \text{ ft}$$

Deflection angles are found by cumulating the individual deflections to full stations from the P.C. to P.T. as follows:

Station	Deflection angles	Chord, ft	Curve data
P.T. 98 + 57.80	52°18.0′		
		57.76	
98	48°49.9′		
		99.82	
97	42°49.9′		
		99.82	$I = 104°36'$
96	36°49.9′		
		99.82	$D_a = 12°00'$
95	30°49.9′		
		99.82	$R = 477.46$ ft
94	24°49.9′		
		99.82	$T = 617.77$ ft
93	18°49.9′		
		99.82	$L = 871.67$ ft
92	12°49.9′		
		99.82	Sta. at P.I. = 96 + 03.90
91	6°49.9′		
		99.82	
90	0°49.9′		
		13.87	
P.C. 89 + 86.13	0°00.0′		

Note that the total deflection angle should equal $I/2$, which provides a positive check on the cumulated deflection angles.

Example 14.2 The directions of two tangents on an existing road are established from the curb lines. The deflection angle to the right is 44°00′00″. Because the existing curbs are to be retained, a curve must be determined for which the tangent distance does not exceed 50 m but is equal to or greater than 45 m. Determine a radius and integral value for D_a to satisfy this specification and calculate the parameters and deflection angles needed to set one-tenth stations on this curve. The station of the P.I. is 8 + 43.892.

SOLUTION A trial value for R can be found using Eq. (14.3):

$$R = \frac{T}{\tan(I/2)} = \frac{45}{\tan 22°} = 111.379 \text{ m}$$

which yields

$$D_a = \frac{5729.578}{111.379} = 51.4°$$

Round this value of D_a to 50°, recompute R, and calculate T:

$$R = \frac{5729.578}{50°} = 114.592$$

$$T = (114.592) \tan 22° = 46.298 \text{ m}$$

This satisfies the stated specification. The balance of the curve computations are included in the sample note form in Fig. 14.10.

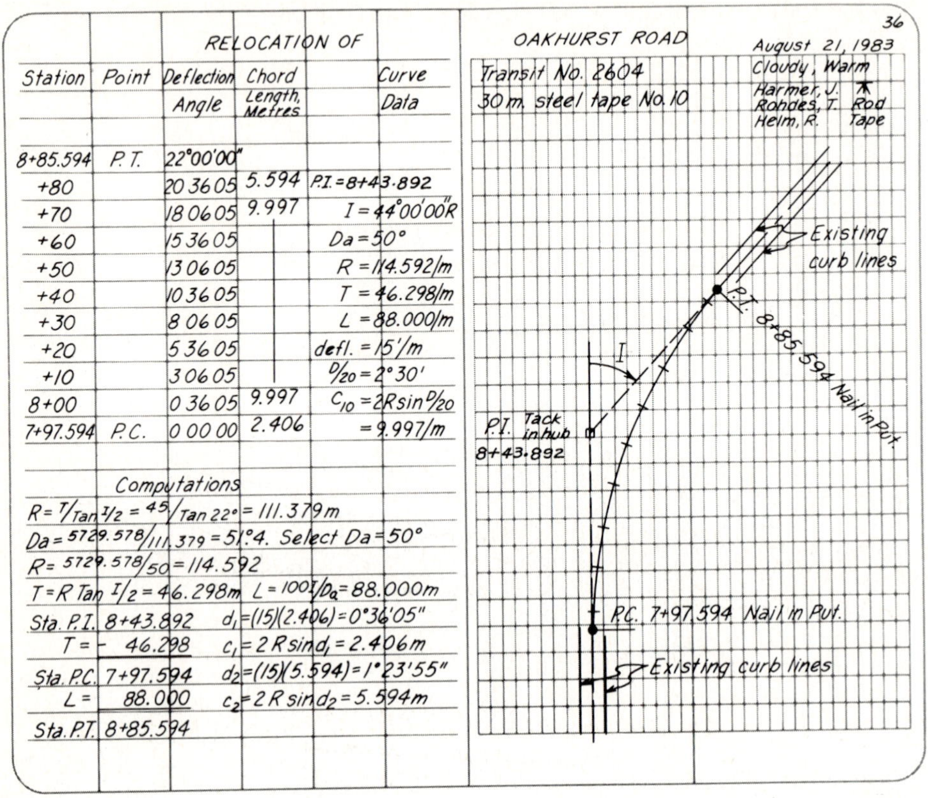

Fig. 14.10 Notes for layout of a simple curve.

14.8. Transit setups on the curve Because of obstacles, great lengths of curve, and so on, often it is impractical or impossible to run all of a given curve with the transit at the P.C. In such cases, one or more setups are required along the curve between P.C. and P.T.

Figure 14.11 illustrates the case where the transit is set up at an intermediate point A. The curve is begun at the P.C. and is located as far as A, where a hub is set. The transit is then set up at A. A backsight (with the A vernier set on 180° and the telescope direct) is taken on the last preceding station at which the transit was set up, in this case the P.C. The upper motion is loosened and the telescope is turned toward B with the angle $\alpha/2 + \beta/2$ being set off on the horizontal circle. The line of sight is thus directed along the chord AB. The angle $(\alpha + \beta)/2$, turned from the line P.C. to A extended, is equal to the deflection angle at the P.C. for point B; therefore, the vernier setting to locate point B from transit station A is the same as that which would have been used had the transit remained at the P.C. According to this method, the following procedure may be used to orient the transit at any point on the curve:

1. Compute deflection angles as for use at the P.C.
2. When set up at any point on the curve, backsight at any preceding transit station with the telescope direct and the A vernier set on 180° + the deflection angle to the station sighted

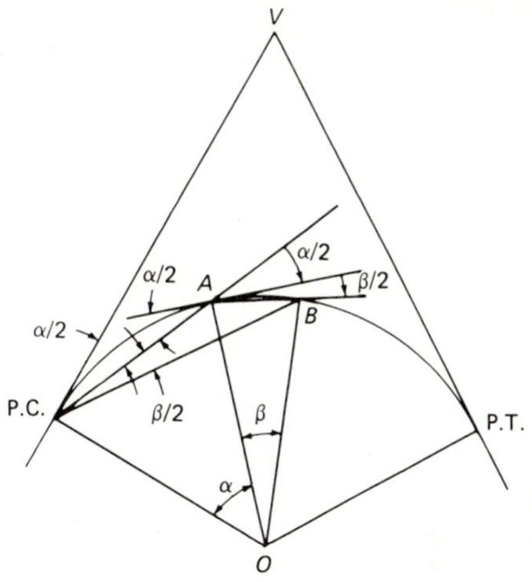

Fig. 14.11 Transit setup on a simple curve.

for curves deflecting to the right and 180° – the deflection angle to the station sighted for curves deflecting to the left.

3. To locate stations in a forward direction along the curve, loosen the upper motion, turn the instrument through 180°, and turn the previously computed deflection angles.

If the instrument is in good adjustment, an alternative procedure is possible. Backsight the preceding station with the telescope reversed and with the vernier set on the deflection angle for the station sighted. To locate subsequent stations, plunge the telescope and use the computed deflection angles. If the backsight is to the P.C., the backsight vernier reading is zero.

14.9. Vertical curves On highways and railways, in order that there may be no abrupt change in the vertical direction of moving vehicles, adjacent segments of differing grade are connected by a curve in the vertical plane called a *vertical curve*. Usually, the vertical curve is the arc of a parabola, as this shape is well adapted to gradual change in direction and as elevations along the curve are readily computed.

Figure 14.12 illustrates crest and sag parabolic and equal tangent vertical curves. Point A is the beginning of the vertical curve labeled BVC or PVC, V is the P.I. or P.V.I. the intersection of the tangents, and B is the end of the vertical curve labeled EVC or PVT. The length of the curve AB is designated L and is measured horizontally. The two grades in the direction of stationing along tangents AV and VB expressed in percent are G_1 and G_2, respectively. When the tangent rises in the direction of stationing, the grade is positive, and when the tangent slopes downward, the grade is negative.

Design of crest and sag vertical curves is a function of the difference in grades of the intersecting tangents, stopping or passing sight distance, which in turn are functions of vehicle design speed and height of the driver's eye above the roadway, and drainage. In addition to these factors, design of a sag vertical curve also depends on headlight beam distance, rider comfort, and appearance. Details governing design of vertical curves are beyond the scope of this text and may be found in route

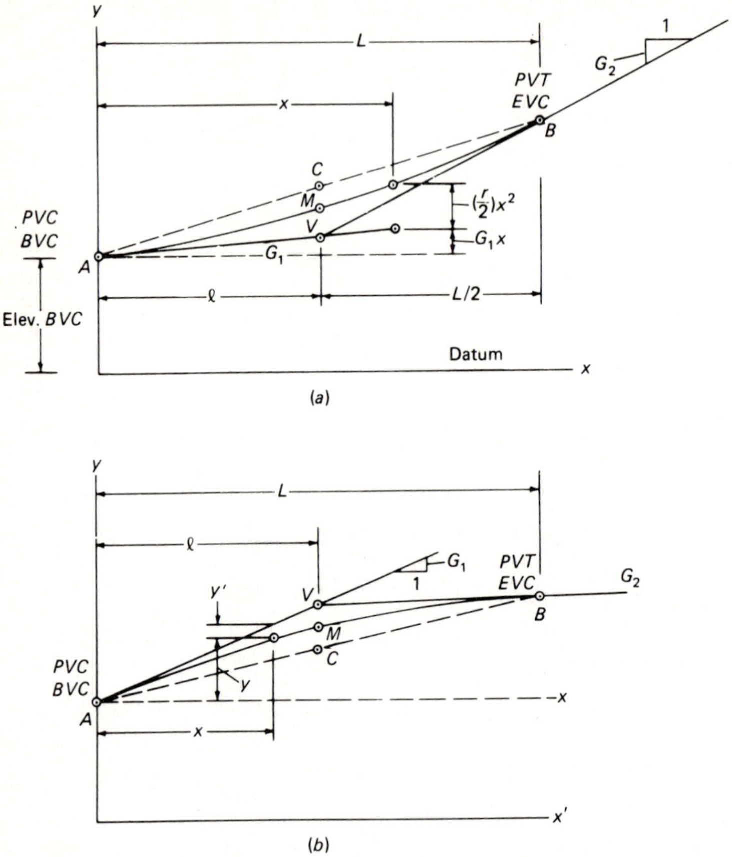

Fig. 14.12 (*a*) Sag vertical curve. (*b*) Crest vertical curve.

surveying textbooks as well as in the AASHO Policies on Rural and Urban Highways (Refs. 1 and 2).

Elevations on vertical parabolic curves can be computed by using the (1) equation of the parabola directly, (2) geometric properties of the parabola to calculate vertical offsets from the tangent, and (3) geometric properties of the parabola as exemplified by the chord gradient method, which is beyond the scope of this textbook.

14.10. Vertical curves by equation of the parabola (Ref. 11) Consider the plane parabolic curve AB of Fig. 14.12a with the y axis passing through the BVC and the x axis corresponding to the datum for elevations. Let L be the length of the vertical curve in stations. The slope of the curve at the BVC is G_1 and the slope at the EVC is G_2. Since the rate of change in slope of a parabola is constant, the second derivative of y with respect to x is a constant, or

$$\frac{d^2y}{dx^2} = \text{constant} = r \tag{14.19}$$

Integration of Eq. (14.19) yields the first derivative or the slope of the parabola,

$$\frac{dy}{dx} = rx + H \tag{14.20}$$

When $x = 0$, the slope $= G_1$ and when $x = L$, the slope $= G_2$. Thus,

$$G_1 = 0 + H \qquad \text{and} \qquad G_2 = rL + H$$

from which

$$r = \frac{G_2 - G_1}{L} \tag{14.21}$$

The value of r is the rate of change of the slope in percent per station. Substitution of Eq. (14.21) into Eq. (14.20) gives

$$\frac{dy}{dx} = \left(\frac{G_2 - G_1}{L}\right)x + G_1 \tag{14.22}$$

To obtain y, integrate Eq. (14.22) to yield

$$y = \left(\frac{G_2 - G_1}{L}\right)\frac{x^2}{2} + G_1x + c \tag{14.23}$$

When $x = 0$, y is the elevation of the BVC so $c =$ elevation of the BVC or y_{BVC} and Eq. (14.23) becomes

$$y = \left(\frac{r}{2}\right)x^2 + G_1x + (\text{elevation of } BVC) \tag{14.24}$$

which is the equation of the parabolic curve AB and can be used to calculate elevations on the curve given G_1, G_2, L, and the elevation of the BVC. Note that the first term, $(r/2)x^2$, is the vertical offset between the curve and a point on the tangent to the curve at a distance x from the BVC. The second term G_1x represents the elevation on the tangent at a distance x from the BVC, and the third term is the elevation of the BVC above the datum (see Fig. 14.12*a*). Any system of units can be employed in Eq. (14.24). The grades G_1 and G_2 are dimensionless ratios and expressed as percentages, but r [Eq. (14.21)] must be determined in units compatible with the units of x and y. If distances are measured in feet, L and x in Eq. (14.21) must be in stations where 100 ft = one station. If distances are in metres, L and x are in metric stations where 100 m = one station.

The high or low point of a vertical curve is frequently of interest for drainage purposes. At the high or low point, the tangent to the vertical curve is zero. Equating the first derivative of y with respect to x [Eq. (14.22)] to zero gives

$$rx + G_1 = 0 \tag{14.25}$$

or

$$x = -\frac{G_1}{r} = \frac{G_1L}{G_1 - G_2} \tag{14.26}$$

by substituting from Eq. (14.21).

Example 14.3 A 200-m equal tangent parabolic vertical curve is to be placed between grades of $G_1 = 1.25$ percent, and $G_2 = -2.75$ percent intersecting at station 18 + 00, which has an elevation of 270.190 m above mean sea level (m.s.l.). Calculate elevations at even 20-m stations on the vertical curve and determine the station and elevation of the high point on the vertical curve. Figure 14.13 shows the geometry involved.

SOLUTION The elevation of the $BVC = 270.190 - (0.0125)(100) = 268.940$ m and the elevation of the $EVC = 270.190 - (0.0275)(100) = 267.440$ m above m.s.l. Next compute the rate of change in grade by Eq. (14.21), where $L = 2$ stations. Thus,

$$r = \frac{-2.75 - 1.25}{2} = -2.000$$

so that

$$\frac{r}{2} = -1.000$$

From Eq. (14.24), the equation of the parabola for this vertical curve is

$$y = -1.000x^2 + 1.25x + 268.940$$

in which y equals the elevation of a point on the vertical curve x stations from the BVC. Calculation of elevations on the vertical curve by this equation is best illustrated by Table 14.1. Note that elevations on the curve for a given station are the sum of columns 3, 4, and 6 of the table.

By Eq. (14.26), the station of the high point is

$$x = -\frac{G_1}{r} = \frac{1.25}{2} = 0.625 \text{ station or } 62.500 \text{ m}$$

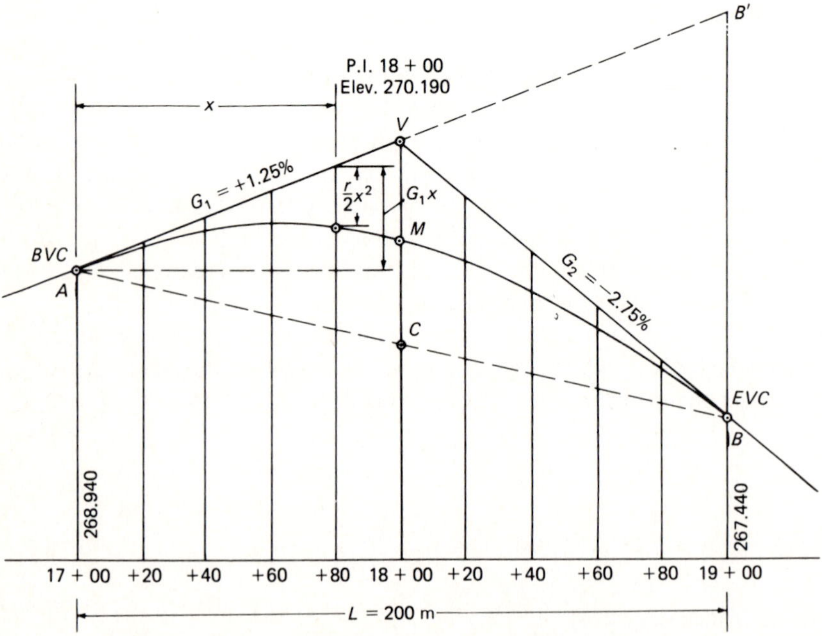

Fig. 14.13 Vertical curve by equation of the parabola and vertical offsets.

Table 14.1 Elevations on a Vertical Curve by Equation of Parabola

Station	x	$(r/2)x^2$	G_1x	$(r/2)x^2 + G_1x$	Elevation *BVC*, m	Elevation on curve, m
BVC 17 + 00	0.	0.	0.	0.	268.940	268.940
+20	0.2	−0.04	0.250	0.210	268.940	269.150
+40	0.4	−0.16	0.500	0.340	268.940	269.280
+60	0.6	−0.36	0.750	0.390	268.940	269.330
+80	0.8	−0.64	1.000	0.360	268.940	269.300
18 + 00	1.0	−1.00	1.250	0.250	268.940	269.190
+20	1.2	−1.44	1.500	0.060	268.940	269.000
+40	1.4	−1.96	1.750	−0.210	268.940	268.730
+60	1.6	−2.56	2.000	−0.560	268.940	268.380
+80	1.8	−3.24	2.250	−0.990	268.940	267.950
EVC 19 + 00	2.0	−4.00	2.500	−1.500	268.940	267.440

so that the

station of high point = (17 + 00) + 62.500 = 17 + 62.500

Using the equation of the parabola, the elevation of this station is

$$\text{elevation at } 17 + 62.500 = (-1.0)(0.625)^2 + (1.25)(0.625) + 268.940$$
$$= 269.331 \text{ m above m.s.l.}$$

14.11. Vertical curves by vertical offsets from the tangent In Fig. 14.12*b*, translate the x coordinate axis so that it passes through *BVC* at *A*. Then by similar triangles,

$$\frac{y - y'}{x} = G_1$$

where y' (measured from the tangent) is positive upward and negative downward or

$$y = G_1 x + y' \tag{14.27}$$

Substituting Eq. (14.27) into Eq. (14.23) yields ($c = 0$)

$$G_1 x + y' = \left(\frac{G_2 - G_1}{2L}\right) x^2 + G_1 x$$

which can be solved for y' to give

$$y' = \left(\frac{G_2 - G_1}{2L}\right) x^2 = \left(\frac{r}{2}\right) x^2 \tag{14.28}$$

or vertical offsets from the tangent vary as the square of the distance from the point of tangency. Now, at point V, the P.I. of the vertical curve, $y' = VM$ and $x = L/2$, so that

$$y' = VM = \left(\frac{G_2 - G_1}{2L}\right)\frac{L^2}{4} = \frac{(G_2 - G_1)L}{8} \tag{14.29}$$

Divide Eq. (14.28) by Eq. (14.29) and solve for y' to give

$$y' = \left(\frac{x^2}{\ell^2}\right) VM \tag{14.30}$$

in which $\ell = L/2$.

In Fig. 14.12*b*, point *C* bisects the chord *AB* and since from geometry, the midpoint *M* on the vertical curve is halfway between *V* and *C*, $VM = \frac{1}{2}VC$. Thus, given the elevations of the *BVC* and *EVC*, *VM* can be calculated. On sag vertical curves, *VM* is positive and on crest curves *VM* is negative, so that y' is positive on sag (Fig. 14.12*a*) and negative on crest (Fig. 14.12*b*) vertical curves.

To determine elevations on the vertical curve by vertical offsets, calculate the elevations for desired stations along the tangent to the vertical curve from the *BVC* to *V* using G_1 and from *V* to the *EVC* using G_2. Next, calculate the vertical offsets [Eq. (14.30)] and apply to the elevations on the tangent. Since an equal tangent parabolic vertical curve is symmetrical, vertical offsets need to be computed for only one-half the curve when stations are equally spaced and symmetrical about the P.I., as in Example 14.3. Then these same vertical offsets can be applied to elevations along the tangent at the *EVC*, allowing determination of elevations on the second half of the curve. Table 14.2 shows the calculations necessary to compute the vertical curve of Example 14.3 by vertical offsets from the tangent.

VM is calculated as follows:

Point	Elevation
BVC	268.940
EVC	267.440
	536.380
C	268.190

$$VM = -\frac{\text{el.}V - \text{el.}C}{2}$$

$$= -\frac{270.190 - 268.190}{2}$$

$$= -1.000$$

Table 14.2 Elevations on a Vertical Curve by Vertical Offsets

Station	Grade along tangent	x in stations	$(x^2/\ell^2)VM^{\dagger}$	Elevation on curve
17 + 00 *BVC*	268.940	0.	0.	268.940
+20	269.190	0.2	−0.040	269.150
+40	269.440	0.4	−0.160	269.280
+60	269.690	0.6	−0.360	269.330
+80	269.940	0.8	−0.640	269.300
18 + 00	270.190	1.0	−1.000	269.190
+20	269.640	0.8	−0.640	269.000
+40	269.090	0.6	−0.360	268.730
+60	268.540	0.4	−0.160	268.380
+80	267.990	0.2	−0.040	267.950
19 + 00 *EVC*	267.440	0.	0.	267.440

† $VM = -1.0$; $\ell = 1$ station.

Elevations along the tangent are computed from BVC to V using G_1. Elevations along the tangent from V to the EVC are calculated with G_2. In this case where points are equally spaced, second differences of the elevations must be equal. This is an important check which can be applied to all vertical curves calculated by the equation of the parabola (as in Example 14.3) or by vertical offsets.

If stations are not equally spaced and symmetrical about the P.I., grades are calculated for the desired stations along the tangent AV extended to B' (Fig. 14.13) and vertical offsets are computed with Eq. (14.30), where x is always measured from the BVC.

14.12. Earthwork operations Earthwork operations consist of the movement of materials in order to establish a predetermined surface for the construction of public and private works and determination of the volume of materials moved. Field work involves acquisition of terrain data (usually profile and cross sections) and setting points (grade stakes and slope stakes) to guide construction work on the site. Office work involves acquisition of terrain data from maps or by photogrammetric methods, processing terrain data, calculating volumes of excavated or embanked materials, and determination of the most economic procedure for performing the excavation and embankment. Earthwork operations and computations are key elements in the overall route location procedure.

For large projects, most of the terrain data can be extracted more accurately and at a lower cost from large-scale photogrammetrically compiled topographic maps or by applying photogrammetric methods to remotely sensed digital data such as can be obtained from aerial photographs. The traditional field methods described in the following sections are still appropriate for projects of small extent, for terrain that is relatively level, and for field conditions where the ground surface is concealed by heavy vegetation.

Although the main emphasis in this chapter is on earthwork operations for route surveying, the field and office methods developed are also applicable for determining volumes of stockpiles of materials, reservoir volumes, and volumes of concrete structures.

14.13. Cross-section levels Frequently, in connection with problems of drainage, irrigation, grading of earthwork, location and construction of buildings, and similar enterprises, the shape of the surface of a parcel of land is desired. This may be obtained by staking out the area into a system of squares and then determining the elevations of the corners and of other points where changes in slope occur. The sides of the squares are usually 100, 50, or 25 ft (30, 15, or 10 m). Directions of the lines may be obtained with either the tape or the transit, distances may be laid off either with the tape or by stadia, and elevations may be determined with either the engineer's level or the hand level, all depending upon the required precision. The elevations are carried forward and the rod readings on ground points are determined as in profile leveling. The data may be employed in the construction of a contour map (see Art. 12.13).

CROSS-SECTION LEVELS BORROW PIT PROPERTY OF A. N. HAWK

Sta.	B.S.	H.I.		IFS	Elev.
BM 10	3.320	127.778			124.458
A-1				3.12	124.66
A-2				3.00	124.78
A-3				2.48	125.30
A-4				2.12	125.66
B-1				2.00	125.78
B-2				2.85	124.93
B-3				2.60	125.18
B-4				2.50	125.28
C-1				2.00	125.78
C-2				2.45	125.33
C-3				3.00	124.78
C-4				3.05	124.73
D-1				0.50	127.28
D-2				1.00	126.78
D-3				1.20	126.58
D-4				1.50	126.28
E-1				1.00	126.78
E-2				2.00	125.78
E-3				2.50	125.28
F-1				0.80	126.98
F-2				0.40	127.38
BM 10			3.320		124.458

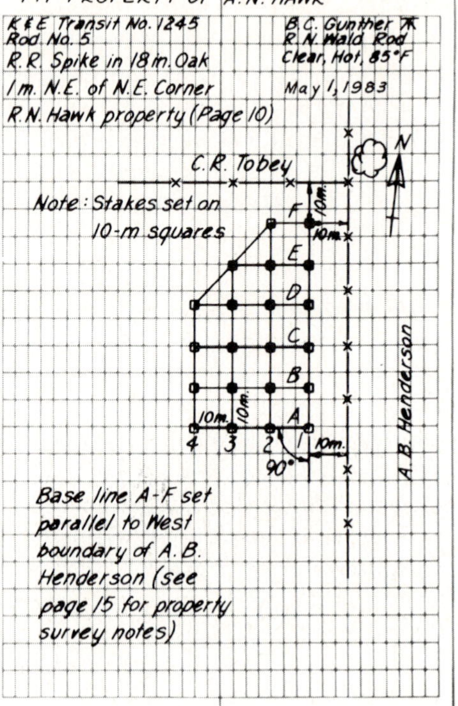

Fig. 14.14 Cross-section (checkerboard) notes.

Figure 14.14 illustrates a suitable form of notes. The sketch on the right-hand page shows the area divided into 10-m squares, the lines running in one direction being numbered and those running the other direction being lettered. The coordinates of a given corner may then be stated as the letter and number of the two lines intersecting at the corner. Thus, *B*2 is a point at the intersection of lines *B* and 2.

Surveyors who own EDM equipment with real-time display of horizontal distance, in particular those who have access to an electronic surveying system (Art. 9.14), should consider optional methods of approaching this problem as outlined in Art. 12.15.

14.14. Route cross sections Surveys for highways, railroads, or canals are often made by establishing a center line along the proposed route. This center line may be a location on paper using a topographic map, may consist of points set on the ground in the field, or may be a line of numerically known positions with respect to a digital terrain model (DTM). In any case, station points (a full station is 100 ft in the English and 100 m in the metric systems) are designated or set at full, half, or one-tenth stations along the route center line. Odd station points are set at breaks in

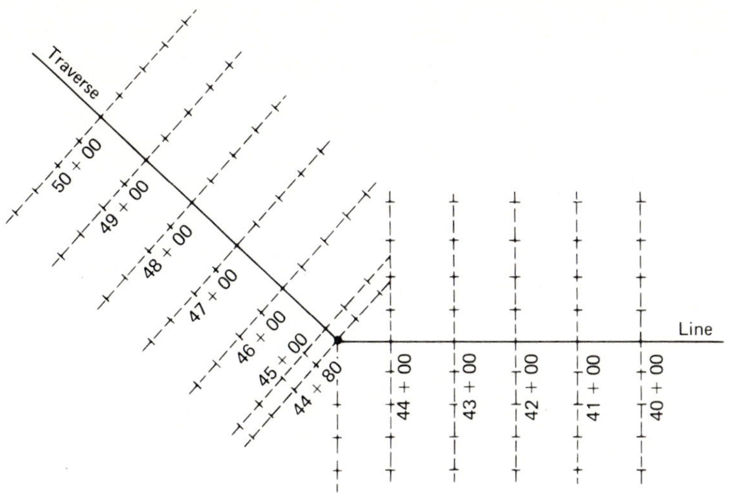

Fig. 14.15 Cross lines for cross sections by the plus and offset method.

the terrain configuration. Then, in order to furnish data for location studies and for estimating volumes of earthwork, a profile is run along the center line (Art. 5.42), and cross sections (Art. 12.12) are taken along lines passing through each station and at right angles to the center line. The elevations along these crosslines and respective distances to the left and right (as one looks toward the next numerically higher station) constitute terrain cross-section data. The solid line in Fig. 14.15 is the center line, the dashed lines represent cross-section lines, and the tick marks are points where elevations are obtained to the left and right of center line, which is usually designated as ℄.

There are four major methods by which terrain cross-section data can be obtained: (1) field surveys; (2) from topographic maps compiled by field surveys or photogrammetric methods, (3) from photogrammetric stereomodels, and (4) from digital terrain models. The data will ultimately be in the form described above in defining the cross section no matter which method is used. The method selected depends on the size and nature of the project under study, and the equipment, personnel, and instrumentation available.

14.15. Terrain cross sections by field methods This method constitutes the traditional approach and resembles the procedure for obtaining data for topographic mapping by the cross-section method (Art. 12.12). The center line is set in the field and a profile is run over the center-line stations by differential leveling (Art. 5.42). The directions of short crosslines are set off by eye; that of long crosslines by transit or right-angle prism. Usually, elevations are determined with an engineer's level and rod in level terrain and by a hand level and rod in rough, irregular country. For each cross section, the height of instrument is established by a backsight on the center-line station. The rod is then held on the crossline at specified distances or at breaks in the surface slope, where rod readings are observed and distances measured with a metallic tape. In very steep, irregular terrain, the use of self-reducing tacheometers and the stadia method (Art. 7.16) or EDM equipment and theodolites to obtain distance and elevation would be advantageous.

A note form for recording cross-section data by the traditional method is illustrated in Fig. 14.16. The elevations for center-line stations are obtained from the profile-level notes. This form of notes would be utilized when profile levels and cross sections are run separately and recorded in separate notebooks. The right-hand page of the notebook represents the traverse line, and to the right or left of this line are recorded the observed distances, rod readings, and the computed elevations. When a second H.I. is required to secure the necessary observations for a given crossline, a second line for that station is shown in the notes; thus station 405 occupies two lines in the notes, and station 406 occupies three. The location of fences, streams, etc., may be indicated by appropriate symbols or abbreviations.

Figure 14.17 shows a note form where profile levels and cross-section data are recorded on the same form, with all readings referenced to the same height of instrument (H.I.). The actual operations of profile and cross sections could be performed concurrently or separately. For example, profile leveling could be run separately, yielding the notes on the left-hand page, and at another time, cross sections could be run with a hand level, tape, and rod as described for the cross section in Art. 12.12. The notes are recorded on the right-hand pages in space left for that purpose when the profile was run.

Since most cross-section data are processed by computer programs, it is frequently advantageous to record field notes on a note form that is directly usable by a key-punch operator for punching terrain data in a specified format. Figure 14.18 illustrates such a form using the data from cross sections given in Fig. 14.16 for stations 405 and 406.

CROSS-SECTIONS
C. & R. EXTENSION

O.H. Ellis, ⊼ No. 2510   Jan. 20, 1984 (24)
C.O. Lord, Rod   Cold, Snow
J.A. Crum, Tape

Sta.	B.S.	H.I.	F.S.	Elev.		Left	℄	Right
405	12.4	633.0		620.6	(Dist.) (Elev.) (Rod)	300 210 12[illegible] 80 632.1 630.8 627.0 626.7 0.9 2.2 6.0 6.3	620.6	
405	0.6	621.2		620.6				50 160 250 350 617.0 612.3 610.0 609.7 4.2 8.9 11.2 11.5
406	12.1	628.9		616.8		90 50 628.2 624.3 0.7 4.6	616.8	
406	11.5	639.7	0.7	628.2		280 200 120 638.3 635.6 632.2 1.4 4.1 7.5		
406	1.9	618.7		616.8				60 100 200 300 615.5 614.2 610.9 609.3 3.2 4.5 7.8 9.4
407	4.7	615.9		611.2		300 180 100 615.6 614.7 612.6 0.3 1.2 3.3	611.2	75 155 270 606.2 604.5 602.9 9.7 11.4 13.0
408	10.6	615.9		605.3		280 200 100 611.6 609.2 607.7 4.3 6.7 8.2	605.3	100 200 300 604.1 603.2 603.0 11.8 12.7 12.9

Fig. 14.16 Route cross-section notes.

PRELIMINARY CROSS SECTIONS
HALF MOON BAY

28

O.H. Ellis, ⊼
C.O. Lord, Rod
J.A. Crum, Tape

June 5, 1983
Fair, Warm
75°F

Sta.	B.S.	H.I.	F.S.	Elev.								
BM24	2.750	213.200		210.450	metres	Level No. 4211 Rod No. 10						
						Brass plate set in concrete. See page 10.						
					(Dist.)	25	16	8	℄	8	16	24
43+00					(Rod)	3.05	3.60	4.40	2.51	2.05	1.45	1.00
					(Elev.)	210.15	209.60	208.80	210.69	211.15	211.75	212.20
						36						35
						3.95						0.30
						209.25						212.90
TP1	0.815	210.263	3.752	209.448								
						25	15	8	℄	7	15	25
44+00						3.42	3.15	2.95	2.25	2.04	1.25	0.95
						206.84	207.11	207.31	208.01	208.22	209.01	209.31
						35						35
						4.01						0.25
						206.25						210.01

Fig. 14.17 Notes for cross sections taken concurrently with profile leveling (metric units).

Cross sections by field methods are applicable on small projects, in areas where the terrain is level, in regions where heavy vegetation inhibits use of photogrammetric methods, and where final cross sections and slope staking are combined.

The use of a direct-elevation rod (see Art. 5.21) in collecting cross-section data yields the required elevation at each data point immediately, thus saving much time in data reduction.

Code 1, ℄ Elevation in col's. 17-24; Code 2, H.I. in col's. 17-24; Offset left is minus, Offset right is plus.

Code	Whole station	Plus	℄ Elevation or H.I.	Offset	Rod	Offset	Rod	Offset	Rod	Offset	Rod
1	405.		620.6	0.	-12.4	-80.	-6.3	-123.	-6.	-210.	-2.2
1	405.		620.6	0.	-12.4	-300.	-0.9	.	.	.	.
1	405.		620.6	0.	-0.6	50.	-4.2	160.	-8.9	250.	-11.2
1	405.		620.6	0.	-0.6	350.	-11.5	.	.	.	.
2	406.		628.9	0.	-12.1	-50.	-4.6	-90.	-0.7	.	.
2	406.		639.7	0.	-11.5	-120.	-7.5	-200.	-4.1	-280.	-1.4
2	406.		618.7	0.	-1.9	60.	-3.2	100.	-4.5	200.	-7.8
2	406.		618.7	0.	-1.9	300.	-9.4	.	.	.	.

Fig. 14.18 Format for field cross sections to be processed by electronic computer.

14.16. Terrain cross sections from topographic maps During the successive stages of highway mapping for route alignment (reconnaissance, project planning, and design), trial locations for proposed highway center lines are plotted on topographic maps for evaluation of grades and earthwork volumes. One method of obtaining cross-section data for evaluating earthwork volumes is to plot the cross-section lines on the topographic map at each full or half-station and all breaks in the terrain. Offset distances from the center line are then scaled from the map and associated elevations are interpolated from the contours. The simplest application of this procedure involves scaling offset distances to contours with an engineer's scale and recording offset distances and elevations as terrain notes.

A portion of a topographic map at a scale of 1 : 480 and with a contour interval of 2 ft is shown in Fig. 14.19 (scale in the figure is reduced). A trial location for a route center line and cross-section lines at full stations are also plotted on the map.

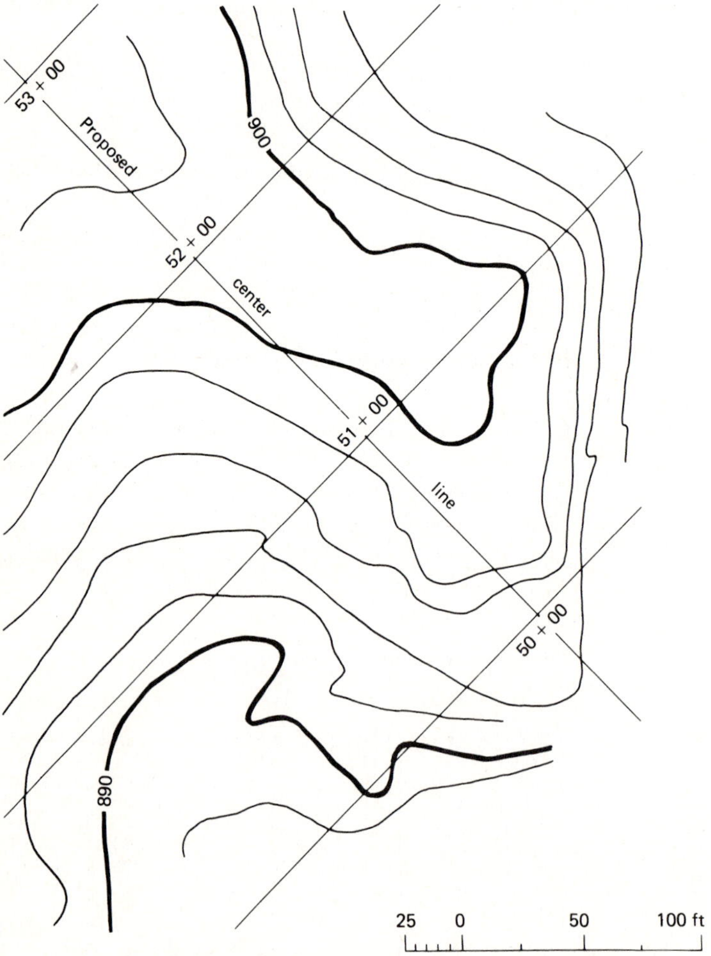

Fig. 14.19 Cross-section data from topographic maps.

150	87	60	35	10	℄	19	91	106	116	128	150
890.8	892.0	894.0	896.0	898.0	898.2	900.0	900.0	898.0	896.0	894.0	892.0

51 + 00

(*a*)

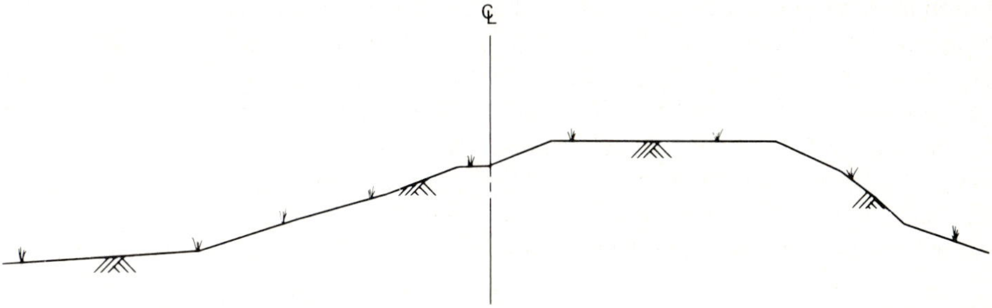

(*b*)

Fig. 14.20 (*a*) and (*b*) Cross-section notes and plotted cross section from a topographic map.

Cross-section data are obtained by scaling from the center line to the contours that intersect the crossline. If the crossline is parallel to the contour, elevations are interpolated at specified intervals left and right of the center line. Cross-section notes determined from station 51 + 00 in Fig. 14.19 are shown in Fig. 14.20*a*. The plotted cross section for station 51 + 00 with a horizontal scale of 1 : 480 and vertical scale of 1 : 12 is illustrated in Fig. 14.20*b* (scale in the figure is reduced). A note form compatible with punching the data on cards for electronic computer processing is shown in Fig. 14.21.

Cross-section data from a topographic map can be observed and recorded automatically by means of an analog-to-digital converter. The basic elements of the system are an electronic unit for storing digital data, a scaling bar with a movable pointer that can be aligned with a cross-section line, a keyboard to permit entering station numbers and elevations into the storage unit, and a card- or tape-punching device or magnetic tape recorder on which digital data in the storage unit can be recorded when a button is pressed.

Terrain Data

Plus distances are right of centerline, minus distances are left of centerline, in the order of increasing stations.

ID Code	Whole station	Pluses	Elev	x-dist	Elev	x-dist	Elev	x-dist	Elev	x-dist
30	51+	.	898.2	0	898.0	-10	896.0	-35	894.0	-60
30	51+	.	892.0	-87	890.8	-150	900.0	+19	900.0	+91
30	51+	.	898.0	+106	896.0	+116	894.0	+128	892.0	+150
30	+	.								

Fig. 14.21 Format for cross-section data taken from a topographic map or photogrammetric stereomodel.

To record data, the operator aligns the scaling bar with the crossline, indexes the bar to zero distance, and enters the station number and center-line elevation into the storage unit using the keyboard. These data are then put on the recording medium by pressing a button. Next, the operator moves the pointer along the scaling bar to a contour or point of desired elevation and introduces this elevation into the storage unit by way of the keyboard. The distance from the center line to the point is automatically cumulated in the storage unit as the pointer is moved along the scaling bar. The operator only needs to indicate the proper sign of + to the right and − to the left of center line and then records the information in the storage unit on the recording medium by pressing the button. This operation is repeated for each point desired on the cross section and for all cross sections along the center line. The procedure is efficient and provides data on any desired input medium and in the necessary format so as to be compatible with the computing equipment available.

14.17. Terrain cross-section data from photogrammetric stereomodels Topographic maps prepared for the planning and design stages of route locations are usually compiled by photogrammetric methods. Consequently, aerial photography of the study area is available almost from the very beginning of an alignment study. In order to compile a topographic map by photogrammetric methods, overlapping aerial photographs are placed in a stereoplotter and are relatively and absolutely oriented according to the methods outlined in Art. 13.15. The stereomodel that results from these operations provides a scaled, three-dimensional model of the photographed terrain. Given the proper instrumentation, it is possible to obtain cross sections directly from the stereomodel. Since fewer operations are involved, cross-section data from the stereomodel are more accurate than those taken from a topographic map compiled from the same stereomodel.

The first step in obtaining cross sections from the stereomodel is to plot the positions of known horizontal control points and the location of the proposed center line on a base sheet using a coordinatograph (Fig. 10.6, Art. 10.10). All center-line stations are also plotted and lines perpendicular to the center line are drawn through each station. The stereomodel is then absolutely oriented with respect to horizontal control points plotted on the base sheet and known vertical control points. At this point, the profile and cross-section data may be obtained by centering the measuring mark in the platen on the tracing table of the stereoplotter over a given center-line station and placing this mark on the apparent surface of the terrain in the stereomodel. The elevation of the terrain is read from the counter on the tracing table and the station and elevation are recorded on a specially prepared form. This procedure is then repeated at each break in the terrain to the left and right of the station along the perpendicular plotted through the center-line station. The offset distance is scaled and offset and elevation are recorded for each point. Figure 14.21 shows cross-section data obtained in this manner. This process is tedious, time-consuming, and requires manual scaling of offset distances. Feasible practical application of this technique requires that the stereoplotter be equipped with a cross-section guide, keying device, and electronic storage unit similar to that described for obtaining data from a topographic map in Art. 14.16.

Figure 14.22 illustrates a stereoplotter with keying device and a digitized tracing table attached to a cross-section guide. This guide, consisting of a channel beam and attachment to the tracing table, can be oriented so that the digitized measuring mark in the tracing table moves in line with a given plotted cross section. First, the operator aligns the guide and measuring mark with a plotted cross-section line. Then the measuring mark is centered over a center-line station and is placed on the apparent terrain surface of the stereomodel. The station number is recorded using the keying device and the digitizing unit is indexed to zero. When the proper button on the keying device is depressed, station number, offset, and terrain elevation are automatically recorded on magnetic tape, punched cards, or punched-paper tape. The operator then moves the floating mark along the crossline to the first break in the terrain, places the floating dot on the apparent terrain, and presses the button again recording elevation and offset distance. This process is repeated for all breaks in terrain and at specified distances from the center line for the entire crossline. The data recorded constitute a cross section of the terrain at the given station. The entire procedure is repeated for all crosslines on the base sheet.

Photogrammetric cross sections have been shown to be more accurate than the usual field methods in rugged, steep terrain where the ground is not concealed by heavy vegetation. Field methods are more accurate and less expensive when the terrain is level.

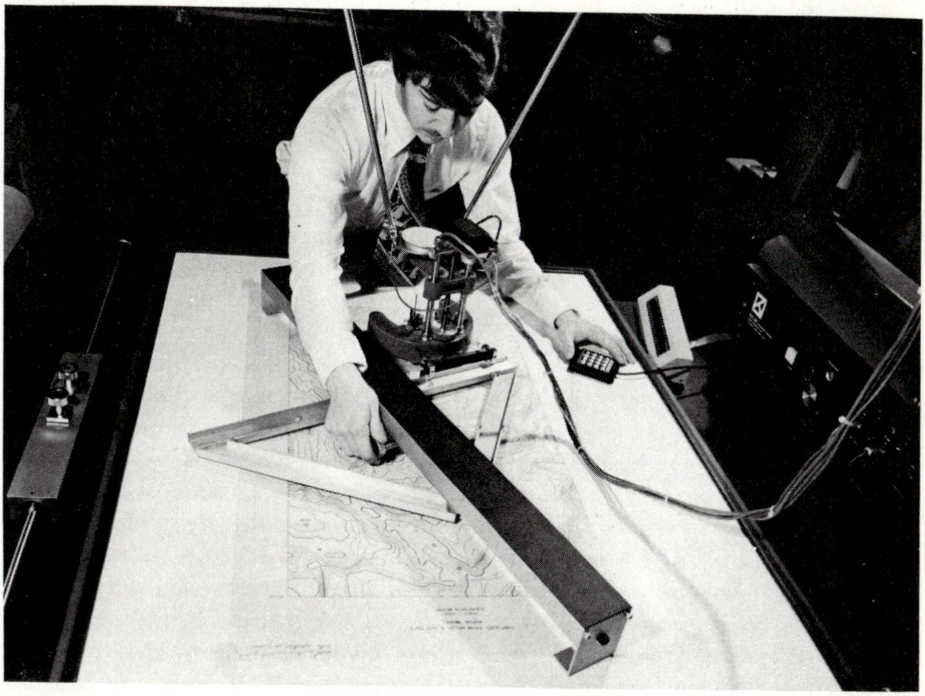

Fig. 14.22 Scaling bar used to obtain cross-section data from stereomodel. (*Courtesy Danko Arlington, Inc.*)

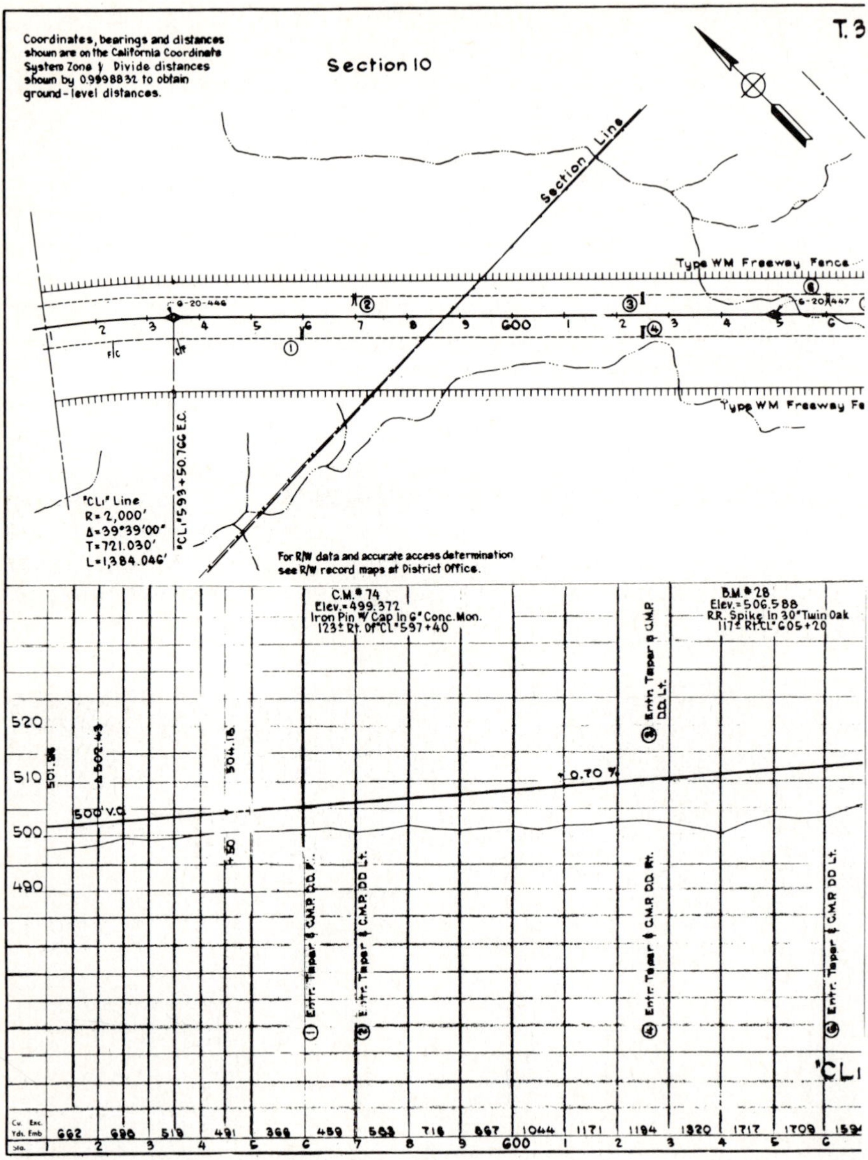

Fig. 14.23 Typical location plan and profile. (*Courtesy California Department of Transportation.*)

14.18. Cross sections and slope stakes Preliminary cross sections are those terrain cross sections obtained by any one of the three methods described in Arts. 14.15 to 14.17, which are then used for studying various alignments and for making preliminary estimates of earthwork.

When a preliminary alignment and grade are being studied, a trial grade is selected and the preliminary grade line is plotted on the plan and profile sheet (Fig.

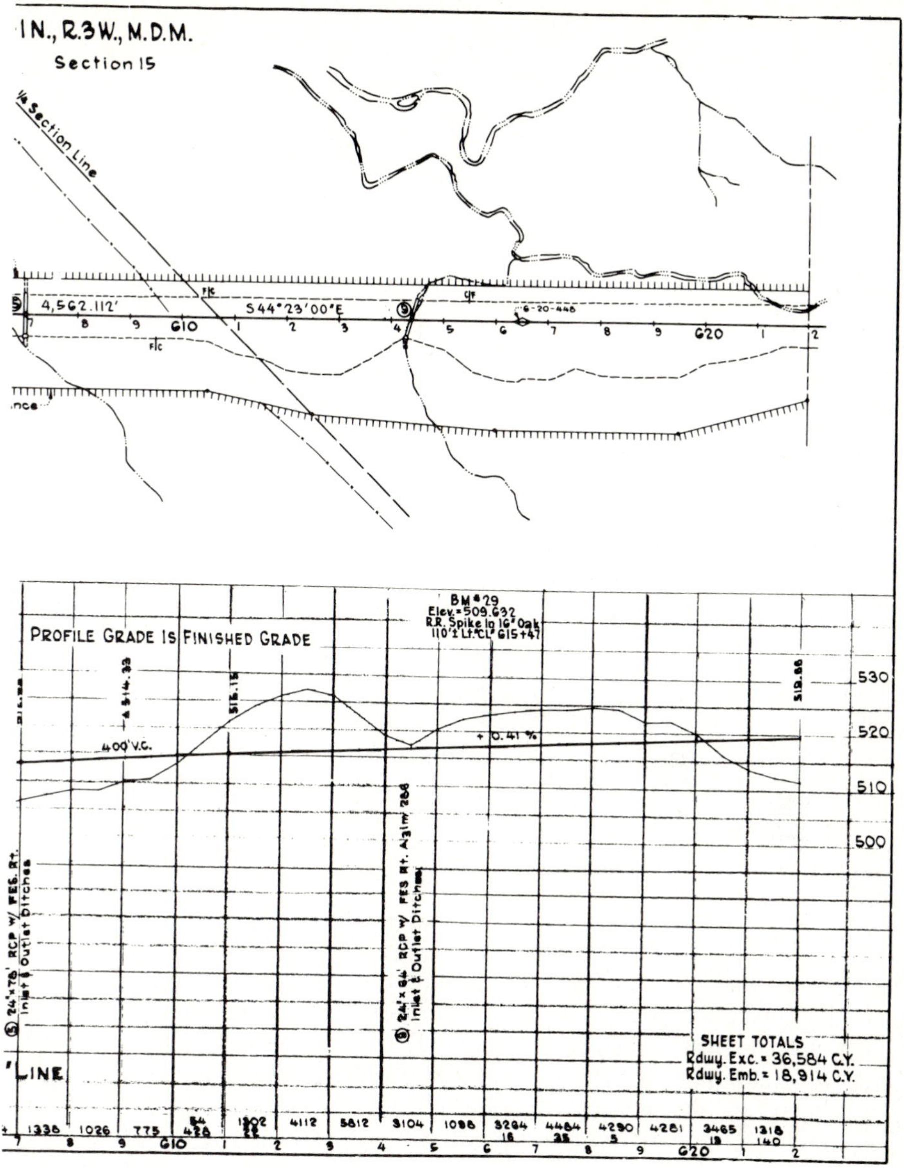

14.23). Details concerning plotting the profile can be found in Art. 5.42. This trial grade is based on maximum permissible rates of grade and elevations at controlling points such as termini, stream, and railway and roadway crossings. A crude effort is also made at this time to balance the amount of cut and fill on the center line. When a grade line is temporarily fixed, grade elevations are computed for all center-line stations and the finished roadbed templets superimposed on the terrain cross

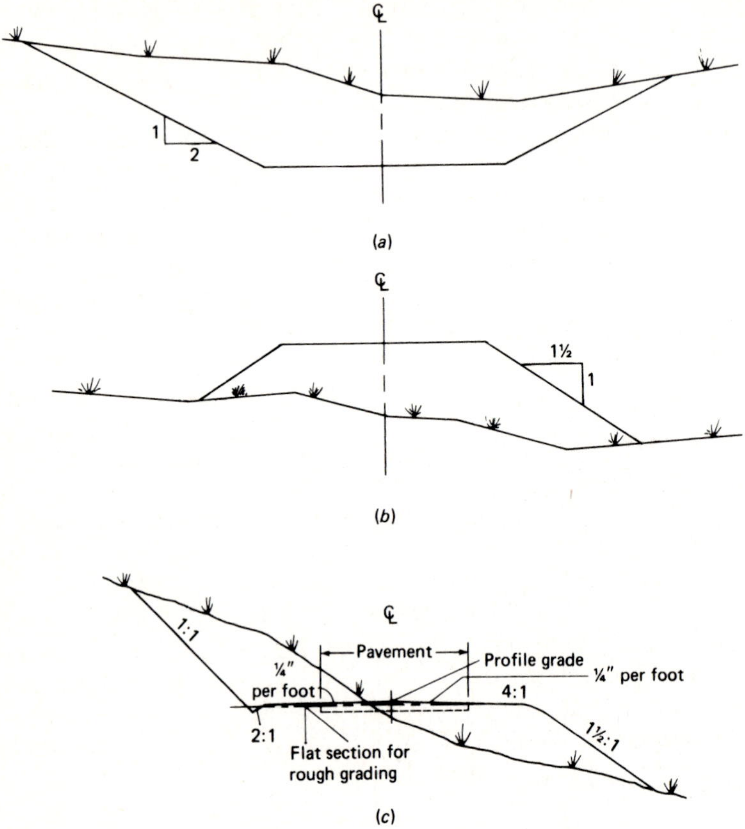

Fig. 14.24 Typical cross sections. (*a*) Cross section in cut. (*b*) Cross section in fill. (*c*) Side-hill cross section.

sections are plotted to a scale of from 1 : 60 to 1 : 120. Figure 14.24 shows typical cross sections in cut, in fill, and a side-hill cross section for a highway.

The roadbed templet consists of a subgrade which is usually a plane surface, transversely level but on highway curves perhaps superelevated, and side slopes at a given rate of slope. On a given road, the subgrade is of one width in cut (Fig. 14.24*a*) with drainage ditches ignored at this stage; of a second and smaller width in fill (Fig. 14.24*b*); and of still a third width where the section is partly in cut and partly in fill (Fig. 14.24*c*). The finished cross section may be sloped in various ways to provide shoulders, drainage, and rounded corners, as illustrated in Fig. 14.24*c*. The rate of the side slope is stated in terms of the number of units measured horizontally to one unit measured vertically; thus a 2 to 1 slope indicates a slope that in a horizontal distance of 2 units rises (or falls) 1 unit. The slope most commonly employed for cuts or fills through ordinary earth is $1\frac{1}{2}$ to 1. For coarse gravel, the slope is often made 1 to 1; for loose rock, $\frac{1}{2}$ to 1; for solid rock, $\frac{1}{4}$ to 1; for soft clay or sand, 2 or 3 to 1.

After study of the tentative profile and the cross-section data, areas of the cross sections are determined for each station according to the methods given in Art. 14.19, and cumulated volumes of earthwork are computed to allow plotting a mass diagram for use in evaluating subsequent earthwork operations.

For major highway alignment projects where earthwork volumes are determined by computation on an electronic computer, cross-sectional areas are not plotted manually. Terrain cross-section data and roadbed templet information, plus alignment and grade parameters, are used as input to a computer program which computes cumulated earthwork volumes, positions for slope stakes and will plot the profile, mass diagram, and the outline of the cross-section areas at specified stations. Computation and analysis of the mass diagram are beyond the scope of this book but are readily available in the references at the end of this chapter.

In either case, earthwork volumes and mass diagram for the tentative alignment are studied and whatever modifications necessary are made to provide the most economic solution and at the same time satisfy all the standards specified for horizontal and vertical alignment for the proposed highway.

Prior to the actual construction, final cross sections are taken and slope stakes marking the intersection of the side slopes with the natural ground surface (also called *catch points*) are set opposite each center-line station.

Final cross-section data, which are used to calculate earthwork quantities for payment, can be determined by field surveys or by photogrammetric methods. Distances to slope stakes can be determined by trial and error in the field concurrently with running the final cross sections, or can be calculated from the photogrammetric cross section. In the latter case, slope stakes are then set in the field during construction layout.

1. Final cross sections and slope stakes by field methods Level readings for final cross sections are generally taken with an engineer's level, and distances to the right or to the left of the center line are measured with a metallic tape, all to tenths of feet or to centimetres. In rough, steep terrain, final cross sections may be taken by measuring slope distances and vertical angles. EDM equipment can be used to advantage in such situations.

Prior to going to the field, the surveyor secures a record of elevations of ground points as obtained from the profile levels, and also the elevation of the established grade at each station. In the field, the instrument is set up in any convenient location, and the H.I. is obtained by a backsight on a bench mark. At each station the rod is held on the ground (sometimes on a short wooden peg driven flush with the ground), a foresight is taken, and the ground elevation is checked against that obtained by profile leveling. Next, the computed cut or fill is marked on the back of the center stake, and a cross line through the station is established as described in Art. 14.15.

Figure 14.25 shows the engineer's level in position for taking rod readings at a section *in cut*. The height of instrument (H.I.) has been determined; the elevation of grade at the particular station is known. The leveler computes the difference between the H.I. and the grade elevation, a difference known as the *grade rod*; that is,

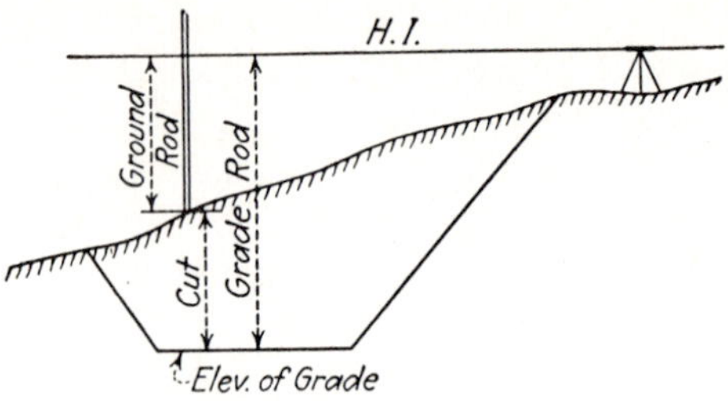

Fig. 14.25 Road cross section in cut.

H.I. – elevation of grade = grade rod. The rod is held at any point for which the cut is desired, and a reading called the *ground rod* is taken. The difference between the grade rod and the ground rod is equal to the cut designated *c*.

Figure 14.26 is a similar illustration for a cross section *in fill*. If the H.I. is *above* grade (as at *A*), the fill is the *difference* between the grade rod and the ground rod; if the H.I. is *below* grade (as at *B*), the fill is the *sum* of the grade rod and the ground rod. Fill is designated by *f*.

If the ground is level in a direction transverse to the center line, the only rod reading necessary is that at the center stake, and the distance to the slope stake can be calculated once the center cut or fill has been determined; such a cross section is called a *level section* (Fig. 14.30*a*). When rod readings are taken at each slope stake in addition to the reading taken at the center, as will normally be done where the ground is sloping, the cross section is called a *three-level section* (Fig. 14.30*b*). When rod readings are taken at the center stake, the slope stakes, and at points on each side of the center at a distance of half the width of the roadbed, the cross section is called a *five-level section* (Fig. 14.30*c*). A cross section for which observations are taken to points between center and slope stakes at irregular intervals is called an *irregular section* (Fig. 14.24*a* or *b*). Where the cross section passes from cut to fill, it is called a *side-hill section* (Fig. 14.24*c*), and an additional observation is made to determine the distance from center to the grade point, that is, the point where the subgrade will intersect the natural ground surface. A peg is usually driven to grade at this point, and its position is indicated by a guard stake marked "grade." In this case, cross sections are also taken at additional plus stations where the center line, the left edge, and the right edge of the roadway pass from cut to fill or from fill to cut, as at stations *C*, *D*, and *F* in Fig. 14.27.

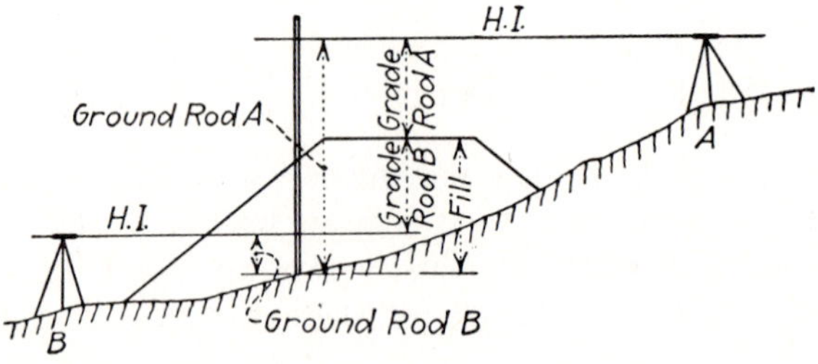

Fig. 14.26 Road cross section in fill.

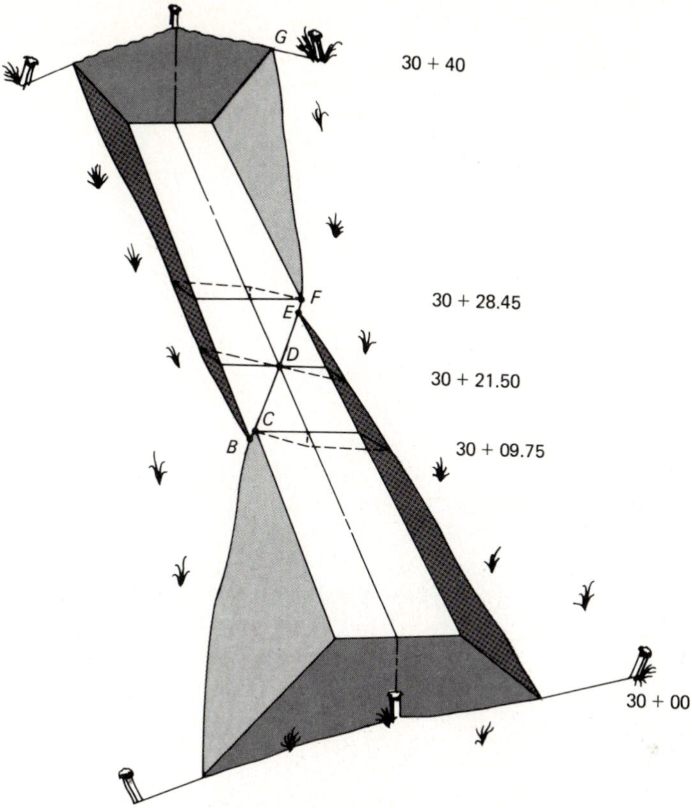

Fig. 14.27 Cross sections in a transition area.

The process of setting the slope stakes requires some additional explanation. If w is the width of roadbed or canal bed, d the measured distance from center line to the slope stake, s the side-slope ratio (ratio of horizontal distance to drop or rise), c the cut (or f the fill) at the slope stake, then from Fig. 14.28, when the slope stake is in the correct position (at C),

$$d = \frac{w}{2} + cs \tag{14.31}$$

and

$$d = \frac{w}{2} + fs \tag{14.32}$$

Setting slope stakes is a trial and error process which progresses rather quickly with an experienced crew.

The following numerical example for a cut illustrates the steps involved in establishing the correct location for a slope stake in the field.

Example 14.4 Let $w = 20$ ft, side slope $s = 1\frac{1}{2}$ to 1, and grade rod = 15.2 ft. Suppose that a slope stake is to be set on the left of the center stake (Fig. 14.28). As a first trial, the rod is held at A; ground rod = 6.6 ft.

c_1 = grade rod − ground rod = 15.2 − 6.6 = 8.6 ft

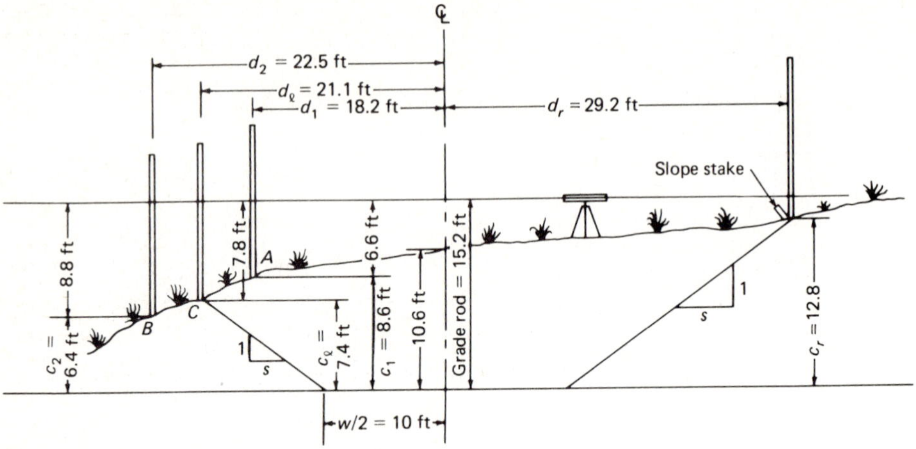

Fig. 14.28 Slope stakes for a cross section in cut.

The computed distance for this value of c_1 by Eq. (14.31) is $w/2 + c_1 s = 10.0 + (8.6)(\frac{3}{2}) = 22.9$ ft. Measurement from the center stake shows d_1 to be 18.2 ft; hence, the rodman should go farther out.

For a second trial, the rod is held at B; ground rod = 8.8 ft; c_2 = grade rod − ground rod = 15.2 − 8.8 = 6.4 ft; $w/2 + (c_2)(s) = 10.0 + 9.6 = 19.6$ ft. The measured value of d_2 is 22.5 ft; hence, the rod is too far out.

Eventually, by trial, the rod will be held at C; ground rod = 7.8 ft; $c = 15.2 - 7.8 = 7.4$ ft. The computed distance for this value of c is $w/2 + cs = 10.0 + (7.4)(\frac{3}{2}) = 21.1$ ft. The measured value of d is also 21.1 ft; hence, this is the correct location for the slope stake. The slope stake on the right, also set by trial and error, is 29.2 ft right and has a cut of 12.8 ft. The notes for this final cross section are recorded as follows, where the symbol c designates a cut:

c7.4	c8.6	c10.6	c12.8
21.1	18.2	0.0	29.2

The reading at 18.2 ft left is recorded because it represents a break in the terrain.

Consider an example of slope stakes for a cross section in fill, as illustrated in Fig. 14.29.

Example 14.5 Let $w = 7$ m; side slopes 2:1, grade elevation at 32 + 00 = 240.36 m; center-line elevation = 239.25 m.

A backsight on a B.M. gives the H.I. of 241.90 m. The rod reading on center line is 2.65 m, which verifies the given center-line elevation. The grade rod = H.I. − grade elevation = 241.90 − 240.36 = 1.54 m. In order to set the right slope stake the rod is held, as a first trial, at point 1, where the ground rod = 3.23. Thus, f_1 = grade rod − ground rod = 1.54 − 3.23 = −1.69 m, where the negative sign indicates a fill. The computed distance for the absolute value of f_1 by Eq. (14.32) is 3.5 + (1.69)(2) = 6.88 m. The measured value for d_1 is 6.2 m, so that the rod must be moved farther from the center line.

A second trial is made at point 2, where the ground rod is 3.62, $f_2 = 1.54 - 3.62 = -2.08$ m, and the computed distance to f_2 is 3.5 + (2.08)(2) = 7.66 m. Since the measured distance d_2 is 8.2 m, the rod is out too far.

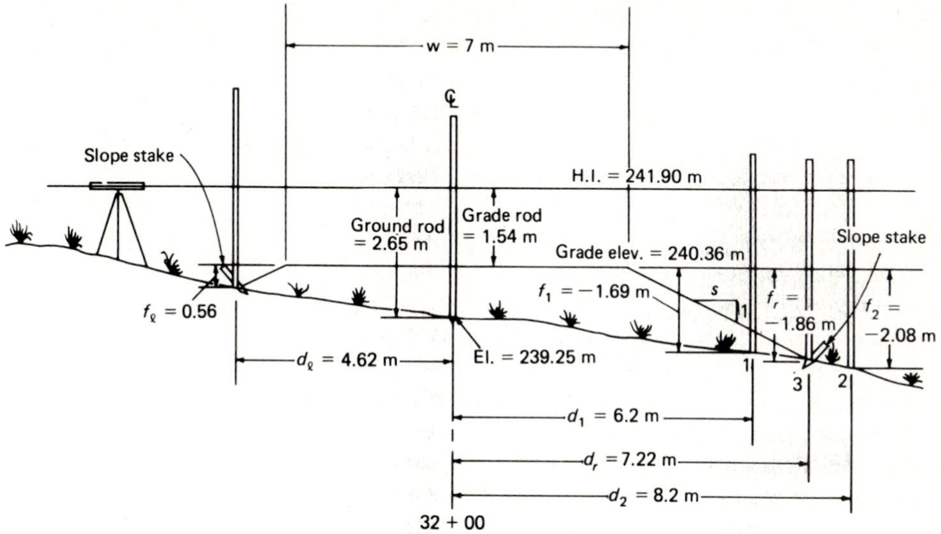

Fig. 14.29 Slope stakes for a cross section in fill.

Eventually, by trial, the rod is held at point 3, where the ground rod = 3.40, $f_r = 1.54 - 3.40 = -1.86$ m, and the computed distance is $3.5 + (1.86)(2) = 7.22$ m. This distance agrees with the measured distance d_r; hence, point 3 is the correct location for the slope stake. The notes for this final cross section are:

H.I. 241.90			
32 + 00	$\frac{f0.56}{4.62}$	$\frac{f1.11}{0.00}$	$\frac{f1.86}{7.22}$

where the left slope stake is also set by trial and error and the symbol *f* designates a fill.

Slope stakes are set with one side to the center line, sloping outward in fill and inward in cut. On the back of the stake the station number is marked with crayon. On the front (side nearest the center line) are marked the cut or fill at the stake and the distance from center line to slope stake. The numbers read down the stake.

Some organizations do not set stakes at the top or bottom of the slope (frequently referred to as the *catch point*) but set a reference point at a fixed distance back from the catch point. This distance is from 10 to 20 ft (3 to 7 m) from the catch point. Figure 14.27 shows this method for stations 30 + 00 and 30 + 40. In this way the stakes are not so likely to be destroyed during roadway construction. Some organizations set stakes at both the catch points and reference points.

2. Final cross sections and slope stakes by photogrammetric methods Final cross sections by photogrammetric methods (Art. 14.17) are feasible when the terrain is steep and fairly free of heavy vegetation or if the right of way has been cleared of vegetation prior to beginning construction. In the latter case, photography is taken specifically for determining final cross sections photogrammetrically. The photogrammetric cross sections are combined with design templet data (roadbed width, side slopes, etc.) for calculating areas of cross sections (Art. 14.19) and volumes of earthwork (Art. 14.20). The computations are performed using an electronic computer program which also provides as output, the data needed for field layout of the slope stakes. These data include roadbed templet notes, side-slope ratios, and distances right and left of center line to the slope stakes.

3. Location of cross sections When measurements are in feet, cross sections are taken at all full or half-stations and at all intermediate stations where significant breaks in the terrain occur. When the metric system is used, stations are generally set at 10- or 20-m intervals (one-tenth and two-tenth station points). If grading is very heavy or rock is present, cross sections should be taken at closer intervals.

Additional cross sections are required when there is a transition from cut to fill on a side-hill location. Figure 14.27 shows a transition from fill to cut from station 30 + 00 to station 30 + 40. Additional cross sections are required at *B*, *C*, *D*, *E*, and *F*. Generally, *B* is so close to *C* and *E* is so close to *F* that these points are omitted. Consequently, three additional cross sections are required: (1) at the fill-base grade point at *C*, (2) at the center-line grade point at *D*, and (3) at the cut-base grade point at *F*.

14.19. Areas of cross sections

14.19. Areas of cross sections Regular cross sections are those for which rod readings are taken at one point on each side of the center line. Level and three-level sections as defined in Art. 14.18 are regular cross sections.

The area of a level cross section (Fig. 14.30*a*) is

$$A = f(w + sf) \tag{14.33}$$

in which f is the center-line fill, w the roadway width, and s the side-slope ratio. If the section is in cut, the center-line cut c is substituted for f in Eq. (14.33).

A three-level section is illustrated in Fig. 14.30*b*. As shown in the figure, this section can be divided into four parts, for which the respective areas are

$$A_1 = \tfrac{1}{2}(f_\ell)\left(\frac{w}{2}\right) \qquad A_2 = \tfrac{1}{2}(f_r)\left(\frac{w}{2}\right) \qquad A_3 = f\left(\frac{d_\ell}{2}\right) \qquad A_4 = f\left(\frac{d_r}{2}\right)$$

in which f_r and f_ℓ are the fills at the right and left catch points, and d_r and d_ℓ are the distances from center line to right and left catch points. Taking the sum of the four areas and collecting terms yields

$$A_{3\ell} = \tfrac{1}{2}f(d_\ell + d_r) + \frac{w}{4}(f_\ell + f_r) \tag{14.34}$$

If the section is in cut, substitute $c = f$, $c_\ell = f_\ell$, and $c_r = f_r$ in Eq. (14.34).

A five-level section is shown in Fig. 14.30*c*. This cross section can be divided into five triangles (where $A_5 = A_6$, so $2A_5 = A_5 + A_6$) as indicated in the figure. The area of each triangle can then be determined as a function of c, $w/2$, c'_ℓ, c'_r, d_ℓ, and d_r. For example, $A_1 = \frac{1}{2}c'_\ell(d_\ell - w/2)$, $A_2 = \frac{1}{2}c'_r(d_r - w/2)$, etc. Taking the sum of these

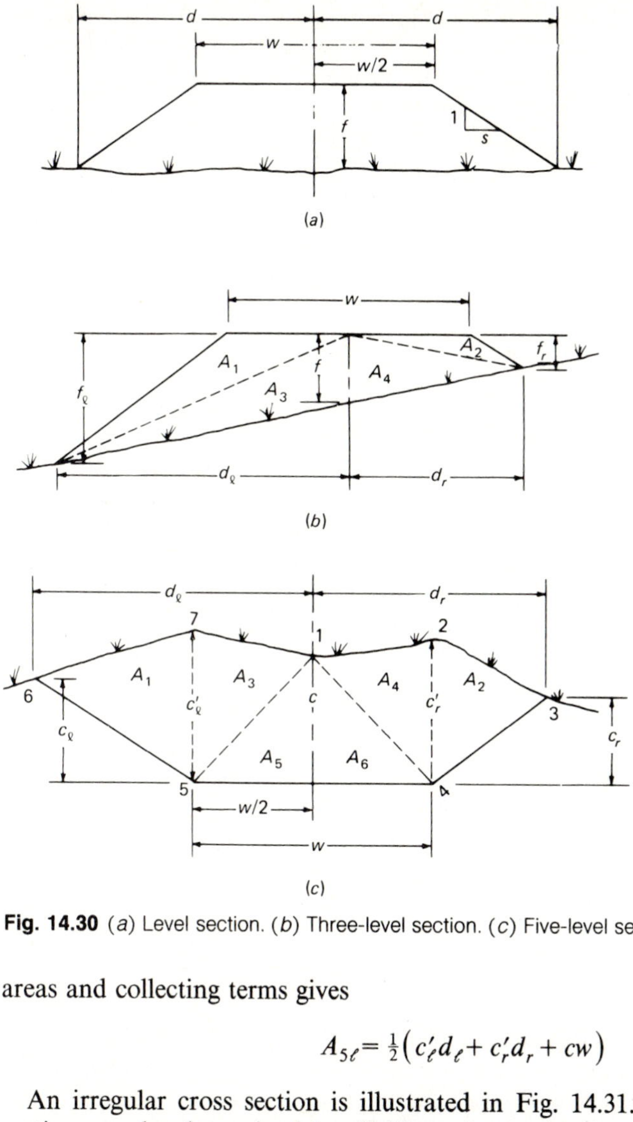

Fig. 14.30 (*a*) Level section. (*b*) Three-level section. (*c*) Five-level section.

areas and collecting terms gives

$$A_{5\ell} = \tfrac{1}{2}\left(c'_\ell d_\ell + c'_r d_r + cw\right) \tag{14.35}$$

An irregular cross section is illustrated in Fig. 14.31. The area of this type of section can be determined by dividing the section into trapezoids and triangles. However, the most efficient procedure is to use the method of coordinates as developed to determine the area within a closed traverse in Eq. (8.16), Art. 8.26. To compute the area of a section by Eq. (8.16), let the center line in the plane of the cross section define the plus Y axis and the finished grade line define the X axis. The coordinates of slope stake positions or catch points, break points, and *hinge points* (hinge point is the angle point between the grade line and side slope) in the cross section are the offset distances from center line (X coordinates) and differences in elevation between the finished grade and the terrain (Y coordinates). Thus, the coordinates of these points can be determined from the cross-section notes (Figs.

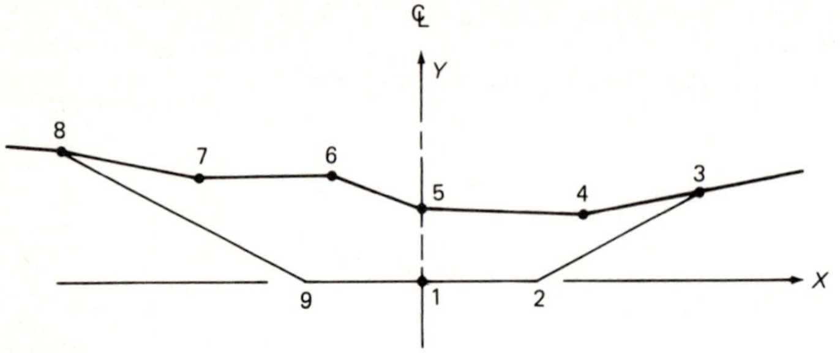

Fig. 14.31 Irregular cross section.

14.17 and 14.20*a*) with due regard for sign of offsets and differences in elevation between roadbed templet and terrain. The area of a cross section having n sides by Eq. (8.16) (Art. 8.26) is

$$A = \tfrac{1}{2}[Y_1(X_2 - X_n) + Y_2(X_3 - X_1) + Y_3(X_4 - X_2) + \cdots + \cdots + Y_{n-1}(X_n - X_{n-2}) + Y_n(X_1 - X_{n-1})]$$

For the cross section shown in Fig. 14.31 with nine sides (sides 1-9 and 2-1 count as two),

$$A = \tfrac{1}{2}[Y_1(X_2 - X_9) + Y_2(X_3 - X_1) + Y_3(X_4 - X_2) + \cdots + \cdots + Y_8(X_9 - X_7) + Y_9(X_1 - X_8)]$$

A convenient way to systematize the calculation of cross-section areas by the coordinate method is to divide the section at the center line, taking the intersection of the grade line and center line (point 1 in Fig. 14.31) as the origin of the coordinate system. Then applying expression (8.19) (Art. 8.26), the area of the right half of the section in Fig. 14.31 is

$$A_R = \frac{1}{2}\left[\frac{Y_1}{X_1} \times \frac{Y_2}{X_2} \times \frac{Y_3}{X_3} \times \cdots \frac{Y_5}{X_5} \times \frac{Y_1}{X_1}\right]$$

where the coordinates are listed for each point with Y in the numerator and X in the denominator starting and ending with the first point and moving in a counterclockwise direction around the loop. Similarly, the area to the left of the center line beginning with point 1 and moving in a clockwise direction around the loop is

$$A_L = \frac{1}{2}\left[\frac{Y_1}{X_1} \times \frac{Y_9}{X_9} \times \frac{Y_8}{X_8} \times \cdots \frac{Y_5}{X_5} \times \frac{Y_1}{X_1}\right]$$

The total area enclosed by the cross section is the sum of the absolute values of A_R and A_L.

Thus, the direction of movement around each loop does not really matter. Using the proper sign for the offset distance ($-$ to the left and $+$ to the right of center line, respectively), the entire cross-sectional area can be found in one operation. Example 14.6 illustrates both techniques.

Example 14.6 Compute the area enclosed by the cross section and roadbed templet of Example 14.5. Use the three-level equation (14.34) and check by the coordinate method. The final cross-section notes are as follows (units are in metres):

$$\frac{f0.56}{4.62} \quad \frac{f1.11}{0.00} \quad \frac{f1.86}{7.22}$$

SOLUTION By Eq. (14.34), the area is

$$A_{3\ell} = \tfrac{1}{2}(1.11)(4.62 + 7.22) + \tfrac{7}{4}(0.56 + 1.86)$$
$$= 10.81 \text{ m}^2$$

Checking by the coordinate method yields

$$A_L = \frac{1}{2}\left[\frac{0.00}{0.00} \quad \frac{1.11}{0.00} \quad \frac{0.56}{4.62} \quad \frac{0.00}{3.50} \quad \frac{0.00}{0.00}\right]$$
$$= \tfrac{1}{2}[(0.00)(0.00) + (1.11)(4.62) + (0.56)(3.50) + (0.00)(0.00)$$
$$-(1.11)(0) - (0.56)(0) - (0)(4.62) - (0)(3.50)]$$
$$= 3.54 \text{ m}^2$$

$$A_R = \frac{1}{2}\left[\frac{0.00}{0.00} \quad \frac{1.11}{0.00} \quad \frac{1.86}{7.22} \quad \frac{0.00}{3.50} \quad \frac{0.00}{0.00}\right]$$
$$= \tfrac{1}{2}[(0)(0) + (1.11)(7.22) + (1.86)(3.50) + (0)(0) - (1.11)(0)$$
$$-(1.86)(0) - (0)(7.22) - (0)(3.50)]$$
$$= 7.26 \text{ m}^2$$

$$A = A_L + A_R = 3.54 + 7.26 = 10.80 \text{ m}^2$$

Calculating the area in one operation yields

$$A = \left|\frac{1}{2}\left[\frac{0.00}{0.00} \quad \frac{0.00}{3.50} \quad \frac{1.86}{7.22} \quad \frac{1.11}{0.00} \quad \frac{0.56}{-4.62} \quad \frac{0.00}{-3.50} \quad \frac{0.00}{0.00}\right]\right| = |-3.54 - 7.26|$$
$$= +10.80 \text{ m}^2$$

14.20. Volumes of earthwork Volumes of earthwork are calculated by a variety of methods, depending upon the nature of the excavation and of the data. If cross sections have been taken along a route, their areas are determined as described in preceding paragraphs, and the volumes of the prismoids between successive cross sections are calculated either by the method of average end areas (Art. 14.21) or by the prismoidal formula (Art. 14.22). Total volumes are expressed in cubic yards or cubic metres.

Because of the repetitive nature of computations for earthwork, calculation by electronic computer is highly desirable.

14.21. Volumes by average end area The common method of determining volumes of excavation along the line of highways, railroads, canals, and similar works is that of *average end areas*. It is assumed that the volume between successive cross sections is the average of their areas multiplied by the distance between them, or

$$V = \frac{\ell}{2}(A_1 + A_2) \tag{14.36}$$

in which V is the volume (cubic feet or cubic metres) of the prismoid of length ℓ (feet or metres) between cross sections having areas (square feet or square metres) A_1 and A_2.

If ℓ is in feet and cross sections at successive stations are in square feet, the volume between the stations in cubic yards, V_y is

$$V_y = \frac{\ell}{54}(A_1 + A_2) \tag{14.37}$$

Formulas (14.36) and (14.37) are valid only when $A_1 = A_2$ but are approximate for $A_1 \neq A_2$. As one of the areas approaches zero, as on running from cut to fill on side-hill work, a maximum error of 50 percent would occur if the formulas were followed literally. In this case, however, the volume is usually calculated as a pyramid; that is,

$$\text{volume} = \tfrac{1}{3}(\text{area of the base cross section})(\text{length})$$

Considering the fact that cross sections are usually a considerable distance apart and that minor inequalities in the surface of the earth between sections are not considered, the method of average end areas is sufficiently precise for ordinary earthwork.

Where heavy cuts or fills occur on sharp curves, the computed volume of earthwork may be corrected for curvature (Art. 14.23), but ordinarily the correction is not large enough to be considered.

14.22. Prismoidal formula It can be shown that the volume of a prismoid is

$$V = \frac{\ell}{6}(A_1 + 4A_m + A_2) \tag{14.38}$$

in which ℓ is the distance between end sections, A_1 and A_2 the areas of the end sections, and A_m the middle area or area halfway between the end sections. A_m is determined by averaging the corresponding linear dimensions of the end sections and *not* by averaging the end areas A_1 and A_2. Equation (14.38) is one application of Simpson's rule (Art. 8.30).

Example 14.7 In Table 14.3 are shown the three-level cross-section notes for two stations 100 ft apart. The width of the roadbed is 20 ft, and the side-slope ratio is $1\frac{1}{2}$ to 1.

Table 14.3 Earthwork Data

	Cross section			Area,	Volume,
Station	L	℄	R	ft^2	yd^3
115	$\frac{c4.0}{16.0}$	$\frac{c6.0}{0}$	$\frac{c12.0}{28.0}$	212	
					575
116	$\frac{c2.0}{13.0}$	$\frac{c3.0}{0}$	$\frac{c8.0}{22.0}$	103	
Midsection	$\frac{c3.0}{14.5}$	$\frac{c4.5}{0}$	$\frac{c10.0}{25.0}$	154	

The volume of earthwork between the two stations is to be computed by the prismoidal formula. Below the regular cross-section notes are shown those for the midsection obtained by averaging the values given for sections at stations 115 and 116. The column headed "Area, ft^2" contains areas of cross sections computed by Eq. (14.34) (Art. 14.19). Then by the prismoidal formula given above,

$$V = \frac{100}{6}[212.0 + (4)(154.0) + 103.0] = 15{,}520 \text{ ft}^3 \text{ or } 575 \text{ yd}^3$$

For the foregoing example, the volume computed by average end areas is 583 yd^3, and the difference between the results obtained by the two methods is 8 yd^3 or about 1.4 percent.

As far as volumes of earthwork are concerned, the use of the prismoidal formula is justified only if cross sections are taken at short intervals, if small surface deviations are observed, and if the areas of successive cross sections differ widely. Usually, it yields smaller values than those computed from average end areas. For excavation under contract, the basis of computation should be understood in advance; otherwise, the contractor will usually claim (and obtain) the benefit of the common method of average end areas.

It can be shown that the difference between the volume computed by the prismoidal formula and that computed by average end areas for the prismoids defined by three-level sections is

$$C_V = \frac{\ell}{12}(c_0 - c_1)(d_0 - d_1) \tag{14.39}$$

where C_V = difference in volume, or correction, for a prismoid ℓ feet or metres in length, in cubic feet or cubic metres

c_0 = center height at one end section, feet or metres

c_1 = center height at the other end section, feet or metres

d_0 = distance between slope stakes at the end section where the center height is c_0, feet or metres

d_1 = distance between slope stakes at the other end section, feet or metres

C_V is know as the *prismoidal correction*; it is *subtracted algebraically* from the volume as determined by the average-end-area method to give the more nearly correct volume as determined by the prismoidal formula.

All units in Eq. (14.39) must be the same. When all distances are in feet, the prismoidal correction in cubic yards is

$$C_{Vy} = \frac{\ell}{324}(c_0 - c_1)(d_0 - d_1) \tag{14.40}$$

Example 14.8 For the prismoid of Example 14.7, the prismoidal correction by Eq. (14.40) is

$$C_{Vy} = \frac{100}{324}(6.0 - 3.0)(44.0 - 35.0) = 8.33 \text{ yd}^3$$

which is consistent with the difference (583 − 575) = 8 yd^3 between the volumes obtained by the average-end-area method and the prismoidal equation, respectively.

14.23. Earthwork curvature correction The average-end-area method for calculating volumes is based on parallel end areas. Volumes computed on curves will have discrepancies if this correction is neglected.

According to a theorem of Pappus, a plane area revolved about an axis generates a volume equal to the product of the revolving area and the path generated by the center of gravity of the revolving area. Let

C_e = curvature correction
= corrected volume − volume by the average-end-area method
R = radius of the highway center line
$\bar{e}$ = eccentricity of the cross-section area

The corrected volume, V_{corr}, is

$$V_{corr} = \frac{\ell}{2}(A_1 + A_2) \pm C_e \tag{14.41}$$

where A_1 and A_2 have respective eccentricities of $\bar{e}_1$ and $\bar{e}_2$, and ℓ is the horizontal center-line distance between stations. Then $V_1 = A_1\ell_1$, where ℓ_1 is the length of arc at $R + \bar{e}_1$, so that by proportion

$$\ell_1 = \ell\left(\frac{R + \bar{e}_1}{R}\right)$$

and

$$V_1 = A_1\left(\frac{R + \bar{e}_1}{R}\right)\ell \tag{14.42a}$$

Similarly,

$$V_2 = A_2\left(\frac{R + \bar{e}_2}{R}\right)\ell \tag{14.42b}$$

Next, assume that

$$V_{corr} = \frac{V_1 + V_2}{2} \tag{14.42c}$$

Substitution of Eqs. (14.42*a*) and (14.42*b*) into Eq. (14.42*c*) and then the result of this manipulation into Eq. (14.41) yields

$$C_e = \frac{\ell}{2R}(\bar{e}_1 A_1 + \bar{e}_2 A_2) \tag{14.43}$$

The curvature correction is positive when the excess area is on the outside of the curve and negative when the excess area is on the inside.

The centroid of the cross section is determined by taking moments about the center line. For example, the three-level section shown in Fig. 14.32*a* is divided into four triangles having respective areas of a_1, a_2, a_3, and a_4, each with a centroid at distances of r_1, r_2, r_3, and r_4 from the center line. Taking moments about the center line of the section yields

$$\bar{e} = \frac{r_1 a_1 + r_2 a_2 + r_3 a_3 + r_4 a_4}{a_1 + a_2 + a_3 + a_4} \tag{14.44}$$

For a three-level section such as that illustrated in Fig. 14.32*a*, the areas of the triangles can be calculated using the cuts or fills and offsets taken from the cross-section notes and the dimensions of the roadbed templet. The respective centers of gravity can be scaled from the cross section plotted at a scale of 1 : 120 (10 ft per 1 in) and recalling that the centroid of any triangle falls at the intersection of its medians.

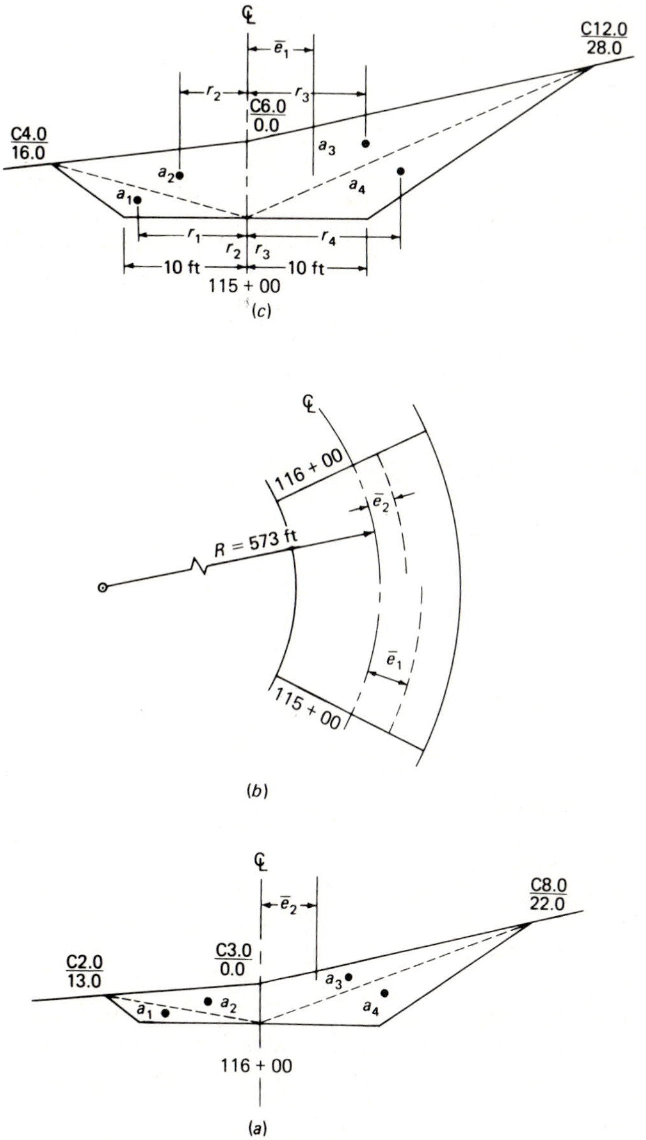

Fig. 14.32 Center of gravity of cross sections. (*a*) 115 + 00. (*b*) Plan view. (*c*) 116 + 00.

Example 14.9 Assume that the center line containing stations 115 and 116 of Example 14.7 is a 10° circular curve (arc definition; see Art. 14.3) which turns to the left. Calculate the curvature correction to the earthwork volume between these two stations.

SOLUTION The cross sections for stations 115 and 116 are plotted in Fig. 14.32. From Fig. 14.32*a* and the cross-section notes of Example 14.7, the centroid for 115 + 00 is found as follows:

$$\left.\begin{array}{ll} a_1 = (\tfrac{1}{2})(10.0)(4.0) = 20\ \text{ft}^2 & r_1 = 8.8\ \text{ft} \\ a_2 = (\tfrac{1}{2})(6.0)(16.0) = 48\ \text{ft}^2 & r_2 = 5.5\ \text{ft} \\ a_3 = (\tfrac{1}{2})(6.0)(28.0) = 84\ \text{ft}^2 & r_3 = 9.9\ \text{ft} \\ a_4 = (\tfrac{1}{2})(10.0)(12.0) = 60\ \text{ft}^2 & r_4 = 12.5\ \text{ft} \end{array}\right\} \text{scaled from cross section}$$

$$\Sigma a_i = 212\ \text{ft}^2$$

$$\bar{e}_1 = \frac{(20)(-8.8) + (48)(-5.5) + (84)(9.9) + (60)(12.5)}{212} = 5.4\ \text{ft}$$

Similarly, for station 116 + 00 (Fig. 14.32*c*),

$$\bar{e}_2 = \frac{-(\tfrac{1}{2})(10)(2.0)(7.8) - (\tfrac{1}{2})(3)(13)(4.4) + (\tfrac{1}{2})(3)(22)(7.6) + (\tfrac{1}{2})(10)(8)(10.2)}{103}$$

$$= 4.8\ \text{ft}$$

From Eq. (14.1*b*) the radius of the center line is

$$R = \frac{5729.58}{10} = 573\ \text{ft}$$

By Eq. (14.43) the curvature correction is

$$C_e = \frac{100}{(2)(573)}[(5.4)(212) + (4.8)(103)] = 143\ \text{ft}^3 = 5\ \text{yd}^3$$

Since the center of gravity falls outside the center line, the correction is positive, so that

$$V = 583\ \text{yd}^3 + 5\ \text{yd}^3 = 588\ \text{yd}^3$$

Note that this correction is only 1 percent of the total. Unless side-hill cuts or fills are very heavy, this correction is frequently neglected.

14.24. Volumes in transitional areas The cross sections required in the transition from cut to fill or fill to cut are discussed in Art. 14.18, section 3, and are illustrated in Fig. 14.27. In that illustration, additional sections are necessary at 30 + 09.75 (point *C*), where the cut runs out; 30 + 21.50 (point *D*), where the grade point is at center line; and 30 + 28.45 (point *F*), where the fill runs out. These cross sections are plotted and shown in Fig. 14.33 together with the sections for 30 + 00 and 30 + 40. Note that these sections are plotted assuming a straight line for the finished grade in cut.

From 30 + 00 to 30 + 09.75, the volume is all fill. From 30 + 09.75 to 30 + 21.50, the volume consists of fill on the right and cut on the left. Note that the volume on

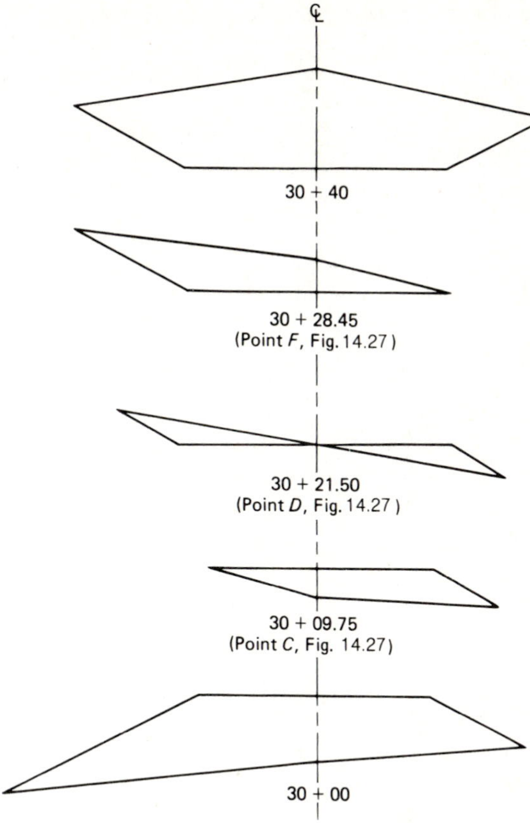

Fig. 14.33 Cross sections in transition from cut to fill.

the left between 30 + 09.75 (where cut runs out) and 30 + 21.50 must be calculated using the equation for volume of a pyramid (Art. 14.21). From 30 + 21.50 to 30 + 28.45, the volume on the left is cut and on the right is fill, which also must be calculated using the equation for a pyramid. In calculating the volumes as pyramids, the bases are the cross-sectional areas at 30 + 21.50 and the altitudes of the pyramids are the distances from 30 + 21.50 to the stations where fill or cut runs out. From 30 + 28.45 to 30 + 40, the volume is all cut.

Example 14.10 Table 14.4 contains cross-section data from which the cross sections of Fig. 14.33 were plotted. Compute the volumes of cut and fill in this transitional area from 30 + 00 to 30 + 40. The units are metres. The roadbed templet is $w = 21$ m in cut and 18 m in fill with 2 : 1 side slopes.

30 + 00: $A = \frac{1}{2}(5.00)(27.20 + 16.20) + \frac{18}{4}(9.10 + 3.60)$

$= 165.65\ \text{m}^2$ fill

30 + 09.75: $A = \frac{1}{2}(2.20)(9.00 + 13.56) + \frac{18}{4}(0 + 2.28)$

$= 35.08\ \text{m}^2$ fill

Table 14.4 Cross-Section Data for Transition Area

Station	Surface elevation	Grade elevation	Left	℄	Right
30 + 00	149.56	154.565	$\frac{f9.10}{27.20}$	$\frac{f5.00}{0.00}$	$\frac{f3.60}{16.20}$
30 + 09.75	152.87	155.073	$\frac{0.00}{9.00}$	$\frac{f2.20}{0.00}$	$\frac{f2.28}{13.56}$
30 + 21.50	155.69	155.686	$\frac{c2.64}{15.78}$	$\frac{0.00}{0.00}$	$\frac{f2.35}{13.70}$
30 + 28.45	158.20	156.048	$\frac{c4.10}{18.70}$	$\frac{c2.15}{0.00}$	$\frac{0.00}{10.50}$
30 + 40	164.10	156.650	$\frac{c4.15}{18.80}$	$\frac{c7.45}{0.00}$	$\frac{c3.84}{18.18}$

30 + 21.50: calculate as triangles, first on the right:

$$A_{\text{right}} = \tfrac{1}{2}(9)(2.35) = 10.58 \text{ m}^2 \qquad \text{fill}$$

$$A_{\text{left}} = (\tfrac{1}{2})(10.5)(2.64) = 13.86 \text{ m}^3 \qquad \text{cut}$$

$$30 + 28.45: \quad A = \tfrac{1}{2}(2.15)(18.70 + 10.50) + \tfrac{21}{4}(4.10 + 0.0) = 52.92 \text{ m}^2 \qquad \text{cut}$$

$$30 + 40: \quad A = \tfrac{1}{2}(7.45)(18.80 + 18.18) + \tfrac{21}{4}(4.15 + 3.84) = 179.70 \text{ m}^2 \qquad \text{cut}$$

These results and the accompanying volumes are listed in Table 14.5.

Volumes are computed as follows: from 30 + 00 to 30 + 09.75 by Eq. (14.36):

$$V = \frac{9.75}{2}(165.65 + 35.08) = 978.6 \text{ m}^3 \qquad \text{fill}$$

From 30 + 09.75 to 30 + 21.50 the volume in fill by Eq. (14.36) is

$$V = \frac{11.75}{2}(35.08 + 10.58) = 268.3 \text{ m}^3 \qquad \text{fill}$$

From 30 + 21.50 to 30 + 28.45 there is a pyramid of fill on the right,

$$V = \tfrac{1}{3}(10.58)(6.95) = 24.5 \text{ m}^3 \qquad \text{fill}$$

Table 14.5 Summary of Areas and Volumes

	Area, m²		Volume, m³	
Station	Cut	Fill	Cut	Fill
30 + 00		165.65		
				978.6
30 + 09.75		35.08		
			54.3	268.3
30 + 21.50	13.86	10.58		
			232.1	24.5
30 + 28.45	52.92			
			1343.4	
30 + 40	179.70			
			1629.8	1271.4

From 30 + 09.75 to 30 + 21.50 there is a pyramid of cut on the left,

$$V = \tfrac{1}{3}(13.86)(11.75) = 54.3 \text{ m}^3 \qquad \text{cut}$$

and the volume of cut is computed by Eq. (14.36) between 30 + 21.50 and 30 + 28.45:

$$V = \frac{6.95}{2}(13.86 + 52.92) = 232.1 \text{ m}^3 \qquad \text{cut}$$

and between 30 + 28.45 and 30 + 40

$$V = \frac{11.55}{2}(52.92 + 179.70) = 1343.4 \text{ m}^3 \qquad \text{cut}$$

The total volumes are 1629.8 m^3 of cut and 1271.4 m^3 of fill.

14.25. Use of digital terrain models (DTM) in route location The digital terrain model [DTM or digital elevation model (DEM)] is described in detail in Art. 13.17. For the purposes of this section, it is sufficient to define the DTM as being a numerical representation of terrain relief and planimetric detail. This numerical representation usually consists of X, Y, Z coordinates in a known system (state plane coordinates and elevations, for example) for a dense array of points within the area to be studied. The data that comprise the DTM are generally recorded on magnetic tape but can be punched on cards or stored in the solid-state memory of an electronic computer.

It is possible to use data from all sources to form the DTM. For example, field survey data (Art. 12.16), data taken from topographic maps (Art. 10.18), and photogrammetrically determined data (Art. 13.17) can be digitized and combined to form a DTM. In this way all existing techniques and information are employed to the fullest extent.

In order to use the DTM for route alignment, it is necessary to acquire or develop electronic computer programs which permit extraction of profiles and cross-section data from the DTM stored on magnetic tape. These extracted data can then be used in the route alignment procedure in the same way as terrain data obtained by traditional methods. Thus, a proposed alignment and many optional locations can be thoroughly studied without access to additional data other than that obtained from the DTM.

The primary requirements for utilization of the DTM in this way are (1) a system for compiling and recording the data used to form the DTM, (2) sufficient ground-control data to permit absolute orientation of the DTM, and (3) a set of electronic computer programs, normally called *software*, which allows the design engineer(s) to extract the information needed for the route location.

At the time of this writing, use of the DTM for route location is still in the experimental stage. However, the potential of the DTM is enormous and extensive use of this source of information for route location and design is inevitable in the near future.

14.26. Highway construction surveys There are two basic approaches to highway construction surveys. These two approaches consist of layout from (1) center-line control, and (2) random control stations. The method chosen depends on terrain configuration, amount of vegetation present, equipment available, construction requirements, and other factors unique to the particular project.

1. Center-line control Usually, just prior to the beginning of construction of a section of highway, the center line is rerun as a base line for construction, missing stakes are replaced, and hubs are referenced. Borrow pits (if necessary) are staked out and cross-sectioned. Lines and grades are staked out for bridges, culverts, and other structures. Slope stakes are set except where clearing is necessary; in that case they are set when the right of way has been cleared. For purposes of clearing, only rough measurements from the center-line stakes are necessary.

The method of setting slope stakes is described in detail in Art. 14.18 for the case where an engineer's level, tape, and rod are used to set the stakes at the catch points. When modern equipment is available, it has become common practice to set slope stakes by measuring slope distances and determining differences in elevation by trigonometric leveling. When this method is employed, the field party is frequently provided with an electronic computer listing which contains roadbed templet notes, slope ratios, and theoretical catch point or slope stake locations.

A typical set of slope stake notes derived from a computer listing for field layout, is shown in Fig. 14.34*a*. The data in the computer listing could have been determined from field notes, topographic map, or from photogrammetric cross sections. In this particular example, the data were derived photogrammetrically. Figure 14.34*b* shows the plotted cross section for station 46 + 00. At this station the grade at center line is 220.48 and the cut at center line is 15.52 ft. The center line of the ditch 35 ft right has a grade elevation of 218.88 ft and the catch point 65.23 ft right has an elevation of 228.95 ft. The slope stake data are given in the column farthest to the right where the $C = (30.23)(0.333) = 10.07$ ft is for a point 30.23 ft to the right of the center line of the ditch (see Fig. 14.34*b*). This catch point is calculated by the program as the intersection of a 3 to 1 slope (0.333) and the terrain surface from the photogrammetric cross section. Similar data are given for the left side of the cross section. The hand-lettered entries in Fig. 14.34*a* are inserted during field operations.

In the field, the center-line station is occupied, the direction of the cross-section line is established, and an offset stake or reference point (R.P.) is set 80.23 ft right of center line or 15 ft from the catch point. In this way the reference point is outside the area of construction and will not be destroyed. The elevation of the reference point is determined trigonometrically as being 227.4 ft. Thus, the catch point is 1.5 ft higher and 15 ft toward the center line from the R.P. (see Fig. 14.34*b*). These data concerning offset distance and elevation of the R.P. and distance and difference in elevation from R.P. are entered on the listing by hand in the field, as shown in Fig. 14.34*a*. The difference in elevation and offset distance between R.P. and catch point $(+1.50/15)$ are also marked on the center-line side of the face of a stake set to guard the R.P. hub. Next, the position of the catch point is set and the elevation is checked against that given on the computer listing. If the elevation of the catch point as determined in the field does not agree with the theoretical location given in the listing, the true position of the catch point is determined by trial according to the

Slope stake left			Catch point left	Center line ditch	Center line	Center line ditch	Catch point right			Slope stake right
	R.P.	R.P.			C15.52			R.P.	R.P.	
C30.16	El. 251.1	-2.0	249.07	218.88	220.48	218.88	228.95	+1.5	El. 227.4	C10.07
@ 90.57										@30.23
s = 0.333	℄ 140.6	15.0	125.57	35.00	0.00	35.00	65.23	15.0	℄ 80.2	s = 0.333
					46 + 00					
					C23.34					
C36.12			254.22	218.06	219.66	218.06	235.52			C17.44
@ 108.47										@ 52.37
s = 0.333			143.47	35.00	0.00	35.0	87.37			s = 0.333
					45 + 50					

(a)

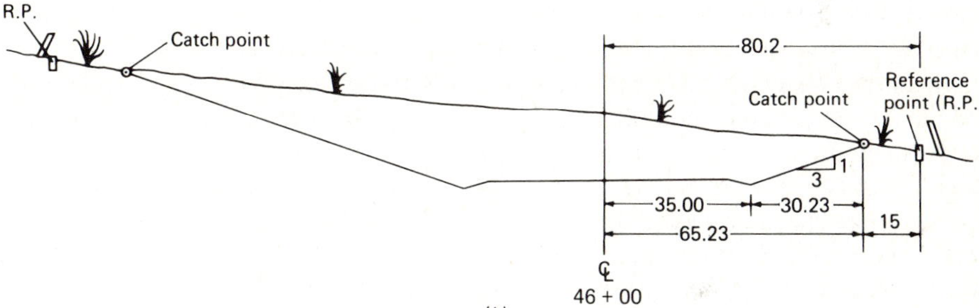

(b)

Fig. 14.34 (a) Listing of slope states derived from computer output for field layout. (b) Plotted cross section for station 46 + 00.

methods given in Art. 14.18 and the reference point and slope stake data are adjusted accordingly. Any changes should be shown directly on the listing.

Where the depth of cuts and fills does not average more than about 3 ft or 1 m, the slope stakes may be omitted; in this case the line and grade for earthwork may be indicated by a line of hubs (with guard stakes) along one side of the road and offset a uniform distance such that they will not be disturbed by the grading operations. Hubs are usually placed on both sides of the road at curves, and may be so placed on tangents; when this is done, measurements for grading purposes may be taken conveniently by sighting across the two hubs or by stretching a line or tape between them.

When rough grading has been completed, a second set of grade stakes is set on both sides of the roadway. It should be understood whether the cuts and fills from the tops of these stakes are to subgrade or to finished grade (final grade). If the slopes of cuts are terraced to provide drainage, finishing stakes are set along the terraces.

In order to give line and grade for the pavement of concrete highways, a line of stakes is set along each side, offset a uniform distance (usually 2 ft or 1 m) from the edge of the pavement. The grade of the top of the pavement, at the edge, is indicated either by the top of the stake or by a nail or line on the side of the stake. The alignment is indicated on one side of the roadway only, by means of a tack in the top of each stake. The distance between stakes in a given line is usually 100 or 50 ft (30 or 10 m) on tangents at uniform grade and half the normal distance on horizontal or vertical curves. The dimensions of the finished subgrade and of the finished pavement are checked by the construction inspector, usually by means of a templet.

2. Random control stations For construction layout from random control stations, three procedures are possible: (*a*) staking directly from random control stations, (*b*) establishing a random line along or near the work area tied into existing random control points, and (*c*) establishing the center line or a line offset from the center line tied to existing random control points.

For each of these procedures, the order of establishing offset stakes for grading is the same as the center-line method described in the first paragraph of this section. It is the method of establishing the positions of the stakes that differs.

As an example, consider staking directly from random control points. The coordinates for all control points, reference points, and catch points are calculated. Inverses between conveniently located random control points and the reference points to the slope stakes are then calculated using an electronic computer program. A listing from the computer giving inverse data from random control point CM 131 to reference points to slope stakes (RPSS) for stations 130 + 00 through 133 + 00 is shown in Fig. 14.35. Given in this listing are the offset distance of the catch point (or slope stake) left or right of the center line, azimuth and zenith angle from the control point to the reference point, slope and horizontal distance from the control point to the reference point in metres and feet, and elevation and coordinates of the reference point. Figure 14.36 shows a plan view of the reference-point layout from control station CM 131 to reference points from slope stakes for stations 131 + 00 through 133 + 00.

Consider as an example the setting of the reference point for station 130 + 00 (140 ft left) using a combined theodolite and EDM. The theodolite is set on station CM 131 and a backsight is taken on station CM 124. An angle of $237°22'34'' - 198°54'19'' = 38°28'15''$ (Fig. 14.35) is turned to the right and a zenith angle of $92°01'30''$ is set on the vertical circle. The reflector is placed on line at the approximate location of the reference point and with the EDM in the tracking mode, the reflector is moved until the EDM registers the slope distance of 734.226 ft (223.793 m) and a hub is set. The measurements are then repeated, including a backsight on CM 124, and the distance and elevation are determined to the hub just set. The field party should also have the slope stake listing (Fig. 14.37) so that the elevation of the catch point can be checked. The catch point is set and checked from the reference point using a tape, hand level, and rod. Note that data for the reference-point location is entered on the computer listing (Fig. 14.37) for station 130 + 00 on the left as RP El. = 1203.0@140 ℄. The catch point is set by measuring 15 ft toward the center line and is found to be −8.7 ft lower than the reference

13 OG 01 01 DATE 04/17/80 PAGE 2

TRANSIT AT	PT	ELEVATION	NORTH COORD	EAST COORD
CM 131	02	1,228.944	550,190.755	1,999,254.589

DESCRIPTION (POINT SIGHTED)	PT	AZIMUTH (NORTH)	ZENITH ANGLE	SLOPE DISTANCE (METERS)	SLOPE DISTANCE (FEET)	HORIZONTAL DISTANCE (METERS)	HORIZONTAL DISTANCE (FEET)	ELEVATION	NORTH COORD	EAST COORD
CM 124	01	198 54 19	90 40 51	288.265	945.751	288.245	945.684	1,217.708	549,296.085	1,998,948.184
CM 134	03	333 15 37	92 04 50	221.642	727.169	221.496	726.690	1,202.545	550,839.733	1,998,927.624
130+00 RPSS 140 LT	04	237 22 34	92 01 30	223.793	734.226	223.653	733.767	1,203.000	549,795.165	1,998,636.590
130+00 RPSS 373 RT	05	213 56 58	126 25 28	95.196	312.321	76.598	251.306	1,043.500	549,982.289	1,999,114.244
130+50 RPSS 145 LT	06	240 39 01	91 59 17	223.551	733.433	223.416	732.992	1,203.500	549,831.487	1,998,615.681
130+50 RPSS 368 RT	07	230 05 53	127 31 15	86.250	282.971	68.407	224.433	1,056.600	550,046.787	1,999,082.417
131+00 RPSS 152 LT	08	243 52 27	92 09 55	225.427	739.589	225.266	739.061	1,201.000	549,865.313	1,998,591.039
131+00 RPSS 370 RT	09	248 41 01	127 45 20	84.100	275.919	66.492	218.149	1,060.000	550,111.454	1,999,051.364
131+50 RPSS 154 LT	10	247 00 46	92 35 38	227.253	745.577	227.020	744.813	1,195.200	549,899.885	1,998,568.920
131+50 RPSS 372 RT	11	266 08 25	125 28 43	88.985	291.944	72.463	237.739	1,059.500	550,174.752	1,999,017.389
132+00 RPSS 133 LT	12	250 30 10	91 23 13	223.446	733.088	223.380	732.873	1,211.200	549,946.149	1,998,563.741
132+00 RPSS 359 RT	13	277 17 02	118 11 00	100.124	328.490	88.253	289.544	1,073.800	550,227.465	1,998,967.381
132+50 RPSS 107 LT	14	254 21 59	89 45 27	220.150	722.274	220.148	722.267	1,232.000	549,996.116	1,998,559.042
132+50 RPSS 365 RT	15	286 42 46	115 37 57	114.667	376.201	103.382	339.178	1,066.200	550,288.294	1,998,929.738
133+00 RPSS 82 LT	16	258 24 48	88 08 56	219.440	719.944	219.325	719.568	1,252.200	550,046.229	1,998,549.684
133+00 RPSS 353 RT	17	290 49 05	111 19 15	132.909	436.052	123.813	406.208	1,070.400	550,335.122	1,998,874.901

ZENITH ANGLES ARE CALCULATED TO A ROD READING = (HI + 0.000 FEET)

TRANSIT AT CM 131 POINT 02 PAGE 2

Fig. 14.35 Computer listing for direction and distances from a random control station to reference points to slope stakes. (*Courtesy California Department of Transportation.*)

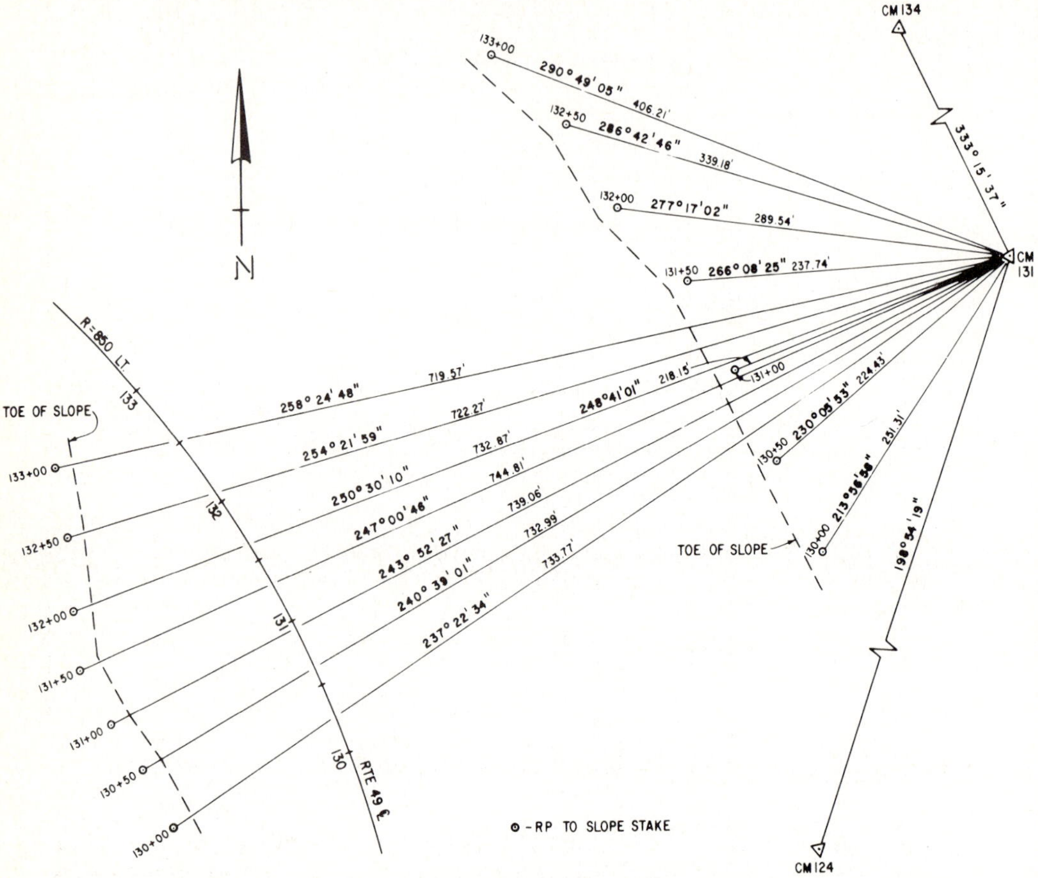

Fig. 14.36 Slope staking from random control points. (*Courtesy California Department of Transportation.*)

point. The field elevation of the catch point is 1203.0 − 8.7 = 1194.3, which checks the precomputed elevation for the catch point also given in the listing. As a rule for earthwork operations, this discrepancy should not exceed 5 ft or about 2 m. Reference points to slope stakes can be checked from a second instrument setup, at random control point CM 134 (Fig. 14.36), for example. However, if the elevations of reference and catch points as determined in the field check the theoretical values in the computer listings (Figs. 14.35 and 14.37, respectively), these checks are assumed sufficient and a second setup is considered unnecessary.

Referring to Fig. 14.37, at station 130 + 50 the field measurements do not check the elevations and offsets in the listing (see the voided entries on the right). Consequently, the correct position of the catch point is determined in the field and adjusted values are entered in the listing as shown.

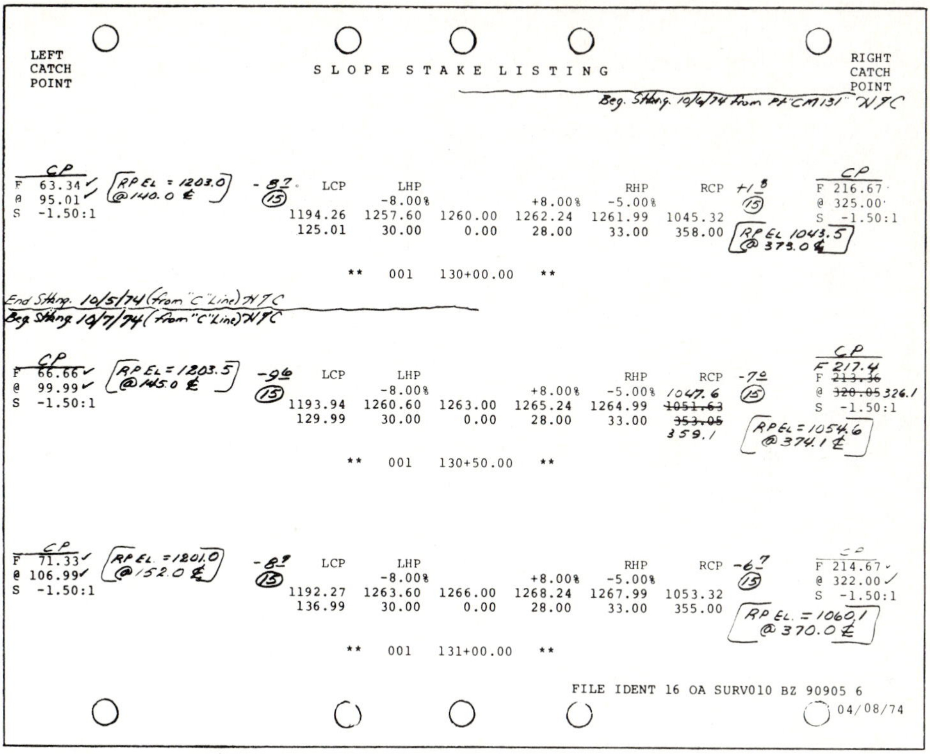

Fig. 14.37 Slope stake listing for field layout. (*Courtesy California Department of Transportation.*)

When a random line is established near the working area, it may be a line connecting two random control points or may be a series of supplementary points established between existing random control points. These supplementary points are used as angle points on the random line. The random line should be located so as to be as close as possible to the points to be set. The relationship between the center line and the random line must be determined and then all reference points to slope stakes are set from the stations on the random line. Figure 14.38 illustrates an example of a random line established to set reference points to slope stakes for a portion of curved center line for stations 129 + 00 through 134 + 00. As in the case of staking directly from random points, this procedure is simplified by electronic computer calculations to determine the distances and directions from the random line to the reference line. Slope stake listings and offset listings are used in the field to assist in the setting of the points. The method is adaptable to the use of combined theodolite EDM instruments and in general involves measuring shorter distances than when staking directly from the control stations.

When excavation is heavy (cuts greater than 30 ft or 10 m), intermediate slope stakes may be required after grading has been partially completed. These stakes can be set from random control points but are more conveniently established from a random offset line or line of constant offset (offset line parallel to the center line).

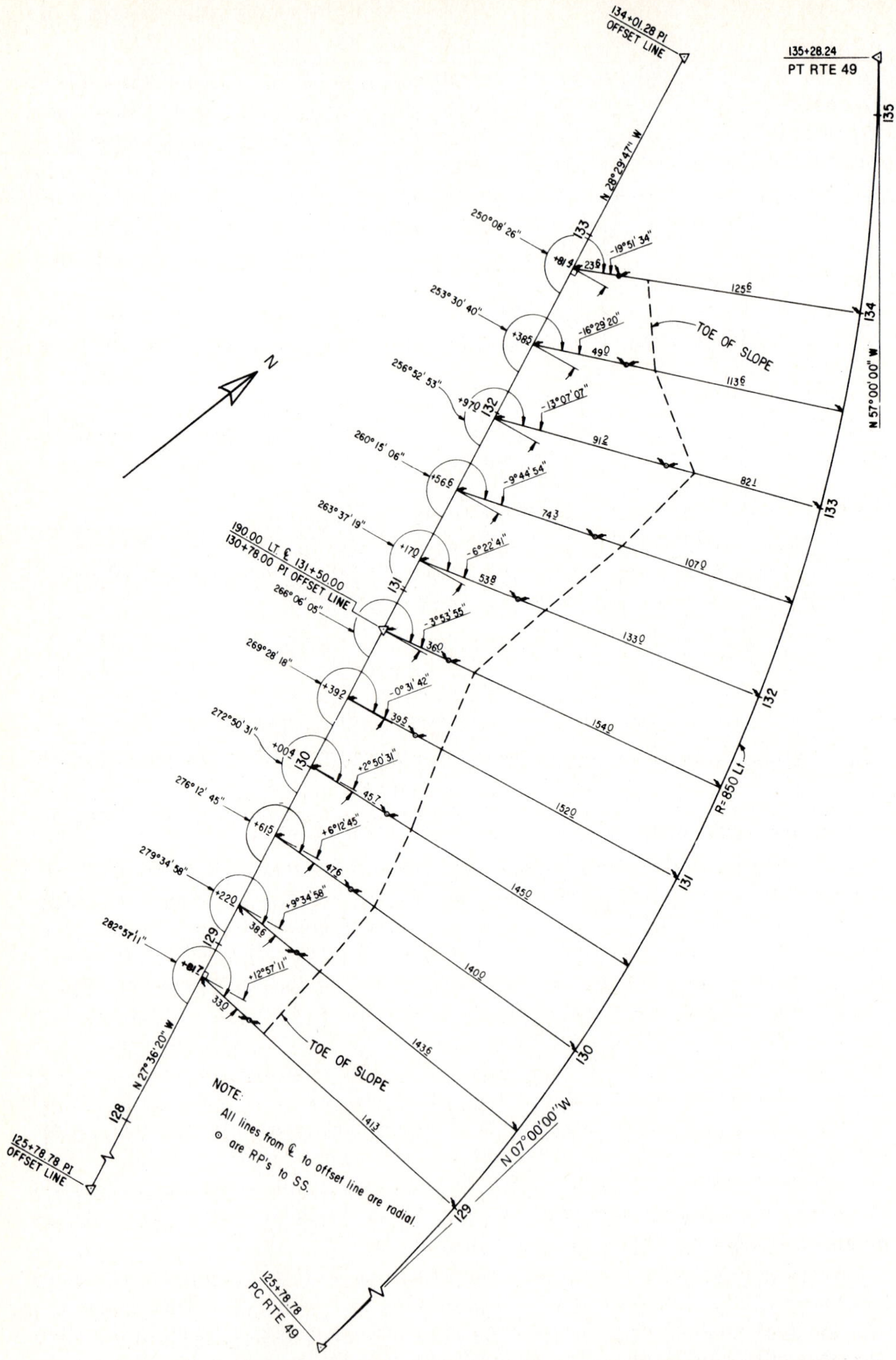

Fig. 14.38 Slope staking from a random offset line. (*Courtesy California Department of Transportation.*)

The center line of the alignment is established from ties to random control points when rough excavation is completed. The balance of the layout work then becomes a combination of staking from random control points and staking from the center line. Prior to setting final grade stakes, bench-mark levels are run throughout the graded alignment and bench marks are set to provide vertical control for all subsequent operations. These bench marks should be established so as to satisfy third-order specifications (Table 5.2, Art. 5.39) and should have a minimum density of three per mile (two per kilometre).

Final grade stakes for subgrade excavation and paving are usually set from the center line. However, grade stakes in the vicinity of an interchange may be more efficiently set from random control points.

As construction proceeds, monthly estimates are made of the work completed to date. A quantity survey is made near the close of each month, and the volumes of earthwork, etc. are classified and summarized as a basis for payment.

14.27. Streets For street construction the procedure of surveying is similar to that just described for highways. Ordinarily, the curb is built first. The line and grade for the top of each curb are indicated by hubs driven just outside the curb line, usually at 50-ft (15-m) intervals.

On alignments containing horizontal and vertical curves, grade stakes should be set at closer intervals commensurate with the curvature of the center line and grade line. Differential levels are run obtaining an elevation on the top of each hub. Cuts or fills are calculated from the top of each hub to finished grade at the top of the curb. These cuts and fills are marked on guard stakes set next to each hub and are also tabulated in the field notebook. Frequently, a separate summary of cuts and fills is prepared for the construction superintendent. A copy of this summary should be filed by the surveyor.

When the curb is completed, levels should be run over the finished curb to check the accuracy of the construction. The grade for the edge of the pavement is then marked on the face of the completed curb; or for a combined curb and gutter it is indicated by the completed gutter. Hubs are set on the center line of the pavement, either at the grade of the finished subgrade (in which case holes are dug when necessary to place hubs below the ground surface) or with the cut or fill indicated on the hub or on an adjacent stake. Where the street is wide, an intermediate row of hubs may be set between center line and each curb. It is usually necessary to reset the hubs after the street is graded. Where driving stakes is impractical because of hard or paved ground, nails or spikes may be driven or marks may be cut or painted on the surface.

The surveys for street location and construction should determine the location of all surface and underground utilities that may affect the project; and notification of necessary changes should be given well in advance.

On paving projects where the rate of grade is constant, the possibility of using a laser level and rod such as are illustrated in Figs. 14.47 and 14.46 should be investigated. The use of laser levels for setting grade stakes is discussed in Art. 14.30. There must be a bench mark within the work area and care must be exercised to avoid sights of excessive lengths. The laser level plane is accurate to within ± 0.03 ft or about 10 mm in a sight of 330 ft (100 m).

The automatic laser guidance equipment described in Art. 14.30 can be used to guide graders in establishing subgrade elevations and paving machines utilized for placing concrete to finished grade elevations.

14.28. Railroads Construction surveys for railroads are similar to those described in Art. 14.26 for highways. Prior to construction, the located center line is rerun, missing stakes are replaced, control station hubs are referenced, borrow pits are staked out, slope stakes are set, and lines and grades for structures are established on the ground. When rough grading is completed, final grade stakes are set to grade at the outer edges of the roadbed, as a guide in trimming the slopes.

The foregoing operations can also be performed from random control points or using a random offset line as described in Art. 14.26.

When the roadbed has been graded, alignment is established precisely by setting tacked stakes along the center line at full stations on tangents and usually at fractional stations on horizontal and vertical curves. Curves are staked out at this time. An additional line of hubs is set on one side of the track and perhaps 3 ft (1 m) from the proposed line of the rail, with the top of the hub usually at the elevation of the top of the rail. Track is usually laid on the subgrade and is lifted into position after the ballast has been placed and compacted around the ties.

14.29. Line and grade Temporary stakes or other markers are usually set at the corners or edges of the proposed structure, as a rough guide for beginning the excavation. Outside the limits of excavation or probable disturbance but close enough to be convenient, are set permanent stations which are established with the precision required for the measurement of the structure itself. These permanent stations should be well referenced (Art. 8.14), with the reference stakes in such number and in such position that the loss of one or two will not invalidate any portion of the location survey. Permanent targets or marks called *foresights* may be erected as convenient means of orienting the transit on the principal lines of the structure and for sighting along these lines by eye.

Stakes or other markers are set on all important lines in order to mark clearly the limits of the work. The number of such markers should be sufficient to avoid the necessity for many measurements by the workmen but should not be so great as to cause confusion. A simple and uniform system of designating the various points, satisfactory to the construction superintendent, should be adopted. Also, the exact points, lines, and planes from which and to which measurements are to be made should be well understood.

Grade elevations are referenced to a system of bench marks established near the structure in convenient locations that will probably not be subject to disturbance, either of the bench mark itself or of the supporting ground. From time to time these bench marks should be checked against one another to detect any disturbance. Every care should be taken to preserve existing bench marks of state and federal surveys; if construction necessitates the removal of such marks, the proper organization should be notified and the marks transferred in accordance with its instructions.

The various grades and elevations are defined on the ground by means of stakes as a guide to the workers. The grade stakes may or may not be the same as the stakes used in giving line. When stakes are used, the vertical measurements may be taken from the top of the stake, from a crayon mark or a nail on the side of the stake, or (for excavation) from the ground surface at the stake; in order to avoid mistakes, only one of these bases for measurement should be used for a given kind of work, and the basis should be made clear at the beginning of construction.

On some jobs, grade stakes consisting of hubs driven flush with the ground are set, elevations are obtained on the tops of the hubs, and cuts or fills to finished grade are calculated in whole and fractions of units from the top of the hub. These hubs are also referred to as grade stakes and are usually guarded by a flat stake.

In many cases, line and grade are given more conveniently by means of *batter boards* than by means of stakes. A batter board is a board (usually 1 by 6 in) nailed to two substantial posts (usually 2 by 4 in) with the board horizontal and its top edge preferably either at grade or at some whole number of feet above or below grade. The alignment is fixed by a nail driven in the top edge of the board. Between two such batter boards a stout cord or wire is stretched to define the line and grade. Batter boards for sewer construction are shown in Fig. 14.39. Note that the string

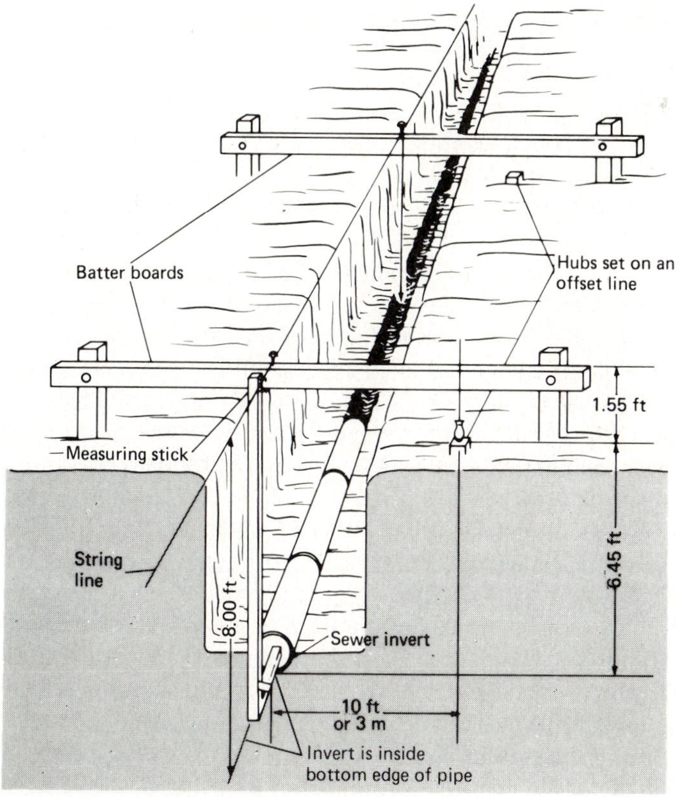

Fig. 14.39 Batter boards for sewer construction.

line provides the reference line for alignment and grade between the surveyed stations.

When batter boards are used, the vertical measurements are taken from the top edge of the board, which is horizontal. The stake or the batter board may be set either at grade or at a fixed whole number of feet (metres) above or below grade.

14.30. Alignment and grade by electronic surveying systems When construction layout is performed using transit or theodolite and tape, alignment is usually established in the field by direct measurement from the highway center line, the exterior dimensions of a structure, or from the property line. This means that center line, side line, or property line must be established prior to setting offset stakes or batter boards. This procedure has definite advantages in that it provides a check on office calculations and furnishes proof that the structure as designed is consistent with basic control points in the field. Also, the process for setting offset stakes or batter boards from center line or building corners is straightforward, involves right-angle offsets or parallel lines, and provides ample visual checks for the validity of the layout. The disadvantages are that the procedure is time-consuming, expensive, and during initial stages of construction (e.g., rough excavation) most of the points set will be lost and will have to be replaced for the final layout.

Use of electronic surveying systems with direct readout of distance and a tracking mode (Art. 9.14) in conjunction with coordinate geometry make construction layout from randomly located control points feasible. In order to perform this operation, horizontal and vertical control points are positioned at locations appropriate for layout. These locations can be outside the area of construction, but must provide good visibility for construction layout. All random control points must be coordinated and be part of a closed survey network. It is also advantageous to have elevations on these control points. The accuracy maintained is a function of the type of work being done.

The points to be staked must also be coordinated. Thus, if stakes on right-angle offsets from the corner of a building or offsets from the catch point in slope staking (Art. 14.18) are to be set, the offset point is coordinated. Next, the inverse solution between the random control point and coordinated offset point is computed to yield distance and direction (azimuth or bearing). Most surveying organizations have standard electronic computer programs for this computation, given the coordinates for both ends of the line. The minimum amount of data provided for field operations consists of azimuth or bearing and horizontal distance from the random control point to the offset point. When offset stakes are to be set for alignment and grade (as for slope stakes in highway excavation), azimuth or vertical angle, slope distance, horizontal distance, and elevation of the offset point are given. Data are usually precalculated and furnished to field personnel in the form of a hard-copy listing.

In the field, the operation consists of occupying the designated random control point, taking a backsight on an intervisible random control station, turning off the direction, and measuring the distance to set the specified offset stake. Ideally, a series of offset stakes are set from a single well-located random control point. To verify calculations and field work, the locations of the stakes as set are checked by measuring distance and direction from another random control point or by measurements between offset stakes on the site.

Obviously, the locations of random control stations must be chosen carefully so as to achieve maximum benefits from this method. In preparation for layout from random stations, enough inverse data must be generated from several control stations so as to provide for checking and to guard against lack of visibility along certain lines. In case insufficient inverses are generated to complete the job, ordinary or programmable pocket calculators provide enough computing power to allow calculation of the necessary inverses on the site.

The choice as to whether construction layout should be performed in the traditional manner or from random controlling points depends on many factors related to characteristics of the job and site. These factors are generally unique for every project. In all probability a combination of both methods will be used on most projects. For example, in highway construction, slope stakes may be set from random control points while the final grade stakes are established from the center line, which of course is set from random control points. Examples of setting slope stakes from random control points were given in Art. 14.26.

14.31. Laser leveling and alignment equipment for construction layout As noted in Art. 4.29, laser beams are used in certain EDM instruments as the carrier beam for modulated light in distance measurement, thus increasing the range of these instruments substantially. Laser light is of a single color or *monochromatic*, is *coherent* (i.e., the light waves are in step with each other), and is highly collimated, so that the beam spreads only very slightly as the distance from generator to target increases. These characteristics make the laser a useful device for surveying instruments to be utilized in various types of construction layout.

As an example of the use of lasers in surveying instruments, consider application of a helium neon laser or gas laser to an ordinary theodolite or transit (Ref. 26). The laser is a separate unit equipped with a portable power supply. This laser can be strapped to the leg of the tripod. The parallel rays of the beam emitted by the laser are focused by an objective lens down to a very small diameter (0.08 mm) for transmission by a flexible fiber optic cable to the telescope of the theodolite equipped with a special eyepiece attachment. Within the telescope the light is focused on a reticle which masks out a cross. The laser beam then passes through a beam-splitting cube which directs the laser along the optical axis of the telescope. The surface of the beam-splitting cube, visible through the eyepiece, contains a second reticle, designed to permit the telescope to be used as usual under good lighting conditions. A filter is installed to absorb any laser light directed toward the eyepiece. The beam directed toward the telescope objective is a narrow beam containing a projected cross which may be brought to a focus on a target by focusing the telescope on the target.

Figure 14.40 shows a theodolite equipped with a gas laser, power supply, and eyepiece attachment. Figure 14.41 is a schematic drawing which illustrates the way in which the laser is projected into the telescope using the attachment shown in Fig. 14.40.

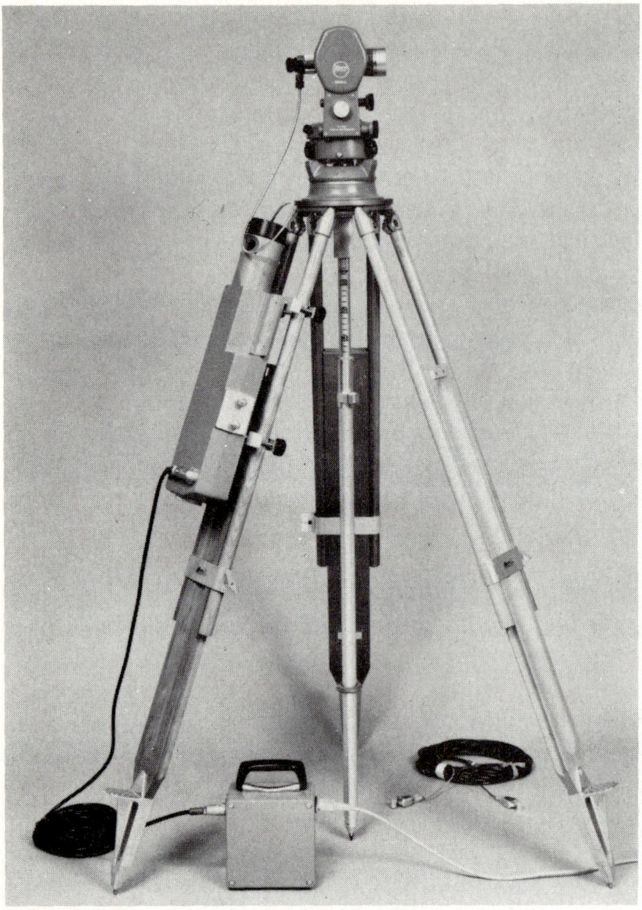

Fig. 14.40 Kern DKM 2 theodolite equipped with a gas laser. (*Courtesy Kern Instruments, Inc.*)

Laser eyepieces, now manufactured as accessories, can also be attached to levels and the laser plummet. A laser plummet which projects a vertical laser beam up and down is illustrated in Fig. 14.42.

The system just described and shown in Fig. 14.40 is a *single-beam* laser that projects a visible "string line" which can be seen on targets under all lighting conditions. Thus, if a sight is taken on a distant point in the conventional manner, alignment can be achieved at any intermediate point by moving a special target about until the projected cross is centered on the target. This alignment can be established at any intermediate point without instructions from an operator behind the instrument.

The laser beam can be projected at any inclination or may be used as a level beam of light. For leveling, the graduated rod would have a movable target equipped with a vernier that is moved up or down until the beam or projected cross is centered on the target. The operator of the rod would then read the rod to the least count of the

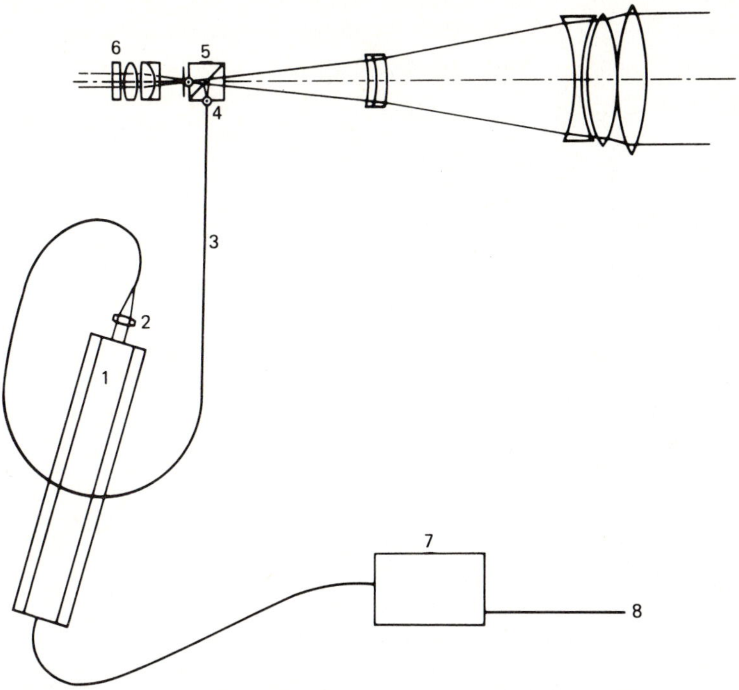

Fig. 14.41 Schematic of the transmission of laser light to the theodolite: (1) laser light source; (2) objective; (3) light tube (glass fiber optics); (4) reticle; (5) light-splitting cube; (6) filter; (7) laser power converter; and (8) power source. (*Courtesy Kern Instruments, Inc.*)

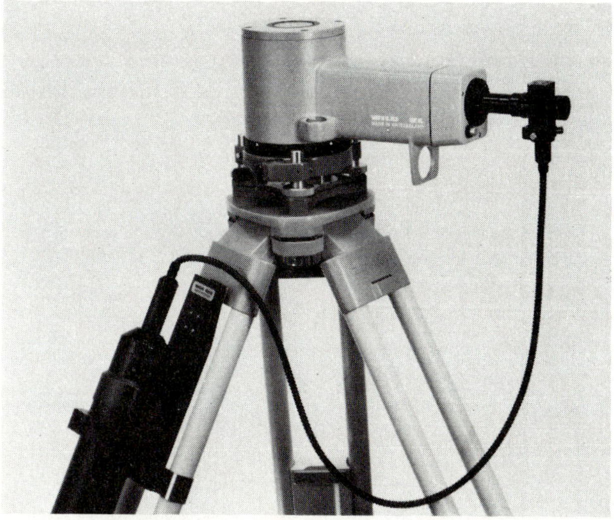

Fig. 14.42 Laser plummet, Wild ZL with eyepiece attachment GLO2. (*Courtesy Wild Heerbrugg Instruments, Inc.*)

scale. The range of the laser attachment shown in Fig. 14.40 is from 10 to 400 m (30 to 1300 ft) with a maximum range of 300 m (1000 ft) under moderate dust conditions as might be common in mining and tunneling operations. Naturally, the range of the laser beam is a function of the power of the laser. Using an output beam with a diameter of 1.7 mm (0.006 ft), the estimated standard deviation in alignment is about 6 to 10 seconds of arc or a displacement of from 6 to 10 mm (0.020 to 0.033 ft) in 200 m ($\sim$ 650 ft). This accuracy is most easily achievable by utilizing a laser beam with a dark spot or cross in the center of the beam for more precise centering of the target. A stable instrument setup is extremely important and therefore whenever possible permanent stands with built-in centering plates or trivets should be used.

The system described is quite elementary but serves to illustrate the principles involved and application of laser technology to existing equipment.

Experiments have been performed using prototypes of laser-equipped theodolites and levels for high-precision alignment and leveling. The reader should consult Refs. 6 and 7 listed at the end of the chapter for details.

As laser technology developed, an impressive array of special purpose laser-equipped instruments has become available. Most of these instruments have been designed for construction layout and can be classified into two general groups:

1. Instruments which project single beams that are visible on targets under all lighting conditions. These devices are usually employed for horizontal and vertical linear alignment. Included in this group are transits or theodolites equipped with lasers (Fig. 14.40), lasers for alignment of tunneling machinery, and lasers for alignment of pipes and drains.

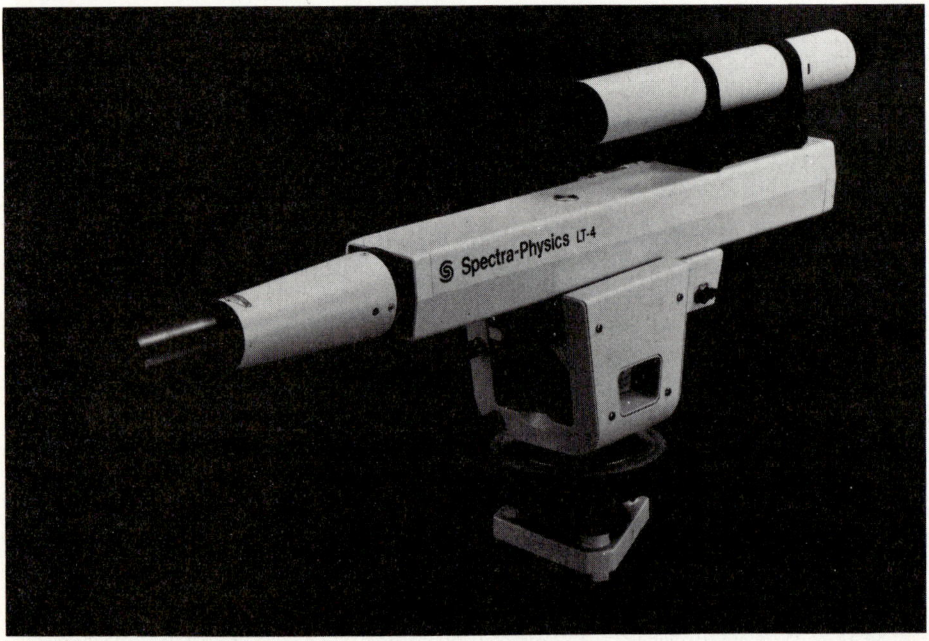

Fig. 14.43 Laser alignment instrument. (*Courtesy Spectra-Physics, Inc.*)

2. Instruments in which the laser beam is rotated by rapidly spinning optics so as to provide a reference plane in space over open areas and trace reference lines which are visible indoors. The speed of rotation can be varied and at zero revolutions per minute a single beam is available.

A single-beam laser alignment instrument is shown in Fig. 14.43. This laser is mounted on a transit-like framework which has horizontal and vertical motions, a spirit level parallel to the axis of the laser, a vertical circle (not visible in figure), and a horizontal circle. Horizontal and vertical circles are graduated in full degrees with double, direct-reading verniers having a least count of 5 minutes. Mounted on top of the laser enclosure is a telescope which allows the operator to sight the location of the transmitted laser spot. The spot diameter of the beam is 0.016 m (0.053 ft), at the laser, and due to spread becomes 0.062 m at 1 km (0.33 ft at 1 mi), and 0.50 m at the

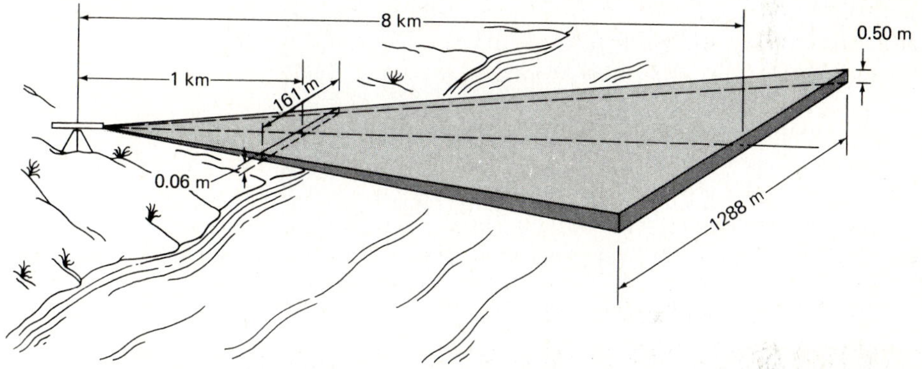

(*a*)

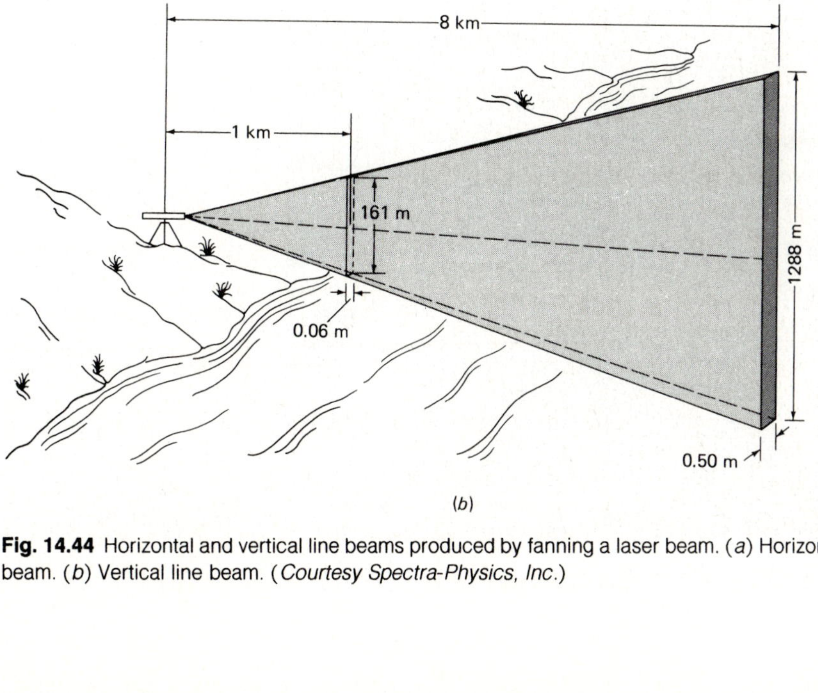

(*b*)

Fig. 14.44 Horizontal and vertical line beams produced by fanning a laser beam. (*a*) Horizontal line beam. (*b*) Vertical line beam. (*Courtesy Spectra-Physics, Inc.*)

maximum range of 8 km (1.67 ft at 5 mi). This range is effective in mild haze and increases somewhat at night.

A fanning lens can be attached to the emission end of the laser housing. This attachment allows the beam spot to be converted into a horizontal or vertical line beam shape (Fig. 14.44). The dimensions of this beam are 0.062 m by 161 m at a distance of 1 km (0.33 ft by 850 ft in 1 mi), increasing to 0.50 m by 1288 m at 8 km

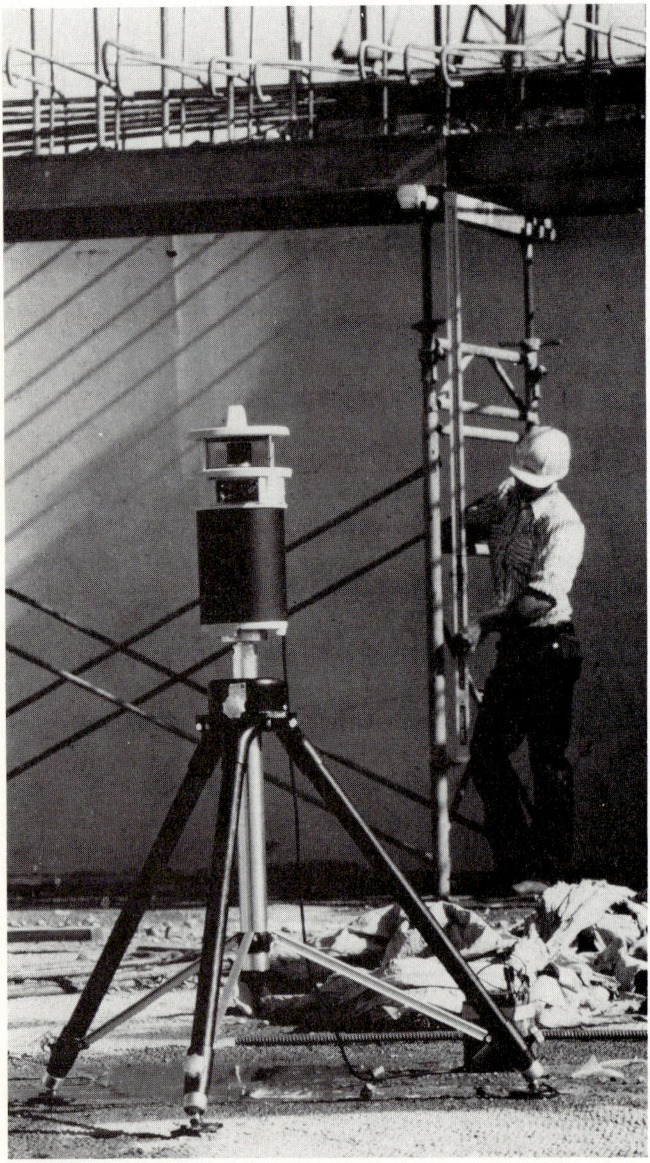

Fig. 14.45 Rotating laser level. (*Courtesy Spectra-Physics, Inc.*)

(1.67 ft by 4250 ft in 5 mi). This instrument can be used for horizontal and vertical control of dredging operations, control of tunneling machines, and azimuth and grade control for pipe-laying operations.

A self-leveling, rotating laser is shown in Fig. 14.45. In this instrument, the laser unit is mounted vertically on a platform containing two orthogonally mounted sensors which act like spirit levels and deviate from center when the platform is not level. The amount of deviation is detected electronically and the consequent electrical impulses drive servomotors which automatically level the base and make the axis of the laser vertical. The laser beam is emitted at an angle of 90° to the axis of the laser by an optical train and the optics rotate to form a horizontal reference plane. This device can also be side-mounted so that the axis of the laser is in a horizontal position and a vertical plane can be formed by the rotating beam. There is also an electronic sensing device parallel to the axis of the laser to allow self-plumbing of the rotating beam. The instrument is self-leveling and self-plumbing within a range of 8°. The tolerance specified for the position of the reference plane with respect to true level or true vertical is 20 seconds of arc. Thus, in a distance of 100 m (330 ft) from the instrument, a deviation of 10 mm (0.03 ft) is possible.

A laser rod equipped with a laser detector (Fig. 14.46) is used in conjunction with the rotating laser. This rod contains a sliding battery-powered sensor on the front face of the rod. When within 0.14 m (0.45 ft) above or below the rotating laser beam, this sensor "locks on" to the beam and emits a "beep" which indicates that a

Fig. 14.46 Laser level rod equipped with a laser detector. (*Courtesy Spectra-Physics, Inc.*)

reading should be taken. The operator then reads the rod directly to the nearest 5 mm (0.01 ft) at an index on the sensor which allows reading the scale on the front of the rod. There are two modes for the sensor. The *lock* mode means the sensor will seek the beam, average the position of the beam, and then lock onto it, giving a beep so that the operator may read the scale. The *float* mode enables the sensor to fix on the laser beam and continue reading the beam as the rod is moved up and down. The lock mode is used for normal leveling and determination of elevations or position. The float mode is useful when forms or stakes must be adjusted. The sensor is controlled by a mode switch at the top of the rod.

Grade and alignment by laser-equipped instruments The basic idea in the application of laser devices to construction layout is that the laser plane replaces the horizontal line of sight of the engineer's level and the laser beam replaces the string line. This plane or beam can be located in space at any intermediate point by using a small plastic target or an electronic laser detector which automatically seeks the laser plane or beam when it is within a specified range. Consequently, no operator must be stationed at the instrument when it is desired to get on line or obtain a rod reading.

When using a horizontal laser beam or rotating laser beam to form a horizontal reference line or plane, the operation of setting grade stakes to a given elevation is the same as described in Art. 14.29 except that there is no need for instructions from the operator of the instrument. The horizontal laser beam is visible as a red dot on a target which slides up and down along the face of the rod. The horizontal laser plane is detected by a sliding sensor on the face of the rod which emits a shrill sound or beep when the sensor index mark coincides with the laser plane. Thus, the operator of the rod must know what rod reading is required to set the grade stake and then can set the stake or form accordingly without assistance from the other end of the line at the instrument. When the laser beam is used as a substitute for the string line, the laser transmitting device is set over a control point at one end of the line and sighted at a control point at the other end of the line. The laser beam then serves the same function as the string line described in Art. 14.29.

Another possibility involves attaching the laser target to the machine which is grading, paving, tunneling, etc. The operator of the machine can observe the target and by keeping the visible red dot centered on the target can maintain the machine on the proper alignment and grade. A more sophisticated approach to this operation consists of attaching an electronic laser detector to the machine. In this way, the machine can be guided automatically in the proper direction.

A simple example consists of the grading operations for an area which is to be graded horizontally at a specified elevation. Assume that the laser level of Fig. 14.45 is set up near the area to be graded, permitted to level, and a reading is taken on a nearby bench mark. This reading establishes the height of the laser level above the datum (H.I.) so that the cut to the given elevation desired for the area can be calculated. Assume that this cut is 6.05 ft (1.84 m), so that the blade of the grader must be adjusted to 6.05 ft below the laser plane. Special graduated devices which support laser detectors and can be attached to various types of machinery are available for this purpose. Figure 14.47 shows a laser level mounted on a tripod and a power grader with laser detectors mounted at each end of the blade. When the laser detector is within about 0.5 ft (0.15 m) of the beam, it will automatically energize display lights in the cab which indicate low, on beam, or high positioning of the blade. There is also a pointer display to show how much adjustment is needed to get the detector on the beam. When the on-beam light is illuminated, the cutting edge of the blade is on grade. The same principles can be used to guide tunneling, paving, and ditch-digging machinery.

There is a precaution (in addition to the usual steps to ensure safety) which must be observed when using any of the laser alignment devices. Theoretically, one of the advantages of these types of systems is that once the laser transmitter is set up and aligned, no operator needs to stand by the instrument. However, laser transmitters are subject to the same disturbances that cause conventional optical alignment telescopes to become misaligned. Settlement, vibration, and accidental bumping are common factors around any construction job which can cause a perfectly aligned laser beam to be deflected. Consequently, periodic checks of the laser-beam alignment are absolutely necessary. Such checks can consist of having a party member check the alignment device at regular intervals or the checking device may be built into the system. The laser level illustrated in Fig. 14.45 shuts off when the laser beam deflects from horizontal. This constitutes a built-in check on the system.

Laser alignment devices and systems have many applications, particularly in construction layout. However, it should be emphasized that many of these devices

Fig. 14.47 Grading machine controlled by laser level and laser detector. (*Courtesy Spectra-Physics, Inc.*)

are not general-purpose surveying instruments having the flexibility in application of an engineer's transit or theodolite (with the exception of the laser attached to a theodolite shown in Fig. 14.40). Consequently, laser systems are more likely to be useful to contractors with a high volume of construction layout operations.

Various applications of laser alignment devices are described in the articles that follow.

14.32. Precision For purposes of excavation only, elevations are normally given to the nearest 0.1 ft (3 cm). For points on the structure, usually elevations to 0.01 ft (3 mm) are sufficiently precise. Alignment to the nearest 0.01 ft (3 mm) will serve the purpose of most construction, but greater precision may be required for prefabricated steel structures or members.

It is still desirable to give dimensions to the workers in feet, inches, and fractions of an inch. Ordinarily, measurements to the nearest $\frac{1}{4}$ or $\frac{1}{8}$ in (6 or 3 mm) are sufficiently precise, but certain of the measurements for the construction of buildings and bridges should be given to the nearest $\frac{1}{16}$ in (1.5 mm). Often, it is convenient to use the relation that $\frac{1}{8}$ in equals approximately 0.01 ft ($\sim$ 3 mm).

14.33. Sewers, pipelines, and culverts The center line for a proposed sewer is located on the ground with stakes or other marks set at 50-ft or 15-m intervals where the grade is uniform and as close as 10 ft or 3 m on vertical curves. At one side of this line, just far enough from it to prevent being disturbed by the excavation, a parallel line of hubs is set, with the hubs at the same intervals as those on the center line. A guard stake is driven beside each hub, with the side to the line; on the side of the guard stake farthest from the line is marked the station number and offset, and on the side nearest the line is marked the cut [to the nearest $\frac{1}{8}$ in (0.01 ft or 3 mm)]. In paved streets or hard roads where it is impossible to drive stakes and pegs, the line and grade are marked with spikes (driven flush), chisel marks, or paint marks.

When the trench has been excavated, batter boards are set across the trench at the intervals used for stationing. The top of the board is set at a fixed whole number of feet or decimetres above the sewer invert (inside surface of bottom of sewer pipe); and a measuring stick or rod of the same length is prepared. A nail is driven in the top edge of each batter board to define the line. As the sewer is being laid, a cord is stretched tightly between these nails, and the free end of each section of pipe is set at the proper distance below the cord as determined by measuring with the rod (see Fig. 14.39).

For pipelines, the procedure is similar to that for sewers, but the interval between grade hubs or batter boards may be greater, and less care need be taken to lay the pipe at the exact grade.

For both sewers and pipelines, the extent of excavation in earth and in rock is measured in the trench, and the volumes of each class of excavation are computed as a basis of payment to the contractor.

The records of the survey should include the location of underground utilities crossed by, or adjacent to, the trench.

A culvert is a drain under a road or embankment. At the intersection of the center line of the culvert with the located survey line of the highway or railway, the angle of intersection is turned off, and a survey line defining the direction of the culvert is projected for a short distance beyond its ends and is well referenced. At (or offset from) each end of the culvert, a line defining the face is turned off and referenced. If excavation is necessary for the channel to and from the culvert, it is staked out in a manner similar to that for a roadway cut. Bench marks are established nearby, and hubs are set for convenient leveling to the culvert. Line and grade are given as required for the particular type of structure.

Laser alignment instruments can be used instead of the batter boards and stringline described above for sewer construction. The simplest approach consists of utilizing a laser-equipped theodolite to establish a laser beam parallel with the flow line of the sewer. This alignment beam would be used first for rough grading to get the ditch excavated and then for placing the pipe on line and at grade in the ditch.

There are also several special-purpose laser alignment devices designed specifically for sewer construction. Figure 14.48 shows such an instrument placed in the invert of a manhole. This type of device can also be mounted on a tripod. These instruments are self-leveling and transmit a laser beam that can be adjusted from a −10 to a +30 percent grade. Instruments such as the laser level (Fig. 14.45) and this one are not useful for general surveying purposes but would be appropriate for a general contractor engaged in substantial amounts of excavation and sewer construction.

Note that the laser alignment instruments serve only to establish a line between previously surveyed control points and bench marks. Thus, the sewer alignment still needs to be established initially as in the traditional approach.

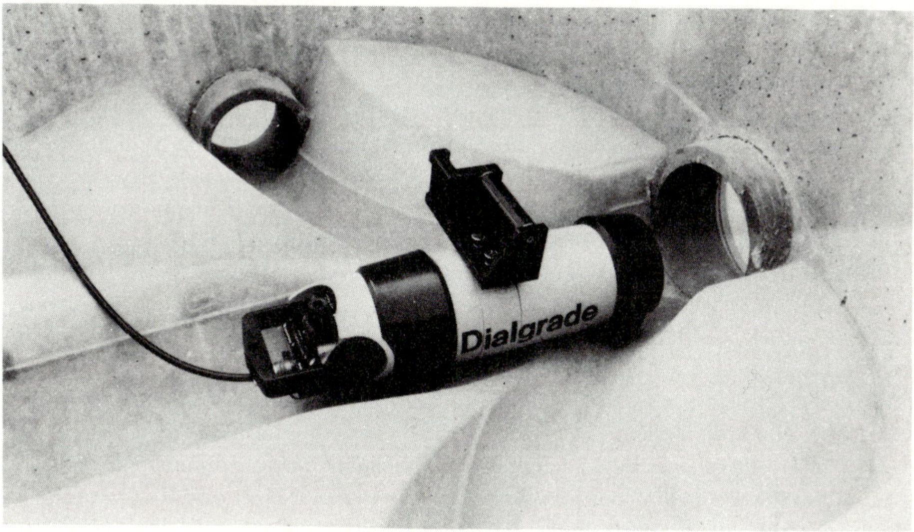

Fig. 14.48 Laser device for sewer alignment. (*Courtesy Spectra-Physics, Inc.*)

14.34. Tunnels Tunnel surveys are run to determine by field measurements and computations the length, direction, and slope of a line connecting given points, and to lay off this line by appropriate field measurement. The methods employed naturally vary somewhat with the purpose of the tunnel and the magnitude of the work. A coordinate system is particularly appropriate for tunnel work.

For a short tunnel such as a highway tunnel through a ridge, a traverse and a line of levels are run between the terminal points; and the length, direction, and grade of the connecting line are computed. Where practical, the surface traverse between the terminals takes the form of a straight line. Outside the tunnel, on the center line at both ends, permanent monuments are established. Additional points are established in convenient surface locations on the center line, to fix the direction of the tunnel on each side of the ridge. As construction proceeds, the line at either end is given by setting up at the permanent monument outside the portal, taking a sight at the fixed point on line, and then setting points along the tunnel, usually in the roof. Grade is given by direct levels taken to points in either the roof or the floor, and distances are measured from the permanent monuments to stations along the tunnel. If the survey line is on the floor of the tunnel, it is usually offset from the center line to a location relatively free from traffic and disturbance; from this line a rough temporary line is given as needed by the construction force.

The dimensions of the tunnel are usually checked by some form of templet transverse to the line of the tunnel, but may be checked by direct measurement with the tape. Photogrammetric methods may also be used to check the cross-sectional dimensions of the tunnel (see Ref. 14).

Railroad and aqueduct tunnels in mountainous country are often several miles in length and are not uniform in either slope or direction. Tunnels of this character are usually driven not only from the ends but also from several intermediate points where shafts are sunk to intersect the center line of the tunnel. The surface surveys for the control of the tunnel work usually consist of a precise triangulation and/or trilateration system tied to monuments at the portals of the main tunnel and at the entrances of shafts and adits, and a precise system of differential levels connecting the same points. With these data as a basis, the length, direction, and slope of each of the several sections of the tunnel are calculated; and construction is controlled by establishing these lines and grades as the work progresses.

Alignment of tunneling machinery was one of the previously mentioned applications of laser-equipped instruments. The procedure for establishing control and the center line of the tunnel is the same as described previously in this article except that a laser beam is projected ahead to control the direction of the tunneling machine (see Refs. 19 and 27). As the center line of the tunnel is produced, a laser-equipped theodolite (Fig. 14.40) is suspended from a support at a center-line station on the roof of the tunnel. The line of sight is produced ahead by double centering from a backsight. The laser transmitter is turned on to create a laser beam which is parallel to the center line of the tunnel. The tunneling machine has two targets placed on the top of its framework. The first target is glass with a center cross and the second is an opaque target with a center mark. The machine is adjusted until the laser beam passes through the center of the first target and strikes the center of the second

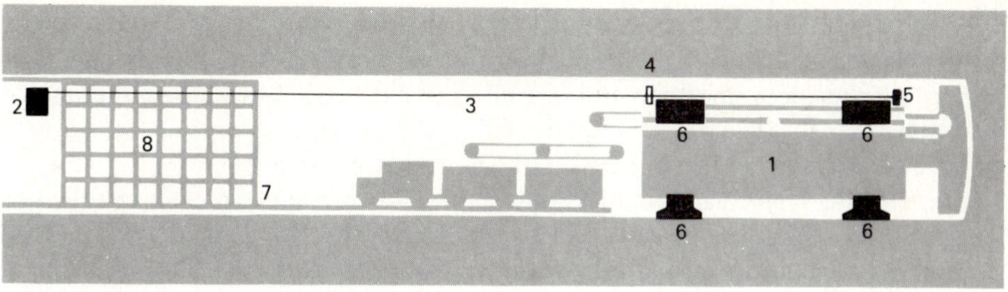

Fig. 14.49 Laser alignment to control a tunneling machine: (1) tunneling machine; (2) laser theodolite station; (3) laser beam; (4) glass target; (5) second target; (6) hydraulic bracing; (7) tunnel invert (lining); (8) concrete forms. (*Courtesy Kern Instruments, Inc.*)

target. When these two conditions are met, the tunneling machine is on line and work can progress. The operator of the machine can detect and correct for any deviation from this alignment. The laser beam is also used to control setting of forms for the tunnel lining. Figure 14.49 illustrates in schematic form the arrangement for control of the tunneling operation.

The range of the laser beam is only about 1000 ft (300 m), owing to the dust created by tunneling operations. Thus, periodically the line is produced ahead by double centering and a target is centered on a new station attached to the tunnel roof and immediately behind the tunnel machine. The target and theodolite are interchanged, the theodolite is centered by forced centering, a backsight is taken on the target, and the line is again produced by double centering. As the tunneling progresses, the backsight should be checked periodically to ensure that the laser beam does not deviate from the correct position.

When the alignment of the tunnel is curved, the machine must be driven ahead along chords on the curve. In order to get directed along a chord from the point of curvature, the targets on top of the machine are shifted laterally by a precomputed amount so as to produce the desired amount of deflection of the cutting heads in the direction of the chord. When the machine has moved ahead enough to provide clearance, the chord point is set on the roof of the tunnel, the theodolite is moved ahead, a backsight is taken on the previous station, and the deflection angle to the next chord point is turned. This procedure establishes the direction of the tunnel until the cutting heads of the machine advance to the next chord point, where the entire procedure is repeated. The machine is driven forward from chord point to chord point until the end of the curve is reached. Obviously, to maintain correct direction of the tunnel, very careful attention must be given to centering the theodolite and producing the laser beam by double centering. Further details concerning use of lasers for tunnel alignment can be found in Refs. 19 and 24.

Electronic distance measuring equipment (EDM) is also very useful in establishing the distances to theodolite stations, points of curvature, and chord points as the tunnel is driven.

To establish absolute directions (azimuth), a gyroscopic theodolite can be used. This instrument consists of a repeating or direction theodolite (Art. 6.30) equipped with a north-seeking gyroscope so that azimuth determination is possible without sighting along a line of known direction of observations on celestial bodies. For further details, see Ref. 9. Periodic determinations of the azimuth of the tunnel center line provide a check on the direction established by producing the line by double centering.

14.35. Bridges Normally, the location survey will provide sufficient information for use in the design of culverts and small bridges; but for long bridges and for grade-separation structures usually a special topographic survey of the site is necessary. This survey should be made as early as possible in order to allow time for design and—in the case of grade crossings or navigable streams—to permit approval of the appropriate governmental agencies to be secured. The site map should show all the data of the location survey, including the line and grade of the roadway and the marking and referencing of all survey stations.

A scale of 1 : 1200 (100 ft/in) with a 5-ft (2-m) contour interval is usually suitable for large bridges in rough terrain. Where the terrain is level and in urban regions, a scale of 1 : 600 with a contour interval of 2 ft (1 m) is desirable.

Where the bridge crosses rivers or other bodies of water, a hydrographic survey which will provide a continuous profile of the bottom of the river or bay at 10- to 25-ft (3- to 8-m) intervals is required. Hydrographic surveys are beyond the scope of this book but Ref. 9 at the end of this chapter can be consulted for additional details.

When planning the horizontal and vertical control for design mapping, due consideration should be given to establishing control points suitably located for the construction survey. Long crossings over water require elaborate triangulation and/or trilateration networks of first-order accuracy (Tables 9.1 and 9.2) which must be planned well in advance of the actual construction. For a short bridge with no offshore piers, first the center line of the roadway is established, the stationing of some governing line such as the abutment face is set on the located line, and the angle of intersection of the face of the abutment with the located line is turned off. This governing crossline may be established by two well-referenced control points at each end of the crossline offset beyond the limits of excavation or, if the face of the abutment is in the stream, by a line parallel to the face of the abutment offset on the shore. Similarly, governing lines for each of the wing walls are established on shore beyond the limits of excavation, with two stations on the line prolonged at one, or preferably both, ends of the wing-wall line. If the faces are battered, usually one line is established for the bottom of the batter and another for the top. Stakes are set as a guide to the excavation and are replaced as necessary. When the foundation concrete has been cast, line is given on the footings for the setting of forms and then by sighting with the theodolite for the top of the forms. As the structure is built, grades are carried up by leveling, with marks on the forms or on the hardened portions of the concrete. Also, the alignment is established on completed portions of the structure. The data are recorded in field books kept especially for the structure, principally by means of sketches.

For long sights or for work of high precision, as in the case of offshore piers, various control points are established on shore and/or on specially built towers by a system of triangulation and/or trilateration, such that favorable intersection angles and checks will be obtained for all parts of the work. Electronic distance measuring (EDM) equipment and first-order theodolites should be employed in establishing this control network. For ordinary river crossings, a single quadrilateral or two adjacent quadrilaterals with the common side coincident with the center line of the bridge are appropriate figures and provide well-located stations for construction layout. For river and bay crossings, vertical control is established by simultaneous, reciprocal leveling (Art. 5.40). For long crossings, special illuminated targets are used on the rod and all measurements are made at night.

To establish the location of offshore structures, simultaneous sights are taken from two coordinated control points which have favorable geometric locations with respect to the point being set. Thus, the points are located by intersection. When combined theodolite and EDM systems are available, points can be located by distance and direction so that triangulation and trilateration are involved. The survey is usually accomplished in stages with the pier locations being established with lower precision than those of the fittings subsequently established on top of the pier.

Laser-equipped instruments can be used for establishing the alignment and vertical positions for cofferdams and pile-driving equipment used in the construction of foundations for piers. The laser alignment device shown in Fig. 14.43 used in conjunction with EDM equipment provides an excellent system for this purpose.

As soon as the pier foundation is placed, survey control is transferred to the foundation so as to verify the location and provide control for construction of the rest of the pier. Elevations are also transferred to the pier at this time. When the pier is completed, the precise locations of anchor bolts and other fastenings to which the bridge structure is connected are set out on the top of the pier. All center-line points, anchor-bolt locations, and control monuments are coordinated in a system referred to a common datum. Thus, it is possible to locate anchor-bolt positions, etc., by triangulation and trilateration from at least two control monuments that have good geometric positions with respect to the points being located. Note that this procedure is an application of location by random control points (Arts. 14.26 and 14.30).

When anchor bolts have been placed, their positions should be verified by triangulation and trilateration from control monuments. In addition, distances between anchor bolts and other pier connections on adjacent piers should be measured using EDM equipment.

On completion of the structure, permanent survey points are established and referenced for use in future surveys to determine the direction and extent of any movement of the piers and bridge structure.

14.36. Buildings and building layout Prior to the layout of the building a property survey must be performed, setting all property corners and lines. At this stage, it is the surveyor's responsibility to be sure that the proposed structure is on the correct lot and is placed according to the client's desires and as set forth on the construction plans. Note that many communities have ordinances establishing *building lines* or *set backs* from the front property line and minimum clearances from side property lines. The surveyor in charge of the work must be familiar with these ordinances and perform the layout so as to avoid any encroachments.

At the beginning of excavation, the corners of the building are marked by stakes, which will of course be lost as excavation proceeds. Sighting lines are established and referenced on each outside building line and line of columns, preferably on the center line of wall or column. A batter board is set at each end of each outside building line, about 3 ft or 1 m outside the excavation. If the ground permits, the tops of all boards are set at the same elevation; in any event the boards at opposite ends of a given line (or portion thereof) are set at the same elevation so that the cord stretched between them will be level. The elevations are chosen at some whole number of feet or metres above the bottom of the excavation, usually that for the floor rather than that for the footings, and are established by holding the level rod on the side posts for the batter board and marking the grade on the post. When the board has been nailed on the posts, a nail is driven in the top edge of the board on the building line, which is given by the transit. Carpenter's lines stretched between opposite batter boards define both the line and the grade, and measurements can be made conveniently by the workers for excavation, setting forms, and aligning masonry and framing. If the distance between batter boards is great enough that the sag of the carpenter's line is appreciable, the grades must be taken as approximate only. Laser alignment and leveling devices provide an excellent substitute for the string line (Art. 14.31) and are adaptable for establishing line and grade. Use of these devices is particularly convenient when obstructions prohibit the erection and use of batter boards.

When excavation is completed, grades for column and wall footings are given by hubs driven in the ground to the elevation of either the top of the footing or the top of the floor. Lines for footings are given by batter boards set in the bottom of the excavation. Column bases and wall plates are set to grade directly by a level and level rod. The position of each column or wall is marked in advance on the footing; and when a concrete form, a steel member, or a first course of masonry has been placed on the footing, its alignment and grade are checked directly.

In setting the form for a concrete wall, the bottom is aligned and fixed in place before the top is aligned.

As before, laser alignment and leveling devices can also be utilized at this stage to establish line and grade.

Similarly, at each floor level the governing lines and grade are set and checked, except that for prefabricated steel framing, the structure as a whole is plumbed by means of the transit, theodolite, or plumbed laser beam at every second or third story level. Notes are kept in a field book used especially for the purpose, principally by means of sketches.

Whenever the elevation of a floor is given, it should be clearly understood whether the value refers to the bottom of the base course, the top of the base course, or the top of the finished floor.

Throughout the construction of large buildings, selected key points are checked by means of stretched wires, plumb lines, optical plummet, plumbed laser beam, plumbed laser plane, transit, theodolite, or level in order to detect settlement, excessive deflection of forms or members, or mistakes. Bench marks are checked to detect any disturbance.

The above instructions are of general nature and can be applied to a great variety of structures ranging from a simple square or rectangular building to complex, multistory structures. In order to illustrate the actual layout procedure, consider an example problem involving the construction of a small dwelling on an urban lot. Figure 14.50 shows the lot and building layout as it would be recorded in the field notebook. Assume that the property survey (Chap. 15) has been previously performed so that all corners are set and known to be in the correct positions and there is a 25-ft building line in this development with a minimum of 15 ft clearance from

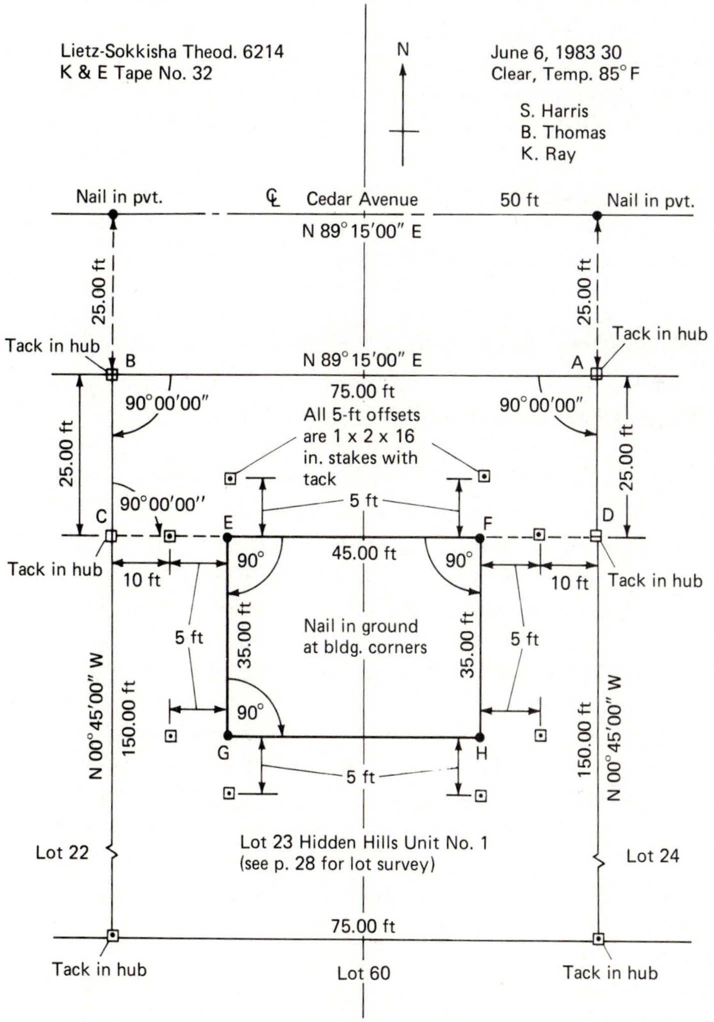

Fig. 14.50 Building layout.

the side property lines. The initial instrument set up is on the NE corner (point A). A sight is taken on the NW corner (point B) and an angle of 90°00′00″ turned to the left to establish the direction of the east property line. A distance of 25.00 ft is measured along this line to set a tack in a hub at the 25-ft building line (point D). Point B is occupied and following a similar procedure, a hub with a tack is set at point C at the 25-ft building line on the west property line. With the building line established, occupy point C, sight B, turn 90°00′00″ right and set the west offset stake with a tack, nails in the ground at building corners E and F, and the east offset stake, all for the front face of the building, checking on point D for line and distance. This last check is *important* and should always be made. If a discrepancy occurs, the cause must be located and rectified before continuing the layout. Next, set the instrument on corners E and F in sequence, setting building corners G and H and setting north and south offset stakes for corners E, G, F, and H, respectively. One final instrument setup is required at G or H to check the layout and set offset stakes east and west of corners G and H. Following the layout, differential levels (Art. 5.31) should be run from the nearest bench mark to obtain elevations on lot corners and building corners as required, and to set elevations on hubs at C and D from which cuts or fills (calculated from construction plans) to the basement elevation can be given for excavation. It would also be advisable to set a temporary bench mark near the site for future use as construction progresses.

The method described is for a very simple project and represents only one of many possible approaches to the given task. The job as outlined indicates the work done by the surveyor hired by the owner of the land to do the property survey and to lay out the building. Placement of batter boards from the given stakes would be the responsibility of the construction contractor's survey engineer.

14.37. Dams Prior to the design of a dam, a topographic survey is made to determine the feasibility of the project, the approximate size of the reservoir, and the optimum location and height of the dam. To provide information for the design, a topographic survey of the site similar in many respects to that for a bridge (Art. 14.35), is made usually by photogrammetric methods supplemented by field surveys where necessary. Extensive soundings and borings are made, and topography is taken in detail sufficient to define not only the dam itself but also the appurtenant structures, necessary construction plant, roads, and perhaps a branch railroad. A property-line survey is made of the area to be covered by, or directly affected by, the proposed reservoir.

When feasibility and location studies are being made for dams in remote areas, due consideration should be given to using doppler positioning and inertial positioning systems (Chap. 9) for establishing the basic control network. Inertial systems can also be used to obtain data for preliminary topographic mapping where heavy ground cover (such as rain forest) combined with bad weather conditions make ground surveys and photogrammetric topographic surveys impractical.

Prior to construction, horizontal control stations, sighting points, and bench marks are permanently established and referenced upstream and downstream from the dam at advantageous locations and elevations for sighting on the various parts of the structure as work proceeds. These reference points are usually established by triangulation and/or trilateration and all points are referred to a system of rectangular coordinates. In the United States, the state plane coordinate systems should be used.

Precise levels (see Ref. 9) should be run to establish vertical control. For major dam projects, first-order specifications should be satisfied in establishing both horizontal and vertical control (Tables 9.1, 9.2, and 5.2). A dam is a vast project in which construction may last for a period of several years. Comprehensive planning of the survey control is essential from the beginning of the project. Careful attention must be given to setting horizontal and vertical control points in locations suitable for subsequent construction layout.

To establish the horizontal location of a point on the dam, as for the purpose of setting concrete forms or of checking the alignment of the dam, simultaneous sights are taken from two theodolites set up at horizontal control points, each instrument being sighted in a direction previously computed from the coordinates of the control station and of the point to be established. The method of staking from random control points as described in Art. 14.26 can be used to advantage in dam construction. Thus, coordinated points on the dam structure are set from at least two coordinated horizontal control points by triangulation using combined theodolite and EDM or an electronic surveying system (Art. 9.14).

Vertical control is usually established by direct leveling. When warranted, trigonometric leveling (Art. 5.5) is used to establish elevation by using measurements obtained with one of the electronic surveying systems from two coordinated vertical control stations.

Wherever possible, laser alignment instruments should be utilized. For example, lasers can be used in controlling tunneling operations in the underground excavation associated with the construction of the dam.

A traverse is run around the reservoir above the proposed shoreline and monuments are set for use in connection with property-line surveys and for future reference. Similarly, bench marks are established at points above the shoreline. The shoreline may be marked out by trace contour leveling (Art. 12.14), with stakes set at intervals along the specified contour line. The area to be cleared is defined with reference to these stakes.

When the dam is completed, permanent horizontal control monuments and bench marks are placed on the dam structure appropriate for subsequent studies related to deformation of the dam. A network of control points suitable for combined triangulation/trilateration (Art. 9.10) provides optimum control for monitoring possible movement of the dam. Observations on this control network should be made with first-order theodolites (Art. 9.6) and EDM equipment (Arts. 4.29 through 4.38). Elevations should be determined by geodetic leveling (see Ref. 9). First-order specifications should govern all horizontal and vertical control surveys (Tables 9.1 and 9.2, Art. 9.2, and Table 5.2, Art. 5.39).

14.38. As-built surveys The *as-built survey* is performed on completion of a construction project for the purpose of reestablishing the principal horizontal and vertical controlling points and to locate the positions of all structures and improvements. As-built surveys should be performed after the construction of highways, railroads, bridges, buildings (including private homes), dams, and underground facilities.

Essentially, the as-built survey consists of rerunning center lines and/or property lines, setting permanent monuments at controlling points, and locating all improvements relative to these lines. These data are then plotted on a map to provide a permanent record or inventory of the work done and its precise location. A record of these data filed in a data bank (magnetic tape, disk storage, or core memory) according to a coordinated position in the state plane coordinate system is perhaps the most efficient method for storing this information. Such a record is easily accessible using a computer program and permits efficient retrieval and automatic plotting of the desired data.

14.39. Problems

14.1 Given: $I = 34°30'$, D (chord basis) $= 3°00'$, and P.C. = station 74 + 30.0. Required: R, L, T, and E; also deflection angles arranged in notebook form for staking out this curve, using 100-ft stations.
14.2 Given: $I = 60°40'$, $E = 125.5$ ft. Required: R, D (arc basis), C, T, M, and L.
14.3 Given: $I = 37°30'$, $D_a = 22°00'$, the station of the P.I. = 14 + 80.000 m. Compute (*a*) T, E, L, the middle ordinate for the entire curve, and the middle ordinate for one full station or 100 m; (*b*) the station of the P.C., the station of the P.T., and deflection angles for staking the curve at even one-tenth station points; (*c*) the true lengths of one-tenth station chords and the true lengths for the subchords at either end of the curve. (*d*) Write the necessary field notes for staking this curve in the form in which they should appear in the field book.
14.4 Given: $I = 92°30'$, $T = 425.00$ ft, and P.C. = station 25 + 10.00. Required: R, D (arc basis), C, E, M, and L; also deflection angles arranged in notebook form for staking out this curve, using 50-ft stations.
14.5 If the curve of Prob. 14.4 represents the center line of a highway curve, suppose it is desired to set alignment stakes along two curves, one of which is to be 10 ft outside, and the other 12 ft inside, the center line. Required: L_1, L_2, D_1, D_2, E_1, E_2, and the deflection angles arranged in notebook form for staking out these curves.
14.6 Given the following finished grade elevations for the center line of a roadway: A at 0 + 00 = 901.200 m, E at 20 + 96.143 = 990.000 m, and D at 41 + 75.470 = 950.000 m, each above mean sea level. An equal tangent parabolic vertical curve with a total length of 150 m and centered about point E is to be calculated for subsequent construction layout. Compute the: (*a*) grade elevations for the *BVC* and *EVC* and at points spaced at 20-m intervals throughout the vertical curve; (*b*) station and elevation of the high (or low) point on the vertical curve. Use the equation of the parabola to solve this problem.
14.7 In Prob. 14.6, do the computations for parts (*a*) and (*b*) using the method of vertical offsets from the tangent.
14.8 On a highway a −6.0 percent grade meets a +4.0 percent grade at station 67 + 50 and at elevation 516.32. Compute the elevations of stations at 50-ft intervals along a 400-ft vertical curve connecting the two grades. Compute the station and elevation of the lowest point on the curve.
14.9 Following are the notes for cross sections at stations 109 and 110. The width of the roadbed is 24 ft, and the side slopes are 2 to 1. Plot these two cross sections at a scale of 1 : 120 (1 in per 10 ft). Compute the areas of the two sections.

Station	Cross section		
109	$\frac{c2.4}{16.8}$	$\frac{c1.2}{0.0}$	$\frac{c0.4}{12.8}$
110	$\frac{c12.2}{36.4}$	$\frac{c9.2}{0.0}$	$\frac{c4.8}{21.6}$

14.10 Following are notes for an irregular cross section. The width of the roadbed is 24 ft and the side slopes are $1\frac{1}{2}$ to 1. Plot the cross section at a scale of 1:120 (1 inch per 10 ft). Compute the cross-sectional area.

$$\frac{c4.2}{18.3} \quad \frac{c6.8}{12.0} \quad \frac{c11.2}{0} \quad \frac{c14.4}{10.0} \quad \frac{c16.8}{25.0} \quad \frac{c18.4}{39.6}$$

14.11 Plot the preliminary cross section for station 44 + 00 in Fig. 14.17 using a horizontal scale of 1:200 and vertical scale of 1:100. Assuming a roadbed width of 10 m, side slopes of 2 to 1, and a center-line grade elevation of 207.300 m, plot the roadbed templet and compute the cross-section area by the coordinate method.

14.12 Using the roadbed templet specifications given in Prob. 14.11, compute the cross-section area in square metres for the data of station 43 + 00 in Fig. 14.17.

14.13 Compute the volume in cubic metres between stations 43 + 00 and 44 + 00 of Probs. 14.11 and 14.12 by the average-end-area method.

14.14 The width of roadbed for a highway is 8 m with side slopes of 2 to 1; the grade elevation at station 48 + 00 is 345.40 m, and the center-line elevation is 344.29 m. The H.I. for obtaining the final cross section at this station is 346.940 m. The rod readings at the right and left slope stakes are 4.60 and 7.10 m, respectively. Compute the grade rod, the center-line cut or fill, and the cut or fill and distance to each slope stake.

14.15 The width of roadbed of a proposed railroad is 24 ft in cut, and the side slopes are $1\frac{1}{2}$ to 1. At a given station the elevation of grade is 515.75. For obtaining the cross section at the station the H.I. of the level is 528.32. The ground rod at the center stake is 6.5. Compute the grade rod and the center cut. The rod reading at the right slope stake is 1.2 and at the left slope stake is 10.9. Compute the cut and the distance out to each slope stake.

14.16 Compute the volume in cubic yards between stations 109 and 110 of Prob. 14.9. Use both the average-end-area method and the prismoidal formula. Note the discrepancy expressed as a percentage between volumes as determined by the two methods.

14.17 What error in volume between station 109 and 110 of Prob. 14.9 would be introduced if the recorded cuts at centers and slope stakes were 0.1 ft too great? What is the error in terms of percentage of the volume by average end areas?

14.18 Assume that the center line for the cross-sectional data in Prob. 14.9 turns to the right with a simple circular curve having $D_a = 10°$ (arc definition). Compute the curvature correction in cubic yards to the earthwork volume between these two stations and the percentage of the total volume of earthwork which this correction constitutes.

14.19 Compute the volume in cut and in fill for the given cross sections between stations 62 and 64. The roadbed is 24 ft wide in cut and 20 ft wide in fill, and the side slopes are $1\frac{1}{2}$ to 1. Tabulate the data in the following form: "Station," "Cross section," "Area," and "Volume." Use the prismoidal formula.

Station	Cross section			
62	$\frac{c2.6}{15.9}$	$\frac{c4.8}{0.0}$	$\frac{c6.4}{21.6}$	
63	$\frac{0.0}{12.0}$	$\frac{c3.1}{0.0}$	$\frac{c4.4}{18.6}$	
63 + 25	$\frac{f4.6}{16.9}$	$\frac{0.0}{0.0}$	$\frac{c2.6}{15.9}$	
64	$\frac{f7.2}{20.8}$	$\frac{f4.8}{0.0}$	$\frac{0.0}{6.0}$	$\frac{c1.8}{14.7}$

14.20 Solve Prob. 14.19 by the method of average end areas, computing the volume of pyramids by the relation, volume = $\frac{1}{3}$ (area of base times length).

14.21 Compute the volume in cut and fill in cubic metres for the given cross sections (all distances and elevations in metres) between stations 10 + 00 and 10 + 65. The roadbed is 8 m wide in cut and 6 m wide in fill and the slopes are 1.5 to 1. Tabulate as directed in Prob. 14.19. Use the method of average end areas.

Station	Cross section			
10 + 00		$\frac{c2.00}{7.00}$	$\frac{c1.50}{0.00}$	$\frac{c0.80}{5.20}$
10 + 30		$\frac{c1.60}{6.40}$	$\frac{c1.00}{0.00}$	$\frac{0.00}{4.00}$
10 + 40		$\frac{c0.85}{5.28}$	$\frac{0.00}{0.00}$	$\frac{f1.40}{5.10}$
10 + 65	$\frac{c1.0}{5.50}$	$\frac{0.00}{2.00}$	$\frac{f1.62}{0.00}$	$\frac{f2.54}{6.81}$

14.40. Field Problem

LOCATION OF A SIMPLE CURVE BY DEFLECTION ANGLES

Object To calculate and stake a simple circular curve in the field by the method of deflection angles (Art. 14.7). Suggested parameters for the curve are as follows: in the metric system $I = 30°00'$, $D_a = 33°$, station at the P.I. = 16 + 34.500; and using the foot as the measuring unit, $I = 30°00'$, $D_a = 10°$, station of the P.I. as given above.

Procedure Before the beginning of the field session, compute (*a*) length of curve; (*b*) radius; (*c*) tangent distance; (*d*) deflection angles from the P.C. to each half-station (one-tenth station for metric units); (*e*) chord lengths between the P.C. and the first curve station, between each half-station (tenth stations when metric units are used) and between the last station on the curve and the P.T. All calculations are to be to the nearest hundredth of a foot (nearest millimetre).

In the field, the instructor will designate the P.I. and a point on the back tangent. Each party will occupy the assigned P.I. and locate the P.C. and P.T. according to the instructions given below. Each party should make one setup on an intermediate station near the midpoint of the curve to complete the curve layout. Consult the text (Art. 14.8) for procedures. The procedure for curve layout is:

1. Set the transit over the P.I. and take a backsight on the P.O.T. Measure the tangent distance from the P.I. and set a hub at the P.C. Measure to the nearest hundredth of a foot (nearest millimetre).
2. Lay off the direction of the forward tangent by turning a double deflection angle according to the method outlined in Art. 6.42.
3. Measure the tangent distance from the P.I. along the forward tangent to set a hub at the P.T. Measure to the nearest hundredth of a foot (millimetre).
4. Occupy the P.C. with the transit. With the *A* vernier or horizontal index mark on zero, sight the P.I.
5. Set stations by deflection angles and chord distances.
6. About halfway along the curve, set a hub to mark the curve station. (This point will be designated by the instructor.) Occupy this point with the transit; set the *A* vernier or horizontal index mark to zero, invert the telescope, and backsight on the P.C. using the lower motion. Replunge the telescope to its direct position and continue setting curve stations.
7. Measure the distance error and offset error at the P.T. Record discrepancies in the field book.
8. Record data for curve and sketch the curve in your field book. Use the note form illustrated in Fig. 14.10.

References

1. American Association of State Highway Officials, *A Policy on Geometric Design of Rural Highways* (Blue book), Washington, D.C., 1966.
2. American Association of State Highway Officials, *A Policy on Design of Urban Highways and Arterial Streets* (Red book), Washington, D.C., 1973.
3. American Society of Civil Engineers Task Committee Report on "Engineering Surveying, Chapters 2, 5, 6, 8, 16, and 20," *Journal of the Surveying and Mapping Division, ASCE*, Vol. 102, No. SU1, December 1976, p. 59.
4. California Division of Highways, *Highway Design Manual of Instructions,* Sacramento, Calif., 1972.
5. California Department of Transportation, *Surveying Manual*, Sacramento, Calif., 1977, Chaps. 7, 11.
6. Chrzanowski, A., Jarzymowski, A., and Kaspar, M., "A Comparison of Precision Alignment Methods," *The Canadian Surveyor*, Vol. 30, No. 2, June 1976.
7. Chrzanowski, A., and Janssen, J.-D., "Use of Laser in Precision Leveling," *The Canadian Surveyor*, Vol. 26, No. 4, September 1972.
8. Cooney, Austin, "Laser Alignment Techniques in Tunneling," *Journal of the Surveying and Mapping Division, ASCE*, Vol. 94, No. SU2, September 1968, p. 203.
9. Davis, R. E., Foote, F. S., Anderson, J. M., and Mikhail, E. M., *Surveying: Theory and Practice,* 6th ed., McGraw-Hill Book Co., New York, 1981.
10. Hickerson, T. F., *Route Location and Design,* 5th ed., McGraw-Hill Book Co., New York, 1967.
11. Hurdle, V. F., "Vertical Curves," unpublished notes, University of California, Institute of Transportation and Traffic Engineering, Berkeley, Calif., 1974.
12. Kahn, Michael, "Lasers, Today's Great Levelers in Construction," *World Construction,* February 1977.
13. Kahn, Michael, "Lasers as Construction Tools," *Constructor,* April 1976.
14. Katibah, G. P., "Photogrammetry in Highway Practice," Unpublished report, California Department of Transportation, 1968
15. Leeming, J. J., "Road Curvature and Superrelevation," *Survey Review,* January 1973.
16. Meyer, C. F. and Gibson D. W., *Route Surveying,* 5th ed., Harper & Row, Publishers, New York, 1980.
17. Moffitt, F. H., and Bouchard, H., *Surveying,* 7th ed., Harper & Row, Publishers, New York, 1982.
18. More, Norman L., and Begell, R. G., "Control Surveys for Major Bridges," *Journal of the Surveying and Mapping Division, ASCE,* Vol. 98, No. SU1, May 1971.
19. Peterson, Edward W., and Frobenius, Peter, "Tunnel Surveying and Tunneling Machine Control," *Journal of the Surveying and Mapping Division, ASCE,* Vol. 99, No. SU1, September 1973, p. 21.
20. Pryor, W. T., "Metrication," *Surveying and Mapping,* Vol. 35, No. 3, September 1979, pp. 229–237.
21. Sanyaolu, A., "New Analytic Formulas and Automatic Checks for Earthwork," *The Canadian Surveyor,* Vol. 32, No. 2, June 1978.
22. State of California, Business and Transportation Agency, Department of Transportation, *Action Plan for Transportation Development,* Sacramento, Calif., 1973.
23. State of California, Business and Transportation Agency, Department of Public Works, Division of Highways, *Surveys Manual,* Sacramento, Calif., 1976.
24. Thompson, Bruce J., "Planning Economical Tunnel Surveys," *Journal of the Surveying and Mapping Division, ASCE,* Vol. 100, No. SU2, November 1974, p. 95.
25. Willis, Maynard J., "Planning Concrete Dam Construction Control," *Journal of the Surveying and Mapping Division, ASCE,* Vol. 98, No. SU1, July 1972, p. 27.
26. "Kern DKM2-A Laser Theodolite," *Kern Bulletin 15,* Kern Instruments Inc., 5001 Aarau, Switzerland.
27. "Laser Control in Tunnel Driving," *Kern Bulletin 15,* Kern Instruments Inc., 5001 Aarau, Switzerland.

CHAPTER 15
Land surveys

15.1. General Land surveying deals with the laying off or the measurement of the lengths and directions of lines forming the boundaries of real or landed property. Land surveys are made for one or more of the following purposes:

1. To secure the necessary data for writing the legal description and for finding the area of a designated tract of land, the boundaries of which are defined by visible objects
2. To reestablish the boundaries of a tract for which a survey has previously been made and for which the description as defined by the previous survey is known
3. To subdivide a tract into two or more smaller units in accordance with a definite plan which predetermines the size, shape, and location of the units

Whenever real estate is conveyed from one owner to another, it is important to know and state the location of the boundaries, particularly if there is a possibility of encroachment by structures, roadways, or adjoining property owners.

The functions of the land surveyor are to carry out field surveys, to calculate dimensions and areas, to prepare maps showing the lengths and directions of boundary lines and areas of lands, and to write descriptions by means of which lands may be legally conveyed, by deed, from one party to another.

The land surveyor must be familiar not only with technical procedures but also with the legal aspects of real property and boundaries. Usually, the land surveyor is required to be licensed by the state, either directly or as a civil engineer.

In this chapter, practices as applied to both rural and urban properties are described, and some of the legal aspects of land surveying are discussed. The United States system of subdividing the public lands is also outlined. Methods of calculating and subdividing areas are discussed in Chap. 8.

15.2. Kinds of land surveys In accordance with the purposes listed in Art. 15.1, land surveys may be classified as follows:

1. *Original surveys*, to measure the unknown lengths and directions of boundaries for the first time. Surveys of this character are usually of rural lands. For example, Adams may purchase from Brown a certain parcel of land bounded or defined by features or objects such as fences, roads, or trees. In order that the deed may contain a definite description of the tract, a survey is necessary.

2. *Resurveys*, to reestablish the boundaries of a tract for which a survey has previously been made. The surveyor is guided by a description of the property based upon the original survey, and by evidence on the ground. The description may be in the form of the original survey notes, an old deed, or a map or plat on which are recorded the measured lengths and bearings of sides and other pertinent data. When land is transferred by deed from one party to another, often a resurvey is made.
3. *Subdivision survey*, to subdivide land into more or less regular tracts in accordance with a prearranged plan. The division of the public lands of the United States into townships, sections, and quarter sections is an example of the subdivision of rural lands into large units. The laying out of blocks and lots in a city addition or subdivision is an example of the subdivision of urban lands. Subdivision surveys are usually original surveys.

15.3. Equipment and methods Land surveys can be run with a transit and tape, transit or theodolite and electronic distance measurement (EDM) device, or electronic surveying system (Art. 9.14). In some states, land surveys have been performed using photogrammetric methods to establish the control points from which property corners have been located and set. Ample evidence exists to indicate that photogrammetric methods provide adequate accuracy for property surveys (Ref. 4). Acceptance of photogrammetry as one of the tools useful for performing property surveys is bound to come as more surveyors become familiar with the procedures (Chap. 13) and as the legality of the method is recognized in the courts.

Directions of lines for property surveys are usually referred to the true or astronomic meridian, and angular measurements are transformed into bearings. Distances are still given in feet and decimal parts of the foot, but with metrication on the horizon, it is recommended that distances be given in feet and in metres. Angles are measured to minutes or fractions thereof. On the U.S. public-land surveys, all distances are in Gunter's chains (66 ft) with 100 links per chain as prescribed by law.

Formerly, the surveyor's compass and 66-ft link chain were used extensively, particularly in rural surveying; and the directions and lengths of lines contained in many old deeds are given in terms of magnetic bearings and Gunter's chains. In retracing old surveys of this character, allowance must be made for change in magnetic declination since the time of the original survey. Also, it must be kept in mind that the compass and link chain used on old surveys were relatively inaccurate instruments and that great precision was not regarded as necessary since usually the land values were low. Further, for many years the U.S. public lands were surveyed under contract, at the low price of a few dollars per mile. Many of the lines and corners established by old surveys are not where they theoretically should be; nevertheless, these boundaries legally remain fixed as they were originally established.

Wherever possible, the field procedure is such that the lengths of boundary lines and the angles between boundaries are determined by direct measurement. Therefore, the land survey is in general a traverse, the stations being at corners of the property, and the traverse lines coinciding with property lines. Where obstacles render direct measurement of boundaries impossible, such as in Fig. 15.1, a traverse is run as near the property lines as practical, and measurements are made from the

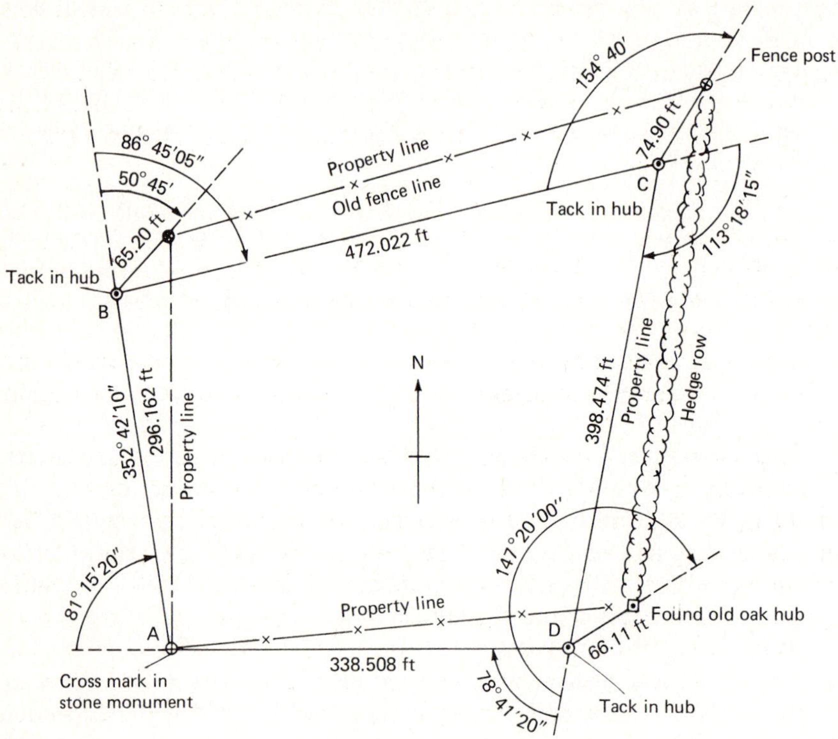

Fig. 15.1 Traverse to locate property lines.

traverse to property corners; the lengths and directions of the property lines are then calculated. Where the boundary is irregular or curved, the traverse is established in a convenient location, and offsets are taken from the traverse line to points on the boundary; the length of the boundary is then calculated.

In general, the required precision of land surveys depends upon the value of the land. The possibility of increase in land values should also be considered.

Wherever possible, the survey utilized to establish property corners should be tied to monuments for which state plane coordinates are available (Arts. 11.16 to 11.20). Use of state plane coordinates integrates all surveys into a common system, facilitates replacing lost corners, and provides a means of parcel identification that is compatible with filing survey data in electronic computer data banks.

15.4. Boundary records Descriptions of the boundaries of real property may be found from deeds, official plats or maps, or notes of original surveys.

Records of the transfer of land from one owner to another are usually kept in the office of the county registry of deeds, exact copies of all deeds of transfer being filed in deed books. These files are open to the public and are a frequent source of information for the land surveyor in search of boundary descriptions when it is inconvenient or impossible to secure permission of the owner to examine the original deed.

Originally, all official records were copied by hand into the record books. The first step toward modernization of this system was to make photocopies of each record, which were still filed in the books. As records proliferated, many recording offices switched to making microfilm copies of new and old records so as to save space. The microfilm copies can be stored on cards (microfiche) and then are filed in a system that permits manual or automatic retrieval.

No matter what system of filing the records is employed, some method for indexing is required. The system that has and still is in most general use in the United States consists of an alphabetic index, usually kept by years, giving in one part the names of the *grantors* or persons selling property and in the other the names of the *grantees* or persons buying property. Thus, if either or both parties to the transfer and the approximate date of the transfer are known, a given deed can be located. Usually, the preceding transfer of the same property is noted on the margin of the deed.

In most cases the deeds of transfer of city lots give only the lot or block number and the name of the addition or the subdivision. The official plat or map showing the dimensions of all lots and the character and location of permanent monuments is on file either in the office of the city clerk or in the county registry of deeds; copies are also on file in the offices of city and county assessors. Filing land transfer records according to these data is called indexing by *lot, block, and parcel,* or more simply indexing by *land parcel.*

Current trends are toward storing legal records related to land transfers in computer data banks (see Arts. 1.11 and 3.15). Such a procedure has the advantage that title searches can be performed more rapidly and at lower cost. Most title companies already have automated or semiautomated search procedures for land records. Electronic computer storage and retrieval methods and interactive computer techniques are utilized in these procedures (Ref. 13). The problems encountered in such systems are mostly related to the fact that the grantor-grantee alphabetic listing is not exactly compatible with efficient computer usage and a unique identifier for each parcel involved in a land transfer is necessary.

15.5. Property descriptions There are two basic methods for describing property: (1) metes and bounds descriptions, and (2) subdivision descriptions.

In the older portions of the United States, nearly all of the original land grants were of irregular shape, many of the boundaries following stream and ridge lines. Also, in the process of subdivision the units were taken without much regard for regularity, and it was thought sufficient if lands were specified as bounded by natural or artificial features of the terrain and if the names of adjacent property owners were given. An example of a description of a tract as recorded in a deed reads:

> Bounded on the north by Bog Brook, bounded on the northeast by the irregular line formed by the southwesterly border of Cedar Swamp of land now or formerly belonging to Benjamin Clark, bounded on the east by a stone wall and land now or formerly belonging to Ezra Pennell, bounded on the south and southeast by the turnpike road from Brunswick to Bath, and bounded on the west by the irregular line formed by the easterly fringe of trees of the wood lot now or formerly belonging to Moses Purington.

As the country developed and land became more valuable, and as many boundaries such as those listed in the preceding description ceased to exist, land litigations became numerous. It then became the general practice to determine the lengths and directions of the boundaries of land by measurements. In describing a tract surveyed in this manner the lengths and bearings of the several courses are given in order, and the objects marking the corners are described; if any boundary follows some prominent feature of the terrain, the fact is stated; and the calculated area of the tract is given. When the bearings and lengths of the sides are thus given, the tract is said to be described by *metes and bounds*. Within the limits of the precision of the original survey, it is possible to relocate the boundaries of a tract if its description by metes and bounds is available, provided that at least one of the original corners can be identified and the true direction of one of the boundaries can be determined.

There are several types of metes and bounds descriptions, the most common of which is the method of *successive bounds*, as discussed above. The type of metes and bounds description in Example 15.1 is sometimes regarded as the true metes and bounds description.

Example 15.1 Alan Tart has sold a portion of his property to Richard Merlin. Tart's land was obtained in a previous conveyance from Biggert to Tart. There is no recorded plan of lots (subdivision), so that a metes and bounds description is necessary for the deed that is the legal document in which the transaction from Tart to Merlin will be recorded. Prior to the sale, the land to be sold is surveyed and the corners are marked as illustrated in Fig. 15.2. The following is one form of a deed description with a metes and bound description by successive bounds:

A parcel of land in the City of _____, County of _______________, State of _______________, being a part of the land conveyed by John Biggert to Alan Tart by deed dated June 14, 1956, and recorded in Book 201, page 131, at the ______________ County Recorder's Office, and described as follows:

Beginning at a concrete monument on the northerly line of Woodlawn Avenue, said monument bearing S47°40′E a distance of 100.00 ft (30.480 m) along the northerly line of

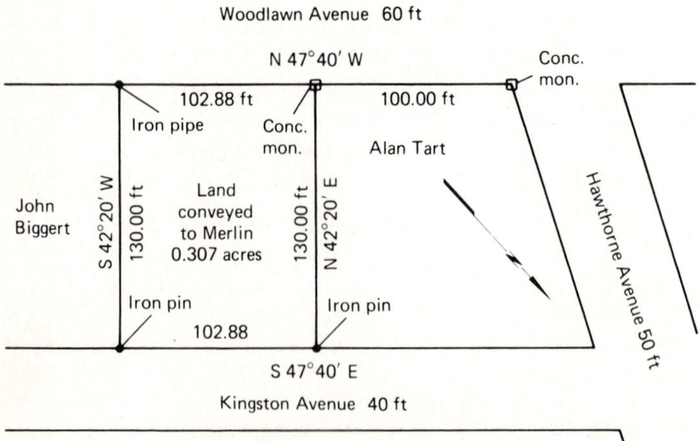

Fig. 15.2 Plan of property described in Example 15.1.

Woodlawn Avenue from the intersection of the northerly line of Woodlawn Avenue with the easterly line of Hawthorne Avenue and running: (1) thence along the northerly line of Woodlawn Avenue S47°40′E a distance of 102.88 ft (31.358 m) to an iron pipe (2 in or 51 mm inside diameter) at the line of the land of John Biggert; (2) thence along said land N42°20′E a distance of 130.00 ft (39.624 m) to an iron pin ($\frac{1}{2}$ in or 13 mm outside diameter) on the southerly line of Kingston Avenue; (3) thence along the southerly line of Kingston Avenue N47°40′W a distance of 102.88 ft (31.358 m) to an iron pin ($\frac{1}{2}$ in or 13 mm outside diameter) at the line of the land of the grantor; (4) thence along the land of the grantor S42°20′W a distance of 130.00 ft (39.624 m) to the point of beginning and containing 0.307 acres, more or less.

All directions are based on the stated bearing of the northerly line of Woodlawn Avenue.

This description was written on July 1, 1978, by _______________, Professional Land Surveyor, Registration No. _______________, State of _______________, and is based on data gathered in the field by _______________, on June 15, 1978.

The deed contains a general statement which indicates the approximate location of the land. The body of the deed includes a specific description of the land conveyed by the method of successive bounds. Note that the description has a point of beginning which is referenced to the block corner. Such a reference is necessary to fix the position of the land being conveyed. Then each course is described in turn and the description ends with a return to the point of beginning.

Following the description are given the basis for bearings, the name and registration number of the surveyor, and the dates when the deed was written and field work performed.

The basis for bearings should always be stated clearly. Thus, bearings should be described as being astronomic, grid, or as being referenced to a stated line. The basis for bearings can be stated indirectly. In Example 15.1, the initial statement "said monument bearing S47°40′E along the northerly line of Woodlawn Avenue..." fixes the basis for bearings along that line. Hence, the statement on basis for bearings, following the body of the deed, is redundant only in situations such as this example.

Several other types of metes and bounds descriptions exist, as follows:

1. *Strip conveyances* are employed to describe a road easement or right of way. The form used is "a right of way for highway purposes over and across a strip of land lying 25 ft (7.620 m) on either side of the following described center line. . . ." Stationing is used to denote distance along the center line.
2. *Conveyance by division line*: "All of lot 62 lying south of Rock Creek" is a description by division line.
3. *Conveyance by distance*: "The westerly 20 m of lot 32" constitutes the very brief description by distance.
4. *Proportional conveyances* are those which describe a fractional part of the whole tract, such as "the eastern 1/8 of lot 11."
5. Land may be conveyed by *exception*, such as "all except the westerly 100 ft of lot 61," or by *acreage*, such as "the west 2 acres of lot 15."

Further details concerning the variations on the true metes and bounds descriptions listed above can be found in Ref. 2.

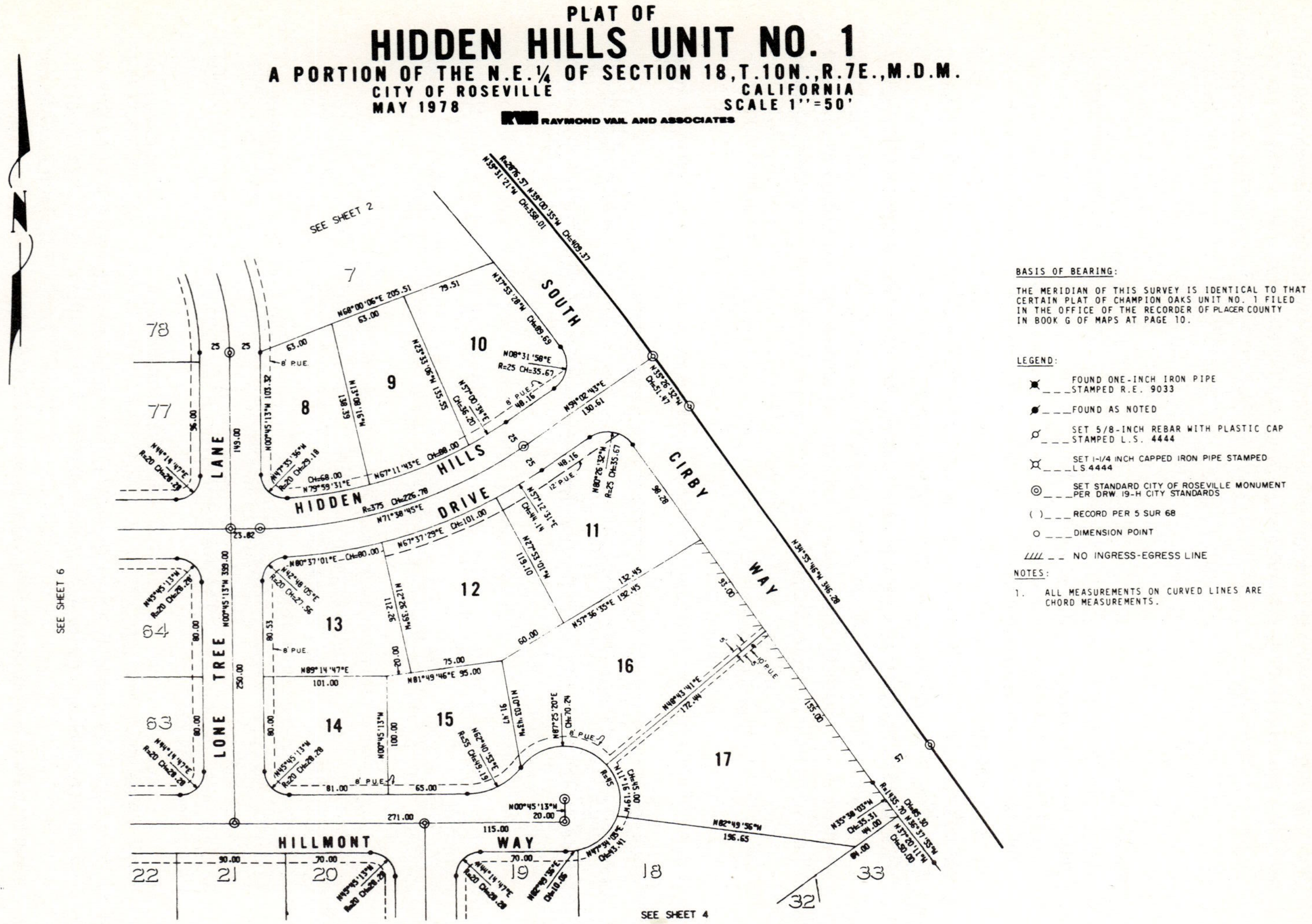

Fig. 15.3 Portion of a subdivision plotted automatically. (Courtesy Raymond Vail and Associates.

When land has been partitioned into lots and a map or plat showing the manner in which the land has been divided and also satisfying state, county, and local laws is placed on record, a subdivision has been created. The plat shows all dimensions, directions, areas, and monuments defining or delineating all lots of the subdivision (see Fig. 15.3). Sectionalized land surveyed according to the Public Land Surveys in the United States (Arts. 15.9 to 15.12) and private subdivisions fall within this definition.

Most states have laws that control subdivision development. In the more densely populated regions, counties and local municipal governments also have laws regulating subdivisions.

When the boundaries of the property to be described coincide exactly with a lot within a subdivision for which there is an official recorded map, the lot may be described legally by a statement giving the lot and block numbers and the name, date, and place of filing the map. Most city property and many lots in urban developments are described in this way. The following are descriptions of this type:

Example 15.2 Lot 29, Block 0, Map of Berkeley Highlands Addition, filed November 20, 1912, in Map Book 8, Page 194, Contra Costa County Records, State of California.

Lot 15 in Block 5 as said lots and blocks are delineated and so designated upon that certain map entitled *Map of Thousand Oaks, Alameda County, California*, filed August 23, 1909, in Liber 25 of Maps, page 2, in the office of the County Recorder of the said County of Alameda.

If the boundaries of a given tract within a subdivision for which there is a recorded map do not conform exactly to boundaries shown on the official map, the tract is described by metes and bounds, with the point of beginning referred to a corner shown on the official map. Also, the numbers of lots of which the tract is composed are given. Following is an example of a description of this kind:

Example 15.3 Beginning at the intersection of the Northern line of Escondido Avenue, with the Eastern boundary line of Lot 16, hereinafter referred to; running thence Northerly along said Eastern boundary line of Lot 16, and the Eastern boundary line of Lot 17, eighty-nine (89) feet; thence at right angles Westerly, fifty-one (51) feet; thence South 12°06′ East, seventy-five (75) feet to the Northern line of Escondido Avenue; thence Easterly along said line of Escondido Avenue, fifty-three and $\frac{13}{100}$ (53.13) feet, more or less, to the point of beginning.

Being a portion of Lots 16 and 17, in Block 5, as said lots and blocks are delineated and so designated upon that certain map entitled *Map of Thousand Oaks, Alameda County, California,* filed August 23, 1909, in Book 25 of Maps, page 2, in the office of the County Recorder of the said County of Alameda.

15.6. Riparian rights An owner of property that borders on a body of water is a riparian proprietor and has riparian rights (pertaining to the use of the shore or of the water) which may be valuable. Because of the difficulties arising from the irregularity of such boundaries, it is important that the surveyor be familiar with the general principles relating to riparian rights and with the statutes and precedents established in the particular state where work is being performed. For example, as regards the ownership of the bed of a navigable river, the two states of Iowa and Illinois bordering on the same river have very different laws. Clark (Ref. 5) states: "It is a rule of property in Illinois, that the interest of the riparian owner of land in that state bordering on the Mississippi River extends to the middle line of the main channel of the river," whereas the Iowa courts hold "that the bed of the Mississippi River and the banks to the high-water mark belong to the state, and that the title of the riparian proprietor extends only to that line."

15.7. Legal authority of the surveyor A resurvey may be run to settle a controversy between owners of adjoining property. The surveyor should understand that, although it is possible to act as an arbiter in such cases, it is not within the power of the surveyor legally to fix boundaries without the mutual consent and authority of all interested parties. In the event of a dispute involving court action, the surveyor may present evidence and argument as to the proper location of a boundary, but has no authority to establish such a boundary against the wishes of either party concerned. A competent surveyor by wise counsel can usually prevent litigation; but if no agreement occurs, the boundaries in dispute become valid and defined only by a decision of the court. In boundary disputes the surveyor is an expert witness, not a judge.

The surveyor is therefore a professional who obtains a license to practice the profession either by examination or by a certain time of practical experience or combination of both.

The right to enter upon property for the purpose of making public surveys is usually provided by law, but there is no similar provision regarding private surveys. The surveyor (or employer, whether public or private) is liable for damage caused by cutting trees, destroying crops or fences, etc.

Permission should be obtained before entering private property, in order to avoid a possible charge of trespassing. If permission is not granted, it may be necessary to traverse around the property or to resort to aerial photographs. In certain cases, as for example, where a boundary is in doubt or dispute, a court order may have to be obtained in order to enter the property.

15.8 Liability of the surveyor It has been held in court decisions that county surveyors and surveyors in private practice are members of a learned profession and may be held liable for incompetent services rendered. Thus, Clark (Ref. 5), quoting from court decisions, states: "If a surveyor is notified of the nature of a building to be erected on a lot, he may be held liable for all damages resulting from an erroneous survey; and he may not plead in his defense that the survey was not guaranteed." Similarly, it has been held that in any case where the surveyor knows the purpose for which the survey is made, he or she is liable for damages resulting from incompetent work.

The general principle invoked in such cases is that the surveyor is bound to exhibit that degree of prudence, judgment, and skill which may reasonably be expected of a member of the profession. Thus, in the following quotations from Clark, a Connecticut court says "the gist of the plaintiff's cause of action was the negligence of the defendant in his employment as a civil engineer. Having accepted that service from the plaintiff, the defendant...was bound to exercise that degree of care which a skilled civil engineer would have exercised under similar circumstances." Also, a Kansas court declared, "reasonable care and skill is the measure of the obligation created by the implied contract of a surgeon, lawyer, or any other professional practitioner." But Ruling Case Law says, "...yet a person undertaking to make a survey does not insure the correctness of his work, nor is absolute correctness the test of the amount of skill the law requires. Reasonable care, honesty, and a reasonable amount of skill are all he is bound to bring to the discharge of his duties."

15.9. U.S. public-lands surveys: general This section deals with the methods of subdividing the public lands of the United States in accordance with regulations imposed by law. The public lands are subdivided into townships, sections, and quarter sections—in early years by private surveyors under contract, later by the Field Surveying Service of the General Land Office, and currently by civil service employees of the Bureau of Land Management which succeeded the General Land Office in 1946. Further subdivision of such lands is made after the lands have passed into the hands of private owners, the work being carried out by surveyors in private practice.

The methods described herein are those now in force, but with minor differences they have been followed in principle since 1785, when the rectangular system of subdivision was inaugurated. Under this system, the public lands of 30 states have been or are in progress of being surveyed. In general, these methods of subdividing land do not apply in the 13 original states and in Hawaii, Kentucky, Maine, Tennessee, Texas, Vermont, and West Virginia. As the progress of the public-land surveys has been from east to west, the details of the methods used earlier, in states east of the Mississippi River, differ somewhat from the details of those used at present. Ohio is unique among the states in that it contains a little of every system of land surveying.

The laws regulating the subdivision of public lands and the surveying methods employed are fully described in the *Manual of Instructions for the Survey of the Public Lands of the United States*, published by the Bureau of Land Management (Ref. 12). From the manual is drawn much of the material for this chapter.

Field notes and plats of the public-land surveys may be examined in the regional offices of the Bureau, and copies may be procured for a nominal fee.

The regulations for the subdivision of public lands have been altered from time to time; hence, the methods employed in surveying various regions of the United States show marked differences, depending upon the dates when the surveys were made. In general principle, however, the system has remained unchanged, the primary unit being the *township*, bounded by meridional and latitudinal lines and as nearly as may be 6 mi (9.66 km) square. The township is divided into 36 secondary units called *sections*, each as nearly as may be 1 mi (1.61 km) square. Because the meridians converge, it is impossible to lay out a square township by such lines; and because the township is not square, not all the 36 sections can be 1 mi square even though all measurements are without error.

Since the time of the earliest surveys, the townships and sections have been located with respect to principal axes passing through an origin called an *initial point*; the north-south axis is a true or astronomic meridian called the *principal meridian*, and the east-west axis is an astronomic parallel of latitude called the *base line* (see Fig. 15.4).

The principal meridian is given a name to which all subdivisions are referred. Thus, the principal meridian which governs the rectangular surveys (in part) of the states of Ohio and Indiana is called the First Principal Meridian. The extent of the surveys which are referred to a given initial point may be found by consulting a map, published by the Bureau of Land Management, entitled "United States, Showing Principal Meridians, Base Lines, and Areas Governed Thereby," or from Ref. 12.

Secondary axes are established at intervals of 24 mi (38.62 km) east or west of the principal meridian and at intervals of 24 mi north or south of the base line, thus dividing the tract being surveyed into quadrangles bounded by astronomic meridians 24 mi long and by true or astronomic parallels, the south boundary of each quadrangle being 24 mi long, and the north boundary being 24 mi long less the convergence of the meridians in that distance. [In some early surveys, these distances were 30 (48.28 km) or 36 (57.94 km) mi.] The secondary parallels are called *standard parallels* or *correction lines*, and each is continuous throughout its length. The secondary meridians are called *guide meridians*, and each is broken at the base line and at each standard parallel.

The principal meridian, base line, standard parallels, and guide meridians are called *standard lines*.

A typical system of principal and secondary axes is shown in Fig. 15.4. The base line and standard parallels, being everywhere perpendicular to the direction of the

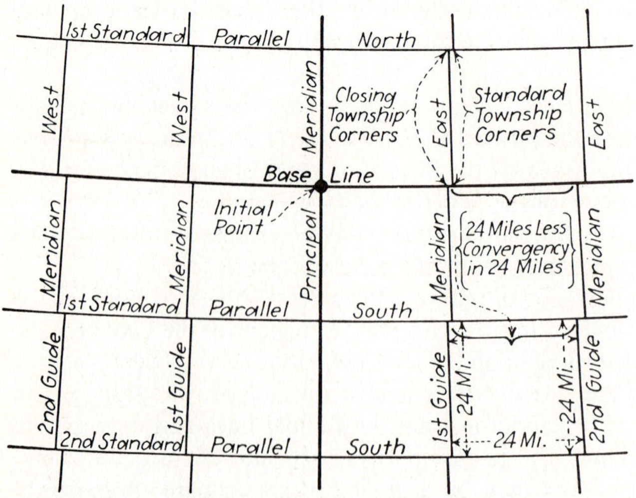

Fig. 15.4 Standard lines.

meridian, are laid out on the ground as curved lines, the rate of curvature depending upon the latitude. The principal meridian and guide meridians, being astronomic north and south lines, are laid out as straight lines but converge toward the north, the rate of convergence depending upon the latitude.

Standard parallels are counted north or south of the base line; thus, the *second standard parallel south* indicates a parallel 48 mi (77.25 km) south of the base line. Guide meridians are counted east or west of the principal meridian; thus the *third guide meridian west* is 72 mi (115.87 km) west of the principal meridian.

The division of the 24-mi (38.62-km) quadrangles into townships is accomplished by laying off astronomic meridional lines called *range lines* at intervals of 6 mi (9.66 km) along each standard parallel, the range line extending north 24 mi to the next standard parallel; and by joining the township corners established at intervals of 6 mi (9.66 km) on the range lines, guide meridians, and principal meridian with latitudinal lines called *township lines.*

The plan of subdivision is illustrated by Fig. 15.5. A row of townships extending north and south is called a *range*; and a row extending east and west is called a *tier*. Ranges are counted east or west of the principal meridian and tiers are counted north or south of the base line. Usually, for purposes of description the word "township" is substituted for "tier." A township is designated by the number of its tier and range and the name of the principal meridian. For example, T7S, R7W (read *township seven south*, *range seven west*) designates a township in the seventh tier south of the base line and the seventh range west of the principal meridian.

The division of townships into sections is performed by establishing, at intervals of 1 mi (1.61 km), "meridional" lines parallel to the east boundary of the township and by joining the section corners established at intervals of 1 mi with straight latitudinal lines. (These lines are not exactly meridional, but are parallel to the east boundary of the township, which is a meridional line.) These lines, called *section lines*, divide each township into 36 sections, as shown in Fig. 15.6. Monuments are set every $\frac{1}{2}$ mi or 40 chains (0.80 km) along the section lines. The sections are numbered consecutively from east to west and from west to east, beginning with No.

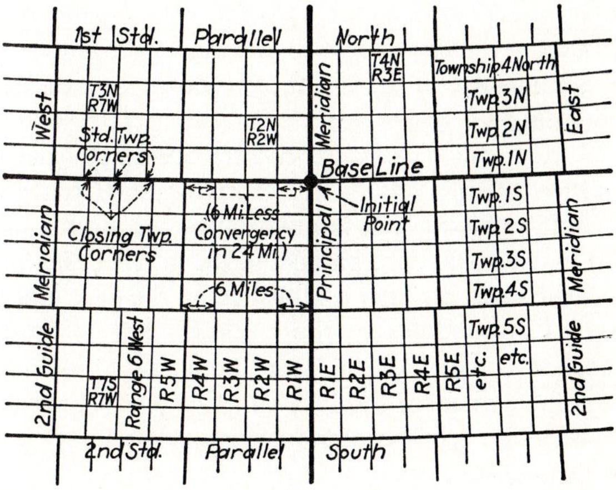

Fig. 15.5 Township and range lines.

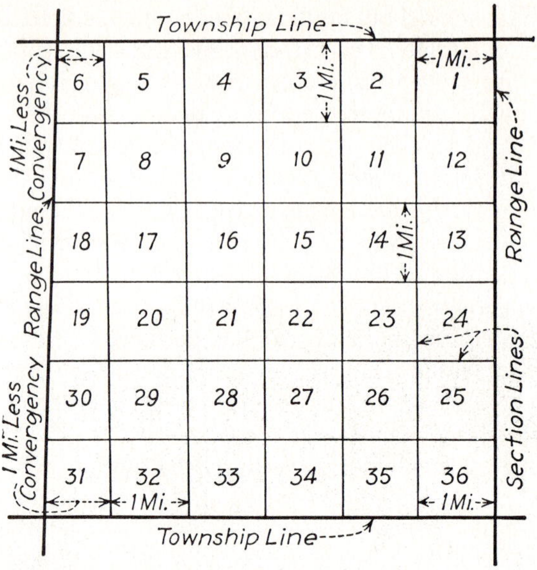

Fig. 15.6 Numbering of sections.

1 in the northeast corner of the township and ending with No. 36 in the southeast corner. Thus section 16 is a section whose center is $3\frac{1}{2}$ mi (5.63 km) north and $3\frac{1}{2}$ mi (5.63 km) west of the southeast corner of a township.

A section is legally described by giving its number, the tier and range of the township, and the name of the principal meridian; for example, Sec. 16, T7S, R7W, of the Third Principal Meridian (usually abbreviated, as 3PM).

On account of the convergence of the range lines (true or astronomic meridians) forming the east and west boundaries of townships, the latitudinal lines forming the north and south boundaries of townships are less than 6 mi (9.66 km) in length, except for the south boundary of townships that lie just north of a standard parallel. As the north-south section lines are run parallel to the *east* boundary of the township, it follows that all sections except those adjacent to the west boundary will be 1 mi (1.61 km) square, but that those adjacent to the west boundary will have a latitudinal dimension less than 1 mi by an amount equal to the convergence of the range lines within the distance from the section to the nearest standard parallel to the south.

Corners called *standard corners* are established on the base line and standard parallels at intervals of 40 chains or 2640 ft (804.67 m); these standard corners govern the meridional subdivision of the land lying between each standard parallel and the next standard parallel to the north. Other corners called *correction corners* or *closing corners* are later established on the base line and standard parallels during the process of subdivision; these corners fall at the intersection of the base line or standard parallel either with the meridional lines projected from the standard township corners of the next standard parallel to the south (see Fig. 15.5) or with the intermediate section and quarter-section lines. Standard parallels are also called *correction lines*.

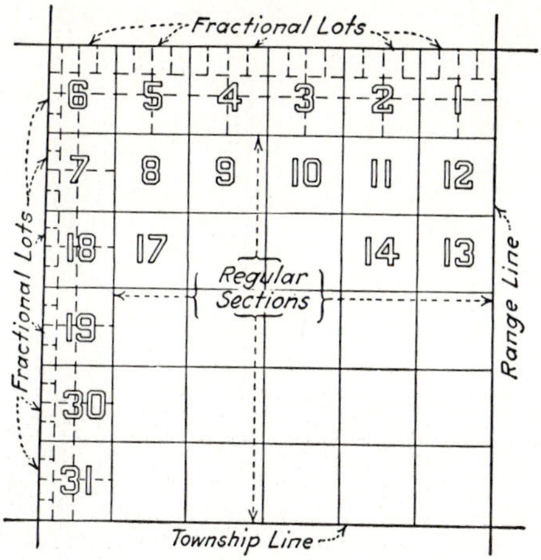

Fig. 15.7 Subdivision of a township.

It should be understood that the plan of subdivision just described is the one which is carried out when conditions allow. Of course, errors of measurement are always present so that the actual lengths and directions established in the field do not entirely agree with the theoretical values. But in addition, conditions met in the field often make it inexpedient or impossible to establish the lines of the survey in exact accordance with the specified plan. The inevitable errors of measurement which do occur are thrown into the west and north row of sections no matter whether it is an excess or deficiency.

Before considering further the methods employed in the subdivision of townships, the legal requirement relative to the rectangular surveys of the public lands should be stated. Of the 36 sections in each normal township (Fig. 15.7), 25 are returned as containing 640 acres (259 ha) each; 10 sections adjacent to the north and west boundaries (comprising sections 1–5, 7, 18, 19, 30, and 31) each containing regular subdivisions totaling 480 acres (194 ha) and in addition 4 *fractional lots* each containing 40 acres (16.2 ha) plus or minus definite differences to be determined in the survey; and 1 section (section 6) in the northwest corner contains regular subdivisions totaling 360 acres (146 ha) and in addition 7 fractional lots each containing 40 acres (16.2 ha) plus or minus certain definite differences to be determined in the survey. The aliquot† parts of 640 acres (259 ha), or regular subdivisions of a section, are the quarter section ($\frac{1}{2}$ mi or 0.805 km square containing 160 acres), the half-quarter or eighth section ($\frac{1}{4}$ by $\frac{1}{2}$ mi or 0.402 by 0.805 km containing 80 acres), and the quarter-quarter or sixteenth section ($\frac{1}{4}$ mi square or 0.402 by 0.402 km), the last containing 40 acres (16.2 ha) and being the legal minimum for purposes of disposal under the general land laws.

15.10. Subdivision of sections Upon the official government township plats the interior boundaries of quarter sections are shown as dashed straight lines connecting

†A part of the area which divides the area without a remainder (Ref. 1, Manual No. 32).

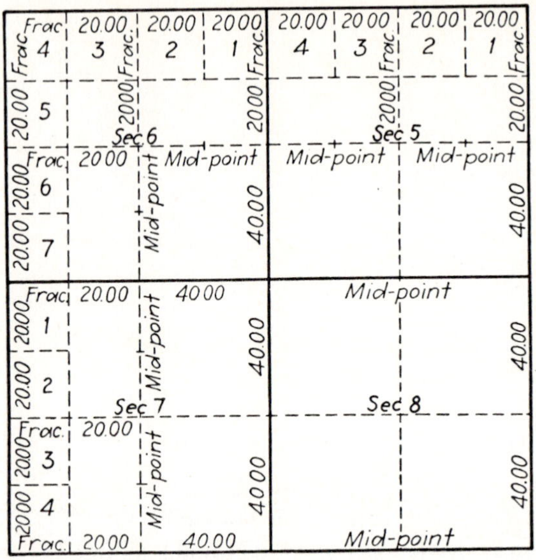

Fig. 15.8 Subdivision of sections.

opposite quarter-section corners. The sections adjacent to the north and west boundaries of a normal township, except section 6, are further subdivided *by protraction* into parts containing two regular half-quarter sections and four *lots*, the latter containing the fractional areas resulting from the plan of subdivision of the normal township. This process of plotting the interior, unsurveyed boundaries on the official plat is called *subdivision by protraction*. Figure 15.8 illustrates the plan of the normal subdivision of sections. The regular half-quarter sections are protracted by laying off a full 20 chains (1320 ft or 402.34 m) from the line joining opposite quarter-section corners. The lines subdividing the fractional half-quarter sections into the fractional lots are protracted from midpoints of the opposite boundaries of the fractional quarter sections.

In section 6 the two quarter-quarter-section corners on the south and east boundaries of the fractional northwest quarter are similarly fixed, one at a point 20 chains north and the other at a point 20 chains west of the center of the section, from which points lines are protracted to corresponding points on the west and north boundaries of the section. Hence, the subdivision of the northwest quarter of section 6 results in one regular quarter-quarter section and three lots.

In all sections bordering on the north boundary the fractional lots are numbered in succession beginning with No. 1 at the east. In all sections bordering on the west boundary the fractional lots are numbered in succession beginning with No. 1 at the north, except section 6 which, being common to both north and west boundaries, has its fractional lots numbered in progression beginning with No. 1 in the northeast corner and ending with No. 7 in the southwest corner, all as illustrated by Fig. 15.8.

Figure 15.9 illustrates a typical plat of section 6 on which the protracted areas are shown. Figure 15.10 is a similar section giving the calculated dimensions of the protracted areas.

In addition to sections made fractional by reason of their being adjacent to the north and west boundaries of a township, there are also sections made fractional on

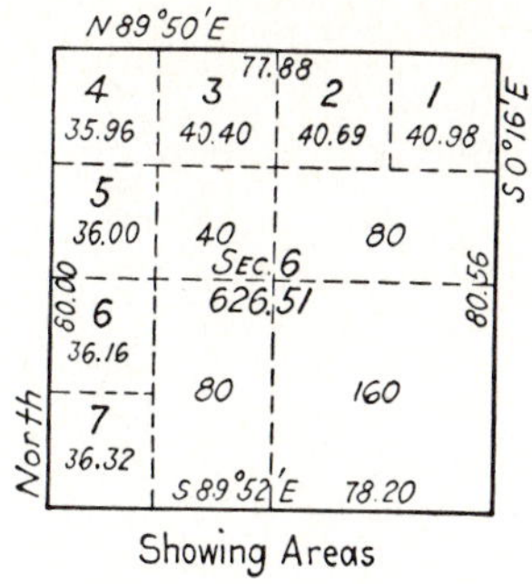

Fig. 15.9 Subdivisional areas of section 6 in acres.

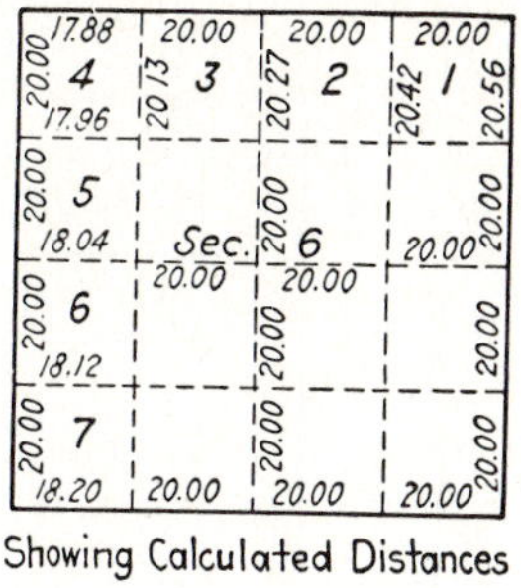

Fig. 15.10 Subdivisional dimensions of section 6 in chains.

account of meanderable bodies of water, mining claims, Indian Reservations, military land grants, and other segregated areas within their limits. Such sections are subdivided by protraction into such regular and fractional parts as are necessary for the entry of the undisposed public lands and to describe these lands separately from the segregated areas.

A major advantage of the public-lands survey system is the ease with which the aliquot parts of a section of land can be legally described and located. In order to visualize this procedure, refer to Fig. 15.11 which shows four quadrangles of 16 townships. The shaded area in the lower right quadrangle of Fig. 15.11 represents one township. Figure 15.12*a* illustrates the township which is shaded in Fig. 15.11. The shaded area in Fig. 15.12*a* is described as Section 23, Township 2 South, Range 4 East of the Second Principal Meridian or abbreviated as Sec. 23, T2S, R4E, 2PM. Section 23 contains 640 acres, more or less. There may be some error in measurement in the laying out of Section 23 so the term "more or less" is used to indicate that the land in question is indeed Section 23 regardless of whether or not it contains exactly 640 acres.

Further subdivision of the section is illustrated in Fig. 15.12*b*. In this figure, tract A is the northeast quarter of Section 23. The description of tract A is NE$\frac{1}{4}$, Sec. 23, T2S, R4E, 2PM. Tract A contains 160 acres. In the same figure, tract B contains 40 acres and is described as the southwest quarter of the northwest quarter of Section 23,..., etc., or SW$\frac{1}{4}$, NW$\frac{1}{4}$, Sec. 23, T2S, R4E, 2PM. Tract J contains 80 acres and is described as E$\frac{1}{2}$, NW$\frac{1}{4}$, Sec. 23, T2S, R4E, 2PM.

As another example, refer to Fig. 15.9, where the tract numbered 4 is described as Lot 4, Sec. 6, T—,..., etc. From the plat (Fig. 15.9), the area of lot 4 is 35.96 acres. The dimensions of lot 4, Sec. 6 are shown in Fig. 15.10.

The public-lands survey system is only the skeleton to which property descriptions are tied. In the case of irregular tracts or tracts which are not aliquot parts of a section, a standard metes and bounds description (Art. 15.5) is written. These descriptions tie the property to one or more corners of an aliquot part of a section. As an example, assume that the property shown in Fig. 15.1 represents an irregular tract in a region subject to the public-lands survey system. To describe this property, a metes and bounds description could be written using station A as the point of

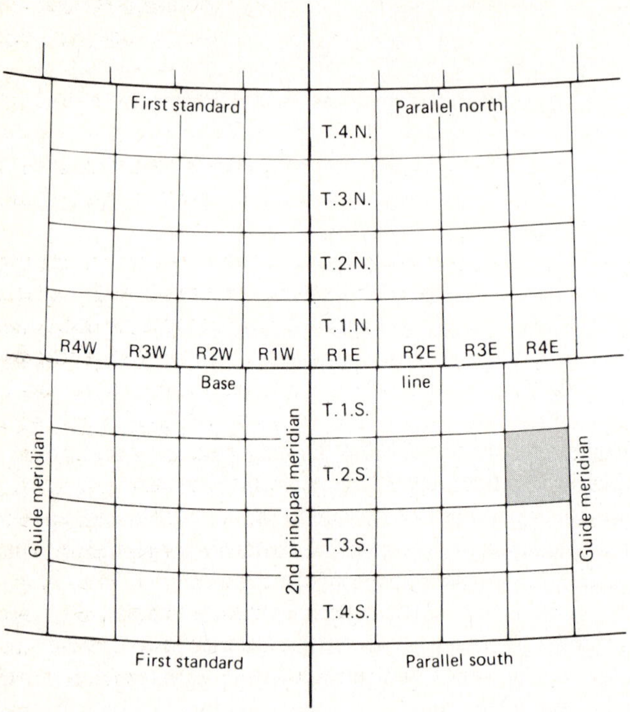

Fig. 15.11 Four quadrangles of 16 townships, each bounded by standard lines.

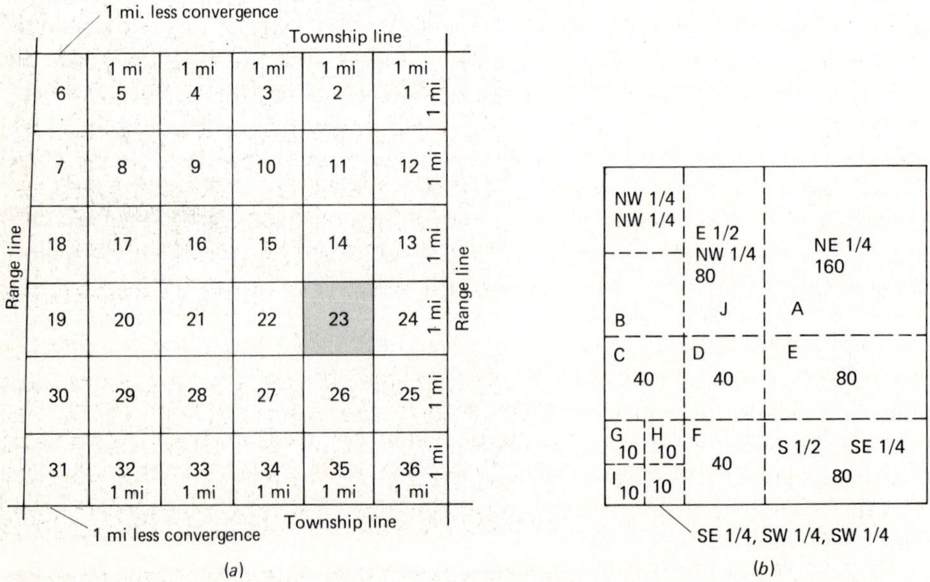

Fig. 15.12 *Reprinted with permission from AVI Publishing Company, Inc., from An Introduction to Agricultural Engineering, 5th Printing, 1982 by Roth, Crow, and Mahoney.*

beginning. Then a bearing and distance to station A from the nearest section or quarter section corner must be measured and included in the description. This procedure is similar to that shown in Example 15.2, where the point of beginning was referenced to the intersection of Woodlawn Avenue and Hawthorne Avenue.

15.11. Meandering In the process of surveying the public lands, all navigable bodies of water and other important rivers and lakes below the line of mean high water are segregated from the lands which are open to private ownership. In the process of subdivision, the regular section lines are run to an intersection with the mean high-water mark of such a body of water, at which intersection corners called *meander corners* are established. The traverse which is run between meander corners, approximately following the margin of a permanent body of water, is called a *meander line*, and the process of establishing such lines is called *meandering*. The mean high-water mark is taken as the line along which vegetation ceases. The fact that an irregular line must be run in tracing the boundary of a reservation does not entitle such a line to be called a meander line except where it follows closely the shore of a lake or the bank of a stream.

Meander lines are not boundaries but are lines run for the purpose of locating the water boundaries approximately, and although the official plats show fractional lots as bounded in part by meander lines, it is an established principle that ownership does not stop at such meander lines.

15.12. Marking lines between corners As a final step in the survey of the public lands, it is the aim to fix the location of the legal lines of subdivision permanently with reference to objects on the surface of the earth. This aim is accomplished (1) by setting monuments at the regular corners, (2) by finding the location of the officially surveyed lines with respect to natural features of the terrain, and (3) by indicating the position of the regular lines through living timber by *blazing* and by *hack marks*.

The last method of fixing the location of the regular subdivisional lines is required by law just as definitely as is the establishment of monuments at the corners. All legal lines of the public-land surveys through timber are marked in this manner. Those trees which lie on the line, called *line trees*, are marked with two horizontal notches, called *hack marks*, on each side of the tree facing the line; and an appropriate number of trees on either side of the line and within 50 links (33 ft or 10.06 m) thereof are marked by flat axe marks, called *blazes*, a single blaze on each of two sides quartering toward the line.

In the subdivision of the public lands as described in the preceding sections, it is required that the U.S. surveyors shall permanently mark the location of the township, section, exterior quarter section, and meander corners, as well as those quarter-quarter-section corners necessary in connection with the subdivision of fractional sections. Monuments of a character specified by regulations of the Bureau of Land Management are employed for this purpose. The location of every such corner monument is, in accordance with definite rule, referred to such nearby objects as are available and suitable for this purpose; and where the corner itself cannot be marked in the ordinary manner an appropriate witness corner is established.

Where a true corner point falls within an unmeandered stream or lake, within a marsh, or in an inaccessible place, a witness corner is established in a convenient location nearby, preferably on one of the surveyed lines leading to the location of the regular corner. Also, where the true point falls within the traveled limits of a road, a cross-marked stone is deposited below the road surface, and a witness corner is placed in a suitable location outside the roadway.

The Bureau of Land Management has adopted a standard iron post for monumenting the public-land surveys. This post is to be used unless exceptional circumstances warrant the use of other material (Ref. 12).

Where the procedure is duly authorized, durable native stone may be substituted for the model iron post described above, provided that the stone is at least 20 in (51 cm) long and at least 6 in (15.2 cm) in its least lateral dimension. Stone may not be used as a monument for a corner whose location is among large quantities of loose rock. The required corner markings are cut with a chisel, and usually the stone is set with about three-fourths of its length in the ground.

For a complete treatment of the system of marking corner monuments established in the survey of the public lands, the reader is referred to the *Manual of Instructions for the Survey of Public Lands of the United States* of the Bureau of Land Management.

15.13. Photogrammetric methods applied to property surveys In Chap. 13 methods for determining positions of photographed objects were developed. Qualitative interpretation of terrain features is also possible with overlapping aerial photographs. These two aspects of photogrammetry, quantitative and qualitative, make it a powerful tool for performing property surveys.

The procedure involved in preparing and executing a photogrammetric property survey is similar in many respects to the methods described for original surveys, resurveys, and subdivision surveys. The primary differences are in the methods employed to determine horizontal positions of points. Generally, the surveyor and photogrammetrist work as a team on a photogrammetric surveying project. Few individuals possess sufficient experience and expertise to do the job alone. Consider, as an example, a subdivision survey for a large tract of land.

As in any property survey, all available maps, aerial photographs, descriptions of the property to be surveyed, descriptions of adjacent lands, descriptions and references to all existing horizontal and vertical control points, and all field notes should be assembled and analyzed prior to any field operations or photogrammetric work. Assume for purposes of this discussion that a topographic map adequate for planning is available for the area and that a tentative plan of streets and lots has been prepared for the subdivision. The objective of the photogrammetric survey is to establish horizontal positions of control points for two purposes: (1) in order to fix the positions of boundaries around the perimeter of the tract; and (2) to provide horizontal control points around the perimeter and throughout the interior of the area with sufficient density to permit locating subdivision monuments and center-line control points for subsequent layout of all street center lines and property corners in the subdivision.

Prior to taking the aerial photography, all ground control points, photogrammetric control points, existing monuments and bench marks, and all potential and identified property corners are marked with targets. These targets should be of a size, configuration, and color suitable for the photogrammetric analysis. In choosing the size of the target, the scale of the photography to be flown must be taken into account.

After all points have been marked, or concurrently with the placing of targets, a traverse is run around the tract. This traverse should be a closed loop tied to at least two monuments in the national control system or the individual sections of the traverse must be closed on monuments of known position. *No point should remain unchecked.* Ties should be made to all U.S. National Ocean Survey, U.S. Geological Survey, and county monuments in the area. Computations should be performed in the state plane coordinate system and all elevations should be referred to the National Geodetic Vertical Datum (NGVD) (Art. 10.2).

When marking of all points and field surveys are completed, the data are reduced and adjusted yielding state plane coordinates and elevations for all ground points. After aerial photography has been acquired, the images of all ground control and photogrammetric control points are measured using a comparator. The measurements and the X, Y coordinates and elevations for known ground control points are utilized in an electronic computer program which performs the calculations required for photogrammetric triangulation. This program yields as output the X, Y coordinates and elevations for all unknown perimeter and interior photogrammetric control points included in the adjustment. These data are retained for future computations.

The control point locations around the perimeter of the tract determined by both field survey and photogrammetric methods are now used to make final determination of boundary locations around the property. Evidence located by stereoscopic examination of aerial photographs permits defining the area of search within rather narrow limits so as to minimize field examination for evidence of property lines. Azimuths and distances are computed from ground and photogrammetric control points to positions of potential property corners and monuments so as to facilitate the search in the field. When all data from examining photography and field searches are consolidated, analyzed, and correlated with the deed description of the property, the exterior boundaries of the tract are fixed and coordinates are computed for each property corner and angle point. Next, coordinates are computed for all center-line and lot-line controlling points and all lot corners within the subdivision. Finally, the azimuths and distances from photogrammetric control points to the nearest subdivision monument or control point are calculated. A listing of these azimuths and distances is prepared for use in the field. It is also possible to plot automatically (Art. 10.10) the positions of photogrammetric control points and the nearest subdivision monuments on work sheets which can also be used in the field.

Field layout surveys consist of occupying the photogrammetric control points, backsighting on another photogrammetric control point, and setting the nearest subdivision monument by angle and distance from the data sheets prepared in the office. When a sufficient number of subdivision monuments have been established, the center lines of streets and the balance of the lot corners should be set so as to provide a check on the layout. No lot corner or subdivision monument should remain unchecked.

As the monumentation and lot survey proceeds, spot checks on the accuracy of the photogrammetric survey can be made. Independent field traverses are run from the perimeter control traverse for this purpose. These traverses should be well distributed throughout the tract so as to provide an adequate sample of check points.

When the monumentation survey and checking have been completed to the satisfaction of the surveyor, the photogrammetrist, and the county engineer or surveyor, the photogrammetric survey is completed. All control points and monuments should be thoroughly referenced for subsequent use and the subdivision plan should be submitted to the proper authorities for approval and recording.

The procedure as outlined for a new subdivision can be applied in essentially the same way to the resurvey of an old subdivision, a group of lots located by metes and bounds, and land subdivided according to the U.S. public-land surveys. To summarize, some of the principal factors to be remembered in any photogrammetric property survey are:

1. The surveyor and photogrammetrist must work as a team. Careful advance planning to correlate field and photogrammetric work is absolutely necessary.
2. Whenever possible, locate all visible evidence of property lines and corners by examination of existing photography and field searches prior to acquiring the aerial photography to be used in the photogrammetric survey.
3. Set targets on all ground control points, photogrammetric control points, and existing monuments and bench marks. Set targets on or near all identified and potential property corners.

Finally, it is necessary to remember that the photogrammetrist can determine the position of a signalized control point with an accuracy more than adequate for most property surveys. However, to locate a property corner properly from this control point requires the services of a surveyor with substantial experience in property surveying and an adequate knowledge of the laws and customs regulating property subdivision in the region where the work is being performed.

References 4 and 8 contain additional details concerning accuracies possible and application of photogrammetric methods to property surveys.

15.14. Problems

15.1 Name and discuss the different types of property or land surveys.

15.2 What is a metes and bounds description? Describe five different types of metes and bounds descriptions.

15.3 Write a metes and bounds description for lot 11 shown in Fig. 15.3.

15.4 Give examples of descriptions by subdivision. What is a significant difference between a description by subdivision and one by metes and bounds?

15.5 Write a lot and block description for lot 9 in the plan shown in Fig. 15.3.

15.6 Describe, briefly, how photogrammetric methods can be used for property surveys.

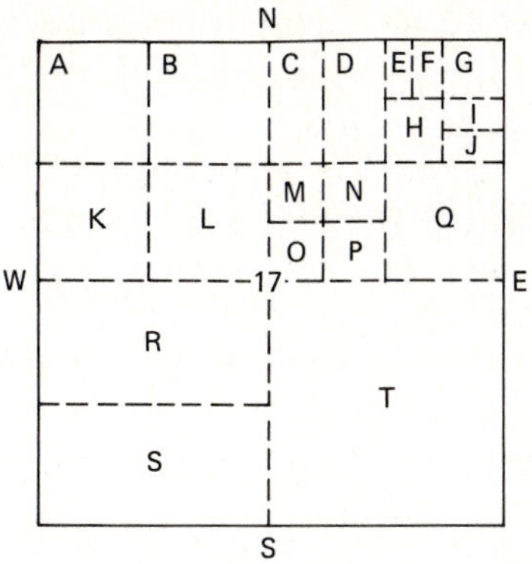

Fig. 15.13 *Reprinted with permission of AVI Publishing Company, Inc. from An Introduction to Agricultural Engineering 5th Printing, 1982, by Roth, Crow, and Mahoney.*

15.7 In performing a property survey using photogrammetric methods, describe the responsibilities of (*a*) the photogrammetric engineer; (*b*) the registered land surveyor. Assume that both are involved in a joint effort in the task.

15.8 Write descriptions for tracts D, E, F, G, H, and I in Fig. 15.12*b*.

15.9 Figure 15.13 represents Sec. 17, T23N, R4W, 2PM. Write the description and find the area for each tract of land indicated.

15.10 Show by a sketch or sketches: E$\frac{1}{2}$, NE$\frac{1}{4}$, NW$\frac{1}{2}$, Sec. 25, T2N, R5W of the Second Principal Meridian. What are its dimensions in chains?

15.11 Write a metes and bounds description for the traverse in Fig. 15.1. Assume station A is the SE corner of Sec. 22, T26N, R3E, 2PM.

15.12 Using the data and under the assumption of Prob. 15.11, write a metes and bounds description for the property shown in Fig. 15.1.

15.13 Refer to the tract of land which is shaded in Fig. 15.14. (*a*) How many acres does the tract contain? (*b*) What are the dimensions of the tract in chains? In rods? In feet? (*c*) In what section does

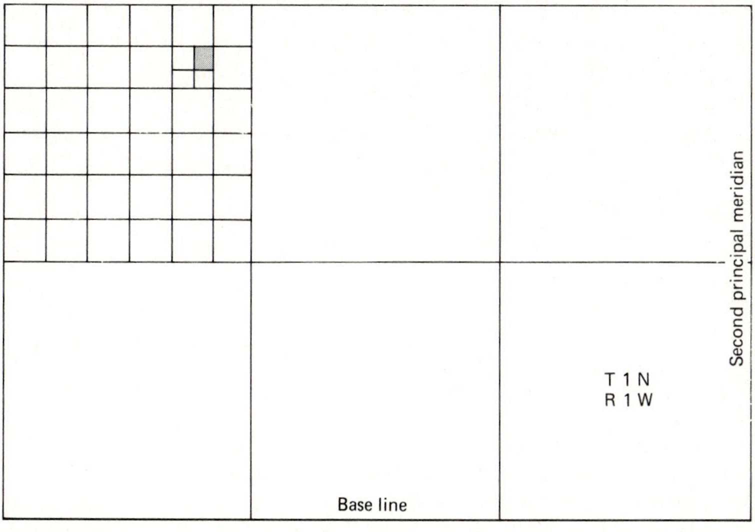

Fig. 15.14

the tract lie? (*d*) In what township and range does the tract lie? (*e*) How far in miles is it from the southwest corner of the tract to the Second Principal Meridian (shortest distance neglecting convergence).

15.15. Office problem

DEED DESCRIPTION

Object To obtain the legal description and other data needed to perform a property survey for a specific lot.

Procedure

1. Go to the county recorder's office (or other public office used for this purpose) and obtain the deed description for the specified property. This property may be the lot on which the house where you live is located or some other appropriate lot.
2. If the deed contains a reference to a recorded plan of lots (usually by plan book volume and page number), locate the plan and get a copy of those portions of the plan you believe necessary to perform the survey.
3. Using the calls from the deed or data from a subdivision map, calculate departures and latitudes for the sides of the property, compute the closure error, and calculate coordinates for the property corners (Art. 8.16). Plot the outline of the property, at an appropriate scale, on an $8\frac{1}{2}$ by 11 in sheet (Art. 10.10).
4. Rewrite the description by the metes and bounds method using coordinates.
5. The final report should consist of copies of the deed description obtained at the recorder's office; plan of lots in which the property lies; map of the lot on the $8\frac{1}{2}$ by 11 in sheet, and the deed written using coordinates.

References

1. American Society of Civil Engineers, Manuals of Engineering Practice, *Technical Procedures for City Surveys*, No. 10, 1963; *Urban Planning Guide*, No. 49, 1969; *Accommodation of Utility Plant within the Rights of Way of Urban Streets and Highways*, No. 14, 1974; and *Definitions of Surveying Mapping and Related Terms*, No. 32, 1978 (rev.).
2. Brown, Curtis M., *Boundary Control and Legal Principles*, 2d ed., John Wiley & Sons, Inc., New York, 1969.
3. Brown, C. M., Robillard, W. G., and Wilson, D. A., *Evidence and Procedures for Boundary Location*, 2d ed., John Wiley & Sons, Inc., New York, 1981.
4. Brown, Duane C., "Accuracies of Analytical Triangulation in Applications to Cadastral Surveys," *Surveying and Mapping*, September 1973.
5. Clark, F. F., *Law of Surveying and Boundaries*, 3d ed., The Bobbs-Merrill Company, Inc., Indianapolis, Ind., 1959.
6. Dean, D. R., and McEntyre, J. G., "Surveyor's Guide to the Use of a Law Library," *Surveying and Mapping*, Vol. 35, No. 3, September 1975.
7. Grimes, J., *Clark on Surveying and Boundaries*, 4th ed., The Bobbs-Merrill Company, Inc., Indianapolis, Ind., 1976.
8. Lafferty, M. E., "Photogrammetric Control for Subdivision Monumentation," *Proceedings, Fall Convention, ACSM*, 1971.
9. McEntyre, J. G., *Land Survey Systems*, John Wiley & Sons, Inc., New York, 1978.
10. Roth, L. O., Crow, F. R., Mahoney, G. W. A., *An Introduction to Agricultural Engineering*, 5th Printing, AVI Publishing Company, Inc., Westport, Conn., 1982.
11. U.S. Bureau of Land Management, *Restoration of Lost or Obliterated Corners and Subdivision of Sections*, Government Printing Office, Washington, D.C., 1974.

12. U.S. Bureau of Land Management, U.S. Department of the Interior, *Manual of Instructions for the Survey of Public Lands of the United States*, 1973, Technical Bulletin No. 6, Government Printing Office, Washington, D.C.

13. Vorhies, J. H., "A Computerized Land Title Plant in Joint Use," *Proceedings Annual Meeting, ACSM*, 1977.

14. Wattles, G. H., *Writing Legal Descriptions*, Gurdon H. Wattles Publications, Inc., Orange, Calif., 1976.

15. Wattles, W. C., *Land Survey Descriptions*, 10th ed., Gurdon H. Wattles Publications, Inc., Orange, Calif., 1974.

APPENDIX A
Error propagation in, and adjustment of, survey measurements

A.1. Error propagation There are two basic types of measurements in surveying: distances and angles. In the field, slope distances are measured along generally inclined terrain using a tape or an electronic distance measurement (EDM) instrument (Chap. 4) and are then reduced to horizontal distances by calculations. Vertical distances are measured with an engineer's level and a rod to yield elevations or differences in elevation (Chap. 5). Angles are measured in horizontal or vertical planes with a transit or theodolite (Chap. 6).

Often, these basic field measurements are used to calculate other quantities which are of direct interest. Such calculated quantities are evaluated through relationships with the measured values. Since the measurements have errors, the calculated quantities will inevitably also have errors. The operation of finding the errors in the calculated quantities from the errors in the measurements is called *error propagation*.

In the treatment of error propagation first consider the simplest form of a specific error, e, such as systematic error. We will then proceed to the more common error, namely, the random error which is reflected by the variance, σ^2. As an example of systematic error, consider a tape which is *too short* by an error e_L. A distance measured to be one length of the tape, L, will in fact be $(L - e_L)$. If we are interested in a distance D, which is five tape lengths, then

$$D = 5L$$

is the relationship needed for error propagation. The error e_D in D will be calculated simply as

$$e_D = 5e_L$$

and the corrected distance becomes, according to Eq. (2.6) (Art. 2.6, Chap. 2),

$$D_c = 5L_c$$

where $L_c = L - e_L$ is the corrected tape length. Then

$$D_c = 5(L - e_L)$$

which is clearly the result one would expect to get.

Carrying this example a bit further, consider a square parcel, each side of which is $S = 2L$, and determine its correct area. First, the area A containing an error e_A is

$$A = S^2 = 4L^2$$

The corrected side S_c is given by

$$S_c = 2L_c = 2(L - e_L)$$

and then the corrected area is

$$A_c = (2L_c)^2 = 4(L - e_L)^2$$

The error in the area, e_A, is the difference between A_c and A. Thus, again from Eq. (2.6).

$$\begin{aligned} e_A &= A - A_c \\ &= 4L^2 - 4(L - e_L)^2 \\ &= 4L^2 - 4L^2 + 8Le_L - 4e_L^2 \end{aligned}$$

or

$$e_A = 8Le_L - 4e_L^2$$

Since e_L is much smaller than L, the term $4e_L^2$ can be neglected and

$$e_A \simeq 8Le_L \tag{A.1}$$

(As a numerical example, L may be 50 m, e_L may be 0.04 m, and thus $8Le_L = 16\ \text{m}^2$ while $4e_L^2 = 0.0064\ \text{m}^2$, which is much smaller. Such a term is usually referred to as *higher-order term*.) The relevant quantities are shown in Fig. A.1. The sum of the areas of the two error rectangles $AA'B_2B$ and $CC'B_1B$ is equal to $8Le_L$, while the

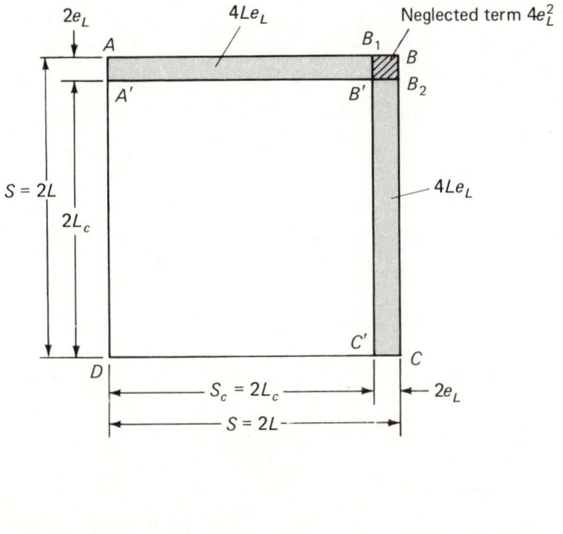

Fig. A.1

area of the small error square B_1BB_2B' is $4e_L^2$. It is the area of this small error square that is neglected. [The instructor will realize that the coefficient $8L$ of e_L in Eq. (A.1) is the first derivative of A with respect to L. Because of the elementary nature of this discussion, calculus is avoided. Interested readers are advised to consult Refs. 5 and 6 at the end of this appendix.]

A.2. Propagation of random error: standard deviations of sums and differences As noted previously, random errors are governed by probability rules and their behavior is expressed by probability distribution. This is quite different from systematic errors (which are called *deterministic*), the natures of which are usually fully known. To see the fundamental difference between these two types of errors, refer to Fig. A.2.

Figure A.2*a* illustrates the case of a tape which is too short; the error is $e_S = 0.04$ m to the left. The corrected distance is 99.96 m. Figure A.2*b* shows the case of a tape too long by $e_S = 0.04$ m. The error is to the right and the corrected distance is again 99.96 m. From these two figures it should be clear that we know the magnitude and direction (sign) of the systematic error. Random error, on the other hand, is never *definitively* known. In Fig. A.2*c*, it is distributed about the end point of the line according to the probability density function (such as the normal distribution). For propagation purposes, the single dispersion measure of variance, σ^2, is used because

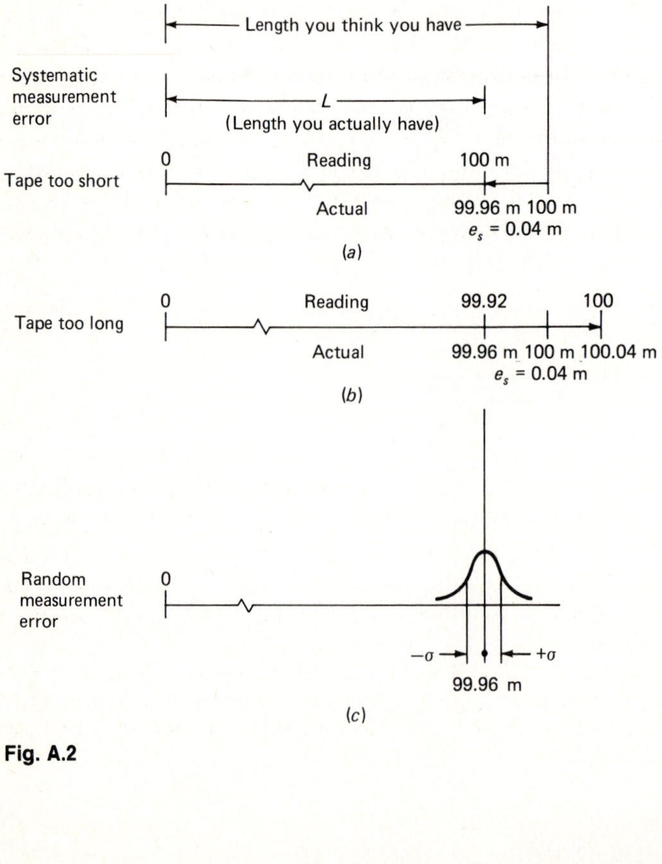

Fig. A.2

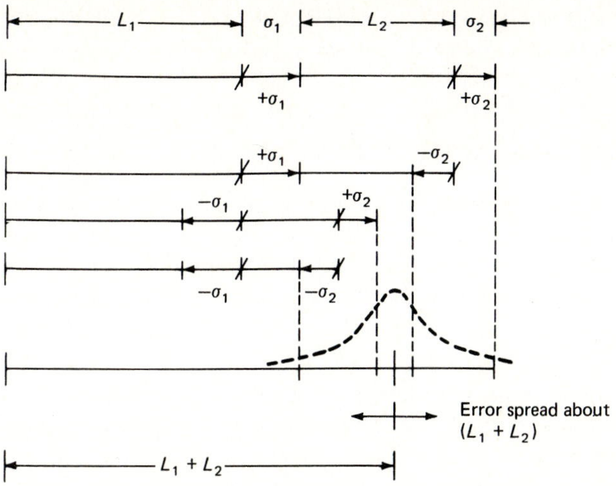

Fig. A.3

as can be seen in the figure the standard error† or standard deviation σ occurs on either side of the end point (σ has no relationship to the systematic error 0.04).

Consider the sum of two distances: L_1, with standard error σ_1, and L_2, with standard error σ_2. Figure A.3 shows four of the essentially infinite number of possibilities. The top line represents the case with $+\sigma_1$ and $+\sigma_2$, the second line $+\sigma_1$ and $-\sigma_2$, the third line $-\sigma_1$ and $+\sigma_2$, and fourth line $-\sigma_1$ and $-\sigma_2$. It is seen from bottom line of Fig. A.3 that because of the random errors, the distribution is spread about the "true" sum $L_1 + L_2$. Recalling that the standard error σ is just one possible measure of dispersion (corresponding to a probability of 0.6826), it should be easy to recognize that other errors of different magnitude and sign would occur. The end result is that a distribution (usually taken as the normal) will exist as schematically depicted in Fig. A.3. From mathematical relations beyond the scope of this text it can be shown that the variance of this distribution σ_S^2 is the sum of σ_1^2 and σ_2^2, or

$$\sigma_S^2 = \sigma_1^2 + \sigma_2^2 \tag{A.2}$$

Thus the standard deviation of the sum is

$$\sigma_S = \sqrt{\sigma_1^2 + \sigma_2^2} \tag{A.3}$$

Suppose that instead of the sum $(L_1 + L_2)$ we are interested in the difference $(L_1 - L_2)$. This relation could be thought of as $L_1 + (-L_2)$. Since we are working with the variance (i.e., square of σ) and since σ occurs on either side of the end point, the relation in Eq. (A.3) applies also for the difference. (The student could draw a figure similar to Fig. A.3 to show the same resulting distribution; see Prob.

† In statistics, the square root of the variance is called the standard deviation and is denoted by σ. An equivalent term is the standard error, which is in common use in surveying. In this text, the same symbol σ will be used to mean *both* the standard deviation and standard error without any distinction in meaning between the two terms.

A.4 at the end of this appendix.) To generalize, then, if

$$L = L_1 \pm L_2 \pm L_3 \pm \cdots \pm L_n \tag{A.4}$$

and all component distances are *uncorrelated* and have standard errors $\sigma_1, \sigma_2, \sigma_3, \ldots, \sigma_n$, respectively, then the standard error σ_L of L is given by

$$\sigma_L = \sqrt{\sigma_1^2 + \sigma_2^2 + \sigma_3^2 + \cdots + \sigma_n^2} \tag{A.5}$$

Correlation is a statistical term used to indicate that two or more random variables do not behave entirely independently from one another. This means that if x and y are two random variables which are correlated, an error e_x in x will cause some error e_y in y. The interaction between the two random variables is reflected by their correlation. This concept is well beyond the level of this text. It is important, however, that the student realize that the relations in Eq. (A.5), and similarly in Eqs. (A.2) and (A.3), do not hold if the random variables involved are correlated. In other words, existence of correlation between random variables will require different propagation rules which are beyond this discussion. The reader may consult Refs. 4 and 5 at the end of this appendix for more information on the subject.

Example A.1 A distance is the sum of several part measurements with respective estimated standard deviations, or standard errors, as follows:

Measurement, m	σ, m
155.204	0.005
500.139	0.008
116.080	0.004
771.423	

Calculate the standard deviation of the total measurement.

SOLUTION By Eq. (A.5):

$$\sigma_{\text{sum}} = \sqrt{(0.005)^2 + (0.008)^2 + (0.004)^2} = 0.010 \text{ m}$$

A.3. Standard deviation of measurement multiple The standard deviation of a multiple of a measurement *cannot* be derived from Eqs. (A.4) and (A.5). For example, if $L' = 3L$, with σ being the standard error of L, then one cannot write

$$L' = L + L + L$$

and apply Eq. (A.5) to it. This is because the three terms on the right-hand side are *perfectly correlated*, being the same quantity. Instead, the multiple case is viewed as a simple scaling of the distribution in which the standard error is enlarged by the multiple. (This can also be shown mathematically, but is beyond the scope of this development.) Thus for

$$L' = kL \tag{A.6}$$

then

$$\sigma' = k\sigma \tag{A.7}$$

$$\sigma'^2 = k^2\sigma^2 \tag{A.8}$$

where σ' and σ are the standard errors of L' and L, respectively.

A.4. Standard deviation of the mean The relations in Eqs. (A.4) and (A.8) can be applied to determine the standard error of the mean of a set of measurements. Referring to Eq. (2.16) in Chap. 2, the mean M of n measurements is

$$M = \frac{m_1 + m_2 + m_3 + \cdots + m_n}{n} \tag{A.9}$$

which may be rewritten as

$$M = \frac{1}{n}m_1 + \frac{1}{n}m_2 + \frac{1}{n}m_3 + \cdots + \frac{1}{n}m_n$$

or

$$M = m_1' + m_2' + m_3' + \cdots + m_n' \tag{A.10}$$

If $\sigma_1, \sigma_2, \sigma_3, \ldots, \sigma_n$ are the standard errors of the individual measurements $m_1, m_2, \ldots, m_n$, then according to Eq. (A.7), the standard errors of $m_1', m_2', \ldots$ are

$$\sigma_1' = \frac{1}{n}\sigma_1 \qquad \sigma_2' = \frac{1}{n}\sigma_2 \qquad \cdots$$

By Eq. (A.5), the standard error of the mean, σ_m, is

$$\sigma_m = \sqrt{\sigma_1'^2 + \sigma_2'^2 + \sigma_3'^2 + \cdots + \sigma_n'^2}$$

or

$$\sigma_m = \sqrt{\frac{\sigma_1^2}{n^2} + \frac{\sigma_2^2}{n^2} + \frac{\sigma_3^2}{n^2} + \cdots + \frac{\sigma_n^2}{n^2}} \tag{A.11}$$

or

$$\sigma_m = \frac{1}{n}\sqrt{\sigma_1^2 + \sigma_2^2 + \sigma_3^2 + \cdots + \sigma_n^2} \tag{A.12}$$

If the individual measurements have the same standard error σ, which means that we have equal confidence in each or that each is of equal weight, then from Eq. (A.11)

$$\sigma_m = \sqrt{\frac{n\sigma^2}{n^2}}$$

or

$$\sigma_m = \frac{\sigma}{\sqrt{n}} \tag{A.13}$$

Note that this equation is a repeat of Eq. (2.19) except that it is in square root form.

Equation (A.13) is important in that it shows that the precision of the mean of a repeated set of measurements is higher than the precision of each individual measurement. Thus, the mean of four measurements of equal weight will have a standard error which is half the standard error of an individual measurement.

Example A.2 A distance has been measured six times with the following results:

Measurement, m	$v_i = x_i - \bar{x}$, m	v_i^2
165.480	0.006	0.000036
165.472	−0.002	4
165.472	−0.002	4
165.475	0.001	1
165.471	−0.003	9
165.474	0.000	0
Σ = 992.844	0.000	Σ = 0.000054

Calculate the mean, the standard deviation of a single measurement, and the standard deviation of the mean.

SOLUTION By Eqs. (2.16) and (2.18), Art. 2.14, Chap. 2,

$$\bar{x} = \frac{992.844}{6} = 165.474$$

$$\sigma_x = \sqrt{\frac{0.000054}{6-1}} = 0.00329 \text{ m} = 0.003 \text{ m}$$

and using Eq. (A.13)

$$\sigma_m = \frac{0.00329}{\sqrt{6}} = 0.0013 \text{ m} \simeq 0.001 \text{ m}$$

A.5. Standard deviation of area Suppose that a rectangular field has dimensions of a by b with σ_a and σ_b the corresponding standard errors. Then the area is given by

$$A = ab$$

In order to evaluate the standard error σ_A of the area A, without using calculus, refer to the discussion in Art. A.1 and write

$$\sigma_A^2 = a^2\sigma_b^2 + b^2\sigma_a^2 \qquad \text{or} \qquad \sigma_A = \sqrt{a^2\sigma_b^2 + b^2\sigma_a^2} \tag{A.14}$$

[The derivation of Eq. (A.14), and in fact of the succeeding equations, are beyond the scope of this book. Interested readers should consult Refs. 4 and 5 at the end of this appendix.]

If the area is of a square with side s, then

$$A = s^2$$

and

$$\sigma_A^2 = 4s^2\sigma_s^2 \qquad \text{or} \qquad \sigma_A = 2s\sigma_s \tag{A.15}$$

If the area is of a circle with radius r, then

$$A = \pi r^2$$

and

$$\sigma_A^2 = 4\pi^2 r^2 \sigma_r^2 \qquad \text{or} \qquad \sigma_A = 2\pi r \sigma_r \tag{A.16}$$

Standard errors of other shapes, simple or composite, can be similarly written, but will require derivations beyond the present scope. The basis behind Eqs. (A.15) and (A.16) is fundamentally different from that behind Eq. (A.14). For the rectangular area, a and b are uncorrelated and one can add two terms, one for the variance of a and the other for the variance of b. It is not possible to do the same when a quantity is squared, such as s^2 or r^2. For example, it would be wrong to consider s^2 as the product $s \cdot s$ and apply Eq. (A.14) because the two multiplicands (i.e., s and s) are here perfectly correlated, being in fact the same quantity. Therefore, Eqs. (A.15) and (A.16) are given for such cases involving squares of measured quantities. For more details on this aspect, which is beyond the scope of this discussion, consult Refs. 4, 5, and 6 at the end of this appendix.

Example A.3 A rectangular field has dimensions of $a = 175.23$ m and $b = 262.54$ m, with respective standard deviations of $\sigma_a = 0.02$ m and $\sigma_b = 0.01$ m. Determine the standard deviation of the area A of the field.

SOLUTION

$$A = ab$$
$$\sigma_A = \sqrt{(175.23)^2(0.01)^2 + (262.54)^2(0.02)^2}$$
$$= 5.54 \text{ m}^2$$

Example A.4 A circle has a radius of $r = 25.00$ m with $\sigma_r = 0.03$ m. Compute the standard deviation for the area A of the circle

SOLUTION

$$A = \pi r^2$$
$$\sigma_A = 2\pi r \sigma_r = (2\pi)(25)(0.03) = 4.71 \text{ m}^2$$

A.6. Linear error versus angular error In plane surveying, the surveyor's field measurements consist mainly of distances and angles. In many cases both types of measurements exist in the same project. Therefore, the effect of errors in each type of measurement, particularly on the location of points, is important to the surveyor.

As before, consider first the case consisting of definitive (or deterministic) errors such as systematic errors. First look at the positive error e_d in the distance d that moves point P to P' as shown in Fig. A.4a. Because P is in the first quadrant of the coordinate system, both the X and Y coordinates are increased because of this error. The total error e_d can be resolved into two components

$$\begin{aligned} e_{d_x} &= e_d \sin A \\ e_{d_y} &= e_d \cos A \end{aligned} \tag{A.17}$$

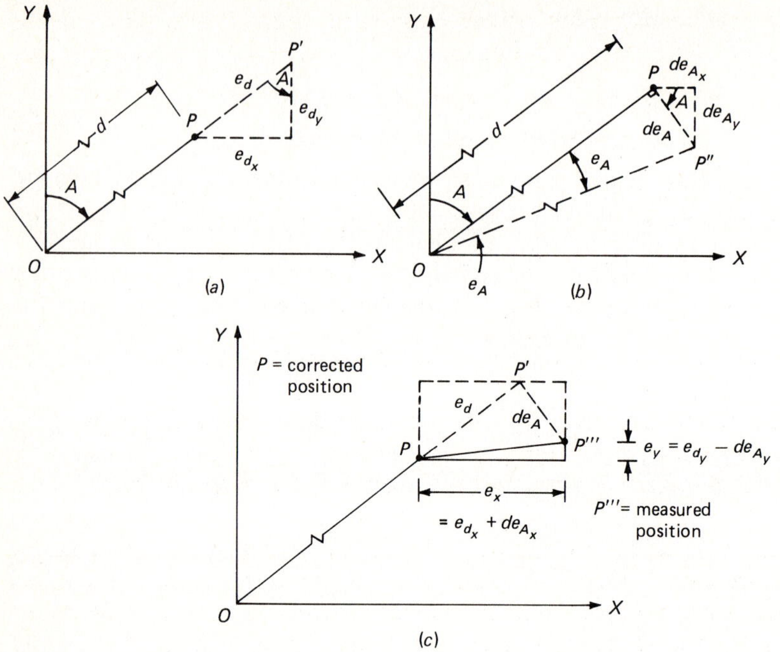

Fig. A.4

in which A is the clockwise angle from the $+Y$ axis to the line OP, or the azimuth of OP.

The effect of the angular error e_A is depicted in Fig. A.4*b*. Point P is shifted perpendicularly† from the line OP to P''. The linear error which corresponds to e_A is the line segment PP'' which is equal to de_A, in which e_A must be in radians. The X and Y components of this linear error are given by

$$de_{A_X} = de_A \cos A$$
$$de_{A_Y} = de_A \sin A \qquad \text{(A.18)}$$

For the specific angular error e_A, the effect on the location of point P is that its X coordinate is increased and its Y coordinate is decreased as shown in Fig. A.4*b*. For situations other than that shown in Fig. A.4*b* with different azimuth values, the signs of the error components will obviously be different.

Figure A.4*c* shows the combined effect of both the linear error e_d and the angular error e_A. It is seen that the final location of point P is at P'''. The final displacement PP''' can be thought of as composed of the displacement PP' followed by the displacement $P'P'''$. (The sequence of adding the two displacements is immaterial, as the same result would be obtained if the displacement due to angular error is used first, then the displacement caused by the linear error.) The effect on the coordinates

† More precisely, P'' is on a circular arc of radius d from point O, but since e_A is a very small angle, it is an acceptable approximation to consider PP'' perpendicular to OP.

is obtained by algebraically summing the component errors. Thus, the combined errors e_X and e_Y in the X and Y coordinates, respectively, as shown in Fig. A.4*c*, are

$$\begin{aligned} e_X &= e_{d_X} + de_{A_X} = e_d \sin A + de_A \cos A \\ e_Y &= e_{d_Y} - de_{A_Y} = e_d \cos A - de_A \sin A \end{aligned} \tag{A.19}$$

Finally, the "corrected" coordinates of the point P are obtained from the given erroneous coordinates of point P''' by subtracting the coordinate errors, or

$$\begin{aligned} X_c &= X_m - e_x = X_m - e_d \sin A - de_A \cos A \\ Y_c &= Y_m - e_y = Y_m - e_d \cos A + de_A \sin A \end{aligned} \tag{A.20}$$

where X_c, Y_c and X_m, Y_m are the corrected and measured coordinates of point P, respectively.

Definitive or deterministic errors as discussed above do not occur as frequently as random errors. Therefore, it is important to the surveyor to study the effects of random linear and angular errors and their interaction. This is demonstrated in Fig. A.5.

Point P is assumed to be determined by the angle A from the Y axis and the distance d from the point of origin O. The distance error is displayed as the standard error σ_d, shown as positive, thus increasing d, or negative, decreasing d. We must understand that σ_d is just one error measure with a given probability (0.683), and other measures could just as easily be considered (see Figs. 2.7 and 2.8, in Chap. 2). The error in distance is in linear units such as metres or feet and acts radially along the line OP. The error in the angle A, by contrast, changes the whole line OP and

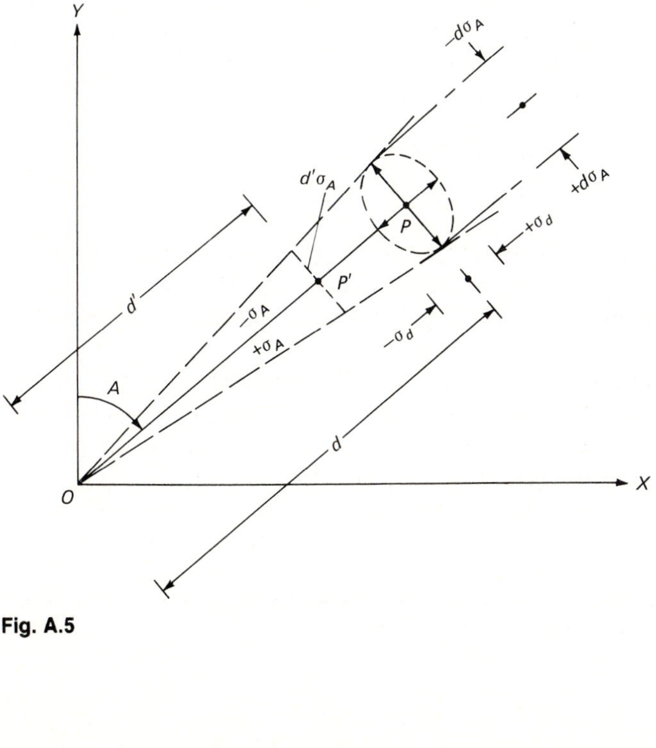

Fig. A.5

thus causes point P to be displaced in a direction which is essentially normal, or perpendicular, to OP. The angular error is represented by the standard error σ_A, which can be either positive, increasing the value of A, or negative, decreasing A. The linear displacement of a point due to angular error depends on how far the point is along the line OP. Thus, as shown in Fig. A.5, the displacement $d'\sigma_A$, of point P', is smaller than the displacement $d\sigma_A$, for point P, because P' is closer to O

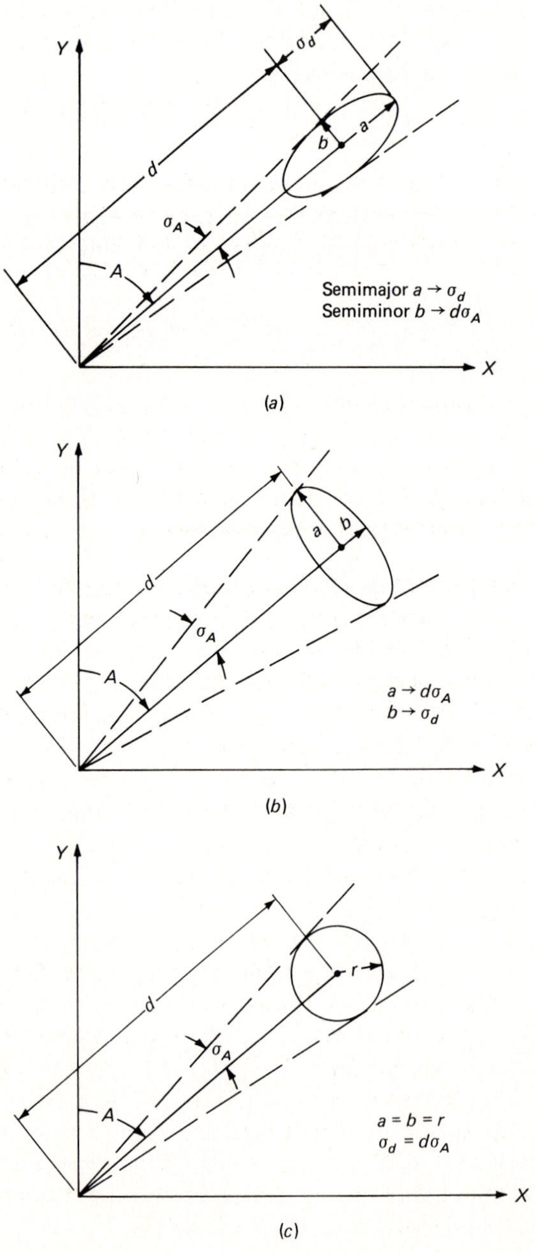

Fig. A.6

than P. Each linear displacement is calculated by multiplying the angular standard error σ_A, *in radians*, by the distance of the point in question (P' or P) to the point O at which the angle A is measured.

The combined effect of both linear and angular random errors is that there is an area of uncertainty around the point to be determined, P. The general shape of this area is an ellipse which is called the *error ellipse*. There is a certain probability associated with an error ellipse of a given size. The larger the size of the ellipse, the higher is the probability and vice versa. The ellipse whose semiaxes are equal to the standard error in distance, σ_d, and the linear displacement due to the angular standard error, $d\sigma_A$, is called the *standard error ellipse*.

The shape of the error ellipse depends on the relative precision of measuring distances and angles. Figure A.6 depicts three cases:

1. In Fig. A.6*a*, the distance measurement is of lower precision than the angle measurement, or $\sigma_d > d\sigma_A$, and therefore the ellipse is elongated in the distance direction.
2. In Fig. A.6*b*, the reverse case of that in Fig. A.6*a* is shown: The distance is measured with higher precision than the angle, or $\sigma_d < d\sigma_A$, and thus the ellipse is elongated in the direction normal to the radial distance.
3. In Fig. A.6*c*, the angle and distance measurements are of equivalent precision, i.e., $\sigma_d = d\sigma_A$ and the error ellipse degenerates to a circle.

The third case, Fig. A.6*c*, is the most important in practice. It is a *balanced* case, where the angular and distance measurements are of equivalent precision. The surveyor should strive to attain this situation by proper planning and choice of instruments and techniques (Chap. 3). It would be uneconomical if one type of measurement were obtained with significantly higher precision than the other type.

A.7. Error propagation in intersection A commonly encountered problem in surveying is that of locating the position of a point P by the intersection of two lines from two known points (see Chap. 9). Frequently, the angles at the two known points, A and B in Fig. A.7, are measured and the two rays thus determined intersect at the desired point P. The accuracy with which point P is located is reflected in the size, shape, and orientation of the error ellipse around it. The ellipse, in turn, depends on the precision with which the two angles A and B are measured. Three different cases are shown in Fig. A.7 so that the student may appreciate the basic geometric concepts. The symmetric case is adopted because it is easier to visualize and demonstrate. Once the concept is understood, the reader can work out other cases, as will be shown subsequently.

In the three parts of Fig. A.7, it is obvious that the intersection angle α at point P is a good indicator of the shape and orientation of the error ellipse. On the other hand, the size of the ellipse depends on the precision σ_A and σ_B of the measured angles A and B, respectively. As shown in Figure A.7*a*, for small values of α ($< 60°$), the semimajor axis of the ellipse is essentially in the Y direction. This means that the X coordinate of point P is determined with a better precision than its Y coordinate.

Figure A.7*b* shows the opposite case to that depicted in Fig. A.7*a*. Here, the intersection angle α is too large ($> 120°$), and the ellipse is therefore oriented such that its major axis is essentially parallel to the X axis. Hence, the X coordinate of point P is determined with less precision than the Y coordinate.

Perhaps the more important case is that shown in Fig. A.7*c*. In this case, the angle of intersection α is at its ideal value of 90°, which gives the strongest geometric intersection of straight lines. Consequently, the ellipse becomes a circle, and the precision of determining point P is uniform in any direction. In practice, one should attempt to get close to this situation as much as possible, and in general keep the intersection angle inside the 60° to 120° interval.

In Fig. A.7 we considered cases where the two angles A and B were basically equal, and thus the axes of the ellipse were essentially parallel to the coordinate axes.

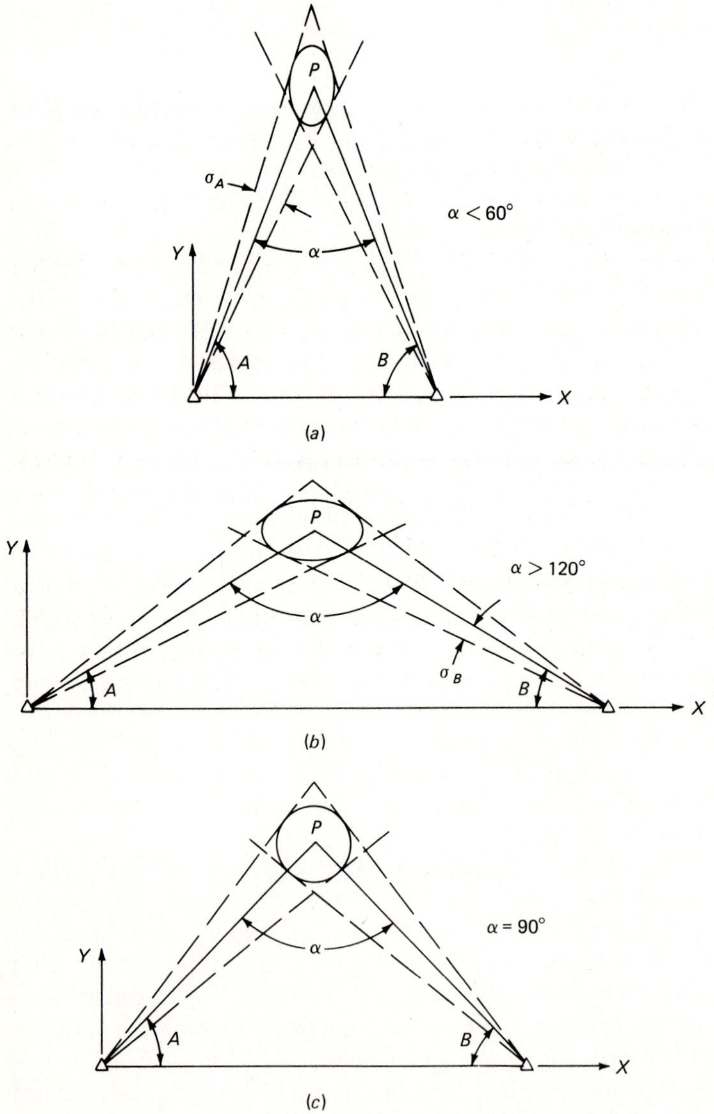

Fig. A.7

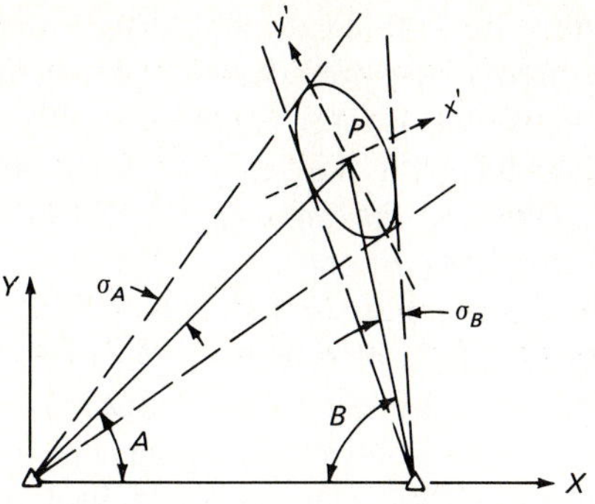

Fig. A.8

In practice this is frequently not the case and instead the general situation depicted in Fig. A.8 may occur. The two angles A and B are unequal and therefore the ellipse axes x', y' are rotated relative to the coordinate axes X, Y.

The unknown point P may also be determined if its *distances* from two known points are measured. This is a very simple case of *trilateration* (see Chap. 9). In Fig. A.9, one case is shown as an example. It is the case which is comparable to that depicted in Fig. A.7*a* except here are involved two measured distances d_1 and d_2, with standard errors σ_1 and σ_2, respectively. It is seen from Fig. A.9 that the ellipse still depends on the intersection angle α. Its size, of course, will depend on the values of the standard errors σ_1 and σ_2. The reader should at this point be able to construct figures which are comparable to Figs. A.7*b*, A.7*c*, and A.8. Needless to say, when both measured angles and distances are involved in locating point P, redundant measurements will exist and it is not that easy to sketch the corresponding error ellipse.

A.8. Adjustment of survey observations When embarking on measurements in the field, the surveyor must have a clear idea how the measurements fit in an overall plan. Consider a very simple task of wanting to find out the distance between two monuments in the field. If the surveyor measures the distance once, the necessary

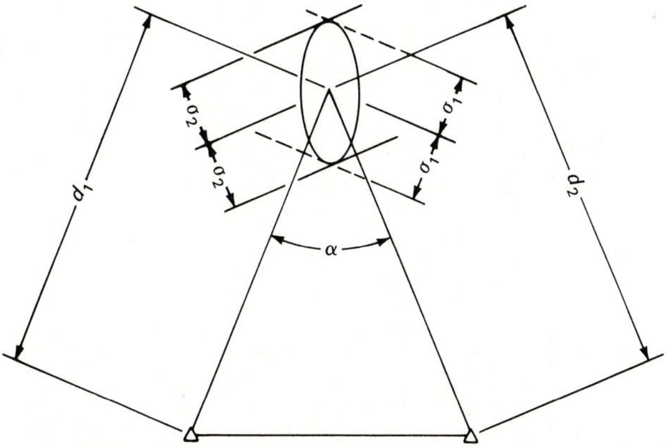

Fig. A.9

information is obtained. However, there may be little confidence in this single measurement. If the distance is measured twice, blunders are detected, and all systematic errors are corrected, the surveyor is still quite likely to have two somewhat different values, such as L_1 and L_2, for the same distance. Now, which one of these measurements should be taken as the final value? Or should a new value, which is a function of both measurements, be calculated? Unless the surveyor has overwhelming evidence that one of the two measurements is totally undependable, the second alternative is much more likely to be exercised. The problem then is how to calculate this new value. The technique used is called *adjustment*. It is necessary in order to overcome the inconsistency between measurements that results from random errors. In performing the adjustment, one must take into account the relative confidence one has in each of the measurements involved.

Figure A.10*a* shows that L_1 is less than L_2 and that the difference d (usually called the *discrepancy*) is

$$d = L_2 - L_1$$

Assuming equal confidence in each observation, or in other words that L_1 and L_2 each have equal weight, then the difference can be equally divided. Thus, the value adopted (which is a better estimate than either L_1 or L_2) is

$$\begin{aligned}\bar{L} &= L_1 + \tfrac{1}{2}d \\ &= L_1 + \tfrac{1}{2}(L_2 - L_1) \\ &= \tfrac{1}{2}(L_1 + L_2)\end{aligned}$$

or

$$\bar{L} = \frac{L_1 + L_2}{2}$$

which is the arithmetic mean of the two measurements. (The same result would be obtained from $L = L_2 - \tfrac{1}{2}d$.) The adjustment residuals (or corrections) are $v_1 = +d/2$

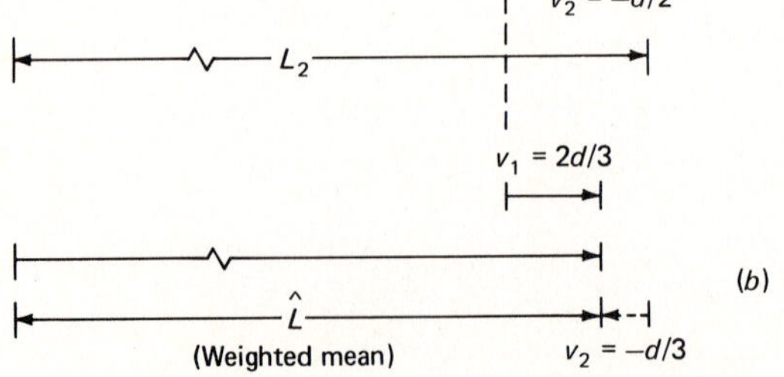

Fig. A.10

for L_1 and $v_2 = -d/2$ for L_2, as shown in Fig. A.10. This means that each observation is affected by the same amount (although their signs are opposite) because of being of equal weight.

The observations may be of unequal weight, meaning that we have more confidence in one observation than the other. Such increased confidence is related to the variance of the measurement. Specifically, it is an inverse relationship between the weight and variance as discussed in Art. 2.17 of Chap. 2. Thus, suppose that the variances of L_1 and L_2 are σ_1^2 and σ_2^2, respectively, and that $\sigma_1^2 = 2\sigma_2^2$. According to Eq. (2.23) the corresponding weights w_1 and w_2 are

$$w_1 = \frac{k}{\sigma_1^2}$$

$$w_2 = \frac{k}{\sigma_2^2}$$

where k is an arbitrary constant of proportionality. Since interest is usually in the ratios between weights, let $w_1 = 1$. This leads to

$$w_1 = \frac{k}{\sigma_1^2} = 1 \qquad \text{or} \qquad k = \sigma_1^2$$

$$w_2 = \frac{k}{\sigma_2^2} = \frac{\sigma_1^2}{\sigma_2^2} = \frac{2\sigma_2^2}{\sigma_2^2} = 2$$

With $w_1 = 1$ and $w_2 = 2$, one is twice as confident in L_2 as in L_1. Therefore, the residual (or change or correction) for L_2 should be smaller than that for L_1. In fact, it is quite reasonable on the basis of weights that $|v_2| = |\frac{1}{2}||v_1|$, or that

$$v_1 = \tfrac{2}{3}d$$

$$v_2 = -\tfrac{1}{3}d$$

Then, the final estimate $\hat{L}$ is

$$\hat{L} = L_1 + \tfrac{2}{3}d = L_1 + \tfrac{2}{3}(L_2 - L_1)$$

$$= \tfrac{1}{3}L_1 + \tfrac{2}{3}L_2$$

or

$$\hat{L} = \frac{(1)L_1 + (2)L_2}{3} = \frac{(1)L_1 + (2)L_2}{(1) + (2)} = \frac{(w_1)L_1 + (w_2)L_2}{(w_1) + (w_2)}$$

The value $\hat{L}$ as given is called the *weighted mean*.

In terms of variances of the observations, $\hat{L}$ may be given as

$$\hat{L} = \frac{(k/\sigma_1^2)L_1 + (k/\sigma_2^2)L_2}{(k/\sigma_1^2) + (k/\sigma_2^2)}$$

and dividing numerator and denominator by k yields

$$\hat{L} = \frac{(L_1/\sigma_1^2) + (L_2/\sigma_2^2)}{(1/\sigma_1^2) + (1/\sigma_2^2)}$$

We can now generalize the arithmetic and weighted means to n repeated measurements of the same quantity x. Thus, for $x_1, x_2, \ldots, x_n$ equally precise (or equally weighted) and uncorrelated measurements, the arithmetic mean $\bar{x}$ is given by Eq. (2.16), Chap. 2, which is repeated here for convenience:

$$\bar{x} = \frac{1}{n} \sum_{i=1}^{n} x_i \tag{A.21}$$

If the measurements are uncorrelated and have unequal weights $w_1, w_2, \ldots, w_n$, then the best estimate is the weighted mean $\hat{x}$, given by

$$\hat{x} = \frac{w_1 x_1 + w_2 x_2 + \cdots + w_n x_n}{w_1 + w_2 + \cdots + w_n} \tag{A.22}$$

or

$$\hat{x} = \frac{\sum_{i=1}^{n} w_i x_i}{\sum_{i=1}^{n} w_i}$$

In terms of variances, the weighted mean is given by

$$\hat{x} = \frac{(x_1/\sigma_1^2) + (x_2/\sigma_2^2) + \cdots + (x_n/\sigma_n^2)}{(1/\sigma_1^2) + 1/\sigma_2^2) + \cdots + (1/\sigma_n^2)} \tag{A.23}$$

or

$$\hat{x} = \frac{\sum_{i=1}^{n} x_i/\sigma_i^2}{\sum_{i=1}^{n} 1/\sigma_i^2}$$

Example A.5 From a sample of 32 measurements it has been determined that $\sigma = 0.006$ m for a single measurement using a particular EDM instrument and experienced operator. Using this EDM device and the same operator, a base line was measured repetitively over a 3-yr period of time, to yield the following three average uncorrelated measurements:

Average measurements value, m	Number of measurements, n	Standard deviation of the mean	Variance $= \sigma^2/n$	$w = \dfrac{k}{\sigma^2/n} = n$
206.139	5	$\sigma/\sqrt{5}$	$\sigma^2/5$	5
206.134	4	$\sigma/\sqrt{4}$	$\sigma^2/4$	4
206.123	6	$\sigma/\sqrt{6}$	$\sigma^2/6$	6

Calculate the weighted mean distance for the base line.

SOLUTION By Eq. (A.13), $\sigma_m = \sigma/\sqrt{n}$. Thus, the standard deviation for each mean value can be computed. Also, by Eq. (2.23), Art. 2.17, $w = k/\sigma^2$, where k is the constant of proportionality between the weight w and the standard deviation of the measurements σ. If k is set equal to $\sigma^2 = (0.006)^2$, then

$$w = \frac{\sigma^2}{(\sigma/\sqrt{n})^2} = n$$

and the weight of the measurement is seen to be proportional to the number of measurements. Thus, the weighted mean by Eq. (A.22) is

$$\hat{x} = \frac{(5)(206.139) + (4)(206.134) + (6)(206.123)}{5 + 4 + 6} = 206.131 \text{ m}$$

Example A.6 Suppose the same sample of measurements and $\sigma = 0.006$ as in Example A.5 are given with standard deviations given for each average value as follows:

Average value, m	σ_{m_i}, m	$w_i = n_i$	w_i'
206.139	0.00268	$(0.006)^2/(0.00268)^2 = 5.0$	1.000
206.134	0.00300	$(0.006)^2/(0.00300)^2 = 4.0$	0.800
206.123	0.00245	$(0.006)^2/(0.00245)^2 = 6.0$	1.200

Assuming uncorrelated measurements, compute the weighted mean.

SOLUTION Using Eq. (2.23), Art. 2.17, and assuming $k = \sigma^2$, $w_i = \sigma^2/\sigma_{m_i}^2$ where $\sigma^2 = (0.006)^2$, we have

$$\hat{x} = \frac{(5)(206.139) + (4)(206.134) + (6)(206.123)}{5 + 4 + 6} = 206.131 \text{ m}$$

Note that the selection of k in determining w_i is somewhat arbitrary. If σ is not given, it is generally chosen to be a reasonable value. For example, the first value may be given a unit weight, or $w_i' = 1$. Thus, $k = \sigma_{m_1}^2$ and the remaining weights will be those given in the last column of the tabulation. Note that the ratio $w_1 : w_2 : w_3$ is the same as $w_1' : w_2' : w_3'$. On the basis of w_i', x' is

$$\hat{x}' = \frac{(1)(206.139) + (0.800)(206.134) + (1.200)(206.123)}{1 + 0.800 + 1.200}$$

$$= 206.131 \text{ m}$$

which is identical to $\hat{x}$.

A.9. Adjustment of simple survey figures The preceding discussion referred to the adjustment of a set of observations which pertains to the same quantity, such as a distance or an angle. There are other survey problems where the observations may refer to a geometric figure such that a relatively simple relationship must be satisfied. The single plane triangle in which the sum of the interior angles must be 180° is one example.

Suppose that the plane triangle shown in Fig. A.11 is to be constructed. Begin by drawing the length a of the base BC, then construct the angles B and C. The two

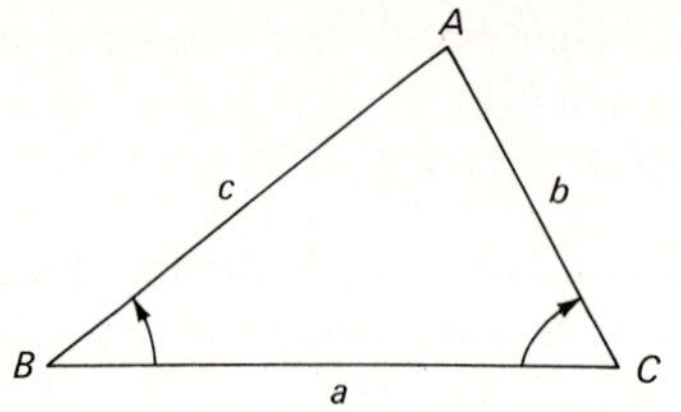

Fig. A.11

lines BA and CA will intersect at point A and the triangle will be complete. Consequently, if plane triangle ABC is to be uniquely determined, three measurements—the base length a and the two angles B and C—are required. Any more measurements would be *redundant* and an adjustment would then be necessary. As an example, if the third angle A is also measured, then it is quite likely that the sum of the three angles will not be equal to 180°. (One can also say that the observations are not consistent with the *model*, the model being the plane triangle. For more details about this topic, consult Refs. 4, 5, and 6 at the end of this appendix.) Thus the angles A, B, C must be *adjusted* such that their sum will be 180°. The amounts by which A, B, C are changed, i.e., the residuals, depend on how much the sum $A + B + C$ differs from 180° and on our relative confidence in the measured values of the angles. Denote by e_c, called *closure error*, the difference between the sum of the angles and 180°, or

$$A + B + C - 180° = e_c \tag{A.24}$$

or

$$A + B + C - e_c = 180° \tag{A.25}$$

The fraction of e_c which must be subtracted from each angle needs to be determined. If the angles are all equally reliable, e_c is distributed equally. Consequently,

$$v_1 = v_2 = v_3 = -\frac{e_c}{3} \tag{A.26}$$

and

$$\hat{A} = A + v_1 = A - \frac{e_c}{3}$$

$$\hat{B} = B + v_2 = B - \frac{e_c}{3}$$

$$\hat{C} = C + v_3 = C - \frac{e_c}{3}$$

where $\hat{A}, \hat{B}, \hat{C}$ are the adjusted angles.

The sum $\hat{A} + \hat{B} + \hat{C}$ must satisfy the *condition* of being 180°. Thus

$$\hat{A} + \hat{B} + \hat{C} = \left(A - \frac{e_c}{3}\right) + \left(B - \frac{e_c}{3}\right) + \left(C - \frac{e_c}{3}\right)$$
$$= A + B + C - e_c$$

or, in view of Eq. (A.25),

$$\hat{A} + \hat{B} + \hat{C} = 180°$$

The single triangle is the simplest plane figure. In general, an s-sided polygon must satisfy the condition that:

$$\text{the sum of the interior angles} = 180° \times (s - 2) \tag{A.27}$$

In a similar manner, the closure error e_c can be divided equally among the s interior angles of the polygon, if all the measured angles are uncorrelated and of equal weight. In this case, the residual for each angle is

$$v_i = -\frac{e_c}{s} \tag{A.28}$$

If the measured angles are uncorrelated but of unequal precision or weight, then each angle would be corrected by a different amount depending on its weight relative to the weights of the other measurements. The measurement with the *highest precision* (i.e., *smallest variance*) or weight will receive the *smallest* correction so as to change the least. Conversely, the angle measured with the smallest precision or weight, i.e., with the largest variance, will change the most and therefore receive the largest correction. Therefore, the corrections or residuals can be calculated as fractions of the closure error e_c in proportion to the individual variances to the sum of the variances. Thus, for the plane triangle, if $\sigma_1^2, \sigma_2^2, \sigma_3^2$ are the variances of the three measured angles, then the residuals are

$$\begin{aligned} v_1 &= -\frac{\sigma_1^2 e_c}{\sigma_1^2 + \sigma_2^2 + \sigma_3^2} \\ v_2 &= -\frac{\sigma_2^2 e_c}{\sigma_1^2 + \sigma_2^2 + \sigma_3^2} \\ v_3 &= -\frac{\sigma_3^2 e_c}{\sigma_1^2 + \sigma_2^2 + \sigma_3^2} \end{aligned} \tag{A.29}$$

In terms of weight, since σ_i^2 is inversely proportional to w_i, then any residual v_i is given by

$$v_i = -\frac{e_c/w_i}{1/w_1 + 1/w_2 + 1/w_3}$$

or

$$v_i = -\frac{e_c/w_i}{\sum_{i=1}^{3} (1/w_i)} \tag{A.30}$$

In a similar fashion, for an s-sided polygon, the residual v_i for the ith interior angle is given by

$$v_i = -\frac{\sigma_i^2 e_c}{\sum_{j=1}^{s} \sigma_j^2} \tag{A.31}$$

in which σ_i^2 is the variance of the ith measured interior angle and the sum of the variances is designated by

$$\sum_{j=1}^{s} \sigma_j^2 = \sigma_1^2 + \sigma_2^2 + \sigma_3^2 + \cdots + \sigma_s^2$$

In terms of weights, one gets:

$$v_i = -\frac{e_c/w_i}{\sum_{j=1}^{s} (1/w_j)} \tag{A.32}$$

Example A.7 Each angle in a triangle was measured six times to give the following results:

Measured angle	Correction	Adjusted angle
$\alpha = 45°25'10''$	$-10''$	$45°25'00''$
$\beta = 65°20'15''$	$-10''$	$65°20'05''$
$\gamma = 69°15'05''$	$-10''$	$69°14'55''$
$\Sigma = 180°00'30''$		$\Sigma = 180°00'00''$

The angles are uncorrelated and since each was turned the same number of times, it is reasonable to assume that they are of equal precision. From Eqs. (A.24) and (A.26)

$$\alpha + \beta + \gamma - 180 = e_c$$

$$e_c = 30''$$

$$v_1 = v_2 = v_3 = \frac{-e_c}{3} = -10''$$

Example A.8 The angles in triangle *ABC* and the number of times each angle was turned are:

Measured angle	n	$1/w_i = 1/n$	v_i	Adjusted angle
$34°33'10''$	12	1/12	$+05''$	$34°33'15''$
$82°46'00''$	6	1/6	$+10''$	$82°46'10''$
$62°40'20''$	4	1/4	$+15''$	$62°40'35''$
$\Sigma = 179°59'30''$		$\Sigma = 1/w_i = 6/12$		$\Sigma = 180°00'00''$

Assume that σ for each angle is 01″ so that $k = 01''^2$ and $w_i = n$ or the weight is proportional to the number of times the angle is turned (refer to Example A.5, where $w_i \propto n_i$). By Eq. (A.24), $e_c = -30''$. Then the corrections by Eq. (A.30) are:

$$v_1 = -\frac{\frac{1}{12}(-30)}{6/12} = +05''$$

$$v_2 = -\frac{\frac{1}{6}(-30)}{6/12} = +10''$$

$$v_3 = -\frac{\frac{1}{4}(-30)}{6/12} = +15''$$

A.10. Concept of least-squares adjustment In Art. A.8, the idea of adjustment was introduced with an indication that adjustment becomes necessary whenever redundant measurements occur. It was shown, essentially intuitively, how discrepancies between measurements can be dealt with, both for cases of weighted and of unweighted measurements. In Art. A.9, the concept was extended, still relying on intuitive sense, to the adjustment of simple survey figures such as a plane triangle or a polygon. Again, measurements with both equal and unequal weights were treated.

In all of the above mentioned cases, a fundamental concept was exploited without mentioning the fact explicitly. This fundamental concept is called the *least-squares* principle. In the case of (uncorrelated) measurements of equal weight, the least-squares principle is that the *sum of the squares of the residuals is a minimum*, or

$$f = v_1^2 + v_2^2 + v_3^2 + \cdots = \text{minimum} \tag{A.33}$$

Let us refer to the discussion in Art. A.8 where L_1 and L_2 are the two measurements of a distance, with $L_2 - L_1 = d$. Note that there were residuals $v_1 = d/2$ and $v_2 = -d/2$. Then for this case

$$f_1 = v_1^2 + v_2^2 = \frac{d^2}{4} + \frac{d^2}{4} = \frac{d^2}{2} \tag{A.34}$$

Suppose now that instead of (intuitively) picking the above values for v_1 and v_2, we had selected the values

$$v_1 = \frac{d}{3} \qquad \text{and} \qquad v_2 = -\frac{2d}{3}$$

Then

$$f_2 = \frac{d^2}{9} + \frac{4d^2}{9} = \frac{5d^2}{9} \tag{A.35}$$

It is clear that f_1 is less than f_2. In fact, the student can try as many combinations as desired and it will be obvious that no value for f can be less than f_1. In other words f_1 is a *minimum*.

The same will apply to other problems with equally weighted observations. For instance, consider the plane triangle discussed in Art. A.9 where Eq. (A.26) gave the residuals as

$$v_1 = v_2 = v_3 = -\frac{e_c}{3}$$

where e_c is the triangle closure error. Then

$$f_1 = v_1^2 + v_2^2 + v_3^2 = \frac{3e_c^2}{9} = \frac{e_c^2}{3} \tag{A.36}$$

Any other values for the residuals will yield a value for f that will always be larger than $e_c^2/3$. For example, for

$$v_1 = -\frac{e_c}{2} \qquad \text{and} \qquad v_2 = v_3 = -\frac{e_c}{4}$$

we have

$$f_2 = \frac{e_c^2}{4} + \frac{2e_c^2}{16} = \frac{3e_c^2}{8} \tag{A.37}$$

It is clear that $f_2 > f_1$.

For weighted observations, the least-squares principle states that the *sum of the weighted squares of the residuals is a minimum*, or

$$g = w_1 v_1^2 + w_2 v_2^2 + w_3 v_3^2 + \cdots = \text{minimum} \tag{A.38}$$

where $w_1, w_2, w_3, \ldots$ are the weights of the (uncorrelated) measurements.

To demonstrate this case, consider the two length measurements: L_1 with $w_1 = 1$ and L_2 with $w_2 = 2$. With $d = L_2 - L_1$ we showed in Art. A.8 that $v_1 = 2d/3$ and $v_2 = -d/3$ and thus

$$\begin{aligned} g_1 &= w_1 v_1^2 + w_2 v_2^2 \\ &= (1)\frac{4d^2}{9} + (2)\frac{d^2}{9} = \frac{6d^2}{9} = \frac{2d^2}{3} \end{aligned} \tag{A.39}$$

Any other combination of residual values will give a value for g which is larger than g_1. For instance, let

$$v_1 = \frac{d}{2} \qquad \text{and} \qquad v_2 = -\frac{d}{2}$$

Then

$$g_2 = (1)\frac{d^2}{4} + (2)\frac{d^2}{4} = \frac{3d^2}{4} \tag{A.40}$$

It is obvious that $g_2 > g_1$. Other examples can be easily done by the student.

Least-squares adjustment is a powerful tool which is used extensively by the surveyor for all kinds of problems. It is unfortunately well beyond the scope of this discussion. The interested reader is advised to consult Refs. 4, 5, and 6 at the end of this appendix.

A.11. Problems

A.1 Figure A.12 shows a field composed of a rectangular area adjacent to a semicircle. The measurements are: $a = 132.12$ m with $\sigma_a = 0.02$ m and $b = 279.22$ m with $\sigma_b = 0.03$ m. Calculate the area A of the field and its standard deviation σ_A.

A.2 Figure A.13 shows two adjacent horizontal angles. Their measured values and the corresponding 90 percent errors are:

Angle	Value	90% error
AOB	23°19′00″	30″
BOC	57°46′40″	20″

What is the value of ∠*AOC* and its standard error?

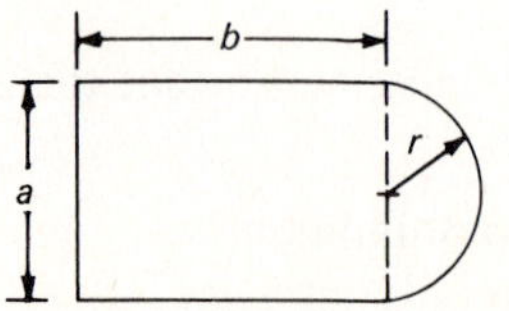

Fig. A.12

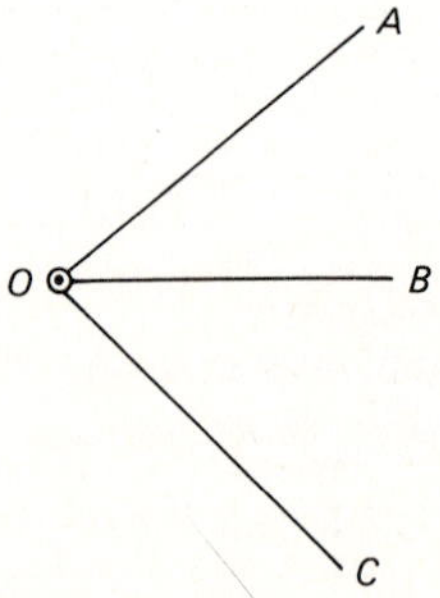

Fig. A.13

A.3 An angle is measured with three different instruments A, B, C. Its values and the corresponding 95 percent errors are:

Instrument	Angle value	90% error
A	58°26′20″	10″
B	58°26′30″	15″
C	58°26′00″	20″

Find the relative weights of these three measurements and their weighted mean.

A.4 Construct a figure similar to Fig. A.3 depicting the case of random error propagation for the difference $L_1 - L_2$ between two measurements L_1 and L_2. Show the four possibilities of combining $\pm\sigma_1$ and $\pm\sigma_2$ and the combined effect.

A.5 The elevation of point A has been determined by leveling over five different lines from five different points, the elevations of which are known. The results are shown below.

Line	Length L, mi	Calculated elevation of point A, m
1	2	642.03
2	4	642.02
3	3	642.04
4	1	642.06
5	4	642.00

(*a*) Find the mean elevation of point A assuming that all lines are of equal weight.

(*b*) Find the mean elevation of point A assuming that the weights of the lines are inversely proportional to their length, i.e., that $w_i = 1/L_i$.

A.6 A line, 1000 ft in length, is measured by three instruments:

An EDM with	$\sigma_E = 0.015$ ft
A steel tape with	$\sigma_S = 0.03$ ft
A Gunter's chain with	$\sigma_G = 0.05$ ft

Calculate the angular error corresponding to each case, and specify a compatible angle-measuring instrument for each.

A.7 An angle is measured by three different instruments:

A theodolite with	$\sigma_t = 10''$
A transit with	$\sigma_r = 30''$
A compass with	$\sigma_c = 10'$

Calculate the corresponding linear errors for a line about 1000 m long and designate a distance-measuring instrument that is compatible for each case.

A.8 Using the data from Probs. A.6 and A.7, sketch the error ellipse for each of the following combinations. Take the azimuth of the line to be 60°.

Case	Angle measured with:	Distance measured with:
A1	Theodolite	EDM
A2	Theodolite	Steel tape
B1	Transit	EDM
B2	Transit	Steel tape
C1	Compass	Steel tape
C2	Compass	Gunter's chain

A.9 Three adjacent angles have the following mean values and standard deviations:

Angle	Mean value	s
1	20°07′30″	20″
2	15°43′00″	15″
3	19°00′45″	10″

Calculate the mean value and standard deviation of the total angle (sum of all three angles).

A.10 A tract of land has a trapezoidal shape with two parallel sides a, b and the distance between them (i.e., height of trapezoid) h. You are given the following data:

$a = 125.12$ m $\quad \sigma_a = 0.10$ m

$b = 150.08$ m $\quad \sigma_b = 0.12$ m

$h = 20.00$ m $\quad \sigma_h = 0.08$ m

The three measurements are uncorrelated. Calculate the area of the tract and its standard deviation.

A.11 When does adjustment become necessary in a survey figure?

A.12 What is the basic objective of a survey adjustment?

A.13 What is the least-squares criterion and why is it needed?

A.14 Why do we need to add residuals to the observations?

References

1. Barry, A., *Engineering Measurements*, John Wiley & Sons, Inc., New York, 1964.
2. Barry, A., *Errors in Practical Measurement in Science, Engineering, and Technology*, John Wiley & Sons, Inc., New York, 1978.
3. Cooper, M. A. R., *Fundamentals of Survey Measurements and Analysis*, Crosby Lookwood, London, 1974.
4. Mikhail, E. M., *Observations and Least Squares*, University Press of America, Lanham, Md., 1976.
5. Mikhail, E. M., and Gracie, G., *Analysis and Adjustment of Survey Measurements*, Van Nostrand Reinhold Company, New York, 1980.
6. Davis, R. E., Foote, F. S., Anderson, J. M., and Mikhail, E. M., *Surveying Theory and Practice*, 6th ed., McGraw-Hill Book Company, New York, 1981.

APPENDIX B
Elementary mathematical concepts

B.1. Angles Angles, such as θ in Fig. B.1*a*, are measured either in degrees, minutes, and seconds, or in radians (abbreviated rad). The full circle in Fig. B.1*a* is divided into 360 *degrees* or 360°. The angle θ is one-twelfth of the circle and is therefore

$$\theta = \frac{360°}{12} = 30°$$

Each degree is divided into 60 parts called *minutes of arc*, or simply minutes, or

$$1 \text{ degree} = 1° = 60 \text{ minutes} = 60'$$

Each minute is divided into 60 parts called *seconds of arc*, or simply seconds, or

$$1 \text{ minute of arc} = 1' = 60 \text{ seconds} = 60''$$

The *radian* measure is a dimensionless quantity. One *radian*, or 1 rad, is an angle whose subtending arc is equal to the radius of that arc, such as the angle ϕ in Fig. B.1*b*. Since π is the ratio of a circumference of a circle to its diameter d, or to twice its radius r,

$$\pi = \frac{\text{circumference}}{\text{diameter}}$$

or

$$\text{circumference} = \pi d = 2\pi r$$

Then the number of radians in a circle is

$$\frac{\text{circumference}}{r} = 2\pi = 6.28 \text{ rad}$$

Thus

$$\begin{aligned} 360° &= 2\pi \text{ rad} \\ 180° &= \pi \text{ rad} \\ 1° &= \frac{\pi}{180°} = 0.0175 \text{ rad} \\ 1 \text{ rad} &= \frac{180°}{\pi} = 57.296° \\ &= 57°17'44.8'' \\ &= 206{,}264.8'' \end{aligned}$$

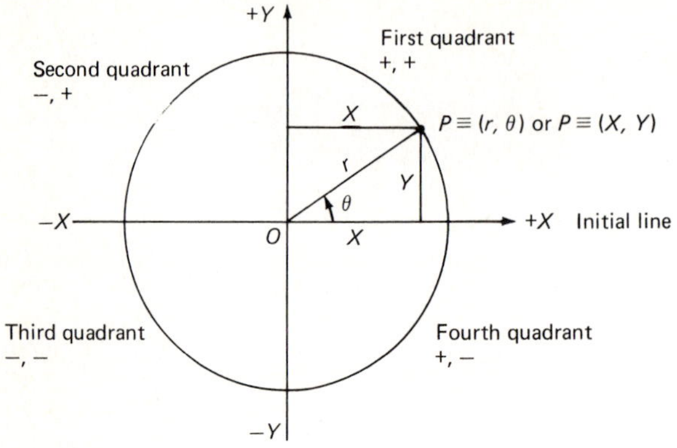

Fig. B.1*a*

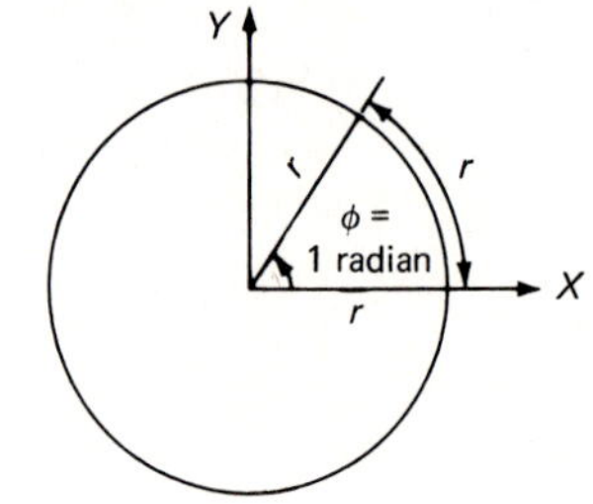

Fig. B.1*b*

The last quantity is referred to by ρ. It is used when one needs to convert from radians to seconds and vice versa. Since $\rho = 206{,}264.8''/\text{rad}$, then

Multiply by ρ when converting radians to seconds.
Divide by ρ when converting seconds to radians.

The degrees, minutes, and seconds (also called the *sexagesimal*) measure is used quite frequently in the United States. In Europe, the *centesimal* system is used where the circle is divided into 400 equal parts, called *grads*, abbreviated$^{(g)}$, each grad having 100 centesimal minutes, abbreviated$^{(c)}$,† and each minute having 100 centesimal seconds, abbreviated$^{(cc)}$‡ (see Art. 1.8, Chap. 1).

B.2. Polar and rectangular (cartesian) coordinates In a plane, any point, such as P in Fig. B.1*a*, is determined by two quantities, called *coordinates*. For *polar* coordinates, a fixed reference point, such as the origin O, and a fixed line, called the *initial line*, are required. The point P is determined by the distance r from O, and the angle θ from the *initial line*. The radius r is *always* positive, and the angle θ takes any value from 0 to 360° and is counterclockwise from the initial line, as shown in Fig. B.1*a*.

A much more common system of coordinates is the *rectangular* or *cartesian* system. The position of any point P is determined by two distances X and Y from two fixed lines which are perpendicular to each other at the origin (point O), as shown in Fig. B.1*a*. The X coordinate is called the *abscissa*, and the Y coordinate is called the *ordinate*. Each coordinate can be either positive or negative. The signs of the coordinates depend on the quadrant in which the point falls. There are four such quadrants, as shown in Table B.1. The data in Table B.1 are also shown in Fig. B.1*a*.

† The $^{(c)}$ stands for the centigrad, or one one-hundredth of a grad.
‡ The $^{(cc)}$ stands for centi-centigrad, or one ten-thousandth of a grad.

Table B.1

Quadrant	Sign of X	Sign of Y	Range of values for θ, deg
First	+	+	0–90
Second	−	+	90–180
Third	−	−	180–270
Fourth	+	−	270–360

B.3. Trigonometric functions For an angle, such as θ in Fig. B.1*a*, there are six functions which are defined as the following ratios:

$$\sin\theta = \frac{Y}{r} \qquad \operatorname{cosec}\theta = \frac{1}{\sin\theta} = \frac{r}{Y}$$

$$\cos\theta = \frac{X}{r} \qquad \sec\theta = \frac{1}{\cos\theta} = \frac{r}{X}$$

$$\tan\theta = \frac{Y}{X} \qquad \cot\theta = \frac{1}{\tan\theta} = \frac{X}{Y}$$

$$\tan\theta = \frac{\sin\theta}{\cos\theta} \qquad \cot\theta = \frac{\cos\theta}{\sin\theta}$$

Each of these trigonometric functions is either positive or negative depending on the signs of X and Y (recall that r is always positive). For example, $\cos 130°$ is negative because in the second quadrant the X value is negative; and $\tan 210°$ is positive since in the third quadrant both X and Y are negative (from Table B.1) and dividing a negative quantity by another negative quantity yields a positive quantity. Table B.2 lists the signs of all six trigonometric functions for angles in each of the four quadrants.

For angles such as 0°, 90°, etc., the functions assume limiting values. These values are listed in Table B.3 for reference.

Table B.2

Quadrant	sin	cosec	cos	sec	tan	cot
First	+	+	+	+	+	+
Second	+	+	−	−	−	−
Third	−	−	−	−	+	+
Fourth	−	−	+	+	−	−

Table B.3

Angle	sin	cosec	cos	sec	tan	cot
0	0	∞	1	1	0	∞
$\frac{\pi}{2} = 90°$	1	1	0	∞	∞	0
$\pi = 180°$	0	∞	−1	−1	0	∞
$\frac{3\pi}{2} = 270°$	−1	−1	0	∞	∞	0
$2\pi = 360°$	0	∞	1	1	0	∞

B.4. Useful relations between trigonometric functions This information is given in the following equations, and in Table B.4.

$$\sin^2\theta + \cos^2\theta = 1 \tag{B.1}$$

$$1 + \tan^2\theta = \sec^2\theta \tag{B.2}$$

$$1 + \cot^2\theta = \operatorname{cosec}^2\theta \tag{B.3}$$

$$\sin(\alpha \pm \beta) = \sin\alpha\cos\beta \pm \cos\alpha\sin\beta \tag{B.4}$$

$$\cos(\alpha \pm \beta) = \cos\alpha\cos\beta \mp \sin\alpha\sin\beta \tag{B.5}$$

$$\tan(\alpha \pm \beta) = \frac{\tan\alpha \pm \tan\beta}{1 \mp \tan\alpha\tan\beta} \tag{B.6}$$

$$\sin 2\theta = 2\sin\theta\cos\theta \tag{B.7}$$

$$\begin{aligned}\cos 2\theta &= \cos^2\theta - \sin^2\theta\\ &= 1 - 2\sin^2\theta\\ &= 2\cos^2\theta - 1\end{aligned} \tag{B.8}$$

$$\sin^2\theta = \tfrac{1}{2} - \tfrac{1}{2}\cos 2\theta \tag{B.9}$$

$$\cos^2\theta = \tfrac{1}{2} + \tfrac{1}{2}\cos 2\theta \tag{B.10}$$

$$\tan 2\theta = \frac{2\tan\theta}{1 - \tan^2\theta} \tag{B.11}$$

$$\sin 3\theta = 3\sin\theta - 4\sin^3\theta \tag{B.12}$$

$$\cos 3\theta = 4\cos^3\theta - 3\cos\theta \tag{B.13}$$

$$\tan 3\theta = \frac{3\tan\theta - \tan^3\theta}{1 - 3\tan^2\theta} \tag{B.14}$$

$$\sin\alpha + \sin\beta = 2\sin\tfrac{1}{2}(\alpha+\beta)\cos\tfrac{1}{2}(\alpha-\beta) \tag{B.15}$$

$$\sin\alpha - \sin\beta = 2\cos\tfrac{1}{2}(\alpha+\beta)\sin\tfrac{1}{2}(\alpha-\beta) \tag{B.16}$$

$$\cos\alpha + \cos\beta = 2\cos\tfrac{1}{2}(\alpha+\beta)\cos\tfrac{1}{2}(\alpha-\beta) \tag{B.17}$$

$$\cos\alpha - \cos\beta = -2\sin\tfrac{1}{2}(\alpha+\beta)\sin\tfrac{1}{2}(\alpha-\beta) \tag{B.18}$$

Table B.4

Angle	Quadrant[†]	sin	cos	tan
$-\theta$	4	$-\sin\theta$	$\cos\theta$	$-\tan\theta$
$+\theta$	1	$\sin\theta$	$\cos\theta$	$\tan\theta$
$90° - \theta = \frac{\pi}{2} - \theta$	1	$\cos\theta$	$\sin\theta$	$\cot\theta$
$90° + \theta = \frac{\pi}{2} + \theta$	2	$\cos\theta$	$-\sin\theta$	$-\cot\theta$
$180° - \theta = \pi - \theta$	2	$\sin\theta$	$-\cos\theta$	$-\tan\theta$
$180° + \theta = \pi + \theta$	3	$-\sin\theta$	$-\cos\theta$	$\tan\theta$
$270° - \theta = \frac{3\pi}{2} - \theta$	3	$-\cos\theta$	$-\sin\theta$	$\cot\theta$
$270° + \theta = \frac{3\pi}{2} + \theta$	4	$-\cos\theta$	$\sin\theta$	$-\cot\theta$
$360° - \theta = 2\pi - \theta$	4	$-\sin\theta$	$\cos\theta$	$-\tan\theta$

† The quadrant is given under the assumption that θ is an acute angle.

$$\sin\alpha\cos\beta = \tfrac{1}{2}[\sin(\alpha+\beta)+\sin(\alpha-\beta)] \tag{B.19}$$
$$\cos\alpha\sin\beta = \tfrac{1}{2}[\sin(\alpha+\beta)-\sin(\alpha-\beta)] \tag{B.20}$$
$$\cos\alpha\cos\beta = \tfrac{1}{2}[\cos(\alpha+\beta)+\cos(\alpha-\beta)] \tag{B.21}$$
$$\sin\alpha\sin\beta = -\tfrac{1}{2}[\cos(\alpha+\beta)-\cos(\alpha-\beta)] \tag{B.22}$$

B.5. Relation between polar and cartesian coordinates Figure B.2 shows the two sets of coordinates, polar (r, θ) and rectangular (X, Y), for any point P. Given r, θ, one can calculate X, Y using

$$\begin{aligned} X &= r\cos\theta \\ Y &= r\sin\theta \end{aligned} \tag{B.23}$$

Conversely, from X, Y the polar coordinates r, θ can be calculated using

$$\begin{aligned} r &= \sqrt{X^2+Y^2} \\ \tan\theta &= \frac{Y}{X} \end{aligned} \tag{B.24}$$

or

$$\theta = \arctan\frac{Y}{X} = \tan^{-1}\frac{Y}{X}$$

As was mentioned earlier, only the positive square root is considered since r is always positive. The symbols "arctan" and "$\tan^{-1}$" indicate "the angle whose tangent is Y/X." These *inverse* relations exist for all trigonometric functions such as $\sin^{-1}$, $\cos^{-1}$, etc. The quadrant in which the calculated angle θ falls is determined by the signs of X, Y, as explained in Art. B.2 and given in Table B.1.

Example B.1 The cartesian coordinates of a point are $X = -6.51$ m and $Y = 17.89$ m. Calculate its polar coordinates.

SOLUTION

$$r = \sqrt{(-6.51)^2 + (17.89)^2} = 19.04 \text{ m}$$

$$\theta = \tan^{-1}\frac{17.89}{-6.51} = \tan^{-1}(-2.7481) = 110°$$

Since X is negative and Y is positive, the angle falls in the second quadrant according to

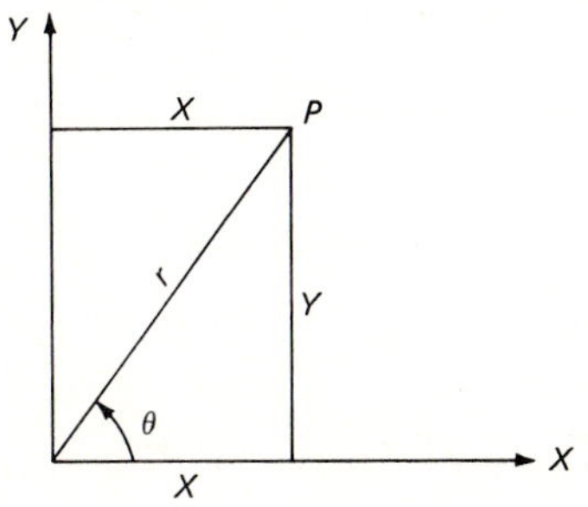

Fig. B.2

Table B.1. No ambiguity exists when one knows the signs of the numerator and denominator.

In some calculators when a negative number is used for the tan, a negative value for the angle is given, $-70°$ in this case. This is because the tan is negative in the second and fourth quadrants. If, as in this case, the angle is known to be in the second quadrant, then the final value of the angle is $180° +$ the value from the calculator, or $180 - 70 = 110°$.

B.6. Distance between two points Although all the relations to be developed in this and following sections are in rectangular coordinates, comparable relations can be derived in polar coordinates. The reason for concentrating on rectangular coordinates is that they are much more common in surveying practice than polar coordinates.

Figure B.3 shows two points P_1 and P_2 with coordinates X_1, Y_1 and X_2, Y_2, respectively. The distance between them, d_{12}, is obtained by employing the Pythagorian relationship to the triangle P_1P_2P or

$$(P_1P_2)^2 = (P_1P)^2 + (P_2P)^2$$

or

$$d_{12}^2 = (X_2 - X_1)^2 + (Y_2 - Y_1)^2$$

Then,

$$d_{12} = \sqrt{(X_2 - X_1)^2 + (Y_2 - Y_1)^2} \tag{B.25}$$

Note that $d_{12} = d_{21}$ since the square of each component is the same regardless whether positive or negative, as for example, $(X_2 - X_1)^2 = (X_1 - X_2)^2$.

B.7. Slope of a line between two points The slope of any line is equal to the tangent of the angle it makes with the X axis. The direction of the line is not important in regard to the slope, but *is important* for the slope angle. The slope angle is the counterclockwise angle from $+X$ to the line in the specified direction. Thus, for the specific case shown in Fig. B.3, the slope angle for the line P_1P_2 is θ_{12}, which is $< 90°$. On the other hand, the slope angle for the line P_2P_1 is θ_{21}, which is $> 180°$. In fact,

$$\theta_{21} = \theta_{12} + 180° \tag{B.26}$$

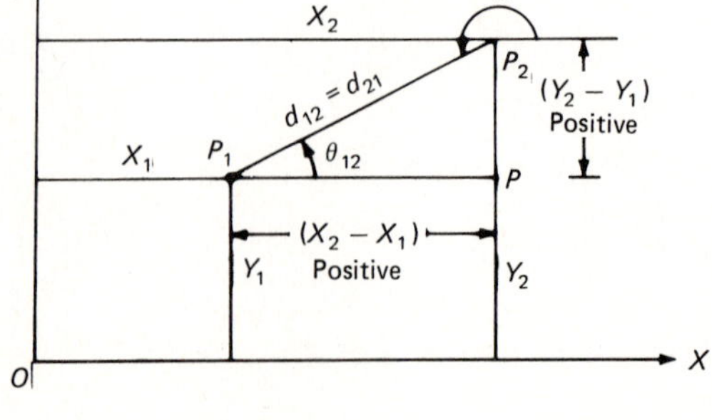

Fig. B.3

The slope m_{12} of the line P_1P_2 is

$$m_{12} = \tan\theta_{12} = \frac{Y_2 - Y_1}{X_2 - X_1} \tag{B.27}$$

and that for the line P_2P_1 is

$$m_{21} = \tan\theta_{21} = \frac{Y_1 - Y_2}{X_1 - X_2} \tag{B.28}$$

Note that $m_{21} = \tan\theta_{21} = \tan(\theta_{12} + 180°) = \tan\theta_{12}$ according to Table B.4, or $m_{21} = m_{12}$, as noted previously.

It does not matter which of the two equations, (B.27) or (B.28), is used. Calculating the coordinate differences so that one gets the correct sign of the slope is important. This is done by remembering that the first coordinates in both the numerator and denominator belong to the same point. For example,

$$m_{AB} = \frac{Y_B - Y_A}{X_B - X_A} = \frac{Y_A - Y_B}{X_A - X_B} \tag{B.29}$$

The slope of the line through points P_1 and P_2, Fig. B.3, is *positive* since θ_{12} and θ_{21} are in the first and third quadrants, respectively, and the tangent is positive in these two quadrants (see Table B.2). The slope of the line through points P_3 and P_4, Fig. B.4, is negative because θ_{34} and θ_{43} are in the second and fourth quadrants, respectively, where the tangent is negative according to Table B.2. Or,

$$\begin{aligned} m_{34} &= \frac{Y_4 - Y_3}{X_4 - X_3} = \frac{+}{-} = \text{negative quantity} \\ m_{43} &= \frac{Y_3 - Y_4}{X_3 - X_4} = \frac{-}{+} = \text{negative quantity} \end{aligned} \tag{B.30}$$

In Eqs. (B.27) through (B.30), one must check to make sure that the denominator is not zero. If it is zero, then the line is parallel to the Y axis, and the slope is infinite.

B.8. Azimuth and bearing Azimuth is always a clockwise angle with a magnitude between 0 and 360°. It is measured either from the North or the $+Y$ axis, or from

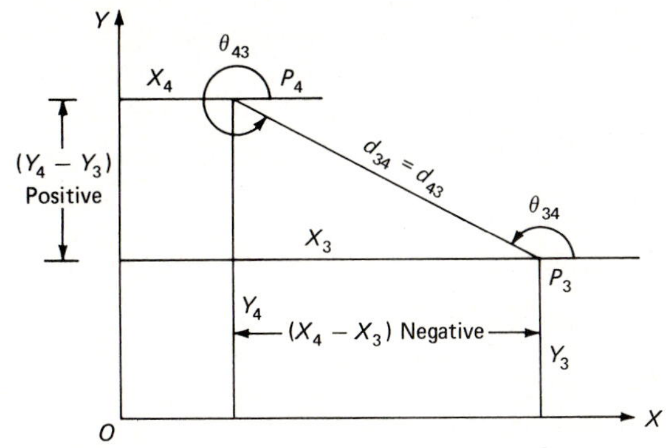

Fig. B.4

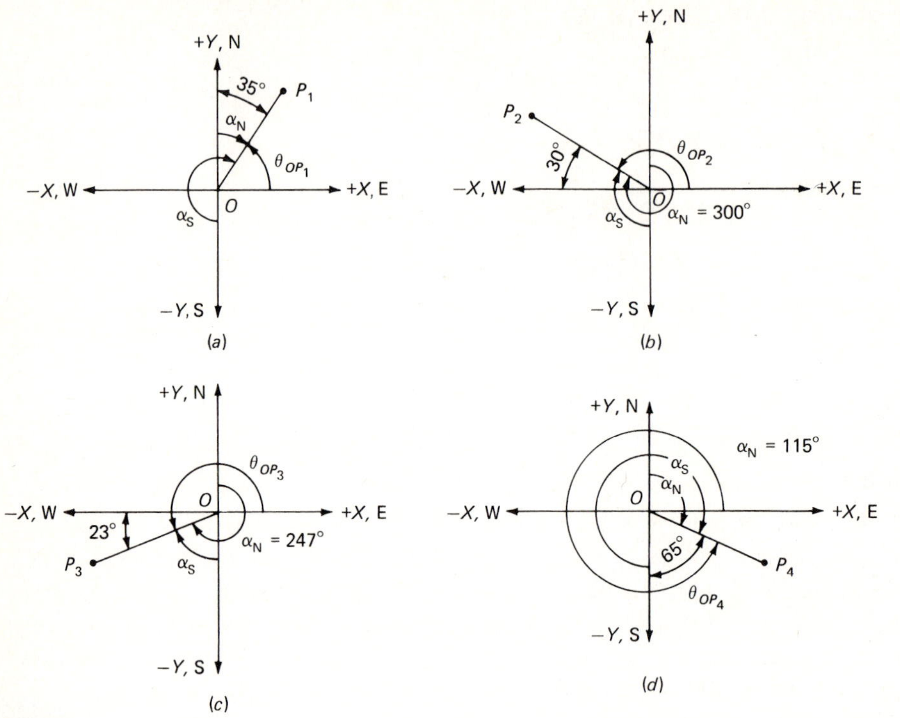

Fig. B.5

the South or the $-Y$ axis. Thus, for any line such as OP_1, in Fig. B.5*a*, α_N is the azimuth from North or $+Y$, and α_S is the azimuth from the South or $-Y$. One azimuth angle is obtained from the other by simply adding 180° and dropping 360° whenever the sum exceeds 360°. Thus,

$$\begin{aligned} \alpha_S &= \alpha_N + 180° \\ \alpha_N &= \alpha_S + 180° \end{aligned} \tag{B.31}$$

Figures B.5*a* to *d* show four lines one in each of the four quadrants. Application of Eq. (B.31) is demonstrated in Table B.5.

Another way of expressing the direction of a line in surveying is the *bearing*. It is always an *acute* angle, with a magnitude between 0 and 90°. The bearing is the angle the line makes with either the $+Y$ or $-Y$ axis. Instead of using $+Y$ and $-Y$, N (for North) and S (for South) are used. The value of the bearing is always a positive

Table B.5

Line	α_N, deg	$\alpha_N + 180°$, deg	α_S, deg	$\alpha_S + 180°$, deg	Figure
OP_1	35	215	215	$395 - 360 = 35$	B.5*a*
OP_2	300	$480 - 360 = 120$	120	300	B.5*b*
OP_3	247	$427 - 360 = 67$	67	247	B.5*c*
OP_4	115	295	295	$475 - 360 = 115$	B.5*d*

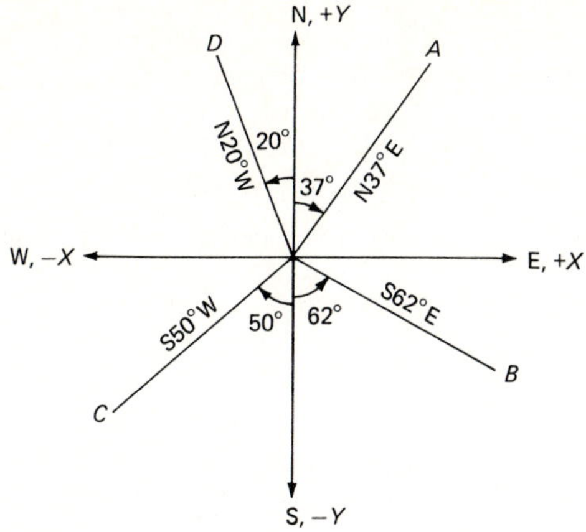

Fig. B.6

Table B.6

Bearing	α_N, deg	α_S, deg
$N\beta°E$	β	$180+\beta$
$S\beta°E$	$180-\beta$	$360-\beta$
$S\beta°W$	$\beta+180$	β
$N\beta°W$	$360-\beta$	$180-\beta$

Table B.7

α_N, deg	α_S, deg	β	Bearing
0 – 90	180 – 270	α_N or $\alpha_S - 180°$	$N\beta°E$
90 – 180	270 – 360	$180° - \alpha_N$ or $360° - \alpha_S$	$S\beta°E$
180 – 270	0 – 90	$\alpha_N - 180°$ or α_S	$S\beta°W$
270 – 360	90 – 180	$360° - \alpha_N$ or $180° - \alpha_S$	$N\beta°W$

quantity. Therefore, the quadrant is indicated by specifying whether the angle is on the east or west side of the meridian (Y axis). Thus, the bearing angle is preceded by either N or S, and succeeded by either E or W. Four examples are given in Fig. B.6.

If β designates a bearing, Table B.6 shows how to convert to azimuth. Table B.7 shows how to convert azimuth to bearing.

Although Tables B.5 through B.7 are useful and cover various cases, it is always very advisable for students to draw a clear sketch and make sure that the angles are in their correct relationship and proper quadrants. Nothing is more helpful than a good sketch of the situation. This is particularly true when angles and lines are added consecutively as in a traverse.

B.9. Relation between azimuth and slope angle Since it is relatively easy to convert from one form of azimuth to the other, consider only the more commonly used azimuth from North, α_N. In Art. B.7, the well-accepted practice in algebra of defining the slope angle of a line as the counterclockwise angle from the $+X$ axis

Table B.8

α_N, deg	$450° - \alpha_N$, deg	θ, deg	$\alpha_N = 450 - \theta$, deg	Figure
35	415 or $55 = \theta_{op_1}$	55	395 or 35	B.5*a*
300	$150 = \theta_{op_2}$	150	300	B.5*b*
247	$203 = \theta_{op_3}$	203	247	B.5*c*
115	$335 = \theta_{op_4}$	335	115	B.5*d*

was used. Given α_N, the slope angle θ is obtained from

$$\theta = 90° - \alpha_N \tag{B.32}$$

Except for $\alpha_N \leq 90°$, the value of θ will be negative. This will cause no problems with trigonometric functions since the proper quadrant will be automatically taken care of. However, if it is preferable to work with positive angles, the student can simply add 360°, or

$$\theta = 450° - \alpha_N \qquad \text{for } \alpha_N > 90° \tag{B.33}$$

For the four cases shown in Fig. B.5, the slope angles are calculated in Table B.8 as an example (see also Table B.5).

The azimuth can also be calculated directly, if the line is fixed by the coordinates of its terminal points. The azimuth of any line *from* a point *to* another, α_{ft}, (where subscript f stands for "from" and t for "to") is given by

$$\alpha_{ft} = \tan^{-1}\frac{X_{\text{to}} - X_{\text{from}}}{Y_{\text{to}} - Y_{\text{from}}} \tag{B.34}$$

The four possible cases, one in each quadrant, are depicted in Fig. B.7. For these cases, the azimuths are given by:

Figure B.7*a*:

$$\alpha_{12} = \tan^{-1}\frac{X_2 - X_1}{Y_2 - Y_1} = \tan^{-1}\left(\frac{+}{+}\right) \qquad \text{first quadrant}$$

$$\alpha_{21} = \tan^{-1}\frac{X_1 - X_2}{Y_1 - Y_2} = \tan^{-1}\left(\frac{-}{-}\right) \qquad \text{third quadrant}$$

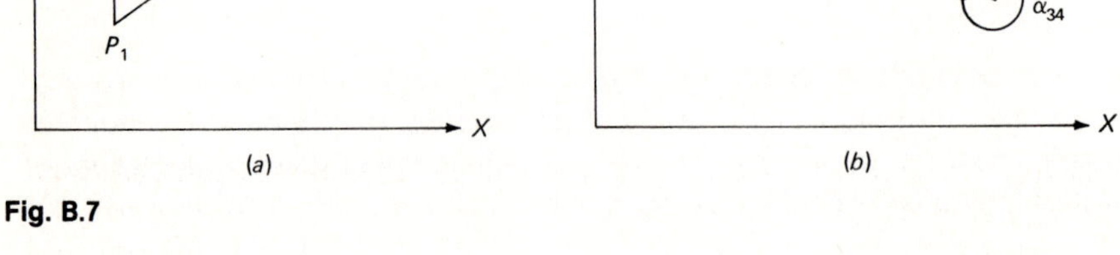

Fig. B.7

Figure B.7*b*:

$$\alpha_{34} = \tan^{-1}\frac{X_4 - X_3}{Y_4 - Y_3} = \tan^{-1}\left(\frac{-}{+}\right) \qquad \text{fourth quadrant}$$

$$\alpha_{43} = \tan^{-1}\frac{X_3 - X_4}{Y_3 - Y_4} = \tan^{-1}\left(\frac{+}{-}\right) \qquad \text{second quadrant}$$

Note that the proper quadrant is determined by the signs of the numerator and denominator. Also, the value of the denominator must first be checked against zero. If it is zero, $\tan\alpha$ is ∞ according to Table B.3. Then the numerator is checked; if it is positive, $\alpha = 90°$, while if it is negative, $\alpha = 270°$.

B.10. Internal and external division of a line segment Given a line segment between two points of known coordinates, it is useful to find the coordinates of either an internal point I or an external point E (Fig. B.8). In Fig. B.8*a*, point I is between P_1 and P_2 and divides the line in the proportion s_1/s_2 or $(P_1I)/(IP_2) = s_1/s_2$. Thus, by proportion in similar triangles P_1Ia and P_1P_2b (see Fig. B.8*a*), the coordinates of point I are given by

$$X_I = \frac{s_1X_2 + s_2X_1}{s_1 + s_2}$$
$$Y_I = \frac{s_1Y_2 + s_2Y_1}{s_1 + s_2} \qquad \text{(B.35)}$$

In Fig. B.8*b*, the division is external, i.e., point E is on the extension of P_1P_2 such that $(P_1E)/(EP_2) = s_1/s_2$. Again by proportion in similar triangles P_1Ef and P_1P_2e (Fig. B.8*b*), the coordinates of point E are given by

$$X_E = \frac{s_1X_2 - s_2X_1}{s_1 - s_2}$$
$$Y_E = \frac{s_1Y_2 - s_2Y_1}{s_1 - s_2} \qquad \text{(B.36)}$$

B.11. Various forms of the equation of a straight line There are many ways of algebraically expressing a straight line in a plane which are quite useful in survey computations. In the following sections, several of these are given.

B.12. Line with known *X* and *Y* intercepts See Fig. B.9.

$$\frac{X}{X_0} + \frac{Y}{Y_0} = 1$$
$$Y_0X + X_0Y = X_0Y_0 \qquad \text{(B.37)}$$

where X_0 and Y_0 are the X and Y intercepts, respectively. This may be referred to as the *two-intercept form* of the line equation.

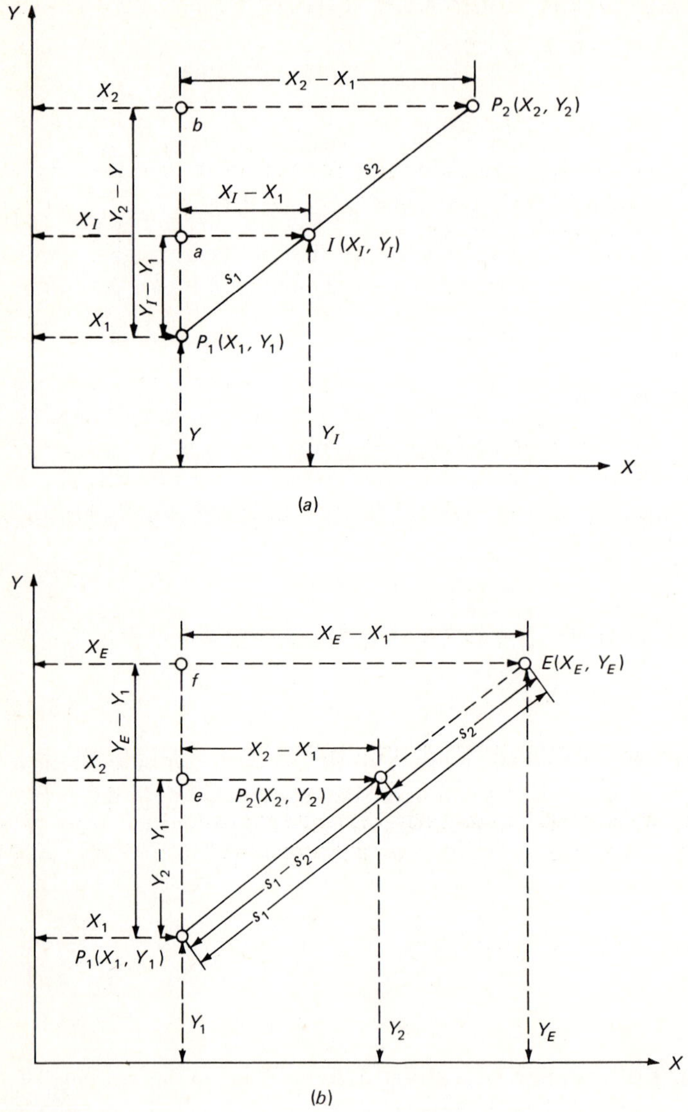

Y
X
$X_2 - X_1$
X_2
$P_2(X_2, Y_2)$
b
s_2
$X_I - X_1$
X_I
$Y_2 - Y$
$I(X_I, Y_I)$
a
s_1
$Y_I - Y_1$
X_1
$P_1(X_1, Y_1)$
Y
Y_I
(a)
$X_E - X_1$
X_E
$E(X_E, Y_E)$
f
$Y_E - Y_1$
s_2
$X_2 - X_1$
X_2
e
$P_2(X_2, Y_2)$
$Y_2 - Y_1$
$s_1 - s_2$
s_1
X_1
$P_1(X_1, Y_1)$
Y_1
Y_2
Y_E
(b)

Fig. B.8

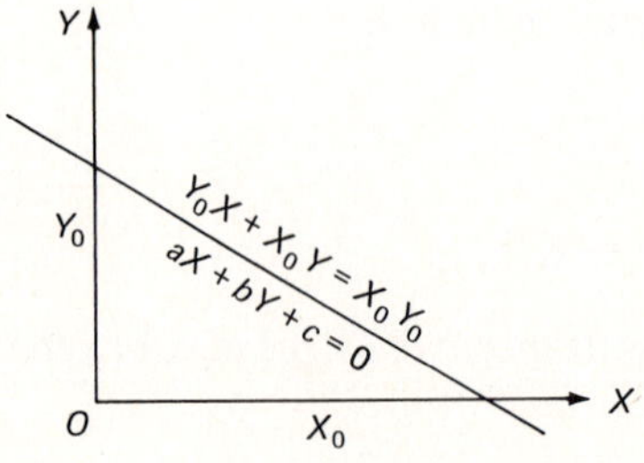

Y
X
O
Y_0
X_0
$Y_0X + X_0Y = X_0Y_0$
$aX + bY + c = 0$

Fig. B.9

B.13. General equation of a line Rearranging Eq. (B.37) yields

$$Y_0 X + X_0 Y - X_0 Y_0 = 0$$

or in general terms

$$aX + bY + c = 0 \tag{B.38}$$

The two intercepts are obtained by substituting zero for each of the two variables X and Y and calculating the value of the other. Thus, with $Y = 0$ and $X = 0$, Eq. (B.38) gives the X intercept and Y intercept, respectively:

$$\begin{aligned} X \text{ intercept} &= -\frac{c}{a} \\ Y \text{ intercept} &= -\frac{c}{b} \end{aligned} \tag{B.39}$$

The slope of the line is given by

$$m = \tan\theta = -\frac{a}{b} \tag{B.40}$$

B.14. Lines parallel to the axes The equation of a line parallel to the X axis is simply

$$Y = k_1 \tag{B.41}$$

where k_1 is the distance of the line from the X axis. Similarly, the equation of a line parallel to the Y axis is

$$X = k_2 \tag{B.42}$$

where k_2 is the distance between the Y axis and the line.

B.15. Line with given slope and *Y* intercept The line shown in Fig. B.10 has a slope $m = \tan\theta$ and a Y intercept $= c$. Let P be any point with coordinates X, Y. Then

$$m = \tan\theta = \frac{Y - c}{X}$$

or

$$Y - c = mX$$

or the equation of the line is

$$Y = mX + c \tag{B.43}$$

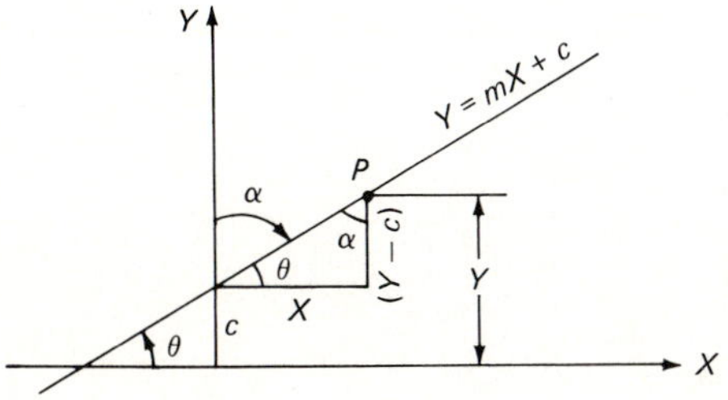

Fig. B.10

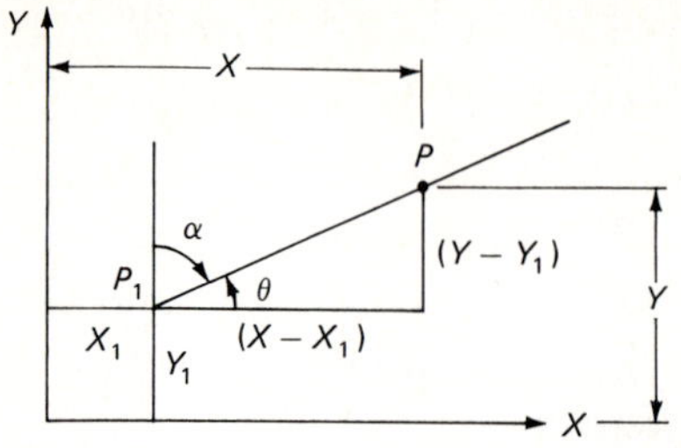

Fig. B.11

In terms of azimuth α, this equation becomes

$$Y = X \cot \alpha + c \tag{B.44}$$

This may be referred to as the *slope intercept* form of the equation of a line.

B.16. Line with a given slope passing through a given point The line shown in Fig. B.11 has a slope $m = \tan\theta$ and passes through point P_1. If P is any point with coordinates X, Y, then

$$m = \tan\theta = \frac{Y - Y_1}{X - X_1}$$

or the equation of the line is

$$Y - Y_1 = m(X - X_1) \tag{B.45}$$

In terms of azimuth, the equation becomes

$$Y - Y_1 = (X - X_1)\cot\alpha \tag{B.46}$$

If the given point is the origin of the coordinate system, then the equations reduce to

$$Y = mX$$

and

$$Y = X \cot \alpha \tag{B.47}$$

This may be called the *point-slope* form of the equation of a line.

B.17. Line joining two points Figure B.12 shows a line passing through two points P_1 and P_2 with coordinates X_1, Y_1 and X_2, Y_2, respectively. With $P(X, Y)$

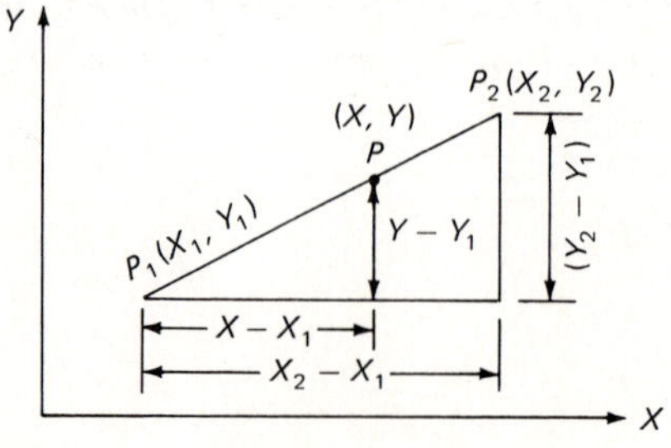

Fig. B.12

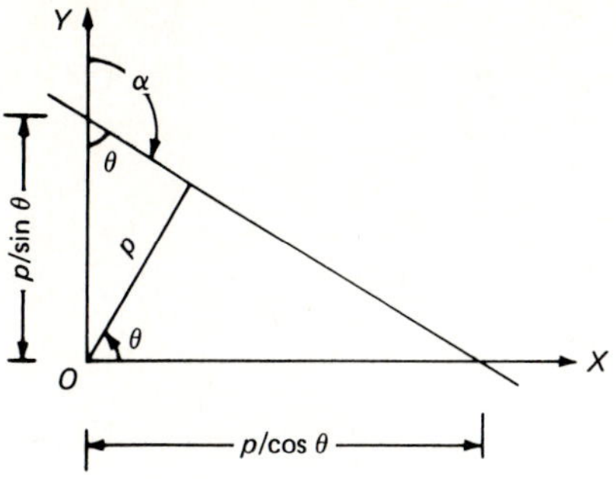

Fig. B.13

being any point on the line, then

$$\frac{Y - Y_1}{X - X_1} = \frac{Y_2 - Y_1}{X_2 - X_1} \tag{B.48}$$

This is called the *two point* form of the equation of a line.

B.18. Line with given length and slope of the perpendicular from origin As shown in Fig. B.13, the perpendicular from the origin O to the line in question has a length p and slope angle θ. As is clear from the figure, the X and Y intercepts are $p/\cos\theta$ and $p/\sin\theta$, respectively. Consequently, according to Eq. (B.37), the equation for the straight line is

$$\frac{X}{p/\cos\theta} + \frac{Y}{p/\sin\theta} = 1$$

or

$$X\cos\theta + Y\sin\theta = p \tag{B.49}$$

This is referred to as the *normal form* of the equation of a line.

B.19. Perpendicular distance from the origin to a line The general form of the equation of a line is [Eq. (B.38)]

$$aX + bY + c = 0$$

or

$$aX + bY = -c \tag{B.50}$$

Comparing Eqs. (B.49) and (B.50), the coefficients of X and Y relate to trigonometric functions of an angle. Figure B.14 shows that the sine and cosine of the slope angle

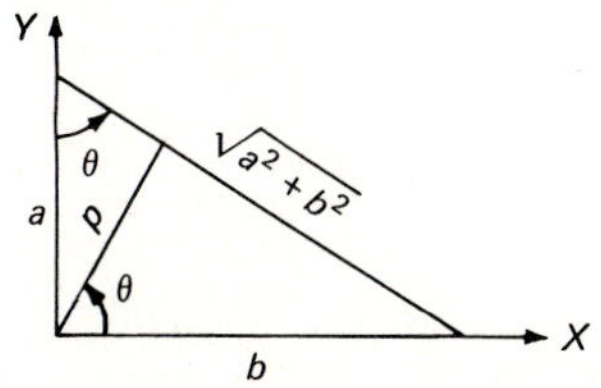

Fig. B.14

of the normal, θ, are given by

$$\cos\theta = \frac{a}{\sqrt{a^2+b^2}} \qquad \sin\theta = \frac{b}{\sqrt{a^2+b^2}} \tag{B.51}$$

Thus, dividing both sides of Eq. (B.50) by $\sqrt{a^2+b^2}$ yields

$$\frac{a}{\sqrt{a^2+b^2}}X + \frac{b}{\sqrt{a^2+b^2}}Y = \frac{-c}{\sqrt{a^2+b^2}}$$

or, in view of Eqs. (B.51),

$$X\cos\theta + Y\sin\theta = \frac{-c}{\sqrt{a^2+b^2}} \tag{B.52}$$

Consequently, from Eqs. (B.49) and (B.52) the perpendicular from the origin to the line with the general equation is

$$p = \frac{-c}{\sqrt{a^2+b^2}} \tag{B.53}$$

The radical is a quantity that represents a length, and so does p. In order to avoid confusion, always take the positive sign of the radical. This means that the sign of p will be either positive or negative depending on whether the sign of c is negative or positive, respectively. Again, to avoid confusion regardless of the resulting sign for p, always take it as a positive length. In other words, Eq. (B.53) may be rewritten in absolute value form, or

$$p = \left|\frac{c}{\sqrt{a^2+b^2}}\right| \tag{B.54}$$

Note that when $c = 0$ the distance p is also zero and the line passes through the origin.

Example B.2 Figure B.15 shows four lines with given equations:

A: $X + Y - 4 = 0$
B: $X - Y + 3 = 0$
C: $X + Y + 5 = 0$
D: $X - Y - 3 = 0$

Calculate the length of the perpendicular from the origin to each of the lines.

SOLUTION By Eq. (B.54)

$$p_A = \left|\frac{-4}{\sqrt{1+1}}\right| = \frac{4}{\sqrt{2}} = \frac{4\sqrt{2}}{2} = 2\sqrt{2}$$

$$p_B = \left|\frac{3}{\sqrt{2}}\right| = 1.5\sqrt{2}$$

$$p_C = \left|\frac{5}{\sqrt{2}}\right| = 2.5\sqrt{2}$$

$$p_D = \left|-\frac{3}{\sqrt{2}}\right| = 1.5\sqrt{2}$$

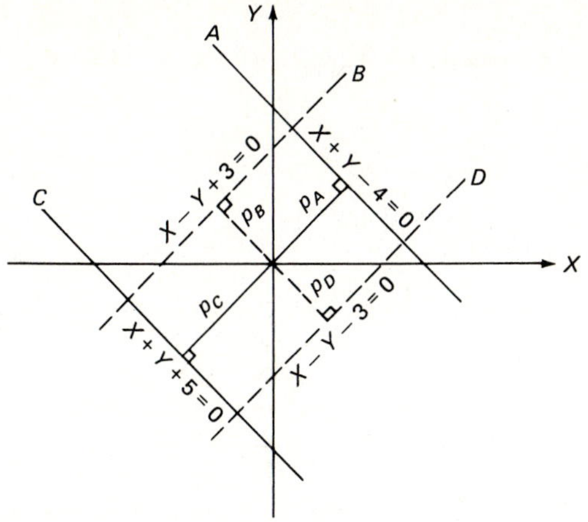

Fig. B.15

Example B.3 Given the length and slope angles of the four perpendiculars:

$$p_A = 2\sqrt{2} \qquad \theta_A = 45°$$

$$p_B = 1.5\sqrt{2} \qquad \theta_B = 135°$$

$$p_C = 2.5\sqrt{2} \qquad \theta_C = 225°$$

$$p_D = 1.5\sqrt{2} \qquad \theta_D = 315°$$

Show that the resulting equations of the four lines are the same as those given in Example B.2 and shown in Fig. B.15.

SOLUTION By Eq. (B.49)

$$A: \quad X\cos 45° + Y\sin 45° = p_A \quad \text{or} \quad \frac{X}{\sqrt{2}} + \frac{Y}{\sqrt{2}} = 2\sqrt{2} \quad \text{or} \quad X + Y - 4 = 0$$

$$B: \quad X\cos 135° + Y\sin 135° = p_B \quad \text{or} \quad -\frac{X}{\sqrt{2}} + \frac{Y}{\sqrt{2}} = 1.5\sqrt{2} \quad \text{or} \quad X - Y + 3 = 0$$

$$C: \quad X\cos 225° + Y\sin 225° = p_C \quad \text{or} \quad -\frac{X}{\sqrt{2}} - \frac{Y}{\sqrt{2}} = 2.5\sqrt{2} \quad \text{or} \quad X + Y + 5 = 0$$

$$D: \quad X\cos 315° + Y\sin 315° = p_D \quad \text{or} \quad \frac{X}{\sqrt{2}} - \frac{Y}{\sqrt{2}} = 1.5\sqrt{2} \quad \text{or} \quad X - Y - 3 = 0$$

B.20. Perpendicular distance from a point to a line Figure B.16 shows a line A with equation $aX + bY + c = 0$ and a point P with coordinates X_1, Y_1. It is required to find the perpendicular distance d of point P from the line A. An alternative form of the equation of line A is $X\cos\theta + Y\sin\theta - p = 0$, where [see Eqs. (B.52) and (B.53)]:

$$\cos\theta = \frac{a}{\sqrt{a^2 + b^2}} \qquad \sin\theta = \frac{b}{\sqrt{a^2 + b^2}}$$
$$p = -\frac{c}{\sqrt{a^2 + b^2}} \tag{B.55}$$

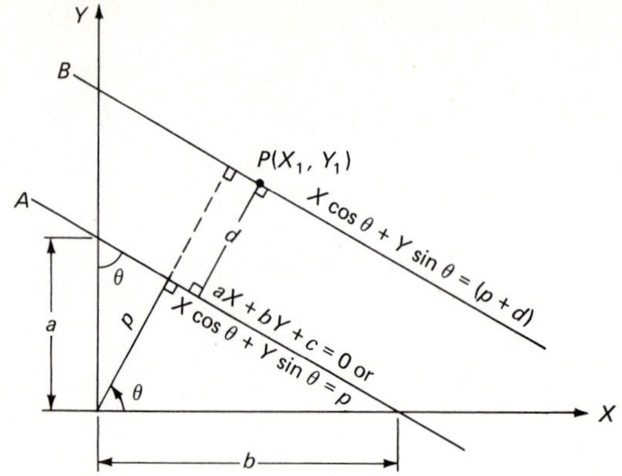

Fig. B.16

The line B, which is parallel to A and passes through point P, has the equation $X\cos\theta + Y\sin\theta = p + d$. Since point P falls on line B, then its coordinates X_1, Y_1 satisfy its equation, or

$$X_1\cos\theta + Y_1\sin\theta = p + d \tag{B.56}$$

Substituting Eq. (B.55) into (B.56) and rearranging, one gets

$$d = \frac{aX_1 + bY_1 + c}{\sqrt{a^2 + b^2}} \tag{B.57}$$

The quantity in the numerator can be either positive or negative because point P can fall on either side of the line A (more explanation of this is given in Art. B.21). Since the radical will always be taken as positive, it is best to take the absolute value; thus

$$d = \left|\frac{aX_1 + bY_1 + c}{\sqrt{a^2 + b^2}}\right| \tag{B.58}$$

Example B.4 Figure B.17 shows a straight line A with X intercept $X_0 = 3$ and Y intercept $Y_0 = -2$. Find the distance d_1 and d_2 of the two points $P_1(4, 4)$ and $P_2(9, 2)$ from the line A.

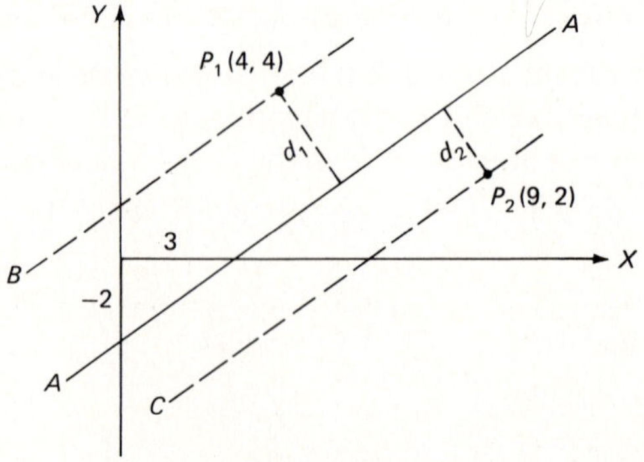

Fig. B.17

SOLUTION By Eq. (B.37) the equation of line A is

$$-2X+3Y=-6$$

or

$$2X-3Y-6=0$$

Next

$$\sqrt{a^2+b^2}=\sqrt{4+9}=3.61$$

For $P_1(4,4)$:

$$aX_1+bY_1+c=2(4)-3(4)-6=-10$$

then

$$d_1=\left|\frac{-10}{3.61}\right|=2.77$$

For $P_2(9,2)$:

$$aX_2+bY_2+c=2(9)-3(2)-6=6$$

then

$$d_2=\left|\frac{6}{3.61}\right|=1.66$$

B.21. Equation of a line through a point and parallel to another line The equation sought here was basically given in Art. B.20. In the normal form it is

$$X\cos\theta+Y\sin\theta=p+d \tag{B.59}$$

where $X\cos\theta+Y\sin\theta=p$ is the equation of the given line, and d is the perpendicular distance between the new line and the given line. The value of d cannot be taken in absolute value as was the case in Art. B.20 where the magnitude was sufficient. Here, the proper sign of d must be determined. Before proceeding, note that the evaluated value of a line equation will always be positive for all points on one side of the line and negative for all points on the other side. (The value is of course zero for points falling on the line.) This fact was demonstrated numerically in Example B.4 (see also Fig. B.17). For the line $2X-3Y-6=0$, the calculated value of d for point P_1 is negative (equal to -10), and is positive for point P_2 (equal to 6). Point P_1 is on the same side of line A as the origin. As a further demonstration, substitute 0, 0, the coordinates of the origin:

$$a(0)+b(0)+c=-6$$

which is also negative as for point P_1.

Using Eqs. (B.55) and (B.57) substituted into Eq. (B.59), one gets

$$\frac{a}{\sqrt{a^2+b^2}}X+\frac{b}{\sqrt{a^2+b^2}}Y=\frac{-c}{\sqrt{a^2+b^2}}+\frac{aX_1+bY_1+c}{\sqrt{a^2+b^2}}$$

which when clearing fractions becomes

$$aX+bY-(aX_1+bY_1)=0 \tag{B.60}$$

that is the equation sought for the line. The coefficients a and b belong to the given line and X_1, Y_1 are the coordinates of the given point.

An alternative procedure is to determine the slope of the given line and use the point-slope form, Art. B.16. Thus, by Eq. (B.40)

$$m = -\frac{a}{b}$$

and then by Eq. (B.45)

$$Y - Y_1 = m(X - X_1) = -\frac{a}{b}(X - X_1)$$

or

$$bY - bY_1 = -aX + aX_1$$

or

$$aX + bY - (aX_1 + bY_1) = 0$$

which is the same as Eq. (B.60).

Example B.5 Write the equations for lines through points P_1 and P_2 and parallel to line A shown in Fig. B.17, using the data in Example B.4.

SOLUTION Line B, through point $P_1(4, 4)$, will have the equation

$$2X - 3Y - [(2)(4) - (3)(4)] = 0$$

or

$$2X - 3Y + 4 = 0$$

Line C, through point $P_2(9, 2)$, has the equation

$$2X - 3Y - [(2)(9) - (3)(2)] = 0$$

or

$$2X - 3Y - 12 = 0$$

B.22. Equation of a line through a point and perpendicular to another line
In Fig. B.18, the given line A has the general equation $aX + bY + c = 0$, and the

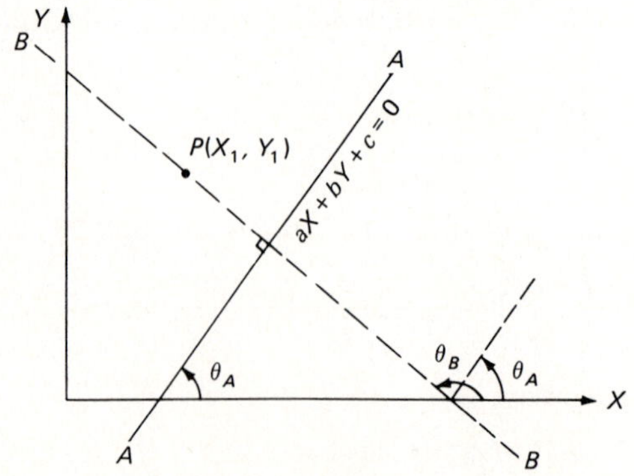

Fig. B.18

given point P has coordinates X_1, Y_1. Required is the equation of line B which passes through point P and is perpendicular to line A. According to Eq. (B.40), the slope of line A is

$$m_A = \tan\theta_A = -\frac{a}{b}$$

And from Fig. B.18, the slope angle θ_B of line B is

$$\theta_B = \theta_A + 90°$$

Then the slope of line B is

$$m_B = \tan\theta_B = \tan(\theta_A + 90) = -\cot\theta_A \qquad \text{(see Table B.4)}$$

or

$$m_B = -\frac{1}{m_A} \tag{B.61}$$

Equation (B.61) indicates that the slope of the perpendicular line B is the negative of the reciprocal of the slope of the given line A. With m_B and point P the equation of line B can be written using the point slope from (Art. B.16), or from Eq. (B.45),

$$Y - Y_1 = m_B(X - X_1) = \frac{b}{a}(X - X_1)$$

or

$$aY - aY_1 = bX - bX_1$$

or

$$bX - aY - (bX_1 - aY_1) = 0 \tag{B.62}$$

B.23. Angle between two lines Figure B.19 shows two lines, the angle between them being designated by γ. It is clear from the figure that

$$\gamma = \theta_2 - \theta_1 \tag{B.63}$$

where θ_1 and θ_2 are the slope angles of line 1 and line 2, respectively. In terms of slope one gets

$$\tan\gamma = \tan(\theta_2 - \theta_1)$$

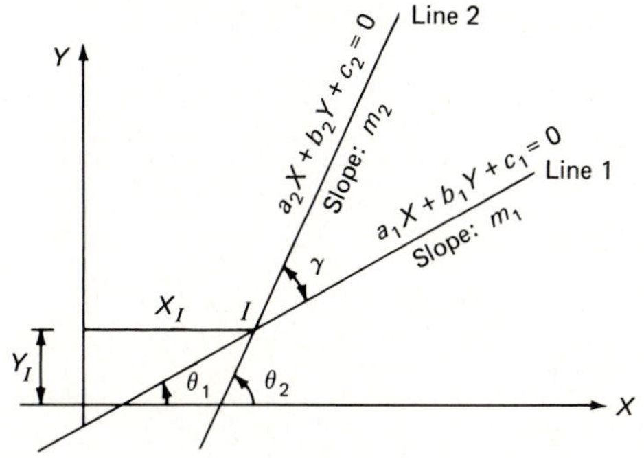

Fig. B.19

which according to Eq. (B.6) becomes

$$\tan\gamma = \frac{\tan\theta_2 - \tan\theta_1}{1 + \tan\theta_1 \tan\theta_2}$$

With slopes $m_1 = \tan\theta_1$ and $m_2 = \tan\theta_2$, of lines 1 and 2, one finally gets

$$\tan\gamma = \frac{m_2 - m_1}{1 + m_1 m_2} \qquad \text{(B.64)}$$

If $\tan\gamma = 0$, or $m_1 = m_2$, the two lines are parallel. On the other hand, if $\tan\gamma = \infty$, or $m_1 m_2 = -1$ [see also Eq. (B.61)], the two lines will be perpendicular to each other. The slopes m_1 and m_2 can be calculated from whatever form is given for the equations of the lines.

B.24. Point of intersection of two lines The two lines in Fig. B.19 intersect at point I. Since this point lies on both lines, its coordinates X_I, Y_I satisfy their equations. Then, to get X_I, Y_I we simultaneously solve the two equations of the lines, or

$$a_1 X_I + b_1 Y_I + c_1 = 0$$
$$a_2 X_I + b_2 Y_I + c_2 = 0$$

Multiply first equation by b_2, the second equation by b_1, and subtract:

$$\begin{array}{c} a_1 b_2 X_I + b_1 b_2 Y_I + c_1 b_2 = 0 \\ \underline{a_2 b_1 X_I + b_1 b_2 Y_I + c_2 b_1 = 0} \\ (a_1 b_2 - a_2 b_1) X_I = c_1 b_2 - c_2 b_1 \end{array}$$

or

$$X_I = \frac{c_1 b_2 - c_2 b_1}{a_1 b_2 - a_2 b_1} \qquad \text{(B.65)}$$

In a similar manner one gets for Y_I

$$Y_I = \frac{c_1 a_2 - c_2 a_1}{a_1 b_2 - a_2 b_1} \qquad \text{(B.66)}$$

B.25. Equation of a circle Figure B.20 shows a circle of radius R, the center of which is at point C with coordinates X_c, Y_c. From Art. B.6, the square of the distance between C and any point on the circle such as $P(X, Y)$ is given by

$$(X - X_c)^2 + (Y - Y_c)^2 = R^2 \qquad \text{(B.67)}$$

This is the equation of a circle when the coordinates of its center and its radius are given. If $X_c = Y_c = 0$, that is, if the circle's center is the origin of the coordinate system, then the equation simplifies to

$$X^2 + Y^2 = R^2 \qquad \text{(B.68)}$$

Equation (B.67) may be expanded to

$$X^2 - 2XX_c + X_c^2 + Y^2 - 2YY_c + Y_c^2 - R^2 = 0$$

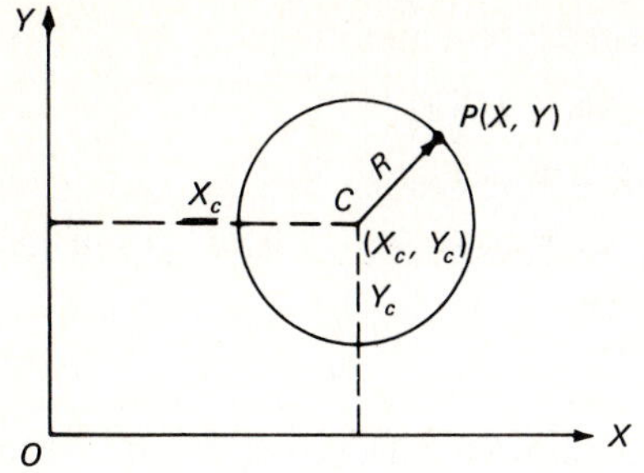

Fig. B.20

or

$$X^2 + Y^2 - 2X_cX - 2Y_cY + (X_c^2 + Y_c^2 - R^2) = 0$$

and then written in the form

$$X^2 + Y^2 + 2dX + 2eY + f = 0 \tag{B.69}$$

Equation (B.69) represents the general form of the equation of a circle. It contains three coefficients d, e, f which represent three geometric elements such as the radius and the two coordinates of its center.

B.26. Intersection of a line and a circle In general, a straight line intersects a circle in two points, as shown in Fig. B.21*a*. Let the equation of the circle be [Eq. (B.69)]

$$X^2 + Y^2 + 2dX + 2eY + f = 0$$

and that of the straight line [Eq. (B.43)]

$$Y = mX + c$$

Substituting Eq. (B.43) into (B.69) yields

$$X^2 + (mX + c)^2 + 2dX + 2e(mX + c) + f = 0$$

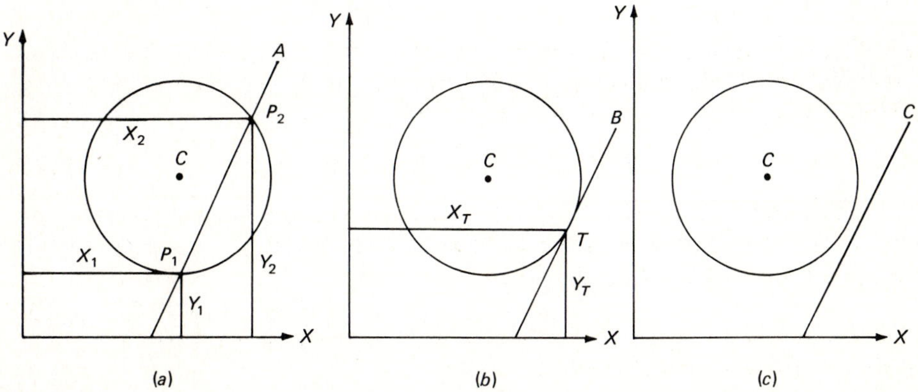

Fig. B.21

which can be reduced to the general *quadratic* equation form

$$AX^2 + BX + C = 0 \tag{B.70}$$

Whatever form the equations for the circle and straight line are, the quadratic form in Eq. (B.70) will always result from the substitution. There are, in general, two values for X obtainable from Eq. (B.70):

$$X_1 = \frac{-B + \sqrt{B^2 - 4AC}}{2A} \quad \text{and} \quad X_2 = \frac{-B - \sqrt{B^2 - 4AC}}{2A} \tag{B.71}$$

and for each of these, a value for Y is obtained from Eq. (B.43). Thus, in general, a straight line intersects a circle in two points such as $P_1(X_1, Y_1)$ and $P_2(X_2, Y_2)$ in Fig. B.21*a*. Two other situations are also possible, depending upon the quantity under the radical $\sqrt{B^2 - 4AC}$. If this quantity is zero, or $B^2 = 4AC$, then $X_1 = X_2$ and the line is tangent to the circle at one point $T(X_T, Y_T)$ as shown in Fig. B.21*b*. In the second case, $B^2 < 4AC$, the quantity is negative, which means that one cannot take the square root. (This is usually referred to as the *imaginary* solution to the quadratic equation.) This case is depicted in Fig. B.21c, where the line misses the circle and thus no intersection takes place.

B.27. Areas

Circle

$$A = \pi R^2 = \frac{\pi D^2}{4} \tag{B.72}$$

where R and D are the radius and diameter, respectively.

Sector

$$A = \tfrac{1}{2}Rs = \frac{\pi R^2 \gamma^\circ}{360^\circ} \tag{B.73}$$

where R is the radius, s is the arc length, and γ is the angle at the center in degrees, as shown in Fig. B.22.

Segment (Less than a semicircle, Fig. B.23)

$$A = \tfrac{1}{2}Rs - \tfrac{1}{2}R^2 \sin\gamma \tag{B.74}$$

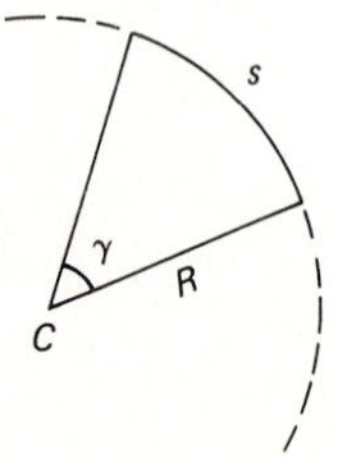

Fig. B.22

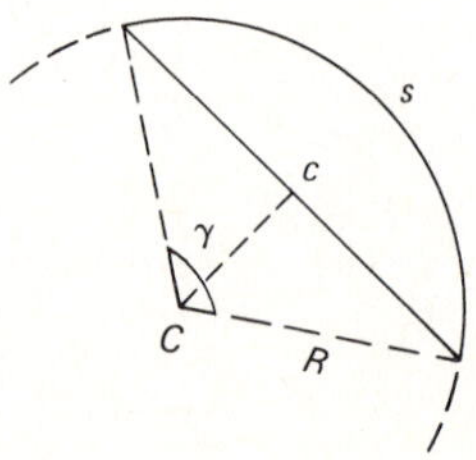

Fig. B.23

or

$$A = \frac{\pi R^2 \gamma°}{360°} - \tfrac{1}{2}R^2 \sin\gamma \tag{B.75}$$

where R, s, and γ are as defined above. If the chord length c is given (Fig. B.23) then

$$\sin\frac{\gamma}{2} = \frac{c}{2R} \tag{B.76}$$

Triangle For a *general* triangle, as shown in Fig. B.24, the area K is given by

$$K = \tfrac{1}{2}bh \qquad (h \text{ is perpendicular to } b) \tag{B.77}$$

or

$$K = \tfrac{1}{2}bc \sin A \tag{B.78}$$

or

$$K = \sqrt{s(s-a)(s-b)(s-c)} \tag{B.79}$$

with

$$s = \tfrac{1}{2}(a+b+c) \tag{B.80}$$

where a, b, c are the lengths of the sides, h is the height of the triangle (which is perpendicular to the base b), and A, B, C are the interior angles as shown in Fig. B.24. Other useful relations for a plane triangle are:

$$A + B + C = 180° \tag{B.81}$$

$$\frac{\sin A}{a} = \frac{\sin B}{b} = \frac{\sin C}{c} \tag{B.82}$$

$$a^2 = b^2 + c^2 - 2bc\cos A \tag{B.83a}$$

$$b^2 = c^2 + a^2 - 2ca\cos B \tag{B.83b}$$

$$c^2 = a^2 + b^2 - 2ab\cos C \tag{B.83c}$$

$$\tan\frac{A}{2} = \frac{1}{s-a}\sqrt{\frac{(s-a)(s-b)(s-c)}{s}} \tag{B.84}$$

$$\cos\frac{A}{2} = \sqrt{\frac{s(s-a)}{bc}} \tag{B.85a}$$

$$\cos\frac{B}{2} = \sqrt{\frac{s(s-b)}{ca}} \tag{B.85b}$$

$$\cos\frac{C}{2} = \sqrt{\frac{s(s-c)}{ab}} \tag{B.85c}$$

$$\frac{a-b}{a+b} = \frac{\tan(A-B)/2}{\tan(A+B)/2} \tag{B.86}$$

$$h = \frac{2}{b}\sqrt{s(s-a)(s-b)(s-c)} \tag{B.87}$$

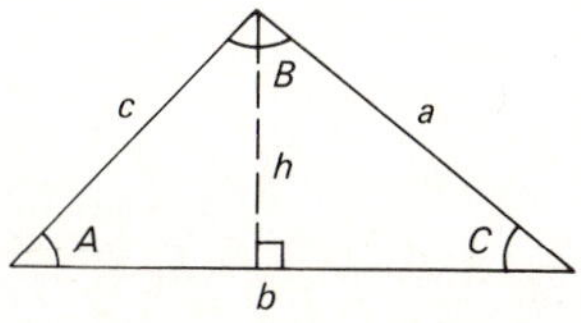

Fig. B.24

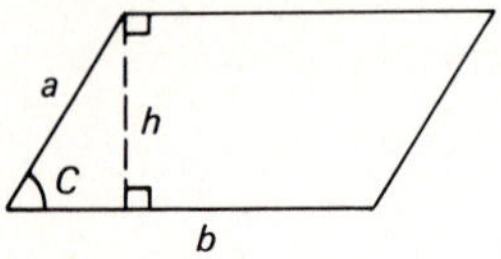

Fig. B.25

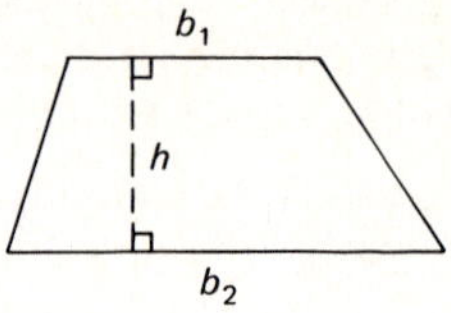

Fig. B.26

For an *equilateral* triangle, where sides $a = b = c$ and angles $A = B = C = 60°$, then the area K becomes

$$K = \frac{a^2\sqrt{3}}{4} \qquad \left(\text{with } h = \frac{a\sqrt{3}}{2}\right) \tag{B.88}$$

Square

$$A = a^2 \tag{B.89}$$

where a is the side length.

Rectangle

$$A = ab \tag{B.90}$$

where a and b are its width and length.

Parallelogram Let a and b be the sides, h the altitude upon side b, C the acute angle, and A the area (see Fig. B.25). Then,

$$A = bh = ab \sin C \tag{B.91}$$

Trapezoid If b_1 and b_2 are the parallel sides, and h is the altitude between them (see Fig. B.26), then the area A is

$$A = \tfrac{1}{2}h(b_1 + b_2) \tag{B.92}$$

For an *isosceles* trapezoid (Fig. B.27), if a is the length of one of the two nonparallel sides, and C is the acute angle between a and b_2, then

$$A = \tfrac{1}{2}a(b_1 + b_2)\sin C \tag{B.93}$$

B.28 Problems

B.1 Find the sin, cos, tan, sec, cosec, and cot of each of these angles: 17°, 135°, 184°, 296°.
B.2 Refer to Fig. B.2; given $X = 6$ ft and $Y = 8$ ft, calculate: r; θ in degrees, minutes, and seconds; θ in radians; $\sin\theta$, $\cos\theta$, $\tan\theta$.
B.3 Given $r = 10$ ft and $Y = 5$ ft, in Fig. B.28, find X, ϕ, θ, $\sin\theta$, $\cos\theta$, $\tan\theta$.

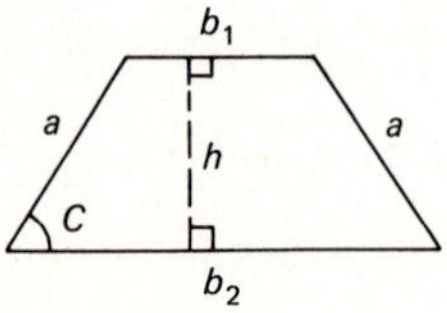

Fig. B.27

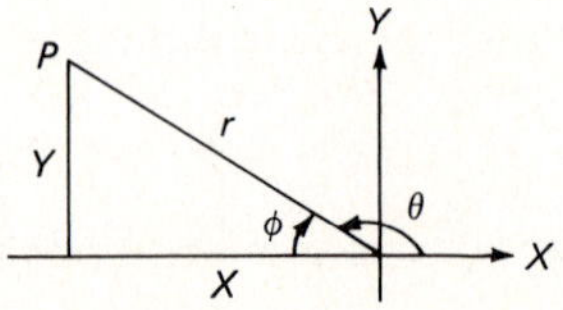

Fig. B.28

B.4 Refer to Fig. B.2; given $r = 10$ ft and $\theta = 50°$, find $X, Y, \sin\theta, \cos\theta, \tan\theta$.
B.5 Refer to Fig. B.28; given $r = 15$ ft and $\theta = 145°$, find $X, Y, \sin\theta, \cos\theta, \tan\theta$.
B.6 Find θ and r for each point, given the following tabulated coordinates:

Point	*X* coordinate, ft	*Y* coordinate, ft
1	+68	−16
2	−28	+61
3	−39	−30
4	+46	+82

B.7 Find the distance between points 1 and 2 and the slope of the line through points 1 and 2, for the following cases:

	Point 1		Point 2	
	X, ft	*Y*, ft	*X*, ft	*Y*, ft
(*a*)	6	8	25	14
(*b*)	3	11	9	4
(*c*)	−7	12	13	−5
(*d*)	−10	−17	33	2
(*e*)	−20	−19	−15	−16
(*f*)	−18	−1	9	−22

B.8 Convert the following azimuths to bearings: 26°, 106°, 216°, 306°.
B.9 Convert the following bearings to azimuths: N6°E, S46°W, S76°E, N16°W.
B.10 Find θ (the slope angle) for the bearings and azimuths in Probs. B.8 and B.9 above.
B.11 Refer to Fig. B.8*a*; find the coordinates of point *I* if the coordinates of point P_1 are $X_1 = 2$ ft, $Y_1 = 1$ ft, and point P_2 are $X_2 = 8$ ft, $Y_2 = 10$ ft, and if $s_1 = 1$ and $s_2 = 2$.
B.12 Refer to Fig. B.8*b*; find the coordinates of point *E* if the coordinates of point P_1 are $X_1 = 3$ ft, $Y_1 = 4$ ft, and point P_2 are $X_2 = 9$ ft, $Y_2 = 7$ ft, and if $s_1 = 3$ and $s_2 = 2$.
B.13 Refer to Fig. B.9; write the equation of the line if $X_0 = 5$ and $Y_0 = 4$.
B.14 Given the equation of a line $2X - Y + 3 = 0$, find the slope m, X intercept, and Y intercept.
B.15 Write the equation of the line that has a *Y* intercept of 4 and a slope of -2.
B.16 Write the equation of the line that passes through point $P(X = +3,\ Y = +7)$ with a slope of $+1$. Find the *X* and *Y* intercepts of this line.
B.17 Write the equation of the line that passes through point P_1 $(X_1 = -3,\ Y_1 = 5)$ and P_2 $(X_2 = 6,\ Y_2 = -1)$. Find its slope and the *X* intercept and *Y* intercept.
B.18 Write the equation of the line in Fig. B.13 if $p = 4$ and $\theta = 60°$
B.19 Refer to Fig. B.15; calculate p for the equation $X + 2Y - 5 = 0$.
B.20 Given the following length and slope angle for the perpendicular p, write the equation of the line (see Fig. B.13): $p = \sqrt{5}$, $\theta = 69°$.
B.21 Find the perpendicular distance from point P $(X = 2,\ Y = 5)$ to the line $X - Y + 1 = 0$.
B.22 Refer to Fig. B.17; if $X_0 = 2$ and $Y_0 = 3$, find the distances d_1 and d_2 of points P_1 $(-3, -4)$ and P_2 $(2, -5)$ from the line.
B.23 Write the equations for lines through points P_1 $(5, 6)$ and P_2 $(8, 1)$ and parallel to line *A* whose equation is $2X - 3Y - 6 = 0$.
B.24 Referring to Fig. B.18, the equation of line *A* is $2X - Y - 6 = 0$. Find the equation of line *B* which passes through point P $(7, 3)$ and is perpendicular to line *A*.
B.25 Find the coordinates X_I and Y_I of the point of intersection of the two lines $3X - Y + 4 = 0$ and $X - Y + 8 = 0$.
B.26 Write the equation of a circle of radius 2 with its center at point C $(X_c = 4, Y_c = 3)$.

B.27 Write the equation of a circle of radius 3 with its center at point $P(5, 4)$. Refer to Fig. B.21, and determine which case applies for the intersection of each of the following lines with the circle. If the line intersects the circle or is tangent to the circle, give the coordinates of the points of intersection or tangency.

Line	Equation
1	$-Y+3=0$
2	$-Y+1=0$
3	$X-1=0$

B.28 Refer to Fig. B.22; find the area of the sector if $\gamma = 33°$ and $R=3$ units.
B.29 Refer to Fig. B.23; find the area of the segment if $\gamma = 111°$ and $R=11$ units. Find the chord length c and the arc s.
B.30 Refer to Fig. B.23; find the area of the segment if $c=8$ units and $R=5$ units. Find the arc lengths s and the angle α.
B.31 Refer to Fig. B.24; find the area of the triangle if the three sides have lengths of 6, 8, and 10 m. Find the value of the angles.

Index

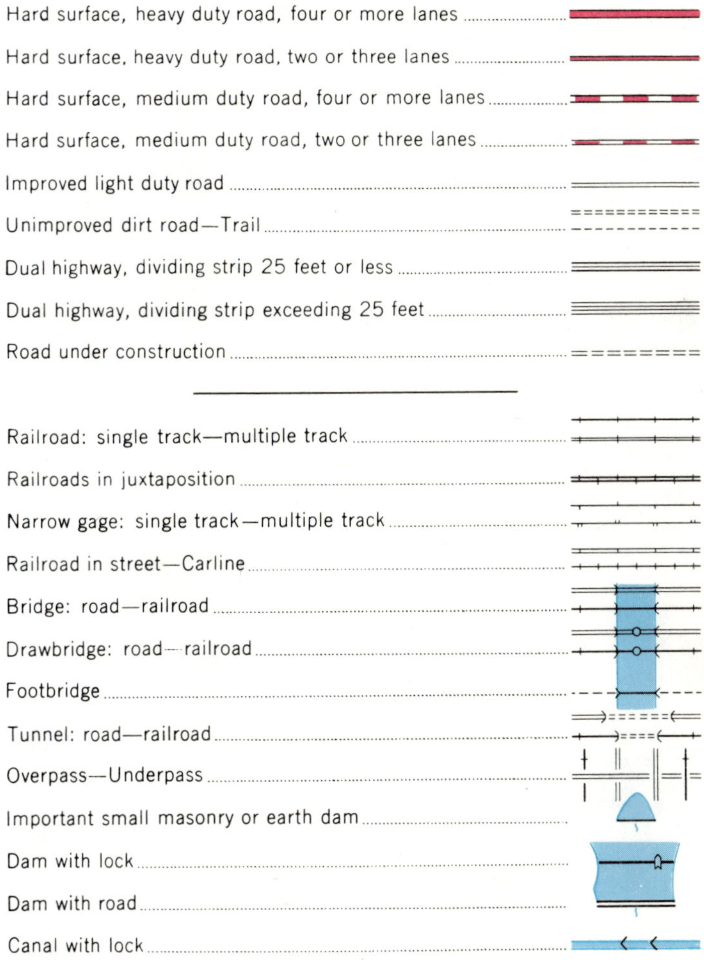

Symbols for roads, railroads, and dams.

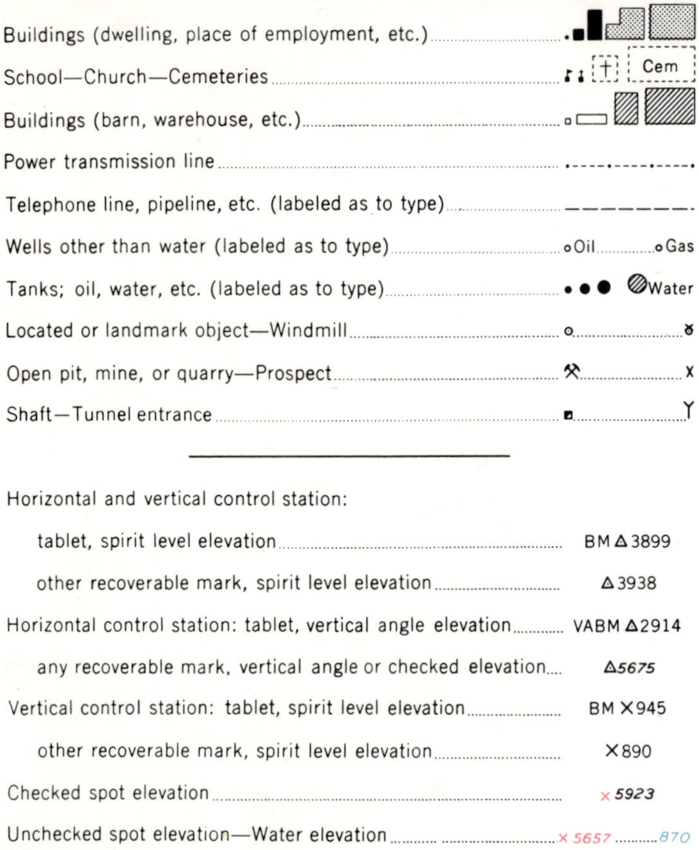

Symbols for structures and stations.

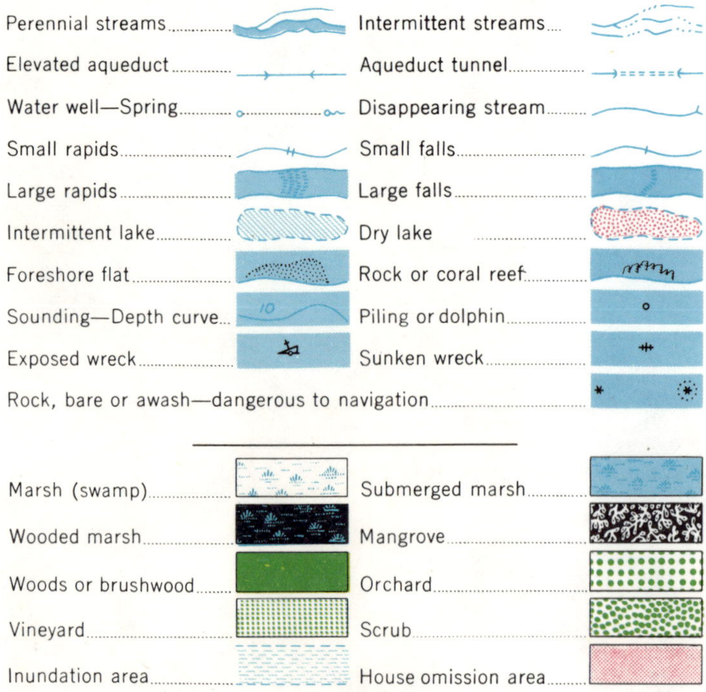

Symbols for hydrography and land classification.